MTBE Remediation Handbook

MTBE Remediation Handbook

Edited by
Ellen E. Moyer
Paul T. Kostecki

Springer Science+Business Media, LLC

ISBN 978-1-4613-4889-4   ISBN 978-1-4615-0021-6 (eBook)
DOI 10.1007/978-1-4615-0021-6

Library of Congress Control Number: 2003100235

© 2003 by Springer Science+Business Media New York
Originally published by Kluwer Academic Publishers in 2003

Acknowledgment

Many, many thanks to Tracy J. Adamski of Tighe & Bond, Inc., and Betty A. Niedzwiecki of the Association for Environmental Health and Sciences for their many hours of assistance in development and coordination of the book.

Preface

The time has come for an *MTBE Remediation Handbook*. There are hundreds of thousands of spills of gasoline containing MTBE in the United States. More than a billion dollars are spent each year to clean up spills of gasoline and manage the risk from existing contamination. Staff of the appropriate regulatory authorities within each state must make decisions to manage these spills on a site-by-site basis. Do they require active cleanup? How much cleanup is necessary? What is the most appropriate technology? What performance should be expected from the available technology? If the state regulators provide good answers to these questions on a site-by-site basis, the money will be well spent.

This handbook is concerned with remediation of MTBE in existing spills. There are a number of myths about MTBE that act as impediments to effective remediation and risk management for MTBE. These myths present MTBE as being qualitatively different from petroleum hydrocarbons. Many still think that benzene is biodegradable in ground water while MTBE is not, that risk management is appropriate for benzene and not appropriate for MTBE, and that drinking water can be treated to remove benzene but not to remove MTBE. These myths have made us reluctant to deal with existing MTBE contamination. As is documented in this *MTBE Remediation Handbook*, we have the technology to clean up MTBE in a rational and economic manner. In general, the same technologies that have worked well for fuel hydrocarbons will work for MTBE contamination.

Experience in the last decade has revealed that prompt source control is key to minimizing impacts and remediation costs. Experience has also shown that remediation of MTBE contamination is much like any other activity. Success depends on the selection of the appropriate technology, on a careful and adequate site characterization, and on sound engineering design and faithful implementation. The *MTBE Remediation Handbook* is a comprehensive and up-to-date compendium of our knowledge. It is my hope that the *MTBE Remediation Handbook* will improve the state of practice for remediation of MTBE.

John T. Wilson, Ph.D.
Ada, Oklahoma

Contents

SECTION I—MTBE HISTORY, PROPERTIES, OCCURRENCE, AND ASSESSMENT

APPENDICES

List of Figures

List of Tables

Section I

MTBE History, Properties, Occurrence, and Assessment

CHAPTER 1

Introduction

Ellen E. Moyer, Ph.D., P.E., Tighe & Bond, Inc.

This book provides comprehensive information on the management of releases of gasoline containing methyl tert butyl ether (MTBE). Our primary focus is MTBE, but other constituents are also discussed because a gasoline release includes several hundred compounds. Other important gasoline constituents include tert butyl alcohol (TBA) and the aromatic compounds benzene, toluene, ethylbenzene, and xylenes (BTEX), among others.

Section I (Chapters 1 through 7) presents information on MTBE history, properties, occurrence, and assessment that is essential to a good understanding of MTBE remediation. Section II (Chapters 8 through 16) covers the various remediation technologies that are applicable to MTBE. Section III (Chapters 17 through 31) presents case studies that demonstrate how these technologies have been applied effectively at a variety of sites.

HISTORY OF MTBE USE

MTBE has been added to gasoline for over 20 years to improve engine performance and make gasoline burn more completely, thereby enhancing air quality. The chronology of use of oxygenates in gasoline is summarized in Table 1-1 (Drogos, 2000). Oil company research on ether additives for gasoline began as early as the 1920s, and alcohols such as ethanol were first added to gasoline to boost octane in the 1930s. Ethanol-blended gasoline was sold in the midwestern United States (U.S.) in the 1930s and 1940s. MTBE was first commercially used in gasoline in Italy in 1973. MTBE was first added to gasoline in the U.S. in the mid-1970s, during the Arab oil embargo, to boost octane and as an extender. In 1979, MTBE, began to be added to gasoline in the U.S. to replace lead and prevent knocking, typically at amounts ranging from less than 1% to 8% by volume. Use of MTBE increased with time. In 1990, the U.S. Clean Air Act Amendments required the use of oxygenates in reformulated gasoline (RFG); the specific choice of oxygenates was left up to the gasoline refiners. This was followed in 1992 by implementation of the winter oxy-

Table 1-1.   **Chronology of Use of Oxygenates in U.S. Gasoline**

Date	Activity
1920s	Oil company research on ether additives for gasoline to boost octane
1930s	Alcohols added to gasoline to boost octane
1930/1940s	Ethanol-blended gasoline sold in the Midwest U.S.
1950s	American Petroleum Institute literature speaks of the applicability of using MTBE in gasoline
1969	TBA was blended into gasoline
1973	The first commercial use of MTBE was in Italy
mid-1970s	MTBE and other ethers were added to gasoline to enhance octane and as extenders during the Arab oil embargo
1978	Gasohol program began, adding 10% ethanol by volume in gasoline
1979	MTBE added to gasoline to boost octane to replace lead, typically at <1% by volume in regular gasoline and 2–8% in premium
1980s	Ether use increases as lead continues to be phased out
1988	Denver, Colorado, implements winter oxygenated fuel program
1989	Southwestern U.S. implements winter oxygenated fuel program
1990	U.S. Clean Air Act Amendments require use of oxygenates in reformulated gasoline
1992	U.S. implements winter oxygenated fuel program, requiring 2.7% oxygen by weight (equivalent to 15% MTBE or 7.3% ethanol by volume) in 40 U.S. metropolitan areas
1995	U.S. implements Reformulated Gasoline Phase I, requiring 2.0% oxygen by weight (equivalent to 11% MTBE or 5.4% ethanol by volume) year-round in 28 U.S. metropolitan areas
1996	California implements California Air Resources Board Phase 2, requiring 2.0% oxygenate by weight state-wide and year-round
2000	U.S. implements Reformulated Gasoline Phase II, still requiring 2.0% oxygen by weight

Source: Drogos, 2000.

genated fuel program, which required 2.7% oxygen by weight in gasoline (equivalent to 15% MTBE or 7.3% ethanol by volume) in 40 U.S. metropolitan areas. Shortly thereafter, in 1995, the U.S. implemented Reformulated Gasoline Phase I, requiring 2.0% oxygen by weight in gasoline year-round (equivalent to 11% MTBE or 5.4% ethanol by volume) in 28 U.S. metropolitan areas shown on Figure 1-1 (USEPA, 2002).

RFG usage accounts for approximately 32% of the total U.S. gasoline market, or approximately 380 million liters (100 million gallons) per day. MTBE is also commonly added to gasoline in non-RFG areas, at lower percentages. Premium gasoline typically contains more MTBE than regular grades of gasoline. In total, MTBE accounts for approximately 4% by volume of all gasoline in the U.S. (OFA, 2002).

MTBE use in Europe has grown in the last decade and is expected to remain stable in the near future (EFOA, 2002). MTBE is used worldwide, and annual world consumption of MTBE is approximately 24 billion liters (6.4

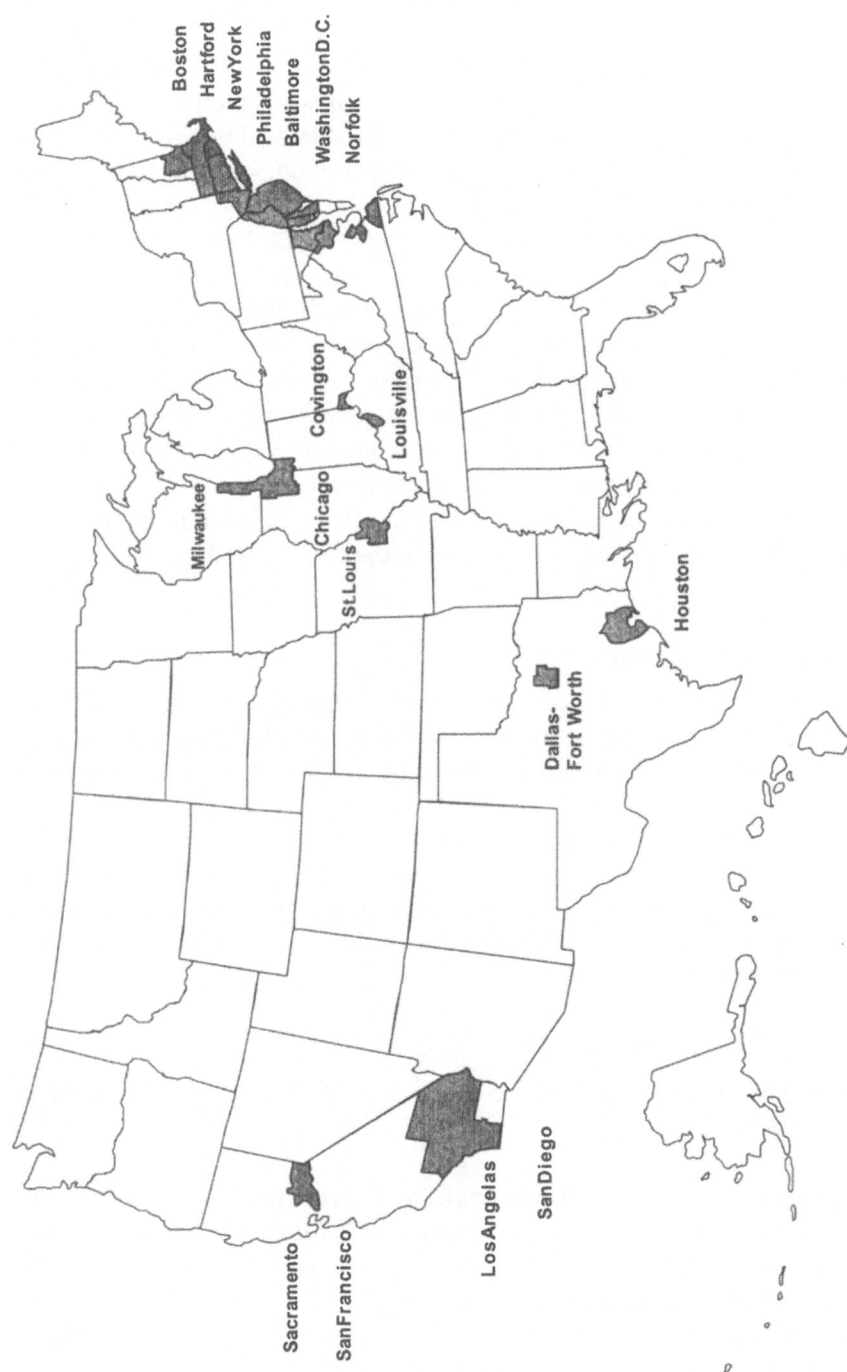

Figure 1-1. Federal Reformulated Gasoline Areas (USEPA, 2002)

billion gallons). The U.S. uses approximately 60% of the MTBE produced; western Europe uses approximately 15%; and the remaining 25% is used in the rest of the world.

Largely because gasoline and other petroleum products are transported in the same pipelines and trucks, MTBE is also present in fuel oil (Robbins *et al.*, 1999), diesel, kerosene, and other middle distillates (Hinchey *et al.*, 2001). MTBE also accumulates in motor oil as it recirculates through the engine (Baker *et al.*, 2000).

There have been other minor uses of MTBE. MTBE has been used to dissolve gallstones in humans, by injecting 2- to 6-cubic centimeter doses of MTBE directly into the gallstones. Hellstern *et al.* (1998) reported on 803 patients in 21 European hospitals that received this treatment and were followed for five years after treatment. There were no toxic injuries reported. MTBE has also been used in analytical laboratories as an extractant; however, other chemicals are preferred, and its use in laboratories is not widespread. MTBE has been used to a minor extent in chemical synthesis, for example, it may be hydrolyzed to make TBA. Other than these small quantities, almost all of the MTBE produced in or imported to the U.S. is used as a gasoline additive.

TERT BUTYL ALCOHOL

TBA is often encountered in association with MTBE. The presence of TBA at a gasoline release site may be due to one more of the following: 1) TBA was blended into the gasoline as a fuel oxygenate in its own right; 2) some unreacted TBA was present in the MTBE that was added to the gasoline (which is possible when MTBE is manufactured from TBA and methanol); and/or 3) MTBE biodegradation or chemical oxidation produced TBA as an intermediate compound. In addition, under certain circumstances, TBA can be generated from MTBE during sample preservation or laboratory analysis (see Chapter 6 for a full discussion).

GASOLINE RELEASES

Environmental releases of gasoline occur in a variety of ways, which are briefly described below.

Underground Storage Tank (UST) Leaks and Overfills.   UST systems have been used to store gasoline and other products for nearly a century. After World War II, the number of USTs in the U.S. increased dramatically with the development of the interstate highway system.

By 1988, there were approximately 2 million UST systems at 700,000 facilities, and many of them were leaking. In 1988, the U.S. Environmental Protection Agency (USEPA) promulgated UST regulations requiring UST removal or upgrade with leak detection and spill and overfill protection systems within ten years. By 1997, many USTs had been removed, reducing the number to 1.2 million USTs at 415,000 facilities. Of those facilities, 195,000 were gas stations and 220,000 were marinas, airports, hospitals, municipalities, and other facilities. By the 1998 deadline, 60% of USTs were estimated to be in compliance with the federal UST regulations.

A recent U. S. Government Accounting Office report (USGAO, 2001) stated that, as of September 2000, 89% of regulated USTs (616,865 USTs out of 693,107) had received the required upgrades, but that 29% of regulated USTs (or 201,001 USTs) were not being operated or maintained properly. The report states that in fiscal year 2000, USEPA and the states confirmed a total of more than 14,500 leaks or releases from USTs subject to federal regulation, although they were uncertain whether the releases occurred before or after the USTs had been upgraded.

Even when UST systems are operated and maintained correctly, the federal UST regulations only require that UST leak detection systems be capable of detecting 0.8 liters (0.2 gallons) per hour or more. This detection limit may seem low, but it is equivalent to 6,632 liters (1,752 gallons) per year. This is a large volume of gasoline as far as environmental contamination is concerned, and this amount could conceivably be released year after year by a compliant UST system. For these reasons, monitoring of groundwater on a regular basis at locations near, and downgradient from, all the components of the UST system, including the tank itself, piping and dispensers, is recommended (Cook *et al.*, 2001). More and more data are becoming available showing that the most leakage prone element of the UST system is the piping. Even a state-of-the-art system like that shown in Figure 1-2 should have environmental monitoring to check for undetected releases of gasoline.

Spills. Spills can occur in a variety of situations, including the following:

- During UST filling operations;
- During vehicle gas tank filling operations, including "drive-aways," in which people accidentally drive away while a gasoline dispenser nozzle is still in the vehicle's gas tank;
- During vehicle repairs and maintenance of lawn mowers, snowmobiles, and other equipment; and
- During vehicle accidents.

Use in Watercraft. Many watercraft engines, such as those for boats and jet skis, have two-stroke engines that release a large percentage of the gasoline, unburned, directly into surface water during operation. Drinking water wells, which ultimately draw on impacted surface water, can become contaminated with gasoline constituents (Baehr and Reilly, 2001). Efforts are cur-

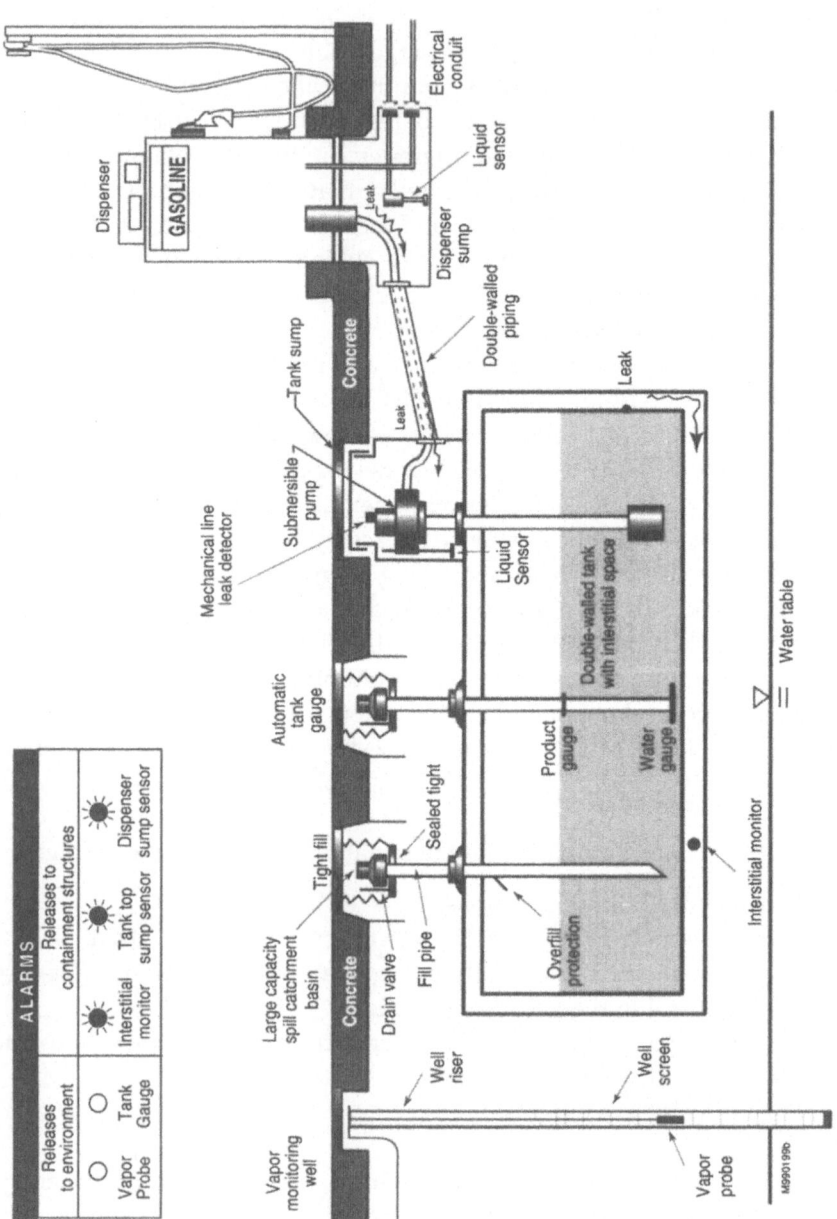

Figure 1-2.   Schematic of Example Double-Walled UST System (Vapor Recovery Not Shown)

rently underway in some states such as California, Massachusetts, and New Hampshire to replace two-cycle engines with less polluting four-cycle engines, and the USEPA is mandating the phase-out of two-stroke engines in 2006.

Volatilization.   Because of high vapor pressures, MTBE and other gasoline constituents are released to the air under a wide variety of situations. Volatilization occurs when gasoline is in contact with air, *e.g.*, during fueling of vehicles and spills.

SUMMARY

MTBE has been used to improve engine performance and enhance air quality in the U.S. and Europe for over 20 years, and it is currently used around the world. Releases of MTBE-blended gasoline to the environment occur in a variety of ways, but releases associated with UST systems are probably of greatest significance. Comprehensive and frequent environmental monitoring adjacent to UST systems is recommended to detect releases early while they are easier to remediate, and also to discover small releases that might otherwise go undetected.

REFERENCES

Baehr, A.L. and Reilly, T.J. 2001. *Water Quality and Occurrence of Methyl Tert-Butyl Ether (MTBE) and Other Fuel-Related Compounds in Lakes and Ground Water at Lakeside Communities in Sussex and Morris Counties, New Jersey, 1998–1999.* Water Resources Investigations Report 01-4149, 86 pp.

Baker, R.J., Best, E.W., and Baehr, A.L. 2000. Used motor oil as a source of methyl tert-butyl ether (MTBE) and gasoline hydrocarbons in ground water. Abstract presented at the *16th Annual International Conference on Contaminated Soils, Sediment and Water,* University of Massachusetts at Amherst. October 16–19, 2000.

Cook, G., David, L., Tulloch, C., and Headlee, C. 2001. Evaluation of MTBE occurrence at operating gasoline UST facilities in Santa Clara County, California, Preliminary findings. In: *Proceedings of the API/NGWA Petroleum Hydrocarbons and Organic Chemicals in Groundwater Conference,* Houston, Texas. November 14–16, 2001. pp. 17–31.

Drogos, D.L. 2000. MTBE v. other oxygenates. In: *Proceedings of Mealey's MTBE Conference,* Marina Del Rey, California. May 11–12, 2000.

EFOA (The European Fuel Oxygenates Association). 2002. *MTBE and Gasoline.* www.efoa.org. Accessed July 2002.

Hellstern, A., Leuschner, U., Benjaminov, A., Ackermann, H., Heine, T., Festi, D., Orsini, M., Roda, E., Northfield, T.C., Jazrawi, R., Kurtz, W., Schmeck-Lindenau, H.J., Stumpf, J., Eidsvoll, B., Aadland, E., Lux, G., Boehnke, E.,

Wurbs, D., Delhaye, M., Cremer, M., Sinn, I., Horing, E., Gaisberg, U., Neubrand, M., Sauerbruch, T., Salamon, V., Swobodnik, W., Sanden, H., Schmitt, W., Kaser, T., Schomerus, H., Wechsler, J.G., Janowitz, P., Lohmann, J., Porst, H., Attili, A.F., Bartels, E., Arnold, W., Strohm, W.D., and Paul, F. 1998. Dissolution of gallbladder stones with methyl tert-butyl ether and stone recurrence: a European survey. *Digestive Diseases and Sciences.* 43(5), 911–920.

Hinchley, E.J., Fox, J.S., and Tayeh, H.C. 2001. Evaluation of MTBE in middle distillate petroleum products in the northeastern United States. Abstract presented at the *17th Annual International Conference on Contaminated Soils, Sediment and Water,* University of Massachusetts at Amherst. October 22–25, 2001.

OFA (Oxygenated Fuels Association). 2002. *MTBE's Role in Reformulated Gasoline.* www.cleanfuels.net. Accessed July 2002.

Robbins, G.A., Henebry, B.J., Schmitt, B.M., Bartolomeo, F.B., Green, A., and Zack, P. 1999. Evidence for MTBE in heating oil. *Ground Water Monitoring and Remediation.* Spring 1999, 65–69.

USEPA (U.S. Environmental Protection Agency). 2002. *Federal Reformulated Gasoline Areas.* www.epa.gov/otaq/rfgmap.jpg. Accessed July 2002.

USGAO (U.S. General Accounting Office). 2001. *Environmental Protection—Improved Inspections and Enforcement Would Better Ensure the Safety of Underground Storage Tanks.* May 2001, Document Number GAO-01-464, 39 pp.

CHAPTER 2

Chemical and Physical Properties

Ellen E. Moyer, Ph.D., P.E., Tighe & Bond, Inc.

INTRODUCTION

It is critical to understand the chemical and physical properties of MTBE and other gasoline constituents because they determine how the chemicals behave in the subsurface environment. These properties also determine which physical and chemical technologies are suitable for remediation. This chapter reviews the chemical and physical properties of a number of gasoline constituents. How these properties affect fate and transport will be discussed in more detail in Chapter 3. The subject of biodegradability is covered in Chapters 3, 12, and 13.

Table 2-1 presents representative values for several important chemical and physical properties of gasoline constituents and related compounds of potential interest. In many cases, different values reported in the literature have been averaged to give single representative values. Compounds include: MTBE, TBA, tert butyl formate (TBF, an intermediate of MTBE's chemical or biological oxidation), ethyl tert butyl ether (ETBE), di-isopropyl ether (DIPE), tert amyl methyl ether (TAME), ethanol, and methanol, all of which are other gasoline oxygenates; BTEX compounds; and 2,2,4-trimethylpentane. The latter is a prevalent naturally-occurring gasoline component that is also a major constituent of isooctane, which some have proposed to add to gasoline to reduce its volatility. The molecular structures of these compounds are shown on Figure 2-1.

Key chemical and physical properties are discussed below. Most of the properties are a function of temperature, and unless otherwise specified, 25°C (77°F) is assumed.

BOILING TEMPERATURE

The boiling temperature is simply the temperature at which a pure substance will boil, changing phase from liquid to vapor. As such, it is an indication of volatility, and an important consideration when applying thermal treatment

Table 2-1.   Representative Values of Chemical and Physical Properties of Various Fuel Components and Degradation Products (at 25 °C unless Otherwise Specified)

Compound	CAS Number	Molecular Weight (gram/mole)	Boiling Temp. (°C)	Specific Gravity (dimensionless)	Water Solubility (mg/l)	Vapor Pressure (mm Hg)	Log K_{ow}	Log K_{oc}	Henry's Law Constant (atm-m3/gram-mole)	Henry's Law Constant (dimensionless)
Methyl tert butyl ether	1634-04-4	88.15	54	0.74	50,000	251	1.2	1.1	1.5E-3	5.5E-2
Tert butyl alcohol	75-65-0	74.12	83	0.79	Infinite	41	0.35	1.6**	1.2E-5	4.9E-4
Tert butyl formate	762-75-4	102.13	82	0.89	~40,000	81*		1.1	2.7E-4	1.1E-2
Ethyl tert butyl ether	637-92-3	102.18	67	0.73	~26,000	152	1.7	1.6	2.7E-3	1.1E-1
Diisopropyl ether	108-20-3	102.18	91	0.74	~9,000*	150*	1.5	1.6	6.9E-3	2.8E-1
Tert amyl methyl ether	994-05-8	102.18	86	0.77	~20,000	68		1.7	1.3E-3	5.2E-2
Ethanol	64-17-5	46.07	79	0.79	Infinite	53	-0.24	0.71	5.9E-6	2.4E-4
Methanol	67-56-1	32.04	65	0.80	Infinite	122	-0.75	0.68	4.4E-6	1.1E-4
Benzene	71-43-2	78.11	80	0.88	1,780	86	2.0	1.9	5.4E-3	2.2E-1
Toluene	108-88-3	92.13	111	0.87	535	28	2.6	1.9	5.9E-3	2.4E-1
Ethylbenzene	100-41-4	106.16	136	0.87	161	10	3.2	2.7	8.4E-3	3.5E-1
m-Xylene	108-38-3	106.16	139	0.88	146	8.3	3.2	2.3	7.7E-3	3.1E-1
o-Xylene	95-47-6	106.16	144	0.88	175	6.6	3.0	1.8	5.1E-3	2.1E-1
p-Xylene	106-42-3	106.17	138	0.86	156	8.7	3.2	2.4	7.7E-3	3.1E-1
2,2,4-Trimethylpentane	540-84-1	114.23	99	0.69	2.4	49	4.1	4.6	3.3E+2	1.3E+4

Notes: * at 20 °C

CAS = Chemical Abstracts Service; mg/l = milligrams per liter; mm Hg = millimeters of mercury; atm−m³ = atmosphere−cubic meter.
Sources for 2,2,4-trimethylpentane: Marchetti, Alfredo A. 2001. Environmental transport and fate of alkylates. Lawrence Livermore National Laboratory Energy and Environmental Directorate. Workshop on the Increased Use of Ethanol and Alkylates in Automotive Fuels in California. Oakland, California, April 10–11, 2001. ABB Environmental Services, Inc. 1990. Compilation of data on the composition, physical characteristics, and water solubility of fuel products. Prepared for the Massachusetts Department of Environmental Protection. Job No. 6042-04. pp. 1–3. Heath, J.S., Koblis, K., Sager, S.L., and Day, C. 1993. Risk assessment for total petroleum hydrocarbons. Calabrese, E.J., and Kostecki, P.T. (eds.), *Hydrocarbon Contaminated Soils—Volume III*. Lewis Publishers, Chelsea, MI, pp. 267–301.
Source for other compounds: National Science and Technology Council Committee on Environmental and Natural Resources. 1997. *Interagency Assessment of Oxygenated Fuels.* pp. 2-51 and 2-52.
**This appears to be the only available published value. It is higher than expected based on the K_{ow} for TBA (0.35) and the relationship between log K_{ow} and log K_{oc} for other compounds. A value of log K_{oc} near 0 would be expected (John T. Wilson, personal communication, December 4, 2002).

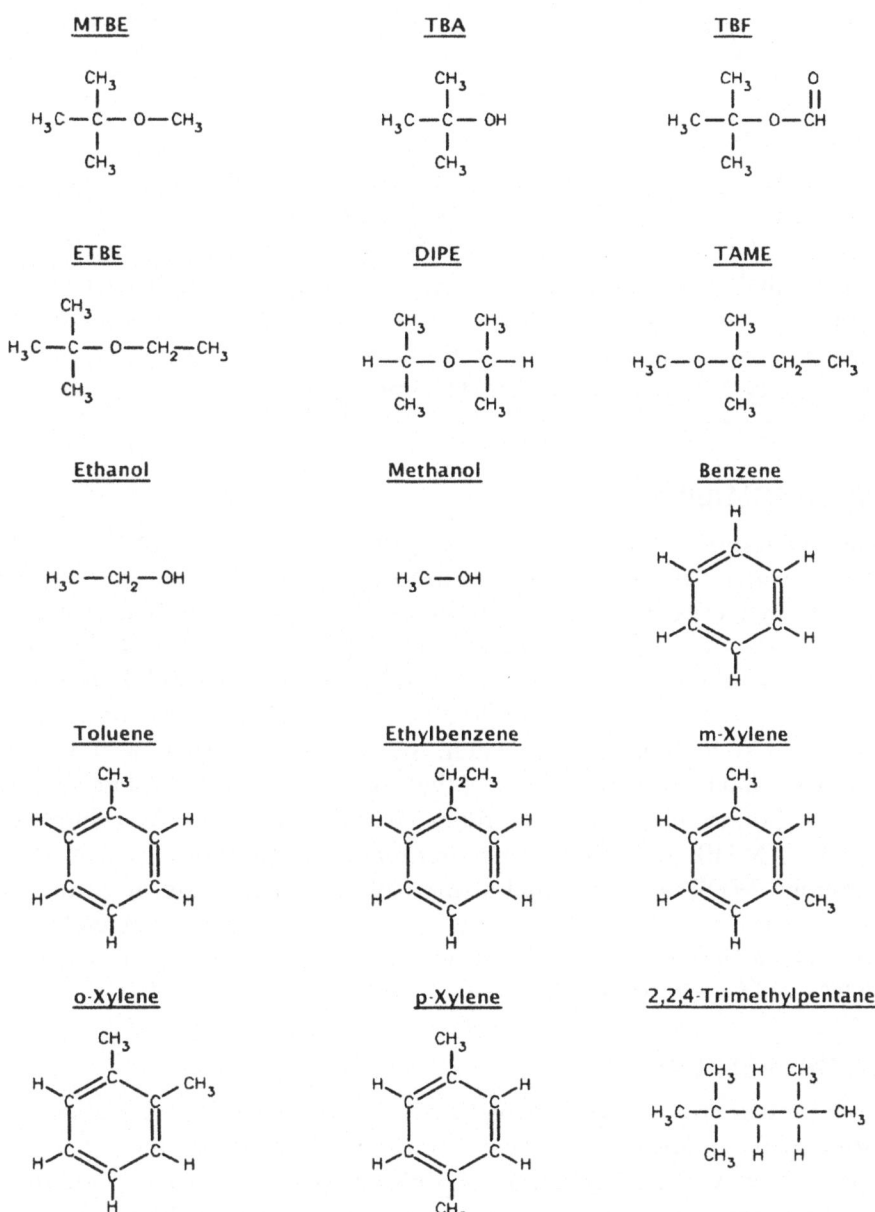

Figure 2-1.   Molecular Structure of Selected Gasoline Constituents

technologies such as thermal desorption or steam stripping. The boiling temperature of water is 100°C (212°F) at one atmosphere absolute pressure. The values in Table 2-1 indicate that all the constituents except toluene, ethylbenzene, and xylenes would volatilize before soil moisture during *in situ* thermal desorption (ISTD) of soil, for example. MTBE has the lowest boiling temperature of the compounds listed (54°C), making it very amenable to thermal treatment technologies.

SPECIFIC GRAVITY

Specific gravity compares the weight of a liquid to that of water. A specific gravity less than one indicates that the liquid will float on a water surface (ground water or surface water), whereas a specific gravity greater than one indicates that the liquid will sink in the water body. The specific gravities of gasoline constituents are less than one, so gasoline in contact with surface water or ground water will float on the water surface. One consequence of this property is that product removal technologies typically need to focus on extraction near the water table.

WATER SOLUBILITY

Water solubility is the extent to which a compound dissolves into water (ground water, surface water, or soil moisture). Solubility increases as temperature increases. The water solubilities of the compounds in Table 2-1 range widely, from 2.4 milligrams per liter (mg/l) for 2,2,4-trimethylpentane, to infinite solubility for TBA, ethanol, and methanol. The solubilities shown in Table 2-1 are for pure compounds at equilibrium. When a mixture such as gasoline is in equilibrium with water, the solubility of each gasoline compound is equal to the mole fraction of that compound in the gasoline times its pure compound solubility, in accordance with Raoult's Law. As a simple example, the MTBE concentration in water that is in equilibrium with gasoline containing MTBE at 1% by moles (mol) (which in the case of gasoline is roughly equivalent to 1% by volume or 1% by mass) would be 500 mg/l, whereas the MTBE concentration in water that is in equilibrium with pure MTBE would be 50,000 mg/l.

VAPOR PRESSURE

Vapor pressure is a measure of volatility. It is the pressure of a vapor in equilibrium with a pure liquid at a given temperature: the higher the vapor pressure, the more volatile the compound. Vapor pressure is relevant to situations in which compounds in contaminated soil volatilize to soil gas (air in the spaces between the soil particles), for example near a leaking UST (LUST), or in which gasoline floating on surface water or ground water volatilizes to overlying air. The Ideal Gas Law and molecular weights can be used to convert vapor pressures to vapor concentrations (for example, milligrams per cubic meter [mg/m^3]). As is the case for water solubility, vapor pressures increase with increasing temperatures, and vapor pressures of constituents in air over a liquid mixture are equal to the mole fractions of those compounds in the liquid mixture times their pure compound vapor pressures. Vapor pressures for the Table 2-1 compounds, all of which are considered volatile, range from 8 millimeters of mercury (mm Hg) (0.3 inches of mercury [in Hg]) for xylenes to 251 mm Hg (10 in Hg) for MTBE. The high vapor pressure of MTBE indicates that soil vapor extraction (SVE) is a very suitable remediation technology for this compound.

VAPOR DENSITY

Vapor density is an indication of whether constituents, once they have volatilized, will sink or rise as vapors in air. Very few compounds have vapor densities less than air. The vapor densities of all gasoline constituents are greater than the density of air, one gram per liter at 25°C (77°F). Therefore, in the absence of pressure or temperature gradients, gasoline vapors will tend to sink and accumulate in low areas such as basements or utility trenches.

ADSORPTION

Adsorption is a measure of the tendency of a compound to adhere to soil. For example, as ground water containing a compound flows through the subsurface, adsorption retards the advance of that compound as it adheres to soil surfaces. The tendency of a compound to adsorb also affects the suitability of granular activated carbon (GAC) or resin adsorption as a treatment technology for that compound. Furthermore, adsorption affects the ability to extract a compound from saturated or unsaturated soil; compounds with a lower tendency to adsorb are more readily extracted. Adsorption decreases with increasing temperature.

Adsorption is quantified in a number of ways, one of which is a soil adsorption coefficient, K_d. This is an equilibrium partitioning coefficient, which is simply the adsorbed concentration in soil, C_s, divided by the concentration in water in equilibrium with the soil, C_w. The adsorption of a compound to soil has been found to be proportional to the amount of organic carbon in the soil, therefore, the soil adsorption coefficient is often expressed as the product of the fraction of organic carbon content in soil, f_{oc}, times an organic carbon partitioning coefficient that is specific to the compound, K_{oc}:

$$K_d = C_s/C_w = f_{oc} K_{oc} \qquad (2\text{-}1)$$

Log K_{oc} values in Table 2-1 range from 0.68 for methanol to 4.6 for 2,2,4-trimethylpentane. The value for MTBE, 1.1, is relatively low, indicating that the movement of MTBE in ground water is not retarded to a significant extent, that MTBE does not adhere as strongly to GAC as some other constituents, and that MTBE is readily extracted from soil, for example through SVE or ground water extraction.

The octanol water partitioning coefficient, K_{ow}, is another way adsorption is quantified, on the theory that octanol is a hydrophobic compound that is generally similar to soil organic carbon. As octanol is one specific compound (as opposed to soil organic carbon), K_{ow} values are more reproducible and easier to determine in the laboratory.

HENRY'S LAW CONSTANT

The Henry's Law constant also reflects a compound's volatility; however, it measures volatilization of the compound from the dissolved phase to air rather than from the pure phase to air. This is relevant to situations such as ground water or surface water with dissolved constituents volatilizing to overlying soil gas or air, or air stripping applications. It is another equilibrium partitioning coefficient, equal to the concentration in air divided by the concentration in water. Henry's Law constant generally increases with increasing temperature. However, this is not always the case because Henry's Law constant quantifies the competing effects of vapor pressure and solubility. It can be estimated for a compound from its vapor pressure, molecular weight (MW), and solubility (Chidgopkar, 1996).

Dimensionless Henry's Law constants in Table 2-1 range over eight orders of magnitude from 1.1×10^{-4} for methanol to 1.3×10^4 for 2,2,4-trimethylpentane (dimensional values range similarly). This indicates that methanol tends to stay dissolved in water and does not readily volatilize to air, and that 2,2,4-trimethylpentane volatilizes from water to air to a great extent. The Henry's Law constant for MTBE is lower than those for BTEX compounds, indicating that it has a weaker tendency to volatilize from water to air. One implication of this is that remediation technologies involving transfer of constituents from the aqueous to the vapor phase, such as air sparging and air stripping, will generally require higher air-to-water ratios for MTBE removal than for BTEX removal.

SUMMARY

Figure 2-2 illustrates how some of the important properties discussed in this chapter dictate the fate and transport of MTBE and benzene at a field site. Benzene is selected for the compound to compare with MTBE because it often drives remediation activities due to its toxicity and carcinogenicity, and thus low required cleanup levels. The lengths of the arrows in Figure 2-2 are proportional to the pure compound values for the two compounds. The arrows are not intended to indicate directionality, just magnitude; the orientation of the arrows does not indicate that vapors will rise or that dissolved constituents will sink.

A LUST is shown with associated residual soil contamination. MTBE will volatilize from product to air to a greater extent than benzene (vapor pressure). Both MTBE and benzene vapors will sink in the building basement (vapor density). Once the released gasoline drains through the soil under the

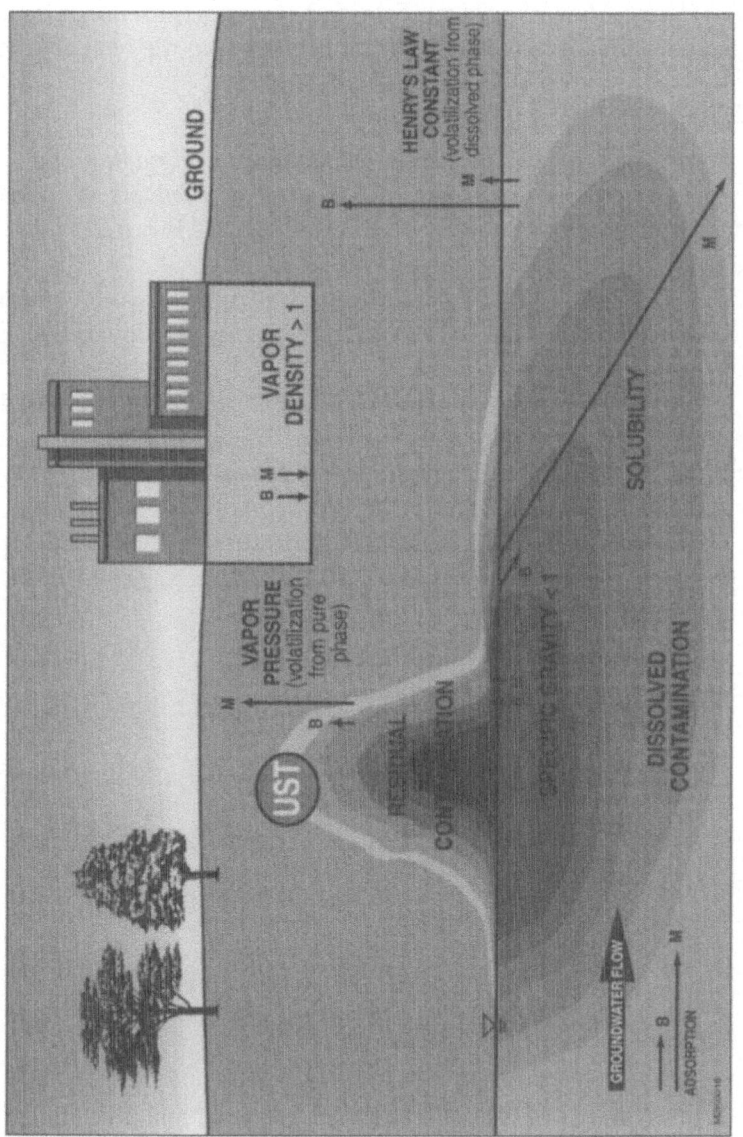

Figure 2-2. Behavior of MTBE and Benzene

force of gravity and reaches the water table, the gasoline (including both benzene and MTBE) will float (specific gravity). More MTBE than benzene will dissolve into the ground water (solubility). MTBE will travel more readily with ground water flow than benzene, whose progress is retarded (adsorption). Benzene will volatilize from the dissolved phase to a greater extent than MTBE (Henry's Law constant). Fate and transport of constituents following a gasoline release are discussed in more detail in the next chapter.

REFERENCES

ABB Environmental Services, Inc. 1990. *Compilation of Data on the Composition, Physical Characteristics, and Water Solubility of Fuel Products*. Prepared for the Massachusetts Department of Environmental Protection. Job No. 6042-04. pp. 1–3.

Chidgopkar, V.R. 1996. Applying Henry's Law to groundwater treatment. *Pollution Engineering*. March 1996. (Available under "back issues" at www.pollutionengineering.com.)

Marchetti, A.A. 2001. Environmental transport and fate of alkylates. Lawrence Livermore National Laboratory Energy and Environmental Directorate. *Workshop on the Increased Use of Ethanol and Alkylates in Automotive Fuels in California*, Oakland, California. April 10–11, 2001.

National Science and Technology Council Committee on Environment and Natural Resources. 1997. *Interagency Assessment of Oxygenated Fuels*. pp. 2-51 and 2-52.

CHAPTER 3

Fate and Transport of MTBE and Other Gasoline Components

John T. Wilson, Ph.D., USEPA National Risk Management Laboratory, Office of Research and Development

Most gasoline releases occur above the water table. As the gasoline drains through the unsaturated zone under the influence of gravity, it leaves behind a residual amount of gasoline held by capillary attraction. Gasoline that drains to the water table will collect in a capillary fringe that is located at the interface between the ground water in the aquifer and the unsaturated zone above. Gasoline is often inaccurately depicted as floating on the water as a light non-aqueous phase liquid (LNAPL). The capillary fringe actually imbibes the gasoline, much as a sponge will soak up water. As a consequence, the gasoline spreads laterally in the capillary fringe much further than it would spread by gravity flow. As the water table moves up and down over time, the gasoline is redistributed in a smear zone. As the vertical position of the water table changes from one time to the next, the position of the smear zone with respect to ground water will change. At any one time most of the smeared gasoline can be above the water table, below the water table, or equally divided between the aquifer and the unsaturated zone above.

The location of the residual gasoline has important consequences for the ability of the MTBE and other constituents in the gasoline to impact ground water. When residual gasoline is present in the unsaturated zone, MTBE and other volatile organic compounds (VOCs) will volatilize from the gasoline and move away from the gasoline as a vapor in soil gas. The vapors of MTBE have strong affinity for water and tend to dissolve in the pore water in the unsaturated zone. Recharge of precipitation can flush the pore water containing MTBE down into the aquifer. When the residual gasoline is below the water table, MTBE can partition directly into the ground water. Transfer of MTBE to ground water is aided by precipitation moving through the gasoline in the smear zone as the precipitation recharges the aquifer.

There are a number of mechanisms that attenuate the concentrations of MTBE and other gasoline constituents in ground water. Once dissolved in ground water, gasoline constituents move along with the natural flow of the ground water. As ground water moves along the flow path, the concentrations of gasoline constituents will be attenuated through dispersion and dilution. However, the contribution of dispersion and dilution is often more apparent than real. For example, plumes can move below the screens of the monitoring wells; this can give a false impression that concentrations are attenuating when in fact they are not. Plumes can change flow direction and sweep contaminated ground water away from a monitoring well that was thought to be downgradient of the plume. Once dissolved in ground water, concentrations of MTBE and other gasoline constituents can also be attenuated through adsorption to aquifer solids and through aerobic and anaerobic biodegradation.

The distribution of MTBE and other gasoline constituents in ground water is controlled by hydraulics and natural attenuation processes acting in ground water. However, the persistence of contamination is usually controlled by the rate of transfer of constituents from the residual gasoline to the moving ground water. Plumes are persistent over time because they are continually regenerated from the residual gasoline in the source area.

These processes and interactions will be discussed in detail in following sections of this chapter.

TRANSPORT AND FATE OF VAPORS OF MTBE IN THE UNSATURATED ZONE

As discussed in Chapter 2, MTBE is volatile; the vapor pressure of pure MTBE is approximately 250 mm Hg (10 in Hg) at 25°C (77°F), equivalent to a saturated vapor concentration of 1.2 grams per liter (Lahvis and Rehmann, 1999). The concentration of MTBE vapor expected in soil gas in equilibrium with RFG (11% MTBE by volume) is 120 mg/l. The dimensionless Henry's constant for MTBE is 0.03, which corresponds to an MTBE concentration of 4,000 mg/l in water in contact with the saturated vapor concentration from RFG.

Figure 3-1 presents a one-dimensional model of MTBE diffusion through soil gas. The model approximates the worst case, a large spill of gasoline under pavement. Under these worst case assumptions for extensive diffusion as a vapor, the concentrations of MTBE in pore water that exceed 10 micrograms per liter (µg/l) extend as much as 11 meters (35 feet) from the edge of the residual gasoline. Ninety percent of the contaminant mass is contained within the first 5 meters (15 feet) from the edge of the residual gasoline. Because more than 97% of the MTBE in the unsaturated zone is dissolved in pore water, flushing of the pore water during recharge by precipitation has a strong influence on the redistribution of MTBE as a vapor. Lahvis and Rehmann (1999) simulated MTBE redistribution from a source in the unsaturated zone with recharge of 20 centimeters (8 inches) per year. The steady

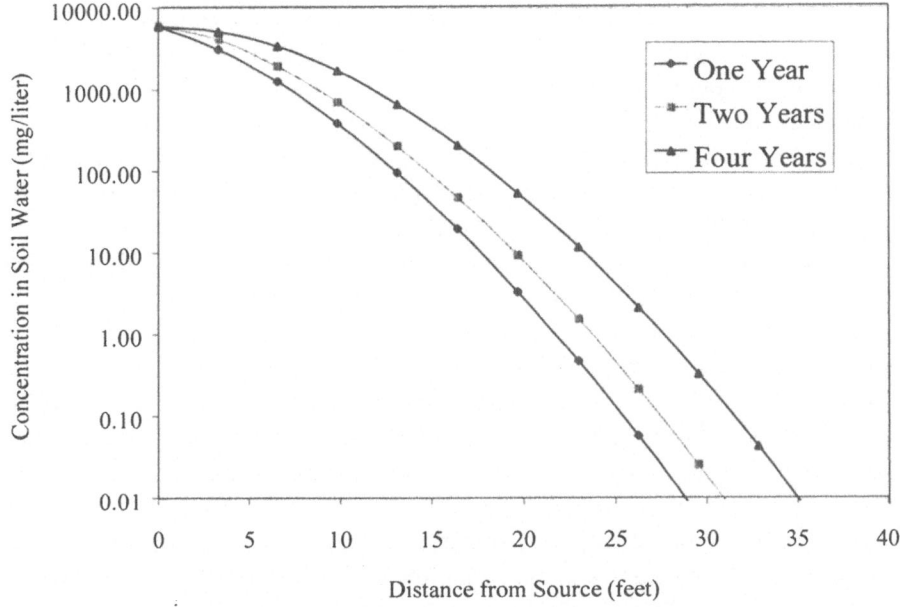

Figure 3-1.   Diffusion of MTBE from Reformulated Gasoline into Soil
Gas over Time. Presented are the results of a simple one-dimensional mathematical model. Concentrations are concentrations of MTBE in soil pore water in equilibrium with
MTBE vapors in soil gas.The model assumes no biodegradation of MTBE. There is no correction for dilution that
would be expected with radial dispersion of vapors from a
point source. The pore water is assumed to be immobile;
the pore water containing MTBE is not displaced and replaced with clean water from precipitation. The model assumes that the residual gasoline in the unsaturated zone
does not weather over time. The calculations assume that
the air-filled porosity is 0.2, the water-filled porosity is 0.2,
that the Henry's constant is 0.024, and that the diffusion coefficient of MTBE in air is 0.0792 cm^2/sec. Diffusion of MTBE
in the pore water is not considered in the calculations.

state concentration of MTBE was reduced by a factor of three at a location
only 0.6 meters (2 feet) from the edge of the residual gasoline.

PARTITIONING OF MTBE FROM GASOLINE
DIRECTLY TO GROUND WATER

As noted in Chapter 2, using Raoult's Law, the equilibrium concentration of
MTBE in water in contact with gasoline can be estimated by multiplying the
solubility of pure MTBE in water by the mole fraction of MTBE in the gaso-

line. As the fraction of MTBE in the gasoline goes up or down, the concentration in the water goes up and down proportionately. The MW of MTBE is 88 daltons (grams per mole), and the average MW of gasoline is near 100 daltons (Squillace *et al.*, 1997). If the gasoline is 2% oxygen by weight (11% MTBE by volume), the mole fraction of MTBE in the gasoline is 0.125 and the expected concentration of MTBE in water at 25°C (77 °F), based on a pure compound aqueous solubility of 50,000 mg/l, is 6,250 mg/l.

Ground water from monitoring wells at MTBE spills very rarely produce concentrations of MTBE as high as would be expected from ground water in equilibrium with RFG. Mace and Choi (1998) presented a frequency distribution of the maximum concentration of MTBE at 609 gasoline spill sites in Texas that had at least one analysis for MTBE in monitoring wells at the site (see Figure 3-2). Of the 609 sites, 93% had MTBE concentrations greater than the detection limit, 85% had concentrations that exceeded 20 µg/l, 60% exceeded 730 µg/l, 50% exceeded 1,600 µg/l, and only about 5% exceeded 100,000 µg/l. Odencrantz (1998) presented a frequency distribution of the maximum MTBE concentrations at gasoline spill sites in Orange County, California in 1997 (see Figure 3-2). In reports from 304 sites, 77% of the sites exceeded 35 µg/l, 47% exceeded 1,000 µg/l, 21% exceeded 10,000 µg/l, and 6% exceeded 100,000 µg/l. The distributions in Southern California and Texas were similar. Kolhatkar *et al.* (2000) collected data from 74 gasoline stations in the eastern U.S., primarily from Pennsylvania. The frequency distribution of the maximum MTBE concentrations at the gasoline spill sites in the eastern U.S. is also presented in Figure 3-2. The median concentration was near 1,000 µg/l, and only 5% of the sites exceeded 100,000 µg/l.

Although water in contact with RFG should theoretically contain 5,000 mg/l or more of MTBE, the concentrations of MTBE in monitoring wells at gasoline spill sites rarely exceed 100 mg/l. Four major factors contribute to this discrepancy between expected and measured concentrations of MTBE in gasoline spills. First, the ground water temperature at many sites is often less than 25°C, and water solubility decreases with decreasing temperature.

The second factor relates to the MTBE content of the gasoline that was spilled. In many of the spills, the MTBE was probably added to the gasoline to enhance the octane rating of the fuel and not to meet an oxygenation criterion for air quality purposes. These gasoline types had a lower content of MTBE, often on the order of 10% of the MTBE required for RFG.

The third factor relates to the effect of partitioning of MTBE from residual gasoline. In the smear zone, gasoline occupies approximately 5% to 10% of the pore space, while the remainder is occupied by ground water. MTBE readily weathers out of the gasoline, depleting its mole fraction in the residual gasoline, which lowers the equilibrium concentration in water (Squillace *et al.*, 1997). In order to evaluate the effect of partitioning in the smear zone and to provide a reasonable upper boundary on concentrations of MTBE that

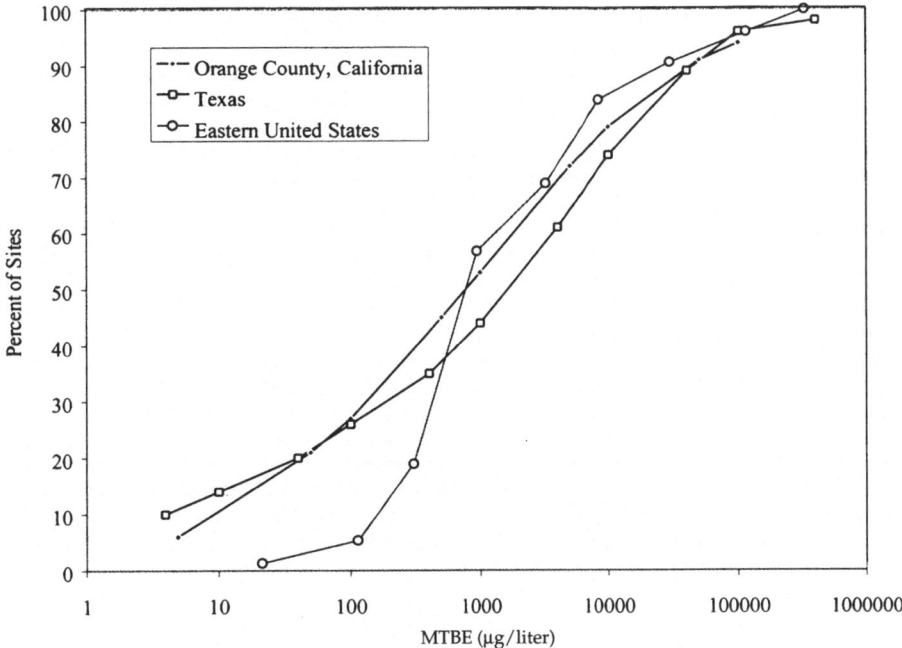

Figure 3-2. **Cumulative Frequency Distribution of the Maximum Concentration of MTBE Measured at Gasoline Service Stations Sites**

should be expected in monitoring wells at gasoline spill sites, the USEPA Office of Research and Development carried out an empirical partitioning experiment. Two samples of regular gasoline and two samples of premium gasoline were amended with MTBE to reach a concentration of 2% oxygen by weight and then equilibrated with water. The ratio of gasoline to ground water was 1 to 10. This represents the quantity of gasoline and water that would be present in a sand aquifer when the total porosity is 30% and the concentration of residual gasoline in a smear zone is 13,000 milligrams per kilogram (mg/kg) as total petroleum hydrocarbons (TPH).

Table 3-1 presents experimental data on the resulting concentrations of MTBE in water. The concentrations of MTBE in water varied from 2,780 mg/l to 4,490 mg/l, somewhat less than the concentration expected from Raoult's Law. Cline *et al.* (1991) developed an empirical relationship to predict the partitioning of fuel components between gasoline and water. Their equation was used to predict the concentration of MTBE in water in Table 3-1. The achieved concentrations were in reasonable agreement with the concentration predicted by Raoult's Law and by partitioning theory.

Table 3-1 also presents data for the resulting concentrations of ETBE,

Table 3-1. **Concentrations of Gasoline Oxygenates to Be Expected in Ground Water**

Gasoline Oxygenates	Water Solubility (mg/liter)	Equilibrium Concentration in Ground Water in Contact with Gasoline (mg/liter)					
		Regular Gasoline #1	Regular Gasoline #2	Premium Gasoline #1	Premium Gasoline #2	Average	Predicted
MTBE	48,000	4,490	4,380	3,410	2,780	3,770	5,240
ETBE	12,000	1,200	1,110	1,180	1,100	1,150	1,370
Ethanol	miscible	4,340	4,450	4,420	4,490	4,430	NA
Methanol	miscible	3,090	3,064	3,080	3,240	3,120	NA
TBA	miscible	6,230	4,400	5,870	5,100	5,400	NA
TAME	10,700	1,530	1,240	1,310	1,560	1,410	1,220
DIPE	6,500	902	812	855	1,060	908	745

Gasoline was extracted with ground water at a ratio of 1 to 10. This ratio simulates conditions in the smear zone of a gasoline spill when the concentration of residual gasoline in the smear zone is 13,000 mg/kg TPH. The concentrations of the oxygenates in the gasoline were sufficient to achieve 2% by volume oxygen in the gasoline.

TAME, DIPE, TBA, ethanol, and methanol when they were added to the gasoline at concentrations to achieve 2% oxygen in the gasoline. The achieved concentrations were near the concentrations achieved for MTBE. As was the case for MTBE, the concentrations of ETBE, TAME, and DIPE could be predicted by the equations of Cline *et al.* (1991).

The fourth factor relates to extent of weathering of the gasoline in the smear zone. The transfer of MTBE from gasoline in the smear zone to ground water is controlled by the three-dimensional relationship between the smear zone and the conductive intervals in the aquifer that actually carry the flow of ground water. In uniform sand aquifers, the smear zone is in direct contact with the aquifer. MTBE readily partitions to the ground water and is carried away as the ground water moves past. In this circumstance, the achieved concentration of MTBE in ground water can be predicted from equilibrium partitioning relationships. MTBE tends to weather from the smear zone uniformly, and the rate of weathering can be predicted from the number of pore volumes of ground water that move through the smear zone. In one-dimensional laboratory columns, an exchange of only 10 pore volumes of ground water past the residual gasoline reduced the concentration of MTBE from 1,000,000 µg/l to 10 µg/l (Rixey and Joshi, 1999, 2000).

In many other landscapes, the smear zone is separated in space from the effective portions of the aquifer that carry the bulk of ground water flow.

There are two important consequences of this separation. Because ground water does not have effective contact with MTBE in the smear zone, the achieved concentration of MTBE in ground water is lower than would be expected from the fraction of MTBE in the fuel that was spilled. Because transfer of MTBE from the smear zone to the moving ground water is not efficient, the smear zone does not weather as rapidly as would be expected, and the MTBE source area persists longer than would be expected.

Figure 3-3 presents data on the vertical distribution of residual fuel, MTBE dissolved in water, and hydraulic conductivity at a fuel spill in a flood plain environment at a site near Elizabeth City, North Carolina (Wilson *et al.*, 2000). The highest concentrations of residual fuel were centered around the average depth of the water table. The smear zone extends to a depth of 3 meters (10 feet). The depth interval from 3 to 5 meters (10 to 15 feet) has relatively low hydraulic conductivity, and the maximum conductivity in the aquifer occurs at a depth of 6 meters (20 feet). At this site, most of the ground water moves below the fuel spill, not through it.

The highest concentrations of MTBE in ground water were associated with the residual fuel, but a concentration gradient of MTBE extended from the smear zone of the fuel down into the aquifer. MTBE occupied only the top half of the aquifer. The concentration gradient of MTBE down into the aquifer is produced by a combination of diffusion and dispersion (see Rixey and Joshi, 2000) and can be regarded as the expression of a mass transfer limitation from the residual gasoline to the flow of ground water in the aquifer.

As a result of the mass transfer limitation, the concentration of MTBE is much less in the ground water monitoring points that are downgradient of the LNAPL. At the site near Elizabeth City, North Carolina, the maximum concentration in contact with the LNAPL is near 20,000 µg/l (Figure 3-3). In contrast, the maximum concentration at any depth in the aquifer downgradient of the LNAPL is 2,160 µg/l. The maximum concentration at any sampling location (averaged over a series of vertical water samples collected with temporary push points) is 1,300 µg/l. The maximum concentration during long-term monitoring at the most contaminated permanent monitoring well varies from 65 to 609 µg/l (Wilson *et al.*, 2000).

SEPARATION OF MTBE FROM BTEX ALONG A FLOW PATH

Frequently, MTBE will appear to move out ahead of BTEX in a plume. This is often explained based on differences in the extent of adsorption to the aquifer. MTBE adsorbs weakly to aquifer material while benzene adsorbs to a slightly greater extent, and the other alkylbenzenes adsorb to an even greater extent. These relationships are illustrated graphically in Figure 3-4.

Table 3-2 compares the retardation ratio of MTBE, benzene, toluene, and xylenes that is expected from the concentration of organic carbon in the

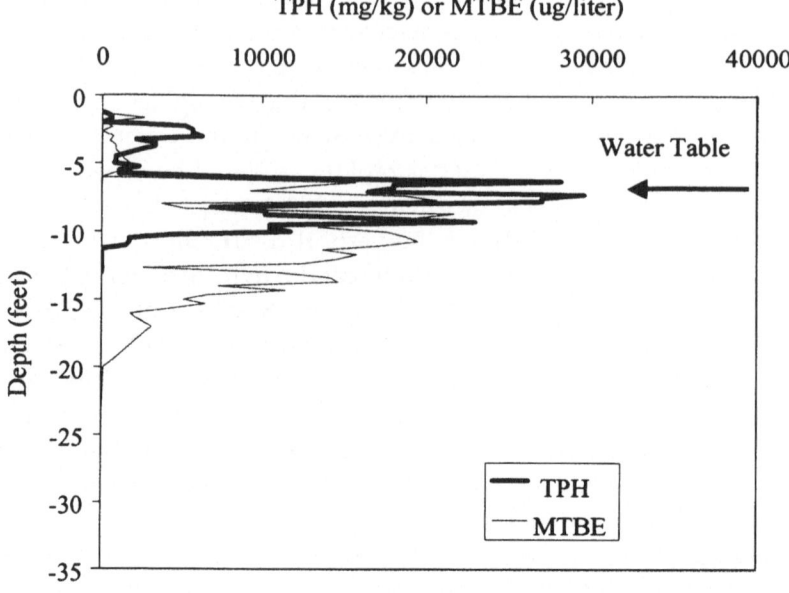

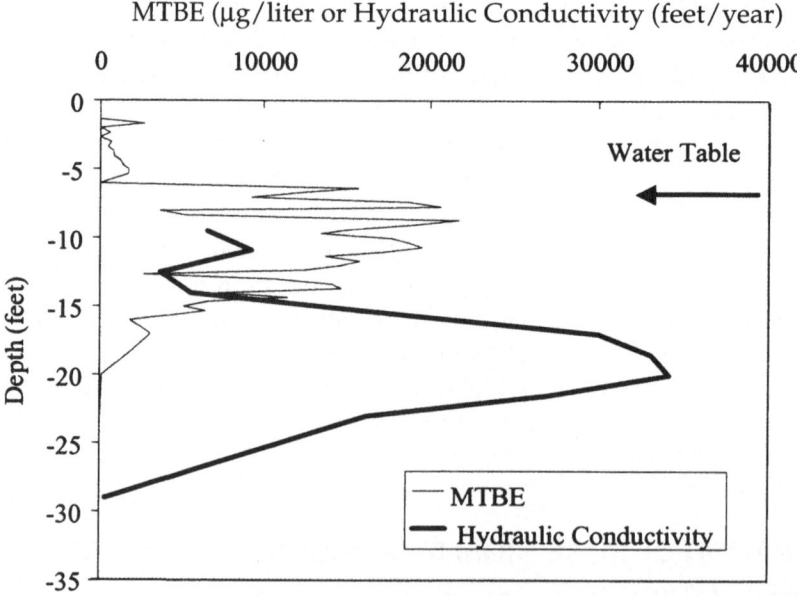

Figure 3-3. **Vertical Distribution of TPH in Sediment, MTBE in Ground Water, and Hydraulic Conductivity at a Fuel Spill in a Flood Plain Landscape. MTBE was extracted from core samples. The data presented are equivalent concentrations in round water, assuming no sorption to aquifer solids, and a porosity of 0.3. Values are corrected for partitioning of MTBE to residual fuel.**

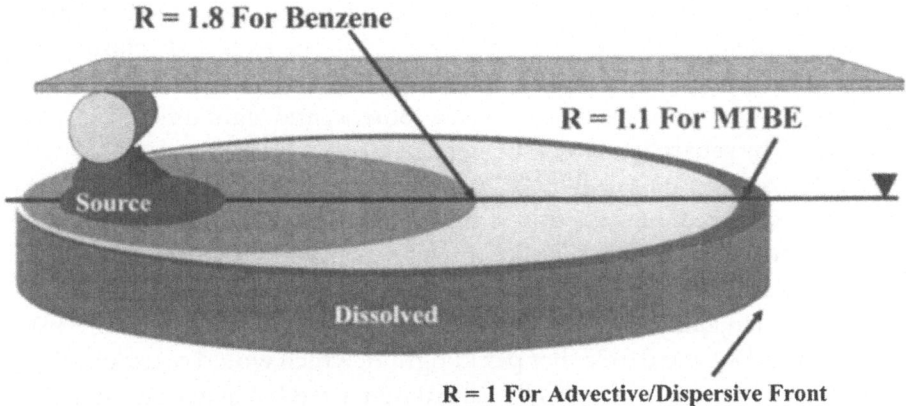

Ground Water Flow Direction ⟶

Figure 3-4.   **Effect of Adsorption on the Retardation of MTBE and Benzene with Respect to the Flow of Water (modified from Small and Weaver, 1999)**

Table 3-2.   **Retardation due to Adsorption That Is Expected from the Organic Matter Content of the Aquifer Solids (Table 3.4, page 145 of Wiedemeier et al., 1999)**

Compound	Fraction of Organic Carbon in Sediment			
	0.001% Low for Aquifers	0.01% Median for Aquifers	0.1% High for Aquifers	1% Typical of soils
MTBE	1.0	1.1	1.6	7
Benzene	1.0	1.2	2.9	20
Toluene	1.1	1.7	7.6	68
Xylene	1.1	2.2	13	120

aquifer. The retardation ratio is simply the rate of movement of water divided by the rate of movement of the contaminants. If the retardation ratio of MTBE is 2.0, then water in the plume moves twice as fast as MTBE. The estimates of the retardation ratio in Table 3-2 are based on equations that predict the adsorption of dissolved organic compounds to native organic material in the aquifer solids (Wiedemeier et al., 1999).

At concentrations of organic carbon in the aquifer solids that are near 0.001%, there should be little retardation of MTBE or any of the BTEX compounds. At concentrations of organic carbon in the aquifer solids near 0.01%, there should be little retardation of MTBE and benzene. However, the separation of toluene and the xylenes compared to MTBE should be noticeable. These concentrations are typical of aquifers with low or average concentrations of organic matter in the aquifer solids. At relatively high concentrations of organic carbon in the aquifer solids near 0.1%, the expected retardation ratio for MTBE would only be 1.6. However, MTBE would be moving almost twice as fast as benzene, five times as fast a toluene, and eight times as fast as the xylenes. The separation would be very distinct.

The retardation ratios in Table 3-2 are calculations and estimates. Shaffer and Uchrin (1997) published experimental data on MTBE adsorption to sandy aquifer material. They measured the adsorption isotherm of MTBE to a sample of the Cohansey Sand from near Chatsworth, New Jersey. The sand they selected had a high content of organic material (1.44% organic carbon). Their measured K_d was 0.0925 liter per kilogram, which would correspond to a retardation ratio of 1.7, whereas the calculations used to generate Table 3-2 predict a retardation ratio of at least 7. Even though the values in Table 3-2 predict little retardation of MTBE, they predicted more adsorption than was seen in a direct experimental measurement.

ROLE OF DILUTION AND DISPERSION

A primary mechanism for attenuation of MTBE in a plume of ground water is dilution and dispersion. In simple terms, the concentration of MTBE is attenuated as the contaminated water is diluted into clean ground water in front of the plume and to the sides of the plume.

Figure 3-5 presents a simple concept of the contribution of dispersion on the spreading of a plume. The parameters D_x, D_y, and D_z are the coefficients of longitudinal, transverse, and vertical dispersion. The coefficients can be measured directly in the field with a tracer test; however, these tests are expensive, time consuming, and often require special regulatory permits. As a consequence, tracer tests to actually measure D_x and D_z are only conducted at a few research sites. The contribution of dilution and dispersion at a particular release is usually extrapolated from the available literature or determined by calibrating a transport and fate model to field data by adjusting the modeled value for dispersion to best fit the existing monitoring data.

Most transport models assume a uniform flow direction and flow velocity. As a result, spreading of the plume due to variations in flow direction and velocity are attributed to dispersion and not to the uncertainty in monitoring data describing the direction of ground water flow.

In some aquifers, the direction and velocity of ground water flow is very stable. In these aquifers, plumes are usually long and narrow. Often the width of the plume far downgradient is no wider than the width of the source area. Transverse and vertical dispersion have minimal contribution, and all the mixing is in the direction of ground water flow. There is little spreading of MTBE to the sides of the plume.

In other plumes, the object in ground water appears to spread laterally as well as longitudinally. This apparent lateral dispersion may be the direct re-

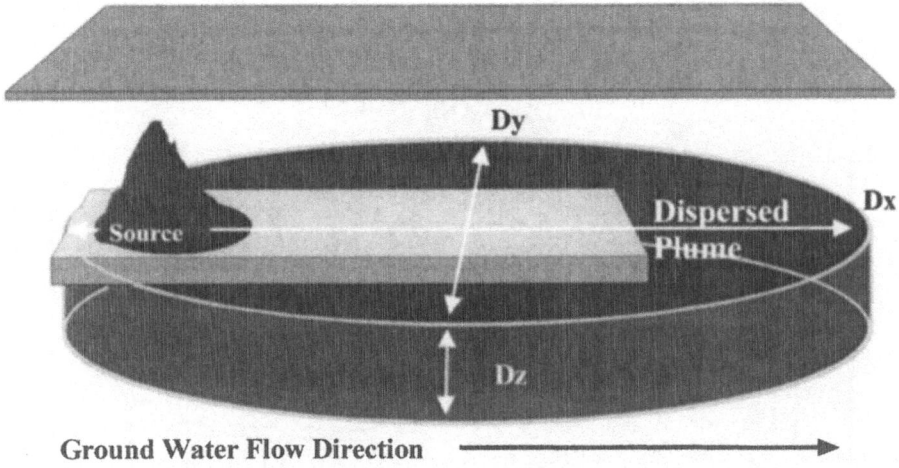

Figure 3-5. **Conceptualization of the Role of Longitudinal and Transverse Dispersion in the Spreading and Dilution of a Plume (modified from Small and Weaver, 1999)**

sult of variations in the direction of ground water flow. What appears to be lateral dispersion is really longitudinal dispersion occurring in different directions. Figure 3-6 presents data on the direction and magnitude of ground water flow at an MTBE site on the U.S. Coast Guard Support Center at Elizabeth City, North Carolina. The site is near the Pasquotank River. The average direction of ground water flow is toward the river; however, the flow at any particular time is sensitive to the stage of the river. Figure 3-6 presents predictions to the direction and velocity of ground water flow from one year of monthly monitoring at the site. Regression analysis was used to fit a plane through the elevation of the water table in the monitoring wells during each month of monitoring (Wilson *et al.*, 2000). An arrow is used to represent the direction and velocity of ground water; the length of the arrow is the distance that ground water would move in a year based on the condition of the water table at that particular month of monitoring. The arrows are given different shades to allow them to be resolved in the figure.

It is apparent that the direction and magnitude of flow vary widely at this site from one month to the next. One round of sampling, or even a few rounds of sampling, would not be adequate to define the direction and magnitude of ground water flow at this site. At this site, the contaminant plume occupies the area encompassed by the variation in the direction of ground water flow.

The standard deviation of the direction of ground water flow over twelve months of sampling, as depicted in Figure 3-6, was 23 degrees. Mace *et al.* (1997) used a similar approach to calculate the variation in the standard deviation of the flow direction from 132 gasoline stations in Texas (Figure 3-7).

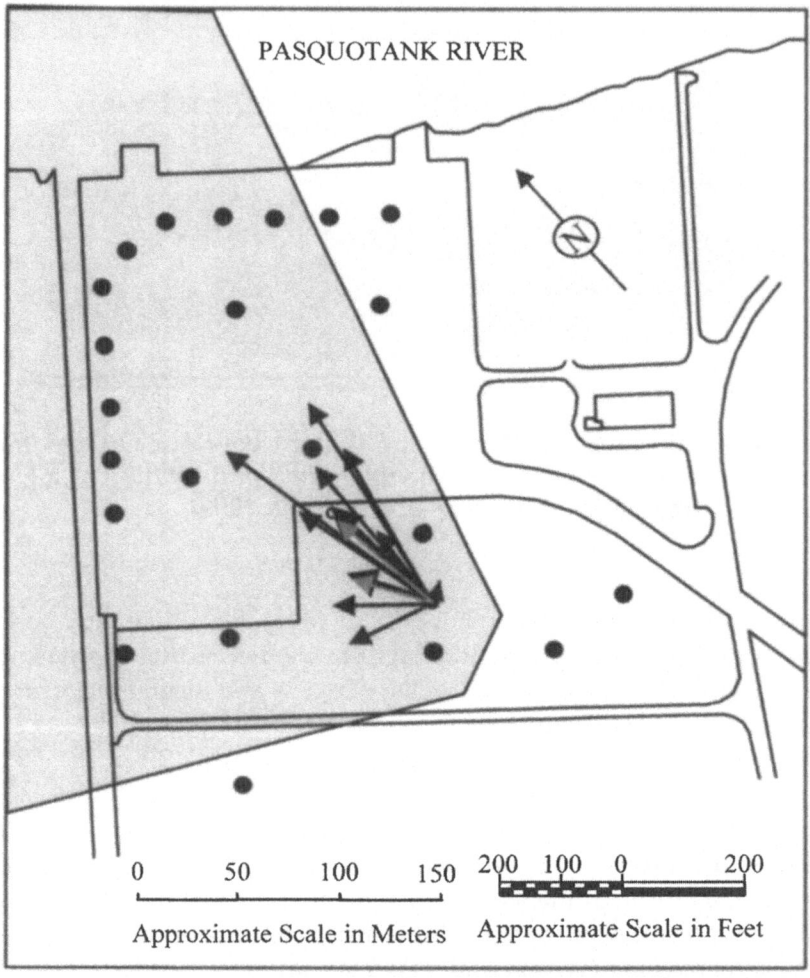

Figure 3-6. **Variation in Direction and Magnitude of Ground Water Flow at a Site in Elizabeth City, North Carolina. Water table elevations were measured each month for a year. The arrows represent the distance that water would move in one year, based on the direction and hydraulic gradient present in a particular round of sampling. The origin of the arrows is the center of the LNAPL source area. The black dots are the locations of temporary push samples for ground water that extend across the vertical extent of the aquifer. The dark shape includes the locations with concentrations of MTBE above 20 μg/l.**

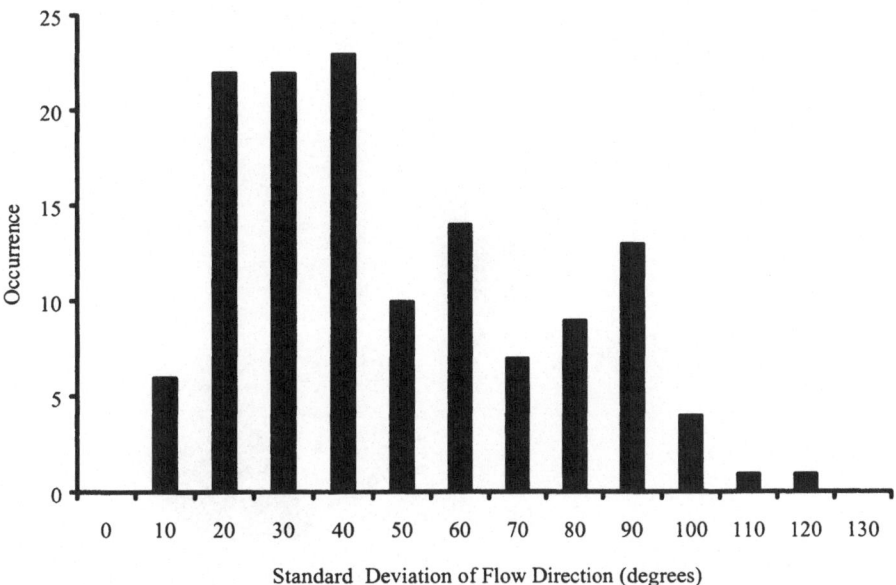

Figure 3-7.   **Variation in the Direction of Ground Water Flow at 132 Gasoline Stations in Texas (data from Figure 14 in Mace *et al.*, 1997)**

They only reported data for the sites in Texas where it was reasonable to represent the shape of the water table as a plane. Most stations in Texas showed more variation than the data presented in Figure 3-6.

Variation in the direction and magnitude of ground water flow is one of the strongest controls on the distribution of contamination moving away from a source area. If the variation in direction is small, the plume tends to look like the tail of comet, moving away from the head. If the variation is very large, the plume tends to appear as a diffuse halo around the source area.

For purposes of comparison, Figure 3-8 summarizes the data in Figure 3-7 by comparing the arcs subtended by one standard deviation in flow direction on either side of the mean direction of flow. A comparison is made between the sites with low deviation, high deviation, and very high deviation. For roughly one-third of the sites in Texas, the direction in ground water flow is highly variable, and the concept of a single flow direction is not the best representation of the behavior of the plume.

ROLE OF BIODEGRADATION OF MTBE

Aerobic and anaerobic biodegradation of MTBE are discussed in detail in Chapters 12 and 13, and natural attenuation is discussed in Chapter 16. Here,

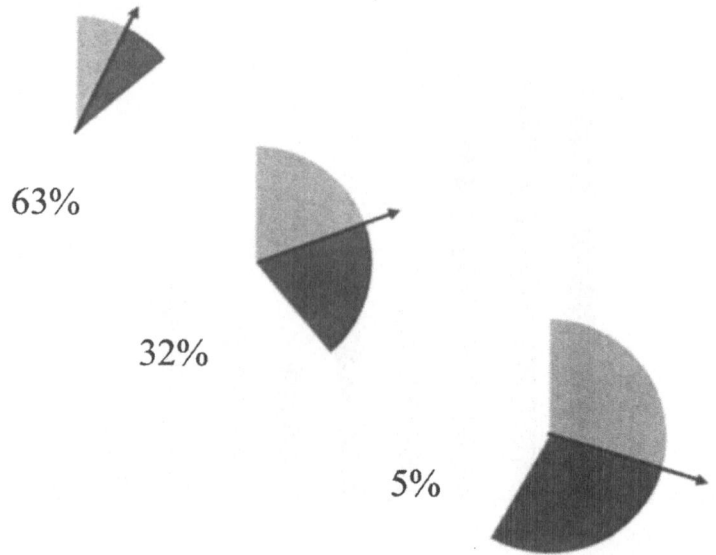

Figure 3-8.   **Effect of the Variation in the Direction of Ground Water Flow. Presented are the arcs subtended by two standard deviations of the direction of ground water flow for the 63% of sites in Texas with low standard deviation, the 32% of sites with high standard deviations, and the 5% of sites with very high standard deviation.**

the discussion will focus on biodegradation rates and the impact of biodegradation on the fate and transport of MTBE and other gasoline components. For a discussion of biodegradation kinetics, see Chapter 12. Tables 3-3 and 3-4 list several rate constants for MTBE biodegradation in ground water or aquifer sediments that are published in the literature. These rates are identified with the 95% confidence interval on the rate. Other rates are extracted from data that were not originally intended for an analysis of the kinetics of biodegradation. They were calculated using the first-order rate law from the initial concentration in the experiment, the final concentration, and the time elapsed. The rates that do not have confidence intervals were extracted in this manner.

The biodegradation of MTBE under aerobic or denitrifying conditions has been documented at a variety of locations (Table 3-3). In general, the rates in laboratory experiments or pilot-scale demonstrations where oxygen is not limiting are rapid. The median rate is 5 per year, corresponding to a half-life of 2 months. Rates that have been documented in the field are much slower, on the order of 0.4 per year or a half-life of 2 years. It is likely that the field-scale rates reflect the rate of reaeration of the plume, as well as the rate of biodegradation when oxygen is available.

Table 3-3.   **Rates of Biodegradation of MTBE under Aerobic and Nitrate-Reducing Conditions**

First-Order Rate Constant (per year)				
Laboratory	Pilot-Scale	Field-Scale	Location	Reference
Oxygen Respiration and Nitrate Respiration				
2 (incomplete)	NA	0.37 ± 0.26	Sampson County, NC	Borden et al., 1997
		0.30 ± 0.15		
		0.0		
5 (incomplete)	NA	0.44	Borden, Ontario	Schirmer et al., 1999
4 (incomplete)				
40	6	NA	Port Hueneme, CA	Salanitro et al., 2000
5	25	NA	Amoco Site, MI	Javanmardian and Glasser, 1997
NA	12	NA	Beaufort, SC	Landmeyer et al., 2001
0.26±0.004	NA	NA	Borden, Ontario	Church et al., 2000
0.40±0.04	NA	NA	Farmington Hill, MI	Church et al., 2000
0.36±0.04	NA	NA	Turnersville, NJ	Church et al., 2000
Sub-oxic				
NA	0.11 to 0.36	NA	Oscota, MI	Barcelona and Jaglowski, 2000

There is one well-documented study of natural MTBE biodegradation under iron-reducing conditions (Landmeyer et al., 1998; see Table 3-4). The removal was statistically significant, but the rate was very slow, on the order of 0.06 per year or a half-life of 12 years. The rate of MTBE biodegradation under iron-reducing conditions may be limited by the supply of biologically available iron in the aquifer sediment. When biologically available iron was added to aquifer sediments, the rate of degradation of MTBE was much faster (Finneran and Lovley, 2001; also Chapter 13).

To date, no one has shown MTBE biodegradation in aquifer sediments under sulfate-reducing conditions (Table 3-4). Amerson and Johnson (2002) used MTBE labeled with a stable carbon isotope to follow MTBE attenuation in the sulfate-reducing portion of a large MTBE plume at Port Hueneme, California. They had no evidence of loss of MTBE mass over the course of one year. (Note that Chapter 26 describes various remediation projects at the Port Hueneme site.) Rates published by Wilson et al. (1999), using data published by Cho et al. (1997), were less than or equal to 0.3 per year, corresponding to a half-life of 2.3 years. These rates were not corrected for dilution and dispersion and should be considered upper boundaries on the rate of biodegradation. In summary, the available data indicate that MTBE biodegradation under sulfate-reducing conditions is either very slow or nonexistent.

Data on MTBE degradation in the laboratory under methanogenic conditions are mixed. Microcosms constructed with material from two sites

Table 3-4.   **Rates of Biodegradation of MTBE under Anaerobic Conditions**

| First1Order Rate Constant (per year) | | | |
Laboratory	Field-Scale	Location	Reference
Iron-Reducing Conditions			
natural rate 0.06	NA	Beaufort, SC	Landmeyer et al., 1998
iron amended 17 6 1.8	NA	Beaufort, SC	Finneran and Lovley, 2001
Mixed			
NA	0.04	Beaufort, SC	Landmeyer et al., 2001
Sulfate-Reducing Conditions			
NA	<0.12 <0.30	Elizabeth City, NC	Wilson et al., 1999 from Cho et al., 1997
NA	0	Port Hueneme, CA	Amerson and Johnson, 2002
0	NA	Blacksburg, VA	Yeh and Novak, 1994
0	NA	Empire, MI	Mormille et al., 1994
Methanogenic Conditions			
2 3.3	NA	Blacksburg, VA	Yeh and Novak, 1994
3.0 ± 0.52	2.7 5.0 to 2.2	Elizabeth City, NC	Wilson et al., 2000
0	NA	Empire, MI	Mormille et al., 1994
0	NA	Norman, OK	Suflita et al., 1993
NA	5.2	Long Island, NY	Kolhatkar et al., 2001
NA	0.41	Philadelphia, PA	Kolhatkar et al., 2000
NA	0.42	Parsippany, NJ	Kolhatkar et al., 2000
NA	0.43	Washington, DC	Kolhatkar et al., 2000

showed the capability to degrade MTBE, while material from two other sites did not show MTBE biodegradation (Table 3-4). When biodegradation occurred, it was rapid. The average rate is near 3 per year, corresponding to a half-life of 3 months. Degradation in the field was rapid at two sites. At three other sites, the rates were an order of magnitude slower.

It is generally considered that the rate of natural bioattenuation of MTBE is much slower than the rate of benzene bioattenuation. Table 3-5 compares rates of natural bioattenuation of benzene in the field as reviewed by Suarez and Rifai (1999), or extracted from the review of Aronson and Howard (1997), to the rates of natural bioattenuation of MTBE at field scale in Tables 3-3 and 3-4. There is a fair amount of uncertainty in this comparison of mean rates of attenuation of benzene and MTBE; however, the available data indicate that the rate of MTBE biodegradation is one-third to one-fourth of the rate for benzene. It is important to remember that this comparison of the rate of MTBE degradation to the rate of benzene degradation only applies to those sites

Table 3-5. **Rate of Natural Degradation in the Field under Anaerobic Conditions**

	First-Order Rate Constant for Degradation (per year)		
	Benzene Suarez and Rifai, 1999	Benzene Aronson and Howard, 1997	MTBE Table 3-3 and 3-4, this chapter
Number of Reported Rates	20	16	10
Mean	3.7	3.9	1.0
Median	NA	1.5	0.41

where MTBE was shown to degrade. There are reported sites where appreciable biodegradation of MTBE was not observed.

The median rate of natural bioattenuation in the field-scale studies listed in Tables 3-3 and 3-4 is 0.41 per year. This rate can best be interpreted in the context of the attenuation required to reach cleanup goals for MTBE, and the time required to reach the cleanup goals. As discussed previously, Figure 3-2 presents the frequency distribution of the maximum concentration of MTBE in monitoring wells in Southern California, Texas, and a sample of sites from the eastern U.S.

For sites in Texas, 26% of the sites had maximum concentrations of MTBE less than 100 µg/l, 44% had maximum concentrations less than 1,000 µg/l, and 74% had maximum concentrations less than 10,000 µg/l. A first-order rate constant of 0.41 per year was used to estimate the time required for natural bioattenuation to bring the maximum concentration of MTBE at each station down to 20 µg/l. This would be an estimate of the survival time of MTBE in a plume after it has moved away from the source area. Results are presented in Figure 3-9.

Fifty percent of the plumes would cleanup in 11 years, 75% would cleanup in 16 years, and 95% would cleanup in 20 years. It is important to not confuse these rates and these time projections with the time required to cleanup a site. These rates presume that the source of contamination to ground water has been controlled, that future contamination of ground water will not occur, and that natural attenuation is acting on the contamination remaining in the ground water. If residual gasoline is left at a site, it can continue to contaminate ground water over time. As a result, the time required for cleanup of a site under natural conditions is often controlled by the rate of natural attenuation of the source area, not the rate of natural attenuation of the plume. If the source of ground water contamination is controlled, as is required by the USEPA directive on monitored natural attenuation (MNA) (USEPA, 1997), ground water at many sites can be restored in less than 20 years.

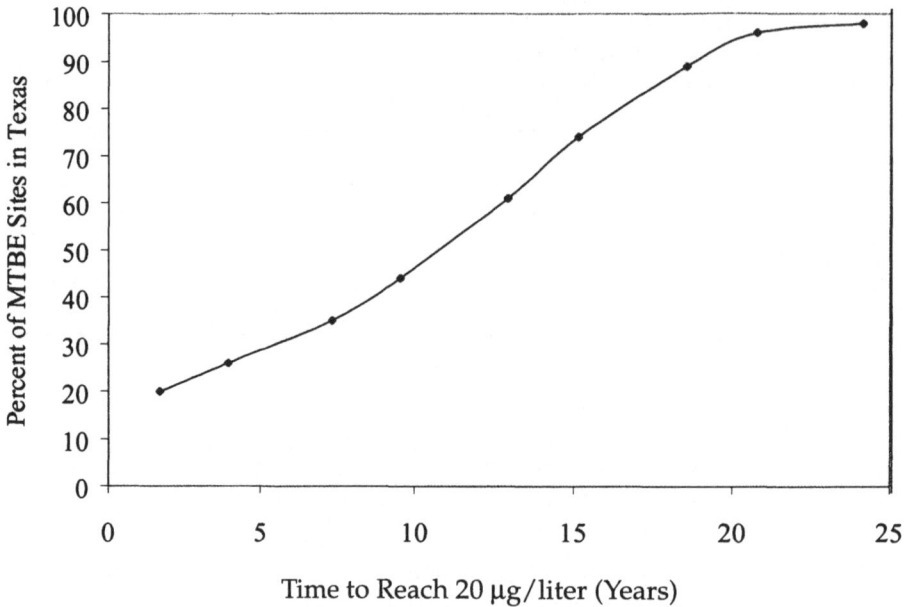

Figure 3-9.   **Effect of Natural Biodegradation of MTBE on the Time Required to Cleanup Ground Water Contamination. The relationship is illustrated with data from Texas (Mace and Choi,
1998) assuming a first-order rate constant for natural
biodegradation of 0.41 per year.**

PRODUCTION AND BIODEGRADATION OF TBA

Kolhatkar *et al.* (2000) reported on the distribution of MTBE and TBA at 74
gasoline stations in the eastern U.S. Figure 3-10 compares the highest concentration of TBA sampled at any well at a particular gasoline service station
with the highest concentration of MTBE sampled at any well at a particular
station. TBA was widely distributed in this sample of stations. Some wells
had high concentrations of TBA with little MTBE, some had high concentrations of MTBE and little TBA, and some wells had high concentrations of
both. In general, TBA was as widely distributed as MTBE, and the concentrations of TBA encountered were equivalent to the concentrations of MTBE.

At many fuel spills, it is impossible to identify the source of TBA present
in ground water (compare Landmeyer *et al.*, 1997) because there are several
plausible sources of TBA in ground water. First, biodegradation of MTBE
may produce TBA as a transformation product. Second, in some areas of the
U.S. TBA has been directly added to fuels as an oxygenate and to enhance octane ratings. Third, commercial MTBE may contain 5% to 10% TBA. Fourth,
TBA may be generated from MTBE during sample preservation or laboratory
analysis (discussed in Chapter 6).

Kramer and Douthit (2000) extracted gasoline from six service stations in New Jersey using a fuel to water ratio of one to four. TBA was detected in extracts of the gasoline from five of the six stations. When detected, the concentrations in the water extract varied from 1,120,000 to 1,690,000 μg/l. These concentrations of TBA in the water extracts would be expected if the MTBE added to the gasoline contained 11% by volume TBA (equivalent to 1.5% by volume in the gasoline).

The lower of the two solid lines in Figure 3-10 is a projection of the concentration of TBA that would be expected if the TBA in the ground water came from TBA that was originally present in the gasoline at a concentration of 10% of the concentration of MTBE, and if all of the TBA and MTBE partitioned from the gasoline to ground water. The higher of the two lines is a projection of the concentration of TBA if all of the TBA partitioned to the ground

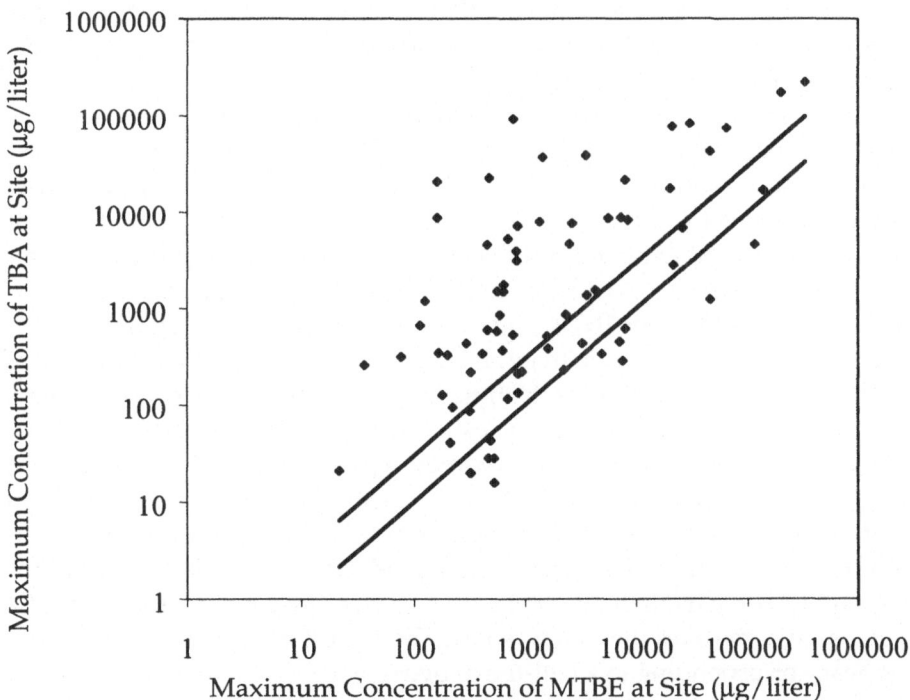

Figure 3-10.   **Distribution of MTBE and TBA in Wells from Gasoline Spills at Selected Service Stations in Pennsylvania, Ohio, Indiana, New York, New Jersey, Maryland, the District of Columbia, and Florida. Technical MTBE used in gasoline often contains approximately 10% TBA. The solid lines are the concentrations of TBA that would be expected if the TBA in ground water was originally present with MTBE in the gasoline.**

water, but two-thirds of the MTBE remained with the gasoline and did not partition to ground water. Data from many of the stations lie near these projections and could be explained by minor concentrations of TBA added to gasoline along with MTBE. However, approximately half of the stations have concentrations of TBA much higher than would be expected if the TBA was a minor component of the oxygenates added to the fuel. These higher concentrations of TBA may result from the use of TBA by itself as the fuel additive, or from biotransformation of MTBE, or possibly as a sampling and analysis artifact.

Kolhatkar *et al.* (2000) examined the distribution of TBA at each of 74 gasoline stations in the eastern U.S. and attempted to extract field-scale rate constants for anaerobic biodegradation. They were able to extract rate constants for TBA degradation that were statistically significant at three sites (see Table 3-6). At the few sites where natural TBA biodegradation could be demonstrated, the rates varied from 0.61 to 13.9 per year, corresponding to half-lives of 1.1 years to 2.4 weeks. These rates are faster than the median rate of natural biodegradation of MTBE

The preponderance of work on biodegradation of TBA under anaerobic conditions has been done in the laboratory of J.T. Novak at Virginia Polytechnic Institute and State University. Novak and his associates used laboratory microcosms to study the potential for anaerobic degradation. Some of their results are summarized in Table 3-6. They originally reported their results as zero-order rate constants. In their experiments, the initial concentration of TBA was 10,000 or 100,000 µg/l. The concentrations of regulatory concern may be on the order of 10 µg/l. To present their data in a form that is more appropriate to determine the time required to reach low concentrations of TBA, their data have been used to estimate first-order rate constants (Table 3-6). This was done by either dividing their reported first-order rate constant by the initial concentration of TBA or by calculating a first-order rate from their figures from the time taken to degrade one-half of the initial concentration of TBA. The rate of anaerobic biodegradation at a variety of locations was rapid, with a median rate of 2 per year, corresponding to a half-life of 4 months. At some sites the rate was low. The lowest rate calculated was 0.16 per year, corresponding to a half-life of over 4 years.

In the microcosm studies of Suflita and Mormile (1993) and Mormile *et al.* (1994), TBA did not degrade in aquifer sediment under anaerobic conditions.

TBA can degrade in ground water if oxygen is available. Salanitro *et al.* (2000) noted that TBA was originally present at concentrations up to 250 µg/l at their demonstration site at Port Hueneme, California. TBA was degraded in a demonstration plot that received oxygen but was not inoculated, but it was degraded more rapidly and to a greater extent in the plot that received oxygen and was inoculated with their culture.

Table 3-6. **Potential for TBA Biodegradation in Anaerobic Ground Water**

Geochemistry	First-Order Rate of Attenuation (per year)		Location	Reference
	Laboratory	Field		
Unknown, anaerobic	0.16 site 1 2.4 site 2 4.6 site 3	NA	Blacksburg, VA, clay deeper vadose zone	Hickman and Novak, 1989
Unknown, anaerobic	0.15 site 1 3.7 site 2	NA	Blacksburg, VA, clay deeper vadose zone	Yeh and Novak, 1994
Unknown, anaerobic	3	NA	Williamport, PA, aerobic sand aquifer	Novak et al., 1985
Unknown, anaerobic	1.3	NA	Williamport, PA, aerobic sand aquifer	Hickman et al., 1989
Unknown, anaerobic	0.5	NA	Wayland, NY, glacial wash and silty clay aquifer	Novak et al., 1985
Unknown, anaerobic	4	NA	Wayland, NY, glacial wash and silty clay aquifer	Hickman et al., 1989
Unknown, anaerobic	0.9	NA	Dumfries, VA, sand and silty clay aquifer	Novak et al., 1985
Unknown, anaerobic	0.4	NA	Dumfries, VA, sand and silty clay aquifer	Hickman et al., 1989
Unknown, anaerobic	2	NA	Newport News, VA, sand and gravel aquifer	Hickman et al., 1989
Unknown, anaerobic	4.6	NA	Newport News, VA, sand aquifer	Hickman et al., 1989
Unknown, anaerobic	1.5	NA	Newport News, VA, sand aquifer	Yeh and Novak, 1994
Methanogenic	NA	13.9	Long Island, NY	Kolhatkar et al., 2000
Methanogenic	NA	0.61	Maryland	Kolhatkar et al., 2000
Methanogenic	NA	5.5	District of Columbia	Kolhatkar et al., 2000
Methanogenic	0	NA	Norman, OK, sandy aquifer	Suflita and Mormile, 1993
Methanogenic	0	NA	NA	Mormile et al., 1994

FALSE ATTENUATION: MISSING THE PLUME WITH MONITORING WELLS

Many gasoline service stations are sited in the flood plains of river valleys. At many sites, the surface material is primarily silts and clays deposited in previous flood events. Beneath the surface silt and clay are sand and gravel deposits associated with previous meanders of the river. The water table is frequently in the surface silt and clay, while the materials with a capacity to carry ground water and transport a plume occur deeper in the aquifer. At many sites the environment is even more complex. There may be several layers of sands and silty clays stacked on top of each other. The plume of contamination tends to move in the conductive layers. The ground water in the shallow silts and clays may be local recharge water; wells that are screened in the non-conductive silts and clays may miss the plume.

This situation is illustrated by the behavior of a plume at the site at Elizabeth City, North Carolina. A release of JP-4 jet fuel produced a plume of BTEX and MTBE in the shallow water table aquifer. The pattern of ground water flow at this site was presented earlier in Figure 3-6. The vertical distribution of residual fuel, MTBE, and hydraulic conductivity in the source area of the plume were presented in Figure 3-3.

Figure 3-11 presents an interpretation of the plume from a contractor's report that was presented to the owner of the site. The interpretation was based on samples from conventional water table monitoring wells. In this interpretation, the plume was less than 200 meters (700 feet) long. The perimeter wells to either side of the plume and in front of the plume were clean. Ground water inside the plume was depleted in oxygen, sulfate, and nitrate, and contained methane and Fe(II). The water in the perimeter wells contained ambient concentrations of oxygen, sulfate, and nitrate; methane and Fe(II) were absent. The ground water velocity at the site averaged 80 meters per year (270 feet per year), and the release was at least 10 years old when the data were collected. The plume should have been at least 820 meters (2,700 feet) long. The report argued that the length of the plume was restricted by natural aerobic biodegradation in the ground water at the margins of the plume.

The conventional monitoring wells that were installed at the site followed the conventional good practice for monitoring wells at the time that they were installed. The wells were 5-centimeter (2-inch) polyvinyl chloride (PVC) wells with a vertical screened interval of either 3 or 6 meters (10 or 15 feet). The screens were set to include the expected variation in elevation of the water table within the screened interval.

The sands below the water table at this site are very uniform in color and appearance. There was no indication from auger cuttings or from examination of core samples that there was a significant contrast in hydraulic conductivity with depth. Adjacent to each permanent well, water was sampled with a push tool in a vertical profile that extended from the water table to a depth of 8 meters (25 feet) below the water table. Water samples were taken every meter (3 feet). In addition, a pumping test was conducted at each depth interval sampled to determine the local hydraulic conductivity (Wilson et al., 1997).

In the well considered to be within the plume (ESM-14 in Figure 3-11), the screen included most of the conductive interval in the water table aquifer (compare Figure 3-12). Well ESM-14 was 76 meters (250 feet) downgradient of the LNAPL source area. Within this distance, the plume had found its way into the most conductive depth interval. The maximum concentration of

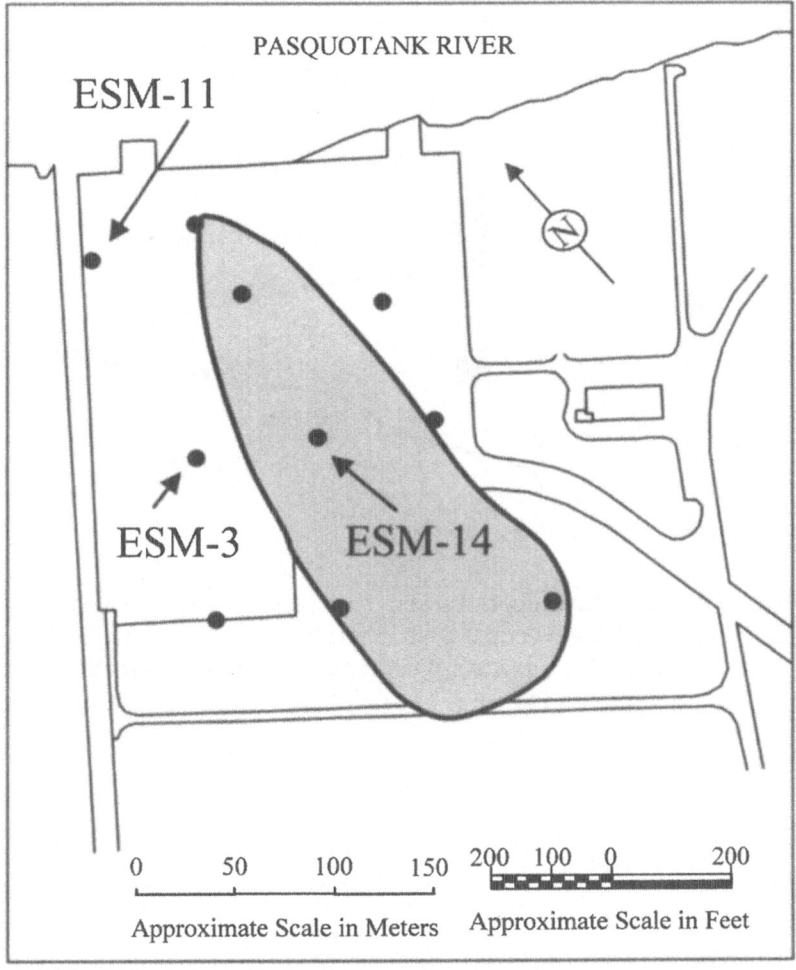

Figure 3-11.   **An Interpretation of the MTBE Plume at Elizabeth City, North Carolina Based on Conventional Wells with Either Ten-Foot Screens Or Fifteen-Foot Screens. The black dots represent the conventional wells. In this interpretation, the plume is contained within the shaded area extending from the source past location ESM-14, but not extending to Pasquotank River. Compare to the interpretation in Figure 2A-6 made from temporary push samples extending vertically across the aquifer.**

MTBE was in the depth interval with the maximum hydraulic conductivity (compare Figure 3-12 and Figure 3-13).

The wells that were considered to be outside the plume, ESM-3 and ESM-11, were screened above the interval with high hydraulic conductivity (Figure 3-14) and above the plume (Figure 3-15). Table 3-7 compares the concen-

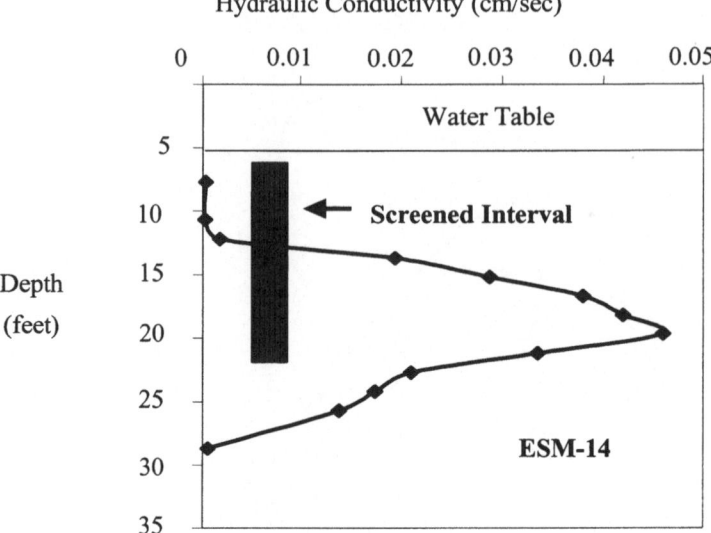

Figure 3-12.   Relationship between the Screened Interval of the Well
and the Vertical Distribution of Hydraulic Conductivity at
ESM-14, a Location That Was Considered to Be Inside the
Plume in Figure 3-11

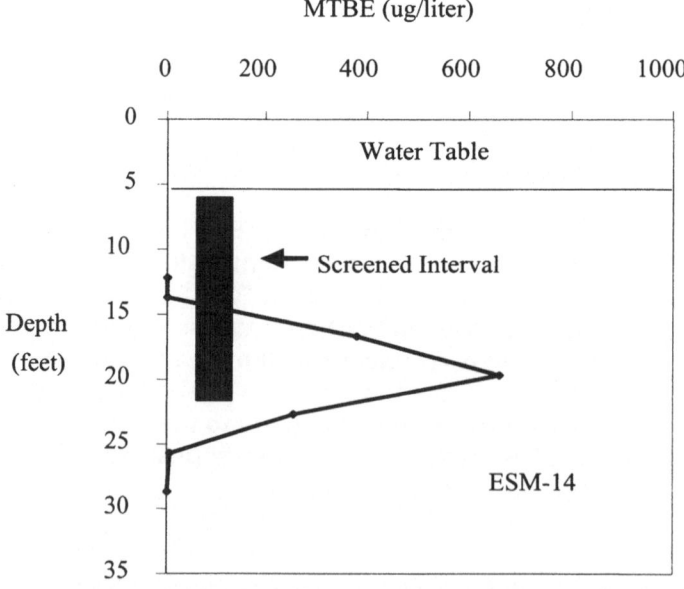

Figure 3-13.   Relationship between the Screened Interval of the Well
and the Vertical Distribution of MTBE at ESM-14, a Location That Was Considered to Be Inside the Plume in Figure
3-11

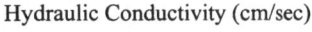

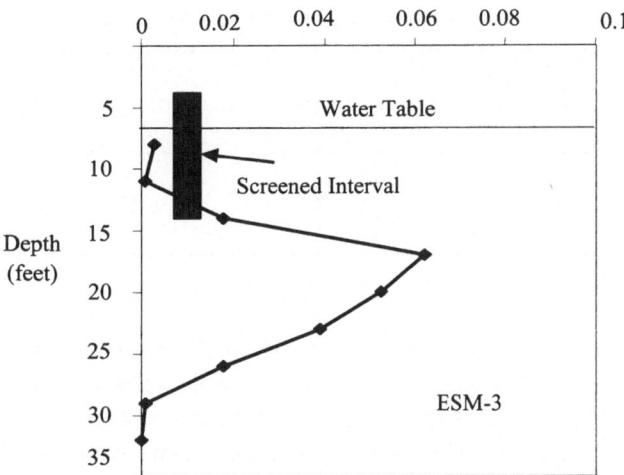

Figure 3-14.   **Relationship between the Screened Interval of the Well and the Vertical Distribution of Hydraulic Conductivity at ESM-3, a Location That Was Considered to Be Outside the Plume in Figure 3-11**

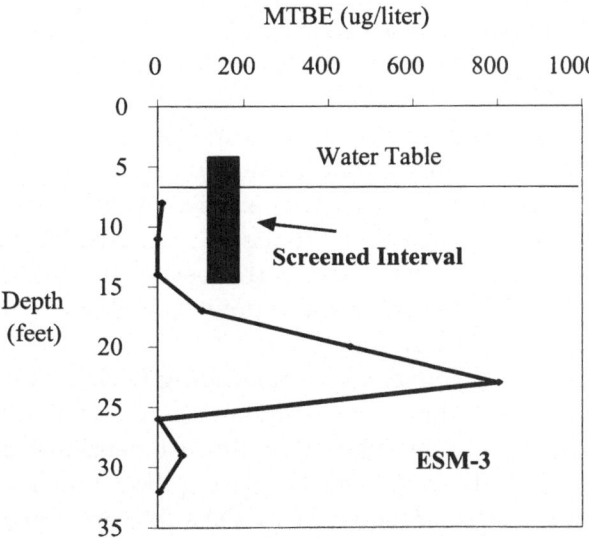

Figure 3-15.   **Relationship between the Screened Interval of the Well and the Vertical Distribution of MTBE at ESN-3, a Location That Was Considered to Be Outside the Plume in Figure 3-11**

Table 3-7. **Comparison of Concentration of MTBE in Water from Conventional Wells Screened across the Water Table to the Concentration from Vertical Push Samples**

Well	Interpretation Based on Conventional Well	Screen Length (feet)	MTBE in Conventional Well (µg/l)	MTBE in Push Samples (µg/l)
ESM-14	in plume	15	353	383
ESM-3	outside plume	10	<1	320
ESM-11	outside plume	10	<1	13.5

trations revealed by the conventional wells and the concentrations revealed by a vertical profile. Because the conventional wells failed to bound the plume, the entire site was characterized with push samples that extended across the entire aquifer. The plume was much more extensive than was revealed by the conventional wells (compare Figure 3-6 showing locations of push samples and Figure 3-11 showing the conventional characterization).

At locations ESM-14 and ESM-3, the highest concentrations of MTBE were almost 6 meters (20 feet) below the top of the water table. The well at ESM-14 detected the plume because it had a 5-meter (15-foot) screen instead of the more conventional 3-meter (10-foot) screen. Conventional criteria require that the screens of monitoring wells should straddle the water table to allow them to detect free product. The screen in the well at ESM-14 was set below the water table. If the screen at location ESM-14 had been set at the same depth interval as the screen at ESM-3, it also would have missed the plume.

In aquifers with significant variation in hydraulic conductivity with depth, it is important to insure that the regions with the highest hydraulic conductivity are sampled. If significant variation is suspected, all the monitoring wells should be tested for their specific capacity. The wells with the highest specific capacity should have the greatest weight in the interpretation of the dimensions of the plume. Data from wells with low specific capacity should be interpreted with caution.

The possibility that conventional monitoring wells will fail to detect a diving plume increases with distance from the source area. If the wells were installed without prior knowledge of the three-dimensional distribution of geological structures, the possibility of missing the plume must be considered. If the geochemistry of the water in wells that are downgradient of a source is similar to wells that are cross-gradient or upgradient of the release, it is probable that they are sampling uncontaminated recharge water and not the plume. The most useful geochemical indicators to distinguish a plume from background water are the accumulation of dissolved inorganic carbon and methane, and the depletion of sulfate and nitrate.

MISSING THE PLUME:
PLUME DIVING BEHAVIOR IN UNIFORM SAND AQUIFERS

In landscapes where the water table is contained within material of uniform texture (for example, many unconsolidated sand aquifers of glacial origin) and where the aquifer is recharged from precipitation, water recharged above the plume tends to push the plume deeper into the aquifer. As a consequence, these plumes move deeper into the aquifer as they move away from their source. Wells with short screens that are only screened across the water table may miss the plume. The effect is well illustrated by the Hagerman Avenue Plume in East Patchogue, New York (Weaver *et al.*, 1999). The plume is in glacial sands on the south side of Long Island. Figure 3-16 depicts the vertical distribution in hydraulic conductivity at the site. There was very little variation with depth. In particular, there are no effective confining layers that would prevent the vertical migration of a plume deeper into the aquifer. The recharge of clean water above the plume can drive the plume deeper into the aquifer.

A plume of MTBE was detected in a private water well in 1994. Investigations by the New York State Department of Environmental Conservation (NYSDEC) tracked the plume back to a gasoline service station more than

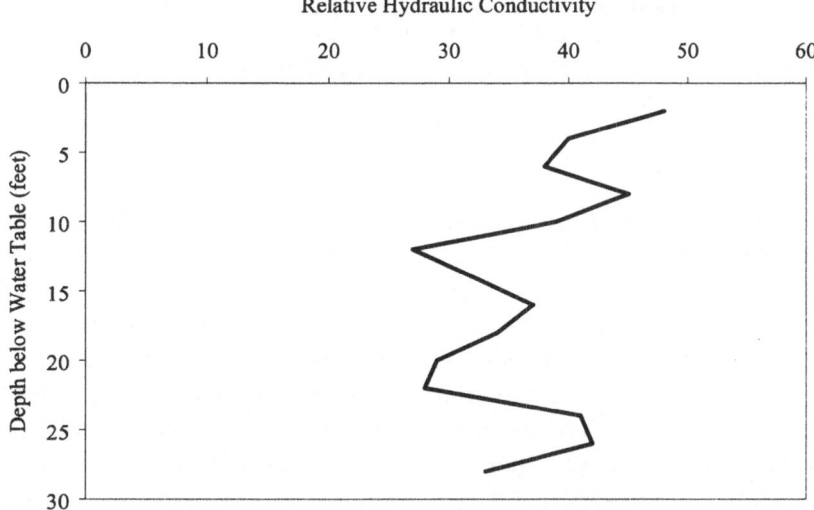

Figure 3-16. **Conditions That Are Conducive to Plume Diving Caused by Recharge of Precipitation. Presented are results of a downhole flow meter test in a fully screened well within a plume on Long Island, New York. Over an interval of 28 vertical feet, there was little variation in the capacity of the aquifer sands to transmit water.**

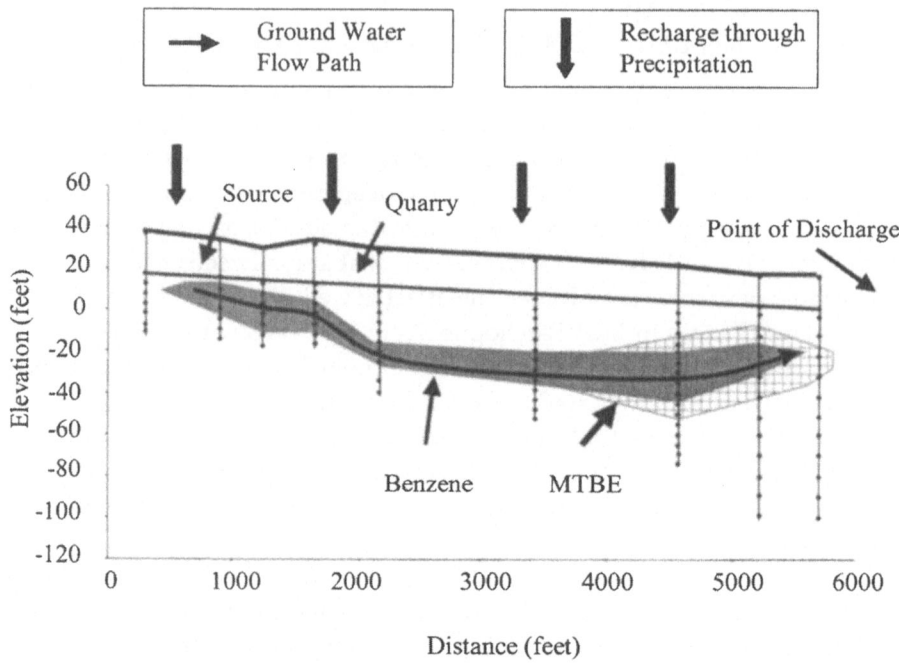

Figure 3-17. **Evidence for "Plume Diving" in the Deep Sand Aquifer on Long Island, New York: A Vertical Transect along the Plume of Benzene at the Hagerman Avenue Site in East Patchogue. The dots along a vertical bar represent the location of discrete sampling wells in a vertical cluster of wells.**

1,200 meters (4,000 feet) upgradient. The USTs had been removed from the service station in 1988. Figure 3-17 depicts the distribution of the plume of benzene and the plume of MTBE in ground water in 1995. By 1995 the MTBE had detached from the source, but the plume of benzene was more or less continuous and revealed the flow path of ground water in the aquifer.

Long Island has substantial rainfall and some snow, and because the surface soil is sandy, much of the rain and snowmelt infiltrate the aquifer. Because the flow of ground water is rapid in this aquifer, the plume was approximately 1,500 meters (5,000 feet) long. Because recharge is important in this aquifer, the top of the plume was depressed from 8 to 9 meters (25 to 30 feet) below the water table. Recharge across a plume may not always be uniform. At the Hagerman Avenue plume, much of the plume diving was associated with a gravel pit that collected storm water and recharged ground water. At many UST sites, conventional wells are screened across the first 3 meters (10 feet) of the aquifer. Figure 3-18 superimposes a 3-meter (10-foot)

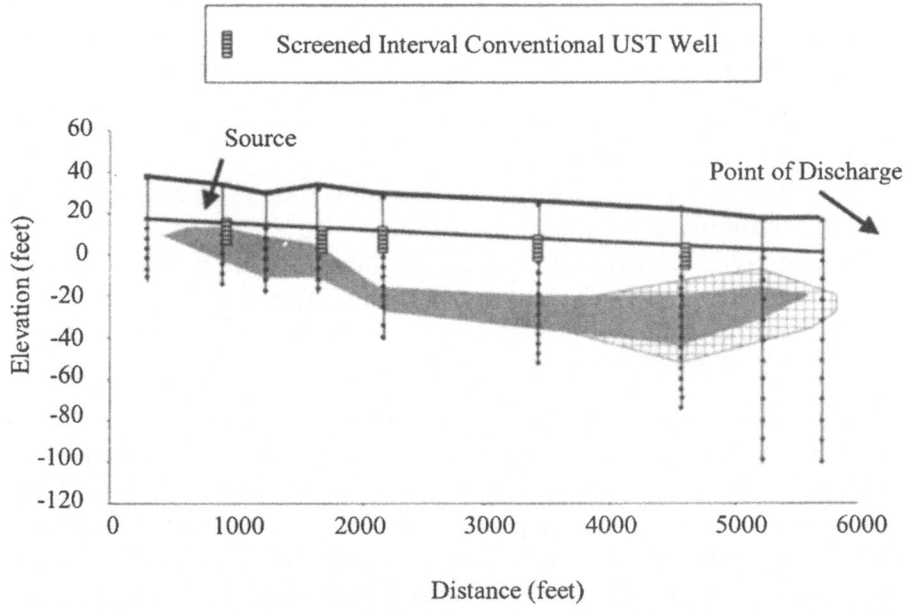

Figure 3-18. **Conventional Wells That Are Screened over the First Ten Feet of the Aquifer Cannot Detect the Downgradient Portion of the Plume at the Hagerman Avenue Site**

well screen over the horizontal cross section. Convention wells with 3-meter (10-foot) screens would have detected the plume up to 600 meters (2,000 feet) from the source. Once the plume moved downgradient of the gravel pit, it was too deep to be sampled by conventional wells with 3-meter (10-foot) screens.

TWO POSSIBLE LIFE CYCLES OF PLUMES

Small and Weaver (1999) present a useful conceptual model of the factors controlling the life cycle of a plume of MTBE in ground water. The following discussion follows their concept.

The size of a plume reflects a balance between the rate of release of the contaminant to the aquifer, the rate of transport of the contaminant away from the source area, and the rates of degradation and dispersion which remove mass or reduce plume concentrations. Depending on the relationship between the rate of dissolution of the contaminant from the fuel spill and the rate of attenuation of the contaminant in ground water, plumes may grow, plumes may come to a stable equilibrium, or plumes may shrink and eventually disappear. When source dissolution and advection dominate over dispersion and biodegradation, plumes expand. Reduction in plume mass occurs when degradation becomes dominant.

Rice *et al.* (1995) proposed a conceptual plume life cycle model to describe the formation of dissolved benzene plumes. The plume life cycle is described with the following four stages:

I. Ground water plume expanding: residual source present, mass flux exceeds attenuation;

II. Ground water plume stable: residual source present, mass flux in "equilibrium" with attenuation;

III. Ground water plume shrinking: residual source nearly exhausted, attenuation exceeds mass flux, reducing plume mass; and

IV. Ground water plume exhausted: average plume concentration low, final stages of source dissolution.

The actual distribution of MTBE in ground water will depend on the relationship between two rates of attenuation: the rate of attenuation of the source and the rate of attenuation along the flow path in the plume. The longevity of the plume will depend on the rate of transfer of MTBE from the fuel spill to the ground water through dissolution and advection. Transfer of MTBE will weather the fuel spill and, over time, will reduce the concentrations in ground water that is in contact with the fuel spill.

The rate of weathering is the rate of natural attenuation of the source. The rate of attenuation of the source determines how long a plume will persist. Biodegradation and dispersion will attenuate concentrations of MTBE as water moves away from the source. This is the rate of attenuation along the flow path. The rate of attenuation along the flow path determines how far a plume will extend away from the source.

The contrasting roles of these two distinct rates of natural attenuation will be illustrated in two case studies. At a release at Elizabeth City, North Carolina, natural anaerobic biodegradation of MTBE has a strong influence on the fate of MTBE. At a release at Port Hueneme, California, natural biodegradation of MTBE is not significant. The fuel spilled at Port Hueneme was gasoline containing 1.2% MTBE and produced maximum concentrations in monitoring wells of 40,000 µg/l. The fuel spilled at Elizabeth City was JP-4 jet fuel containing 0.02% MTBE and produced maximum concentrations in monitoring wells of 609 µg/l.

The plumes are otherwise very similar. At both sites, the smear zone of fuel is centered around the average elevation of the water table. At both sites, the smear zone is 1 to 2 meters (3 to 7 feet) thick. Both occupy shallow semi-confined aquifers in unconsolidated sand. The conductive interval at Elizabeth City is approximately 6 meters (20 feet) thick while the conductive interval at Port Hueneme varies from 3 to 6 meters (10 to 20 feet) in thickness. At both sites, fuel in the smear zone occupies finer grained material just above the conductive portion of the aquifer. The releases are approximately the same age. The release at Port Hueneme occurred in 1984 and 1985. The release at Elizabeth City was controlled in 1991. Free product recovery began in 1990. The ground water seepage velocities are similar. The ground water seepage velocity is 80 meters (270 feet) per year at the Elizabeth City site, and 85 meters (280 feet) per year at Port Hueneme.

Due in part to the contribution of natural anaerobic biodegradation, the MTBE plume at Elizabeth City has reached a steady state. In the absence of significant natural anaerobic biodegradation, the MTBE plume at Port Hueneme has continued to expand, and in the year 2000 extended 1,200 meters (3,900 feet) past the non-aqueous phase liquid (NAPL) source area.

THE PLUME COMES TO STEADY STATE, THEN RECEDES BACK TO THE LNAPL

Figure 3-19 illustrates the behavior expected when dissolution is slow and attenuation along the flow path is rapid. This behavior is to be expected for sites where natural biodegradation of MTBE is rapid, or where the flow of water past the spill is slow, or where the fuel spill is trapped in silts and clays and does not have good contact with ground water flow. After a time, the plume reaches its maximum extent, and then goes into a slow decline. The entire plume attenuates concurrently, and the "hot spot" stays near the source area.

This behavior is illustrated at a site in Elizabeth City, North Carolina (Wilson et al., 2000). Other properties of the plume have been discussed previously (see Figures 3-6 and 3-11). Figure 3-20 presents the transect of wells used to estimate the rate of attenuation of MTBE and benzene in ground water. Figure 3-21 plots the concentration of MTBE, benzene, and methane along the transect. The average plume velocity at this site is 2.5×10^{-4} centimeters per second (cm/s) (0.7 feet per day [ft/day]). Figure 3-21 presents the distribution of MTBE and benzene in 1998, 10 years after the site was no longer used for storage of fuels. The plume of MTBE and benzene extends 200 meters (700 feet), equivalent to 3 years of travel time.

Ambient ground water in the aquifer contains oxygen and is free of methane. Methane is produced in the source area of the plume by fermentation of the gasoline constituents. As ground water moves through the LNAPL source area, the concentration of methane increases (Figure 3-21). After ground water moves past the source area, concentrations of methane change little out to a distance of 200 meters (700 feet). At this site, the attenuation of methane is most likely due to dilution and dispersion. Methane provides a convenient "footprint" for the plume. In contrast to the behavior of methane, the concentrations of MTBE and benzene decline approximately fifty-fold over a distance of 200 meters (700 feet). Microcosm studies at this site established that concentrations of both MTBE and benzene at this site can be attenuated by natural anaerobic biodegradation.

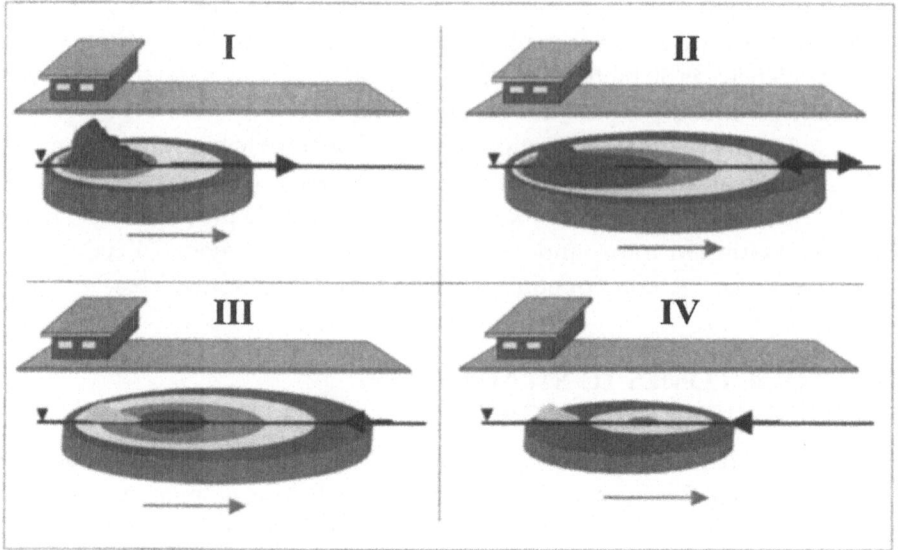

Figure 3-19. **MTBE Plume Life Cycle Where Attenuation of the Source is Slow Compared to Attenuation in the Ground Water (Modified from Small and Weaver, 1999). The gray arrow is the direction of ground water flow. The black arrow is the expected direction of movement of the leading edge of the plume.**

The rate of attenuation along the flow path in Figure 3-21 was $1.8 +/- 1.1$ per year, equivalent to a half-life of 5 months. The rate of attenuation of the source due to natural weathering was 0.06 per year, equivalent to a half-life of 12 years (Wilson *et al.*, 2000). Although natural biodegradation can destroy MTBE and benzene in this plume within a five-year residence time, the plume is being continuously regenerated at its source. At a rate of attenuation of the source of 0.06 per year, the plume can be expected to last for another 60 years.

THE PLUME FAILS TO COME TO STEADY STATE, AND THE HOT SPOT MOVES DOWNGRADIENT

Figure 3-22 illustrates the second possible life cycle, where weathering of the source is fast and attenuation along the flow path is slow. This behavior is to be expected at sites where significant natural biodegradation of MTBE does not occur and the only attenuation mechanism is dispersion. It can also be expected where the flow of water past the spill is rapid, and there is good contact between the flowing ground water and the fuel spill. Concentrations of MTBE in the flow path are maintained, and the length of the plume increases

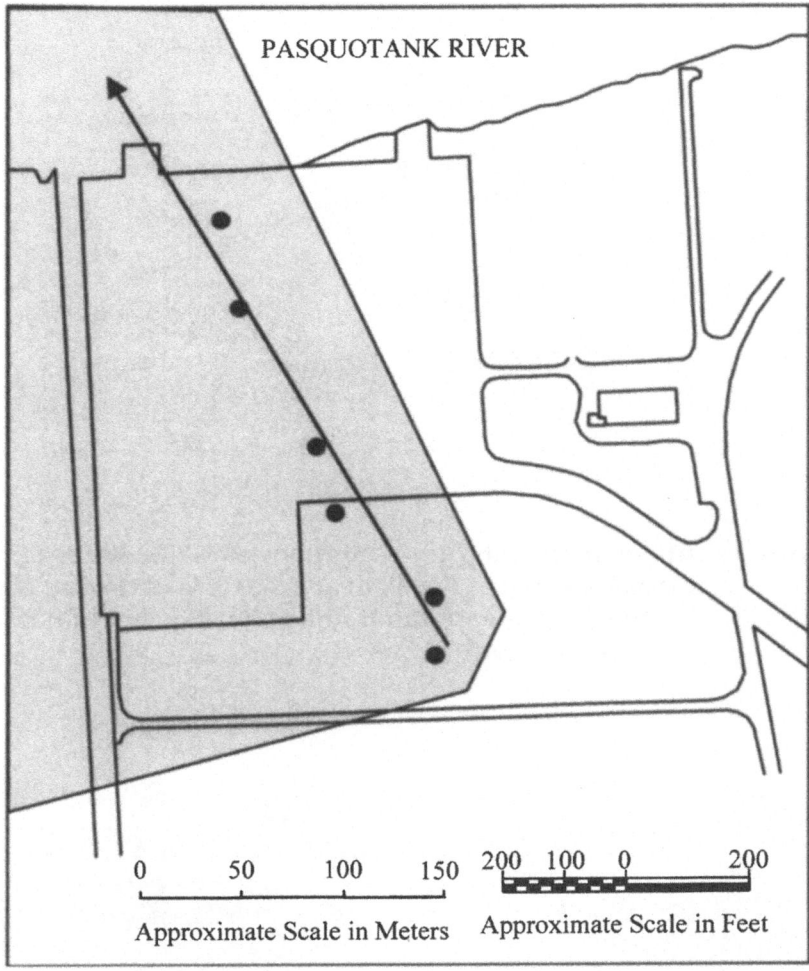

PASQUOTANK RIVER

| 0 | 50 | 100 | 150 | 200 | 100 | 0 | 200 |

Approximate Scale in Meters   Approximate Scale in Feet

Figure 3-20.   **Sampling Locations Used to Estimate the Rate of Attenuation of MTBE, Benzene, and Methane along a Flow Path at a Site (Elizabeth City, North Carolina) Where Natural Anaerobic Biodegradation of MTBE Is Important**

over time. As the source weathers, the concentrations at the source decline over time; however, concentrations at the leading edge of the plume will increase. The "hot spot" moves downgradient of the source over time.

The second case study to illustrate this life cycle is a site in Port Hueneme, California (see also Chapter 26). Between September 1984 and March 1985, there was a release of approximately 42,000 liters (11,000 gallons) of leaded and unleaded gasoline containing MTBE. By 2000, the plume of BTEX was reportedly less than 400 meters (1,300 feet) long, but the plume of MTBE

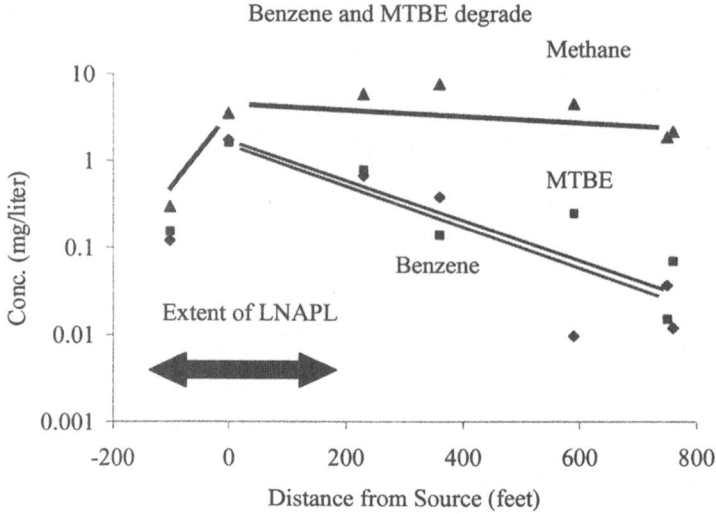

Figure 3-21.   **Attenuation of the Concentrations of MTBE, Benzene, and Methane Along a Flow Path at a Site (Elizabeth City, North Carolina) Where Natural Anaerobic Biodegradation of MTBE Is Important**

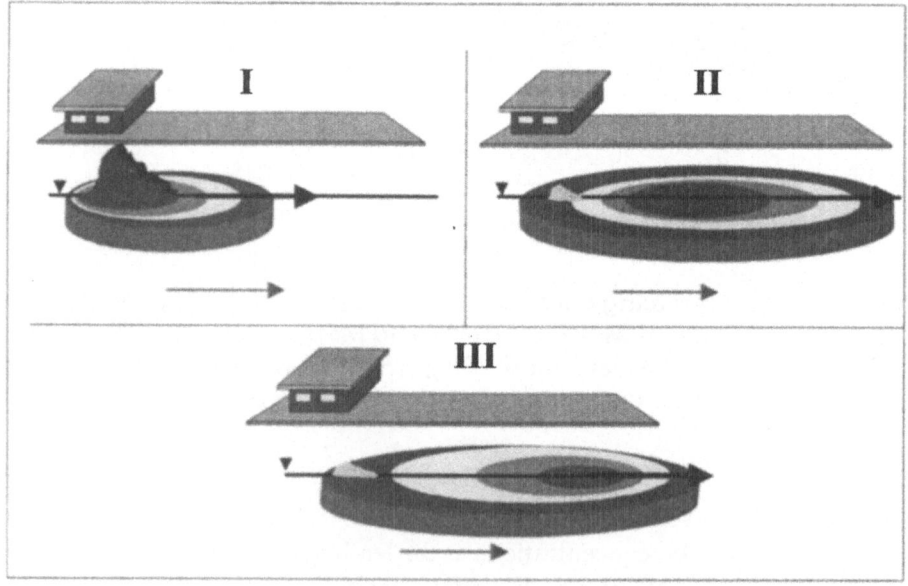

Figure 3-22.   **MTBE Plume Life Cycle Where Attenuation of the Source Is Fast Compared to Attenuation in the Ground Water (Modified from Small and Weaver, 1999). The gray arrow is the direction of ground water flow. The black arrow is the expected direction of movement of the leading edge of the plume.**

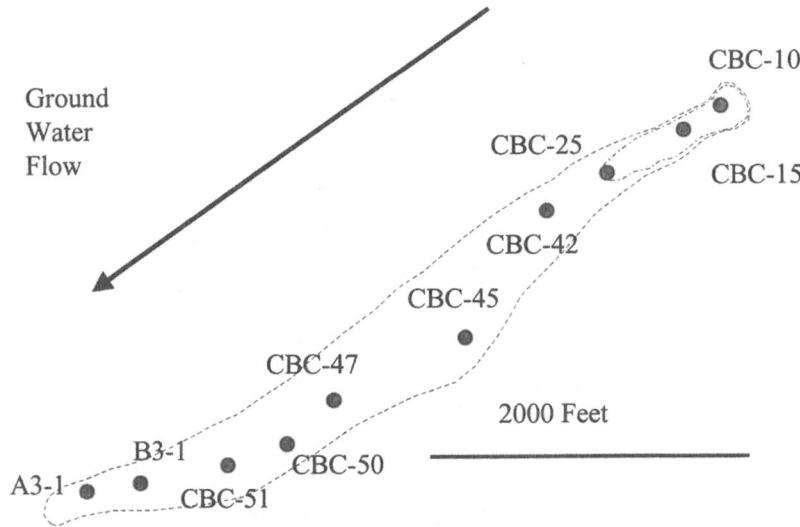

Figure 3-23. **Sampling Locations Used to Estimate the Rate of Attenuation of MTBE, Benzene, and Methane along a Flow Path at a Site (Port Hueneme, California) Where Natural Anaerobic Biodegradation of MTBE Is Not Important, and Dispersion Is the Primary Mechanism of Natural Attenuation.**

was over 1,200 meters (4,000 feet) long (Figure 3-23). This was the maximum extent of the MTBE plume as it developed under natural conditions. The plume is currently being remediated.

Figure 3-24 presents the distribution of MTBE, benzene, and methane along the plume centerline in 1997, which was 13 years after the spill occurred. Compare the data from Port Hueneme in Figure 3-24 to data from Elizabeth City in Figure 3-21. Note that the length of the plume at Port Hueneme is almost 10 times the length at Elizabeth City, and the axes of the figures are scaled appropriately.

The rate of attenuation of MTBE along the flow path at Port Hueneme was 0.37 +/− 0.21 per year, equivalent to a half-life of 22 months. The rate of attenuation of methane as a conservative tracer was slightly faster, 0.55 +/− 0.15 per year, indicating that the apparent attenuation of MTBE could be entirely explained by dispersion. The rate of attenuation of benzene was rapid, at least 2.0 per year. The rate of benzene degradation in the plume at Port Hueneme is similar to the rate of benzene degradation at the site in Elizabeth City, North Carolina (2.4 per year).

The plume was 15 years old in 2000, but it had not yet reached a steady state. Although concentrations of MTBE started to decline in the source area after 1997, the leading edge of the plume continued to advance along the flow

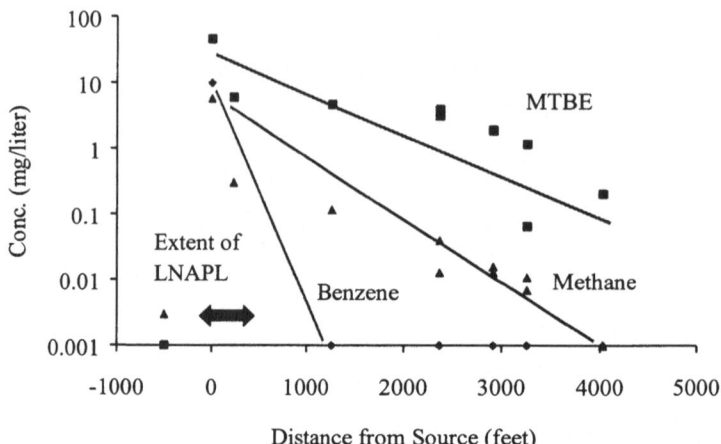

Figure 3-24. **Attenuation of the Concentrations of MTBE, Benzene, and Methane along a Flow Path at a Site (Port Hueneme, California) Where the Primary Mechanism of Natural Attenuation Is Dispersion and Natural Anaerobic Biodegradation of MTBE Is Not Important.**

path (Figure 3-25). If the plume had not been captured, it would have continued to advance until dispersion brought the concentration of MTBE below the action level. This would have required an additional six years and extended the plume an additional 500 meters (1,600 feet).

Figure 3-26 compares the rate of attenuation of MTBE in a well near the leading edge of the LNAPL and a well near the downgradient edge (see Figure 3-23 for the location of the wells). The concentration of MTBE at the leading edge had been fairly stable from the time of the spill until 1998, when the concentration at monitoring well CBC-10 started to attenuate rapidly. Once attenuation began in 1998, the rate of attenuation was 0.0072 per day, or 2.6 per year, equivalent to a half-life of 3 months. Based on accumulation of floating gasoline in monitoring wells, monitoring well CBC-10 was between 20 and 55 meters (70 and 180 feet) from the leading edge of the LNAPL. The estimated seepage velocity of ground water is 2.7×10^{-4} cm/s (0.77 ft/day). In the 15 years between the time of the spill and the time when rapid weathering became apparent, between 23 and 59 pore volumes of ground water had been exchanged along the flow path. This behavior is in good agreement with the laboratory data of Rixey and Joshi (2000), who achieved equivalent removals after the exchange of 10 pore volumes.

The source area is not weathering uniformly. As ground water sweeps

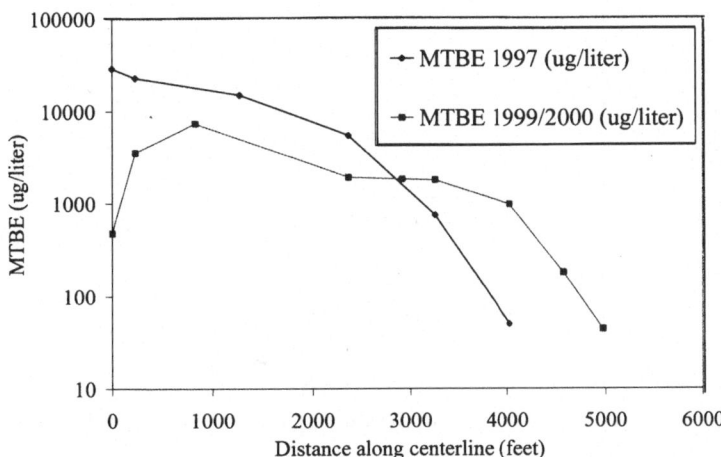

Concentrations along Plume Centerline

Figure 3-25. **Two Time Intervals in the Life Cycle of a Plume (Port Hueneme, California) That Has Not Come to a Steady State. Although the concentration of MTBE in the source area declined, the plume continued to advance along the flow
path.**

into the LNAPL, there may be accelerated weathering at the leading edge of
the fuel spill. If this is the case, there may be a weathering front moving across
the spill, and it took from 1984 when the spill occurred until 1997 for the
weathering front to reach monitoring well CBC-10. As a consequence, the hot
spot has appeared to move downgradient. In contrast, the concentration of
MTBE in monitoring well CBC-25 at the downgradient edge of the LNAPL
has declined slowly over time (Figure 3-26). The well is approximately 300
meters (1,000 feet) from the leading edge of the LNAPL. In the 16 years since
the release, approximately 4.3 pore volumes of ground water have been exchanged along the flow path to well CBC-25. In the time since 1993 when the
concentration of MTBE reached its peak, the rate of attenuation of MTBE over
time at CBC-25 is 0.38 per year, equivalent to a half-life of 22 months. Based
on the rate of attenuation at location CBC-25, the source area can be expected
to last for an additional 30 years.

Many, if not most, MTBE plumes will fall into a pattern where the source
area in the LNAPL is persistent over time, due in part to mass transfer limitations. Peargin (2000, 2001) extracted the rate of natural attenuation of MTBE
in 22 wells in the smear zone of 15 gasoline stations in the eastern U.S. The
fastest rate of attenuation of the source was 0.7 per year, equivalent to a half-

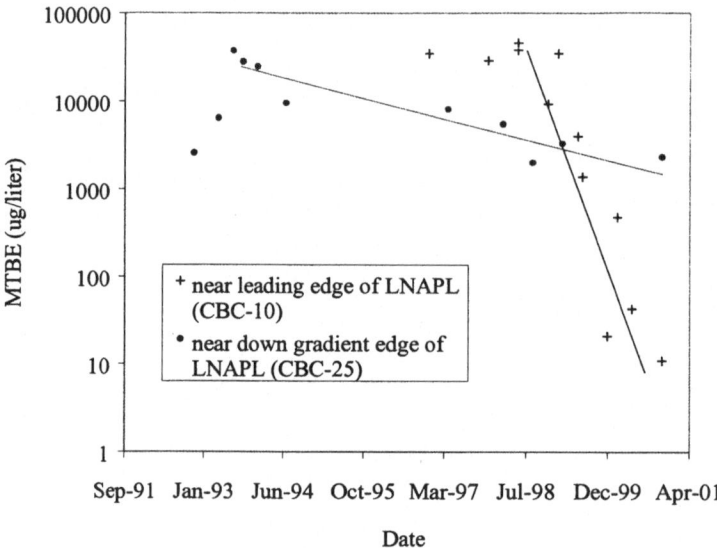

<figure>
Figure 3-26.   **Contrasting Behavior in the Attenuation of MTBE over Time in Two Monitoring Wells at a Site (Port Hueneme, California)**
</figure>

life of 1 year. The mean rate of attenuation was 0.04 per year, equivalent to a half-life of 17 years.

OVERVIEW OF FACTORS THAT LEAD TO LONG MTBE PLUMES

There are three site-specific factors that can produce long plumes of MTBE in ground water. The first important factor is the effective velocity of ground water in the plume. Obviously, longer plumes are expected when the hydraulic conductivity of the aquifer and the hydraulic gradient are higher. Effective velocity is also higher when there is little change in plume direction. Simply stated, plumes that spend more time moving in the same direction tend to move further than plumes that change direction. Effective velocity can be higher when there is heterogeneity in the hydraulic conductivity of materials in the aquifer. Plumes tend to find their way into the most conductive materials. Stringers of sands and gravels can project a plume much faster than would be expected from the average flow of ground water (as predicted using Darcy's Law based on a slug test of a monitoring well).

The second important factor is a failure of MTBE to biodegrade at significant rates. This is most likely in old spills and large spills where biodegradation of the petroleum hydrocarbons has depleted oxygen in the ground water, and none is available for degradation of MTBE. Failure to degrade significantly is also likely in plumes that are sulfate-reducing but are not methanogenic.

The most important factor is the fraction of MTBE in the fuel. MTBE in ground water adjacent to a spill of RFG with 11% MTBE by volume can reach concentrations in the range of 1,000,000 µg/l. These higher concentrations in ground water must be attenuated on the order of 100,000-fold to reach cleanup goals relevant to drinking water. MTBE in ground water adjacent to a spill of gasoline containing 1% MTBE for octane enhancement can reach concentrations in the range of 100,000 µg/l. These intermediate concentrations must be attenuated on the order of 10,000-fold to reach cleanup goals relevant to drinking water. MTBE in ground water adjacent to a spill of gasoline or other petroleum hydrocarbons with incidental concentrations of MTBE (0.1% or less) can reach concentrations in the range of 10,000 µg/l. These lower concentrations in ground water must be attenuated on the order of 1,000-fold. Higher starting concentrations require longer travel distances to be attenuated by dilution and dispersion, or longer residence times to be attenuated by natural biodegradation.

DISCLAIMER

The USEPA through its Office of Research and Development funded the research described here under in-house task 5857 (Natural Attenuation of MTBE) at the Ground Water and Ecosystems Restoration Division. It has not been subjected to Agency review and therefore does not necessarily reflect the views of the Agency. No official endorsement should be inferred.

REFERENCES

Amerson, I. and Johnson, R.L. 2002. A natural gradient tracer test to evaluate natural attenuation of MTBE under anaerobic conditions. *Ground Water Monitoring and Remediation*. In press.

Aronson, D. and Howard, P.H. 1997. *Anaerobic Biodegradation of Organic Chemicals in Groundwater: A Summary of Field and Laboratory Studies*. Final Report prepared by the Syracuse Research Corporation for the American Petroleum Institute, the Chemical Manufacturer's Association, the National Council of the Paper Industry for Air and Stream Improvement, the Edison Electric Institute, and the American Forest and Paper Association.

Barcelona, M.J. and Jaglowski, D.R. 2000. Subsurface fate and transport of MTBE in a controlled reactive tracer experiment. In: *Proceedings of the 2000 Petroleum Hydrocarbons and Organic Chemicals in Ground Water: Prevention, Detection, and Remediation, API and MGWA Conference and Exposition*, Houston, Texas. November 17–19, 2000. pp. 123–137.

Borden, R.C., Daniel, R.A., LeBrun IV, L.E., and Davis, C.W. 1997. Intrinsic biodegradation of MTBE and BTEX in a gasoline-contaminated aquifer. *Water Resources Research*. 33(5), 1105–1115.

Cho, J.S., Wilson, J.T., DiGiulio, D.C., Vardy, J.A., and Choi, W. 1997. Implementation of natural attenuation at a JP-4 jet fuel release after active remediation. *Biodegradation.* 8, 265–273.

Church, C.D., Pankow, J.F., and Tratnyek, P.G. 2000. Effects of environmental conditions on MTBE degradation in model column aquifers: II Kinetics. In: *Preprints of Extended Abstracts,* 40(1), pp. 238-240. Symposia papers presented before the Division of Environmental Chemistry American Chemical Society, San Francisco, California. March 26–30, 2000.

Cline, P.V., Delfino, J.J., and Rao, P.S.C. 1991. Partitioning of aromatic constituents into water from gasoline and other complex solvent mixtures. *Environmental Science and Technology.* 25(5), 914–920.

Finneran, K.T. and Lovley, D.R. 2001. Anaerobic degradation of methyl tert-butyl ether (MTBE) and tert-butyl alcohol (TBA). *Environmental Science and Technology.* 35(9), 1785–1790.

Hickman, G.T. and Novak, J.T. 1989. Relationship between subsurface biodegradation rates and microbial density. *Environmental Science and Technology.* 23(5), 525–532.

Hickman, G.T., Novak, J.T., Morris, M.S., and Rebhun, M. 1989. Effects of site variations on subsurface biodegradation potential. *Research Journal of the Water Pollution Control Federation.* 61(9), 1564–1575.

Javanmardian, M. and Glasser, H.A. 1997. In-situ biodegradation of MTBE using biosparging. In: *Preprints of Extended Abstracts,* Volume 37(1), p. 424, April 1997. American Chemical Society, Division of Environmental Chemistry.

Kolhatkar, R., Wilson, J., and Dunlap, L.E. 2000. Evaluating natural biodegradation of MTBE at multiple UST sites. In: *Proceedings of the Petroleum Hydrocarbons and Organic Chemicals in Ground Water: Prevention, Detection, and Remediation Conference, API, NGWA, STEP Conference and Exposition,* Anaheim, California. November 15–17, 2000. pp. 32–49.

Kolhatkar, R., Wilson, J., and Hinshalwood, G. 2001. Natural biodegradation of MTBE on Long Island, NY. In: *Bioremediation of MTBE, Alcohols, and Ethers, The Sixth International In Situ and On-Site Bioremediation Symposium,* San Diego, California. June 4–7, 2001. pp. 43–50. (Magar, V.S, Gibbs, J.T, O'Reilly, K.T., Hyman, M.R., and Leeson, A., Eds.).

Kramer, W.H. and Douthit, T.L. 2000. Water soluble phase oxygenates in gasoline from five New Jersey service stations. In: *Proceedings of the Petroleum Hydrocarbons and Organic Chemicals in Ground Water: Prevention, Detection, and Remediation, API and NGWA Conference and Exposition,* Anaheim, California. November 15–17, 2000. pp. 283–295. (Stanley, A., Ed.).

Lahvis, M.A. and Rehmann, L.C. 1999. Simulation of Methyl Tert-Butyl Ether (MTBE) Transport to Ground Water From Immobile Sources of Gasoline in the Vadose Zone. In: *Proceedings of the 2000 Petroleum Hydrocarbons and Organic Chemicals in Ground Water: Prevention, Detection, and Remediation, API, NGWA, STEP Conference and Exposition,* Houston, Texas. November 17–19, 1999. pp. 247–259.

Landmeyer, J.E., Chapelle, F.H., Bradley, P.M., Pankow, J.F., Church, C.D., and Tratnyek, P.G. 1998. Fate of MTBE relative to benzene in a gasoline-contaminated aquifer (1993–98), *Ground Water Monitoring and Remediation.* Fall 1998, 93–102.

Landmeyer, J.E., Chapelle, F.H., Herlong, H.H., and Bradley, P.M. 2001. Methyl tert-butyl ether biodegradation by indigenous aquifer microorganisms under natural and artificial oxic conditions. *Environmental Science and Technology.* 35(6), 1118–1126.

Landmeyer, J.E., Pankow, J.F., and Church, C.D. 1997. Occurrence of MTBE and tert-butyl alcohol in a gasoline-contaminated aquifer. In: *Preprints of Symposia Papers,* Volume 37(1), pp. 413–415. Presented before the Division of Environmental Chemistry, American Chemical Society, San Francisco, California. April 13–17, 1997.

Mace, R.E. and Choi, W. 1998. The size and behavior of MTBE plumes in Texas. In: *Proceedings of the Petroleum Hydrocarbons and Organic Chemicals in Ground Water: Prevention, Detection, and Remediation Conference, API and NGWA Conference and Exposition,* Houston, Texas. November 11–13, 1998. pp. 1–11.

Mace, R.E., Fisher, R.S., Welch, D.M., and Para, S.P. 1997. *Extent, Mass, and Duration of Hydrocarbon Plumes from Leaking Petroleum Storage Tank Sites in Texas.* Geological Circular 97-1. Bureau of Economic Geology, The University of Texas at Austin.

Mormille, M.R., Liu, S., and Suflita, J.M. 1994. Anaerobic biodegradation of gasoline oxygenates: Extrapolation of information to multiple sites and redox conditions. *Environmental Science and Technology.* 28(9), 1727–1732.

Novak, J.T., Goldsmith, C.D., Benoit, R.E., and O'Brien, J.H. 1985. Biodegradation of methanol and tertiary butyl alcohol in subsurface systems. *Water Science Technology.* 17(9), Copenhagen, 71–85.

Odencrantz, J.E. 1998. Implication of MTBE for intrinsic remediation of underground fuel tank sites. In: *Proceedings of the Petroleum Hydrocarbons and Organic Chemicals in Ground Water: Prevention, Detection, and Remediation Conference, API, NGWA, STEP Conference and Exposition.* Houston, Texas. November 11–13, 1998. pp. 571–579.

Peargin, T.R. 2000. Relative depletion rates of MTBE, benzene, and xylene from smear zone NAPL. In: *Proceedings of the 2000 Petroleum Hydrocarbons and Organic Chemicals in Ground Water: Prevention, Detection, and Remediation. Special Focus: Natural Attenuation and Gasoline Oxygenates, API, NGWA, STEP Conference and Exposition,* Anaheim, California. November 15–17, 2000. pp. 207–212.

Peargin, T.R. 2001. Relative depletion rates of MTBE, benzene, and xylene from smear zone non-aqueous phase liquid. In: *Bioremediation of MTBE, Alcohols, and Ethers. The Sixth International In Situ and On-Site Bioremediation Symposium,* San Diego, California. June 4–7, 2001. pp. 67–74. (Magar, V.S., Gibbs, J.T., O'Reilly, K.T., Hyman, M.R., and Leeson, A., Eds.).

Rice, D.W., Grose, R.D., Michaelsen, J.C., Dooher, B.P., MacQueen, D.H., Cullen, S.J., Kastenberg, W.E., Everett, L.G., and Marino, M.A. 1995. *California Leaking Underground Fuel Tank (LUFT) Historical Case Analyses.* Environmental Protection Department, Environmental Restoration Division, Lawrence Livermore Laboratories. UCRL-AR-122207.

Rixey, W.G. and Joshi, S. 1999. Dissolution characteristics of MTBE, BTEX and 1,2,4-Trimethylbenzene from a residually trapped gasoline source. In: *Proceedings of the 2000 Petroleum Hydrocarbons and Organic Chemicals in Ground Water: Prevention, Detection, and Remediation Conference, API, NGWA, STEP Conference and Exposition,* Houston, Texas. November 17–19, 1999. pp. 107–122.

Rixey, W.G. and Joshi, S. 2000. *Dissolution of MTBE from a Residually Trapped Gasoline Source: A Summary of Research Results. A Summary of Research Results from AI's Soil and Groundwater Technical Task Force,* No. 13, American Petroleum Institute. http://www.api.org/ehs/ehss&gw.htm or http://www.api.org/ehs/bulletin13.pfd.

Salanitro, J.P., Johnson, P.C., Spinnler, G.E., Maner, P.M., Wisniewski, H.L., and Bruce, C. 2000. Field-scale demonstration of enhanced MTBE bioremediation through aquifer bioaugmentation and oxygenation. *Environmental Science and Technology.* 34(19), 4152–4162.

Schirmer, M., Butler, B.J., Barker, J.F., Church, C.D., and Schirmer, K. 1999. Evaluation of biodegradation and dispersion as natural attenuation processes of MTBE and benzene at the Borden field site. *Physics and Chemistry of the Earth, Part B: Hydrology, Oceans and Atmosphere.* 24(6), 557–560.

Shaffer, K.L. and Uchrin, C.G. 1997. Uptake of methyl tertiary butyl ether (MTBE) by groundwater solids. *Bulletin of Environmental Contamination and Toxicology.* 59, 744–749.

Small, M.C. and Weaver, J. 1999. An updated conceptual model for subsurface fate and transport of MTBE and benzene. In: *Proceedings of the Petroleum Hydrocarbons and Organic Chemicals in Ground Water: Prevention, Detection, and Remediation Conference, API, NGWA, STEP Conference and Exposition,* Houston, Texas. November 17–19, 1999. pp. 209–220.

Squillace, P.J., Pankow, J.F., Korte, N.E., and Zorgorski, J.S. 1997. Review of the environmental behavior and fate of methyl *tert*-buty ether. *Environmental Toxicology and Chemistry.* 16(9), 1836–1844.

Suarez, M.P. and Rifai, H.S. 1999. Biodegradation rates for fuel hydrocarbons and chlorinated solvents in groundwater. *Bioremediation Journal.* 3(4), 337–362.

Suflita, J.M. and Mormile, M.R. 1993. Anaerobic biodegradation of known and potential gasoline oxygenates in the terrestrial subsurface. *Environmental Science and Technology.* 27(5), 976–978.

USEPA. 1997. *Use of Monitored Natural Attenuation at Superfund, RCRA Correc-*

tive Action, and Underground Storage Tank Sites. Office of Solid Waste and Emergency Response. Directive 9200.4-17.

Weaver, J.W., Haas, J.E., and Sosik, C. 1999. Characteristics of gasoline releases in the water table aquifer of Long Island. In: *Proceedings of the Petroleum Hydrocarbons and Organic Chemicals in Ground Water: Prevention, Detection, and Remediation Conference, API, NGWA, STEP Conference and Exposition*, Houston, Texas. November 17–19, 1999. pp. 260–261.

Wiedemeier, T.H., Rafai, H.S., Newell, C.J., and Wilson, J.T. 1999. *Natural Attenuation of Fuels and Chlorinated Solvents in the Subsurface.* New York, John Wiley & Sons, Inc.

Wilson, J.T., Cho, J.S., Beck, F.P., and Vardy, J.A. 1997. Field estimation of hydaulic conductivity for assessments of natural attenuation. In: *In Situ and On-Site Bioremediation: Volume 2, Papers from the Fourth International In-Situ and On-site Bioremediation Symposium*, New Orleans, Louisiana. April 28–May 1, 1997, pp. 309–314.

Wilson, B.H., Shen, H., Cho, J., and Vardy, J. 1999. Use of BIOSCREEN to evaluate natural attenuation of MTBE. In: *Natural Attenuation of Chlorinated Solvents, Petroleum Hydrocarbons, and Other Organic Compounds 5. Proceedings of the Fifth International In Situ and On-Site Bioremediation Symposium*, San Diego, California. April 19–22, 1999. Volume 5(1). pp. 115–120. (Alleman, B.C. and Leeson, A., Eds.).

Wilson, J.T., Cho, J.S., Wilson, B.H., and Vardy, J.A. 2000. *Natural Attenuation of MTBE in the Subsurface under Methanogenic Conditions.* EPA/600/R-00/006. www.epa.gov/ada/kerrcenter.html.

Yeh, C.K. and Novak, J.T. 1994. Anaerobic biodegradation of gasoline oxygenates in soils. *Water Environment Research.* 66(5), 744–752.

Chapter 4

MTBE Occurrence in Surface and Ground Water

James A. M. Thomson and James W. McKinley, Applied Hydrology Associates, Inc.
Robert C. Harris and Alwyn J. Hart, Ph.D., Environmental Agency
Peter Hicks and David K. Ramsden, Ph.D., URS Corporation
Barbara Wilson, Dynamac Corporation

INTRODUCTION

This chapter summarizes the occurrence of MTBE in surface and ground water as reported in several studies. Information reviewed includes the last ten years of sampling results for the United States Geological Survey's (USGS) National Water Quality Assessment (NAWQA) Program and the findings of other MTBE sampling programs in the U.S. In addition, this chapter describes a study of MTBE occurrence performed by the Environment Agency of England and Wales; this provides a valuable counterpoint to the situation in the U.S. Most of the studies consider only MTBE occurrence. The end of the chapter summarizes additional studies comparing MTBE and benzene plume lengths performed in Texas, Florida, California, and South Carolina.

Because of the extensive nature of these data, this chapter presents an abbreviated outline of the major studies and findings. More complete details are presented in Appendix A.

MTBE AND THE USGS NAWQA PROGRAM

In 1991, the USGS initiated the NAWQA program, a systematic water quality survey conducted in 50 river basins and aquifer systems in nearly all 50 states (USGS, 2001). The program attempted to answer the questions of how and why water quality varies across the nation. Over the last ten years, the USGS has collected and interpreted data concerning water chemistry, hydrology, land use, stream habitat, and aquatic life, and analyzed ground and surface water samples for pesticides, nutrients, VOCs, trace elements, and aquatic ecology. Of the 4,023 ground water samples collected, 388 (10%) contained detectable concentrations of MTBE. The average MTBE concentration for the

388 samples was 280 μg/l, while the maximum concentration detected was 23,000 μg/l. Of the 94 springs sampled, MTBE was detected in only 3 (3%). Of the 1,515 surface water samples collected, 463 (30%) contained detectable concentrations of MTBE.

Most of the locations were not sampled repeatedly, but for those that were, MTBE concentrations generally decreased over time. From 1993 through 1999, the ratio of detections to analyses remained relatively constant. In 2000 and 2001, there were no reported detections; however, minimal reported data exist for those years. Spatially, only three river basins (South Platte River Basin, Connecticut/Housatonic/Thames River Basins [Unit MA-100], and the Coastal New Jersey and Long Island River Basin) account for approximately 66% of the total number of MTBE detections, but only 20% of the total number of MTBE analyses. Other basins across the U.S. exhibited a relatively low detection to analyses ratio.

The NAWQA program is a useful study for determining background MTBE concentrations; however, it does have some shortcomings. As part of the well selection criteria, wells where MTBE has been previously detected, such as LUST, Resource Conservation and Recovery Act (RCRA), and Superfund sites, are avoided. Similarly, wells near roads and highways, where one would expect to find a relatively high occurrence of gasoline components, are avoided (Lapham *et al.*, 1995). Since wells where MTBE already has been found or is likely to be found are avoided, it is hard to conclude that the NAWQA MTBE results are truly representative of the nationwide occurrence of MTBE. The long time-scale of the program limits the ability to identify a new contaminant of concern and take quick remedial action. Small-scale, more focused studies will be needed to accomplish these goals.

NATIONAL MTBE SURVEY AND THE NORTHEASTERN AND MID-ATLANTIC STATES STUDY

The USGS has also been looking at MTBE occurrence on a regional basis. Working with the Metropolitan Water District of Southern California and the Oregon Graduate Institute, the USGS conducted a National MTBE Survey for the American Water Works Association Research Foundation (Clawges *et al.*, 2000). The survey randomly selected 954 community water supplies (CWSs) consisting of 579 wells, 171 rivers, and 204 reservoirs within all 50 states and Puerto Rico. A public water supply (PWS) is defined as a system that serves piped water to at least 25 people or 15 service connections for at least 60 days a year. A CWS is a PWS that serves people year round in their homes (USEPA, 2001). The initial findings of the survey, reported on June 20, 2001, indicated that MTBE was the second most commonly detected VOC after chloroform. MTBE was detected in 9% of all sources sampled (14% of surface water sources and 5% of ground water sources), 4% of CWSs serving 10,000 people or less, and 15% of CWSs serving 50,000 people or more (Hirsch, 2001).

Working with the USEPA's Office of Ground Water and Drinking Water, the USGS conducted a Northeastern and Mid-Atlantic States Study, which summarized the occurrence of MTBE in Connecticut, Maine, Maryland, Massachusetts, New Hampshire, New Jersey, New York, Rhode Island, Vermont, and Virginia. Data were provided for 10,479 CWSs within those states; USGS randomly selected 2,110 of these CWSs for data review. Data on MTBE sampling and analysis were provided for 1,194 of the 2,110 CWSs selected. MTBE was detected in only 9% of the CWSs with available MTBE analysis data. MTBE concentrations were above the USEPA advisory level of 20 µg/l in less than 1% of those CWSs impacted by MTBE (Grady and Casey, 1999). Based on the occurrence percentages, the USGS extrapolated that an additional 180 CWSs in the study area could contain MTBE at concentrations above 5 µg/l, and that MTBE concentrations in 80 CWSs could exceed 20 µg/l.

NORTHEAST STATES FOR COORDINATED AIR USE MANAGEMENT (NESCAUM)

On November 9, 1998, as a follow-up to the Northeastern and Mid-Atlantic States USGS report, New Hampshire Governor Jeanne Shaheen asked NESCAUM to assess what steps the northeastern states should take to maximize air quality without sacrificing water quality. In response to the request, NESCAUM compiled a series of technical papers that summarized the occurrence of MTBE in Connecticut, Maine, Massachusetts, New Hampshire, New Jersey, New York, Rhode Island, and Vermont (NESCAUM, 1999). The papers concluded that MTBE detected in the study area was typically at concentrations below the USEPA advisory level, but that LUST sites could pose a serious health risk due to MTBE's high solubility and low adsorption potential. BTEX and MTBE occurrence data were compared; BTEX compounds were detected at only 12% of the sites where MTBE was detected. Overall, BTEX was detected infrequently.

MIDWESTERN STATES STUDY

In October 2001, ENSR International and Applied Hydrology Associates, Inc. (AHA) reported on the occurrence of gasoline components in the ground water of seven Midwestern states (ENSR and AHA, 2001). The report analyzed 77 ground water samples from 29 PWS systems in Colorado, Illinois, Minnesota, and Nebraska. In these states and Indiana, Kansas, and Wisconsin, 231 ground water samples were collected at 70 LUST sites, and 35 samples were collected at intermediate locations between 6 LUST sites and the nearest PWSs. Samples from only one PWS site (two wells in Hyannis, Nebraska containing 19 and 170 µg/l benzene) exceeded standards for any of the gasoline constituents measured (BTEX, ethanol, MTBE, and TBA). Of the LUST sites that were sampled, BTEX and MTBE were detected at 86% and 70% of the sites respectively, and BTEX concentrations exceeded MTBE plus TBA concentrations at 64% of the sites. The study concluded that gasoline releases have had minimal impact on study area PWSs and that remediation efforts and natural attenuation have been sufficient in reducing LUST impacts to PWSs.

INDIVIDUAL STATE STUDIES

In addition to regional MTBE studies, many states have conducted their own investigations of MTBE occurrence. Summaries of these individual state studies are presented in Appendix A. In general, MTBE detection frequencies are low in most areas of the U.S. In those areas where detection frequencies are relatively high compared to the national background, MTBE concentrations are often below the USEPA advisory level of 20 µg/l. Most of the individual state reports did not include data on the occurrence frequency or average concentrations of BTEX components; therefore, a comparison of MBTE to BTEX was not possible.

MTBE OCCURRENCE IN ENGLAND AND WALES

Most of the unleaded fuel in England and Wales contains MTBE, but at proportions lower than fuels used in the U.S. Ground water accounts for approximately 35% of PWSs in England and Wales, and almost 75% in the densely populated southeastern area of England.

In 2000, the England and Wales Environment Agency collected data on the occurrence of ethers in ground water from all available data sources and archives, which included approximately 800 site investigations and 3,000 water samples from PWSs and monitoring wells. Additionally, major oil companies provided data on 2,069 retail gas stations, transfer depots, and oil terminal sites where gasoline releases were suspected. Of the 2,069 sites, 837 had data on ethers, and MTBE was detected at 25% of these sites. Of 940 PWS and observation wells, 255 were tested for ethers; MTBE was detected in 32 wells (12%), although only 3 wells contained MTBE at concentrations above 5 µg/l. Based on the detection frequencies listed above, the Agency constructed a model (Environment Agency, 2000) that would extrapolate the potential number of MTBE impacted wells among the 1,944 PWS wells in England and Wales. The model concluded that 203 (10%) of the PWS wells could contain MTBE, but only 6 would have MTBE concentrations exceeding 5 µg/l. As a result, the Agency concluded that MTBE currently does not pose an environmental or human health threat in England and Wales.

PLUME LENGTH STUDIES

The above sections have dealt primarily with the frequency of MTBE detections in ground and surface water across the U.S., England, and Wales. However, there have also been a number of studies that summarize the movement and behavior of MTBE in the subsurface. These plume studies, in Texas, Florida, California, and South Carolina, were conducted to evaluate the fate and transport of MTBE compared with other gasoline components, especially BTEX.

The Texas plume study (Mace and Choi, 1998) compiled MTBE and benzene data from 609 LUST sites; the compiled data allowed for the comparison of plumes at 79 sites. For these 79 sites, the average MTBE plume length at a concentration of 10 µg/l was 55 meters (182 feet), 8 meters (27 feet) longer than the average benzene plume length of 47 meters (155 feet). At the 609 LUST sites, MTBE concentrations were non-detect at 24% of the monitoring wells, decreasing at 9% of the wells, stable at 50%, increasing at 7%, and erratic at 10%.

Data from 149 British Petroleum (BP) sites were reviewed as part of the Florida plume study (Reid et al., 1999), which allowed for the comparison of plumes at 55 sites. In contrast to the Texas study, average plume lengths were determined in the Florida study by taking into account the state benzene drinking water standard and the USEPA MTBE advisory level (1 µg/l and 20 µg/l respectively). At these concentrations, 96% of the MTBE plumes were equal to or shorter in length than the corresponding benzene plume. Of all the benzene and MTBE plumes, 4.4% were increasing in size, 6.6% were stable, and 89% were decreasing in size.

The California plume study (Happel et al., 1998) examined 236 LUST sites, which allowed plume comparisons at 63 sites. For the average plume lengths, the California study also used 1 µg/l and 20 µg/l for contouring benzene and MTBE, respectively. Using these concentrations, 81% of MTBE plumes were equal to or shorter in length than the corresponding benzene plumes. When taking into account the USEPA advisory level for MTBE and the maximum contaminant level (MCL) for benzene in drinking water, benzene was more often the driver for remediation.

The South Carolina plume study (Wilson et al., 2001) evaluated MTBE and BTEX concentration data from 212 UST cleanup sites. At 171 of the 121 UST sites, the BTEX plume length was equal to or greater than the corresponding MTBE plume length. Plumes were also evaluated by soil type (sand, silt, or clay); the median plume lengths and areas for MTBE and BTEX were larger in silt formations, but the mean lengths and areas did not vary significantly with soil type. Wilson's findings were comparable with the results from the Texas, Florida, and California plume studies described above.

HISTORY IN CALIFORNIA

Over recent years, California has received extensive media attention concerning MTBE, much of it negative. Initial studies reported in 1995 assumed that detection frequencies would either stay stable or increase as more widespread testing was performed. As a result, there was an initial feeling that MTBE would become a much greater problem than has actually been observed in the last six years.

At the end of 1995, MTBE was detected in 9 of 11 of Santa Monica's high volume production drinking water wells, which were subsequently shut down (Cal/EPA, 1999). In the Charnock wellfield, MTBE concentrations reached highs of 610 μg/l in production wells, 17,000 μg/l in regional monitoring wells, and 230,000 μg/l in LUST site monitoring wells (Blue Ribbon Panel, 1999). In South Lake Tahoe in 1996, 13 of the District's 34 drinking water wells were shut down because of the existing or potential threat of MTBE contamination (Dernbach, 2000). In 2000 as a result of the shutdown, the District limited the percentage of MTBE in gasoline to 0.6% by volume (Anonymous, 2001).

After these initial scares, the occurrence of MTBE in California does not appear as widespread as initially thought. On September 5, 2000, the California Department of Health Services (CDHS) estimated that only 0.8% of drinking water sources and 1.9% of PWSs sampled in California in 2000 contained detectable levels of MTBE; MTBE concentrations in less than 1% of these sources exceeded the state primary MCL of 13 μg/l. Over the last six years, MTBE detection frequencies averaged 1.3% for all samples. The highest frequency was in 1995 (approximately 5%), but in the last five years, the frequency has dropped and stayed below 2%. Between 1994 and 2000, no MTBE was detected in 95% of California drinking water supplies. Of the remaining 5% of California drinking water supplies in which MTBE was detected, 73% of samples and 86% of sources contained concentrations below the state's MCL of 13 μg/l (Williams and Sheehan, 2001; Williams *et al.*, 2000).

Because of concerns about MTBE impact in California, Governor Gray Davis established a panel of University of California (UC) professors and researchers to prepare a comprehensive overview of the human health and environmental effects of MTBE use (Keller *et al.*, 1998; Keller, 1999). The results were released in a November 1998 report entitled "Health and Environmental Assessment of MTBE," which influenced Davis' decision to phase out the use of MTBE in California by December 31, 2001.

Malcolm Pirnie (Malcolm Pirnie, 2001) prepared a commentary on the UC report for the Methanol Institute. According to this commentary, current evidence did not support the negative stand on MTBE taken by the UC panel. The UC report had extrapolated that 60 to 340 additional PWSs in California could be impacted by MTBE in the future. However, Malcolm Pirnie's commentary accounted for the lower MTBE detection frequencies in 2000 and 2001, and estimated the number of potential future MTBE-impacted PWSs closer to 16.

CONCLUSIONS

The studies described above represent a very large body of data. Drawing general conclusions needs to be approached with caution because of variations between studies. An important variable is the purpose, which can contain inherent biases towards a particular sample set. Specific program objectives could include one or more of the following:

1. To statistically represent water quality conditions in an aquifer system;
2. To detect previously unknown VOC plumes and attempt to identify their sources;
3. To characterize the status of LUST site management;
4. To determine the effectiveness of the UST upgrade program, remedial response, or natural attenuation in reducing MTBE occurrence, concentration, or plume size;
5. To identify which chemical should "drive" remediation and the oversight program activities in an administrative area; and
6. To compare MTBE fate and transport with that of benzene or BTEX.

Other factors that often vary from study to study and from state to state are listed below.

1. Is the study area a designated RFG area or a non-RFG area?
2. If the study area is an RFG area, when was oxygenated fuel introduced, and what oxygenates have been used?
3. If a non-RFG area, are oxygenates used for reasons other than RFG designation?
4. When was MTBE changed from an optional to a required monitoring analyte?
5. Is the CWS or PWS drawn from a ground water or a surface water source? MTBE breakdown is more rapid and detection frequency is lower in surface water.
6. Are wells to be sampled selected only from known LUST sites? This biases detection frequencies upwards.
7. Alternatively, are known LUST sites excluded from the sampled population? This biases detection frequencies downwards.
8. Are "randomly" chosen wells actually randomly distributed? For example, wells may follow the position of a specific aquifer formation, which in turn may be associated with areas of urban development and LUST sites.
9. What type of well is sampled? The type of well sampled affects its catchment area: samples from large CWS wells represent water drawn from a larger radius than smaller, private domestic supply wells, which in turn sample a larger area than unpumped monitoring wells. However, monitoring wells are typically installed in urban areas, often around gas stations, and, therefore, have more probability of being in or near a LUST plume.
10. What is the vertical permeability of the unsaturated zone? This will affect the rate of plume development.

11. What is the lateral permeability and ground water flow rate? This affects the rate of plume growth.
12. What are the local standards for BTEX and MTBE, and what are the variations in these standards over time?
13. Are plume lengths defined by computer or hand-drawn plume contours?
14. What is the defining contour for plume length (*e.g.*, 1 µg/l, 20 µg/l, 200 µg/l)?
15. How complete is the monitoring network for plume length characterization? (Was vertical delineation complete?)
16. What detection limits have been used for MTBE and other constituents?

The broadest conclusions that may be drawn from the data reviewed herein are that the threat to water quality posed by MTBE has not proved to be as widespread, persistent, and intractable as initially forecast, and in general it is associated with other gasoline constituents. Typically, the environmental concern is a LUST problem rather than an MTBE problem.

Overall, MTBE was detected at the highest frequencies and concentrations between 1995 and 1998. Detection frequencies and concentrations have reduced since 1998, and in most cases, continue to drop from year to year. If the 1995 to 1998 sampling was truly representative, and if MTBE plumes were rapidly growing, then continued sampling would be expected to result in stable or increasing MTBE detection frequencies and concentrations. However, since 1998, results have shown the opposite trend. This suggests that: (1) the 1995 to 1998 sampling was biased towards sites with higher MTBE concentrations; (2) the incidence of oxygenated fuel releases is decreasing due to better USTs and improved monitoring; and (3) remedial efforts and natural attenuation are mitigating MTBE impacts to ground water.

REFERENCES

Anonymous. 2001. *Some FAQ's About MTBE in South Tahoe.* http://tahoe. ceres.ca.gov/stpud/faqsmtbe.html.

Blue Ribbon Panel on Oxygenates in Gasoline. 1999. *Achieving Clean Air and Clean Water: The Report of the Blue Ribbon Panel on Oxygenates in Gasoline.* EPA-A20-R-99-021. September 15, 1999.

Cal/EPA (California Environmental Protection Agency). 1999. *Public Health Goal for Methyl Tertiary Butyl Ether (MTBE) in Drinking Water.* Pesticide and Environmental Toxicology Section, Office of Environmental Health Hazard Assessment, California Environmental Protection Agency, Sacramento, California. March 1999.

Clawges, R.M., Zogorski, J.S., and Bender, D. 2000. *Abstract: Key MTBE Findings Based on National Water-Quality Monitoring.* U.S. Geological Survey, Rapid City, South Dakota.

Dernbach, L.S. 2000. The complicated challenge of MTBE cleanups. *Environmental Science and Technology*. 34(23), 516A–521A.

ENSR International and Applied Hydrology Associates, Inc. 2001. *Investigation of Selected Gasoline Constituents in Groundwater in Seven States*. ENSR Document #04373-006-200. October 2001.

Environment Agency. 2000. *A Review of Current MTBE Usage and Occurrence in Groundwater in England and Wales*. TSO ISBN 0 11 310181 3.

Grady, S.J. and Casey, G.D. 1999. *Occurrence and Distribution of Methyl tert-Butyl Ether and Other Volatile Organic Compounds in Drinking Water in the Northeast and Mid-Atlantic Regions of the United States, 1993-1998*. U.S. Geological Survey Water-Resources Investigation Report, 00-4228.

Happel, A.M., Beckenbach, E.H., and Halden, R.U. 1998. *An Evaluation of MTBE Impacts to California Groundwater Resources*. Lawrence Livermore National Laboratory, University of California, Livermore, California. UCRL-AR-130897. June 11, 1998.

Hirsch, R.M. 2001. *Statement Before the United States House of Representatives, Committee on Energy and Commerce, Subcommittee on Oversight and Investigations*. Associate Director for Water, U.S. Geological Survey, U.S. Department of the Interior. November 1, 2001. http://sd.water.usgs.gov/nawqa/vocns/USGS_MTBE_testimony.html.

Keller, A., Froines, J., Koshland, C., Reuter, J., Suffet, I., and Last, J. 1998. Health and environmental assessment of MTBE, summary and recommendations. In: *Health and Environmental Assessment of MTBE*, Volume 1. University of California Toxics Research and Teaching Program, University of California Davis. November 1998.

Keller, A. 1999. Health and environmental assessment of MTBE—The California perspective. In: *Proceedings of the American Water Works Association Annual Conference*, Chicago, Illinois. June 20–24, 1999.

Lapham, W.W., Wilde, F.D., and Koterba, M.T. 1995. *Ground-Water Data-Collection Protocols and Procedures for the National Water-Quality Assessment Program: Selection, Installation, and Documentation of Wells, and Collection of Related Data*. U.S. Geological Survey Open-File Report 95-398.

Mace, R. E. and Choi, W. 1998. The size and behavior of MTBE plumes in Texas. In: *Proceedings of the Petroleum Hydrocarbons and Organic Chemicals in Ground Water: Prevention, Detection, and Remediation Conference, API and NGWA Conference and Exposition*, Houston, Texas. November 11–13, 1998. pp. 1–11. (Stanley, A., Ed.).

Malcolm Pirnie, Inc. 2001. *Water Quality Impacts of MTBE: An Update Since the Release of the UC Report*. Prepared for the Methanol Institute, Oakland, California. August 2001.

NESCAUM. 1999. *RFG/MTBE Findings and Recommendations*. Northeast States for Coordinated Air Use Management, Boston, Massachusetts. August 1999. http//www.nescaum.org/RFG/RFGPh2.shtml.

Reid, J.B., Reisinger, H.J. II, Bartholomae, P.G., Gray, J.C., and Hullman, A.S.

1999. A comparative assessment of the long-term behavior of MTBE and benzene plumes in Florida, USA. In: *Natural Attenuation of Chlorinated Solvents, Petroleum Hydrocarbons, and Other Organic Compounds, Proceedings of the Fifth International In Situ and Onsite Bioremediation Symposium*, Volume 1, San Diego, California. April 19–22, 1999. pp. 97–102. (Alleman, B.C. and Lesson, A., Eds.). Columbus, Ohio, Battelle Publications.

USEPA (U.S. Environmental Protection Agency). 2001. *Where Does My Drinking Water Come From?* www.epa.gov/OGWDW/wot/wheredoes.html.

USGS (U.S. Geological Survey). 2001. *The National Water-Quality Assessment Program-Informing Water-Resource Management and Protection Decisions.* http://water.usgs.gov/nawqa/docs/xrel/external.relevance.pdf.

Williams, P.R.D., Scott, P.K., Sheehan, P.J., and Paustenbach, D.J. 2000. A probabilistic assessment of household exposures to MTBE in California drinking water. *Human and Ecological Risk Assessment.* 6(5), 827–849.

Williams, P.R.D. and Sheehan, P.J. 2001. A better perspective on the incidence and implications of MTBE in California's drinking water. *Contaminated Soil Sediment and Water.* Spring (Special Issue), 23–28.

Wilson, B.H., Shen, H., Pope, D., and Schemelling, S. 2001. Cost of MTBE remediation. In: *Bioremediation of MTBE, Alcohols, and Ethers. The Sixth International In Situ and On-Site Bioremediation Symposium.* San Diego, California. June 4–7, 2001. pp. 129–136. (Magar, V.S., Gibbs, J.T., O'Reilly, K.T., Hyman, M.R., and Leeson, A., Eds.). Columbus, Ohio, Battelle Press.

CHAPTER 5

Site Assessment

Nancy E. Milkey, P.G., Tighe & Bond, Inc.

Following the identification of a release or suspected release on a site, a site assessment is conducted to identify the nature and extent of contamination, potential sources, sensitive receptors, and migration pathways. A well-designed and implemented site assessment provides a sound basis for risk characterization and subsequent remediation. The heterogeneity of the subsurface environment and ground water flow regimes encountered at sites necessitate the need for careful evaluation of the underlying stratigraphy, affected media, and potential migration pathways.

A site assessment usually progresses through a series of phases including a historical records review; initial subsurface investigation to determine the source(s) of contamination, as well as the nature and general extent of the release; a detailed assessment to fully delineate the release and migration pathways; and, if necessary, the collection of additional data needed for remediation design. The information obtained during each phase of the investigation is used to focus the subsequent work and create a comprehensive understanding of the site.

Due to the volatile nature of the majority of compounds involved in gasoline releases, the presence or absence of MTBE does not significantly impact how a site assessment is conducted. This chapter highlights some of the most common methods and approaches used in the assessment of gasoline release sites.

HISTORICAL ASSESSMENT

Initially, the history of the site and surrounding area is researched to determine current and previous property uses; the size, age, contents, construction, and location of USTs; proximity to drinking water source(s); types of sewage disposal systems (municipal versus private); known releases at the site and neighboring properties; and hazardous materials usage. Part of the assessment includes interviews with both current and previous occupants or owners of the site, and a review of municipal files and/or interviews with municipal officials including those of the building, assessors, planning, fire, health, sewer, and water departments. This information is used to focus the subsurface investigation on sensitive receptors or areas of the site that are likely to have been impacted by the release.

The historical assessment also includes a review of available mapping including site plans, and topographic and geologic maps for the site vicinity. Based on the review of published mapping, regional ground water flow direction can often be estimated.

Identification of Receptors.   During the historical and municipal records review and interviews, the presence of sensitive receptors at, and near, the site are identified. Some examples of receptors that should be noted include: private/municipal water supply wells; building basements; wetlands/surface water bodies; daycare facilities/schools; and residences (particularly focusing on homes where children and/or elderly reside).

INITIAL SUBSURFACE INVESTIGATION

The first phase of the subsurface investigation typically includes the installation of three to four borings that are completed as water table monitoring wells (well screens intersect the water table). The number of wells is dependent on the size and complexity of the site and the anticipated extent of the release(s). The locations of the monitoring wells are based on the location of potential sources and receptors and the estimated direction of ground water flow. Wells are usually installed:

- Upgradient of the suspected source to provide a monitoring point to determine the quality of ground water migrating onto the site;
- At the location of sensitive receptors;
- At or adjacent to the suspected source(s); and
- Downgradient of the suspected source(s) to provide initial information on the general extent of the plume at the water table.

Utility Clearance.   By law, prior to installing monitoring wells or advancing borings, a utility locating service must be notified. Marking companies use paint, stakes, or flags to identify the locations of utilities. Colors differentiate the type of utility; for example, in New England electric lines are marked in red and potable water pipes in blue.

Boring Advancement.   The purpose of characterizing the subsurface soil conditions is to identify features that affect the flow of ground water and the migration of contaminants (API, 2000). Subsurface conditions, particularly in previously glaciated terrain, are rarely homogeneous and it is particularly useful to identify highly permeable strata that may serve as preferential contaminant migration pathways.

During boring advancement, soil samples are traditionally collected at a minimum of 1.5-meter (5-foot) intervals. However, it is recommended that during an initial site investigation, samples be collected continuously to identify changes in stratigraphy that may be missed by discontinuous sampling. Samples are examined and logged, and field notes taken documenting as much detail as possible on the moisture content, color, soil texture, and odor for each sample. Of particular importance is the depth at which changes in the stratigraphy occur. The field notes are used to draft boring logs that provide a description of the geologic formations and thickness of each stratum encountered as a function of depth (Todd, 1980).

As soil samples are collected they are typically screened in the field for fuel components such as VOCs and TPH. The screening is used to direct the subsurface investigation in the field and to provide a basis for selecting samples for laboratory analysis.

Soil Screening Methods. A variety of meters and test kits are available to assist in the identification and delineation of releases in the field; however, the biases and limitations of each technique must be understood to select the best method for a given situation and effectively utilize the data generated. The selection of the appropriate piece of equipment includes an evaluation of the objectives of the field screening, anticipated analytes and concentrations, potential matrix interferences, and the end use of the data.

Field screening is useful to determine if contamination may be present in a sample. However, due to the general nature of field screening, it is often not possible to determine specific contaminant identities and concentrations. For example, photoionization detectors (PIDs) detect VOCs with ionization potentials less than that of the instrument's lamp. The majority of PIDs are equipped with either a 10.2 or 10.6 electron volt (eV) lamp with which the majority of chlorinated solvents, BTEX compounds, and MTBE would be detected. A positive response indicates that a volatile compound is present, but it would not be possible to identify or quantify the specific compound. Also, if a VOC with an ionization potential higher than that of the lamp is present, it would not be detected.

Due to the high vapor pressure of MTBE and other gasoline constituents, dual samples should be collected when field screening is required. One sample jar can be used for the screening. The second jar remains sealed and is reserved for potential laboratory analysis. To minimize the loss of VOCs prior to submittal of the samples for laboratory analysis, soil samples need to be collected in compliance with USEPA Method 5035. The method's most significant element is the preservation of soil samples with MeOH (MADEP, 1999).

Several of the most common soil screening tools are discussed below (MADEP, 2001 and 1996).

- PID—Excellent screening tool for gasoline releases (but less useful for products with lower volatilities). The instrument responses can be negatively affected by high humidity or moisture, and the instrument can provide non-linear responses for concentrations in excess of 150 parts per million (ppm). PIDs are calibrated to an isobutylene standard that is in the middle of the range of the PID sensitivity (Robbins, 1996).
- Flame Ionization Detector (FID)—FIDs are typically used for the same purpose as PIDs, although ionization within an FID is caused by a hydrogen flame. Compounds with an ionization potential of 15.4 eV or less can be detected with an FID. FIDs are less susceptible to the effects of humidity than PIDs, and by calibrating with methane (its most sensitive constituent), the FID is more sensitive than the PID for field screening for gasoline. However, if methane is present (from the degradation of organic materials such as in septic systems) a positive reading will be obtained. The presence of methane can be identified by using a carbon filter attached to the FID; duplicate FID readings with and without the filter can determine if other VOCs are present (Robbins, 1996).
- Immunoassay Test Kits—Used to detect either specific compounds or groups of compounds, such as polynuclear aromatic hydrocarbons (PAHs), gasoline, or aviation fuel. The kits do not directly quantify carbon fractions, individual compounds, or TPH.
- Emulsion-Based TPH Methods—A solvent is used to extract hydrocarbons from a soil sample. Suspended materials are then filtered from the extract to prevent interference. A developing solution is added, and the soil extract develops a response in proportion to the concentration of hydrocarbon contained in the soil sample. The sample is then placed into an analyzer, which provides a digital readout in ppm. Interference is possible in soils containing a high organic content or clay.

Well Development.   Following completion of the soil borings as monitoring wells, the wells are developed to remove the fine particles from the permeable zone(s) adjacent to the well screen. Well development increases the production (specific capacity) of the well. There are numerous methods commonly used for development including overpumping (pumping at a high rate), backwashing (causing backflow into the formation to break down sand bridges that result when water flows in only one direction), and mechanical and air surging (Driscoll, 1986).

Ground Water Sample Collection. Ground water samples are typically collected after at least seven days have elapsed since the monitoring wells were installed. Additional information on ground water sampling techniques is included later in this chapter.

As during soil sample collection, field observations are recorded, including any color or odor associated with ground water evacuated from each well. The color of ground water can provide information on potential contaminants. For example, ground water containing elevated concentrations of iron frequently has a reddish color. In addition, the presence of any sheens, silt, or free product must be noted. Personnel in the field are the eyes and ears of the project managers and design professionals who will be using their data. Consequently, comprehensive and detailed field notes are essential to the success of the investigation.

Once the well has been purged, ground water samples can be collected. The samples are placed directly into containers specifically prepared for the analyses to be conducted and then placed on ice in preparation for timely transport to an analytical laboratory.

Determination of Ground Water Flow Direction. Site-specific ground water flow direction is an important parameter to determine early in an assessment. Once monitoring wells have been installed on a site, the locations and elevations of the wells are surveyed relative to a benchmark. The depth to water is measured in all monitoring wells located on the site. The measurements can then be used in conjunction with the survey data to calculate the relative ground water elevations. The data are used to determine ground water elevation gradients and, in conjunction with hydraulic conductivity and porosity, groundwater velocities at the site.

The initial ground water flow direction, determined from a minimum of three wells and based on one sampling event, is useful to provide a general sense of the flow regime at the site. However, as discussed in Chapter 3, variations in ground water flow are common based on seasonal changes, rainfall events, and interactions with surface water. Consequently, it is important to define these variations; ground water depth measurements should be collected on a seasonal basis throughout the year to determine the cycle of annual fluctuations. It is also important to keep in mind that ground water flow paths are not always straight lines.

METHODS OF SOIL AND GROUND WATER SAMPLE COLLECTION

The primary goal of any site assessment is to collect data that are representative of conditions at the site. The selection of particular sampling methods or equipment will be based on the objectives of the sampling program, the contaminants and concentrations of concern, the geologic setting, and the physical constraints of the site.

Drilling and Soil Sample Collection. Numerous methods are available for collecting soil samples and installing wells. Several conventional methods include the following.

Traditional Drilling Methods. A drilling method is selected based on the purpose of the boring or well, the depth to ground water, and geologic conditions at the site. Frequently, a particular drilling technique is common in a localized region because it is most effective in the geologic conditions present in that area. In general, the driller is best qualified to determine which drilling method should be used for a particular geologic condition and purpose (Driscoll, 1986).

- Hollow Stem Auger (HSA)—The HSA method is commonly used for the installation of monitoring wells in unconsolidated materials. It is a fast and efficient technique for completing wells to moderate depths. The use of HSA is particularly advantageous for obtaining accurate depth samples (Driscoll, 1986). Soil samples are usually collected using 0.6-meter (2-foot) long split spoon samplers that are driven ahead of the augers to collect *in situ* samples from selected depths. Once the boring has reached a desired depth, a monitoring well can be installed without the use of casing or drilling fluids.
- Air rotary—Air rotary is an option for installing wells in unconsolidated materials that do not require a fluid to support the walls against caving. Drilling depths can exceed 150 meters (490 feet) under favorable conditions (Todd, 1980). The use of this method is not recommended when the collection of soil samples is necessary.
- Cable Tool—The cable tool percussion method was the earliest drilling method developed and has been in use for about 4,000 years. Cable tool drilling is accomplished by the repeated percussion of lifting and dropping a string of heavy drilling tools. The drill bit breaks or crushes consolidated rock into small fragments, which are removed in a slurry of water and rock fragments (Driscoll, 1986). This method is least useful in unconsolidated sand and gravel because the overburden slumps into the borehole and caves in around the bit. This technique can be used for the installation of monitoring wells at depth even through boulders and fractured or fissured rocks that other methods may not be able to penetrate (Todd, 1980).
- Sonic—Sonic drilling is also known as rotasonic, rotosonic, sonicore, vibratory, or resonantsonic drilling. It employs the use of high frequency mechanical vibration (50 to 150 hertz or cycles per second) to collect continuous core samples of overburden and "soft" bedrock formations, such as sandstone, shale, or slate. Harder bedrock may require adaptation of the rig for diamond wire line or air hammer

drilling (Boart Longyear, 1997). One of the advantages of sonic drilling is that very few drill cuttings are generated during boring advancement (Global Environment and Technology Foundation, 2002). However, it is frequently an expensive option as compared to other drilling methods.

- Coring—Coring can be used with a variety of drilling methods to obtain samples of bedrock prior to disturbing the material with a drill head or bit. Cores are useful during site assessment to evaluate fracturing and determine the competency of the rock underlying the site.

Direct Push Methods. Direct push methods are more recent innovations that provide a rapid means of collecting soil samples without generating excess cuttings. Soil samples can be collected continuously or the probe can be advanced to a discrete depth for sample collection. The versatility of direct push methods makes them ideal for assessment activities where a variety of media need to be sampled; tools attached at the end of steel rods can be used to collect soil, ground water, and soil gas samples. Small diameter wells and permanent soil gas points can be installed if additional rounds of sampling are required.

In addition, a membrane interface probe (MIP) has been developed for direct push pneumatic hammer equipment. MIPs provide depth-discrete detection and quantitation of dissolved phase organic constituents. As the MIP is advanced into the subsurface either a PID, FID, or both instruments are connected in series to detect VOCs. In addition, an electrical log is created of the formation. The electrical log, in conjunction with the sensor data, provides information on contaminant distribution and migration pathways (Geoprobe®, 2002). The MIP has also been adapted for use with a standard geophysical cone penetrometer tip (USDOE, 2002).

Direct push equipment can be mounted on a variety of vehicles including pickup trucks, all-terrain vehicles, and skid rigs. However, the technology does not work at sites with gravel, shallow bedrock, or numerous boulders.

Cone Penetrometer Technology (CPT). CPT provides "real-time" data for use in characterizing subsurface conditions. A steel cone is hydraulically advanced into the ground, and sensors at the tip and sidewalls of the cone collect data that can be used to classify soil type and detect the presence of a variety of contaminants. CPT can also be used in conjunction with ground water sampling equipment to extract ground water grab samples. The application of CPT can be limited at sites with high soil densities or numerous boulders. The data collected using CPT are primarily used as a screening tool to provide initial site characterization data, and to identify the underlying site geology and zones of high and low hydraulic conductivity. Additional samples need to be collected and analyzed by other methods to confirm soil textures and contaminant concentrations.

The U.S. Army Engineer Waterways Experiment System has developed the Site Characterization and Analysis Penetrometer System (SCAPS). The system provides a rapid and cost-effective means to characterize soil conditions. A SCAPS platform is used to push a cone penetrometer into the ground. Data are collected continuously and can be used to determine soil stratigraphy and identify the presence of contaminants. Sensors and sampling systems, attached to the end of the penetrometer, provide real-time detection of petroleum products, explosives, VOCs, solvents and gamma emitting radionuclides (USACE, 1998).

Ground Water Sample Collection.   The most common method of sample collection includes purging a well using either relatively high flow pumping or bailing until a minimum number of well volumes (usually three) has been evacuated. Concerns, including the entrainment of particles and the costs of disposing of large volumes of contaminated purge water, have led to the development of alternative methods for purging and sampling (USEPA, 1993).

Low-Flow (Minimal Drawdown) Ground Water Sampling.   Water quality data collected using low-flow procedures are believed to be more representative of actual ground water conditions than data obtained using high flow purging or bailing techniques. Low-flow methodology requires that purging and sampling be conducted in a manner that minimizes chemical and hydrological disturbance in and around the well. In addition, the completion of purging is based on the stabilization of water quality parameters rather than the removal of a fixed number of well volumes. According to USEPA, the advantages of low-flow sampling include the collection of more representative samples, the minimization of contaminated purge water volumes, and better spatial resolution in the sampling (USEPA, 1993).

Passive Diffusion Bag (PDB) Samplers.   PDB sampling employs a low-density polyethylene bag filled with deionized water that acts as a semipermeable membrane when lowered into the ground water. VOCs diffuse across the bag until equilibrium between the concentration in the PDB and ground water is reached. PDB samplers are disposable and require only a small amount of field equipment, minimizing the need to decontaminate equipment between holes and saving time. The results of laboratory testing indicate that PDB samplers should not be used for MTBE, acetone, and most semi-volatile organic compounds (SVOCs). Studies indicate that these compounds are transmitted through the bag, but the resulting concentrations are lower than in the surrounding ground water (ITRC, 2002).

Multi-Level Monitoring Systems. To completely characterize the vertical extent of contamination at sites with MTBE releases, the collection of depth-specific samples are frequently necessary (API, 2000). Multi-level monitoring can be accomplished by installing multiple wells to different depths; however, this is time-consuming and expensive.

Alternative multi-level monitoring systems using a single well have recently been developed. One system, employing multi-channel tubing, uses either inflatable packers or bentonite seals to separate individual screened sections of the well (*In situ*, 2002).

A second technology utilizes a low profile collapsible bag sampler. Ground water samples are collected from immediately above a check valve that prevents water from entering the sampler while it is lowered into the well. It has similar advantages to the PDB sampler in that no purge water is generated, but unlike the PDB it can successfully be used at sites with MTBE because it relies on advection rather than diffusion for sample collection. Multi-level sampling can be accomplished by stacking several samplers in the well to provide a vertical contaminant profile (Cordry, 2002). (Long well screens should always be used with caution due to the possibility of penetrating aquicludes and cross-contaminating aquifers.)

Ground Water Field Screening Equipment. Many parameters are most accurately measured in the field at the time of sample collection, and a variety of portable test kits and meters are available for field screening. Oxygen-reduction potential, temperature, pH, dissolved oxygen, and ferrous iron concentrations are commonly measured in the field. Several of these parameters can be measured by one piece of equipment, minimizing the number of meters needed in the field.

Table 5-1 lists types of test kits and meters that can be used for the field analysis of a number of ground water parameters and constituents. The screening equipment can be obtained through environment equipment retailers or rental dealers. For some parameters, select samples should be submitted for laboratory analysis to provide a baseline for comparison with the screening results. Exceptions might include pH, dissolved oxygen, temperature, and ferrous iron concentrations, which can change quickly with time (*i.e.*, during transport to the laboratory).

Down-hole Probes. Numerous probes are available that provide down-hole monitoring capability. The probes are lowered into the ground water in a well and can provide multi-parameter monitoring information including water level, temperature, pressure, dissolved oxygen, conductivity, pH, oxygen-reduction potential, total dissolved solids, resistivity, salinity, nitrate, chloride, ammonium, ammonia, and turbidity (In-situ, 2002). The parameters the probe detects are dependent on the selected sensors placed in the probe. The

Table 5-1.   **Ground Water Field Screening Methods**

Ground water Parameter	Meter	Test Kits
pH	X	X
Temperature	X	
Dissolved oxygen	X	X
Methane	X	
Carbon Dioxide	X	X
Sulfate		X
Nitrate/Nitrite	X	X
Oxygen Reduction Potential	X	
Ferrous/Ferric Iron		X
Sodium	X	
Turbidity	X	X
Total Dissolved Solids	X	
Phosphate		X

probes have a datalogging capability that interfaces through submersible cables with either a personal data assistant or laptop computer to facilitate downloading of the probe data. Different size probes are available to monitor hydraulic properties within a variety of monitoring well and microwell borehole annulus.

Soil and Ground Water Analytical Methods.   Both soil and ground water samples are typically submitted for laboratory analysis for constituents of concern at the site as determined during the historical records review and field screening. Laboratory analysis is covered in detail in Chapter 6.

At a site where a gasoline release has occurred, analytical samples, at a minimum, are submitted for analysis of VOCs, including BTEX and MTBE. Initial analyses should also include lead and TBA. To date, samples have not commonly been submitted for analysis of other oxygenates such as DIPE, ETBE, and TAME. However, analysis of these additional components is advisable at least in the initial subsurface investigation and is becoming increasingly required. In addition to petroleum compounds, samples should be collected and submitted for laboratory analyses of chlorinated VOCs if the site was used for vehicle repair. Based on the results of the initial laboratory analysis, the list of analytes can be narrowed for subsequent phases of the investigation.

Petroleum compounds are a mixture of hundreds of components, and the composition of an individual petroleum product is a function of the origin and chemistry of the crude oil, the refining and blending processes, and the use of additives (MADEP, 2001). Consequently, methods have been developed that report the concentrations of specific carbon ranges typically associated with petroleum products. For example, Massachusetts has implemented the volatile and extractable petroleum hydrocarbon (VPH/EPH) methods. The VPH and EPH methods were developed based on three broad observations:

- Petroleum products are comprised mainly of aliphatic/alicyclic and aromatic hydrocarbon compounds;
- Aromatic hydrocarbons appear to be more toxic than aliphatic compounds; and
- The toxicity of aliphatic compounds appears to be related to their carbon number/MWs (MADEP, 2001).

The VPH and EPH methods report six aliphatic and aromatic carbon fractions: three each for VPH (C_5-C_8 aliphatics, C_9-C_{12} aliphatics, and C_9-C_{10} aromatics) and EPH (C_9-C_{18} aliphatics, C_{19}-C_{36} aliphatics, and C_{11}-C_{22} aromatics). The methods also include the identification and quantification of target analytes including BTEX, naphthalene, and MTBE in the VPH analysis and 17 PAHs with the EPH analysis. The VPH fractions are most applicable to gasoline, whereas the EPH fractions are more relevant to fuel oils, diesel, waste oils, and other heavier petroleum products.

Analytical Detection Limits.   Understanding detection limits is imperative to properly evaluating analytical data. In samples that have a high concentration of a particular compound, the detection limits of other compounds present at lower concentrations may be elevated, sometimes giving the appearance that the other compounds are not present ("not detected").

Gasoline composition varies widely, but based on a typical RFG that contains 11% by volume of MTBE, the maximum approximate MTBE and BTEX concentrations expected in water in contact with freshly released RFG (assuming ground water is at 25°C [77°F]) are presented in Table 5-2. The concentrations were calculated by multiplying the pure compound solubilities (*e.g.*, 50,000 mg/l for MTBE) by the mole fractions of the compounds in the gasoline (roughly equal to the percent by volume, which for MTBE is 11%).

In this example, there is over 300 times as much MTBE in the water as benzene. Due to the high concentration of MTBE and the need to dilute the sample, the resulting detection limit for benzene in this sample would be approximately 50 µg/l or more, and benzene would be reported as not detected even though it is present at a concentration that exceeds the federal MCL of 5 µg/l. Often a sample must be run several times, at different dilutions, to detect all of the constituents at concentrations of interest.

Data Analysis.   An issue paper prepared by USEPA (2001) presents an interesting discussion on the use of "effective data." The paper discusses the concept that analytical methods generate "definitive data," whereas field screening methods only provide "screening data," which are considered to be inferior and not legally defensible. The concept of effective data relies on the

Table 5-2.   **Maximum MTBE and BTEX Concentrations in Water in Contact with Freshly Released Reformulated Gasoline at 25°C**

Compound	Solubility	Volume/Volume (%)	Concentration in Water (mg/l)
MTBE	50,000	11	5,500
Benzene	1,780	1	18
Toluene	535	11	59
Ethylbenzene	161	2	3
Xylenes	175	11	19

principle that data (field screening or laboratory analytical) quality depends heavily on the combination of sampling and analytical design, and the intended end use of the data, not just laboratory quality assurance/quality control. When the interactions between these factors are understood, the data generated by screening methods can be used in making important and defensible project decisions.

The paper argues that a "judicious blending" of both screening and laboratory analytical data should be used to optimize data collection in a cost-effective manner as long as the distinctions between them are understood. This topic will be of increasing importance in the near future as regulators, consultants, and property owners try to balance the frequently competing issues of accuracy, thoroughness, and affordability of site assessments.

Geophysics.   There are two general types of geophysical methods—active and passive. Active methods measure the subsurface response to electromagnetic, seismic, and electrical energy. Passive methods measure the earth's existing magnetic, electrical, and gravitation fields. Geophysical methods can be employed to provide information on the subsurface geology, aquifer properties, locations of buried objects, and occasionally identify residual or floating product (USEPA, 1997).

Geophysics is commonly used in conjunction with other information (laboratory analytical data, historical maps, boring logs, etc.) to assist in interpretation of the data. A variety of data sources is often necessary to resolve anomalies detected by the geophysical methods (USEPA, 1997).

The most common use of geophysics in site assessments is to determine the presence and orientation of buried objects, particularly USTs and utilities. Several methods typically employed in site assessments are presented below.

Ground-Penetrating Radar (GPR).   High frequency electromagnetic waves are emitted from an antenna into the subsurface. If an interface such as a UST, void, pipe, or bedrock is encountered, a portion of the energy is reflected back to the antenna. GPR is commonly used to cover large areas to determine if USTs or buried utilities may be present.

Seismic Refraction. Seismic surveys measure characteristics (depth, thickness, and attitude) of stratigraphic layers based on the velocity with which seismic waves move through each layer. Seismic waves are generated using a wave source (*i.e.*, a sledgehammer). The waves are tracked by sensors called geophones and recorded on a seismograph. The principle of refraction is based on the contrast in velocity between two different media or stratigraphic layers. The data can be used to determine the saturated thickness of aquifers and the location of faults and joints (Reynolds, 1997).

Electrical Resistivity. Electrical resistivity surveys use two pairs of electrodes: current and potential. An electrical current is passed through the current electrodes and the drop in potential is measured across a pair of potential electrodes placed at a measured distance from the current electrode pair. For each electrode spacing, apparent resistivity is calculated using the potential drop, applied current, and electrode spacing. The data can be used to determine the thickness of sand and gravel aquifers which overly bedrock, the location of saltwater-freshwater interfaces, and the extent of leachate migration at landfills (Freeze and Cherry, 1979).

Geomagnetic Survey. Geomagnetic methods are typically utilized in site assessments to locate buried metal objects including drums, pipes, and cables. A magnetometer, an instrument that measures the earth's magnetic fields, is used to survey an area in a grid pattern to identify anomalies at the site. The location of any anomalies are marked and subsequently investigated by other usually intrusive methods (*e.g.*, excavation) (Reynolds, 1997).

DETAILED ASSESSMENT

Once an initial assessment has been conducted, a more detailed evaluation of the nature and extent of the release is usually undertaken. Site assessment is an iterative process that is data dependent. The data collected during the initial phases of the investigation are evaluated, and data gaps are identified and used to direct subsequent phases of the investigation. The information will subsequently be used in a risk characterization, to determine if and where remediation is required, and in the selection and design of appropriate remedial measures. The quality of the data collected will be directly reflected in the effectiveness of the remediation.

Additional monitoring wells, test pits, or soil borings may be required to complement the data collected during the initial assessment. In addition to soil and ground water sample collection and analysis, more sophisticated methods may be employed to provide information on the site. Several of those methods are discussed below.

Tracers. Tracers can be used in many applications in both water and air to determine fluid velocity, flow patterns, hydraulic conductivity, and contaminant dispersion. Types of tracers include dyes, salts, radioisotopes, isotopes, fluorinated organic acids, halocarbons, and bio-organisms (Marley, 1993). Commonly used tracers include sulfur hexafluoride (SF_6), bromide (Br), helium (He), Rhodomine WT (a red fluorescent dye), and methane.

Tracers are chosen based on detectability and the ability to travel through the medium undeterred. Tracers should not be hazardous, and they must be conservative, *i.e.*, they are not consumed by chemical or biological reactions and do not adsorb to soil particles. For example, use of bromide as a tracer should be evaluated on a site-specific basis, as it may be toxic to certain aquatic species, and in areas where it may come into contact with chlorine, it may form a more toxic bromine compound (Vujević *et al.*, 2000).

Aquifer Tests. The hydraulic conductivity of an aquifer, or the capacity of a porous medium to transmit water, can be determined by a number of methods. The two most common methods are discussed below.

Slug Tests. These tests are conducted using a "slug" of known volume (often a bailer filled with water or a solid cylindrical slug) to displace water in the well. Water level measurements are recorded at predetermined intervals throughout the tests using either a water level meter or, preferably, a downhole pressure transducer connected to an electronic data logging system. Hydraulic conductivity values are usually calculated from the field data using the Bouwer and Rice method (Bouwer and Rice, 1976).

Pumping Tests. A constant-rate pumping test is the most accurate way to determine aquifer parameters such as hydraulic conductivity, transmissivity, and storativity. The well is pumped at a constant rate for a predetermined period of time. Throughout the test, drawdown measurements are recorded at observation wells installed at varying distances from the pumping well. Usually, the recovery of the aquifer is monitored once the pumping well has been turned off. The recovery data are used to confirm the pumping test data. In order to obtain the aquifer parameters, the drawdown data are plotted versus time or distance from the pumped well (Driscoll, 1986).

Evaluation of Soil Gas and Indoor Air Migration Pathways. As part of a comprehensive assessment following a release of gasoline containing MTBE, the evaluation of soil gas and indoor air within and adjacent to nearby structures may be warranted. Due to the volatile nature of gasoline components, including MTBE, the migration of these contaminants to indoor air must be assessed if the release may have impacted an occupied building. Typically, an assessment of a residential structure is initiated with the installation of between three and five soil gas probes. Soil probes are advanced either through the basement or first story floor or adjacent to the building foundation. Sampling probes are generally installed to enable the collection of a soil gas sample from immediately beneath the lowest floor of the building (MADEP, 2001). Samples may need to be collected at several times throughout the year to evaluate the effect of seasonal variations, including changing ground water depths, on vapor concentrations and migration.

If contaminants are detected at concentrations of potential concern in soil gas samples, indoor air samples may need to be collected. Multiple samples, including one to represent background, should be collected from each building in locations that reflect the areas frequented by residents or employees.

Several different means of collecting samples for laboratory analysis are available including Summa® canisters and Tedlar bags. Tedlar bags are simple to use and transport, and multiple analyses (screening, dilutions, and reanalysis) can be conducted from a single bag. However, there is a short holding time (24 to 48 hours), and the bags are easy to puncture. Summa® canisters have the same advantages as Tedlar bags, but the holding time on samples is longer (14 days), and the canisters are sturdy and easy to use.

A variety of methods are available for analysis of air samples. Samples collected with Summa® canisters for analysis of VOCs, including MTBE, are analyzed by USEPA Method TO-14 (USEPA, 1988). Massachusetts has developed a specific method, air petroleum hydrocarbons (APH), to complement the VPH/EPH methodology used for soil and ground water samples. This method reports air sampling results for three carbon fractions (C_5-C_8 aliphatics, C_9-C_{12} aliphatics, and C_9-C_{10} aromatics), plus BTEX, MTBE, and naphthalene (MADEP, 2000).

Care should be taken during collection of indoor air samples due to the numerous potential sources that can create false positives. For example, containers of paints, solvents, and cleaning supplies, or cars idling in adjacent driveways or streets beside an open window can impact the analytical results. A careful inventory of potential interferences should be taken prior to the collection of samples. (It is because of this problem of false positives that initial analysis of soil gas beneath the building is recommended.)

Carbon Isotope Analysis. Isotopes of certain elements have the same atomic number, but different atomic weights due to varying numbers of neutrons in the nucleus. Isotopes are divided into two categories: stable and radioactive. Stable isotopes are primarily used to identify contaminant sources or monitor natural attenuation, whereas radioactive isotopes are typically employed to determine the age of ground water. Site assessments rarely require the use of radioactive isotopes for age dating.

MTBE is manufactured using one of three methods. The isotopic composition of the starting material and any isotope effects associated with its manufacture determine the isotopic composition of the MTBE. Stable carbon isotopes exist as a mixture of ^{12}C and ^{13}C, and the ratios of these two isotopes are expressed relative to a standard in delta notation ($\delta^{13}C=[R_{sample}/R_{standard}-1] \times$ 1,000 (Smallwood *et al.*, 2001). R_{sample} is the ratio of $^{12}C/^{13}C$ of the sample; $R_{standard}$ is the ratio of $^{12}C/^{13}C$ of the international standard, Vienna Peedee Belemnite (VPDB) (see Chapter 30).

If the carbon ratio of the MTBE within a plume can be determined it may be able to be matched with a similar ratio from a source. For comparison, a sample collected from the upgradient portion of the plume must be analyzed because biodegradation occurring downgradient may affect the carbon ratio, thus eliminating the possibility of source identification. The carbon ratio in that sample is compared to a sample of virgin product from each of the suspected sources. If the carbon ratio is similar to one source and different from others, the correlation identifies the source of the MTBE (Smallwood *et al.*, 2001). Carbon isotope analysis is most effective when used in conjunction with other lines of evidence developed from gas chromatography/mass spectrometry (GC/MS) to validate the results (Mansuy *et al.*, 1997).

Compound-specific isotope ratios can also be used for monitoring *in situ* biodegradation. As biodegradation occurs, a parent compound such as MTBE becomes enriched in the heavy isotopes (^{13}C) because of a preference of microorganisms for the lighter molecules. In contrast, physical processes (volatilization, dispersion, or absorption) are not accompanied by isotopic fractionization. Consequently, carbon isotope analysis can be used to demonstrate that the observed reduction in contaminant concentration is from biodegradation and not from physical dispersion or absorption (Hunkeler *et al.*, 2001). This can be instrumental in demonstrating MNA.

IDENTIFYING MIGRATION PATHWAYS

Stratified surficial geology is frequently encountered on sites located in flood plains. The combination of high energy flood events and river meanders deposit a variety of both fine and coarse overburden materials which complicate the assessment and identification of preferential migration pathways. This type of stratified aquifer influences the migration of contaminants as the plume tends to migrate through the more conductive strata, even if they are located at depth.

An example of this is a site in Chicopee, Massachusetts where petroleum from LUSTs is migrating toward several surface water bodies. The overburden at the site can be demarcated into three distinctive intervals (Figure 5-1). The intervals are generally described as upper layers of fine to medium sand and silt above a layer ranging between fine to coarse sand to gravel and cobbles. The deepest interval ranges from a silty clay to a silt with some sand interlayering. Geotechnical analysis of representative samples from each layer

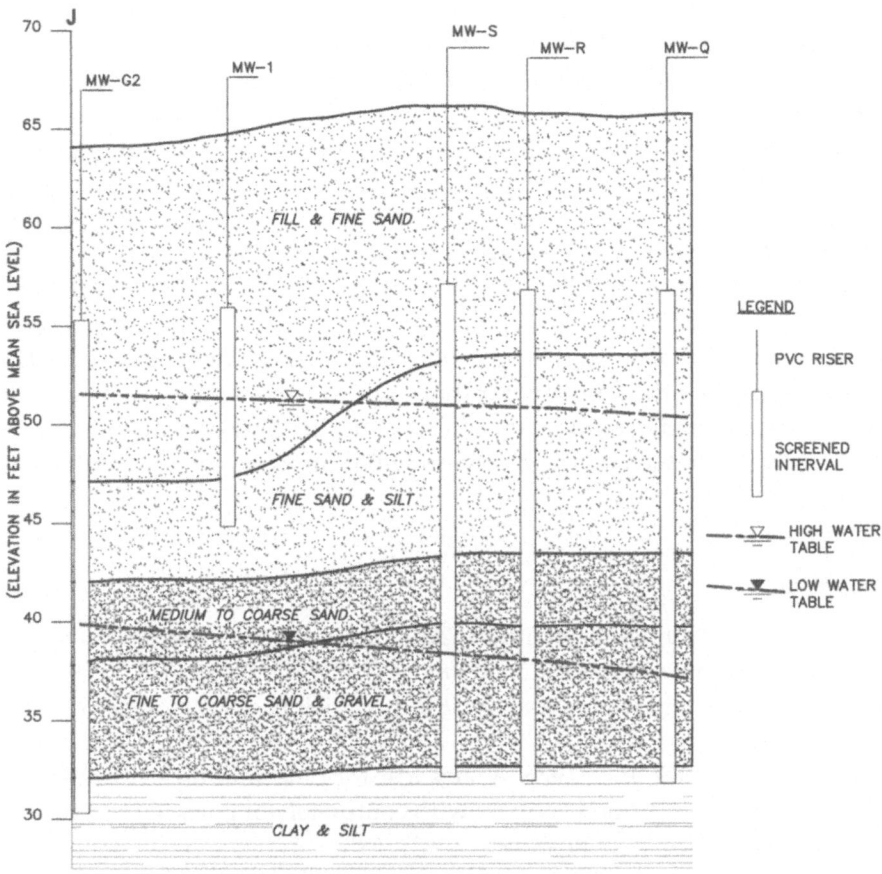

Figure 5-1.   **Cross-Section Identifying the Stratigraphy Underlying an Industrial Site in Chicopee, Massachusetts**

indicate that the sand and gravel hydraulic conductivity values range between 10^{-2} and 10^{-3} cm/s (10^0 and 10^1 ft/day). By contrast, the hydraulic conductivity of the stratum immediately above the sand and gravel is 10^{-5} cm/s (10^{-2} ft/day). Even more significant, the hydraulic conductivity of the underlying stratum ranges between 10^{-5} and 10^{-7} cm/s (10^{-2} and 10^{-4} ft/day), many orders of magnitude less than the sand and gravel.

Laboratory analysis of clay and silt samples collected at varying depths beneath the sand and gravel indicate that petroleum hydrocarbons have only been detected at relatively minor concentrations. Twenty-six out of the 32 samples had no positive detections, whereas extensive contamination was identified in the vast majority of samples collected from the sand and gravel unit. These results confirm that the more conductive sand and gravel layer is the pathway for migration of free-phase petroleum product even though it is well below the water table for the majority of the year.

At sites where a preferential migration pathway exists at depth, the installation of conventional water table monitoring wells will not be sufficient to characterize the plume. As illustrated in Figure 5-1, if a conventional monitoring well (MW-1) were installed to span the water table, the well would have missed the plume and incorrectly assessed the extent of contamination. Whereas, if the monitoring well was installed to the aquitard (silt and clay layer), through the sand and gravel, significant contamination would be encountered. In addition, the collection of ground water samples from discrete stratigraphic intervals would have facilitated the assessment of the site. This example illustrates the necessity for thoroughly characterizing a site in three dimensions and understanding the effects of the geology on contaminant migration prior to implementing costly remedial measures or concluding that none are needed.

REFERENCES

API (American Petroleum Institute). 2000. *Strategies For Characterizing Subsurface Releases of Gasoline Containing MTBE.* Publication No. 4699.

Boart Longyear Environmental Drilling Division. 1997. *Sonic Drilling.* Brochure.

Bouwer, H. and Rice, R.C. 1976. A slug test for determining hydraulic conductivity of unconfined aquifers with completely or partially penetrating wells. *Water Resources Research.* 12, 423–428.

Cordry, K. 2002. HydraSleeve—A new discreet interval, no-purge, MTBE ground-water sampler. Presented at: *2002 NGWA Conference on MTBE: Assessment, Remediation, and Public Policy*, Orange, California. June 6–7, 2002.

Driscoll, F.G. 1986. *Groundwater and Wells.* 2nd Edition. pp. 168, 203, 268, and 313. St. Paul Minnesota, Johnson Filtration Company, Inc.

Freeze, R.A. and Cherry, J.A. 1979. *Groundwater.* pp. 308-309. Englewood Cliffs, New Jersey. Prentice-Hall, Inc.

Geoprobe® Systems. 2002. www.geoprobe.com. Accessed September 23, 2002.

Global Environment and Technology Foundation. *Innovative Technology Summary—ResonantSonic Drilling.* www.gnet.org/archive/4645.html. Accessed July 12, 2002.

Hunkeler, D., Butler, B., Aravena, R., and Barker, J.F. 2001. Monitoring biodegradation of MTBE using compound-specific carbon isotope analysis. *Environmental Science and Technology.* 35, 676–681.

In-Situ, Inc. 2002. www.in-situ.com. Accessed July 31, 2002.

Marley, M.C. 1993. Evaluation of vadose zone air flow pathways utilizing tracer gases and the subsequent implication on vapor extraction system design. In: *Hydrocarbon Contaminated Soils*, Volume III. pp. 385. (Calabrese, E. J. and Kostecki, P.T. Eds.). Boca Raton, Lewis Publishers.

Mansuy, L., Philp, R.P., and Allen, J. 1997. Source identification of oil spills based on the isotopic composition of individual components in weathered oil samples. *Environmental Science and Technology*. 31, 3417–3425.

MADEP (Massachusetts Department of Environmental Protection). 2001. *Characterizing Risks Posed by Petroleum Contaminated Sites: Implementation of the MADEP VPH/EPH Approach*. Final Draft, June 2001.

MADEP (Massachusetts Department of Environmental Protection). 2000. *Method for the Determination of Air Phase Petroleum Hydrocarbons (APH)*. Public Comment Draft, February 2000.

MADEP (Massachusetts Department of Environmental Protection). 1999. *Preservation Techniques for Volatile Organic Compound (VOC) Soil Sample Analysis*. WSC #99-415.

MADEP (Massachusetts Department of Environmental Protection). 1996. *Commonwealth of Massachusetts Underground Storage Tank Closure Assessment Manual*. April 1996.

Interstate Technology Regulatory Council (ITRC). 2002. *Passive Diffusion Bag (PDB) Samplers.www.itrcweb.org*. Accessed July 12, 2002.

Reynolds, J.M. 1997. *An Introduction to Applied and Environmental Geophysics*. pp. 116–207, 276–320. Chichester, England, John Wiley & Sons.

Robbins, G.A. 1996. Recommended Guidelines for Applying Field Screening Methods in Conducting Expedited Site Investigations at Underground Storage Tank Sites in Connecticut. Developed for LUST Trust Fund Program. Bureau of Waste Management, Connecticut Department of Environmental Protection. Department of Geology & Geophysics, University of Connecticut.

Smallwood, B.J., Philp, R.P. and Burgoyne, T.W. 2001. The use of stable isotopes to differentiate specific source markers for MTBE. *Environmental Science and Technology*. 2, 215–221.

Todd, D.K. 1980. *Groundwater Hydrology*. 2nd Edition. pp. 179–183. New York, John Wiley & Sons.

USACE (U.S. Army Corps of Engineers). 1998. *Site Characterization and Analysis Penetrometer System (SCAPS) Technology Development/Application*. www.wes.army.mil/el/scaps.html. Last updated March 4, 1998.

USDOE (U.S. Department of Energy). *Characterization, Monitoring and Sensor Technology Crosscutting Program. www.cmst.org*. Accessed September 23, 2002.

USEPA (U.S. Environmental Protection Agency). 1988. Determination of volatile organic compounds (VOCs) in ambient air using Summa® passivated canister sampling and gas chromatographic analysis. In: *Compendium of Methods for the Determination of Toxic Organic Compounds in Ambient Air*. EPA-600-4-89-017.

USEPA (U.S. Environmental Protection Agency). 1993. *Ground Water Sampling—A Workshop Summary*, Dallas, Texas. November 30–December 2, 1993. Robert S. Kerr Environmental Research Laboratory, Office of Research and Development, Ada, OK. EPA/600/R94-205.

USEPA (U.S. Environmental Protection Agency). 1997. *Expedited Site Assessment Tools for UST Sites*. EPA 510-B-97-001.

USEPA (U.S. Environmental Protection Agency). 2001. *Applying the Concept of Effective Data to Environmental Analyses for Contaminated Sites*. EPA 542-R-01-013.

Vujević, M., Vidaković-Cifrek, Z., Tkalec, M., Tomić, M., and Regula, I. 2000. Calcium chloride and calcium bromide aqueous solutions of technical and analytical grade in Lemna bioassay. *Chemosphere*. 41, 1535–1542.

Additional Web Sites:

American Petroleum Institute: www.api.org
ASTM: www.astm.org
USEPA technology links: www.clu-in.org
USEPA: www.epa.gov

CHAPTER 6

Laboratory Analysis of Oxygenated Gasoline Constituents

Robert J. Pirkle, Ph.D. and Patrick W. McLoughlin, Ph.D., Microseeps, Inc.

INTRODUCTION

Assessment and remediation of oxygenated gasoline contamination in environmental samples requires that environmental laboratories perform speciation and quantification of each component accurately, reliably, and routinely. Reviews (Rhodes and Verstuyft, 2001; Uhler *et al.*, 2000) of the available methodologies have been previously published. Concerns remain regarding the applicability of these methodologies, particularly to samples containing oxygenates in the form of ethers and alcohols. Standard methods for the preservation, preparation, and measurement of hydrocarbon components of gasoline in environmental samples have been thought to be adequate for the analyses of samples that also contain fuel oxygenates such as MTBE. We are beginning to learn that this is not the case and the reasons why. It is our purpose here to critically discuss these methodologies and associated problems with respect to speciation and quantification of fuel oxygenates in environmental samples and to suggest best practices from the available choices.

To facilitate the discussion, we define "analysis" as the process that includes the preservation, preparation, and measurement of the analytes of interest in an environmental sample. It will be the judicious combination of selected components of analysis, which will form the best option for speciation and quantification of the analytes of interest.

The oxygenated gasoline sample can be in either an ambient air, NAPL, contaminated soil gas, soil, or ground water matrix. For analysis of ambient air, a set of relatively complex components, as used in USEPA Method TO-14 (USEPA, 2002) must be used. Discussion of such a set of complex components is beyond the scope of this work. Much has been done in analyzing pure hydrocarbon samples. That work is also beyond the scope of this chapter. For further discussion of the analysis options for NAPL the reader is referred to ASTM D 5599 for oxygenates, and ASTM D 5769, ASTM D 4815 or ASTM D 5134 for all potential gasoline compounds. A soil gas or SVE sample can be easily analyzed by the methodologies discussed herein and, though the com-plications associated with sample preservation and preparation are mini-mized, the problems associated with measurement by a gas chromatography (GC) methodology still must be accounted. Through washing, a soil sample can be treated as a ground water sample. Thus, the focus in this work is the analysis of ground water. The starting assumption is that a water sample has been taken using proper sampling procedures and placed in a 40 milliliter volatile organic analysis (VOA) vial, without headspace, in preparation for shipment to the laboratory. (For some example protocols, see USGS, 2002.)

It is further assumed that this water sample will be preserved and pre-pared for a measurement method involving GC. There are options for both preservation and preparation of the sample for introduction into the GC for measurement. Preservation options include cooling, or the addition of acid or base. Preparation options include purge and trap, headspace equilibration, direct aqueous injection (DAI) and solid phase microextraction (SPME). Fur-ther, there are options for analyte measurement including PIDs, FIDs, and mass spectral detectors (MSDs). In the succeeding sections of this chapter these options will be discussed and evaluated to enable the selection of the optimum analysis method for environmental samples containing volatile hy-drocarbon and fuel oxygenate contaminants of concern in environmental samples.

PROPERTIES OF OXYGENATED GASOLINE COMPONENTS

The analysis of environmental samples containing oxygenates is a chemical process, so it is appropriate for us to begin by considering the chemical com-pounds in the oxygenated gasoline that are of concern in environmental sam-ples. The compounds consist of three major groups: hydrocarbons, ethers, and alcohols. The hydrocarbons make up the bulk of gasoline. They consist of alkanes (*e.g.*, pentane, hexane, and heptane) and aromatics (*e.g.*, BTEX).

The second and third groups of compounds have very different physical and chemical properties from those of the hydrocarbons. These properties make standard preservation, preparation, and measurement methodologies less applicable. The specific compounds in these groups are:

- Ethers: MTBE, ETBE, DIPE, and TAME; and
- Alcohols: TBA, tert amyl alcohol (TAA), isopropyl alcohol (IPA), ethanol, and methanol.

Compared to hydrocarbons, ethers and alcohols are much more soluble in water, are much more polar, and have lower Henry's Law constants (*i.e.*, a greater affinity for water). These differences are greatest for the alcohols.

The ethers and alcohols also have very different chemistry than the hy-drocarbons. It is this chemistry that can make their analysis using standard methodologies problematic. While the hydrocarbons are generally unaffected by acid, the ethers can react with water in the presence of acid through a process known as hydrolysis. As an example, MTBE can hydrolyze to TBA and methanol:

$$MTBE + H_2O \xrightarrow{H^+} TBA + MeOH \qquad (6\text{-}1)$$

This hydrolysis can take place in acidified waters at elevated temperatures, or over a highly polar resin (O'Reilly et al., 2001a, 2001b).

SAMPLE PRESERVATION METHODS

It is standard procedure to preserve a ground water sample before shipment to the laboratory if that sample is to be analyzed for VOCs. The purpose of this preservation is to minimize the continued degradative activity of the indigenous microbial population. The use of non-oxidative acids such as hydrochloric acid in combination with cooling and maintaining the sample at 4°C (39°F) has been shown to be an effective preservation methodology in the case of many ground water samples containing volatiles such as BTEX.

For fuel oxygenates, however, the use of acid preservatives, while effective as a microbial inhibitor, introduces another problem: the potential hydrolysis of the ethers. In samples preserved with acids, such as hydrochloric acid, MTBE may chemically degrade in the sample prior to, or during, analysis.

Hydrolysis of MTBE during analysis produces an analytical result that understates the concentration of MTBE and overstates the concentration of TBA originally in the sample, since the hydrolysis of MTBE produces TBA. TBA is also a primary biodegradation product of MTBE in ground water. Therefore, the result could erroneously be interpreted to suggest the occurrence of biodegradation in ground water. Alternately, the observation of artificially elevated TBA levels could lead to a costly overestimate of the burden required for an active remediation system.

It is beneficial to consider the following set of data derived from samples preserved with hydrochloric acid and analyzed by a heated headspace-GC/MS method (discussed below). Because of the high levels of MTBE, some of the samples required the additional analysis of a diluted sample. The TBA concentrations were determined from both the diluted (10X) and undiluted (1X) samples. Typically, when the results are corrected for the dilution, agreement to within 10% is expected. The results are displayed in Figure 6-1.

As the sample is diluted, the hydrochloric acid that was used as a sample preservative is diluted as well. In a diluted sample, less hydrochloric acid is present to promote hydrolysis; therefore, less TBA is produced by hydrolysis. This is represented in Figure 6-1.

A second example is derived from a study of potential natural and stim-

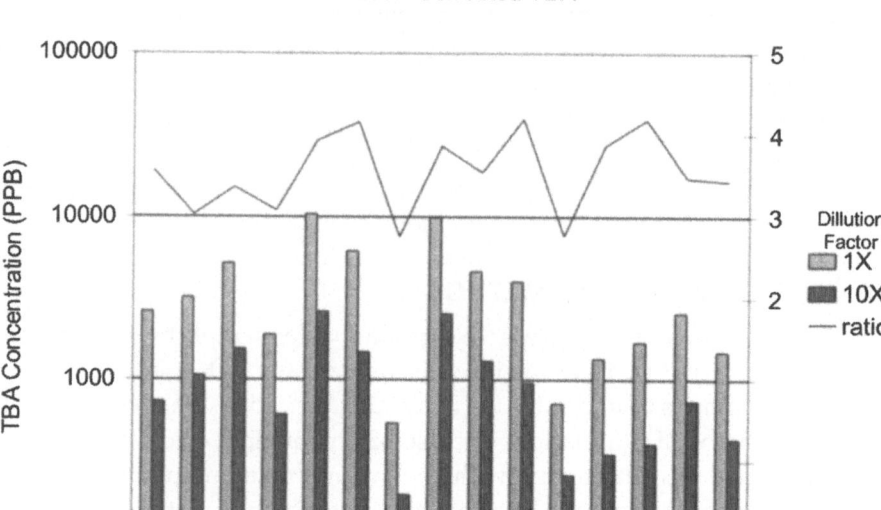

Figure 6-1.   **Dilution Corrected Results from a Set of Samples that Were Run Undiluted (1X) and Diluted (10X). For the readers benefit, the ratio of the results (1X/10X) is also displayed (right axis).**

ulated biodegradation of MTBE. Typical results that had been observed over several sampling events during this study are shown in Figure 6-2. These samples were collected in May, 2001 and had been preserved with hydrochloric acid. They were analyzed for both MTBE and TBA via a heated headspace- GC/FID technique. As shown in Figure 6-2, substantial amounts of MTBE and TBA were detected. After reviewing the results of the study portrayed Figure 6-1, it was decided to see if hydrolysis played a role in the data shown in the results in Figure 6-2.

To accomplish that, a set of samples was collected from the same area in January 2002, and those samples were preserved with a basic salt, tri-sodium phosphate. Those samples were then analyzed using the heated-headspace GC/FID technique and the results are shown in Figure 6-3. The general concentration of the detected MTBE has not changed, but there is now no TBA detected at concentrations greater than 5 µg/l. This is in stark contrast to the May 2001 results of Figure 6-2 and suggests that the TBA observed in those analyses may have been produced during the analysis. To confirm that, several of the duplicate samples collected in the January 2002 study were acidified with hydrochloric acid to a pH less than 1.5 and analyzed using the heated headspace-GC/FID technique. The results of those analyses, shown

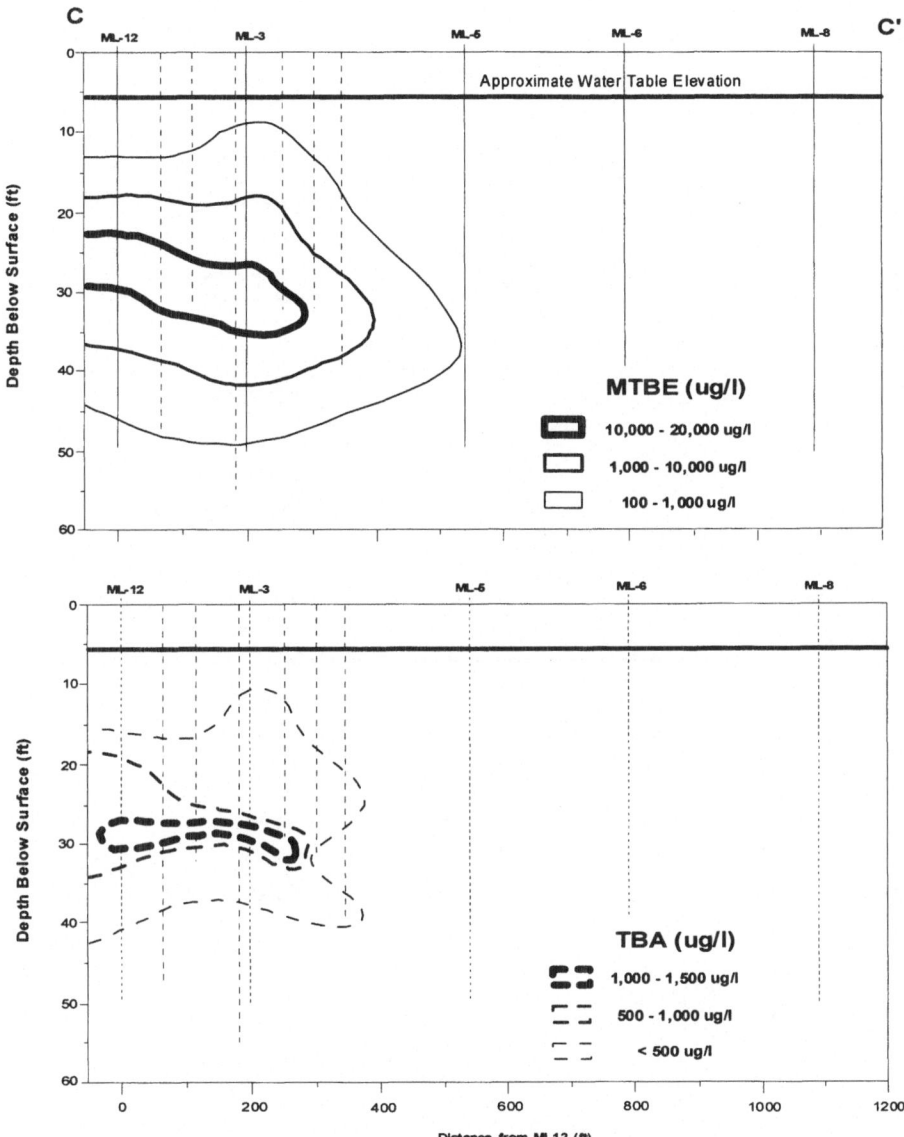

Figure 6-2.   **Cross Sectional Plot of Results from the Analysis of Hydrochloric Acid Preserved Samples**

in Figure 6-4, reveal a substantial concentration of TBA. The only conclusion that could be drawn was that the TBA observed in May 2001 was, most likely, not from the ground water, but from the analysis, and the actual ground water concentrations are similar to those in Figure 6-3.

The results in Figures 6-1 through 6-4 naturally lead to the question "is

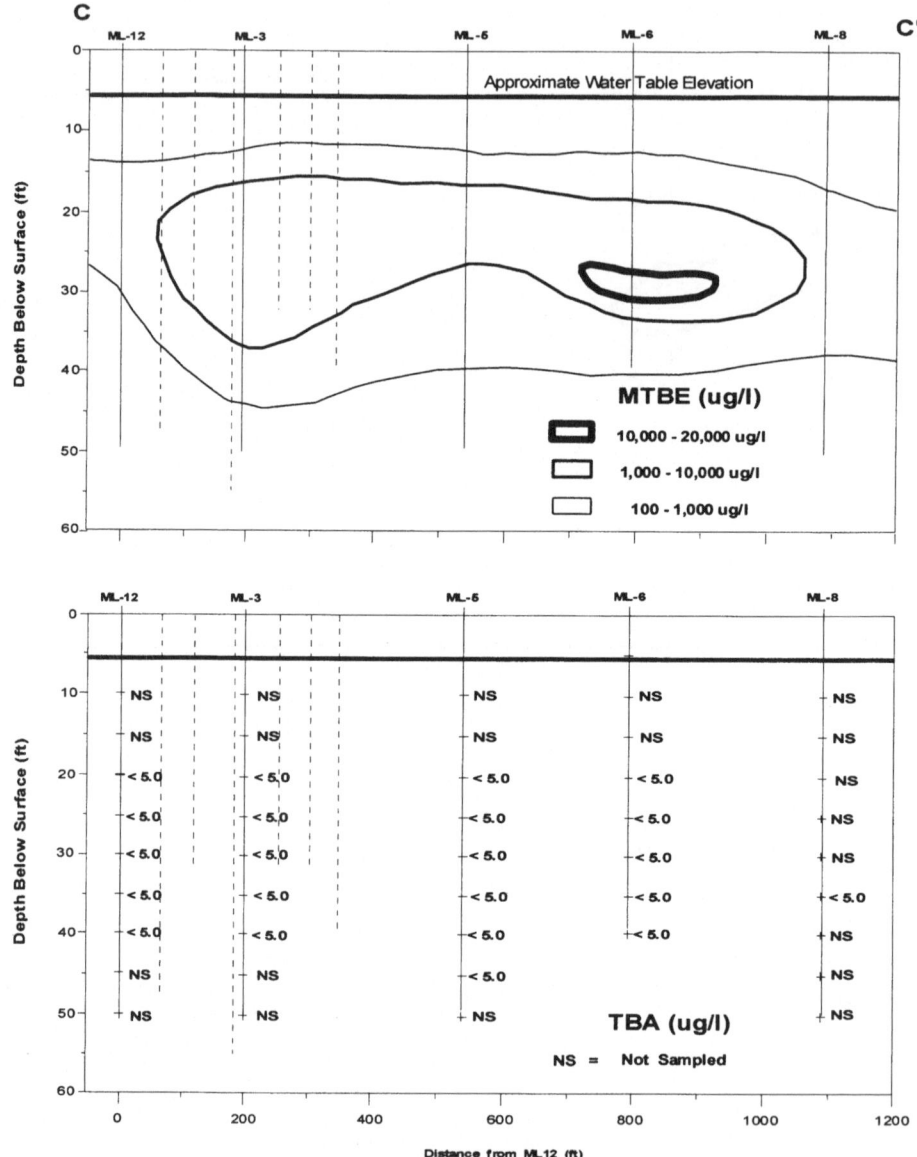

Figure 6-3.   **Cross Sectional Plot of Results from the Analysis of Tri-Sodium Phosphate Preserved Samples**

hydrolysis a problem at my site?" Once the data is collected, it is difficult, and often impossible, to answer this question with certainty. However, the more important question might be "given the potential for hydrolysis, what can I do about it?"

To minimize the potential for hydrolysis, maximize the quality control

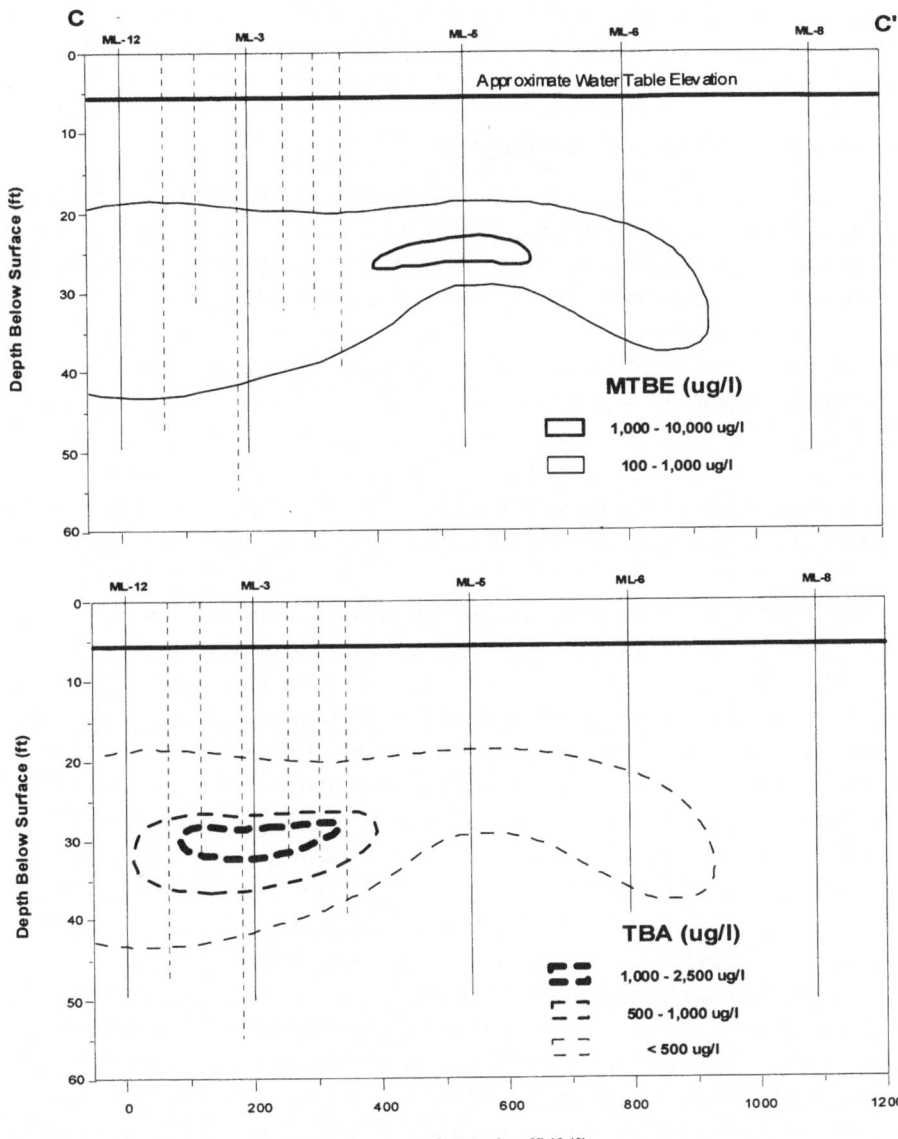

Figure 6-4.   **Cross Sectional Plot of Results from the Analysis of Acidified, Tri-Sodium Phosphate Preserved Samples**

that enables an evaluation of the integrity of the results, and control cost, the
investigator must carefully choose the proper analysis. This chapter is intended to empower the reader to make those judicious choices. Through a
discussion of the components of analysis, the options available for each component, and the pros and cons of each option, we present the information

needed for an analytical method to be chosen. Further, we will discuss the quality assurance techniques that can be implemented to assure that the oxygenates data collected through that analysis is reliable.

SAMPLE PREPARATION METHODS

Unless the sample matrix is gaseous, the sample must be prepared for measurement. A soil sample is readily converted to a slurry, which can be treated as an aqueous sample. In general, measurement methodologies for VOCs involving GC function optimally when the VOCs have been separated from aqueous solution. As a further preparation, in some procedures the VOCs, once separated from the aqueous phase, are concentrated. In the following discussion the separation and concentration steps will be discussed individually.

Separation of Volatiles from Aqueous Solution. There are a variety of ways that the separation of volatile organics from water can be accomplished.

Purge. The most widely used method for separation of VOCs from ground water samples is the "purge" of purge-and-trap. Purge-and-trap by SW846-5030 (USEPA, 1986) is the sample preparation method in USEPA Methods 502.2 and 524.2 for drinking water and USEPA Methods 601/602 and 624 for wastewater. In wastewater purge-and-trap, a 5 milliliter ground water sample is purged with helium, and the volatiles in that sample are entrained in the purge gas flow. While this is the standard method for separation from aqueous solution for hydrocarbons such as BTEX and works adequately for the fuel oxygenate ethers, it does not work well for the much more water-soluble alcohols. Typical detection limits for ethers purged from aqueous solution are 1 to 5 µg/l, while for the alcohols they are typically 50 µg/l.

Lawrence Livermore National Laboratory conducted a study (Happel *et al.*, 1998) that used a purge-and-trap technique (SW846-5030/8260) to measure a detection limit for oxygenates in gasoline-impacted ground water. They calculated a detection limit of 35 µg/l for TBA using a 20°C (68°F) purge vessel. By heating the purge vessel to 40°C (104°F) and doubling the volume of sample purged, they were able to produce a detection limit of 5 µg/l.

The hydrolysis rate increases with temperature. Using the activation energy for aqueous hydrolysis of MTBE reported by O'Reilly *et al.* (2001a), the increase in hydrolysis rate with temperature was calculated and the data are displayed in Figure 6-5. The data presented in Figure 6-1 were recorded after raising the temperature of the samples to 80°C (176°F). This results in a 3,000-fold increase of the hydrolysis rate as compared to the rate in a 20°C (68°F) purge vessel.

The achievement of the 5 µg/l detection limit noted above by Lawrence

Livermore National Laboratory involved increasing the temperature to 40°C (104°F) and doubling the sample volume from 5 to 10 milliliters. The temperature increase from 20 to 40°C (68 to 104°F) increases the rate of hydrolysis by a factor of 20 as shown in Figure 6-5. While a factor of 20 sounds acceptable compared to 3,000, it must be realized that doubling the sample volume will decrease the working range of the analytical equipment, in turn increasing the need to analyze samples at multiple dilutions. To avoid this decrease in working range, many laboratories will be inclined to further increase the purge vessel temperature rather than doubling the sample volume. The data shown in Figure 6-5 indicate that the hydrolysis rate is very sensitive to temperature increases, and extreme care must be used when heating the purge vessel to avoid conversion of MTBE to TBA during analysis.

Headspace Equilibration. Another method for separation of VOCs from aqueous solution is the method of headspace equilibration. This method is described in SW846-5021. Typically, a sample may be comprised of a 13-milliliter volume of water and an 8-milliliter headspace in a vial sealed with a crimped septum. VOCs in the water partition into the headspace according to Henry's Law. This works very well with analytes whose aqueous solubility is low (high Henry's Law constant), but less well for more soluble analytes. These solubility problems may be minimized by either heating the sample vial, thereby decreasing the aqueous solubility of the species, or by modifying the aqueous solubility by addition of a matrix modifier such as chloride. The combination of these two modifications can yield acceptable detection limits for all species of interest in fuel oxygenate samples. For ex-

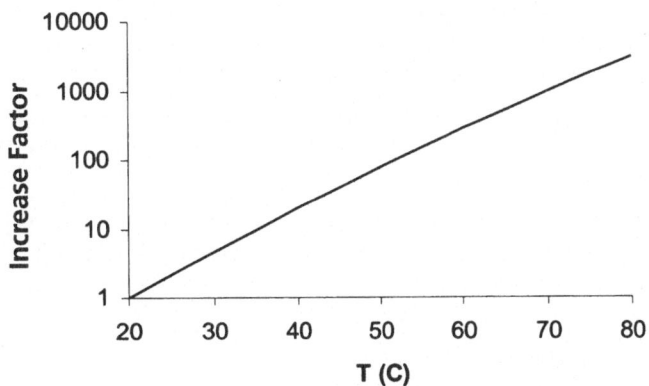

Hydrolysis Rate Increase

Figure 6-5.   **The Increase in the Rate of MTBE Hydrolysis in Water. The increase is calculated relative to the rate at 20°C.**

ample, the USEPA R.S. Kerr Research Laboratory reports both MTBE and TBA to 1 µg/l using a heated headspace method and an ion trap GC/MS measurement methodology; Microseeps reports both MTBE and TBA to 5 µg/l using heated headspace and FID detection.

The choice of a matrix modifier deserves special care. The matrix modifier recommended in SW846-5021 is sodium chloride. Increasing the ionic strength of the matrix modifier can maximize the salting-out effect. This increase can be achieved either through increasing the concentration of the modifier or changing the modifier used. The choice and concentration of a modifier depends upon the required detection limits. To enhance sensitivity, the matrix modifier may be changed, but care should be taken that this does not result in acidification of the sample.

Direct Aqueous Injection (DAI).   In order to maximize the probability of measurement of the range of ethers, alcohols, and hydrocarbons in a typical oxygenated gasoline sample, some have developed methods involving DAI of the sample onto the GC (Church, *et al.*, 1997).

In one of the more recent applications of this technique, Church *et al.* (1997) utilized a 10 microliter DAI volume and a uniquely configured injection port/liner that provided some condensation of the vaporized water and yielded satisfactory chromatography on a highly polar polyethylene glycol column. They obtained detection limits of 10 µg/l for both MTBE and TBA using a standard mass scan from 41 to 150 m/z (m/z is the mass to charge ratio). The use of selective ion monitoring (SIM) for quantification reduced detection limits to less than 0.1 µg/l for all analytes of interest.

Solid Phase Microextraction (SPME).   In the technique known as SPME (Pawliszyn, 1997; Black and Fine, 2001), a fiber is held immediately above, or below, the surface of the water that is being sampled. The analytes preferentially adsorb onto the fiber. To enhance that adsorption, the sample may be heated, and a matrix modifier may be added to the sample. The fiber is removed from the solution and placed into a specialized GC injector port. In that port the fiber is rapidly heated and the analytes desorb from the fiber and are swept onto the GC column.

SPME relies on the use of a solid-phase fiber, and such solid phases, particularly the more polar ones used to more efficiently adsorb alcohols, can catalyze hydrolysis (O'Reilly *et al.* 2001a). The heating of the fiber that occurs when the fiber is placed into the injector for analyte desorption may also elevate the potential for hydrolysis.

Concentration of Separated Volatiles.   Once VOCs are separated from aqueous solution, a means to concentrate the separated VOCs may be required to achieve the needed detection limits. The extent to which the separated VOCs must be concentrated varies with the selected method of volatiles separation from the ground water and the sensitivity goals of the measurement. If the VOCs are purged out of the water, they may be purged into a volume of purge gas of over 400 milliliters. Since this volume is much larger than can be accommodated by a single injection or handled by a modern analytical column, concentration is necessary.

Concentration is most typically performed via the "trap" of purge-and-trap. As the volatiles are purged out of the water, they are "trapped" onto a low efficiency, high capacity column. That column is then backflushed and rapidly heated to remove the volatiles quickly. Relatively less polar traps, which work well for hydrocarbons and ethers such as MTBE, do not work nearly as well for the more polar alcohols such as TBA and ethanol. Increasing the polarity of some, or all, of the trap can increase the trapping efficiency for TBA. However, one must exercise caution when choosing a trap because highly polar traps may act in a similar way to the acids in aqueous solution, catalyzing the decomposition of MTBE into TBA (O'Reilly *et al.*, 2001a). The potential for hydrolysis may be even greater at the elevated temperatures that occur during desorption. The flexibility herein discussed in the choice of a trap is, in practice, limited by the commercially available selection.

Alternately, in a matrix-modified heated-headspace preparative methodology, VOCs are transferred into a roughly 8 milliliter headspace, and a substantial portion of that volume can readily be injected onto the chromatographic column. Using this methodology, adequate sensitivity (1 to 10 µg/l detection levels, as discussed above) is achievable without concentration of the VOCs.

In the DAI and SPME methodologies, preparation and concentration are combined into a single step.

MEASUREMENT METHODS

Although there may well be additional methods for the measurement of fuel oxygenates from ground water samples, we have chosen here to consider only those methods that utilize GC. There are two fundamental components of GC that need to be discussed: first, the choice of analytical column used to separate the volatile compounds found in ground water samples contaminated by oxygenated gasoline; and, second, the choice of detector that is used to detect the eluting species at the end of the analytical column.

In modern GC of VOCs, the use of capillary columns is near universal. Additionally, considerable advances have been made in the so-called "stationary phase" or lining of columns. When combined with choices of column length and column diameter, the column can now provide much of the selectivity previously required from the detector (Agilent, 2002).

Perhaps the most distinguishing characteristic of GC measurement methods is the choice of detector. There are three detectors that will be discussed here: FID, PID, and MSD. There are two types of MSDs in general commercial use today, ion traps and quadrupoles. In addition, quadrupole mass spectrometers can be operated in either full mass scan mode or SIM mode.

Rhodes and Verstuyft (2001) have pointed out that one area of important difference between these three detectors is that of selectivity. A specific example common to many columns is the potential for MTBE to coelute with two compounds often found in gasoline: 2-methyl pentane and 3-methyl pentane. The FID is relatively nonselective, responding with more or less equal sensitivity to all of the components of oxygenated gasoline. The PID responds to all of the ethers and alcohols, the aromatic hydrocarbons, and some saturated hydrocarbons, particularly the branched-chain saturated hydrocarbons such as 2-methyl pentane. Identification of analytes using either the FID or PID rests on the similarity of retention times of known solutions or standards and the retention times of peaks in unknown ground water samples. The MSD offers significantly better selectivity from the evaluation of mass fragmentation patterns that are uniquely characteristic of each analyte. Laboratory costs for GC/MS are higher than those for GC/PID or GC/FID, and quadrupole MS may be less sensitive than other detectors, but it is chosen for its selectivity. However, advances in GC columns can move much of the burden of selectivity off of the detector and onto the column.

This idea is supported by the realization that FIDs are routinely used to characterize complex mixtures of VOCs such as fuels with good specificity. Using modern capillary columns these analyses achieve near-baseline separation of hundreds of compounds in complex mixtures. The problem is significantly reduced in ground water, since most of the saturated hydrocarbons that are abundantly present in fuel mixtures have minimal solubility in ground water.

For oxygenated gasoline components, the sensitivity of the FID is greater than, or equal to, that of the quadrupole MSD operated in full mass scan mode. The sensitivity of the PID is comparable to the FID for most of the analytes of interest. Sensitivity of the quadrupole MSD can be enhanced significantly by the judicious use of SIM, which integrates the signal of selected ions over a great number of scans at selected m/z's. Increases in detection levels using SIM are in the range of 10X to 100X that of full mass scan mode. Additionally, the ion trap MSD, used by some laboratories, offers enhanced sensitivity over the use of quadrupole MSD operated in full mass scan mode. Ion trap MSDs, however, are not in general use in commercial environmental laboratories.

Another important consideration between the FID and the MSD or PID is working concentration range. A typical working range for the measurement of MTBE using an FID in a heated headspace method is 5 to 4,000 parts per billion (ppb); using an MSD and SW846 8260, a typical working range for measurement of MTBE is 5 to 200 ppb. As discussed above, decreases in the working range lead to increases in the number of required dilutions and re-analyses, and this results in a greater possibility for analytical error and either an elevation of detection levels or an increase in laboratory expenses. Thus, there is a significant advantage in terms of working concentration range using the FID that translates to real benefits to the data user.

OPTIMUM METHODS FOR ANALYSIS OF FUEL OXYGENATES IN GROUND WATER

There are a number of choices that can be made in selecting an analytical technique for oxygenated gasoline environmental samples; however, most choices require an explanation of potential associated problems. Much work is still being done to discern these potential pitfalls and the extent to which they impact the analytical result. It the intent of this work to describe the options as they are currently understood and to recommend a set of those options that is proven to work.

In terms of sample preservation, the use of tri-sodium phosphate instead of hydrochloric acid is preferable because it universally eliminates the issue of preservative catalyzed hydrolysis of fuel oxygenate ethers. Further, the basic conditions of tri-sodium phosphate preserved environmental samples are unlikely to have negative effects on other compound groups of interest in oxygenated gasoline environmental samples. This preservative has been recommended by Kovacs and Kampbell (1999) of USEPA for use with hydrocarbon VOCs and is in current use by the Robert S. Kerr Laboratories of the USEPA for oxygenated gasoline contaminated environmental samples (Wilson, 2001).

For sample preparation, the use of headspace equilibration, SW846-5021, coupled with heating and matrix modification is preferred for several reasons:

1. It is a relatively simple analytical process that is easily automated and readily adaptable to FID, PID, or MSD measurement methods;
2. Reduction of the solubility of polar organics is easily accommodated through matrix modification by addition of inorganic salts;
3. A sample prepared in this method requires no further concentration to achieve detection limits of 1 to 10 µg/l for all components of oxygenated gasoline, including ethanol;
4. The use of a solid substrate, such as SPME fibers or traps that can catalyze hydrolysis, is not required; and
5. A sample preserved with tri-sodium phosphate has minimal chance of degradation using the headspace method of sample preparation.

Any measurement methodology, *i.e.*, FID, PID, or MSD, may be used in conjunction with the preservation and preparative methodologies mentioned above. The choices have already been described, and it is likely that economics and regulatory concerns will have as much to do with the practitioner's decision as technical issues.

We believe that the use of GC/FID coupled with tri-sodium phosphate preservation, matrix modification, and heated headspace equilibration offers an outstanding compromise for analysis of oxygenated gasoline compounds in water samples. In samples with high hydrocarbon concentrations where coelution of hydrocarbons with MTBE is a cause for concern, the use of SW846-8260 on one or more samples preserved with tri-sodium phosphate can be used to verify the identity of analytes of concern (Rhodes and Verstuyft, 2001). Such analyses should be necessary only during initial assessment of the site, not during later rounds of routine remedial monitoring.

Rhodes and Verstuyft (2001) have pointed out that the FID does not offer ultimate species selectivity and that the potential for coelution of multiple analytes is a serious liability of this choice of methodology. It is felt here that this problem, while real, is 1) minimized by advances in separation capability of GC columns, and 2) not as serious as the problems of purge-and-trap preparation that must be used when the MSD using full mass scan mode is used for analyte measurement.

Analysis of fuel oxygenates using heated headspace preparative methodology coupled with either an ion trap MSD or a quadrupole MSD in SIM mode perhaps offers the ultimate in preparative methodology with the most selective measurement methodology. However, neither ion trap MSD nor SIM mode is used routinely in commercial environmental laboratories. Both of these MSD measurement methods have a considerably limited working concentration range as compared to measurement with the FID.

Recently, at the request of USEPA, three laboratories (USEPA R.S. Kerr Environmental Research Center, the Calgary Laboratory of Ecotest Laboratories, and Microseeps, Inc.) participated in a study to demonstrate the applicability of the combination of preservation with tri-sodium phosphate and a headspace sample preparation technique for ground water samples containing oxygenated gasoline compounds. In this "tri-lab" study, each of the laboratories used different measurement techniques, which are summarized in Table 6-1. Correspondingly, each of the laboratories had different reporting limits for each of the compounds analyzed and those reporting limits are presented in Table 6-2.

Twenty-four samples were collected from Site A, with duplicate samples collected and sent to each of the participating laboratories. In those samples, only ethers were detected. The results from each laboratory for ETBE, TAME, and MTBE are shown in Figures 6-6 through 6-8, respectively.

It was an objective of this study to test this analytical program on all three groups of compounds of interest in oxygenated gasoline: hydrocarbons, ethers, and alcohols. Historical data had shown the presence of TBA in sam-

Table 6-1. **Measurement Methods for the Laboratories Participating in the Tri-Lab Study**

R.S. Kerr	ETL	Microseeps
Ion trap MSD	Quadrupole MSD (full scan)	FID

Table 6-2. **Reporting Limits for the Laboratories Participating in the Tri-Lab Study**
All concentrations are specified in µg/l.

Analyte	R.S. Kerr[1]	ETL	Microseeps
Ethanol	14	300	10
TBA	2.4	100	5
MTBE	0.2	5	5
ETBE	0.2	10	5
DIPE	0.2	10	5
TAME	0.2	10	5
TAA	1	100	5
Benzene	0.2	1	5
Toluene	0.4	1	5
Ethylbenzene	0.3	1	5
m&p-Xylenes	0.2	1	10
o-Xylene	0.3	1	5

[1] The limits for R.S. Kerr are method detection limits, for the other laboratories reporting limits are given (typically reporting limits are 2 to 5 times higher than method detection limits).

ples from Site A. However, it was found that these data were produced from the analysis of samples that were preserved with hydrochloric acid. Twenty-three samples were collected at Site B, and the results from the analyses for benzene, MTBE, and TBA are shown in Figures 6-9 through 6-11, respectively.

For all data shown, the results of the tri-lab study are in excellent agreement. However, precision is only a part of what is needed—accuracy is similarly important. The only way to judge the accuracy of analysis is to spike a known amount of a compound into a sample and to see if subsequent analysis of that sample quantifies the concentration of analyte that was spiked into the solution.

Ideally, an analysis should be capable of measuring the exact amount of the compound spiked into the sample. The amount actually measured is normalized by the amount that was truly spiked and the normalized result, expressed as a percent, is called the recovery. A perfect recovery is 100%. Variations in the instrument response, or in the response of a particular sample matrix, to the analysis as caused by interference from the sample, may raise

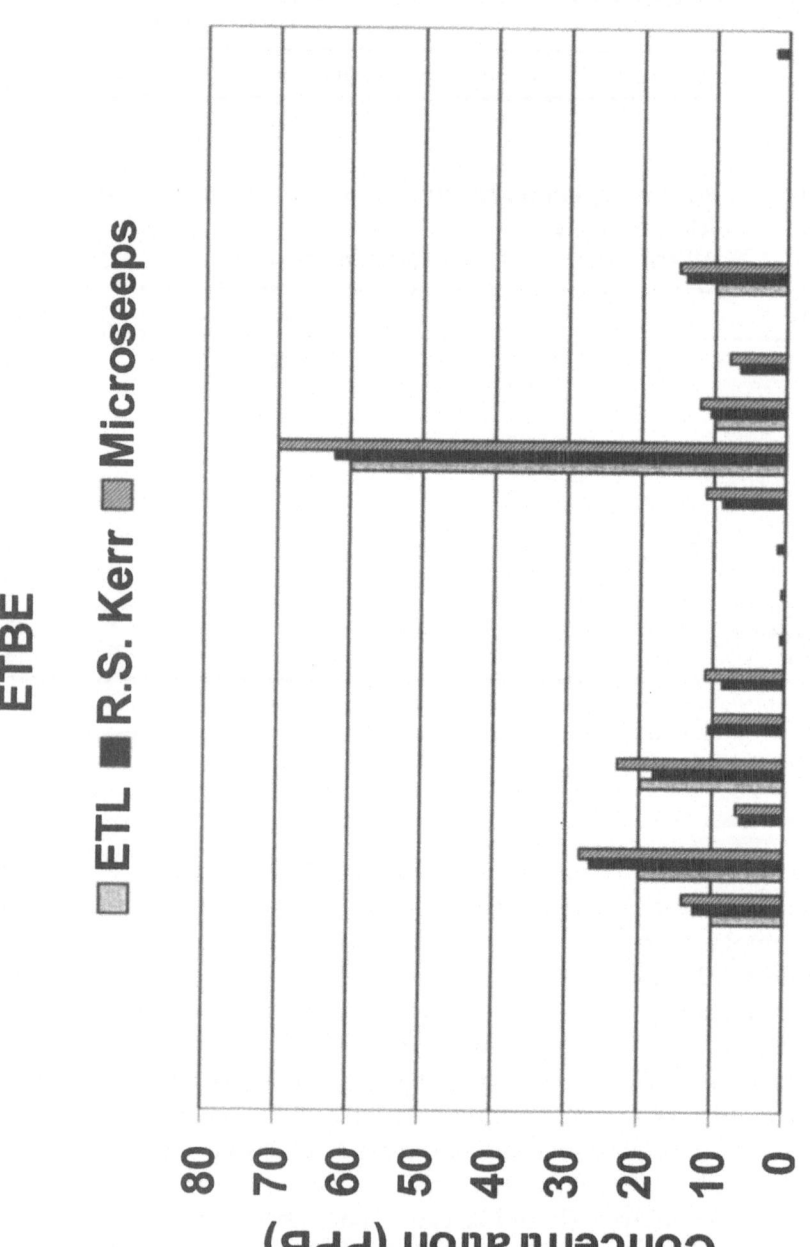

Figure 6-6.   Results of Tri-Lab Study for ETBE at Site A

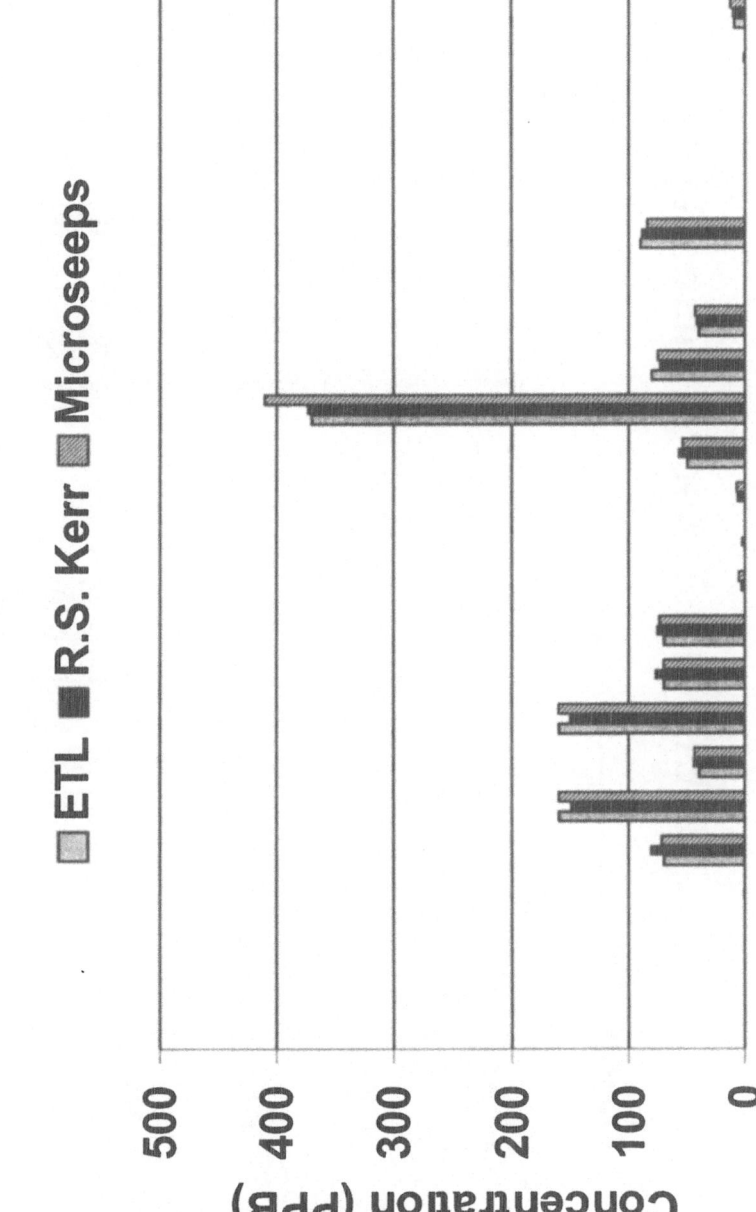

Figure 6-7. Results of Tri-Lab Study for TAME at Site A

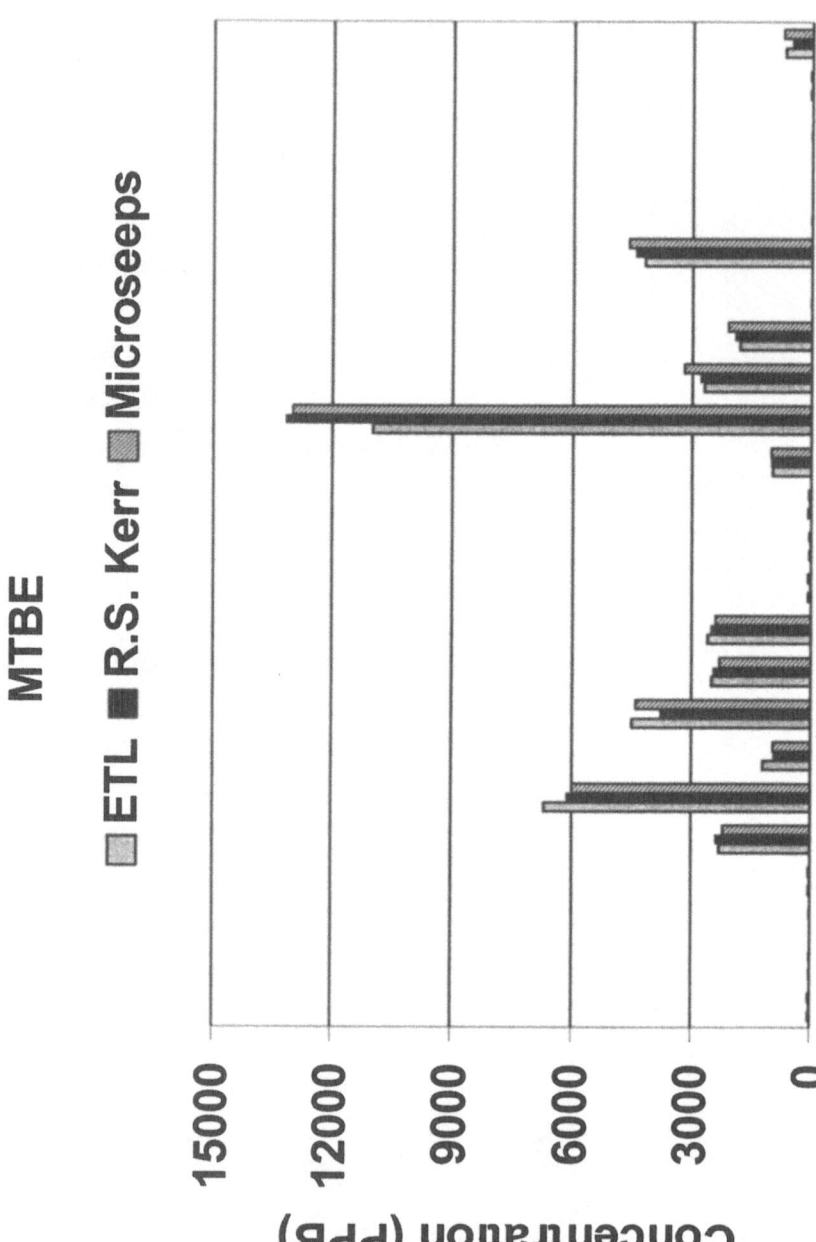

Figure 6-8.   Results of Tri-Lab Study for MTBE at Site A

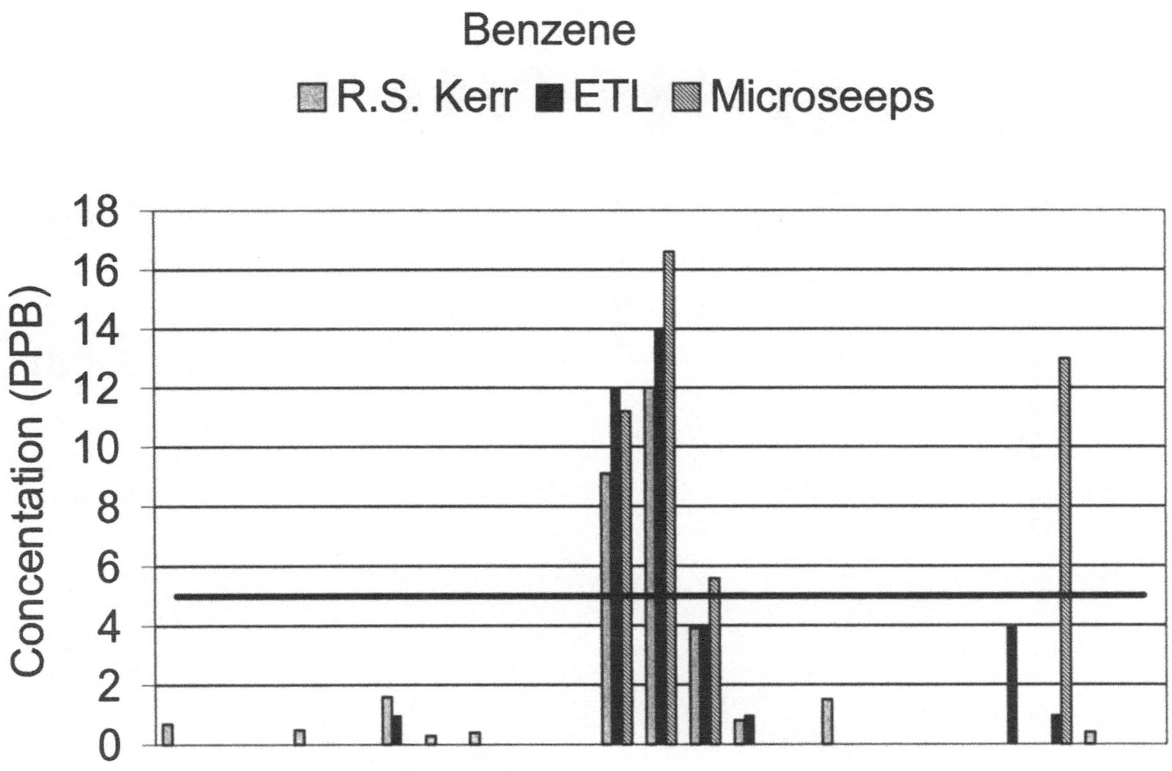

Figure 6-9. **Results of Tri-Lab Study for Benzene at Site B Microseeps has a 5 PPB reporting limit for benzene.**

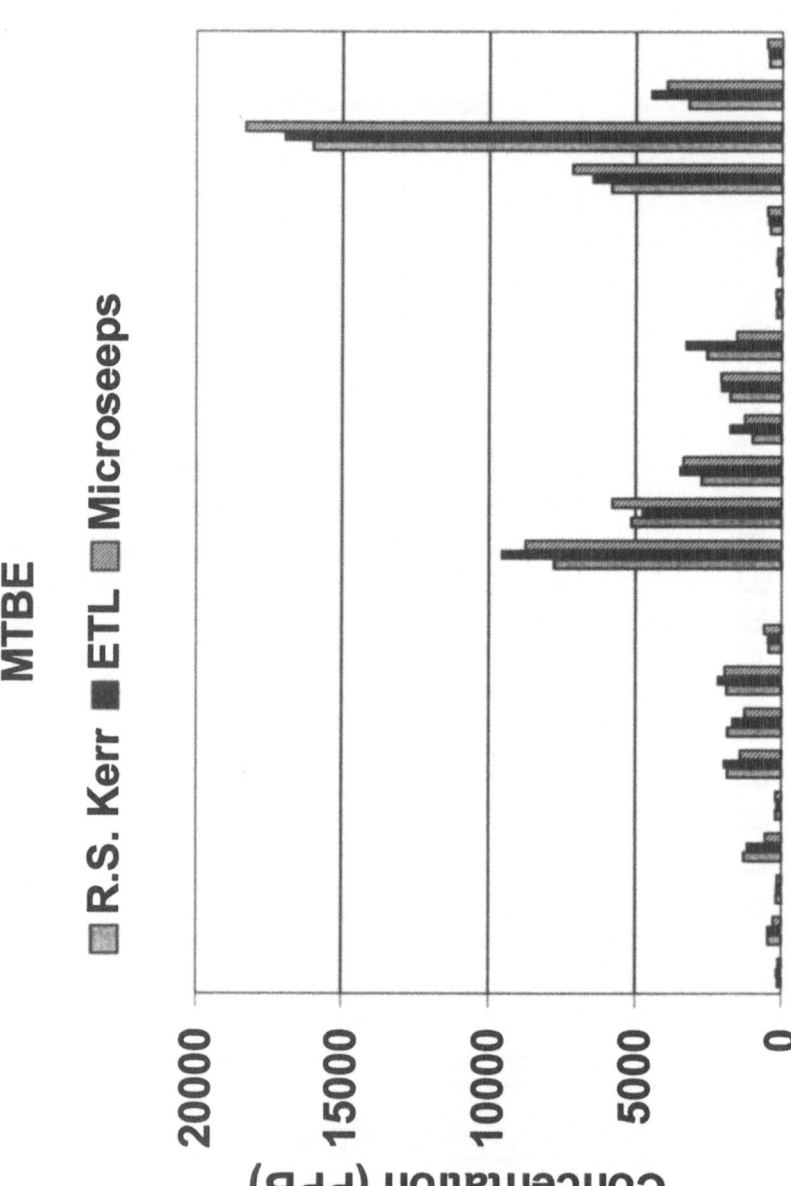

Figure 6-10. Results of Tri-Lab Study for MTBE at Site B

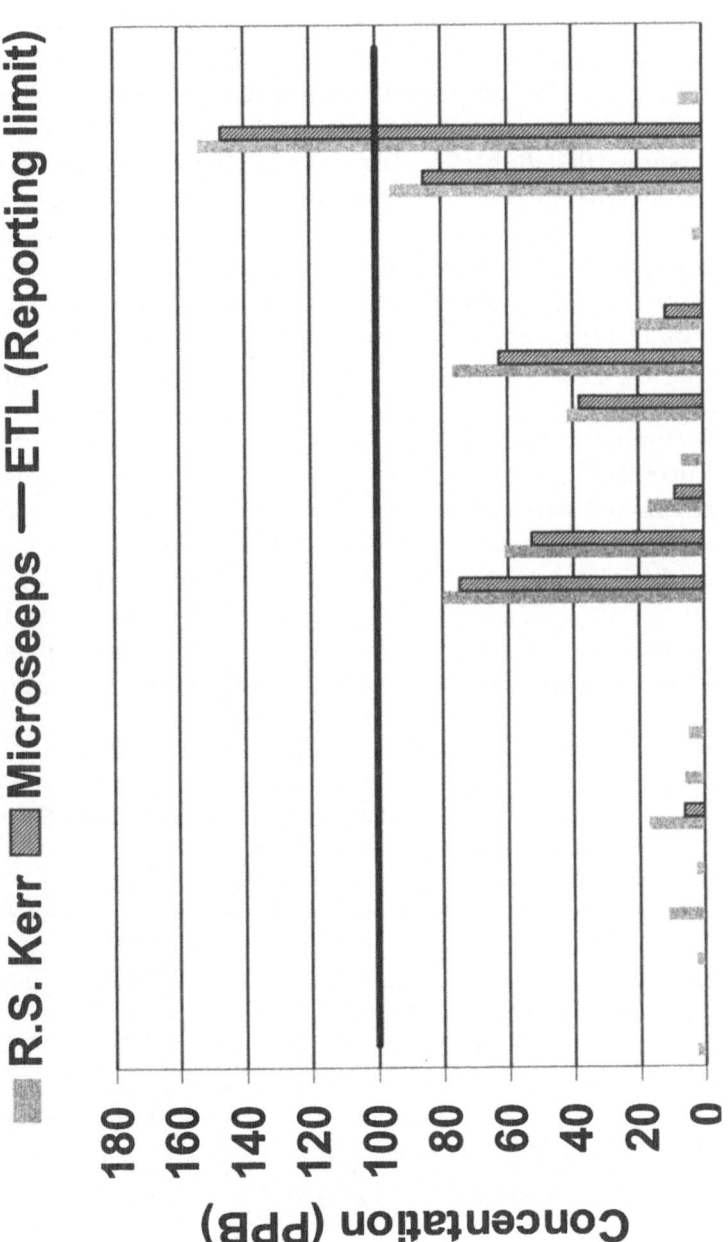

Figure 6-11.   Results of Tri-Lab Study for TBA at Site B

or lower the measured recovery. For VOC analyses, recoveries between 80% and 120% are considered good.

There are two kinds of spikes that apply to field samples. If the compound that was spiked into the sample is extremely unlikely to be in the unspiked sample, the spike is considered a surrogate spike. If the compound that was spiked into the sample is likely to be in the sample already, the spike is considered a matrix spike.

It is assumed that the background concentration of surrogate compounds is zero, so there is no need for analysis of an unspiked sample. Surrogate spikes can be, and are, added to every sample. Since the compounds that are in a matrix spike may be in the sample even before the addition of the spike, it is necessary to do an analysis of the unspiked sample as well as of the spiked sample. Because of this, it is not practical to do a matrix spike on every sample. Matrix spikes are typically done at a minimum frequency of once per 20 samples.

The compounds that were used as surrogate spikes in the tri-lab study are not a useful indicator of accuracy for ethers and alcohols because they are hydrocarbons and do not behave like ethers and alcohols. The most useful indicators of accuracy are the recoveries from the matrix-spiked samples. Only two of the three laboratories reported matrix spike recoveries. For Site A, the laboratory specific results of the original, unspiked samples used for matrix spikes are presented in Table 6-3; the concentration of the applied spike is given in Table 6-4; and the spike recoveries are summarized in Figure 6-12.

Table 6-3.   **The Laboratory Specific Results of the Original Samples used for Matrix Spikes for Site A**
All concentrations are given in µg/l.

Analyte	R.S. Kerr ML-5C	R.S. Kerr ML-12C	Microseeps ML-12C	Microseeps ML-8D
Benzene	<0.2	<0.2	< 5.0	< 5.0
DIPE	<0.2	<0.2	< 5.0	< 5.0
Ethanol	15.2*	<14	< 10	< 10
Ethylbenzene	<0.3	<0.3	< 5.0	< 5.0
ETBE	<0.2	<0.2	< 5.0	<5.0
m&p-Xylenes	<0.2	<0.2	< 10	< 10
MTBE	47.6	0.7*	< 5.0	67
o-Xylene	<0.3	< 0.3	< 5.0	< 5.0
TAA	<1.0	<1.0	< 5.0	< 5.0
TAME	<0.2	<0.2	< 5.0	5.5
TBA	<2.4	<2.4	< 5.0	< 5.0
Toluene	<0.4	<0.4	< 5.0	< 5.0

*The indicated value is less than that of the least concentrated standard (100 µg/l for ethanol, 1 µg/l for MTBE).

Table 6-4. **The Concentration of the Spikes Used for the Recovery Studies**
All concentrations are given in the sample concentration in µg/l. All compounds are in the spike mix at the same concentration except for ethanol in the R.S. Kerr mix (10-fold greater concentration) and m&p-xylene in the Microseeps mix (2-fold greater concentration).

R.S. Kerr ML-5C	R.S. Kerr ML-12C	Microseeps ML-12C	Microseeps ML-8D
50	10	400	400

For Site B, the laboratory specific results of the original, unspiked samples used for matrix spikes are presented in Table 6-5; the concentration of the applied spike is given in Table 6-6; and the spike recoveries are summarized in Figure 6-13.

The spike recoveries were all acceptable, except for MTBE in Microseeps spike of MW-3. As can be seen in Table 6-5, the original sample contained 3,400 µg/l of MTBE. A spike of 400 µg/l is only a 12% increase in the concentration. That is a small change and could easily result in a poor matrix spike recovery. The spike recovery of the TBA would be much greater than 100% if conversion of MTBE to TBA occurred through hydrolysis. The TBA recovery of 109% is not significantly greater than 100% and the recovery of all of the other spiked compounds is very good, so it is likely that the poor recovery of the MTBE was due to the high concentration of MTBE in the original sample.

The spike recovery data here validate the idea that the results from the tri-lab study were accurate as well as precise. As this example shows, spike recovery data can be very helpful when assessing MTBE analytical results. Unfortunately, when applying this analysis to a previously obtained data set, even if spike recovery data were collected, MTBE may not have been one of the spiked compounds. If this is the case, the spike recovery data are of very limited use for evaluating the accuracy of the MTBE analytical results.

CONCLUSIONS

The example provided by the tri-lab study shows that, irrespective of the detector that is used, a combination of tri-sodium phosphate preservation and headspace preparation produces an accurate and precise analysis. Additionally, that example shows the utility of matrix spike analyses and of including both the target ethers and their corresponding alcohols in the matrix spike mix. Finally, that example should be useful not only in choosing analytical

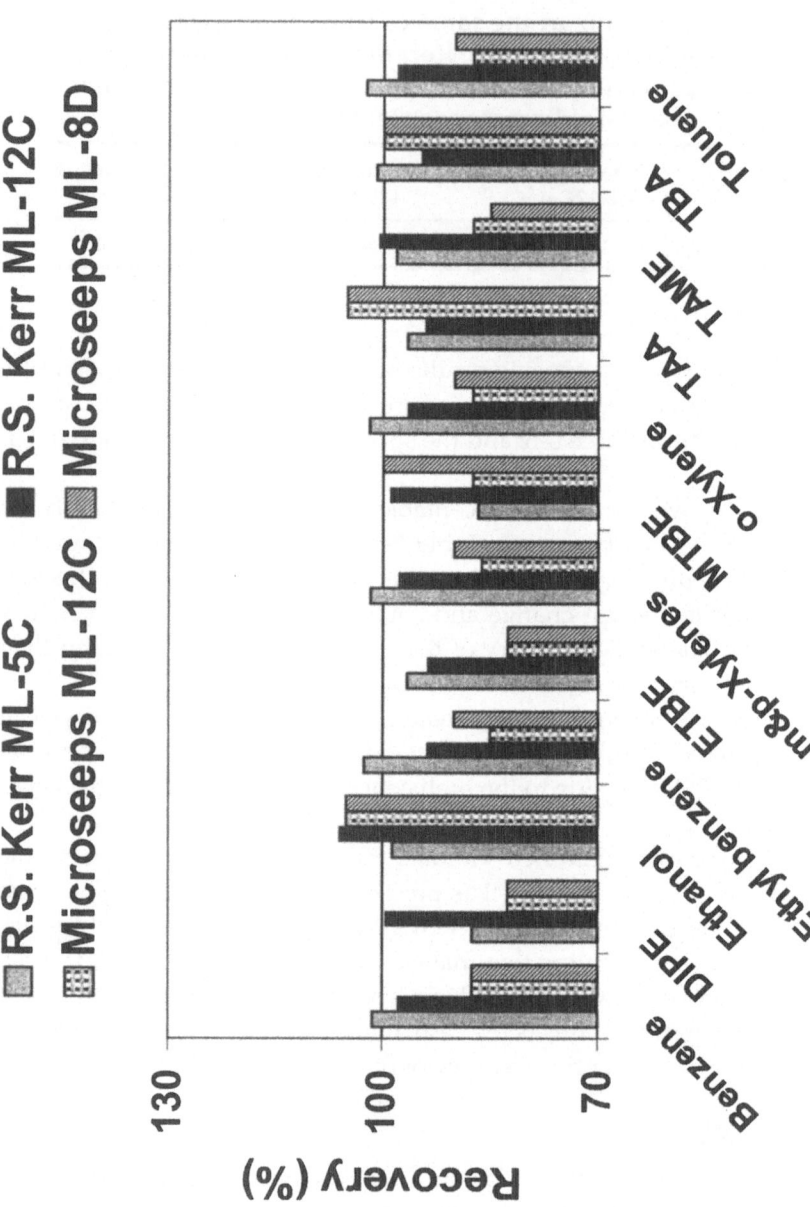

Figure 6-12.   Matrix Spike Recoveries for Site A

Table 6-5.   The Laboratory Specific Results of the Original Samples
             Used for Matrix Spikes for Site B
             All concentrations are given in µg/l.

Analyte	R.S. Kerr MW-13	R.S. Kerr ML-10B	Microseeps MW-3
Benzene	0.7*	<0.2	< 5.0
DIPE	<0.2	<0.2	< 5.0
Ethanol	<14	<14	< 10
Ethylbenzene	1.7	<0.3	< 5.0
ETBE	<0.2	<0.2	< 5.0
m&p-Xylenes	2.0	<0.2	< 10
MTBE	130	2.5	3400
o-Xylene	3.1	<0.3	8.7
TAA	<1.0	<1.0	< 5.0
TAME	1.9	<0.2	49
TBA	2.7	<2.4	63
Toluene	6.2	<0.4	< 5.0

*The indicated value is less than that of the least concentrated standard (1 µg/l for Benzene).

Table 6-6.   The Concentration of the Spikes Used for
             the Recovery Studies
             All concentrations are given in the sample
             concentration in µg/l. All compounds are
             in the spike mix at the same concentration
             except for ethanol in the R.S. Kerr mix
             (10-fold greater concentration) and
             m&p-xylene in the Microseeps mix
             (2-fold greater concentration).

R.S. Kerr ML-10B	R.S. Kerr MW-13	Microseeps MW-3
10	100	400

methods to be used for future analyses, but in presenting tools to be used to evaluate analyses that have already been done.

The USEPA is currently revising methods SW846-5021 and SW846-8015 to reflect the techniques used in the tri-lab study. They are also issuing a guidance document that discusses the hydrolysis issue and recommends the use of a tri-sodium phosphate preservative if water or soil samples are to be analyzed for fuel oxygenates. Both the method revisions and the guidance document are currently in review and should be available in 2002. Details should be posted on the SW846 website.

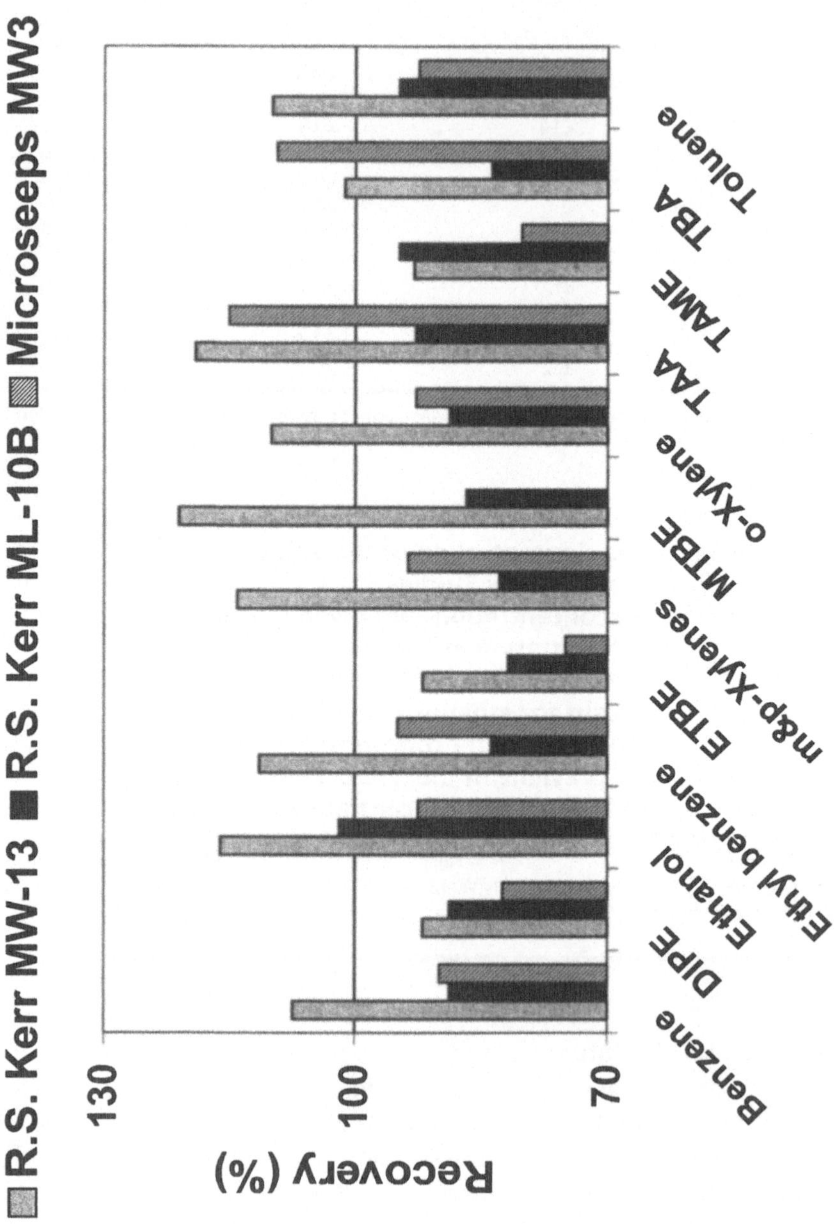

Figure 6-13.   Matrix Spike Recoveries for Site B

REFERENCES

Agilent Technologies. 2002. *Positive Identification of Methyl tert-Butyl Ether (MTBE)*. GC Petrochemicals Application Note #P5. www.chem.agilent.com/cag/cabu/pdf/b-0248.pdf. Accessed July 2002.

Black, L. and Fine, D. 2001. High levels of monoaromatic compounds limit the use of solid-phase microextraction of methyl tert-butyl ether and tert-butyl alcohol. *Environmental Science and Technology.* 35, 3190.

Church, C.D., Isabelle, L.M., Pankow, J.F., Rose, D.L., and Tratnyek, P.G. 1997. Method for determination of methyl tert-butyl ether and its degradation products in water. *Environmental Science and Technology.* 31, 3723.

Happel, A.M., Beckenbach, E.H., and Halden, R.U. 1998. *An Evaluation of MTBE Impacts to California Groundwater Resources.* Lawrence Livermore National Laboratory, UCRL-AR-130897.

Kovacs, D.A. and Kampbell, D.H. 1999. Improved method for the storage of groundwater samples containing volatile organic analytes. *Archives of Environmental Contamination and Toxicology.* 36, 242.

O'Reilly, K., Moir, M.E., Taylor, C.D., Smith, C.A., and Hyman. M.R. 2001a. Hydrolysis of tert-Butyl Methyl Ether (MTBE) in dilute aqueous solutions. *Environmental Science and Technology.* 35, 3954.

O'Reilly, K., Moir, M.E., Taylor, C.D., and Hyman, M.R. 2001b. Hydrolysis of MTBE: implications for anaerobic and abiotic natural attenuation. Redox conditions at fuel oxygenate release sites. In: *Proceedings of The Sixth International In Situ and On-Site Bioremediation Symposium.* Columbus, Ohio, Battelle Press.

Pawliszyn, J. 1997. *Solid Phase Microextraction: Theory and Practice.* New York, Wiley-VCH.

Rhodes, I.A.L. and Verstuyft, A.W. 2001. Selecting analytical methods for the determination of oxygenates in environmental samples and gasoline. *Environmental Testing and Analysis.* 10, 24.

Uhler, A.D., Stout, S.A., Uhler, R.M., and McCarthy, K.J. 2000. Considerations for the accurate chemical analysis of MTBE and other gasoline oxygenates. *Soil, Sediment & Groundwater.* Special MTBE Issue, 70.

USEPA (U.S. Environmental Protection Agency). 1986. *Test Methods for Evaluating Solid Waste (SW-846),* 3rd edition, Office of Solid Waste and Emergency Response, Washington, DC.

USEPA (U.S. Environmental Protection Agency). 2002. Technology Transfer Network Ambient Monitoring Technology Information Center. www.epa.gov/ttn/amtic/airtox.html. Last updated July 19, 2002.

USGS (U.S. Geologic Survey). 2002. National Water-Quality Assessment (NAWQA) Method, Sampling, and Analytical Protocols. water.usgs.gov/nawqa/protocols.doc_list.html. Last updated July 23, 2002.

Wilson, J.T. 2001. Personal communication.

CHAPTER 7

Risk Assessment

Pamela R.D. Williams, Sc.D., and Patrick J. Sheehan, Ph.D., Exponent

INTRODUCTION

Risk assessment is the process of determining the nature and extent of risks to human health and the environment. In a widely cited report, *Risk Assessment in the Federal Government: Managing the Process* (often referred to as the "Red Book"), risk assessments for human health are broadly defined as "the characterization of the potential adverse health effects of human exposures to environmental hazards" (NRC, 1983). The Red Book also details the main elements of the risk assessment process, including a description of potential adverse health effects (based on epidemiological, clinical, toxicological, and environmental research); extrapolation from available data to predict the type and estimate the extent of health effects in humans under certain exposure conditions; evaluation of the number and characteristics of persons exposed at various intensities and durations; and characterization of the overall magnitude of the public health problem and any uncertainties inherent in the process of inferring risk (NRC, 1983).

Current approaches for characterizing human health risks can be traced to the fields of occupational health, food safety, and radiation exposure (Kolluru, 1996; McClellan, 1998; Paustenbach, 1995). Over the last several decades, however, risk assessments have been performed for a wide variety of environmental hazards related to water contaminants, hazardous waste sites, air contaminants, occupational hazards, consumer hazards, and risks to wildlife (Paustenbach, 1989a, 2002). Many administrative and regulatory agencies in the U.S. also routinely conduct or review risk assessments, including the USEPA, the Food and Drug Administration (USFDA), the Occupational Safety and Health Administration (OSHA), the Department of Energy (USDOE), and the Consumer Product Safety Commission (USCPSC). The widespread use and general acceptance of health risk assessments in the U.S. has led to the publication of numerous reference materials on how to conduct risk assessments, and interpret and utilize the results (Bolger *et al.*,

1996; CRAM, 1997; Gargas *et al.*, 1999; Graham, 1995; Graham and Wiener, 1995; Graham *et al.*, 1988; Kolluru, 1996; Masters, 1998; NRC, 1983, 1994, 1996; Paustenbach, 1989b, 1995, 2002; USEPA, 1984, 1986a, 1986b, 1986c, 1989, 1992a, 1995a, 1995b, 1995c, 1996a, 1999b, 2000a).

The fuel oxygenate MTBE has generated substantial controversy in the U.S., and a comprehensive evaluation under the classical risk assessment paradigm would help resolve many contentious issues. Several attempts have been made to assess MTBE exposures or health effects in specific regions of the U.S., but these assessments have been limited in scope and do not provide an adequate characterization of the potential risks posed by MTBE. For example, although the UC study, *Health and Environmental Assessment of MTBE* (UC, 1998), gave the impression that exposures to MTBE in California posed a public health risk, this study relied on limited data sets and did not provide a quantitative risk characterization of MTBE to support its conclusions. Similarly, the NESCAUM (1999) report entitled, *RFG/MTBE Findings and Recommendations*, concluded that MTBE exposures in the northeast U.S. may exceed a health-protective threshold level under certain exposure conditions, but these findings were based on unrealistic exposure conditions and a misrepresentation of the available data. Although more comprehensive risk characterizations have been conducted for MTBE in the U.S., the primary focus of these evaluations has been on drinking water-related exposures only (Stern and Tardiff, 1997; Williams *et al.*, 2000a).

The purpose of this chapter is to provide a broad overview of the risk assessment process, particularly as conducted for chemical risks and human health in the U.S. Specific examples for and references to MTBE are also provided, where available. The information presented here should give the reader a general understanding of the essential components of a risk assessment, as well as the current state of knowledge about MTBE exposures, toxicity, and risk. Note that this chapter only provides a summary of the available literature on MTBE and does not represent an attempt to present a new or thorough risk characterization for MTBE; therefore, only tentative conclusions can be drawn about the potential health risks posed from MTBE air and drinking water exposures. As mentioned, a comprehensive evaluation of MTBE under the classical risk assessment paradigm has not yet been conducted, but would aid in resolving contentious issues and in future decision-making about MTBE.

The remaining sections of this chapter include: (1) a description of the risk assessment process for evaluating human health risks, (2) a brief description of the risk assessment process for evaluating ecological risks, (3) an overview of a recent European risk assessment of MTBE, (4) a discussion of the broader risk analysis framework, and (5) concluding thoughts about the risk assessment process and MTBE.

EVALUATING HUMAN HEALTH RISKS

Human health risks typically are evaluated by following the four steps of the risk assessment process: hazard identification, dose-response assessment, exposure assessment, and risk characterization (see Figure 7-1). In the first step, hazardous situations or agents are identified. In the second and third steps, which can be performed in any order, likely human exposures to a hazard and the health effects arising from these exposures are quantified where possible. In the final step, exposure and toxicity assessments are integrated to characterize the overall risk of a hazard. Risk assessment differs from risk management in that the former entails quantifying and characterizing risks, while the later involves making decisions about what to do about these risks (NRC, 1983). Each step of the risk assessment process, including specific applications for MTBE, is discussed in greater detail below.

Hazard Identification.   Hazard identification is the process of determining whether exposure to an agent can cause an increase in the incidence of a specified health effect (NRC, 1983; Masters, 1998). Identifying hazards involves characterizing the nature and strength of evidence of causation. The question of whether a substance causes an adverse health effect is theoretically a "yes-no" question, but few chemicals have adequate human data to reach such a definitive conclusion. Therefore, the hazard identification process is often based on an evaluation of data from animal studies or other test systems. An adverse effect is one that affects the organism's ability to survive, and adverse effects typically are thought of in terms of non-cancer and cancer effects. Non-carcinogenic effects are often systemic in nature (*i.e.*, affect the body's organ systems) and can include immunological, lymphoreticular, neurological, reproductive, or developmental effects. Carcinogenic effects, on the other hand, are defined as an increase in malignant tumors in animals or humans.

Over the last few years, a number of studies have assessed the potential health effects of MTBE on rodents, and a few have evaluated acute responses in humans based on testing of volunteers. Many of these investigations, including agency reviews, have been published in the literature (HEI, 1996; Clary, 1997; Mennear, 1997; NSTC, 1997; USEPA, 1997a; ATSDR, 1998; IPCS, 1998; UC, 1998; IARC, 1999; CA-OEHHA, 1999). Subchronic and chronic animal toxicity studies have also been performed for TBA, one of the primary byproducts of MTBE metabolism (Cirvello *et al.*, 1995; NTP, 1995). In general, the acute toxicity of MTBE and TBA in exposed animals and humans is expected to be low, but questions have been raised concerning the potential human risk of developing cancer from long-term exposure to MTBE (and/or its metabolites) based on animal data.

Systemic (Non-Carcinogenic) Toxicity.   Subchronic animal studies suggest that

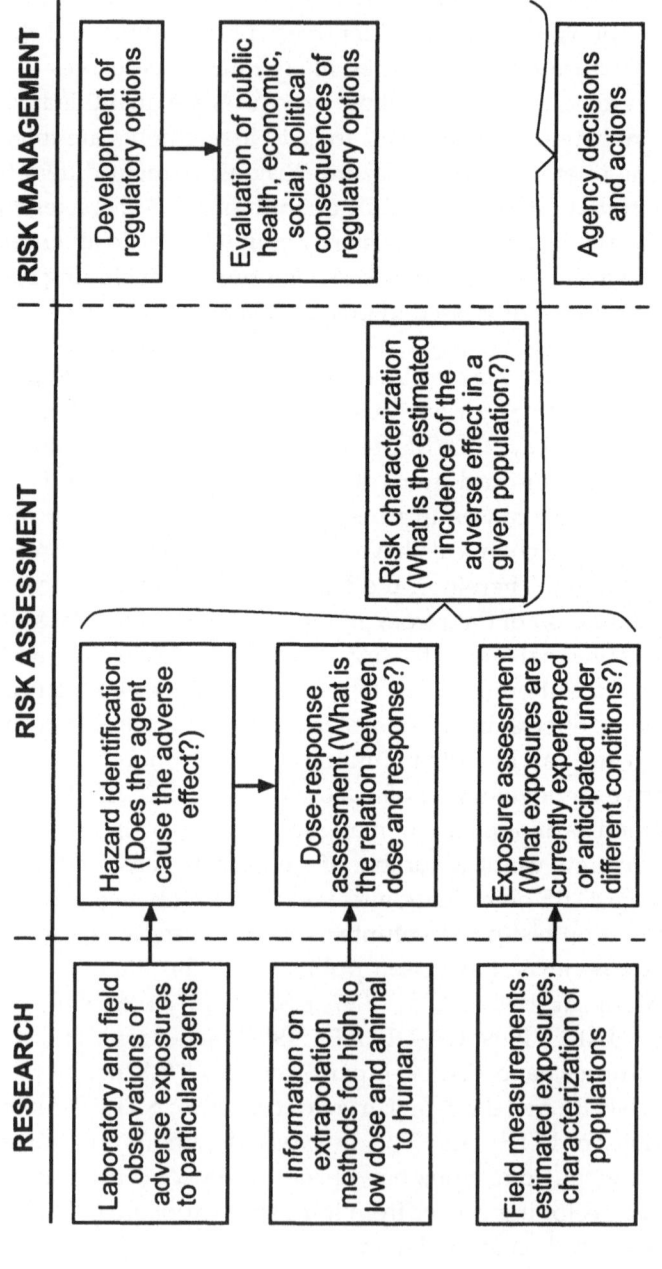

Figure 7-1.   Traditional Risk Assessment Framework. Source: NRC, 1983.

the kidney and liver are the most sensitive organs to MTBE exposure. The highest administered dose that failed to elicit an adverse effect—*i.e.*, the no-observed-adverse-effect level (NOAEL)—for increased kidney weight associated with oral MTBE exposures have been reported by Johnson *et al.* (1992) and Robinson *et al.* (1990) as 90 and 100 milligrams per kilogram per day (mg/kg-day), respectively. Similar systemic effects have been observed in MTBE inhalation exposure studies; for example, Chun *et al.* (1992) reported a NOAEL for renal effects associated with inhalation exposures of 400 ppm. In more recent studies, where the toxicity of MTBE was evaluated in rodents following repeated exposures, MTBE enhanced cell proliferation in mice (Bird *et al.*, 1997) and caused hepatocellular hypertrophy and kidney lesions in rats (Williams *et al.*, 2000b). Exposure to MTBE also produced reversible effects on the central nervous system, including sedation, hypoactivity, ataxia, and anesthesia in rodents at higher concentrations (Chun and Kintigh, 1993).

Similar transient effects on the central nervous system have also been reported from acute inhalation exposures to TBA (NTP, 1997). Subchronic oral toxicity studies of TBA indicate that in rodents the urinary tract is a primary target system, and males are more sensitive to TBA toxicity than females (CA-OEHHA, 1999; NTP, 1995). Decreased survival rates, body weights, liver weights, and kidney weights were observed in male and female rats and mice given TBA (greater than 99% pure) in drinking water for 13 weeks (NTP, 1995). Chronic oral exposures to TBA have also been associated with adverse effects on the kidney, urinary bladder, and thyroid gland, although systemic toxicity was generally only observed at the highest dose levels (NTP, 1995; MSDS, 2001).

MTBE exposures have not produced reproductive or developmental effects in animals at concentrations below those that are toxic to the reproducing adults (IARC, 1999). Similar findings were observed for TBA (CA-OEHHA, 1999; NTP, 1995). In contrast, moderate and long-term oral and inhalation exposure of animals to benzene, another gasoline constituent, resulted in adverse effects including bone marrow damage, changes in blood cells, developmental and reproductive effects, and alterations in immune response (CA-OEHHA, 2001). The levels of benzene exposure producing adverse effects are well below the NOAELs for MTBE, indicating that benzene is substantially more toxic to animals.

A limited number of human studies have evaluated the health effects associated with acute exposures to MTBE in gasoline vapors, and these studies have been reviewed and summarized by Borak *et al.* (1998). Self-reported symptoms from MTBE inhalation exposure include headache, eye irritation, burning of the nose and throat, cough, nausea, and dizziness (Mohr *et al.*, 1994; Anderson *et al.*, 1995). Although there is anecdotal evidence that inhaled MTBE may cause short-term effects, the available studies have not shown a significant correlation between the reported effects and MTBE exposure experienced by these individuals (NSTC, 1997; Balter, 1997; USEPA, 1997a; ATSDR, 1998). In a more recent controlled inhalation exposure study, Fiedler *et al.* (2000) found that self-reported sensitive individuals showed no difference in neurobehavioral or psychophysiologic responses when exposed to vapors of gasoline with 15% MTBE versus gasoline containing no MTBE, and symptoms historically associated with MTBE were not found. Therefore, study authors were unable to establish a dose-response relationship for MTBE exposures (Fiedler *et al.*, 2000). These researchers did find that individuals who had self-reported heightened sensitivity to MTBE drove their vehicles more often and fueled their vehicles more frequently than asymptomatic individuals (Opiekun *et al.*, 2001).

Mehlman (1998) reported that ingestion exposures to water contaminated with gasoline containing MTBE produced short-term neurotoxic and respiratory effects in some individuals, but the interpretation of such studies is confounded by the fact that exposures have largely been to gasoline containing MTBE, rather than to MTBE by itself. At present, no epidemiological data are available on the specific long-term systemic (or carcinogenic) effects of MTBE or TBA on humans (ATSDR, 1998; IARC, 1999; CA-OEHHA, 1999).

Carcinogenicity. The carcinogenicity of MTBE has been investigated in both inhalation and gavage long-term animal bioassay studies. Specifically, Chun *et al.* (1992) and Bird *et al.* (1997) evaluated the carcinogenicity in male and female Fisher 344 rats, and Burleigh-Flayer *et al.* (1992) and Bird *et al.* (1997) evaluated the carcinogenicity in CD-1 mice exposed to MTBE via inhalation. Belpoggi *et al.* (1995, 1997, 1998) also evaluated the carcinogenicity of MTBE administered to male and female Sprague-Dawley rats by oral gavage. In test animals exposed to MTBE by inhalation, statistically significant increases in renal tumors and Leydig cell tumors of the testes were observed in male rats, and a statistically significant increase in the incidence of liver tumors was observed in female and male mice. In test animals exposed by oral gavage, a statistically significant increase in Leydig cell tumors of the testes was observed in male rats, and a statistically significant increase in combined lymphomas and leukemias was observed in female rats.

Although these studies suggest that MTBE is carcinogenic to laboratory animals, the available bioassay data have several important limitations with respect to understanding the carcinogenic potential of MTBE in humans. First, the significant increase in observed tumors occurred primarily in animals exposed at the highest dose levels, which were generally systemically toxic to these test animals (Mennear, 1997). Second, some of the observed tumor sites (*e.g.*, Leydig cell of the testes) are those that have historically had a high background rate of tumor occurrence in rodents, making it difficult to assess the statistical or biological significance of such findings (Mennear, 1997). Third, limited pathology reporting makes it difficult to interpret the increased incidence of combined lymphomas and leukemias reported by Belpoggi *et al.* (1995).

Recent investigations into the mode of action by which MTBE causes species- and tissue-specific tumor responses have confirmed that MTBE is a non-genotoxic carcinogen in animals, but has the ability to cause tumors in rodents at high exposure concentrations (Borghoff and Williams, 2000). With regard to kidney tumors in male rats, Borghoff and Williams (2000) conclude that MTBE induces α2u-globulin nephropathy. Because this response to MTBE is mild compared to responses to other chemicals, a quantitative evaluation of the correlation between α2u-globulin accumulation and renal cell proliferation is necessary to demonstrate that this mechanism alone is responsible for the low incidence of kidney tumors that develop in male rats following chronic exposure to MTBE. With regard to liver tumors in female mice, Borghoff and Williams (2000) conclude that the lack of tumor-promoting ability of MTBE does not rule out the possibility that a high level of exposure to MTBE produces an increase in liver tumors by interfering with estrogen hormone function. These authors also note that the concentration of MTBE that produced mouse liver tumors is toxic to the test animals and is well above the concentration to which humans are likely to be exposed. Clegg *et al.* (1997) also suggest that non-genotoxic chemicals that cause Leydig cell tumors in rats operate through pathways that are not relevant to humans. In a recent study of the role of endocrine effects in inducing Leydig cell tumors in rats, the authors conclude that MTBE causes mild perturbations in serum triiodothyronine and prolactin; however, the changes in testosterone and luteinizing hormone levels did not fit the pattern caused by known Leydig cell tumorigens (Williams *et al.*, 2000b).

Chronic toxicity studies of TBA suggest "some evidence" of carcinogenicity in male rats based on increased incidences of renal tubule adenoma or carcinoma (combined), and in female mice based on increased incidences of follicular cell adenoma of the thyroid gland (Cirvello *et al.*, 1995; NTP, 1995). These studies showed "no evidence" of carcinogenicity in female rats (receiving 2.5, 5, or 10 mg/ml TBA in drinking water) and "equivocal" evidence in male mice based on marginally increased incidences of follicular cell adenoma or carcinoma (combined) of the thyroid gland (Cirvello *et al.*, 1995; NTP, 1995). Overall, USEPA (1997a) states that the National Toxicology Program (NTP) studies show no clear evidence of TBA carcinogenicity in either test species. TBA has also been tested for induction of genetic damage both in vitro and in vivo, and all genetic toxicity results have proven negative (NTP, 1995, 2002).

Like MTBE, TBA has been found to interact with α2u-globulin (*i.e.*, is a mild inducer of α2u nephropathy), but it appears that only the parent and not the metabolites of TBA are associated with α2u, unlike some other chemicals (Williams and Borghoff, 2001). In addition, because TBA causes α2u accumulation and renal cell proliferation in male, but not female rats, this implies that TBA interacts with α2u in vivo. The finding that TBA appears to be operating through the α2u-mediated mechanism suggests that the male rat kidney tumors (associated with α2u nephropathy only) should not be used for human cancer hazard identification (Williams and Borghoff, 2001).

No information specifically regarding MTBE's or TBA's cancer potential in humans was found in recent reviews (NTP, 1995; USEPA, 1997a; ATSDR, 1998; IPCS, 1998; IARC, 1999; CA-OEHHA, 1999). Neither MTBE nor TBA was listed as a human carcinogen in the NTP's Ninth Report on Carcinogens (NTP, 1999). The stated reason for not listing MTBE was "rodent cancer data not sufficient for listing in the Report as reasonably anticipated to be a human carcinogen" (NTP, 1999). To date, no national or international regulatory agency has classified MTBE or TBA as a known human carcinogen. In contrast, benzene, another gasoline constituent, is internationally recognized as a known human carcinogen (USEPA, 1979; IARC, 1982). Based on the available data, USEPA (1997a) concluded that the weight of evidence indicates that MTBE is an animal carcinogen and poses a carcinogenic potential to humans. The California Office of Environmental Health Hazard Assessment (CA-OEHHA) also considers MTBE to be an animal carcinogen and a possible human carcinogen (CA-OEHHA, 1999). On the other hand, International Agency for Research on Cancer (IARC, 1999) concluded that there is limited evidence for the carcinogenicity of MTBE in experimental animals and there is inadequate evidence in humans. Similarly, the World Health Organization concluded that data on MTBE are currently inadequate for use in human carcinogenic risk assessment (IPCS, 1998). Neither the USEPA nor IARC have evaluated TBA for carcinogenicity (CARB, 1997).

Dose-Response Assessment. While all substances are toxic at some dose level, a primary objective of toxicologists is to distinguish between the dose that is toxic and a lower dose that is not. Dose-response assessment is the process that characterizes the relation between the dose of an agent received and the incidence of the health effect (NRC, 1983; USEPA, 1986a, 1996a, 1999b). Because animal bioassays are the most common source of toxicity data, characterizing the dose-response relationship for humans requires two extrapolations (USEPA, 1989). First, the adverse effects usually are observed only under high exposure conditions; therefore, low-dose models are necessary to estimate effects in the test species at the much lower exposure levels that are likely to be encountered by humans in the environment. Second, effects in test animals are used to predict responses in humans. The process of characterizing the dose-response relationship depends on whether the agent exhibits a threshold (non-carcinogenic) or apparent non-threshold (carcinogenic) effect.

Threshold (Non-Cancer) Effects.   For non-carcinogens, it is generally assumed that there is some dose level below which no adverse effects are likely to occur. A two-step process is used to identify and quantify these threshold values (USEPA, 1989). First, the highest administered dose that fails to elicit an adverse effect, called the NOAEL, is determined. Second, the NOAEL (or lowest-observable-adverse-effect level [LOAEL] or benchmark dose/concentration) is divided by uncertainty or modifying factors to account for data gaps or uncertainty in the dose-response assessment, such as extrapolation from animal data to humans, and intra-species variability (Barnes and Dourson, 1988; Dourson *et al.*, 1996; Renwick and Lazarus, 1998).

The USEPA refers to the threshold value for ingestion as the Reference Dose (RfD), and the threshold airborne concentration for inhalation as the Reference Concentration (RfC). These values represent doses to which it is believed that humans can be exposed over a specified period of time without experiencing adverse effects (USEPA, 1999a). Specifically, the chronic RfD is defined as "an estimate (with uncertainty spanning perhaps an order of magnitude) of a daily oral exposure to the human population (including sensitive subgroups) that is likely to be without an appreciable risk of deleterious effects during a lifetime" (USEPA, 1999a). The chronic RfC has the same definition, except, instead of daily oral exposure, it refers to "continuous inhalation exposure" over a lifetime (USEPA, 1999a). The Agency for Toxic Substances and Disease Registry (ATSDR) also develops minimal risk levels (MRLs), which are defined as the "estimate of the daily human exposure to a hazardous substance that is likely to be without appreciable risk of adverse non-cancer health effects over a specified duration of exposure" (ATSDR, 2000a). These MRLs are similar in nature to USEPA's RfD/RfC values, in that they specify health guidance or acceptable exposure levels, but they differ in that they are based on the most sensitive substance-induced endpoint considered to be of relevance to humans, which may be less severe than those considered under the USEPA approach.

The USEPA has not yet developed an RfD for oral exposures to MTBE, but an RfC of 3 mg/m^3 has been developed for inhalation of MTBE (USEPA, 2000b). Extrapolation from the USEPA RfC is deemed acceptable because the non-cancer effects (*i.e.*, kidney and liver toxicity) are similar for both routes of exposure, and comparative toxicokinetic data are available for both routes. This extrapolation yields an equivalent oral dose of approximately 1.0 mg/kg-day (Dourson and Felter, 1997). ASTDR (1998) has also established an MRL of 0.3 mg/kg-day for intermediate-duration oral exposures to MTBE. Individual states may also establish their own reference guidance values; for example, the Pennsylvania Department of Environmental Protection (PADEP) has proposed an RfD for MTBE of 0.857 mg/kg-day (PADEP, 1997). The Texas Natural Resource Conservation Commission (TNRCC) uses California Environmental Protection Agency (Cal/EPA) values of 0.01 mg/kg-day for an RfD for MTBE and 3.0 mg/m³ for an RfC for MTBE. In addition, TNRCC has developed an RfC of 0.3 mg/m³ for TBA and an RfD of 0.09 mg/kg-day for TBA (TNRCC, 2002). Neither the USEPA nor CA-OEHHA have developed non-cancer health standards for acute or chronic exposures to TBA (CARB, 1997).

Table 7-1 compares non-cancer threshold values established by USEPA and ATSDR for MTBE and other selected gasoline constituents. As indicated in this table, the threshold values for MTBE are up to 10 times greater than that for other gasoline constituents, suggesting that, in general, MTBE has a much lower non-cancer toxicity than other common gasoline constituents.

Non-Threshold (Cancer) Effects. For carcinogens, it is generally assumed that any amount of an agent will result in some probability of an adverse effect, unless data to the contrary are available (USEPA, 1986a, 1996a, 1999b). The primary step in assessing the dose-response relationship for carcinogens is estimating the agent's cancer potency, which represents the slope of the dose-response curve at low doses and specifies the rate of increase in risk as a function of increasing dose. Cancer potencies can be expressed as a cancer slope factor (CSF), which is defined as an "upper bound, approximating a 95% confidence limit, on the increased cancer risk from a lifetime exposure to an agent," or as a unit risk, which is defined as the "upper bound excess lifetime cancer risk estimated to result from continuous exposure to an agent at a concentration of 1 µg/l in water, or 1 µg/m³ in air" (USEPA, 1999a). Because animal bioassay data generally provide the only information on an agent's dose-response, mathematical models are used to extrapolate from the high doses tested to the low doses of interest (USEPA, 1989). Note that the use of different types of extrapolation models (*e.g.*, tolerance distribution models,

Table 7-1. **Summary of Threshold Doses for Selected Gasoline Constituents**

Constituent	Chronic Oral RfD (mg/kg-day)	Chronic Inhalation RfC (mg/m³)	Acute Oral MRL (mg/kg-day)	Intermediate Oral MRL (mg/kg-day)	Chronic Oral MRL (mg/kg-day)
Toluene	0.2	0.4	0.8	0.02	NA
Ethylbenzene	0.1	1	NA	NA	NA
Xylene	2	NA	1	0.2–0.6	NA
MTBE	1*	3	0.4	0.3	NA

Source: USEPA (1988a, 1991, 1992b, 2000b) and ATSDR (1995b, 1998, 1999, 2000b).
*Based on extrapolation from chronic RfC.
Threshold Doses have not been established for TBA and ethanol.
NA = not available; RfD = reference dose; RfC = reference concentration; MRL = minimal risk level

mechanistic models, and time-to-tumor models) can introduce significant uncertainty in the dose-response assessment and exert a powerful influence on the resulting risk estimate (Paustenbach, 1989b; Graham *et al.*, 1988).

Various adjustments are also made to account for differences between the test animals and humans. These may include scaling factors based on body weight or surface area, used to normalize equivalent doses in animals and humans, or physiologically based pharmacokinetic (PBPK) models for animals and humans for determining target tissue dose estimates for risk assessment. These latter types of models, which involve dividing the animal into relevant compartments and using chemical and species-specific information to estimate the movement and behavior of the chemical within the body, can account for the many metabolic and other pharmacokinetic differences among species, so that a better estimate of the risk to humans at various doses can be identified (Leung, 1991, 2000; Leung and Paustenbach, 1995; Clewall, 1995; Clewell et al, 1995; McDougal, 1996; Gargas *et al.*, 1999).

Although USEPA (1997a, 2001) and IARC (1999) have identified MTBE as an animal carcinogen, neither agency has proposed a human cancer potency factor for MTBE. Similarly, no data were found in the literature on a cancer potency factor for TBA. However, based on the presumption that MTBE poses a carcinogenic risk to humans, CA-OEHHA (1999) recently derived a CSF for MTBE of 1.8×10^{-3} (mg/kg-day)$^{-1}$, as part of their Public Health Goal for drinking water. Note that the CA-OEHHA (2001) Public Health Goal for benzene assumes a CSF of 1.0×10^{-1} (mg/kg-day)$^{-1}$, indicating that California expects oral exposures to benzene to be about 50 times more potent that oral exposures to MTBE. USEPA (2000c) has also established a cancer potency for benzene, which ranges from 1.5×10^{-2} (mg/kg-day)$^{-1}$ to 5.5×10^{-2} (mg/kg-day)$^{-1}$, a value that is 8 to 30 times greater than the CSF for MTBE established by CA-OEHHA (1999).

CA-OEHHA's estimate for MTBE is based on the geometric mean of three potency estimates obtained from Chun *et al.* (1992) and Belpoggi *et al.* (1995), for which tumors were observed at multiple target sites and under inhalation and gavage MTBE dosing regimes (see Tables 7-2 and 7-3). CA-OEHHA (1999) also used a modified PBPK model to estimate the absorbed dose of MTBE in animals (not in humans), although several aspects of CA-OEHHA's approach have been questioned. These debatable components include (1) the use of an unvalidated PBPK model, which was created as a hybrid of two validated PBPK models; (2) the use of a rat PBPK model, but not a human PBPK model; and (3) reliance on animal tumor data that may not be relevant to humans. Note that a PBPK model is not necessary for evaluating the dose-response relationship of benzene, because the cancer potency for benzene is based on epidemiological (human) data, thus diminishing the need to extrapolate from animals to humans.

Aside from CA-OEHHA's modified PBPK assessment, three other PBPK

Table 7-2. **Chronic Animal Bioassay Data Used to Derive Cancer Potency for MTBE in California**

| Study | Tissue/Tumor Type | Dose Level (mg/kg-day)* | | | |
		Control	Low	Mid	High
Sprague-Dawley female rats; oral gavage (Belpoggi *et al.* 1995, 1998)	Hemolymphoreticular tissues • lymphoblastic lymphomas • lymphoblastic leukemias • lymphoimmunoblastic lymphoma)	0 (2/58)	250 (7/51)	NA	1,000 (12/47)
Sprague-Dawley male rats; oral gavage (Belpoggi *et al.* 1995, 1998	Testes • leydig cell adenoma	0 (3/26)	250 (5/25)	NA	1,000 (11/32)
Fischer 334 male rats; inhalation (Chun *et al.* 1992)	Kidney • renal tubular adenoma • carcinoma combined	0 (1/35)	400 (0/32)	3,000 (8/31)	8,000 (3/20)

Source: CA-OEHHA, 1999.
*Observed number of responses noted in parentheses.
NA = not available

models are currently available for MTBE (Borghoff *et al.*, 1996; Rao and Ginsberg, 1997; Licata *et al.*, 2001). In the initial model developed by Borghoff *et al.* (1996), in vivo rate constants were measured from gas uptake studies, and the rodent model was scaled to humans by incorporating human physiological and anatomical parameters and allometrically scaling metabolic parameters. Even though this model was able to predict the human MTBE blood levels calculated from the rat studies during exposure, it tended to under-predict post-exposure levels. Allometric scaling between rodents and humans for metabolic rate constants also appears to be insufficient to describe post-exposure levels of MTBE in the blood. A more refined PBPK model developed by Licata *et al.* (2001) overcomes this limitation by incorporating an understanding of the key metabolic processes that affect MTBE in rodents and humans; therefore, it is better able to estimate target tissue doses for risk assessment. It is likely that any future estimates of the carcinogenic potency of MTBE will be based on this refined PBPK model and on any new toxicity data collected.

Exposure Assessment.   Exposure is defined as "the event during which a person comes into contact with an agent;" therefore, exposure represents the joint occurrence (in space and time) of a person and an event (Ott, 1985). Exposure assessment is the process of measuring or estimating the intensity, frequency, and duration of human exposure to an agent currently present in the

Table 7-3. **Estimated Cancer Slope Factors for MTBE Based on Rat Oral and Inhalation Studies**

Route	Sex	Tumor Site and Type	q_1* $(mg/kg\text{-}day)^{-1}$	LED_{10} $(mg/kg\text{-}day)$	CSF $(mg/kg\text{-}day)^{-1}$
Inhalation (Chun et al., 1992)	Male	Renal tubular cell adenoma and carcinoma	1.9×10^{-3}	55	1.8×10^{-3}
Inhalation (Chun et al., 1992)	Male	Testicular interstitial cell tumors	9.2×10^{-3}	11	8.7×10^{-3}
Gavage (Belpoggi et al., 1995, 1998)	Male	Leydig cell tumors			
		Original 1995 report	1.4×10^{-3}	76	1.4×10^{-3}
		Revised 1998 data	1.6×10^{-3}	64	1.6×10^{-3}
Gavage (Belpoggi et al., 1995, 1998)	Male	Leukemia/lymphoma			
		Original 1995 report	2.1×10^{-3}	49	2.0×10^{-3}
		Revised 1998 data	2.2×10^{-3}	48	2.1×10^{-3}

Source: CA-OEHHA, 1999.

q_1* = 95% upper bound on linear slope at low dose using the linearized multistage (LMS) model.

LED_{10} = 95% lower bound on dose predicted to give a 10% tumor incidence.

CSF = Cancer slope factor based on LED_{10} (i.e., $0.1/LED_{10}$).

environment, or of estimating hypothetical exposures that might arise from the release of a new agent into the environment (NRC, 1983; USEPA, 1989). All available information pertaining to a particular exposure condition or scenario, including the exposure route, magnitude, duration, population affected, and data uncertainties, is presented in the exposure assessment (USEPA, 1988b; Paustenbach, 2000).

Exposure Pathways. A primary component of the exposure assessment process is to determine all of the ways in which people may be exposed to an agent or chemical of concern (USEPA, 1992a; ATSDR, 1995a). For example, exposures can occur from inhalation of toxic substances that are airborne; ingestion of contaminated food, water, dust, or soil; or dermal (skin) contact with a contamination source (see Figure 7-2). Because exposures can occur via multiple routes, and various transformations can occur as chemicals travel from one place or medium to another, understanding the fate and transport of environmental contaminants is important (Masters, 1998; USEPA, 1989).

Despite documented concerns over MTBE's impact on water quality, the primary route of human exposure to MTBE for most persons is the ambient air. MTBE levels can be detected in nearly any indoor or outdoor location, due to the widespread use of MTBE as a fuel oxygenate and the resultant exhaust

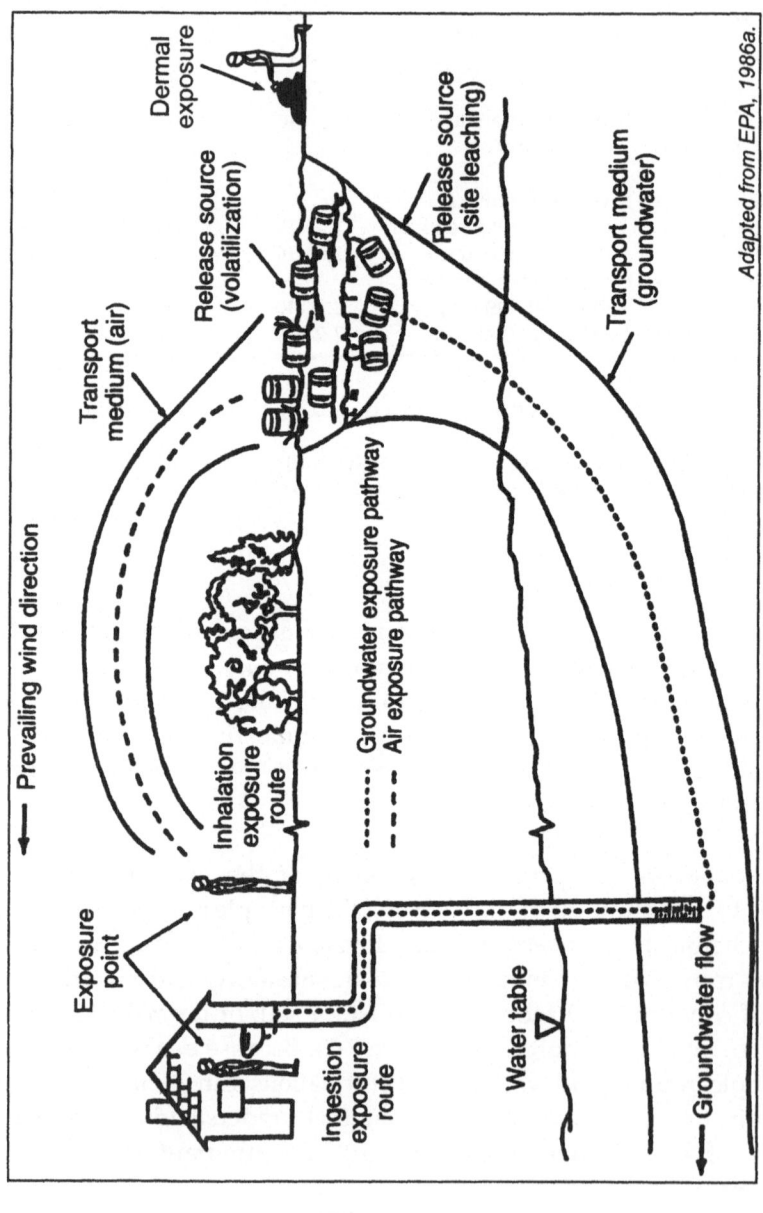

Figure 7-2.   Multi-Media Exposure Pathways. Source: USEPA, 1986a.

and evaporative emissions. During 1996 to 1997, outdoor air concentrations of MTBE were estimated to range from 1 to 48 $\mu g/m^3$ in urban regions of California (UC, 1998), although evaluation of the entire air monitoring data set during this period reveals that MTBE levels were about 8 microgram per cubic meter ($\mu g/m^3$) and 24 $\mu g/m^3$ at the 50th and 95th percentiles, respectively (BAAQMD, 1998; CARB, 1998). In the Northeast and nationwide, MTBE outdoor air concentrations have been estimated to range from about 1 to 3 $\mu g/m^3$ for the general population (Brown, 1997; NESCAUM, 1999). The highest airborne concentrations of MTBE are encountered during refueling and driving. For example, median MTBE air concentrations during refueling (with or without a stage II vapor recovery system) have been estimated to range from about 400 to 5,000 $\mu g/m^3$ (HEI, 1996; API, 1993; Cook and Kovein, 1994; Lioy et al., 1994; Johnson, 1993), while average MTBE levels during commuting have been estimated to range from about 60 to 140 $\mu g/m^3$ (Lioy et al., 1994; Brown, 1997).

Another, less likely, route of human exposure to MTBE is from the consumption of contaminated drinking water. Although few data are available on the extent of MTBE drinking water contamination nationwide or over time, assessments performed in several regions suggest that the majority of drinking water supplies have not been affected by MTBE or contain MTBE at very low levels. For example, a comprehensive review of California's water quality monitoring data indicated that only a small fraction (about 1 to 3%) of sampled drinking water sources from 1995 to 2000 contained detectable levels of MTBE. Of those drinking water sources with at least one MTBE detection, about 86% had concentrations below the state's primary (health-based) standard of 13 $\mu g/l$ (Williams, 2001). Slightly higher detection frequencies (about 5 to 16%) were observed in the Northeast during 1997 through 1999, but detected concentrations of MTBE were still typically below 2 $\mu g/l$, and only 0.5 to 1.5% of drinking water supplies were found to contain MTBE levels above 35 ug/l (NESCAUM, 1999). The most recent evaluation by the USGS indicates that MTBE was detected in about 9% of the CWSs in selected Northeast and Mid-Atlantic regions from 1993 through 1998, and that most MTBE concentrations were less than 5 $\mu g/l$ (less than 1% of CWSs were found to have MTBE concentrations at or above 20 $\mu g/l$, the USEPA's lower bound advisory level for MTBE in drinking water) (Grady and Casey, 2001). These findings suggest that the consumption of MTBE-containing drinking water is not expected to represent a significant exposure route for most persons in the U.S. (NSTC, 1997).

Note that, besides the consumption of MTBE-containing drinking water, other water-related activities in a household (e.g., showering) can result in MTBE exposures from inhalation of volatilized vapors or dermal contact with contaminated water (Brown, 1997; Stern and Tardiff, 1997). Although these latter water-related exposures typically result in a lower MTBE dose than the direct ingestion of MTBE in drinking water, their relative contribution to total dose can be high under certain exposure conditions (Williams *et al.*, 2000a). In general, dermal exposures are not expected to represent a significant pathway for MTBE compared to inhalation or ingestion exposures (UC, 1998).

Exposed Population.   Another important component of the exposure assessment process is to determine what segments of the population are exposed (USEPA, 1988b, 1997b). This may include the entire U.S. population, persons living in a specific region of the U.S., or subgroups within a specified population. Identifying exposed population groups is particularly important if a certain segment of the population is especially vulnerable or susceptible to an agent's effects, such as children, the elderly, pregnant women, or persons with impaired health (USEPA, 1988b; NRC, 1983; ATSDR, 1995a).

As mentioned, the general population may be exposed to MTBE due to airborne vehicle and fuel emissions. Other potentially more highly exposed population groups include persons who consume contaminated drinking water, persons living near a gasoline station, service station attendants, road tank drivers, or other occupational workers involved in the manufacturing or distribution of MTBE (UC, 1998; Brown, 1997). Currently, there is no clear evidence that certain segments of the population, such as children or the elderly, could potentially be more susceptible to the toxic effects of MTBE than the general population, although some agencies still caution against this possibility (ATSDR, 1998).

Estimating Dose.   Perhaps the most important aspect of the exposure assessment process is to quantify human exposure to an agent, using either direct or indirect methods (Paustenbach, 2000). The former approach uses monitoring data to directly measure an agent's concentration in the environment, while the latter approach uses mathematical models to predict human exposures (Ott, 1985). Exposure estimates are typically expressed as an average daily dose (ADD) or a lifetime average daily dose (LADD), which represent the amount of a substance to which an individual will be exposed on a daily basis or over a specified exposure period, respectively. Only the LADD can be used to characterize longer-term (chronic) health risks (see Equation 7-1). Dose metrics require information on environmental concentrations, absorption rates in the body, and details about the exposed individual (*e.g.*, body weight, intake rate, exposure duration). Population-specific estimates are generally used for these latter values when available, otherwise default assumptions that correspond to a typical or average individual in the population can be used (Finley *et al.*, 1994; USEPA, 1997b). Note that for all of these exposure input parameters, the use of quantitative probabilistic modeling techniques (such as Monte Carlo analysis) is becoming increasingly common in order to incorporate and evaluate the key sources of variability and uncertainty in exposure and risk assessments (Finley and Paustenbach, 1994; Thompson and Graham, 1996; USEPA, 1997c; Cullen and Frey, 1999; Paustenbach, 2000).

Estimating Lifetime Average Daily Dose (LADD)

$$\text{LADD (mg/kg-day)} =$$

$$\frac{\text{Concentration (C)} \times \text{Intake Rate (IR)} \times \text{Exposure Duration (E)} \times \text{Absorption Factor (A)}}{\text{Averaging Time (AT)} \times \text{Body Weight (BW)}} \quad (7\text{-}1)$$

where:

C — Amount of agent measured or modeled in environment; units are mg/kg (food), mg/l (water), or mg/m^3 (air)

IR — Rate of inhalation, ingestion, or dermal contact; units are milligram per day (mg/day) (food), liter per day (l/day) (water), or cubic meter per day (m^3/day) (air)

E — Number of days exposed over a lifetime; units are days

A — Fraction of agent absorbed into body; unitless

BW — Body weight of individual; units are kilograms

AT — Number of days in lifetime; units are days

Much of the available information for evaluating MTBE exposures is based on direct monitoring data. For example, air and drinking water data on MTBE in California are routinely collected or maintained by the Air Resources Board (CARB) and the CDHS, respectively. Limited information is also available on how much MTBE is absorbed into the body following exposure; the absorption rate for inhaled MTBE has been found to be 50% or less (Johanson et al., 1995; Pekari et al., 1996), while the absorption rate for ingested MTBE is generally assumed to be 100% (ATSDR, 1998; CA-OEHHA, 1999). Default values or established distributions from the literature can be used to estimate other input parameters, such as exposure frequency, averaging time, intake rate, and body weight (USEPA, 1989; Cal/EPA, 1994; Finley et al., 1994; Finley et al., 1993).

Using these types of data, MTBE doses have been estimated for specific population groups in several studies, although these evaluations have been based on alternative end points and differing assumptions about selected input parameters and exposure scenarios. For example, the UC study (UC, 1998) provided a range of estimated ADDs for MTBE for the general population and different subgroups in California based on both air and drinking water exposures (see Table 7-4). Williams et al. (2000a) also estimated the distribution of ADDs (and LADDs) for MTBE for different population groups in California, but these were for water-related activities only (see Figures 7-3a,

Table 7-4.   **Estimated Average Daily Dose from MTBE Air and**
            Drinking Water Exposures for Different
            Population Groups in California

Population Group	Average Daily Dose (µg/kg-day)	
	Minimum	Maximum
General population: adults		
Los Angeles/Burbank	0.6	4.8
San Francisco/Bay Area	0.3	5.2
Central Valley	0.3	2.8
South Coast Area	0.4	3.6
General population: children (10 yrs)		
Los Angeles/Burbank	—	4.8
San Francisco/Bay Area	—	5.7
Central Valley	—	3.0
South Coast Area	—	3.8
Road tank driver (LA/Burbank)	—	170
Service station attendant (LA/Burbank)	—	105
Persons living near service station (LA/Burbank)*	2.2	5.1
Persons consuming MTBE in drinking water at 35 µg/l (LA/Burbank)	1.6	5.5

Source: UC, 1998.
*Based on ambient air concentrations of MTBE obtained from service station
 perimeter monitoring data.

7-3b). ADDs of MTBE have been estimated in the Northeast by NESCAUM (1999), based on different assumptions about MTBE drinking water levels and the presence of stage II vapor recovery systems during refueling at gas stations (see Table 7-5). On a national level, Brown (1997) has estimated ADDs from exposure to MTBE in tap water and LADDs from exposure to MTBE in ambient air (see Tables 7-6 and 7-7). These studies suggest that MTBE exposures in the U.S. from all routes are likely to range from less than 1 µg/kg-day to 5 µg/kg-day for the general population, but can be somewhat greater for more highly exposed population groups, such as occupational workers.

Risk Characterization.   Risk characterization is the final, and perhaps most important, step of the overall risk assessment process - it involves integrating the information developed in the dose-response and exposure assessments to provide qualitative and quantitative estimates of risk (Williams and Paustenbach, 2002). Risk characterizations should also discuss all data, models, and statistical uncertainties; explain the rationale for using certain dose-response and exposure assessments; and express the level of confidence in exposure

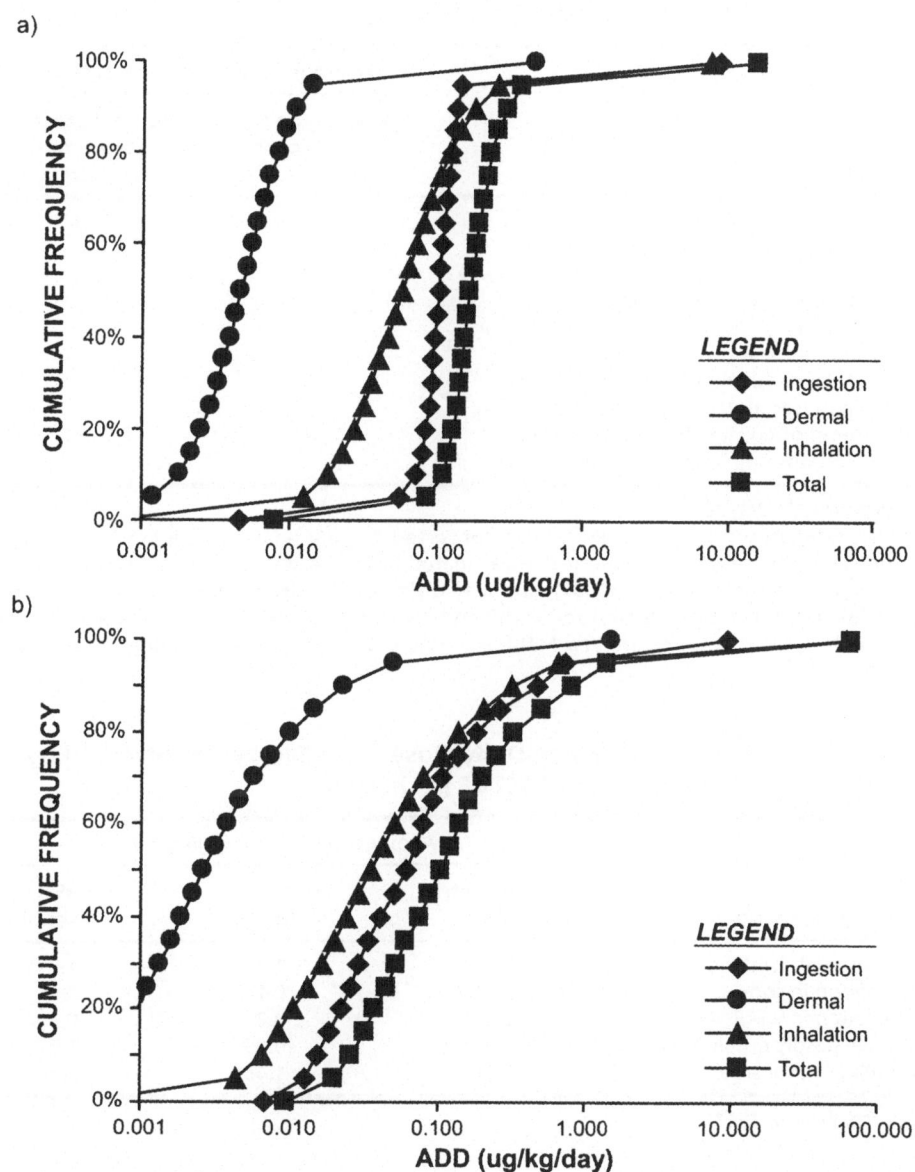

Figure 7-3.   **Distribution of Estimated Average Daily Dose for MTBE By Exposure Route in California (a) General Population (b) Households With Contaminated Drinking Water. Source: Williams et al., 2000a.**

Table 7-5.   **Estimated Average Daily Dose from MTBE Air and Drinking Water Exposures in the Northeast Based on Different Exposure Scenarios**

	Average Daily Dose (µg/kg-day)*	
Scenario	With Stage II Vapor Recovery System	Without Stage II Vapor Recovery System
MTBE level of 0 µg/l in drinking water; low outdoor air levels, no attached garage	5.3	8.3
MTBE level of 35 µg/l in drinking water; low outdoor air levels, no attached garage	6.5	9.4
MTBE level of 35 µg/l in drinking water; high outdoor air levels, attached garage	8.4	11.3
MTBE level of 100 µg/l in drinking water; low outdoor air levels, no attached garage	8.5	11.5
MTBE level of 100 µg/l in drinking water; high outdoor air levels, attached garage	13.4	16.4

Source: NESCAUM, 1999.
*Note: Estimates in units of ug/day were converted to µg/kg-day by dividing by a 70-kg adult. Exposure estimates were further calculated based on stated data sets and assumptions provided in NESCAUM (1999), and may differ slightly from values observed in the published figures. Data related to "with" and "without" stage II vapor recovery system are only applicable for estimating MTBE airborne exposures during refueling at gas station.

Table 7-6.   **Estimated Average Daily Dose from Tap Water Affected by Leaks and Spills of MTBE in the U.S.**

	Average Daily Dose (µg/kg-day)*		
Exposure Scenario	Geometric Mean	Arithmetic Mean	95th Percentile
Ingestion of drinking water	0.008	1.2	1.5
Inhalation in shower	0.0006	0.1	0.1
Inhalation in whole house	0.001	0.2	0.3
Dermal absorption in bath	0.00003	0.05	0.05
Combined exposure	0.01	1.4	2.0

Source: Brown, 1997.

and toxicity estimates (NRC, 1983, 1996; USEPA, 1989, 1995a, 1995b, 1995c, 2000a). Depending on its purpose, a risk characterization may also contain information on the effectiveness of alternative risk management options or the risk of competing or substitute chemicals (NRC, 1989, 1996; USEPA, 1995c, 2000a; Graham and Hartwell, 1997). The most relevant findings and conclusions about risk are summarized in the risk characterization, which in

Table 7-7. **Estimated Lifetime Average Daily Dose for Populations Exposed to MTBE in Air in the U.S.**

Population Group	Number of Persons	Lifetime Average Daily Dose (µg/kg-day)*	
		Geometric Mean	Arithmetic Mean
MTBE manufacturing workers	883	3.0	27.0
Gasoline blending workers	1,800	9.9	90.0
Gasoline transportation workers	1,489	8.6	255
Gasoline distribution workers	7,705	4.7	46.2
Gasoline station workers	150,000	12.4	59.1
Mechanics, etc.	300,000	3.0	11.9
Professional drivers	900,000	0.2	0.6
Commuters	30,000,000	0.1	0.2
Other drivers	22,800,000	0.3	0.4
Gasoline station customers	52,800,000	0.2	0.3
MTBE manufacturing and gasoline blending neighbors	815,000	0.1	0.4
Gasoline station and storage neighbors	11,700,000	0.1	1.1
General public	60,500,000	0.4	1.1

Source: Brown, 1997.
*Lifetime average daily dose incorporates data on atmospheric MTBE concentrations, exposure duration, number of exposure events per year, number of years exposed, and body weight. Number of persons exposed is based on national statistics adjusted for the 30% of the U.S. population living in areas where MTBE is required in gasoline (estimates for manufacturing, blending, transportation, and distribution workers based on Hinton, 1993).
Arithmetic mean = Sum of all observations in a data set divided by the total number of measurements, or the average of all observations.
Geometric mean = Product of all observations in data set taken to the nth root, where n is the sample size, or the antilogarithm of the arithmetic mean of the logarithms of the values. Note: the geometric mean is less affected by extreme values than the arithmetic mean.

turn is used to inform risk managers and decision-makers. In short, the risk characterization process attempts to make sense of the available scientific information and describe what it means to a broad audience.

Non-Cancer Effects.   In assessing the risk of non-carcinogens, estimated exposures are typically compared to an agent's safety level or threshold (Barnes and Dourson, 1988; USEPA, 1989; Masters, 1998). Estimates of non-cancer risk are based on the assumption that there is a level of exposure below which it is unlikely to experience adverse health effects. Common methods of evaluating non-cancer risks are to generate a hazard quotient (HQ), which represents the ratio of exposure to toxicity (usually the RfD), or to use a margin of exposure (MOE) approach, which represents the ratio of toxicity (usually the NOAEL) to exposure (see Equation 7-2).

Estimating Non-Cancer Risks (Assumes Threshold at Low Doses)

$$\text{Hazard Quotient (HQ)} = \frac{\text{Exposure (E)}}{\text{Reference Dose (RfD)}} \qquad (7\text{-}2a)$$

where:

E   Estimated exposure level or intake; units are mg/kg-day

RfD   Estimated non-cancer reference dose; units are mg/kg-day

$$\text{Margin of Exposure (MOE)} = \frac{\text{No-Observed-Adverse-Effect Level (NOAEL)}}{\text{Exposure (E)}} \qquad (7\text{-}2b)$$

where:

NOAEL   Estimated no-observed-adverse-effect-level; units are mg/kg-day

E   Estimated exposure level or intake; units are mg/kg-day.

HQs less than one indicate that exposures are unlikely to result in any adverse health effects, while HQs greater than one suggest that there may be concern for potential non-cancer effects (USEPA, 1989). However, since the HQ ratio is not interpreted as a statistical probability, the level of concern does not increase linearly as the RfD is approached or exceeded. Like carcinogens, risks from simultaneous exposure to more than one non-carcinogenic substance, or from multiple exposure pathways, are generally assumed to be additive, and can be evaluated by summing the individual estimated HQs. The resulting hazard index (HI) assumes that the magnitude of an adverse health effect is directly proportional to the sum of the individual HQ ratios. Similarly, MOE values that exceed 100 generally imply an acceptable level of exposure, while values less than 100 have traditionally been used by regulatory agencies as flags for requiring further evaluation (Barnes and Dourson, 1988; Klaassen, 2001).

Despite the availability of sufficient data on non-cancer effects, neither the UC (1998) study nor the NESCAUM (1999) report provided a quantitative risk characterization of MTBE for non-cancer effects. However, evaluations performed by Stern and Tardiff (1997) and Williams *et al.* (2000a), suggest that MTBE drinking water exposures are unlikely to pose a significant non-cancer health risk, because estimated HQs are well below one (see Table 7-8), and the estimated MOE is well above 100 (see Table 7-9).

Health Standards and Advisories.   A less formal approach to evaluating health risks, particularly for non-carcinogens, is to compare measured environmental exposure levels to available health standards or advisories. For example, the USEPA and many individual states have established enforceable and/or non-enforceable health standards for a variety of drinking water contaminants. Maximum contaminant level goals (MCLGs) are generally non-en-

Table 7-8. **Estimated Distribution of Hazard Quotients (HQs) and Hazard Indices (HI) from MTBE Drinking Water Exposures for Different Population Groups in California**

Population Group and Exposure Route	Distribution*	
	50th Percentile	95th Percentile
General population		
Inhalation (HQ)	6.5×10^{-5}	3.0×10^{-4}
Ingestion (HQ)	3.4×10^{-4}	4.8×10^{-4}
Dermal (HQ)	1.6×10^{-5}	4.6×10^{-5}
Total all routes (HI)	*4.4×10^{-4}*	*7.1×10^{-4}*
Households with contaminated drinking water		
Inhalation (HQ)	4.0×10^{-5}	7.7×10^{-4}
Ingestion (HQ)	2.0×10^{-4}	2.4×10^{-3}
Dermal (HQ)	8.7×10^{-5}	1.6×10^{-4}
Total all routes (HI)	*2.6×10^{-4}*	*3.2×10^{-3}*

Source: Williams *et al.*, 2000a.
*Based on annual average MTBE concentrations in drinking water ranging from 0.6 to 6.4 µg/l for the general population and 5 to 58 µg/l for households with contaminated drinking water.

Table 7-9. **Estimated Margin of Exposure (MOE) from MTBE in Drinking Water Based on Alternative Exposure and Toxicity Estimates**

Source and Estimate of Exposure	Human Chronic or Subchronic NOAELs (mg/kg-day)		
	2-Year Inhalation (1 mg/kg-day)	90-Day Inhalation (0.3 mg/kg-day)	90-Day Oral Gavage (0.1 mg/kg-day)
Atmospheric deposition*			
Geometric mean ADD (7.3×10^{-6} mg/kg-day)	140,000	41,000	14,000
Maximum ADD (6×10^{-5} mg/kg-day)	17,000	5,000	1,700
Leaks and spills*			
Geometric mean ADD (1×10^{-5} mg/kg-day)	—	300,000	100,000
Arithmetic mean ADD (1.6×10^{-3} mg/kg-day)	—	1,900	600
95 percentile ADD (2.0×10^{-3} mg/kg-day)	—	1,500	500

Source: Stern and Tardiff, 1997.
*Estimated MOEs for atmospheric deposition are based on chronic human NOAELs, while estimated MOEs for leaks and spills are based on subchronic human NOAELs, because such exposures are expected to occur for a short period of time.
ADD = Average daily dose. NOAEL = No-observed-adverse-effect-level. MOE = Margin of exposure.

forceable and represent a level at which no known or anticipated adverse health effects are expected to occur, while MCLs are enforceable and represent the maximum permissible level of a contaminant in water delivered to users of a public water system (USEPA, 2000d). In practice, MCLs are set as close to the MCLG as possible, but they allow feasibility and economic issues to be considered. Secondary drinking water regulations (SDWRs) are also non-enforceable guidelines that relate to cosmetic effects (*e.g.*, tooth or skin discoloration) or aesthetic effects (*e.g.*, taste, odor, or color) of drinking water. In addition, drinking water health advisories may provide information on contaminants that can cause human health effects and are known or anticipated to occur in drinking water (USEPA, 2000e). Note that these advisories typically represent guidance values that are based on non-cancer health effects for different durations of exposure (*e.g.*, 1-day, 10-day, and lifetime).

The USEPA (1997a) has established an advisory guidance level for MTBE in drinking water of 20 to 40 µg/l based on taste and odor effects, which is considered protective of sensitive members of the population. Although this advisory level is based on aesthetics, rather than health risk, USEPA (1997a) notes that the 20- to 40-µg/l range is about 20,000 to 100,000 (or more) times lower than the range of exposure levels in which cancer or non-cancer effects have been observed in rodent tests. To date, no national health-based advisory level for MTBE in drinking water has been established by USEPA, although promulgation of a secondary drinking water standard is expected in the near future (USEPA, 2002a). MTBE is currently listed as part of the Unregulated Contaminant Monitoring Rule, which means that additional information is required before a decision can be made on whether to regulate MTBE (USEPA, 2002b). A number of individual states have established their own primary (health-based) or secondary (aesthetic-based) standards for MTBE in drinking water. For example, California set a primary MCL of 13 µg/l (effective May 2000) to be protective of cancer effects, and a secondary MCL of 5 µg/l (effective January 1999) to address taste and odor concerns (CA-OEHHA, 1999; CDHS, 2001). Many states have also set uniform or site-specific action levels and/or cleanup levels for MTBE in contaminated ground water (see Table 7-10).

Interestingly, TBA is classified by CDHS (2002) as an "unregulated" chemical in that it lacks a formal drinking water standard (*i.e.*, MCL). However, monitoring of TBA in drinking water is required according to Title 22 of the California Code of Regulations, which became effective January 2001, and such monitoring was requested a year prior by the California State Water Resources Control Board (CASWRCB, 2000). California has also established a drinking water advisory action level for TBA of 12 ppb, although it is not specified how this action level was determined. In general, action levels are intended to represent health-based advisory levels that are established for chemicals for which primary drinking water standards have not been estab-

Table 7-10.   **Drinking Water and Ground Water Standards for MTBE in Various U.S. Regions**

State	Allowable MTBE Concentration (µg/l)		
	Primary MCL	Action Level*	Cleanup Level
Alabama	—	20	Site-specific
California	13	—	—
Delaware	—	180*	—
District of Columbia	—	Site-specific	Site-specific
Florida	—	—	50
Hawaii	—	20 - 202,000	Site-specific
Massachusetts	70	—	70 - 50,000
Maine	35	25	35
Michigan	—	40	40
Missouri	—	—	40,000 - 400,0000
Montana	—	30,000	—
New Hampshire	13	Site-specific	70
New Jersey	70	—	70
New Mexico	—	100,000	—
New York	50	10	10 or site-specific
North Carolina	—	200	—
Oregon	—	—	Site-specific
Pennsylvania	—	20	—
Rhode Island	—	—	40 - 5,000
South Carolina	—	40	Site-specific
Texas	240	—	—
Vermont	—	40	Site-specific
Utah	—	200	—
Washington	—	—	20
Wisconsin	—	—	60

Source: Simmons et al., 2001 and Stephenson, 2002.
*Represents specified action level or other water quality standard.

lished (CASWRCB, 2000). Note that from 1984 to 2001, TBA was detected only once out of over 2,000 drinking water sources sampled during this period (CDHS, 2002).

TNRCC has established protective concentration limits (PCLs) for TBA in both soil and ground water. In residential areas over aquifers that supply or have the potential to supply drinking water, the ground water PCL is 2.2 mg/l; the soil PCL (assuming combined human health exposure pathways and a 30 acre source area) is 2,100 mg/kg (TNRCC, 2002). A combined human health exposure pathway includes ingestion, inhalation, dermal, and vegetable consumption pathways.

The New Jersey Department of Environmental Protection (NJDEP) has developed ground water remediation interim criterion for TBA of 100 µg/l. This criterion has not yet been promulgated as a regulation; therefore, it is considered guidance and is not enforceable (NJDEP, 2002).

The USEPA website contains information on groundwater cleanup levels for MTBE, TBA, and other fuel oxygenates for the U.S., which is compiled by Delta Environmental Consultants and updated periodically. To access this information, go to www.epa.gov/swerust1/mtbe/index.htm, scroll down to "Essential Information about MTBE" and click on the links in the fifth bullet.

Cancer Effects.   To evaluate the risk posed by suspected human carcinogens, estimated exposures are combined with information on an agent's ability to increase the rate of cancer risk (USEPA, 1986a, 1996a, 1999b). Specifically, risks are expressed as the probability of an individual suffering an adverse effect and are estimated as the incremental or excess individual lifetime cancer risk (above the background rate) based on a specified exposure (*e.g.*, absorbed dose). Estimates of cancer risk are often based on the simplifying assumption that the dose-response relationship is linear at low doses (USEPA, 1989). Under this assumption, risk is directly related to intake and can be calculated by multiplying the ADD over a lifetime by the cancer potency factor (see Equation 7-3). As mentioned, the CSF usually represents the 95% upper confidence limit (UCL) of the probability of response based on experimental animal data used in the multistage model; therefore, the CSF represents an "upper bound" or "plausible upper limit" value (Anderson, 1983; USEPA, 1986a, 1989). In reality, the "true risk" is not expected to exceed this value and, in fact, may be substantially less or equal to zero, either because the compound is not a human carcinogen or the dose-response relationship has an effective threshold above the level of exposure. Note that risks from multiple chemicals or exposure pathways are generally assumed to be additive; *i.e.*, no interactions or synergies between different agents or chemicals are incorporated.

When there are sufficient data to support an assumption of nonlinearity for carcinogens, other types of analyses can be used (USEPA, 1989, 1996a, 1999b). In an MOE analysis, the risk is not extrapolated as a probability of an effect at low doses, but rather represents the toxicity point of departure (*i.e.*, the beginning of the extrapolation) divided by the environmental exposure level (USEPA, 1996a). Alternatively, if there is evidence of a biological threshold and sufficient data on the cancer end point, an RfD may be calculated for carcinogens (USEPA, 1999b). As mentioned, more sophisticated PBPK models may be appropriate if sufficient information is available on a chemical's mechanism of action, in which the administered dose in an animal study is converted to an equivalent administered dose in humans. Discussions about "hormesis"—*i.e.*, dose-response relationship where there is a stimulatory response at low doses, but an inhibitory response at high doses, resulting in a U-shaped or inverted U-shaped dose-response curve—may also become an increasingly important aspect of risk characterizations in the future (Calabrese, 2001).

Estimating Cancer Risks (Assumes No Threshold at Low Doses)

Risk = Lifetime Average Daily Dose (LADD) × Cancer Slope Factor (CSF)   (7-3)

where:

LADD Estimated dose from lifetime exposures; units are mg/kg-day
CSF Estimate cancer potency from lifetime exposures; units are $(mg/kg-day)^{-1}$

To date, a comprehensive cancer risk assessment has not been conducted for MTBE. The UC (1998) study estimated human exposures to MTBE in California and presented an adequate overview of the toxicology literature, but it did not provide a quantitative risk characterization of MTBE based on potential cancer effects. The NESCAUM (1999) report also attempted to evaluate MTBE cancer risks by establishing a chronic health-protective threshold concentration for MTBE, but it relied on a questionable approach and a data set that may not be appropriate for MTBE. In their evaluation of MTBE, Stern and Tardiff (1997) did not estimate potential cancer risks due to limited data on MTBE's carcinogenic potential in humans and the finding that MTBE tumor responses are likely to exhibit a toxicity-associated threshold.

The only known estimates of potential MTBE cancer risk that have been published in the peer-reviewed literature are those by Williams *et al.* (2000a) for water-related exposures in California. In this study, individual lifetime cancer risks at the 95th percentile of MTBE drinking water exposures were estimated to be less than one per million for both the general population and households with contaminated drinking water, assuming a 5-, 8-, or 13-year exposure duration (see Table 7-11). This study, however, did not include am-

Table 7-11. **Estimated Distribution of Lifetime Cancer Risk from MTBE Drinking Water Exposures for Different Population Groups and Exposure Durations in California**

Population Group	Individual Lifetime Cancer Risk (per Million)	
	50th Percentile	95th Percentile
General population		
5 years (1995–1998)	0.02	0.05
8 years (1995–2002)	0.03	0.08
13 years (1995–2008)	0.06	0.1
Households with contaminated drinking water		
5 years (1995–1998)	0.01	0.2
8 years (1995–2002)	0.02	0.3
13 years (1995–2008)	0.03	0.5

Source: Williams *et al.*, 2000a.

bient air inhalation exposures, and it used the CSF established by CA-OEHHA (1999) to characterize MTBE cancer risks. Estimated cancer risks based on both air and drinking water MTBE exposures have also been presented by Williams *et al.* (2000c) for different population groups, but these estimates do not include the most up-to-date exposure data on MTBE and are based on the CA-OEHHA (1999) potency value for MTBE (see Table 7-12).

EVALUATING ECOLOGICAL RISKS

Ecological risk assessment is defined as "the process that evaluates the likelihood that adverse ecological effects may occur or are occurring as a result of exposure to one or more stressors" (USEPA, 1992c, 1998). Environmental stressors may be chemical, physical, or biological, and multiple stressors may be considered in an ecological risk assessment. Environmental receptors include plant, invertebrate, and fish communities. The general process of evaluating ecological risks is very similar to that for evaluating human health risks, with a few important differences (see Figure 7-4). For example, a "problem formulation" component is incorporated in the beginning of the ecological risk assessment process to determine the focus and scope of the assessment. In addition, the hazard identification and dose-response assessments are combined in an "ecological effects assessment" phase. Finally, ecological risk assessments use the term "stressor-response" instead of "dose-response" to highlight the fact that physical changes (which are not measured as doses), as well as chemical contamination, can stress ecosystems (USEPA, 1996b, 1997d). There is also a difference in the biological level of organization of interest in ecological and human health risk assessments. In an ecological risk assessment, risks to populations, communities, and the ecosystem are of primary interest, while in human health risk assessments the risk to individual persons (or a subset of the population) is of primary concern. As is the case for human health risk assessments, a number of guidance materials have been published on how to conduct or interpret the results of ecological risk

Table 7-12.   **Estimated Distribution of Lifetime Cancer Risk from MTBE Air and Drinking Water Exposures for Different Population Groups in California**

Population Group*	Individual Lifetime Cancer Risk (per Million)*	
	50th Percentile	95th Percentile
General population: adults	0.2	0.5
General population: children	0.4	0.9
Service station workers	7.7	67

Source: Williams *et al.*, 2000c.
*Exposure duration is assumed to be eight years for all population groups.

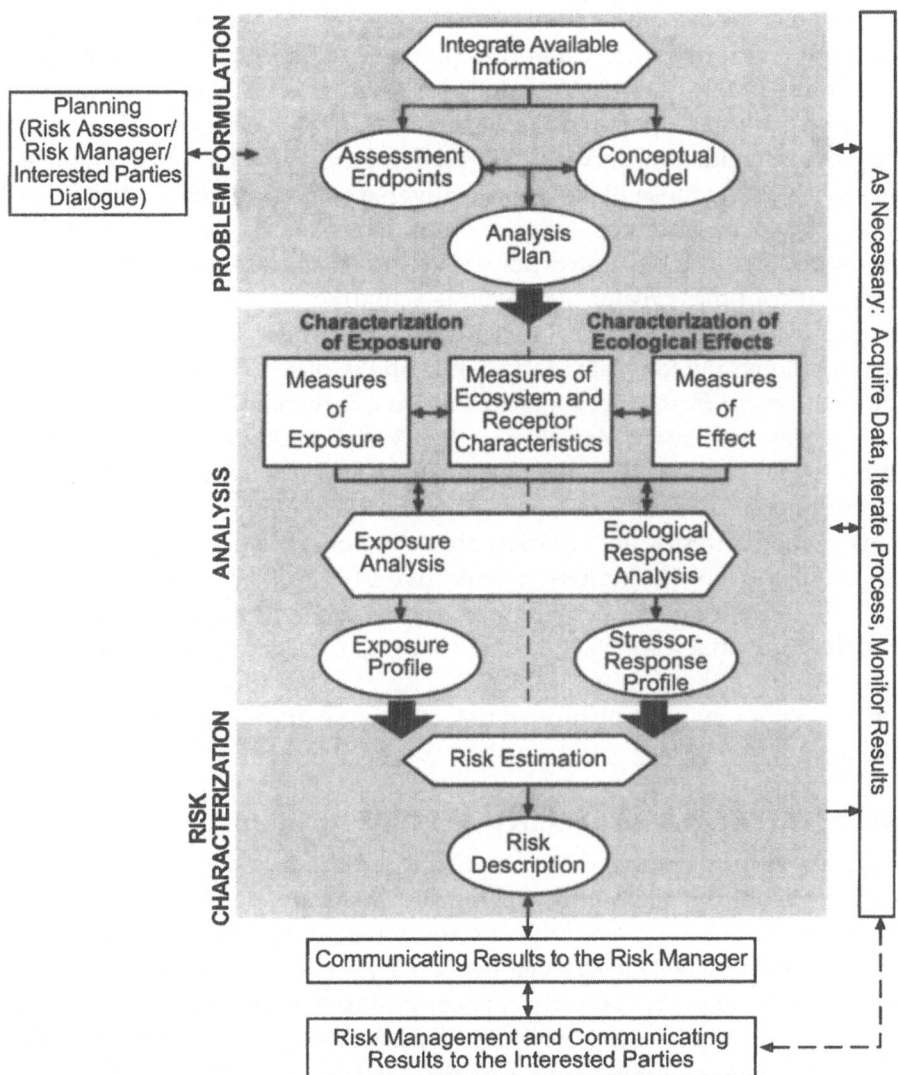

Figure 7-4. **Ecological Risk Assessment Framework. Source: USEPA, 1998.**

assessments (Bartell *et al.*, 1992; USEPA, 1992c, 1993, 1996b, 1997d, 1998, 2001; Freedman, 1989; Paustenbach, 1989a; Suter, 1993.)

Although ecological systems, such as those found in lakes or other surface waters, are potentially at risk from MTBE releases, a review of the available bioassay data suggests that commonly observed environmental exposure levels of MTBE are not likely to be toxic to aquatic life (Werner and Hinton, 1998). The toxicity of MTBE to fish and aquatic organisms was also found to be very low, based on a screening-level ecological risk assessment for aquatic ecosystems in California (Johnson, 1998). Specifically, the most conservative toxicity reference value calculated was 7,000 µg/l for rainbow trout, which, when compared to measured MTBE exposure levels in two surface waters, yielded conservative HQs ranging from 0.001 to 0.006. These levels are far below those likely to pose adverse ecological effects. In fact, adverse effects on rainbow trout are not expected to occur until MTBE concentrations in the water column reach about 4,600 µg/l, a level that is thousands of times greater than what is actually measured in surface water systems. Most of the historical toxicity data for fish are based on short-term investigations, and few data are available on the risks to benthic invertebrate communities in freshwater or estuarine environments (Johnson, 1998). In a more recent evaluation, Mancini *et al.* (2002) evaluated existing aquatic toxicity data and conducted 19 new freshwater and marine tests in order to calculate national ambient water quality criteria for MTBE. Preliminary calculations yielded a freshwater criterion protective of acute exposures of 151,000 µg/l, and a freshwater criterion protective of chronic exposures of 51,000 µg/l. Preliminary marine criteria determined to be protective of acute and chronic exposures were 53,000 and 18,000 µg/l, respectively. These criteria values indicate that current ambient MTBE concentrations detected in U.S. surface waters do not constitute a risk to aquatic organisms (Mancini *et al.*, 2002).

EUROPEAN RISK ASSESSMENT OF MTBE

There are several notable differences between the U.S. and other countries with respect to the manner in which health risks are evaluated and characterized (Paustenbach, 1995; Williams and Paustenbach, 2002). For example, countries such as the United Kingdom (U.K.) and Germany rely primarily on expert judgment (*e.g.*, expert advisory committees) in the risk assessment and decision-making process, rather than a regulatory agency. The types of information presented by other countries also tend to be broader than that contained in most assessments conducted in the U.S., particularly in the risk characterization step. In particular, risk assessment and risk management processes appear to be much more intermingled in some other countries (*e.g.*, U.K., Netherlands, Australia) than in the U.S., and risk characterizations often include discussions about broader cost-benefit, cost-effectiveness, and other resource evaluation issues (Department of the Environment, 1991; El Saadi and Langley, 1991; Walker, 1992; Ministry of Housing, 1994; Ginjaar, 1996). The "precautionary principle," which is based on societal preferences for future environmental protection and safety in situations where very little is known about a chemical or agent, is also a movement that began in Europe in the early 1990s. This "better safe than sorry" approach is often criticized by the U.S. and other countries, because when precautions are taken to protect public health or the environment in the absence of clear evidence of harm, the benefits of such actions cannot be assessed and the costs are often substantial (Hickey and Walker, 1995; Cross, 1996; EEA, 2001; Weiner, 2002).

With regard to MTBE, recent evaluations by the European Union (EU) appear to have been driven more by scientific analyses than precautionary measures, because they rely predominantly on the available scientific data rather than public risk perceptions. As a result, major European countries and policy-makers generally support the use of MTBE and have concluded that the risks to human health or the environment from MTBE releases are not expected to occur or have already been mitigated by implementation of various risk reduction measures (Hery, 2001). In fact, the European *MTBE Health Risk Characterisation* concluded that MTBE does not pose a concern for human health with regard to current occupational and consumer exposures (ECETOC, 1997). A more recent European report, *An Environmental Risk Assessment of MTBE Use in Europe*, also concluded that the environmental risk of using MTBE as a fuel additive is low, based on a life-cycle analysis of the production, formulation, processing, and use of MTBE (Ahlberg *et al.*, 2001). In addition, ground water assessments based on review of the available water monitoring data in England and Wales have concluded that MTBE does not present a significant risk to ground water supplies (Environment Agency, 2000). Finally, a recent study by Arthur D. Little (ADL, 2001) concluded that the risk of ground water contamination in Europe is low, given important differences between the U.S. and EU in the requirements for the construction and operation of USTs.

An important limitation of these evaluations is the apparent lack of a routine monitoring program for MTBE releases in the environment; therefore, the available data in these countries may not be sufficient to make an accurate assessment of the current risk posed by MTBE (Environment Agency, 1999). Recent air and water quality legislation in the U.K. and Europe may result in the collection and analysis of additional MTBE data in the future.

RISK ANALYSIS FRAMEWORK

Risk analysis has been broadly defined as a systematic framework for understanding and managing diverse risks (Ruckelshaus, 1985). The risk analysis framework includes three essential elements: risk assessment, risk management, and risk communication. As discussed above, the risk assessment process entails quantifying and characterizing risks, while the risk management process entails deciding what to do about these risks (NRC, 1983, 1994, 1996). Risk communication, on the other hand, involves conveying information about risks or risk management decisions to different groups (Slovic, 1986; Sandman, 1987, 1993; Plough and Krimsky, 1987; NRC, 1989).

The elements of risk analysis have historically been conducted in isolation from one another. In fact, most regulatory guidance documents and published papers specify that risk assessments and characterizations should remain separate from risk management and other considerations (NRC, 1983; USEPA, 1995a, 1995b). The most recent guidance by the National Research Council (NRC, 1996), suggests that risk characterizations should be more than a mere summary of scientific information and, instead, should be an integral part of the entire decision-making process. As its scope continues to evolve, risk assessments may begin to include quantitative (and qualitative) analyses of other considerations, such as the costs and benefits of alternative public policies (OMB, 2001).

Such broader analyses would be useful in evaluating chemicals like MTBE, which have both risks and benefits associated with their use, and whose substitutes also have countervailing risks and benefits (Graham and Weiner, 1995). For example, the use of RFG containing MTBE has been associated with considerable air quality benefits, which may more than outweigh the potential risks associated with low-level drinking water contamination (USEPA, 1999c; NESCAUM, 1998; Cal/EPA, 1997; NSTC, 1997; Spitzer, 1997). Preliminary assessments also suggest that substitution of MTBE with an alternative, such as ethanol, may result in even greater risks to human health or the environment due to increased evaporative air emissions or impacts on water quality by other gasoline constituents (Malcolm Pirnie, 1998; UC, 1998; NESCAUM, 1999; Rice and Cannon, 1999). In addition, a recent evaluation of the social costs associated with banning MTBE in California concluded that replacement of MTBE with ethanol or non-oxygenated RFG would cost (on average) an additional $1.24 billion and $0.92 billion annually, respectively (Rausser et al., 2001). When compared to other chemicals commonly detected in drinking water, MTBE is also found to pose a substantially lower health risk, suggesting that the intense efforts to regulate or ban MTBE may be misguided from a public health perspective (Williams et al., 2002). A more thorough evaluation of these types of risk and cost-benefit tradeoff issues would undoubtedly aid in future decision-making about MTBE.

SUMMARY

Standardized risk assessment approaches have been used for nearly 20 years to evaluate the significance of potential chemical exposures and health risks (NRC, 1994; Paustenbach, 1989a). Risk assessment, which includes both quantitative and qualitative expressions of risk, represents the first step toward collecting, analyzing, and interpreting data to better inform decision-makers and the public about a particular issue. Recent efforts to improve risk characterizations have focused on the use of more formal quantitative uncertainty analyses, such as Monte Carlo techniques, and methods to improve tissue dose estimates in humans, such as PBPK models (Williams and Paustenbach, 2002). The most obvious limitations to risk assessments are the lack of appropriate toxicity and exposure data and a failure to adequately describe the uncertainties and level of confidence in reported risk estimates, which can result in poorly characterized risks (USEPA, 1995a, 1995b, 1995c).

Such limitations are readily apparent when evaluating potential human health and environmental risks posed by MTBE. Not only has the USEPA not developed an oral RfD for MTBE ingestion exposures, but the inadequacy of the available animal cancer bioassays and mechanistic studies have left the question of whether MTBE should be treated as a human carcinogen unanswered. Greater attention in recent years to the potential toxicity of TBA, the major metabolite of MTBE, also appears to be unfounded based on the available animal toxicity data. That is, there is no clear evidence of acute or chronic toxicity (including carcinogenicity) associated with TBA at likely exposure levels. The State of California has attempted to develop a cancer potency factor for MTBE (CA-OEHHA, 1999), but their assessment suffers from several important limitations, including the lack of an appropriate PBPK model. A more refined PBPK model has been developed recently by Licata *et al.* (2001), but it has yet to be applied to MTBE to yield an alternative cancer potency factor or non-cancer RfD.

Similarly, there are currently no uniform or national databases with sufficient data on airborne or drinking water MTBE concentrations to support a comprehensive quantitative characterization of MTBE exposures and risk in the U.S. (NSTC, 1997). Region-specific analyses, such as those conducted in California (UC, 1998) and the Northeast (NESCAUM, 1999), have also tended to mischaracterize the available air and water data, as well as other important exposure or toxicity parameters. More thorough risk evaluations that have been conducted suggest that MTBE is unlikely to pose a significant public health risk, at least from drinking water exposures (Stern and Tardiff, 1997; Williams *et al.*, 2000a), but the lack of routine sampling for MTBE hinders the ability to provide more accurate estimates of drinking water-related exposures and risk (Williams, 2001; Grady and Casey, 2001). Beginning in 2001, USEPA (2002a) will require all large drinking water systems, and a representative sample of small systems, to monitor and report the presence of MTBE. As mentioned, monitoring of TBA in drinking water is also now required in California, and perhaps other regions. The USGS also routinely monitors ground water and drinking water sources for MTBE and other VOCs in several regions of the country (Squillace *et al.*, 1999; Grady and Casey, 2001). All of these efforts may provide useful data for future analyses.

Despite the data limitations and lack of a clear environmental or health threat, several efforts have been made to ban the use of MTBE in the U.S. due to perceived impacts on water quality (CARB, 1999; USEPA, 2000f). In contrast, it has been determined that the use of MTBE in gasoline does not pose a threat in Europe (ECETOC, 1997; Environment Agency, 2000; Ahlberg *et al.*, 2001). Regardless of these policy decisions, the available data suggest that current environmental exposure levels of MTBE in the U.S. and abroad are not acutely toxic to humans and are unlikely to pose chronic health effects from longer-term exposures. That is, MTBE is rarely detected in public drinking water supplies at notable concentrations; estimated exposures from MTBE in air or drinking water are typically very low for the general population; and predicted toxicity levels (for humans or ecological receptors) are well above those typically encountered in the environment. It is obvious that resolution of the remaining questions on MTBE's carcinogenic potential to humans and additional data on MTBE concentrations in air and drinking water will allow for a more thorough risk characterization of MTBE and greater confidence that human health and ecological risks in MTBE fuel use areas are negligible. It is also likely that decision-makers will continue to seek more information about these issues, particularly on the costs, risks, and benefits of MTBE versus other alternatives in achieving national environmental and air quality goals.

REFERENCES

ADL (Arthur D. Little). 2001. *MTBE and the Requirements for Underground Storage Tank Construction and Operation in Member States: A Report to the European Commission.* Cambridge, Massachusetts. ENV.D.1/ETU/2000/0089R.

Ahlberg, R., Gennart, J.-P., Mitchell, R.E., Schulte-Koerne, Thomas, M.E., Vahervuori, H., Vrijhof, H., and Watts, C.D. 2001. *An Environmental Risk Assessment of MTBE Use in Europe.* ECETOC/EFOA Task Force on Environmental Risk Assessment of MTBE, European Centre for Ecotoxicology and Toxicology of Chemicals, European Fuel Oxygenate Association.

Anderson, E.L. 1983. Quantitative approaches in use to assess cancer risks. *Risk Analysis.* 3, 227–295.

Anderson, H.A., Hanrahan, L., Goldring J., and Delaney, B. 1995. *An Investigation of Health Concerns Attributed to Reformulated Gasoline (RFG) Use in Southeastern Wisconsin.* Wisconsin Department of Health and Social Services, Division of Health, Bureau of Public Health, Section of Environmental Epidemiology and Prevention. Milwaukee, Wisconsin. Final Report, May 1995.

API (American Petroleum Institute). 1993. *Gasoline Vapor Exposure Assessment at Service Stations.* API Publication Number 4553. Washington, DC.

ATSDR (Agency for Toxic Substances and Disease Registry). 1995a. *Public Health Assessment Guidance Manual.* U.S. Department of Health and Human Services, Ann Arbor, Michigan, Lewis Publishers.

ATSDR (Agency for Toxic Substances and Disease Registry). 1995b. *Toxicological Profile for Xylenes.* U.S. Department of Health and Human Services, Atlanta, Georgia.

ATSDR (Agency for Toxic Substances and Disease Registry). 1998. *Toxicologi-

cal Profile for Methyl Tert-Butyl Ether. U.S. Department of Health and Human Services, Atlanta, Georgia.

ATSDR (Agency for Toxic Substances and Disease Registry). 1999. *Toxicological Profile for Ethylbenzene.* U.S. Department of Health and Human Services, Atlanta, Georgia.

ATSDR (Agency for Toxic Substances and Disease Registry). 2000a. *Minimal Risk Levels (MRLs) for Hazardous Substances.* www.atsdr.cdc.gov/mrls. html. Last updated February 17, 2000.

ATSDR (Agency for Toxic Substances and Disease Registry). 2000b. *Toxicological Profile for Toluene.* U.S. Department of Health and Human Services, Atlanta, Georgia.

BAAQMD (Bay Area Air Quality Management District). 1998. *Annual Report—Toxic Air Contaminant Control Program.* Bay Area Air Quality Management District, San Francisco.

Balter, N.J. 1997. Casualty assessment of the acute health complaints reported in association with oxygenated fuels. *Risk Analysis.* 17, 705–715.

Barnes, D.G. and Dourson M. 1988. Reference dose (RfD): description and use in health risk assessments. *Regulatory Toxicology and Pharmacology.* 8, 471–486.

Bartell, S.M., Gardner, R.H., and O'Neill, R.V. 1992. *Ecological Risk Estimation.* New York, New York, Lewis Publishers.

Belpoggi, F., Soffritti, M., and Maltoni, C. 1995. Methyl-tertiary-butyl ether (MTBE)-a gasoline additive-causes testicular and lympho-haematopietic cancers in rats. *Toxicology and Industrial Health.* 11, 119–149.

Belpoggi, F., Soffritti M., Filippini, F., and Maltoni, C. 1997. Results of long-term experimental studies on the carcinogenicity of methyl tert-butyl ether. *Annals of the New York Academy of Sciences.* 837, 77–95.

Belpoggi, F., Soffritti, M., and Maltoni, C. 1998. Pathological characterization of testicular tumours and lymphomas-leukemias, and their precursors observed in Sprague-Dawley rats exposed to methyl tertiary butyl-ether (MTBE). *European Journal of Oncology.* 3, 201–206.

Bird, M.G., Burleigh-Flayer, H.D., Chun, J.S., Douglas, J.F., Kneiss, J.J., and Andrew, L.S. 1997. Oncogenicity studies of inhaled methyl tertiary-butyl ether (MTBE) in CD-1 mice and F-344 rats. *Journal of Applied Toxicology.* 17, S45–S55.

Bolger, P. M., C.D. Carrington, and Henry, S.H. 1996. Risk assessment for risk management and regulatory decision-making at the U.S. Food and Drug Administration. In: *Toxicology and Risk Assessment: Principles, Methods and Applications*, pp. 791–798. (Fan, A.M. and Chang, L.W., Eds.). New York, New York, Marcel Dekker Inc.

Borak, J., Pastides, H., Van Ert, M., Russi, M., and Herastein, J. 1998. Exposure to MTBE and acute health effects: a critical review. *Human and Ecological Risk Assessment.* 4, 177–200.

Borghoff, S.J, Murphy, J.E., and Medinsky, M.A. 1996. Development of a physiologically based pharmacokinetic model for methyl tertiary-butyl ether and tertiary-butanol in male Fischer-344 rats. *Fundamental and Applied Toxicology.* 30, 246–275.

Borghoff, S.J. and Williams, T.M. 2000. Species-specific tumor responses following exposure to methyl tert-butyl ether. *CIIT Activities.* 20(2), 1–9.

Brown, S. L. 1997. Atmospheric and potable water exposures to methyl tert-butyl ether (MTBE). *Regulatory Toxicology and Pharmacology.* 25, 256–276.

Burleigh-Flayer, H. D., Chun, J. S., and Kintigh, W. J. 1992. *Methyl Tertiary Butyl Ether: Vapor Inhalation Oncogenicity Study in CD-1 Mice.* Union Carbide Chemicals and Plastics Company, Inc., Export, Pennsylvania. Bushy Run Research Center Report No. 91N0013A.

Calabrese, E.J. 2001. The frequency of U-shaped dose responses in the toxicological literature. *Toxicological Sciences.* 62, 330–338.

Cal/EPA (California Environmental Protection Agency). 1994. *CalTOX: A Multimedia Total Exposure Model for Hazardous Waste Sites.* Department of Toxic Substances Control. Sacramento, California. Spreadsheet user's guide, version 1.5.

Cal/EPA (California Environmental Protection Agency). 1997. *MTBE Briefing Paper.* Sacramento, California.

CA-OEHHA (California Office of Environmental Health Hazard Assessment). 1999. *Public Health Goal for Methyl Tertiary Butyl Ether (MTBE) in Drinking Water.* California Environmental Protection Agency, Pesticide and Environmental Toxicology Section, Office of Environmental Health Hazard Assessment. March 1999.

CA-OEHHA (California Office of Environmental Health Hazard Assessment). 2001. *Public Health Goal for Benzene in Drinking Water.* California Environmental Protection Agency, Pesticide and Environmental Toxicology Section, Office of Environmental Health Hazard Assessment. June, 2001.

CARB (California Air Resources Board). 1997. *Tert-Butyl Alcohol.* Toxic air contaminant identification list summaries. September, 1997. pp. 159–161.

CARB (California Air Resources Board). 1998. *California Ambient Air Quality Data.* California Environmental Protection Agency. CD Rom. December 1998.

CARB (California Air Resources Board). 1999. *ARB Bans MTBE and Modifies Rules for Cleaner Burning Gasoline.* News Release. California Environmental Protection Agency.

CASWRCB (California State Water Resources Control Board). 2000. *Regional Water Quality Control Board UST Program Managers and Local Oversight Program Managers.* Letter from Elizabeth L. Haven, Manager of Underground Storage Tank Programs, Division of Clean Water Programs. April 12, 2000.

CDHS (California Department of Health Services). 2001. *MTBE in California Drinking Water*. California Department of Health Services. www.dhs.ca. gov/ps/ddwem/chemicals/MTBE/mtbeindex.htm. Last updated December 4, 2001.

CDHS (California Department of Health Services). 2002. *Drinking Water Standards: Unregulated Chemicals Requiring Monitoring*. www.dhs.ca.glv/ps/ddwem/chemicals/unregulated/index.htm. Last updated February 8, 2002.

Chun, J. S., Burleigh-Flayer H. D., and Kintigh W. J. 1992. *Methyl Tertiary Butyl-Ether: Vapor Inhalation Oncogenicity Study in Fischer 244 Rats*. Union Carbide Chemicals and Plastics Company, Inc., Export, Pennsylvania. Bushy Run Research Center Report No. 91N0013B.

Chun, J.S. and Kintigh, W.J. 1993. *Methyl Tertiary Butyl Ether: Twenty-Eight Day Vapor Inhalation Study in Rats and Mice*. Bushy Run Research Center, Export, Pennsylvania. Laboratory project ID 93N1241.

Cirvello, J.D., Radovsky, A., Heath, J.E., Farnell, D.R., Linamood, C. 1995. Toxicity and carcinogenicity of t-Butyl Alcohol in rats and mice following chronic exposure in drinking water. *Toxicology and Industrial Health*. 11, 151–165.

Clary, J.J. 1997. Methyl tertiary butyl ether systemic toxicity. *Risk Analysis*. 17, 661–672.

Clegg, E.D., Cook, J.C., Chapin, R.E., Foster, P.M., and Daston, G.P. 1997. Leydig cell hyperplasia and adenoma formation: mechanisms and relevance to humans. *Reproductive Toxicology*. 11, 107–121.

Clewall, H.J. 1995. The application of physiologically based pharmacokinetics modeling in human health risk assessment of hazardous substances. *Toxicology Letters*. 79, 207–217.

Clewell, H.J., Gentry, P.R., Gearhart, J.M., Allen, B.C., and Andersen, M.E. 1995. Considering pharmaxokinetics and mechanistic information in cancer risk assessments for environmental contaminants: examples with vinyl chloride and trichloroethylene. *Chemosphere*. 31, 2561–2578.

Cook, C.K. and Kovein, R.J.. 1994. *NIOSH Health Hazard Evaluation Report*. National Institute for Occupational Safety and Health, Division of Surveillance, Hazard Evaluations and Field Studies, Cincinnati, Ohio. HETA 94-0220-2526.

CRAM (Commission on Risk Assessment and Risk Management). 1997. *Risk Assessment and Risk Management in Regulatory Decision-Making*. Presidential/Congressional Commission on Risk Assessment and Risk Management, Washington, DC. Final Report, Volume 2.

Cross, F.B. 1996. Paradoxical perils of the precautionary principle. *Washington and Lee Law Review*. 53, 851–925.

Cullen, A.C. and H.C. Frey. 1999. Probabilistic techniques in exposure assessment. *Risk Analysis*. 14, 389–393.

Department of the Environment. 1991. *Policy Appraisal and the Environment: A Guide for Government Departments.* Norwich, United Kingdom.

Dourson, M. L. and Felter, S. P. 1997. Route-to-route extrapolation of the toxic potency of MTBE. *Risk Analysis.* 17, 717–725.

Dourson, M.L., Felter, S.P., and Robinson, D. 1996. Evolution of science-based uncertainty factors in noncancer risk assessment. *Regulatory Toxicology and Pharmacology.* 24, 108–120.

ECETOC (European Centre for Exotoxicology and Toxicology of Chemicals). 1997. *MTBE*

Health Risk Characterisation. Brussels, Belgium. Technical Report No. 72, pp. 1–67.

EEA (European Environment Agency). 2001. *Late Lessons From Early Warnings: The Precautionary Principle 1986–2000.* Denmark. Environmental Issue Report No. 22.

El Saadi, O. and Langley, A. 1991. *The Health Assessment and Management of Contaminated Sites.* South Australian Health Commission, Adelaide, Australia.

Environment Agency (1999). *The Fuel Additive MTBE—A Groundwater Protection Issue?* National Groundwater and Contaminated Land Centre.

Environment Agency (2000). *A Review of Current MTBE Usage and Occurrence in Groundwater in England and Wales.* The Institute of Petroleum. R&D Publication 97.

Fiedler, N., Kelly-McNeil, K., Mohr, S., Lehrer, P., Opiekun, R., Lee, C.W., Wainman, T., Hamer, R., Weisel, C., Edelberg, R., and Lioy, P. 2000. Controlled human exposure to methyl tiertiary butyl ether in gasoline: symptoms, psychopsysiologic and neurobehavioral responses of self-reported sensitive persons. *Environmental Health Perspectives.* 108 (8), 753–763.

Finley, B., Scott P., and Paustenbach, D. 1993. Evaluating the adequacy of maximum contaminant levels as health-protective cleanup goals: an analysis based on Monte Carlo techniques. *Regulatory Toxicology and Pharmacology.* 18, 438–455.

Finley, B., Proctor, D., Scott, P., Harrington, N., Paustenbach, D., and Price, P. 1994. Recommended distributions for exposure factors frequently used in health risk assessment. *Risk Analysis.* 14, 533–553.

Finley, B.L. and Paustenbach, D.J. 1994. The benefits of probabilistic exposure assessment: three case studies involving contaminated air, water, and soil. *Risk Analysis.* 14, 53–73.

Freedman, B. 1989. *Environmental Ecology: The Impacts of Pollution and Other Stresses on Ecosystem Structure and Function.* New York, New York, Academic Press.

Gargas, M.L., Finley, B.L., Paustenbach, D.J., and T.F. Long. 1999. Environmental health risk assessment: theory and practice. In: *General and Applied Toxicology*, Volume 3, 2nd edition, pp. 1749–1809. (Ballantyne, B., Marrs, T., and Syversen, T. Eds.). London, Macmillan.

Ginjaar, L. 1996. *Risk is More Than a Number: Reflections on the Development of the Environmental Risk Management Approach.* Health Council of the Netherlands, Committee on Risk Measures and Risk Assessment, The Ministry of Health, Welfare and Sports, Netherlands.

Grady, S.J. and Casey, G.D. 2001. *Occurrence and Distribution of Methyl tert-Butyl Ether and Other Volatile Organic Compounds in Drinking Water in the Northeast and Mid-Atlantic Regions of the United States, 1993–98.* U.S. Department of the Interior, U.S. Geological Survey. Denver, CO. Water-Resources Investigations Report 00-4228.

Graham, J.D., Green, L., and Roberts, M.J. 1988. *In Search of Safety: Chemicals and Cancer Risks.* Cambridge, Massachusetts, Harvard University Press.

Graham, J.D. 1995. Historical perspective on risk assessment in the federal government. *Toxicology.* 102, 29–52.

Graham, J.D. and Weiner, J.B. 1995. *Risk vs. Risk: Tradeoffs in Protecting Health and the Environment.* Cambridge, Massachusetts, Harvard University Press.

Graham, J. D. and Hartwell, J.K. 1997. *The Greening of Industry: A Risk Management Approach.* Cambridge, Massachusetts, Harvard University Press.

HEI (Health Effects Institute). 1996. *The Potential Health Effects of Oxygenates Added to Gasoline—A Review of the Current Literature.* Cambridge, Massachusetts.

Hery, B. 2001. MTBE in Europe: from sound science to a reliable component. European Fuel Oxygenates Association. Presented to *DeWitt MTBE/Oxygenates & Methanol Conference*, Houston, Texas. October 16–18, 2001.

Hickey, J.E. Jr. and Walker, V.R. 1995. Refining the precautionary principle in international environmental law. *Virginia Environmental Law Journal.* 14, 423–454.

Hinton, J. 1993. Occupational exposures—MTBE. In: *Procedures, Conference on MTBE and other Oxygenates: A Research Update.*

IARC. 1982. Benzene. IARC Monographs on the evaluation of carcinogenic risks of chemicals to man. *Some Industrial Chemicals and Dyestuffs.* 29, IARC Lyon, France.

IARC. 1999. IARC monographs on the evaluation of carcinogenic risks to humans. *Some Chemicals that Cause Tomours of the Kidney or urinary Bladder in rodents and Some Other Substances.* 73, IARC, Lyon, France.

IPCS. 1998. *Environmental Health Criteria 206. Methyl Tertiary-butyl Ether.* International Programme on Chemical Safety, World Health Organization, Geneva, Switzerland.

Johanson, G., Nihlen, A. and Lof, A. 1995. Toxicokinetics and acute effects of MTBE and ETBE in male volunteers. *Toxicology Letters.* 82/83, 713–718.

Johnson, W.D., Findlay, J., and Boynce, R.A. 1992. *28-Day Oral (Gavage) Toxicity Study of Methyl Tert-Butyl Ether in Rats.* Prepared for Amoco. ITT Research Institute, Chicago, Illinois. ITT Research Institute Project No. L08100.

Johnson, M.L. 1998. *Ecological Risk of MTBE in Surface Waters. Volume III: Air Quality and Ecological Effects.* Health and Environmental Assessment of MTBE, University of California.

Johnson, T. 1993. Service station exposures. In: *Proceedings of the Conference on MTBE and Other Oxygenates: A Research Update.* U.S. Environmental Protection Agency, National Center for Environmental Assessment. EPA/600/R-95/134.

Klaassen, C.D. 2001. *Casarett and Doull's Toxicology; The Basic Sciences of Poisons.* (C.D. Klaassen, Ed.). New York, New York, McGraw-Hill.

Kolluru, R.V. 1996. Risk assessment and management: a unified approach. In: *Risk Assessment and Management Handbook,* pp. 1.3–1.41. (Kolluru, R.V., Bartell, S.M., Pitblado, R.M., and Stricoff, R.S., Eds.). New York, New York, McGraw Hill Inc.

Leung, H.W. 1991. Development and utilization of physiologically based pharmacokinetic models for toxicological applications. *Journal of Toxicology and Environmental Health.* 32, 247–267.

Leung, H.W. and Paustenbach, DJ. 1995. Physiologically based pharmacokinetic and pharmacodynamic modeling in health risk assessment and characterization of hazardous substances. *Toxicology Letters.* 78, 55–65.

Leung, H.W. 2000. Physiologically based pharmacokinetic modeling. In: *General and Applied Toxicology,* Volume I, pp. 141–154. (Ballantyne, B., Marrs, T. and Syversen, T., Eds.). London, United Kingdom, Macmillan Publishing. .

Licata, A.C., DeKant, W., Smith, C.E., and Borghoff, S.J. 2001. A physiologically based pharmacokinetic model for methyl tert-butyl ether in humans: implementing sensitivity and variability analyses. *Toxicological Sciences.* 62, 191–204.

Lioy, P.J., Weisel, C.P., Wan-Kuen, J., Pellizzari, E., and Raymer, J.H. 1994. Microenvironmental and personal measurements of methyl-tertiary butyl ether (MTBE) associated with automobile use activities. *Journal of Exposure Analysis and Environmental Epidemiology.* 4, 427–441.

Malcolm Pirnie. 1998. *Evaluation of the Fate and Transport of Ethanol in the Environment. Prepared for the American Methanol Institute.* Oakland, California. November 1998.

Mancini, E.R., Steen, A., Rausina, G.A., Wong, D.C.L., Arnold, W.R., Gostomski, F.E., Davies, T., Hockett, J.R., Stubblefield, W.A., Drottar, K.R., Springer, T.A., and Errico, P. 2002. MTBE ambient water quality criteria development: a public/private partnership. *Environmental Science and Technology.* 36, 125–29.

Masters, G.M. 1998. Chapter 4 Risk assessment. In: *Introduction to Environment Engineering and Science.* pp. 117–162. Englewood Cliffs, New Jersey, Prentice Hall.

McClellan, R.O. 1998. Chapter 127 Risk assessment. In: *Environmental and Occupational Medicine.* (Rom W.N., Ed.). Philadelphia, Lippincott-Raven Publishers.

McDougal, J.N. 1996. Physiologically based pharmacokinetic modeling. In: *Dermatoxicology*, pp. 37–60. (Marzulli, F.N. and Maibach, H.I. Eds.). London, Taylor and Francis.

Mehlman, M.A. 1998. Dangerous and cancer-causing properties of products and chemicals in the oil-refining and petrochemical industry: Part XXV, neurotoxic, allergic, and respiratory effects in humans form water and air contaminated by methyl tertiary butyl ether in gasoline. *International Journal of Occupational Medicine and Toxicology.* 7, 65–87.

Mennear, J.H. 1997. Carcinogenicity studies on methyl tertiary butyl ether (MTBE): critical review and interpretation. *Risk Analysis.* 17, 73–681.

Ministry of Housing. 1994. *Environmental Quality Objectives in the Netherlands.* Spatial Planning and Environment, Hague, Netherlands.

Mohr, S., Fiedler, N., Weisel, C., and Kelly-McNeill, K. 1994. Health effects of MTBE among New Jersey garage workers. *Inhalation Toxicology.* 6, 553–562.

MSDS (Material Safety Data Sheet). 2001. *Tebol (TM) 93 Alcohol.* Prepared by Lyondell Chemical Company. MSDS No: BE136, Variant: USA-EN, Version No: 1.2, Validation Data: December 14, 2001, pp. 1–9.

NESCAUM (Northeast States for Coordinated Air Use Management). 1998. *Relative Cancer Risk of Reformulated Gasoline and Conventional Gasoline Sold in the Northeast.* August 1998.

NESCAUM (Northeast States for Coordinated Air Use Management). 1999. *RFG/MTBE: Findings and Recommendations.* August 1999.

NJDEP (New Jersey Department of Environmental Protection). 2002. *Interim Specific and Generic Ground Water Quality Criteria.* . Last updated January 30, 2002.

NRC (National Research Council). 1983. *Risk Assessment in the Federal Government: Managing the Process.* Washington, DC, National Academy Press.

NRC (National Research Council). 1989. *Improving Risk Communication.* Committee on Risk Perceptions and Communication. Washington, DC, National Academy Press.

NRC (National Research Council). 1994. *Science and Judgment in Risk Assessment.* Washington, DC, National Academy Press.

NRC (National Research Council). 1996. *Understanding Risk: Informing Decisions in a Democratic Society.* Washington, DC, National Academy Press.

NSTC (National Science and Technology Committee). 1997. *Interagency Assessment of Oxygenated Fuels.* National Science and Technology Committee on Environment and Natural Resources (CENR) and Interagency Oxygenated Fuels Assessment Steering Committee; White House Office of Science and Technology Policy (OSTP) through the CENR of the Executive Office of the President, Washington, DC.

NTP (National Toxicology Program). 1995. *TR 436: Toxicology and Carcinogenesis Studies of t-Butyl Alcohol (CAS No. 75-65-0) in F344/N Rats and B6C3F₁ Mice (Drinking Water Studies).* ntp-server.niehs.nih.gov/htdocs/LT-Studies/TR436.html. Report Data: May 1995.

NTP (National Toxicology Program). 1997. *Toxicity Studies of T-Butyl Alcohol (CAS No. 75-65-0) Administered by Inhalation to F334/N Rats and B6C3F₁ Mice. Tox-53.* National Toxicology Program, Research Triangle Park, NC. NTIS# PB98-108905. ntp-server.niehs.nih.gov/htdocs/ST-Studies/TOX053.html. Report Data: July 1997.

NTP (National Toxicology Program). 1999. *Ninth Report on Carcinogens.* National Toxicology Program, Research Triangle Park, North Carolina.

NTP (National Toxicology Program). 2002. *Tert-Butyl Alcohol: Testing Status.* ntp-server.niehs.nih.gov/htdocs/Results_Status/Resstatb/10402-N. Html. Last Update: February 11, 2002.

OMB (Office of Management and Budget). 2001. *Guidelines for Ensuring and Maximizing the Quality, Objectivity, Utility, and Integrity of Information Disseminated by Federal Agencies.* Office of Management and Budget, Executive Office of the President. October 1, 2001.

Opiekun, R.E., Freeman, N., Kelley-McNeil, K., Fiedler, N.L., and Lioy, P.J. 2001. Effect of vehicle use and maintenance patterns of a self-described group of sensitive individuals and nonsensitive individuals to methyl tertiary-butyl ether in gasoline. *Journal of Exposure Analysis and Environmental Epidemiology.* 11, 79–85.

Ott, W.R. 1985. Total human exposure. *Environmental Science and Technology.* 19, 880–886.

PADEP (Pennsylvania Department of Environmental Protection). 1997. *Administration of the Land Recycling Program (Act 2).* 25 PA. Code CH. 250.

Paustenbach, D.J. 1989a. *The Risk Assessment of Environmental Hazards: A Textbook of Case Studies.* (Paustenbach, D.J. Ed.). John Wiley & Sons.

Paustenbach, D.J. 1989b. Health risk assessments: opportunities and pitfalls. *Columbia Journal of Environmental Law.* 41, 379–410.

Paustenbach, D.J. 1995. The practice of health risk assessment in the United States (1975–1995): how the U.S. and other countries can benefit from that experience. *Human and Ecological Risk Assessment.* 1, 29–79.

Paustenbach, D.J. 2000. The practice of exposure assessment: a state-of-the-art review. *Journal of Toxicology and Environmental Health (Part B).* 3, 179–291.

Paustenbach, D.J. 2002. *Human and Ecological Risk Assessment: Theory and Practice.* (Paustenbach, D.J. Ed.). John Wiley & Sons.

Pekari, K., Riihmaki, V., Vainiotalo, S., Teravainen, E., and Aitio, A. 1996. Experimental exposure to methyl-tert-butyl ether (MTBE) and methyl-tert-amyl ether (MTAE). *International Symposium on Biological Monitoring in Occupational and Environmental Health, Finnish Institute of Occupational and Environmental Health*, Helsinki, Finland. FIN-00250.

Plough, A. and Krimsky, S. 1987. The emergence of risk communication studies: social and political context. *Science, Technology, and Human Values*. 12, 4–10.

Rao, H.V. and Ginsberg, G.L. 1997. A physiologically-based pharmacokinetic model assessment of methyl t-butyl ether in groundwater for a bathing and showering determination. *Risk Analysis*. 17(5), 583–598.

Rausser, G.C., Adams, G.D., Montgomery, W.D., and Smith, A.E. 2001. *The Social Costs of an MTBE Ban in California*. Charles River Associates Report.

Renwick, A.G. and Lazarus, N.R. 1998. Human variability and non-cancer risk assessment: an analysis of default uncertainty factors. *Regulatory Toxicology and Pharmacology*. 27, 3–120.

Rice, D. and Cannon, G. 1999. *Health and Environmental Assessment of the Use of Ethanol as a Fuel Oxygenate*. Report to the California Environmental Policy Council in Response to Executive Order D-55-99. Volume I. Executive Summary. Air Resources Board, Office of Environmental Hazard Assessment, and Lawrence Livermore National Laboratory.

Robinson, M., Bruner, R.H., and Olson, G.R. 1990. Fourteen- and ninety-day oral toxicity studies of methyl tertiary-butyl ether (MTBE) in Sprague-Dawley rats. *Journal of the American College of Toxicology*. 9, 525–540.

Ruckelshaus, W. D. 1985. Science, risk and public policy. *Science*. 221, 1026–1028.

Sandman PM. 1993. *Responding to Community Outrage: Strategies for Effective Risk Communication*. American Industrial Hygiene Association, Fairfax, Virginia.

Sandman PM. 1987. Explaining risk to non-experts. *Emergency Preparedness Digest*. Oct/Dec, 25–29.

Simmons, K., Click, D., Kostecki, P., and Calabrese, E. 2001. AEHS's 2000 survey of states' soil and groundwater cleanup standards. *Contaminated Soil Sediment and Water*. February, 22–76.

Slovic P. 1986. Informing and educating the public about risk. *Risk Analysis*. 6, 403–415.

Spitzer, H. L. 1997. An analysis of the health benefits associated with the use of MTBE reformulated gasoline and oxygenated fuels in reducing atmospheric concentrations of selected volatile organic compounds. *Risk Analysis*, 17, 683–691.

Squillace, P., Moran, M., Lapam, W., Price, C., Clawges, R. and Zogorski, J. 1999. Volatile organic compounds in untreated ambient groundwater in the United States, 1985–1995. *Environmental Science and Technology*. 33, 4176–4187.

Stephenson, J.B. 2002. Testimony before the Committee on Energy and Commerce, Subcommittee on Environment and Hazardous Materials. MTBE contamination in ground water: Identifying and addressing the problem. Director of Environmental Issues, U.S. Government Accounting Office, Washington, D.C., May 21, 2002.

Stern, B.R. and Tardiff, R.G. 1997. Risk characterization of methyl tertiary butyl ether (MTBE) in tap water. *Risk Analysis.* 17, 727–743.

Suter, G.W. II. 1993. *Ecological Risk Assessment.* Ann Arbor, MI, Lewis Publishers.

Thompson, K.M. and Graham, J.D. 1996. Going beyond the single number: using probabilistic risk assessment to improve risk management. *Human and Ecological Risk Assessment.* 2, 1008–1034.

TNRCC (Texas Natural Resource Conservation Commission. 2002. *PCL Tables.* www.tnrcc.state.tx.us/permitting/trrp.htm. Last updated March 28, 2002.

USEPA (U.S. Environmental Protection Agency). 1979. *Final Report on Population Risk to Ambient Benzene Exposures.* Springfield, Virginia. NTIS report number PB82-227372.

USEPA (U.S. Environmental Protection Agency). 1984. *Approaches to Risk Assessment of Multiple Chemical Exposures.* Washington, DC. EPA 600/9-84-008.

USEPA (U.S. Environmental Protection Agency). 1986a. Guidelines for carcinogen risk assessment. *Federal Register.* 51, 33992, September 24, 1986.

USEPA (U.S. Environmental Protection Agency). 1986b. Guidelines for health risk assessment of chemical mixtures. *Federal Register.* 51, 34014, September 24, 1986.

USEPA (U.S. Environmental Protection Agency). 1986c. Guidelines for the health assessment of suspect developmental toxicants. *Federal Register.* 51, 34028, September 24, 1986.

USEPA (U.S. Environmental Protection Agency). 1988a. *Xylenes.* Integrated Risk Information System. Washington, DC.

USEPA (U.S. Environmental Protection Agency). 1988b. Proposed guidelines for exposure-related measurements. *Federal Register.* 53 (232), 48830–48853.

USEPA (U.S. Environmental Protection Agency). 1989. Chapter 8: Risk characterization. In: *Risk Assessment Guidance for Superfund (RAGS). Volume I: Human health evaluation manual (HHEM),* Part A, Interim Final. (Office of Emergency and Remedial Response). Washington, DC. EPA/540/1-89/002.

USEPA (U.S. Environmental Protection Agency). 1991. *Ethylbenzene.* Integrated Risk Information System. Washington, DC.

USEPA (U.S. Environmental Protection Agency). 1992a. *Guidance on Risk Characterization for Risk Managers and Risk Assessors.* Washington, DC.

USEPA (U.S. Environmental Protection Agency). 1992b. *Toluene*. Integrated Risk Information System. Washington, DC.

USEPA (U.S. Environmental Protection Agency). 1992c. *Framework for Ecological Risk Assessment*. Risk Assessment Forum. Washington, DC. EPA/630 /R-02/011.

USEPA (U.S. Environmental Protection Agency). 1993. *Wildlife Exposure Factors Handbook*. Volume I and II. Office of Research and Development, Washington, DC. EPA/600/R-93/187.

USEPA (U.S. Environmental Protection Agency). 1995a. *Policy for Risk Characterization*. Science Policy Council, Washington, DC. www.epa.gov/ ordntrn/ORD/spc/rcpolicy.htm.

USEPA (U.S. Environmental Protection Agency). 1995b. *Elements to Consider When Drafting EPA Risk Characterizations*. Science Policy Council, Washington, DC. www.epa.gov/ordntrn/ORD/spc/rcelemen.htm.

USEPA (U.S. Environmental Protection Agency). 1995c. *Guidance for Risk Characterization*. Science Policy Council, Washington, DC. www.epa.gov /ordntrn/ORD/spc/rcguide.htm.

USEPA (U.S. Environmental Protection Agency). 1996a. *Proposed Guidelines for Carcinogen Risk Assessment*. Office of Research and Development, Washington, DC. EPA/600/P-92/003C.

USEPA (U.S. Environmental Protection Agency). 1996b. *Ecological Risk Assessment Guidance for Superfund: Process for Designing and Conducting Ecological Risk Assessments*. Environmental Response Team, Edison, New Jersey. Internal EPA Review Draft.

USEPA (U.S. Environmental Protection Agency). 1997a. *Drinking Water Advisory: Consumer Acceptability Advise and Health Effects Analysis on Methyl Tertiary-Butyl Ether (MTBE)*. Office of Water, Washington, DC. EPA/822 /F-97/008.

USEPA (U.S. Environmental Protection Agency). 1997b. *Exposure Factors Handbook*. Office of Health and Environmental Assessment, Washington, DC.

USEPA (U.S. Environmental Protection Agency). 1997c. *Guiding Principles for Monte Carlo Analysis*. Office of Research and Development, Washington, DC. EPA/630/R-97/001.

USEPA (U.S. Environmental Protection Agency). 1997d. *Ecological Risk Assessment Guidance for Superfund: Process for Designing and Conducting Ecological Risk Assessments*. Solid Waste and Emergency Response, Washington, DC. EPA/540/R-97/0065. Interim Final.

USEPA (U.S. Environmental Protection Agency). 1998. Guidelines for ecological risk assessment. Risk Assessment Forum. Office of Research and Development, National Center for Environmental Assessment. Washington, DC. EPA/630/R-95/002F. *Federal Register*. 63(93), 26846–26924.

USEPA (U.S. Environmental Protection Agency). 1999a. *EPA Glossary of IRIS*

Terms. Integrated Risk Information System. Washington, DC. www.epa. gov/ngispgm3/iris/gloss8.htm.

USEPA (U.S. Environmental Protection Agency). 1999b. *Guidelines for Carcinogen Risk Assessment.* Washington, DC. . SAB review copy, July 1999. www.epa.gov/ncea/raf/crasab.htm

USEPA (U.S. Environmental Protection Agency). 1999c. *Achieving Clean Air and Clean Water: The Report of the Blue Ribbon Panel on Oxygenates in Gasoline.* EPA 420-R-99-021. September 15, 1999.

USEPA (U.S. Environmental Protection Agency). 2000a. *Science Policy Council Handbook: Risk Characterization.* Office of Science Policy, Office of Research and Development, Washington, DC. EPA/100/B/00/002.

USEPA (U.S. Environmental Protection Agency). 2000b. *Methyl Tertiary-Butyl Ether (MTBE).* Integrated Risk Information System. Washington, DC.

USEPA (U.S. Environmental Protection Agency). 2000c. *Benzene.* Integrated Risk Information System. Washington, DC.

USEPA (U.S. Environmental Protection Agency). 2000d. *Drinking Water Health Advisories.* Office of Water, Washington, DC. www.epa.gov/ost /drinking/.

USEPA (U.S. Environmental Protection Agency). 2000e. *Consumption Advisories.* Office of Water. Washington, DC. www.epa.gov/ost/fish/.

USEPA (U.S. Environmental Protection Agency). 2000f. *Clinton-Gore Administration Acts to Eliminate MTBE, Boost Ethanol.* U.S. Environmental Protection Agency. Headquarters Press Release. March 30, 2000.

USEPA (U.S. Environmental Protection Agency). 2001. *ECO Update.* Office of Solid Waste and Emergency Response, Washington, DC. EPA 540/F-01/014.

USEPA (U.S. Environmental Protection Agency). 2002a. *Methyl Tertiary Butyl Ether (MTBE).* MTBE FAQs Drinking Water. Washington, DC. www.epa. gov/mtbe/water.htm.

USEPA (U.S. Environmental Protection Agency). 2002b. *Approved Methods for Unregulated Contaminants.* Washington, D.C. www.epa.gov/safewater/ methods/unregtbl.html#1st2aero. Last updated March 28, 2002.

UC (University of California). 1998. *Health and Environmental Assessment of MTBE.* Report to the Governor and Legislature of the State of California sponsored by SB521. Five Volumes, 874 pp. Submitted through the University of California (UC) Toxic Substances Research and Teaching Program SB521 MTBE Research Program. Davis, California. November 12, 1998.

Walker, K. 1992. *Australian Environmental Policy: Ten Case Studies.* Kensington, Australia, New South Wales University Press.

Weiner, J.B. 2002. Precaution in a multi risk world. In: *Human and Ecological Risk Assessment: Theory and Practice.* (Paustenbach, D.J. Ed.). New York, New York, John Wiley and Sons.

Werner, I. and Hinton, D.E. 1998. *Toxicity of MTBE to Freshwater Organisms. Volume III: Air Quality and Ecological Effects.* Health and Environmental Assessment of MTBE, University of California.

Williams, P.R.D. 2001. MTBE in California drinking water: an analysis of patterns and trends. *Journal of Environmental Forensics.* 2, 75–85.

Williams, P.R.D. and Paustenbach, D.J. 2002. Risk characterization: principles and practice. *Journal of Toxicology and Environmental Health (Part B: Critical Reviews).* 5, 337–406.

Williams, P.R.D., Benton, L., Warmerdam, J., and Sheehan, P. 2002. A comparative risk analysis of six volatile organic compounds in California drinking water. *Environmental Science and Technology.* 36, 4721–4728.

Williams, P.R.D., Scott, P.K., Sheehan, P.J., and Paustenbach, D.J. 2000a. A probabilistic assessment of household exposures to MTBE from drinking water. *Human and Ecological Risk Assessment.* 6, 827–849.

Williams, P.R.D., Sheehan, P.J., and Paustenbach, D.J. 2000c. MTBE ambient air and drinking water exposures in California. Presentation at the *International Society of Exposure Analysis Annual Conference.* Monterey, California. October 24–27, 2000.

Williams, T.M and Borghoff, S.J. 2001. Characterization of tert-Butyl Alcohol binding to α2u-globulin in F-344 Rats. *Toxicological Sciences.* 62, 228–235.

Williams, T.M., Cattley, R.C., and Borghoff, S.J. 2000b. Alterations in endrocrine response in male Sprague-Dawley rats following oral administration of methyl tert-butyl ether. *Toxicological Sciences.* 54, 168–176.

Section II

Applicable Remediation Technologies

CHAPTER 8

Receptor Protection

Jonathan Greene, P.E., Malcolm Pirnie
Theodore R. Davis, P.E., Southwestern Environmental, Inc.
David K. Ramsden, Ph.D., URS Corporation

INTRODUCTION AND MAJOR PHASES

In its broadest sense, receptor protection includes the entire sequence of re-mediation activities following a release of gasoline containing MTBE. In this broad sense, receptor protection only ends when the release is remediated to concentrations that are protective of all receptors, considering all possible pathways that may impact those receptors. In this chapter, receptor protection will be considered in a narrower sense as one of the four major phases of response to a release of gasoline containing MTBE.

Responses to gasoline releases, with or without MTBE, can be considered to include four major phases: receptor protection, source control, remediation of residual and dissolved contamination, and MNA of soil and water. Figure 8-1 illustrates these phases.

Receptor protection for the purposes of this chapter consists of immedi-ate response activities designed to ensure the temporary protection of any re-ceptors that might be exposed to concentrations of gasoline components. It is described in detail in this chapter. Source control consists of the steps taken to remediate the highly impacted areas that are the source of the fuel that is spreading away from the site. The various approaches for source control are described in Chapter 9. Remediation of residual and dissolved contamina-tion includes the active remediation efforts that are required to reduce the contamination levels from those that remain after source control is completed down to levels that are appropriate for MNA to take over. Chapters 10 through 15 describe some of the approaches commonly used for this phase of the overall program. The final phase at most sites is a period of MNA that brings the site to the required cleanup goals. Chapter 16 describes this ap-proach in detail.

The first phase is an immediate or near-term response. The last three

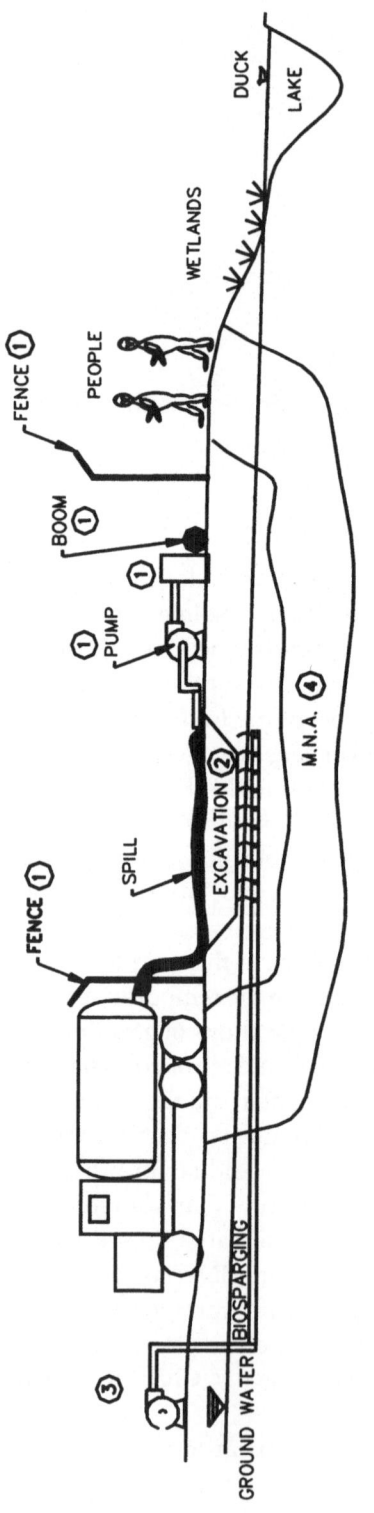

Figure 8-1. **Technology, Sequencing, and Phases**

phases by comparison are longer-term responses to the presence of the gasoline components in the environment. In most instances the initial efforts provide temporary receptor protection until the longer-term remedies can be studied, selected, designed, and sequentially implemented to reduce receptor risk to acceptable levels. These initial responses typically occur soon after the release is discovered but can occur at any time during the full response, because new contamination and pathways posing a more immediate threat to receptors can be discovered even after initiation of source control and long-term remediation.

Conceptualizing the overall remediation effort in four phases helps the site manager avoid some common problems at the site. A frequent error is to imagine that there is one remedy for each site. In this mistaken approach, selecting the best remedy is a search for the one "silver bullet" that will solve all of the environmental problems at the site for the least cost.

Properly considering the four major phases helps avoid that oversimplification and helps to properly sequence various technologies and approaches to most effectively solve the various types of environmental issues at the lowest overall cost. In addition to potentially using different approaches for the four different phases, it is sometimes useful to use a sequence of technologies within one phase. (A good example is the sequencing of technologies to treat the off-gas from an SVE system. Sometimes it makes sense to use a simple incinerator or burner for the highly concentrated stream at the beginning of the remediation period. As the concentrations decline, the burner can be swapped for a catalytic burner and then finally replaced by activated carbon.)

Another advantage of using the four phases is that the site manager keeps in mind that it usually makes sense to allow natural attenuation to polish off the site once the concentrations have been reduced to appropriate levels by more active remediation. MNA is non-intrusive and does not interfere with many potential land use scenarios.

There may be overlap between the phases. For example, ISTD or in-situ chemical oxidation (ISCO) may be appropriate as receptor strategies at one site, while at another site they are not rapid enough and would instead be considered source control technologies. Nonetheless, it is useful to think in terms of the four phases. Many of the technologies described in this chapter are covered in more detail in subsequent chapters (*e.g.*, 10, 11, and 15).

Receptors. Receptors include anything that is valued and potentially affected by the contaminants of concern. Receptors traditionally have a hierarchy. Usually the most valued receptors are humans, with ecosystems, water bodies, plants, or animals falling lower on the hierarchy, except when they are pathways for exposure of humans. The same or similar technologies or techniques can be used to protect all receptors, with the caveat that receptors' sensitivities to contaminants can be variable and that some technologies, such as excavation, that would protect people might destroy the very eco-logical receptor that also was to be protected. Figure 8-2 illustrates some of the human receptors that may be impacted by a release of gasoline compo-nents.

Receptor Threat.   A receptor threat is any circumstance where contamina-tion is impacting or about to impact a susceptible receptor at a concentration that could affect that receptor acutely or chronically.

Receptor Protection.   Receptor protection means actions taken quickly or immediately when a release is found to be impacting, or threatening to im-minently impact, a receptor, and the likely or current long-term remedial ap-proach cannot be designed and implemented in time to stop or control the short-term impact. Protection of receptors may involve a series of actions re-lated to the elements of a risk exposure scenario. Exposure scenarios involve a contaminant at a significant concentration reaching a sensitive receptor by at least one identified pathway. Protection involves removing one or more of these elements: removing the contaminant, preventing significant concentra-tions from reaching the receptor, blocking the pathway to the receptor, or in extreme cases, removing the receptor.

For a contaminant to threaten or impact a receptor, the contaminant must move or be transported to the receptor, or the receptor must move to come in contact with the contaminated matrix. Essentially there are three types of nat-ural matrices that can be contaminant transporters or holders: air (vapor), soil (or other soil-like solids), and water. In addition, free-phase materials in releases on the surface or in the ground are sources of exposure as well as sources to the natural matrices—by volatilization, dissolution in water, or smearing on soils during ground water elevation fluctuations. Because free-phase material in the ground is usually low in mobility and cannot be rapidly recovered except under unique circumstances, it is usually the focus of source control activities rather than rapid response to protect receptors. The derivative threats that come from the free-phase material are usually the focus of receptor protection scenarios.

TECHNOLOGIES

General.   In general two broad technological strategies can be taken for the management of receptor exposure (vapor, water, and soil). These strategies are management at the point of receptor exposure and management before the pathway brings the contaminant to the receptor.

Schedule, rapid implementation, cost, and speedy effect are generally priorities in selecting those technologies or techniques that are most appro-priate for receptor protection in a particular situation. Relatively slow acting,

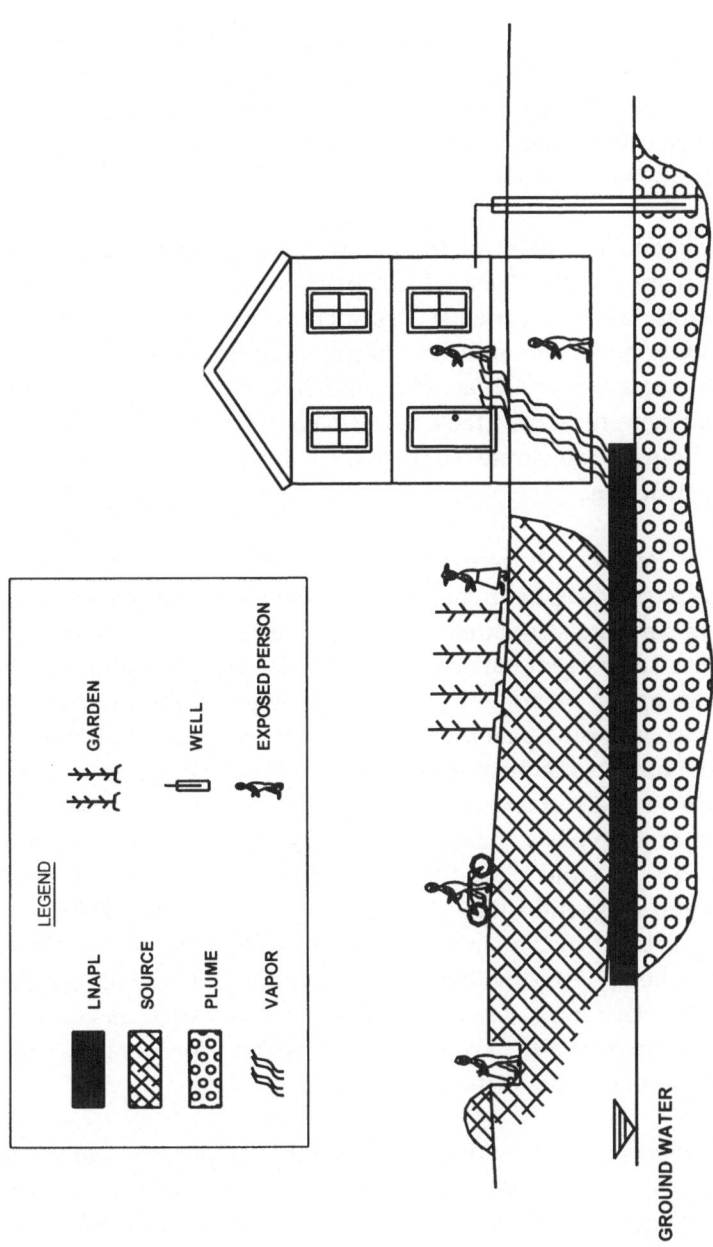

LEGEND

LNAPL

SOURCE

PLUME

VAPOR

GARDEN

WELL

EXPOSED PERSON

GROUND WATER

Figure 8-2.   Human Receptor Exposure

long-term technologies such as phytoremediation and bioremediation have inherent lag times in reducing the exposure concentrations of the chemical of concern or its intermediates of concern. Generally they are inappropriate in the context of rapid response. Passive and active barriers, chemical oxidation, thermal desorption, containment, and removal are potential approaches that may quickly lower exposures by blocking, removing, or limiting access to the contaminated matrices. Tables 8-1, 8-2, and 8-3 list technologies by contaminant carrier matrix (vapor, water, soil) for interim receptor protection from current or imminent exposures.

Vapor Management. Most of the components of gasoline that pose the greatest risks are volatile. Released to the atmosphere they usually only persist and accumulate in enclosed spaces such as basements, low areas, or excavations. Odor may be an indicator of potential gasoline vapors in a structure, but only a vapor survey can determine if contaminated vapor is present at significant concentrations. Table 8-1 lists vapor management technologies for MTBE. Figure 8-3 illustrates some of the components involved in the protection of a structure.

Passive Vapor Barriers. A general management approach for vapors is to block access to the receptor. Intact concrete slabs or walls, as well as coated, below-grade concrete structures may resist passage of gasoline and MTBE vapors. Retrofitting vapor barriers inside or outside of existing structures is sometimes an option. Vapor entry is often through penetrations in the structure such as cracks, pipe penetrations, and exposed soil areas. Sealing these penetrations may be adequate for protection. Products for sealing wall surfaces include a variety of coating products such as epoxys, novolacs, phenolics, and other epoxy-based polymers. Plugging penetrations can also be accomplished with similar products as well as epoxy polysulfides, fluoroelastomers, and vinyl emulsions. Applicability for gasoline, MTBE, and other components present in the vapor stream as well as the user's site-specific needs should be reviewed with the system vendors. Coatings are designed systems, and selection is based on multiple factors such as contaminants, abrasion and fire resistance, costs, moisture, surface preparation, and VOC emissions from the products themselves.

Active Vapor Barrier. Technologies are available to prevent soil vapor from reaching the receptor by breaking or disrupting the exposure pathway leading to the receptor. For soil vapor this can be an open trench for passive dispersal, turbine-vent-capped wells for shallow, low-impact removals, or active SVE systems. Active SVE systems may include vacuum pumps or blowers to extract the soil gas and, sometimes, passive clean air injection wells (horizontal or vertical well arrays or trenches) (USACE, 1999).

Active Vapor Removal. High-volume ventilation can remove or lower vapor concentrations below applicable guidelines in enclosed spaces. The appropriate ventilation rate is calculated with the measured contaminant infiltration rate and the airflow needed to lower concentrations to acceptable concentrations. Ventilation equipment manufacturers can be located at the Home Ventilation Institute website (www.HVI.org/purpose). In a related technology, room or space air can be drawn through an activated carbon filter or similar adsorptive system to reduce the ambient concentrations in the space air to protective or low risk concentrations. As with many adsorption technologies, good maintenance and timely changeout of carbon when it nears breakthrough are essential for continuing protection.

Passive Vapor Removal. For short duration exposures, contaminated vapor management can consist of personal protective equipment (PPE) selected based on the contaminant(s), routes of exposure, and concentrations. To be protective, this requires personnel with formal training who have an understanding of the conditions, risks, and potential for exposure associated with this approach. Such an approach is virtually never appropriate for the general public or for administrative personnel at a site or facility for any duration other than emergency evacuation.

Water Management. Gasoline releases can pose a risk by entering surface water or well water that can come in contact with receptors. Table 8-2 lists water management technologies for MTBE.

Passive Water Management Technologies. Passive water barriers are systems designed to allow the receptor to avoid the contaminated plumes, without necessarily treating the contamination. Passive systems include slurry or specialized sealable sheet pile walls keyed into an underlying impermeable soil layer that create in-ground dams, redirecting ground water around a particular location. Systems that are based on holding back the flow of ground water have a potential flaw in that ground water may still be drawn around the wall if the receptor is a sink, such as a ground water production well. Ground water extraction is often necessary in conjunction with barriers.

Another barrier is an interceptor system. These systems can be ditches, trenches, or French drains that capture a ground water plume and direct it to an appropriate management system. Management includes treatment (as required) and discharge to a sewer, river, or other surface water body. These barriers can use pumping systems to lift and direct the water to a management system. They capture ground water along the existing natural flow path without depressing the water table upstream of the capture zone.

For private water wells, short- and long-term protection can be achieved by connecting to public supplies via pipeline. In the short-term, bottled water

Table 8-1. Receptor Protection—Technologies for Vapor

Technology	Purpose	Applications	Issues
Vapor Impermeable Coatings	Block entry of vapor to receptor area	• Outside or inside surfaces • Epoxy, emulsion, or similar coatings	• Difficult to install on outside of existing structures, below ground • Excavation can lead to structural instability of buildings • Diversion of vapor to another receptor by barrier
SVE	Remove vapor before it reaches receptor	• Well array or trench SVE barrier against intruding contaminant vapor Around or between subsurface rooms, spaces, or basements and vapor plume • Treat extracted vapor by GAC, resin, biofilter, catalytic oxidation, thermal oxidation, and other technologies	• Requires adequate vapor capture for protection • Extracted vapor must be treated or dispersed/diluted prior to release away from receptors • Upwelling and ground water fluctuations can impact operation and maintenance of system • Short-circuiting and preferential flow paths
Slurry, HDPE, Other Barrier Wall	Block or divert vapor from reaching receptor exposure area	• Around large buildings or below ground areas with inaccessible walls for coatings	• Expensive compared to simple coating barriers Selection of compatible impermeable materials • Selection of appropriate placement or installation technologies • Diversion of vapor to another receptor by barrier

Method	Purpose	Description	Considerations
High Volume Ventilation	Dilute vapor in potential receptor exposure area	• Dilution of intruding vapors in workspaces or building spaces • High outside airflow and mixing to lower exposure concentration to protective levels	• May need to be fire code safe with potential high concentrations of hydrocarbon vapors • May need dispersal vent/stack • May require permit in some localities • Potentially increased heating and/or cooling costs
Carbon or Other Filters	Remove contaminant from air in vapor exposure area	• Adsorb BTEX vapors using air filtration to lower exposure concentrations to protective levels	Inadequate filter monitoring may allow breakthrough More effective for other gasoline components such as BTEX Selection of appropriate filter for vapor
PPE	Protect essential personnel without treating entire atmosphere	• Buildings and other confined spaces such as excavations • Personal protection in areas where vapor is present	• Restricts personnel activity • PPE performance and inappropriate selection and use issues could lead to exposure • Excessive vapor could be physical health hazard due to fire or explosion hazard even for PPE protected personnel
Personnel Relocation	Remove receptor from potential contact with contaminant	• Usually non-industrial settings • Usually short-term solution	• Restriction on personnel activity • Public relations • Loss of productivity • Consequential damages claims

Notes: GAC = granular activated carbon; HDPE = high density polyethylene; PPE = personal protective equipment; SVE = soil vapor extraction

Table 8-2. Receptor Protection—Technologies for Water

Technology	Purpose	Applications	Issues
Slurry Wall	Prevent contaminated water from entering receptor area	• Passive barrier between ground water plume and receptor • Receptors can be sensitive aquifer, building/structure, well, spring, or surface water	• Some locations will have many underground obstructions hindering installation • Slurry wall barriers can have "holes" allowing contaminant penetration at localized points • Barriers tend to reroute ground water creating possible new exposures or eventually are end-run by contaminant plume • Change of ground water flow patterns can place other receptors at risk • Ground water withdrawal upgradient of wall is often required
Sheet Pile Wall	Prevent contaminated water from entering receptor area	• Passive barrier • Multitude of receptors • Easy to install • Low permeability situation	• Requires joint seating • Requires positive hydraulic control • Corrosion • Not applicable to productive aquifers
Interceptor System	Capture and reroute contaminated water before it can enter receptor area		
Hydraulic Control/Pump-and-Treat	Prevent contaminated water from entering receptor area	• Well or trench array • Provides GW capture barrier between plume and receptor • Receptors can be sensitive aquifer, building/structure, well, spring, or surface water	• Change of flow pattern can place other receptors at risk • Usually requires treatment for captured water • Extended time for ground water treatment • Lower cost to install but long-term operating costs • Potential discharge to POTW or sewer if concentrations treatable in receiving system

	Objective	Description	Disadvantages
Replacement Water Source	Stop consumption or other exposures to impacted water	• Replace contaminated drinking/bathing water • Deeper, acceptable aquifer and well • Alternative local natural source from surface water • Water purchased from neighboring water systems • Bottled or piped in water	• Cost • Politics • Resource taking • Contaminated source may be sole-source aquifer, lake, spring, or river drinking water source
High Efficiency Sparging and SVE Barrier	Remove contaminants from water before water reaches receptor	• Active barrier between ground water plume and receptor • Receptors can be sensitive aquifer, building/structure, well, spring, or surface water	• Sparging can change water and/or air flow and create new receptor • Disposal of contaminants
Carbon, Resin, or Other Filter	Remove contaminants from water before water reaches receptor	• Contaminated drinking water	• Disposal of contaminated carbon • Potentially high carbon usage for MTBE in standard GAC filter designs • Design system to maximize % MTBE saturation on carbon while operating continuously and minimizing chance of unguarded breakthrough

Notes: POTW = Publicly Owned Treatment Works

Table 8-3.   Receptor Protection—Technologies for Soils

Technology	Purpose	Applications	Issues
Excavation	Remove contaminated soils to eliminate exposure for potential receptors	• Best for easy access areas • Rapid removal of contamination	• Disposal of excavated material and potential long-term liability • Safety of personnel involved in the excavation • Underground obstructions and utilities • Overlying or included sensitive natural, historical, or archeological areas
In Situ Thermal Desorption	Removes contaminants to eliminate or minimize risk to exposed receptor	• Desorption adequate for MTBE and BTEX • Best for shallow applications • Less water is better	• Control moisture in soils if practical; heating more water is more expensive • Must capture and treat off-gas • Off-gas treatment with burner may require permit
In Situ Chemical Oxidation	Destroys contaminants in soils to eliminate or reduce risk to exposed receptor	• Shallow soils best for lance or Geoprobe® applications • Applications for saturated and unsaturated soils • Select oxidizer appropriate to contaminant	• Messy and requires stringent health and safety considerations • Reactive • Can be expensive • May require repeat treatments • Low permeability soils may require close spacing of injections • Oxidizer selections • Selection of injection method; air lances, water lances, well points, direct push technologies (e.g., Geoprobe®)
Capping, Landscaping & Recontouring	Prevent direct soil contact by potential receptors	• Containment and barrier • Prevents direct contact, soil slump, and/or erosion and washout	• Maintenance • Loss of property use • Overlying or included sensitive natural, historical, or archeological areas

Concrete Pads or Walls	Prevent direct soil contact by potential receptors	• Containment and/or barrier • Prevents direct contact, soil slump, and/or erosion and washout • High foot and vehicle traffic areas	• Concrete is not water impermeable; moisture accumulation may allow migration of contaminants into/through concrete • Requires patching if traffic area potholes develop • Bearing strength of underlying soils relative to traffic
Asphalt Cover	Prevent direct soil contact by potential receptors	• Containment and barrier • Prevents direct contact, soil slump, and/or erosion and washout • Moderate foot and vehicle traffic and load areas	• Gasoline components may act as solvents, softening or dissolving asphalt • Requires patching if traffic area • Bearing strength of underlying soils relative to traffic
Geosynthetic Liners	Prevent direct soil contact by potential receptors	• Containment and barrier to prevent direct contact, soil slump, and/or erosion and washout • Low foot traffic areas • No vehicle traffic areas	• Must be protected from mechanical penetration • Compatibility with gasoline liquid and vapor must be determined • Durability under ultraviolet light, temperature extremes, flexing, etc. must be appropriate • Low durability under traffic with or without protective soil cover
Deed Restriction	Prevent direct soil contact by potential receptors by restricting site uses and activities and warning of presence of residual soil contamination	• No/low use properties • Non-residential properties	• Restriction on use of property • Dependent on compliance by property users and/or owners and communication of restriction through time • Loss or limitation of property use and decline in property value
Access Restrictions	Prevent direct soil contact by potential receptors	• Isolation of contaminated area • Fencing and signage • Control or limit time in area to limit exposure to safe levels	• Restriction on personnel activity • Loss of property use and value • Public relations (e.g., from signage)

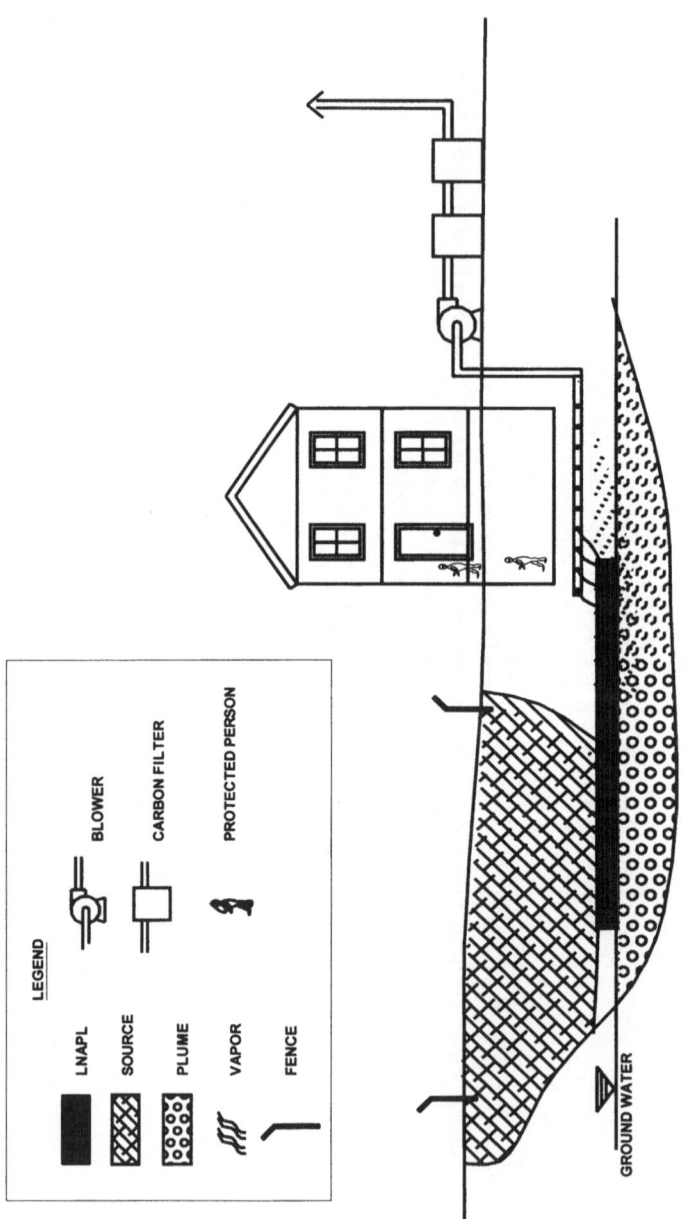

Figure 8-3. **Protect the Structure**

is sometimes used for consumption. A new well, located upgradient of the contaminant source or in an unimpacted and hydraulically isolated aquifer, can also be installed.

Active Water Management Technologies.   Active water management technologies treat the impacted water before it comes in contact with a receptor. An air sparging well array, essentially *in situ* stripping, can be installed to volatilize the contaminants. Such systems often include SVE to capture and manage the off-gas, preventing risk from vapor exposure. Another approach is to treat the contaminated water after it is produced from the well, but before it comes in contact with any receptors. A treatment system for a number of houses is a major project and is usually not considered receptor protection in the context of rapid response. The devices that are typically used for receptor protection are point of entry and point of use treatment systems in individual homes.

Point of use systems are installed on each tap where water is used. They are normally used to control odor and taste in otherwise safe water. Due to the large number of devices required and the difficulty in providing maintenance, they are not normally used for water that, if used without treatment, could actually harm a user.

A point of entry system treats the water where it enters the house and before it is distributed throughout the house. Point of entry systems are appropriate for VOCs, where ingestion, inhalation of vapor from showers or baths, and skin absorption present potential health risks. GAC is often used in point of entry systems treating VOCs such as BTEX and MTBE.

Other possibilities include reverse osmosis (RO), advanced oxidation, small household air stripping tanks, or resin adsorption systems. Information on activated carbon and resin adsorption systems for MTBE treatment is available from GAC venders and from several publications (California MTBE Research Partnership, 1999a, 1999b, 2001). Operation and maintenance are keys to limiting exposure; neglecting to change the carbon can result in exposure. Activated carbon and resin systems must be routinely monitored for saturation and breakthrough.

RO is a high-pressure filter that uses semi-permeable membranes to allow water through and reject the target contaminants. Some RO membranes require a pre-filter to extend the life of the membrane. Various RO units are offered by many household and commercial suppliers that can be found via the internet.

Soil Management.   Contamination of soil often occurs as a result of leaking USTs, tanks, or piping, and also from surface spills. Table 8-3 provides a list of potential treatment systems for soils. Exposures to soil contaminants are best prevented by excavation and contaminant removal technologies, barriers, and engineering or administrative management.

Active Soil Management Technologies.   Excavation permanently removes contaminated soils and eliminates exposure at the original site. Often, excavation requires treatment of excavated materials for disposal or beneficial reuse. Complementary treatment technologies for gasoline in excavated soils include *ex situ* low temperature thermal desorption, SVE of piled soils, and biovault remediation.

In situ, soils can be actively treated with SVE and/or soil venting (USACE, 1999), with the added benefit of controlling possible exposures during the remediation by controlling vapor migration. ISTD processes can also be used with SVE or venting to enhance the volatilization of the contaminants in unsaturated soils.

Passive Soil Management Technologies.   Passive technologies include barriers and engineering or administrative management. Barriers used for engineering controls include:

- enclosing the soils with slurry or other walls to stop migration of vapors within the soil (see vapor control section above);
- capping the soils with asphalt, concrete, liners, clean soils, or combinations of these to prevent water infiltration, direct contact, and vapors coming to the surface; and
- sealing buildings and similar structures to prevent contact with the underlying soils.

Administrative controls include: restricting access, fencing, restricting excavation, and excavating only with prescribed PPE and training.

CONCLUSIONS

There are proven technologies effective in all three media to protect receptors from gasoline components, including MTBE, during the initial phases of the project. The presence of MTBE in the gasoline does not change the range of technologies available to the designer, but may impact details such as equipment sizing and carbon usage rates.

REFERENCES

California MTBE Research Partnership. 1999a. *Treatment Technologies for Removal of Methyl Tertiary Butyl Ether (MTBE) from Drinking Water: Air Stripping, Advanced Oxidation Processes, Granular Activated Carbon, Synthetic Resin Sorbents.* December 1999. National Water Research Institute, Fountain Valley, California. (714-378-3278).

California MTBE Research Partnership. 1999b. *Evaluation of the Applicability of Synthetic Resin Sorbents for MTBE Removal from Water.* December 1999. National Water Research Institute, Fountain Valley, California. (714-378-3278).

California MTBE Research Partnership. 2001. *Treating MTBE—Impacted Drinking Water Using Granular Activated Carbon.* December 2001. National Water Research Institute, Fountain Valley, California. (714-378-3278).

USACE (U.S. Army Corps of Engineers). 1999. *Soil Vapor Extraction and Bioventing.* EM 1110-1-4001. June 1999. www.usace.army.mil/inet/usace-docs/eng-manuals/em.htm.

CHAPTER 9

Source Control

Theodore R. Davis, P.E., Southwestern Environmental, Inc.
Jonathan Greene, P.E., Malcolm Pirnie
David K. Ramsden, Ph.D., URS Corporation

INTRODUCTION

The classical paradigm for successful contaminant management is "where possible, remediate the most concentrated area first." The most contaminated area is commonly called the source area for a release. This applies to successfully managing releases of gasoline containing MTBE. The presence of MTBE as a component in modern gasoline accentuates the need to remove the source area in a timely fashion because of the mobility of MTBE in the underground environment.

SOURCES

Response to a fuel release, whether or not it contains MTBE, may include four stages: receptor protection, source control, remediation of residuals and dissolved plumes, and natural attenuation of soil and water. Following this model, source control focuses on the highly impacted part of the site that is the origin of the plume. For a fuel release, sources can include the tankhold, unsaturated soils, LNAPL, and saturated soils. Typically, replacement, repair, or removal of the UST (or aboveground storage tank [AST]) system, where applicable, is standard following a release to eliminate any ongoing or continuing release. (The UST system includes the underground tanks, dispensers, connecting piping, and associated instrumentation and valving.)

The controls selected for a given source depend on the source type, source volume, source location, threats to receptors, stakeholder issues, regulatory requirements, and company policies. Figure 9-1 illustrates the simple relationship of a release to the source and to the ground water plume. The high source concentrations are from gasoline trapped in the pore spaces of the soil, whether saturated or unsaturated.

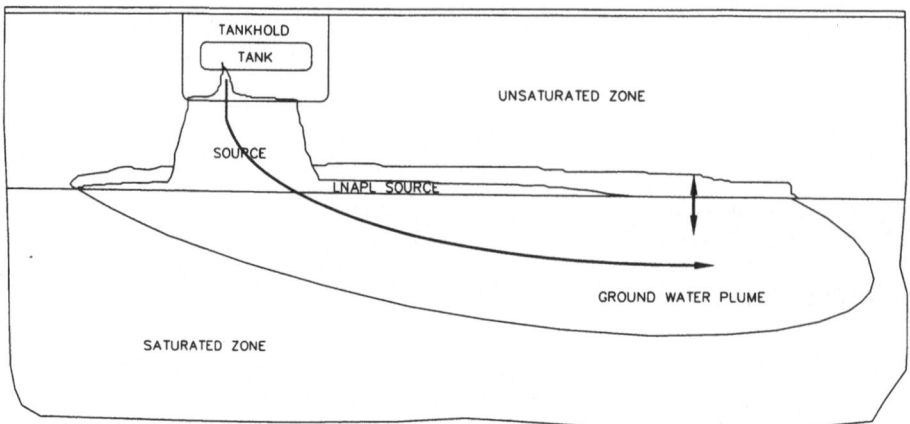

Figure 9-1. **Sources**

Tankhold.   Many MTBE-blended gasoline releases are from UST systems. The tankhold is the excavation made for the USTs and contains the tanks and some of the interconnecting piping. The tankhold is usually backfilled with pea gravel or a mix of pea gravel and sand. These materials are relatively easy to remove for maintenance or remediation when compared to cohesive soils. Where possible, tankholds are placed with the bottom above the water table. In most cases, the underlying, undisturbed material beneath the tankhold is of lower permeability than the backfill and much of the leaking material may be retained in the hold. This often facilitates recovery. Tankholds normally are considered a specialized case of unsaturated soil because of the presence of the relatively permeable, homogenous, non-cohesive backfill.

Unsaturated Soils.   Unsaturated soils are vertically infiltrated by the gasoline release due to gravity. The extent and rate of infiltration depends on the soil type and associated properties such as permeability. Normally the unsaturated soils are less porous than the tankhold backfill material, so the free-phase material may enter soil pore spaces as fingering ganglia or droplets. The free-phase eventually flows to ground water if the volume of the release is great enough. It leaves adsorbed components on the unsaturated soil organic fraction, ganglia or droplets trapped within pores, and dissolved components in the unsaturated soil's moisture content. Large releases can saturate the soil. These materials, and particularly the ganglia, are a substantial contaminant source to infiltrating surface water or ground water.

LNAPL.   Large releases that saturate the unsaturated soils with gasoline will eventually migrate to ground water and may form LNAPL at the interface between ground water and the unsaturated soils. As ground water elevation fluctuates, LNAPL may smear in underlying and overlying soil.

Saturated Soils.   LNAPL may become trapped in saturated soils when the water table rises.

REMEDIATION TECHNOLOGIES

There are many remedial technologies and even more permutations of techniques and practices for each technology. Discussed below are those established remedial technologies for managing sources in the four different areas discussed above.

Tankhold.   Tankhold sources can often be easily and effectively remediated. In general, the sooner after the release action is taken, the easier it is to remediate the release. The tankhold backfill may be easily recovered; liquids may be readily recoverable from the permeable backfill; and the backfill is often dissimilar enough from the underlying and surrounding native soils to limit rapid infiltration due to the capillary break phenomenon (USACE, 1983; USACE, 1998). Tankhold remediation technologies are essentially the same as the technologies used for unsaturated soils as listed in Table 9-1. These may include removal of the tankhold materials and recovery of any free-phase or aqueous liquids pooled in the tankhold using vacuum trucks or pumps and tanks. Underlying, unsaturated soils may be excavated as well. For minimal releases such as those from a small one-time overfill, SVE can be used in the unexcavated tankhold to remove contamination (USACE, 2002).

Unsaturated Soils.   MTBE remediation technologies in unsaturated native soils include excavation of contaminated soil (USACE, 1983; USACE, 1998), SVE, thermal desorption (USEPA, 1995c), chemical oxidation (ITRC, 2001), and flushing techniques (USEPA, 1995b). Table 9-1 lists common remedial options, application factors, and related issues for those options.

LNAPL.   Remediation technologies for LNAPL focus on hydraulic recovery and ways to mobilize more LNAPL mass. There is an expectation that, without physically excavating the encompassing soil, LNAPL recovery will be incomplete and the rate and efficiency of recovery will progressively diminish with time of operation. Many sites will reach the point where further LNAPL recovery is impractical, but remaining LNAPL may still create a ground water plume requiring continued ground water remediation.

LNAPL can be removed from wells by hand bailing or by vacuum truck. More productive wells or trenches can be fitted with skimmer pumps. Water removal and vacuum (USACE, 1999), thermal (API, 2001), or surfactant

Table 9-1. Source Control Technologies for Unsaturated Soils and Tankholds

Technology	Application	Issues
Excavation* (USACE, 1983; USACE, 1998; USACE, 1987)	• High risk, high concentration sites • Short schedule requirements • Low permeability sites • Simple compact sites	• May not remove all liquid accumulation in excavation • Overcomes rebound by contaminated soil removal • May need techniques for vapor control/ventilation during excavation • Need other technologies for ultimate disposal such as: vapor extraction, ex situ chemical oxidation, biovault treatment
In Situ Thermal Desorption and Other Thermal Methods* (API, 2001; USEPA, 1995c) Chapter 10	• Steam, 6-phase, resistance, hot air, or other heating systems • Trend is to low temperature, applicable to VOCs	• Control and recovery of vapor with SVE to avoid air emissions • Not appropriate near active USTs
Ex Situ Thermal Desorption* Figure 9-2	• Rotary kiln, heated pile or other as applicable to setting and contaminant(s)	• Control of off-gas vapor with other technology such as catalytic incinerator, condenser or other • Air emissions
Soil Vapor Extraction* (USACE, 2002; USACE, 1997; USACE, 2001) Figure 9-3 Chapter 10	• Can be enhanced with in situ thermal desorption	• Treatment of vapor with other technologies such as GAC, catalytic incineration, etc. to avoid air emissions • May require surface barrier to prevent air short-circuiting in shallow or highly permeable settings
In Situ Chemical Oxidation (ITRC, 2001) Chapter 11	• Fenton's reagent (hydrogen peroxide and iron) • Other as applicable to contaminant(s) • Injection by wells, water or air lances	• MTBE mineralized by some but not all oxidants • Reaction non-specific • Produces significant reaction and requires health and safety measures appropriate for oxidant and setting • Requires evaluation of spreading of any underlying plume with injected liquids

	• Best for accessible, small, high concentration areas	• Usually requires repeat applications to reach low concentration goals • May require close spacing in unsaturated, low permeability, or highly contaminated soils • Not appropriate near active USTs
Surfactant Flushing* (USEPA, 1995a, 1995b)	• Vertical infiltration of water and surfactant • Driven more by other gasoline components than soluble, non-adsorptive MTBE	• Requires evaluation of spreading of plume with injected liquids • Surfactant and cosolvent volumes may be large • High costs • Usually requires cosolvent and water injection • Must consider toxicity, biodegradability of solvent/surfactant • Requires ground water management system • Depends on vertical permeability for infiltration • Must separate surfactant, cosolvent from contaminant(s) and treat or treat all combined
Soil Flushing* (ITRC, 1998)	• Infiltration of water • Possible amendment with nutrients	• Requires evaluation of spreading of plume with injected liquids • Good for MTBE due to solubility, poorer for BTEX, poorest for other gasoline components • Need to know soil burden — possible formation of LNAPL by vertical water flood • Requires established ground water management system

Notes: * = Technology usually requires accompanying complementary technology to polish waste or manage new waste stream.

Figure 9-2.   **Ex Situ Thermal Desorption**
From left to right: elevated control room, rotating drum,
soil feed hopper; in background; emissions control trailer
and stack.

(USEPA, 1995b) enhancement are sometimes used to increase the rate of recovery. Eventually the LNAPL will be decreased to the level of field residual
saturation, and hydraulic recovery will no longer be feasible. The field residual saturation varies with the soil type and the LNAPL properties. It is defined as the percent of the pore volume occupied by the LNAPL that will not
drain out under gravity. It has been documented to range from below 10% to
above 50% (ITRC, 1996; API, 2001). Since pore volume in soils typically
ranges from 30% to 50% of the soil volume (ITRC, 1996; API, 2001), the residual LNAPL can be between 3% and 25% of the soil volume. (LNAPL is lighter
than soil; therefore, the percent by weight would be less.)

Once the mobile LNAPL is removed, the residual can be remediated by
gas-based technologies (USACE, 2002; USACE, 1997; USACE, 1999), *in situ*
chemical treatment (ITRC, 2001), biological treatment (ITRC, 1998), or other
technologies.

Effective technologies for LNAPL removal are limited. Factors considered in selection are depth to LNAPL, LNAPL volume, physical characteristics of LNAPL and soil, ground water elevation fluctuations, cost, and potential and time for constituents to impact sensitive receptors. Table 9-2 lists

Table 9-2. Source Control Technologies for LNAPL

Technology	Application	Issues
Selective Hydrocarbon Phase Extraction*	• Skimmer pump	• Slow LNAPL recovery • Poor LNAPL spread control • Poor or no ground water control • Minimizes water recovery • Minimizes drawdown • Minimizes biofouling
Surfactant Flushing* (USEPA, 1995b)	• Horizontal and/or vertical infiltration of water and surfactant • Driven more by other gasoline components than soluble, non-adsorptive MTBE	• Requires consideration of spreading of plume with injected liquids • Usually requires cosolvent and water injection • Requires established ground water management system • Expensive • Toxicity of surfactants and cosolvent
Multi-phase Extraction* (USACE, 1999)	• Extraction of ground water and LNAPL	• Contaminated water management • Biofouling • Requires phase separation and treatment of soil gas, water, and hydrocarbons
Bioslurping* (USACE, 1999) Figure 9-4	• A multi-phase extraction method • Extraction tube at interface down well(s) • May be accompanied by amendment injection/infiltration • Takes advantage of slurping aeration effect	• Requires phase separation and treatment of soil gas, water, and hydrocarbons • Some hydrocarbons (e.g., diesel) cause foaming, more stable emulsions • Biofouling
Excavation* (USACE, 1983)	• Shallow soils	• Personnel protection • Excavated material will require further treatment • May involve long-term liability

Table 9.2 Continued

Technology	Application	Issues
In Situ Thermal Desorption and other Thermal Methods* (API, 2001)	• Many applications 　— Steam 　— 6-phase 　— Resistance 　— Hot air 　— Radio frequency • Low temperature is trend	• Temperature below boiling point of water is most economical • Air emissions must be controlled by SVE
In Situ Chemical Oxidation (ITRC, 2001) Chapter 11	• Injection 　— Wells 　— Water lance 　— Air lance • Best for 　— Accessible areas 　— Small areas 　— High concentrations	• Requires consideration of spreading of plume with injected liquids • Usually requires repeat applications to reach low concentration goals • May require close spacing in saturated setting • May require close spacing in low permeability setting

Notes: * = Technology usually requires accompanying complementary technology to polish waste or manage new waste stream.

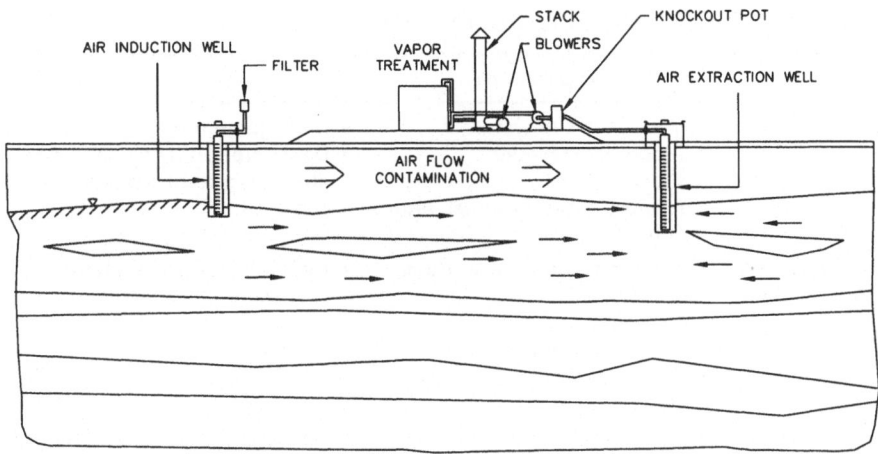

Figure 9-3. **Soil Vapor Extraction**

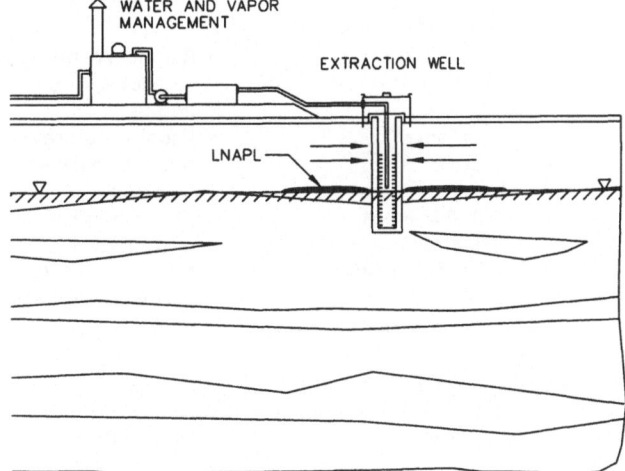

Figure 9-4. **Bioslurping**

LNAPL removal and recovery technologies. These include single-phase recovery, multi-phase extraction (MPE) methods (including bioslurping), and surfactant flushing.

Saturated Soils.   Saturated soil source control technologies are summarized in Table 9-3. These include excavation (USACE, 1983), pump-and-treat, ISCO (ITRC, 2001), and surfactant flushing (USEPA, 1995b). Technologies used for unsaturated soils can be used when it is practical to dewater (USACE, 1983) the soils during treatment.

Table 9-3. **Source Control Technologies for Saturated Soils**

Technology	Application	Issues
Dewatering*	• Manage *in situ* water • Supports excavation • Supports SVE • Supports *in situ* thermal desorption	• Management of extracted water • Dewatered soils can be remediated using technologies in Table 9-1
Pump-and-Treat* (USACE, 2001) Figure 9-5	• Good for soluble, low adsorption MTBE • Well understood • Limited by permeability • BTEX may drive time of remediation	• Hydraulic control is a benefit • Treatment of extracted water • Reinjection/discharge and permitting
In Situ Thermal Desorption and other Thermal Methods* (API, 2001)	• Many Applications — Steam — 6-phase — Resistance — Hot air — Radio frequency • Low temperature is trend	• Temperature below boiling point of water is most economical • Air emissions must be controlled by SVE
In Situ Chemical Oxidation (ITRC, 2001) Chapter 11	• Injection — Wells — DPTs — Water lance — Air lance • Best for — Accessible areas — Small areas — High concentrations	• Requires consideration of spreading of plume with injected liquids • Usually requires repeat applications to reach low concentration goals • May require close spacing in saturated setting • May require close spacing in low permeability setting
Surfactant Flushing* (USEPA, 1995b)	• Smear zones with high concentrations • Driven more by other gasoline components than soluble, non--adsorptive MTBE	• Requires consideration of spreading of plume with injected liquids • Usually requires cosolvent and water injection • Requires established ground water management system • Expensive • Toxicity of surfactants and cosolvent

Notes: * = Technology usually requires accompanying complementary technology to polish waste or manage new waste stream.

CONCLUSIONS

Technologies are available to control sources located in the various areas discussed. In general, shallower sources of all types have the greatest selection of technologies, lowest costs, and greatest success of remediation. This em-

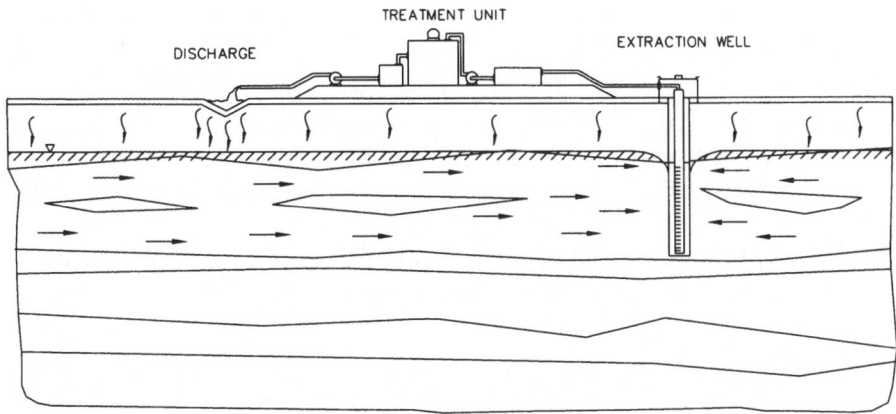

Figure 9-5. **Pump-and-Treat**

phasizes the need for early detection and rapid response to releases, before the released contaminants can potentially migrate to greater depths.

REFERENCES

API (American Petroleum Institute). 2001. *Evaluating Hydrocarbon Removal From Source Zones: Tools to Assess Concentration Reduction.* January 2001 Edition.

ITRC (The Interstate Technology and Regulatory Cooperation). 1996. *Technical Requirements for On-Site Low Temperature Thermal Treatment of Non-Hazardous Soils Contaminated with Petroleum/Coal Tar/Gas Plant Wastes.* Low Temperature Thermal Desorption Task Group. Final, May 1996.

ITRC (The Interstate Technology and Regulatory Cooperation). 1998. *General Protocol for Demonstration of In Situ Bioremediation Technologies.*

ITRC (The Interstate Technology and Regulatory Cooperation). 2001. *Technical and Regulatory Guidance for In Situ Chemical Oxidation of Contaminated Soil and Groundwater.*

USACE (U.S. Army Corps of Engineers). 1983. *Dewatering and Groundwater Control.* Joint Departments of Army, Air Force and Navy. USACE TM 5-818-5.

USACE (U.S. Army Corps of Engineers). 1987. *Bituminous Pavements.* Departments of Army and Airforce. USACE TM 5-822-8.

USACE (U.S. Army Corps of Engineers). 1997. *In-Situ Air Sparging.* CEMP-RT. USACE EM 1110-1-4005.

USACE (U.S. Army Corps of Engineers). 1998. *Removal of Underground Storage Tanks (USTs).* USACE EM 1110-1-4006.

USACE (U.S. Army Corps of Engineers). 1999. *Multi-Phase Extraction.* CEMP-R. EM 1110-1-4010.

USACE (U.S. Army Corps of Engineers). 2001. *Adsorption Design Guide.* CECW-E. USACE DG 1110-1-2.

USACE (U.S. Army Corps of Engineers). 2002. *Soil Vapor Extraction and Bioventing.* CEMP-ET. USACE EM 1110-1-4001.

USEPA (U.S. Environmental Protection Agency). 1995a. *Cosolvents.* Office of Solid Waste and Emergency Response, Technology Innovation Office. EPA542-K-94-006.

USEPA (U.S. Environmental Protection Agency). 1995b. *Surfactant Enhancements.* Office of Solid Waste and Emergency Response, Technology Innovation Office. EPA542-K-94-003.

USEPA (U.S. Environmental Protection Agency). 1995c. *Thermal Enhancements.* Office of Solid Waste and Emergency Response, Technology Innovation Office. EPA542-K-94-009.

CHAPTER 10
Soil Vapor Extraction, Bioventing, and Air Sparging

Brian D. Symons, P.E., The RETEC Group, Inc.
Jonathan Greene, P.E., Malcolm Pirnie

SVE, air sparging, bioventing, and biosparging are the gas-based technologies used to volatilize and/or biodegrade VOCs in the subsurface. Before the introduction of MTBE to gasoline, these gas-based technologies were frequently the technologies of choice for managing gasoline releases. These technologies are still useful solutions for remediating gasoline releases containing MTBE; however, a designer must change some of the details of equipment and airflow sizing to accommodate the differences between MTBE and the other components of gasoline.

This chapter describes each of these technologies. It also describes the different factors a designer must consider in selecting the best technology or combination of technologies for a site. Some of the factors that are crucial for design decisions are the concentrations of gasoline constituents present, the location of the mass of contaminant relative to the water table, and soil characteristics and geology of the site.

GAS-BASED TECHNOLOGIES

There are three major gas-based technologies: SVE, bioventing/biosparging, and air sparging. These three systems are described below.

Soil Vapor Extraction. SVE is one of the most effective and cost-efficient methods of removing VOCs (including gasoline constituents such as MTBE) from unsaturated soils. A basic SVE system consists of one or more extraction points screened in the unsaturated zone and connected to blowers or vacuum pumps that extract air. The extraction points are usually wells or trenches. A more complex SVE system often adds air injection wells, a low-permeability cover at the ground surface, air/water separator, and an off-gas treatment system (USACE, 1995) (Figure 10-1). Airflow is induced in the unsaturated zone when the system withdraws (vacuums) air from wells or trenches in the

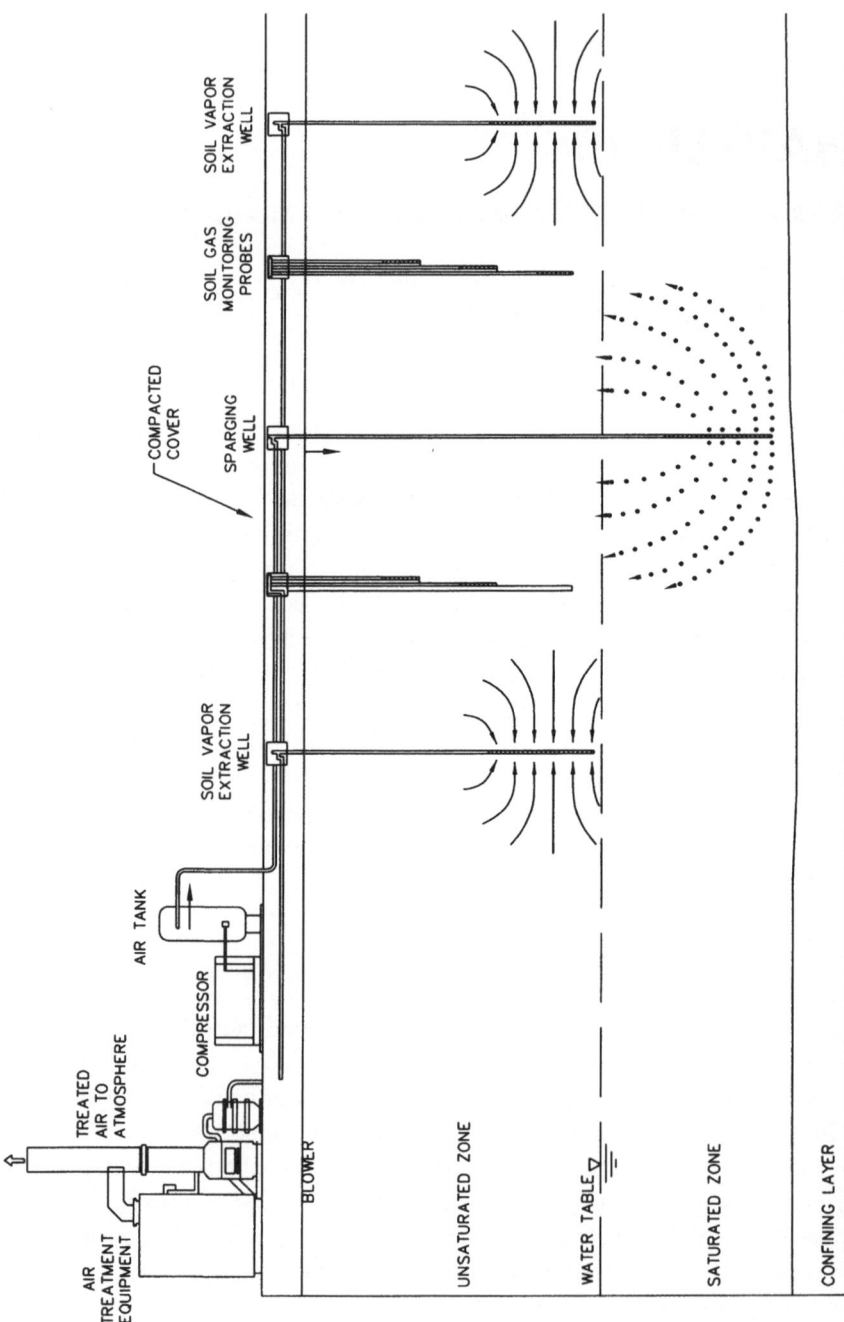

Figure 10-1. Vertical Air Sparging and Vertical Soil Vapor Extraction Wells

subsurface, creating a pressure gradient. SVE systems always withdraw air from the subsurface for discharge to the atmosphere, often after treatment by adsorption on GAC, catalytic or thermal oxidation, or biofiltration. The gas (*i.e.*, air) flow created by the SVE system enhances evaporation of NAPL and soil moisture, volatilization of contaminants dissolved in pore water, desorption of contaminants from the surfaces of soil particles, and biodegradation.

Air Sparging.   Air sparging is the process of injecting air into the saturated subsurface to treat impacted soil and ground water. Air sparging mechanisms partition volatile contaminants from the aqueous and NAPL phases to the vapor phase (stripping) for their subsequent transfer to and removal from the unsaturated zone. Air sparging also transfers oxygen from the injected air to the aqueous phase to enhance aerobic microbial degradation of contaminants (USACE, 1997). It also increases oxygen levels and bioremediation in the unsaturated zone. An important advantage of an air sparging system is that it can effectively strip volatile chemicals from both the saturated and the unsaturated zone. Because it is able to strip VOCs, air sparging is often accompanied with an SVE system to capture and treat the vapors generated before they are discharged to the atmosphere or to a nearby receptor. The SVE system is not needed with air sparging if VOCs are minimal and if potentially affected receptors are far from the air sparging system.

These three gas-based technologies may be implemented alone or in combination at a specific site depending on the chemical properties and concentrations of the contaminants present, site characteristics (*i.e.*, subsurface stratigraphy, depth to ground water, site structures, preferential airflow pathways), age of the impacts, vertical and horizontal extent of impacts, receptors, and required treatment time. The effects of these site considerations on the technologies are further explained in the following sections. Tables 10-1 and 10-2 summarize conditions that influence the feasibility of these gas-based technologies for given site conditions. These tables present generalizations and rules of thumb. Each site is unique, and the designer must carefully consider site conditions before selecting and applying a technology. A number of possible gas-based technology configurations are shown on Figures 10-1 through 10-4.

Bioventing/Biosparging.   Bioventing/biosparging is similar to both SVE and air sparging in that air is caused to flow in the subsurface environment (Leeson and Hinchee, 1996). The essential difference is that treatment of the contaminants takes place *in situ* rather than aboveground, thereby reducing remediation costs (AFCEE, 1996; King *et al.*, 1992). In the case of aerobic mineralization, microorganisms in the unsaturated zone biodegrade contaminants to carbon dioxide, water, and biomass.

All three air-based technologies increase oxygen levels and encourage

Table 10-1. **Rules of Thumb for Applicability of Soil Vapor Extraction**

Soil Vapor Extracor Applicability Factor	Well Suited	Limited Effectiveness
Depth to ground water	>10 feet	<3 feet
Contaminant type	Gasoline, diesel, halogenated solvents, MTBE	Weathered fuel, lubrication oil
Vapor pressure of contaminants	>10 mm Hg	<0.5 mm Hg
Solubility of contaminants	Low	High
Permeability	>10^{-6} cm^2	<10^{-10} cm^2
Unsaturated zone thickness at high water level	> 10 feet	< 3 feet
Soil type	Uniform sand and gravel	Clays, high organic soils
Presence of LNAPL	None or thin layer	Thick layer
Bedrock contamination	Highly fractured bedrock	Unfractured bedrock
Contaminant phase	Residual LNAPL	Dissolved phase

Source: Guide for Conducting Treatability Studies under CERCLA: Soil Vapor Extraction (USEPA, 1990)

Table 10-2. **Rules of Thumb for Applicability of Air Sparging**

Air Sparging Applicability Factor	Well Suited	Limited Effectiveness
Depth to ground water	>10 feet	<3 feet
Contaminant type	Gasoline, diesel, halogenated solvents, MTBE	Weathered fuel, lubrication oil
Volatility of contaminants	High	Low
Solubility of contaminants	Low	High
Hydraulic conductivity	>10^{-4} cm/sec	<10^{-4} cm/sec
Aquifer type	Unconfined	Confined
Soil type	Uniform sand and gravel	Clays, high organic soils
Presence of LNAPL	None or thin layer	Thick layer
Bedrock aquifer contamination	Highly fractured bedrock	Unfractured bedrock
Contaminant phase	Dissolved phase	Free Product

Source: A Technology Assessment of Soil Vapor Extraction and Air Sparging (USEPA, 1992)

aerobic bioremediation. The bioventing/biosparging systems are unique because they are not designed to volatilize gasoline constituents for removal, but to biodegrade them *in situ*. In general, a bioventing/biosparging system has a lower airflow rate then a similar air sparging or SVE system. The bioventing/biosparging systems are designed to supply oxygen to the mi-

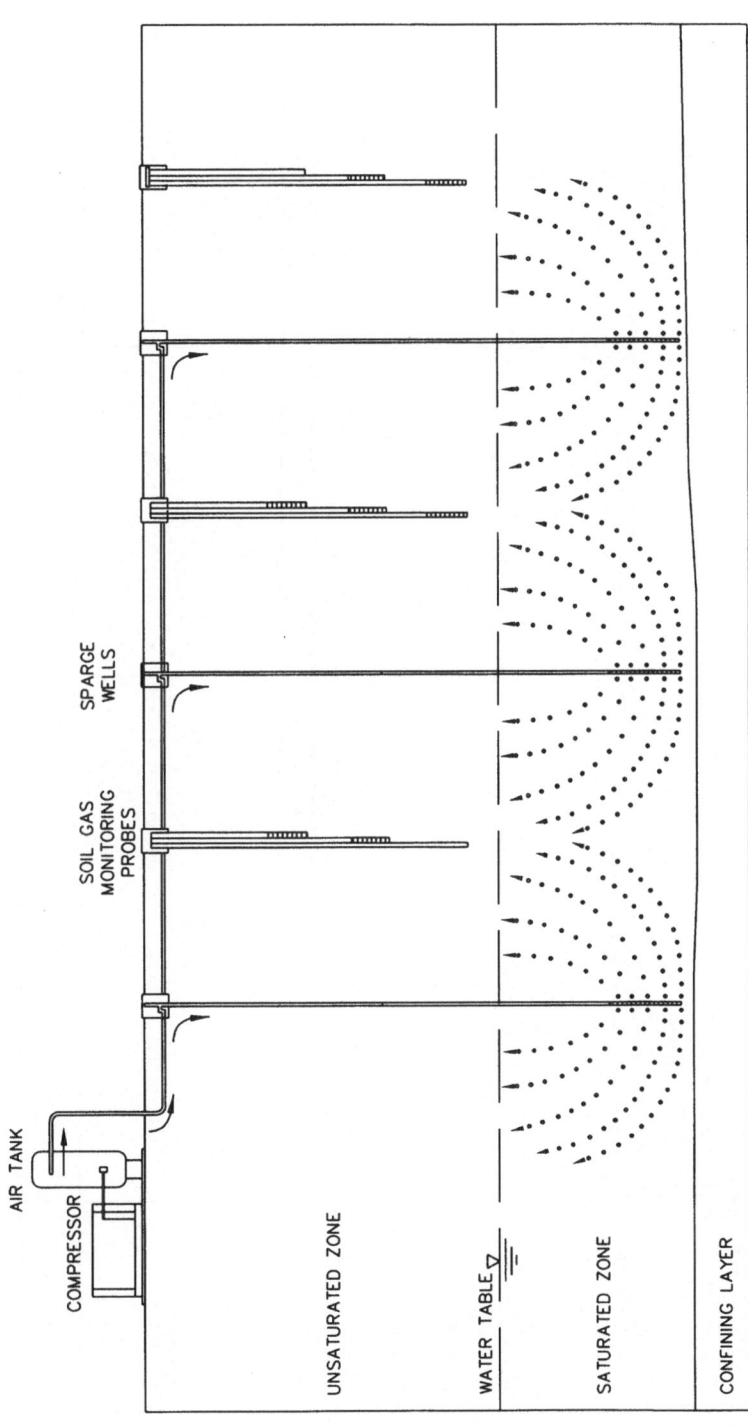

Figure 10-2. Vertical Air Sparging with No Recovery

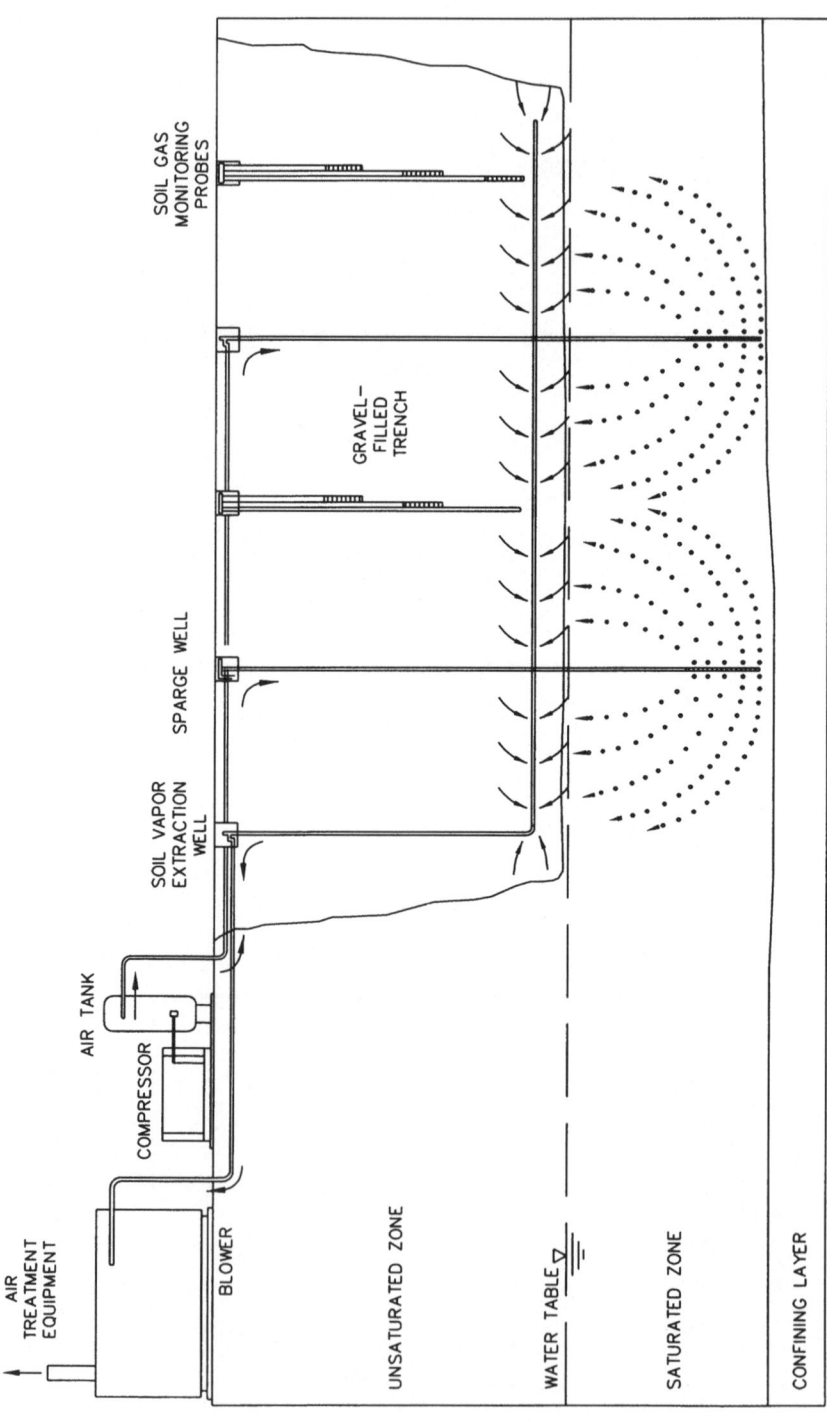

Figure 10-3.   Vertical Air Sparging and Horizontal Trenched Soil Vapor Extraction Wells

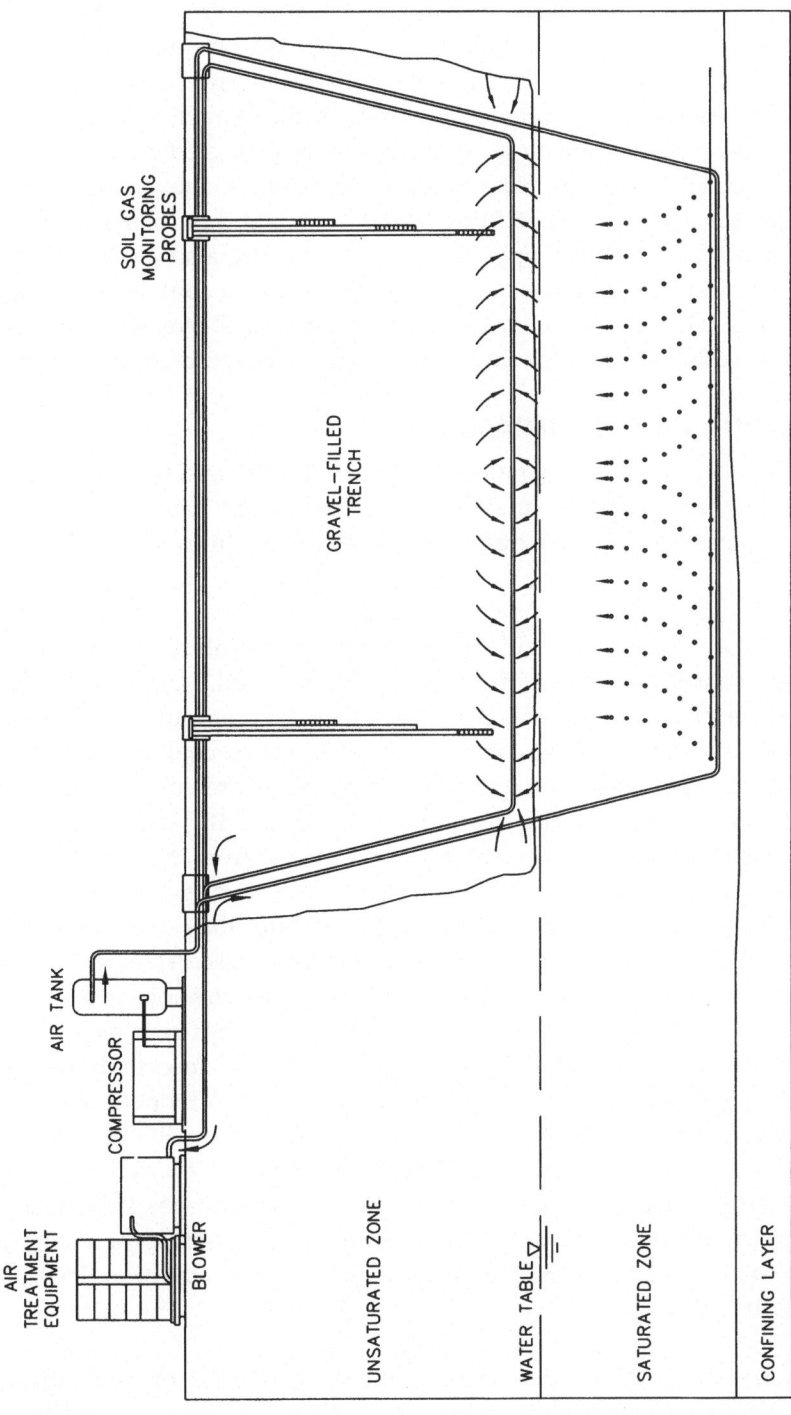

Figure 10-4.   Continual Trenched Horizontal Air Sparging and Soil Vapor Extraction Wells

croorganisms, support biological activity, and minimize volatilization. These systems may be operated in either the saturated zone (biosparging) or unsaturated zone (bioventing). Biosparging systems inject air into saturated zone and the bioventing systems inject air into, or extract air from, the unsaturated zone. The injection mode is often preferred over the extraction mode. The injection mode can focus the air at the center of an impacted area, where it is most needed. This mode also eliminates recovery and management of an off-gas air stream potentially containing VOCs, water, and particulates. The injection mode flow rate and operational efficiency are also less sensitive to fluctuations in water table elevation and soil moisture content than the extraction mode. The extraction mode offers better control over subsurface air-flows and is often selected for sites with sensitive receptors located nearby.

CONTAMINANT CONSIDERATIONS

Primary contaminant considerations include volatility and biodegradability, as these are the two primary mechanisms employed by gas-based technologies to remove gasoline constituents such as MTBE from soil and ground water.

Volatility. A number of specific chemical properties influence the ability of gas-based technologies to volatize VOCs present in soil, NAPL, and ground water. Some of these chemical properties are: the compound's solubility, vapor pressure, Henry's Law constant, and adsorption coefficient (K_{OC}). All gasoline constituents, including MTBE and benzene, are volatile enough for effective SVE and air sparging treatment. Because MTBE has a higher vapor pressure than benzene, it will volatilize more readily from the separate product phase or residual phase. MTBE is more soluble in ground water than benzene. SVE will be most effective if applied soon after the release occurs and before MTBE significantly dissolves into the water phase. At one site, MTBE concentrations were 41% of the total constituent concentrations in the off-gas of the SVE system. Since MTBE does not normally comprise more than 10% to 15% of gasoline, this may indicate that MTBE was preferentially removed due to its higher vapor pressure. After five months of SVE operation, MTBE concentrations in the off-gas dropped to 26%. As the water table elevation decreased, new gasoline containing MTBE was exposed to SVE operations causing the percentage contribution of MTBE in the off-gas to increase to 89%. At another site where only SVE was employed, significant declines in ground water concentrations of MTBE due to SVE were observed (Davidson and Parsons, 1996).

Biodegradability. When determining the effectiveness of bioventing or biosparging, aerobic biodegradability is another key factor in addition to volatility. Aerobic biodegradation kinetics depend upon oxygen concentrations. MTBE is aerobically biodegradable, but sometimes at slower rates than for some of the other gasoline constituents such as benzene. Adding significant amounts of oxygen is usually a major enhancement to biodegrading MTBE. Aerobic biodegradation of MTBE and other gasoline constituents is discussed more fully in Chapter 12. Given appropriate environmental conditions and sufficient time, biodegradation is a significant fate mechanism for MTBE. Stoichiometrically, approximately three grams of oxygen are necessary to biodegrade one gram of gasoline constituents, including MTBE.

SOIL CONSIDERATIONS

Soil properties that control the effectiveness of gas-based technologies include permeability, water saturation, and NAPL saturation. Each of these three soil properties impacts the performance of the three gas-based technologies in different ways.

Soil Permeability.   Soil permeability is the most significant soil property affecting the success of gas-based technologies. Low permeability (less than 10^{-5} cm/s [10^{-2} ft/day]) silt and clay may preclude use of gas-based technologies altogether. This is more of a concern with air sparging and biosparging than it is with SVE and bioventing because it is much easier to apply a high vacuum (660 to 711 mm Hg [26 to 28 in Hg]), than it is to apply a high pressure to the soil without negative consequences. Application of a high pressure to soil can result in pneumatic fracturing of the soil, excessive short-circuiting along the well annulus, or even well failure. For these reasons, air injection pressures should always be below the overburden pressure, which is the sum of the overlying soil weight and hydrostatic head.

Water Saturation.   Water saturation and humidity are important considerations for SVE, bioventing, and biosparging, particularly for MTBE remediation (USEPA, 1995). Both low humidity and high humidity can be of concern.

Air with low humidity is typically encountered in dry climates or during dry periods in more humid climates. Air with low humidity will gain humidity as it flows through deeper, moist subsurface soil, reducing the water content of the deep soil. This condition can have two important impacts on system performance. First, effective bioventing and biosparging systems rely on maintaining sufficient moisture to encourage microbial growth (ideally between 50 and 80% of the soil field capacity). As moisture levels fall below these ideal values, microbial activity is reduced or ceases altogether until soil moisture is restored. In an extreme case where in order to enhance volatilization, hot dry air is introduced into the subsurface from an external heat source, the soil microorganisms could be sterilized near the injection point. This is a consideration when using heat to enhance the system, although biological activity rebounds after high temperature thermal treatments. Fine-grained soils will shrink in volume as moisture is depleted, creating cracks that act as preferential flow pathways for air. This short-circuiting prevents effective volatilization of chemicals to be treated and reduces oxygenation of the entire treatment zone.

High soil moisture content may also reduce the efficiency of an SVE or bioventing system. Unusually high amounts of precipitation, spring snow melts, or other surface events may temporarily increase the moisture content of the soil, thus reducing the air-filled porosity of the soil. This increases the amount of vacuum required to achieve the same air extraction rate than under lower moisture conditions.

As the soil water content approaches 100%, movement of air through the soil is significantly inhibited. A rising water table may temporarily submerge the unsaturated zone. Furthermore, excessive percolation of water through the unsaturated zone may leach chemicals from this zone into the ground water. At higher soil moisture contents, more MTBE, which is highly soluble, will dissolve into the water phase. MTBE partitioning to air from the dissolved phase is relative. This effect is most pronounced near the water table at the capillary fringe. For this reason and to expand the vertical thickness of the treatment zone, it may be beneficial to dewater the area to be vented to reduce water saturation in the impacted zone.

Excessive water also causes problems with the equipment used in extractive systems. Water may condense in the cooler system components. This water is collected into the SVE water separator. If the separator does not have a continuous water transfer system, the system may shut down, or water may overflow into other system piping and components. As water production increases, it may reach a large enough flow to require a continuous water treatment system.

The ground water response to the application of soil vacuum is a rise in the water level, or upwelling. The degree of upwelling increases with increasing applied vacuum. Upwelling may become a significant problem if the length of exposed well screen above the water is short (less than a couple of meters), or if the soil to be treated is near the water table (which is often the case with gasoline). Upwelling is more pronounced in fine-textured soils.

NAPL Saturation.   NAPL saturation is an important consideration for gas-based technologies. Historical experience shows that the gas-based technologies are much more effective at sites where little free or mobile NAPL exists. Normally mobile NAPL is removed before gas-based technologies are deployed.

GEOLOGIC CONSIDERATIONS

It is rare in nature that the soil properties (and thus airflow distribution) are uniform throughout an impacted soil volume. In soils, airflow resulting from gas-based technologies occurs in discrete pore-scale or larger-scale channels, rather than as uniform flow. Both soil stratigraphy and heterogeneity have a profound influence on air channel location and density.

Fine-Grained Lenses. Continuous or discontinuous lenses of finer-grained soils will reduce the effectiveness of all gas-based technologies. In general, the greater is the degree of layering, the poorer the system effectiveness. As air passes through soil and ground water, contaminants that volatilize are carried along with the bulk movement of the air through more permeable regions by advection. Lenses of finer-grained soils may retain MTBE and other VOCs to create diffusion-limited conditions as remediation progresses. Due to these diffusion-limited conditions, recovery of contaminants may plateau. This may prevent the system from reaching the ground water or soil remediation criteria as coarser zones are remediated, but finer-grained zones continue to hold or release mass.

Diversion of Airflow. As supported by a number of sand box visualization studies, lower permeability lenses act as barriers to vertical airflow. This effect is most important for air sparging, as vertical airflow is the desired pathway for air to escape from the saturated to the unsaturated zone. Diversion of airflow will isolate some impacted zones from airflow and may also result in the buildup of air pockets and pressure (to the point where artesian ground water flow conditions may exist in shallow monitoring wells). Nested air extraction/injection wells can help to minimize diversion of airflow, but this can become infeasible for air sparging in multi-layered geologic strata. A more effective method for alleviating the layering problem for air sparging systems is to install air sparging wells in a continuous gravel-filled trench that cuts across the layered system (Figure 10-4).

Diversion of airflow is somewhat less of a problem for SVE and bioventing systems than for injection systems because they are typically screened over a wider interval and they rely more on horizontal airflow. Screens in these systems may extend from the water table to within a meter of the ground surface. Air sparging wells are typically screened 3 to 4.6 meters (10 to 15 feet) or more below the water table and have a small 0.3- to 0.9-meter (1- to 3-foot) screened interval.

Heterogeneous Soils. Heterogeneous soils create preferential airflow paths and a potential for diffusion-limited conditions. All or most of the airflow may pass through cracks and macropores, greatly reducing oxygen transfer and volatilization from finer textured soil. The existence of these conditions is most effectively evaluated during site characterization and subsequent field pilot testing.

AIRFLOW CONSIDERATIONS

Airflow must be designed to create the desired subsurface conditions. The best airflow rate is a balance between two competing considerations. Low airflow rates help reduce off-gas volumes and maximize contaminant concentrations in the off-gas. High airflow rates maximize the zone of influence around air extraction and injection points, thereby reducing the number of required wells and trenches. The designer must optimize these competing considerations.

Maximizing Biodegradation.   For bioventing and biosparging, the design target for biodegradation is to maintain a specified minimum oxygen concentration in the unsaturated and saturated zones. Unsaturated zone oxygen concentrations are normally maintained at levels above 5%. (Oxygen content of ambient air is 20.9%.) Saturated-zone dissolved oxygen levels are usually maintained at levels above 1 to 2 mg/l. Dissolved oxygen levels in unimpacted natural ground water will generally be around 3 to 6 mg/l. Dissolved oxygen is not typically present in highly impacted ground water because it has been depleted by the native bacteria. With an excess of air, saturated dissolved oxygen levels may be as high as 8 to 10 mg/l. Dissolved oxygen levels generally will not exceed 13 mg/l for air-based injection systems (USACE, 1997). Much higher saturated dissolved oxygen levels can be achieved if pure oxygen is injected instead of air.

Average airflow rates are adjusted to economically maintain minimum oxygen concentrations. One way to accomplish this is to use a continuously low flow rate. For bioventing, a rule of thumb is to set the flow rate for 0.25 to 1.0 pore volume exchange per day. The zone of influence is not as great at low bioventing/biosparging flow rates compared to the higher air sparging/soil vapor extraction (AS/SVE) flow rates. To compensate for this, wells may need to be spaced more closely. This closer spacing has advantages. It may more effectively distribute oxygen throughout the area to be treated, thereby reducing the total airflow rate, volatilization, and operational cost. On the other hand, this must be balanced against the higher cost of additional wells. Another technique that is often used to maintain the desired minimum oxygen concentrations is to pulse the airflow on and off. Bioventing cycles may be on the order of a day or two. Biosparging cycles are usually on the order of 15 minutes to a few hours.

Maximizing Volatilization.   For SVE and air sparging, maximizing removal by volatilization requires an approach that concentrates on the mass of VOCs stripped and removed for aboveground handling. The mass of VOCs volatized is described by the equation $M = Q*C$ where M equals volatized mass, Q equals flow rate, and C equals concentration of MTBE and/or other

VOCs in the off-gas. To achieve the optimum M for a system, the designer must select the optimum Q. This task is difficult because Q and C are inversely related. In general, the higher the Q, the lower the C. There is an upper limit to the M, since it is limited by the release of VOCs from the residual NAPL droplets or from water containing dissolved VOCs. The relationship between Q and C is often one of the important findings of a long-term field pilot test. The relationship is then further refined during the initial stages of full-scale system operation. The practical impact of the complex relationship between Q and C is the need to design the system with considerable operational flexibility.

For air sparging, a similar relationship in terms of air to water ratios applies. Data from packed tower strippers (PTS) suggest that the air to water ratio required to achieve the same concentration reduction for MTBE as compared to benzene may be 5 to 10 times higher. In both cases, removals of greater than 95% are possible. Air injection rates that optimize benzene volatilization in air sparging systems generally range from 0.1 to 0.7 standard cubic meters per minute (scmm) (5 to 25 standard cubic feet per minute [scfm]) per well for injection depths of 3 to 4.6 meters (10 to 15 feet) below the water table. If MTBE is present, air injection rates should usually be selected at or beyond the upper end of the range for benzene.

AIRFLOW AND PRESSURE RELATIONSHIPS

Airflow and pressure are directly related in gas-based remediation systems. Typically, as air pressure (for injection systems) or air vacuum (for extraction systems) increase, airflow will increase. A discussion of detailed airflow and pressure relationships, the basis for airflow modeling, is beyond the scope of this discussion (see USEPA, 2001). Before a full-scale system is designed, airflow and pressure relationships at a specific site must be understood to ensure that the individual pieces of equipment are appropriately selected and sized. Table 10-3 summarizes some of the advantages and disadvantages of some available gas-based technology equipment.

ZONE OF INFLUENCE AND WELL SPACING

The zone of influence is an important design parameter for gas-based technologies because it directly determines the well layout and spacing. The zone of influence is the horizontal distance from the well or trench at which the desired airflow is achieved. A working definition of the zone of influence for air sparging is the volume of the saturated zone over which air-filled channels are relatively closely spaced (e.g., total void space contains greater than 10% air). The effective radius of the zone of influence of a shallow air sparging system depends on the soil type, but is likely to average somewhere around 4.6 meters (15 feet) (USACE, 1997).

Table 10-3. Equipment for Gas-Based Remediation Technologies

Type of Equipment	Advantages	Disadvantages
AIR EXTRACTION		
Regenerative Blower	Low cost, low noise level	Low vacuum
Positive Displacement Pump	Moderate cost, moderate vacuum	High noise level
Liquid Ring Pump	Moderate cost, high vacuum	Moderate noise level, water needed
AIR INJECTION		
Regenerative Blower	Low cost, low noise level	Low pressure
Positive Displacement Pump	Moderate cost, moderate pressure	High noise level
Liquid Ring Pump	Moderate cost, moderate pressure	Moderate noise level, water needed
Rotary Screw Compressor	High pressure	High cost, moderate noise level
Oilless Compressor	High pressure, oilless	Very high cost
SVE/BIOVENTING		
Vertical	Low cost for deep wells	Multiple well headers, high vacuum potentially ineffective for shallow wells
Horizontal	Low cost for shallow wells	Single well header (less operational flexibility)
	Lower vacuum required Treatment under buildings	High cost for deep wells
AIR SPARGING/BIOSPARGING		
Vertical (Hollow Stem Auger)	Conventional method	Difficult to install in deep sand aquifers; generates waste soil
Vertical (Well Point)	Rapid and inexpensive	Cannot install in cobbles/boulders
Vertical (Cone Penetrometer/Geoprobe®)	Little or no drill cuttings	Small diameter wells (≤1 inch)
Vertical (Biopolymer Trench)	Creates a uniform medium	Expensive, guar gum residual, waste soil
Vertical (Caissons)	Creates a uniform medium	Expensive, generates waste soil
Horizontal Drilled	Fewer wells, uniform air dispersion	Expensive, special equipment required
Horizontal in Continuous Trench	Uniform medium and air dispersion	Specialized equipment required. Generates large quantities of waste soil
PIPING		
PVC, Schedule 40	Low cost Locally available	Must be buried/shielded for pressure applications, low strength
PVC, Schedule 80	Moderate strength	Moderate cost, must be buried/shielded for pressure applications
HPDE, corrugated	Low cost, flexible	Only large diameters available, possible leaks through joints
Polyethylene tubing	Very low cost, flexible Long lengths without joints	Only available in small diameters, high head loss for high flow/long length

Table 10-3. **Continued**

Type of Equipment	Advantages	Disadvantages
PIPING (continued)		
Galvanized metal	Very high strength, easy to obtain	Difficult to cut, high cost
ChemAir™	Moderate strength Doesn't need to be buried/shielded	High cost, not locally available
VAPOR TREATMENT		
Thermal Oxidizer	Good for high VOC concentrations (>2,000 ppmv)	High fuel cost for low VOC concentrations
Catalytic Oxidizer	Good for medium VOC concentrations 100 to 2,000 ppmv	High fuel cost for low VOC concentrations, possible poisoning of catalyst with metals/H_2S
Biofilter	Good for low to moderate VOC concentrations (100–5,000 mg/m^3)	Initial lag time to reach stable microbial population
Granular Activated Carbon	Very reliable. Good for very low influent concentrations <35 to 100 ppmv).	Very expensive for high VOC concentrations
Condensation/Recycling	Good for high VOC concentrations. Beneficial reuse of hydrocarbons.	Expensive, need buyer for recovered hydrocarbons
None	No cost. No complex recovery/treatment equipment.	Air treatment may be required by regulations

In general, the zone of influence will increase in the system as flow rate and pressure/vacuum increase. However, a point will be reached where increases in flow rate, pressure, and vacuum no longer significantly increase the zone of influence. It is the flow rate and pressure/vacuum combination near the zone of influence inflection point that is normally the initial target for operating gas-based technology systems. However, this flow rate, normally established during pilot testing, must also be balanced with the desired flow rate to maximize biodegradation or volatilization and minimize off-gas treatment costs (if applicable). In air sparging systems, the potential for off-gas migration to undesired locations must also be considered.

For extraction systems, a good approach to well layout is to design an air velocity that exceeds some minimum rate everywhere within the contaminated zone. This translates to a rate of air exchange (pore volume per time). It may possibly require a closer well spacing than was traditionally used, but the closer spacing will likely lead to cleanup to target levels in a shorter time.

Areas of stagnation often develop at sites with multiple wells. This stagnation can be avoided by varying the flow rate at nearby wells to move the stagnation point with time or by the use of air injection wells. Computer models can be used to estimate the effect of passive or active air injection wells.

Table 10-4 shows some common ranges for well spacing for gas-based technologies. It is common practice to space sparge wells densely (3 to 5 meters [10 to 16 feet]) for air sparging systems, particularly for MTBE remediation (Fields *et al.*, 2001).

MODELING AND PILOT TESTING

Modeling and pilot testing are often conducted to evaluate the feasibility and potential effectiveness of gas-based technology systems. A detailed discussion or summary of available mathematical models or pilot test procedures is beyond the scope of this discussion but can be found in USACE (1995, 1997). The modeling and pilot testing procedures are the same for MTBE as they are for the other gasoline constituents such as benzene.

Summary of Extraction System Effectiveness.   Three key chemical properties influence the relative effectiveness of SVE/BV for treatment of BTEX in gasoline compared to MTBE: vapor pressure, Henry's Law constant, and biodegradability. MTBE has a higher vapor pressure than BTEX compounds. SVE off-gas concentrations and removal rates will be higher for MTBE than BTEX (Davidson and Parsons, 1996). MTBE has a lower Henry's Law constant; it will tend to partition into the water phase more readily than BTEX. This is

Table 10-4.   Typical Design Parameters for Gas-Based Systems (One Vertical Well in Permeable Soil)

Parmeter	SVE Typical Range	Bioventing Typical Range	Air Sparging/Biosparging Typical Range
Well Diameter	2 to 4 inches	2 to 4 inches	0.5 to 2 inches
Well Screen Length	5 to 20 feet	5 to 20 feet	1 to 3 feet
Depth of Top of Well Screen	>5 feet BELOW ground surface but 5 feet ABOVE water table	>5 feet BELOW ground surface but 5 feet ABOVE water table	5 to 20 feet BELOW water table
Depth of Bottom of Well Screen	0 to 10 feet BELOW water table	0 to 10 feet BELOW water table	ABOVE confining layer
Air Flow Rate Per Well	50 to 150 scfm	10 to 50 scfm	1 to 20 scfm
Air Pressure/Vacuum	1 to 15 in Hg vacuum	1 to 10 in Hg vacuum (extraction)	
5 to 15 psi pressure (injection)	5 to 15 psi pressure		
Zone of Influence Radius	30 to 100 feet[1]	15 to 60 feet[1]	5 to 15 feet

[1] Water table greater than 10 feet below ground surface.

of secondary importance compared to MTBE's higher volatility, particularly if the soil moisture content is moderate to low. MTBE will generally biodegrade more slowly than BTEX. Longer bioremediation times should be expected for MTBE compared to BTEX if concentrations in the soil are similar.

Summary of Injection System Effectiveness.   Billings and Griswold (2001) evaluated the effectiveness of air sparging at six sites in New Mexico. Dissolved concentrations of MTBE ranged from 4.1 to 44.4 mg/l. Treatment times ranged from 3 to 24 months (generally 12 to 24 months). Under these conditions, dissolved-phase MTBE concentrations were reduced by between 38 and 100%. Fields *et al.* (2001) reported similar results: 90% mass removal of BTEX and over 75% removal of MTBE after 18 months of air sparging operation. The removal of MTBE required 5 to 10 times more airflow than would have been used for BTEX.

DESIGN CONSIDERATIONS

The first design considerations for a gas-based remediation system are the selection of the overall system, off-gas treatment technology, and enhancements. These considerations are discussed below.

Technology Selection.   The first decision is whether to apply SVE, air sparging, bioventing, biosparging, or a combination of these technologies at a gasoline release site. This decision will largely depend on the soil characteristics and on the contaminant concentrations in the saturated and unsaturated zones. The key issue for a site is to determine whether the greatest mass of the contaminant of concern is in the saturated zone or the unsaturated zone. If it is in the saturated zone, air sparging or biosparging are likely candidates. If it is in the unsaturated zone, SVE or bioventing are likely candidates. The choice between the physical removal technologies (air sparging and SVE) and the biological technologies (bioventing and biosparging) often comes down to the concentrations of the contaminants of concern. The physical systems usually remove concentrated contamination faster while the biological systems are often more effective at lower contaminant levels.

Off-Gas Treatment.   To accommodate reductions in soil vapor concentrations with time, four modes of off-gas treatment are typically considered (Fields *et al.*, 2001): (1) thermal oxidation for inlet VOC concentrations greater than about 2,000 ppm by volume (ppmv); (2) catalytic oxidation for inlet VOC concentrations between about 100 and 2,000 ppmv; (3) GAC for inlet VOC concentrations between 35 and 100 ppmv; and (4) direct discharge for inlet VOC concentrations less than 35 ppmv (if permitted based on applicable regulations).

It is usually not practical to apply all of these technologies at a site. For instance, catalytic oxidation may be the first treatment technology applied (accepting the fact that initial inlet VOC concentrations may need to be diluted with ambient air). Also, depending on local air emissions control requirements, direct discharge may be allowed in lieu of using GAC once the lower catalytic oxidation threshold is reached. This may be particularly appropriate at MTBE-impacted sites because MTBE breaks through GAC relatively quickly when compared to other gasoline constituents. An economical GAC design for MTBE often requires the use of several vessels in series so that the capacity of the upstream vessels can be exhausted, even after some MTBE breakthrough, as one or more downstream vessels continue to remove any MTBE not recovered earlier.

Biofiltration is a newer off-gas treatment technology that is a good selection for MTBE. Biofilters can be an economical method of treating low to moderate concentrations of VOCs such as BTEX and MTBE (up to a few thousand ppm). Biofiltration is a well-established air pollution control technology in several European countries and is becoming popular in the U.S. The objective of the biofilter media (typically granular particles or soil, but other options are available) is to temporarily adsorb the VOCs passing through them and to provide a suitable environment for biological growth. Efficient biofilters are relatively maintenance free and self-sustaining, with some periodic attention needed to maintain proper nutrients, moisture, and population density in the media.

Air sparging without SVE is an example of an *in situ* biofiltration system. As organic vapors are emitted from the water table surface, they may pass through a zone of unsaturated soil before exiting at the ground surface. This unsaturated zone soil, containing microorganisms and other naturally occurring organic carbon, adsorbs organic vapors and provides a surface and environment for microbial growth. At many air sparging sites, soil vapor concentrations have been shown to decrease as they approach the ground surface. Biofiltration of MTBE may require larger systems or longer residence times relative to BTEX.

Enhancements.   There are several techniques that may be used to enhance a gas-based remediation system: pulsed operation, injecting gases other than air, and heat. These enhancements are described below.

Pulsed Injection.   Turning the air sparging system alternately on and off (pulsing) is a method of increasing air/water/soil contact and ground water mixing. Due to the induced mixing, pulsed injection is most effective for mobile, dissolved-phase contaminants (such as MTBE). Pulsed injection is designed to take advantage of a recurring expansion phase, as airflow is reinitiated after an off period, during which the zone of influence is larger than during the steady state air sparging. Pulsed injection also allows the designer to specify smaller equipment, as only part of the site is active at any one time.

Injecting Gases Other than Air.   Injecting gases other than air (*e.g.*, pure oxygen, ozone, gaseous nutrients such as triethylphosphate or cometabolic substrates such as butane or propane) may enhance the speed at which bioremediation proceeds or alter the conditions under which it occurs.

Adding Heat (Thermal).   Thermal enhancement is potentially so powerful that it is often considered a completely different technology. The addition of heat to the subsurface environment helps SVE remove the gasoline components by changing the physical properties of the chemicals. Solubility and vapor pressure increase with increasing temperature, while adsorption decreases with increasing temperature. These effects enhance the extraction of contaminants from the subsurface when the temperature is raised.

Thermal mobilization technologies typically use less energy than thermal destruction technologies (the latter of which are overkill for VOCs). There are several types of *in situ* thermal mobilization technologies including steam injection, 6-phase/3-phase electrical heating, and thermal conduction electrical heating.

Steam injection involves heating water to create steam and then injecting the steam into the subsurface. This remediation process has dual-purpose benefits. By injecting steam, there is a stripping component, such as one would get with air sparging, and a heating component. An SVE or MPE system is used with this system to capture vapors and liquids being mobilized from the contaminated media.

The energy input for steam injection is not as high as for thermal destruction technology, but it is higher than for 3- and 6-phase heating. This is because there is some energy used to heat the water until steam is generated and then also in injecting the steam to heat the subsurface. The 3- and 6-phase heating process uses the subsurface soil as the source of the heating by passing electrical current directly into the ground. Electrical resistance heats the soils as the current goes through it. Energy efficiency is maximized because all of the electrical energy is being used for heating the contaminated media.

The 6-phase electrical heating process involves splitting standard 3-phase electrical power into six phases and supplying this energy to six electrode wells that are installed in a hexagonal pattern in the target area and equidistant from each other and from a central neutral electrode well. A soil vapor collection well is also installed adjacent to the neutral well. Current is supplied to one of the six "hot" electrodes while all other electrodes are switched to neutral. The electrical current and wells are spaced at 60 degrees and only one well is "hot" at any given time, such that the electricity passes from the "hot" electrode to the three adjacent neutral electrodes (a central electrode and one on each adjacent side of the hexagon). Electrical resistance in the subsurface causes the target area to heat up and eventually generate steam once the appropriate temperature is reached. This steam helps strip the now vaporized contaminants out of pore spaces.

Three-phase electrical heating involves a series of electrode wells connected to a power source controller. The power source controller uses the electrodes in pairs of hot and neutral, and passes current from the hot to the neutral electrode. The electrodes are placed in the ground based on lithology, contaminant concentrations, and immovable infrastructure, as well as other considerations. The lateral spacing of the electrodes is site-specific, but is typically 5 to 9 meters (15 to 30 feet). SVE or MPE wells are installed in between the electrode wells to capture organic vapors as the target area is being heated. As with 6-phase electrical heating, the resistance of the electricity passing through the subsurface causes the soil to heat up and mobilizes the contaminants.

The thermal conductivity of soil is fairly low and only differs by approximately a factor of four between different soil types. Subsurface temperature is not easily increased, but once heated, holds the heat for a relatively long time.

Thermal mobilization technologies benefit from the heat conducting properties of soils in that as the ground is heated, the heat moves away from the source in a relatively uniform manner.

CONCLUSIONS

Gas-based technologies were very frequently the technology of choice for the management of gasoline releases before the introduction of MTBE. The addition of MTBE as another constituent in gasoline changes some of the details of equipment design, but does not change the usefulness of these technologies for the management of gasoline releases containing MTBE.

REFERENCES

AFCEE (Air Force Center for Environmental Excellence). 1996. *A General Evaluation of Bioventing for Removal Actions at Air Force/Department of Defense Installations Nationwide General Engineering Evaluation/Cost Analysis EE/CA.* AFCEE, Technology Transfer Division, Brooks Air Force Base, San Antonio, Texas.

Billings, B.G. and Griswold, J.E. 2001. Field data regarding air-based remediation of MTBE. In: *In Situ MTBE Biodegradation, Bioremediation of MTBE, Alcohols, and Ethers, The Sixth International In Situ and On-Site Bioremediation Symposium,* San Diego, California. June 4–7, 2001. Volume 1, pp. 115–119. (Magar, V.S., Gibbs, J.T., O'Reilly, K.T., Hyman, M.R., and Leeson, A., Eds.). Columbus, Ohio, Battelle Press.

Davidson, J.M. and Parsons, R. 1996. Remediating MTBE with current and emerging technologies. In: *Proceedings of the Petroleum Hydrocarbons and Organic Chemicals in Groundwater — Prevention, Detection, and Remediation Conference,* Houston, Texas. November 13–15, 1996. pp. 15–29. Dublin, Ohio, National Ground Water Association.

Fields, Keith A., Zwick, T.C., Leeson, A., Wickramanayake, G.B., Doughty, H., and Sahagun, T. 2001. Air sparging life cycle design for an MTBE and BTEX site. In: *Enhanced Aerobic Restoration, In Situ Aeration Aerobic Remediation, the Sixth International In Situ and On-Site Bioremediation Symposium,* San Diego, California. June 4–7, 2001. Volume 6, pp. 1–8. (Leeson, A., Johnson, P.C., Hinchee, R.E., Semprini, L., and Magar, V.S., Eds.). Columbus, Ohio, Battelle Press.

King, B.R., Long, G.M., and Sheldon, J.K. 1992. *Practical Environmental Bioremediation.* Boca Raton, Florida, Lewis Publishers.

Leeson, A. and Hinchee, R. *Principles and Practices of Bioventing.* Battelle, September 1996.

USACE (United States Army Corps of Engineers). 1995. *USACE Engineering Manual: SVE and Bioventing.* EM 1110-1-4001. 1995.

USACE (United States Army Corps of Engineers). 1997. *USACE Engineering Manual: In situ Air Sparging.* EM 1110-1-4005. 1997.

USEPA (United Stated Environmental Protection Agency). 1990. *Guide for Conducting Treatability Studies under CERCLA: Soil Vapor Extraction.* EPA/540/2-91/019B.

USEPA (United Stated Environmental Protection Agency). 1992. *A Technology Assessment of Soil Vapor Extraction and Air Sparging.* EPA/660/R-92/173.

USEPA (United Stated Environmental Protection Agency). 1995. *Manual — Bioventing Principles and Practice.* Volume I. Bioventing Principles. EPA/540/R-95/534a.

USEPA (United Stated Environmental Protection Agency). 2001. *Development of Recommendations and Methods to Support Assessment of Soil Venting Performance and Closure.* EPA 600-R-01-070.

CHAPTER 11

In Situ Chemical Oxidation

Kara L. Kelley, Michael C. Marley, and Kenneth L. Sperry, P.E.,
Xpert Design & Diagnostics, LLC

INTRODUCTION

ISCO can be a cost-effective method for the remediation of MTBE *in situ*. Compared to aerobic bioremediation, ISCO is less constrained by constituent concentrations. Aerobic bioremediation is typically effective only on lower MTBE concentrations or where free product is not present (Leetham, 2001). Several ISCO processes have been tested successfully under laboratory conditions, and a number have proven successful when tested in the field for the degradation of MTBE. This chapter reviews the state of the art with respect to MTBE oxidation for several common oxidants and advanced oxidation processes (AOPs), which are oxidants or oxidant combinations characterized by the production of the hydroxyl radical (OH•). The production of OH• drastically increases the oxidative capabilities of the ISCO system (Liang *et al.*, 2001; Mitani *et al.*, 2001; Acero *et al.*, 2001). Five frequently used oxidants are reviewed in this chapter and include: hydrogen peroxide (H_2O_2), ozone ($O_{3(g)}$), permanganate (MnO_4^-), persulfate ($S_2O_8^{2-}$), and ultrasound.

When choosing an oxidant for use in a specific remediation project, tradeoffs exist between oxidant strength and stability in the subsurface. Typically, the more aggressive oxidants are less stable, have shorter half-lives, and prove more difficult to transport in the subsurface (Clayton *et al.*, 2000). The standard oxidation-reduction potential ($E°$), a measure of oxidant strength, is the electromotive force in units of volts (V) of the oxidation/reduction reaction. Larger, positive $E°$ values indicate a greater potential for the reaction to proceed as written. According to $E°$ values, the oxidants $O_{3(g)}$, $S_2O_8^{2-}$, H_2O_2, and MnO_4^- are listed in order of decreasing strength (Huang *et al.*, unpublished).

$$O_{3(g)} + 2H^+ + 2e^- \rightarrow O_{2(g)} + H_2O \qquad E° = 2.07 \text{ V} \qquad (11\text{-}1a)$$
$$S_2O_8^{2-} + 2e^- \rightarrow 2SO_4^{2-} \qquad E° = 2.01 \text{ V} \qquad (11\text{-}1b)$$

$$H_2O_2 + 2H^+ + 2e^- \rightarrow 2H_2O \qquad\qquad E^\circ = 1.78\ V \qquad (11\text{-}1c)$$
$$MnO_4^- + 4H^+ + 3e^- \rightarrow MnO_{2(s)} + 2H_2O \qquad E^\circ = 1.70\ V \qquad (11\text{-}1d)$$

When implementing an ISCO process, several water quality parameters may influence the initiation and the effectiveness of the reaction. These parameters include pH, alkalinity, natural organic matter, and the concentration of interfering compounds (Acero *et al.*, 2001). Soil oxidant demand (SOD) from natural soil organics, inorganics, and co-contaminants can significantly increase the amount of oxidant required to treat the target constituent. An example of a matrix requiring a substantial oxidant demand would be organic clay that contains NAPL and has a high organic carbon content. This matrix may require hundreds of grams of oxidant per kilogram of soil (Clayton *et al.*, 2000), whereas other soils or subsurface conditions typically require 3 to 20 grams per kilogram (g/kg) of SOD.

Oxidation end products are an important consideration in the selection of an oxidant, as not all oxidants have proven successful in complete mineralization of MTBE. TBF and TBA are the major intermediate products in the oxidative reactions of MTBE, although it has been postulated that because TBF and TBA may be partially oxidized, the intermediates may be more susceptible to biological degradation and natural attenuation (Leetham, 2001; Mitani *et al.*, 2001). It is critical to design the oxidant injection system to maximize contact between the oxidant and contaminant to facilitate complete oxidation. Benign end products of complete mineralization of MTBE include carbon dioxide and water (Wagler and Malley, 1994).

Other factors, including regulatory restrictions, need to be considered when choosing an oxidant for a specific application. ISCO has gained regulatory acceptance, thereby facilitating permitting issues associated with ISCO implementation (Oberle and Schroder, 2000). This chapter will highlight the chemistry of the oxidant/MTBE reaction, successes and limitations observed under laboratory and field conditions, practical design advice when employing the oxidants, and general cost information where available.

A matrix of the oxidants discussed in this chapter and their considerations for use are presented in Table 11-1. The matrix includes strengths of each oxidant such as the formation of OH• and successful field application (as indicated by the completion of a field test). Table 11-1 also includes weaknesses of oxidant application (*e.g.*, dependency of the reaction upon system pH, which may require system buffering).

HYDROGEN PEROXIDE

Description of Process. Hydrogen peroxide is injected into the subsurface to promote either constituent volatilization or radical oxidation. To promote volatilization, concentrated H_2O_2 solutions (35 to 50% by weight) are injected into the subsurface. The resulting rapid decomposition of the H_2O_2 solution

Table 11-1. **Matrix of Oxidants and Their Considerations for Use**

Oxidant	OH• Formation	pH Dependant	Cost Info	Field Test
H_2O_2	✓	✓	✓	✓
O_3	✓	✓*		✓
MnO_4^-			✓	✓
$S_2O_8^{2-}$	✓	✓	✓	
Ultrasound	✓	✓*	✓	✓

* If used with H_2O_2 or Fenton's reagent

produces heat, O_2, CO_2, and possibly salts from non-hydrocarbon constituents.

To promote radical oxidation, less concentrated solutions of H_2O_2 (*i.e.*, 12% by weight) are injected with a stabilizer to slow the decomposition of H_2O_2, and an acidified iron salt or chelated form of ferrous iron (a Fenton's or Fenton's-type reaction) is injected to promote OH• formation (Kakarla, 2002). The OH• cleaves the carbon-carbon bonds, oxidizing organic compounds to water and carbon dioxide. On a stoichiometric basis, 2.62 kilograms (5.78 pounds) of H_2O_2 are required for every 0.45 kilogram (1 pound) of MTBE oxidized. Equation 11-2 below is the stoichiometric equation for the reaction of MTBE and H_2O_2 in the presence of ferrous iron:

$$C_5H_{12}O + 15H_2O_2Fe \rightarrow 5CO_2 + 21H_2O \qquad (11\text{-}2)$$

Fenton's reaction is capable of oxidizing a wide range of organic compounds under optimal pH conditions of 3 to 5 pH units. Although aquifers can be pretreated with an acid solution to lower the pH, acidic conditions can be hard to maintain depending on the buffering capacity of the soil. This pretreatment can be costly and ineffective (Leetham, 2001; Nyer and Vance, 1999; Oberle and Schroder, 2000; Yeh and Novak, 1995). Hydrogen peroxide can be injected with a chelated catalyst promoting radical formation at a higher pH (*i.e.*, 6 pH units), avoiding the need for pretreatment (Kakarla, 2002).

During a Fenton's or a Fenton's-type reaction, a combined chemical-biological reaction sequence can occur where the biological degradation of intermediates (*e.g.*, TBA, acetone) generated from the chemical oxidation of MTBE depletes the system of oxygen, creating an environment suitable for iron reduction. The reduced iron further catalyzes the reaction of available H_2O_2 producing additional OH•; thus, an aerobic-anaerobic cycle can develop where MTBE is degraded through chemical oxidation followed by microbial degradation of the intermediate products. It remains questionable as to whether this cycle could be produced *in situ* (Yeh and Novak, 1995). Deeb *et al.* (2000), in an examination of past and current studies, found that intermediate products TBF and TBA are readily biodegraded once the majority of the MTBE has been removed.

Proven Effectiveness in Field or Laboratory. Kealy *et al.* (2001) presented an *in situ* remediation strategy for MTBE using OxyVac® technology. Prior to full-scale injection, a batch laboratory study was conducted using water and soil from an MTBE-blended gasoline spill site in Massachusetts. The highest initial soil and ground water MTBE concentrations were 992 micrograms per kilogram ($\mu g/kg$) and 3,400 $\mu g/l$ respectively. The plume dimensions were 18 meters by 18 meters (60 feet by 60 feet) by 2 meters (7 feet) thick. The site stratigraphy consisted of a silty sand with seasonal variability in depth to ground water from at-grade to 2.4 meters (8 feet) below grade. Soil and ground water were collected from both unsaturated and saturated portions of the aquifer.

During the batch study, MTBE concentrations were reduced by 54% (unsaturated soil) and 67% (saturated soil) after three applications of 5% H_2O_2 solution over a one-hour reaction time. MTBE concentrations were reduced by 80% (unsaturated soil) and 46% (saturated soil) after similar treatment with a 10% H_2O_2 solution. No information was given as to the mechanism of the oxidation, but the author noted that the addition of ferrous sulfate to the batch reactions did little to increase MTBE concentration reductions (Kealy *et al.*, 2001).

Using the information from the batch study, the full-scale design concentration of H_2O_2 solution was determined. The treatment solution was injected using an OxyVac® system (see system description below) since contamination included other VOCs as well as MTBE and BTEX. Twenty 1.3-centimeter (0.5-inch) diameter steel injection points were installed to a depth of 2.4 meters (8 feet) below grade and screened from 1.2 to 2.4 meters (4 to 8 feet) below grade. SVE wells were installed to collect VOC laden off-gas. Over the course of a year, eight (300- to 1,900-liter [80- to 500-gallon]) batches of a 10% H_2O_2 and 5% $FeSO_4$ solution were injected into a contaminant plume area. The total volume of Fenton's reagent injected was 6,000 liters (1,600 gallons). Although it was not discussed whether the reduction in MTBE was the result of volatilization or oxidation, the OxyVac® system successfully remediated the site, with reductions of MTBE concentrations in ground water ranging from 93 to 98%, with a simultaneous reduction of BTEX of 100% (Kealy *et al.*, 2001).

In an anaerobic laboratory study, using a molar ratio of Fenton's reagent: MTBE of 100:1 and a pH of 3.6, MTBE degradation was greater than 99%. However, complete mineralization was not achieved, as seen by the significantly high concentrations of the intermediate products TBA, TBF, acetone, and methyl acetate (Burbano *et al.*, 2001).

At an active gas station in north Texas, a gasoline spill from a UST collected in the sand and gravel tankhold. Water and product were pumped from the tankhold, but the residual ground water contained MTBE concentrations ranging from 411 to 475 mg/l, as well as high concentrations of benzene (14.4 to 15.8 mg/l), toluene (27.9 to 28.0 mg/l), ethylbenzene, and xylenes (both 1.45 to 2.25 mg/l). A high-pressure lance was used to bore into the subsurface, create micro-fractures in the soils of low permeability that surrounded the tankhold backfill, and inject the liquid amendments, including ferrous sulfate, to promote the Fenton's reaction. After the first *in situ* chemical treatment, the average reduction in MTBE concentration measured was 83.4%, with final concentrations of MTBE in monitoring wells located at each corner of the tankhold ranging between 26.8 and 68.4 mg/l (Leetham, 2001). Although the mechanism of the reaction was not discussed, there was evidence of a vigorous reaction between the H_2O_2, ferrous sulfate, and subsurface organics.

Yeh and Novak (1995) conducted a batch laboratory study of the degradation of MTBE using H_2O_2 with and without an iron catalyst. They found that MTBE concentrations did not decrease in a solution of H_2O_2 and distilled water. When ferrous iron was added to the solution, oxidation of MTBE was rapid. After addition of H_2O_2, in the presence of ferrous iron, the color of the solution changed from a light yellow to a yellowish brown, indicating oxidation and precipitation of iron had occurred. Although sufficient iron was available to initially catalyze the oxidation of MTBE, the iron was quickly oxidized to Fe^{3+}, which was no longer capable of catalyzing the formation of Fenton's reagent (Yeh and Novak, 1995).

Yeh and Novak (1995) found that MTBE was chemically oxidized to TBA and acetone and that the oxidation was influenced primarily by H_2O_2 concentration and pH. Yeh and Novak (1995) determined from the study that oxidation at a pH that results in the formation of iron hydroxides and oxides (as evidenced by the presence of brown colored solution) will stop the degradation of MTBE.

Practical Design Considerations.   If an insufficient amount of naturally occurring ferrous iron exists in the target aquifer, the reaction can be catalyzed by the addition of ferrous sulfate ($FeSO_4$) in the treatment solution. An acid solution can be used to reduce the pH of the soil matrix if needed, and a buffer (*i.e.*, phosphate, especially monophosphate) solution may help to stabilize H_2O_2 in the subsurface (Kealy *et al.*, 2001; Yeh and Novak, 1995). Yeh and Novak (1995) suggest that the presence of phosphate provided stabilization of the H_2O_2 and did not inhibit the chemical oxidation of MTBE.

Health and safety issues relating to ISCO using H_2O_2 are significant. There is more than one undocumented account concerning the application or misapplication of highly concentrated H_2O_2 and liquid amendments. The oxidation reaction is exothermic and is capable of producing extreme heat, and H_2O_2 concentrations as low as 11% can cause ground water to boil (Oberle and Schroder, 2000). Only experienced persons should handle, mix, and inject a concentrated H_2O_2 solution into the subsurface. One undocumented account describes workers having to vacate a site an hour after injecting H_2O_2

beneath a 20-centimeter (8-inch) thick concrete floor when heat could be felt through not only the floor but also the laborers' work boots. Another H_2O_2 injection at a site located in a parking lot caused wells to release gas and the asphalt to rise. A worker poked a screwdriver through the asphalt and was blown off his feet, and a large portion of the asphalt was destroyed (Nyer and Vance, 1999).

TBF and TBA have been identified as stable intermediate products in the ISCO of MTBE; research indicates that these intermediate products can be oxidized under optimal conditions with additional H_2O_2 (Burbano *et al.*, 2001).

OZONE

Description of Process. Ozone may react either directly or indirectly with organic compounds. Direct ozonation is the reaction of compounds with molecular O_3, and indirect ozonation involves the reaction of compounds with OH• produced during O_3 decay. The direct oxidation of MTBE via O_3 is slower than oxidation via OH• (Acero *et al.*, 2001; Mitani *et al.*, 2001).

Ozone, when used in conjunction with other oxidation processes (*i.e.*, O_3 combined with H_2O_2 [peroxone] or a process that combines air stripping with O_3), provides efficient oxidation of MTBE via OH•. The combination of ultrasound and O_3 is presented later in the chapter. Figure 11-1 shows the indirect (via OH•) degradation of MTBE and formation of degradation products as taken from Acero *et al.* (2001).

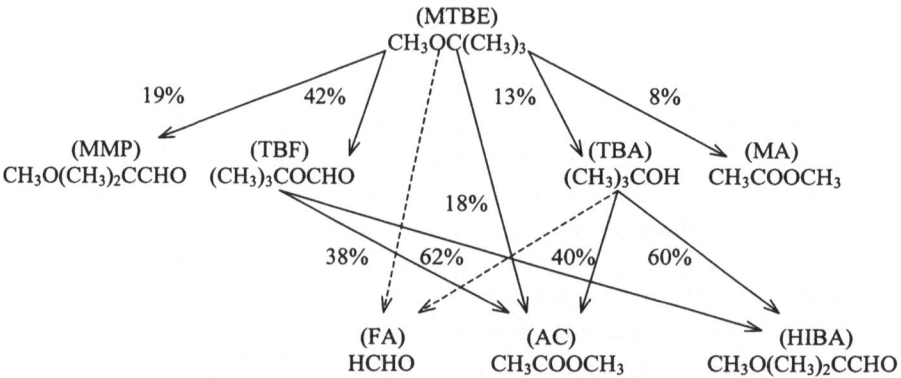

Figure 11-1.   **Indirect (via OH•) Degradation of MTBE and Formation of Degradation Products by Acero *et al.* (2001)**
The percentages given represent the fraction of MTBE oxidized to the degradation product. AC = acetone; FA = formaldehyde; HIBA = hydroxyisobutyraldehyde; MA = methyl acetate; MMP = 2-methoxy-2-methyl propionaldehyde

The initial OH• attack on MTBE by H-abstraction occurs at either methoxy group or any of the three methyl groups. The O-H bond energy is higher than that of the C-H bond of an organic compound, resulting in OH• indiscriminately abstracting hydrogen from organic compounds (Mitani *et al.*, 2001). The likely reaction is via the methoxy group since H-abstraction occurs most easily at the alpha position to the ether function, thus generating TBF, TBA, acetone, and formaldehyde. The other pathway, which includes an OH• attack on the methyl groups, leads to the formation of 2-methoxy-2-methyl propionaldehyde (MMP), methyl acetate, acetone, and formaldehyde. Significant degradation of MTBE daughter products occurs only after significant MTBE degradation has occurred (Acero *et al.*, 2001).

Proven Effectiveness in Field or Laboratory. Ozone is typically combined with another oxidant and or remediation technique such as microsparging to increase efficiency of oxidation. Mitani *et al.* (2001) determined in a laboratory study that if O_3 alone were used to remediate MTBE, then increased residence time, temperature, or O_3 concentration was necessary to completely oxidize MTBE to carbon dioxide.

Ozone with Hydrogen Peroxide. A variety of batch experiments were conducted comparing MTBE oxidation with O_3 and O_3 in conjunction with H_2O_2. Generally, the results of the studies indicate that O_3 in a H_2O_2 solution degraded MTBE faster and more efficiently. With the addition of O_3 to a H_2O_2 solution, OH• was determined to be the predominant degradation reactant (Acero *et al.*, 2001). Table 11-2 compares the O_3 and the OH• rate constants for the degradation of MTBE. As presented in Table 11-2, the rate constants for OH• are nine orders of magnitude faster than the rate constants for O_3 alone. Liang *et al.* (2001) compared the percent removal of MTBE by O_3 alone to a peroxone treatment for low (200 µg/l) and high (2,000 µg/l) initial MTBE concentrations. The results are tabulated in Table 11-3 and show that peroxone was more effective in oxidizing MTBE than O_3 alone, especially at higher

Table 11-2. Rate Constants for Reactions of MTBE with O_3 and OH•

MTBE Concentration	k_{O_3}	$k_{OH•}$	Reference	Notes
1×10^{-4} M	0.14/M-s	1.9×10^9/M-s	Acero *et al.* (2001)	pH 7, O_3 (2 to 10 mg/l), H_2O_2 (2×10^{-4} M)
2 mM	NI	1.6×10^9/M-s	Karpel Vel Leitner (1994)	pH 8, O_3 (11.0 mM/h), O_3/H_2O_2 (2.1 M/M)
NI	<1/M-s*	1.6×10^9/M-s*	Liang *et al.* (2001)	NI

NI = No information was given
* Indicates a secondary reference Buxton et al. (1988).

Table 11-3.   **Percentage MTBE Removal via O_3 and Peroxone**
               by Liang *et al.* (2001)

Ozone (mg/l)	MTBE Removal (%)	Peroxone (O_3/H_2O_2) (mg/l)/(mg/l)	MTBE Removal (%)
"Low" Initial MTBE Concentration (200 µg/l)			
1	19	1/1	41
2	25	2/2	49
4	69	4/4	72
6	83	6/6	88
8	92	8/8	94
10	97	10/10	97
"High" Initial MTBE Concentration (2,000 µg/l)			
NI	NI	1/1	39
NI	NI	2/2	56
4	8	4/4	80
6	25	6/6	92
8	33	8/8	95
10	53	10/10	98

NI = No information was given

initial MTBE concentrations. At lower initial MTBE concentrations, the difference in the percentage of MTBE removed was less dramatic, and in one case (O_3 at 10 mg/l and O_3:peroxone ratio of 1:1) the percentage removal was the same. Karpel vel Leitner *et al.* (1994) found the ratio of ozone/peroxide to ozone for MTBE removal at 1.7, only about 70% more efficient with peroxide added when bubbling is continuous.

Ozone Microsparging.   Direct bubbling of ozone gas into soil solutions under pressure serves to accelerate and stabilize reactions (Shia *et al.*, 2001). These may represent secondary Fenton's-type reactions. The reactions often decay MTBE rapidly with first-order half-lives of a few days (5 to 7 days). The rapid field removal rates of MTBE seem to have a common stoichiometry, independently verified by Kerfoot (2001a), Voci (2001), and Wheeler (2001), approaching a theoretical ratio of 3 mole ozone to 1 mole MTBE, described by Karpel vel Leitner *et al.* (1994). Spargepoints®, which receive both ozone gas and liquid hydroperoxides creating Perozone™ microbubbles, have been developed (Kerfoot, 2001b).

A pilot study was conducted at a gasoline service station on Long Island, New York. A BTEX and MTBE plume had contaminated an aquifer volume of over 500,000 cubic meters (17,000,000 cubic feet). The MTBE plume had spread to 700 meters (2,400 feet) downgradient from the site. Two Spargepoints® (O_3/air) were installed at different depths in a single borehole. Monitoring wells were placed 3.6 and 8.5 meters (12 and 28 feet) downgradient of the injection well. The radius of influence for the system was 8.5 meters (28 feet). After sparging for four weeks there was a measured 76% and 73% reduction in MTBE concentrations from 45 μg/l to 11 μg/l and 6,300 μg/l to 1,700 μg/l, 3.6 and 8.5 meters (12 and 28 feet) downgradient of the injection well, respectively. Three weeks later, the MTBE concentration had dropped to 2 μg/l and 79 μg/l, resulting in reductions in MTBE concentrations of 96% and 99% at 3.6 and 8.5 meters (12 and 28 feet) downgradient of the injection well, respectively. MTBE oxidation was determined to occur in the gas-phase bubbles and in the aqueous phase via direct O_3 and OH• oxidation. There was a measured spike in BTEX concentrations two weeks into the O_3 sparging, but no further information was available concerning the BTEX degradation after this spike was detected (Nichols and Voci, 2001).

Practical Design Considerations.   Bromate is a chemical that is formed during the ozonation of natural waters where bromide is present. In environments with a high bromide content such as seawater, bromide concentrations are approximately 65 mg/l (Acero *et al.*, 2001; Chang and Yen, 2000; Liang *et al.*, 2001; West Hertfordshire Health Authority, 2001). The USEPA has set the MCL for bromate at 0.010 mg/l because toxicology studies have shown bromate to be carcinogenic in laboratory animals (USEPA, 1998). Consideration of bromate formation should be part of the evaluation process for oxidant selection.

During laboratory studies, Liang *et al.* (2001) found that O_3 alone caused an increase in bromate concentration, but with peroxide to O_3 ratios of greater than 1.0, a decrease in bromate concentration was realized. Liang *et al.* (2001) determined that the H_2O_2 in the peroxone treatment reduced aqueous bromine to bromide, resulting in low yields of bromate. In a similar study, Acero *et al.* (2001) surmised that, even at concentrated levels, H_2O_2 cannot completely suppress bromate formation, but the presence of H_2O_2 leads to reduced hypobromous acid concentrations. Hypobromous acid is an important intermediate in bromate formation. In summary, laboratory studies have shown that a H_2O_2 to O_3 ratio of 1:1 minimizes bromate and maximizes production of OH• (Acero *et al.*, 2001; Mitani *et al.*, 2001).

Prior to a full-scale application, a pilot test with a complete monitoring program is essential to determine the site-specific, optimum ozone or AOP dosing that maximizes MTBE oxidation.

PERMANGANATE

Description of Process.   Potassium permanganate oxidizes organic compounds with carbon-carbon double bonds, aldehyde groups, or hydroxyl groups very efficiently (Oberle and Schroder, 2000). Suthersan (1997) states that oxidation with permanganate can occur via electron abstraction, hydrogen atom abstraction, hydride ion abstraction, and direct donation of oxygen to the organics in the substrate.

Ground water pH affects the kinetics of the permanganate reaction and determines whether the oxidation will involve one, three, or five electron exchanges (Damm *et al.*, unpublished). Most aquifers have a pH within the 3.5 to 12 pH unit range; therefore, it can be expected that three electrons will be exchanged during *in situ* oxidation. The effect of lower pH on E° is shown on Table 11-4. Under acidic conditions (pH 3.5 to 7), E° is higher when compared to alkaline conditions (pH 7 to 12). Manganese dioxide is the product of either reaction in which three electrons are transferred as shown on Table 11-4.

Complete MTBE oxidation via permanganate is described by Equation 11-3.

$$21\ MnO_4^- + 2\ C_5\ H_{12}O \rightarrow 21\ MnO_2 + 10\ CO_2 + 24\ OH^- \qquad (11\text{-}3)$$

Note that 21 moles of permanganate are required to completely mineralize 2 moles of MTBE.

Proven Effectiveness in Field or Laboratory.   Batch laboratory tests performed by Damm *et al.* (unpublished) show that the oxidation of MTBE by permanganate follows a second-order reaction overall, and a first-order reaction with respect to the individual constituents.

Although Damm *et al.* (unpublished) found that the effects of pH on the batch tests were minimal, which is typical of $KMnO_4$ reaction kinetics under typical ground water conditions, results from the batch study showed that oxidation of MTBE was favored under more acidic or alkaline conditions than under neutral conditions. However, the change in reaction rate was so small that a pH adjustment of the aquifer before permanganate treatment would be unnecessary.

The rate of MTBE oxidation via the permanganate ion in the Damm et al. (unpublished) batch study was 1.4×10^{-6} milligram per liter per hour (mg/l-h) with half-lives ranging from 55 to 495 hours. This is a slower reaction rate than the literature rates for permanganate oxidation of chlorinated organic compounds, which in the Huang *et al.* (unpublished) study was four hours. The slower reaction is attributed to the tertiary structure of the MTBE

Table 11-4.   The Influence of pH on Oxidation Potential by Damm et al. (unpublished)

Conditions	pH	Equation	Potential E° (Volts)
Acidic	3.5 to 7	$MnO_4^- + 4\ H^+ + 3\ e^- \rightarrow MnO_2 + 2\ H_2O$	+1.70
Alkaline	7 to 12	$MnO_4^- + 2\ H_2O + 3\ e^- \rightarrow MnO_2 + 4\ OH^-$	+0.59

molecule, which requires higher activation energy to initiate a reaction than double-bonded organic compounds (*i.e.*, trichlolorethylene [TCE] and dichloroethene [DCE]).

In the Damm *et al.* (unpublished) batch study, complete oxidation of MTBE did not occur. Intermediate and end products included TBF and TBA, and the measured molar consumption of permanganate and MTBE was 2:1 (permanganate:MTBE), which correlates with the theoretical chemical equation for the oxidation of MTBE to TBA as shown below in Equation 11-4.

$$2\,MnO_4^- + 1\,C_5H_{12}O \rightarrow 2\,MnO_2 + C_4H_{10}O + 1\,CO_2 + 2\,OH^- \quad (11\text{-}4)$$

The discussion for the Damm *et al.* (unpublished) batch study did not include information on the mineralization of TBF and TBA in the presence of excess permanganate.

Clayton *et al.* (2000), in a multi-site field evaluation of ISCO, concluded that MTBE could be successfully oxidized to TBA with permanganate. Treatment efficiencies of greater than 99% were reported for both low-level dissolved and high-level dissolved/adsorbed MTBE (Clayton *et al.*, 2000).

Practical Design Considerations. Ground water geochemistry undergoes changes during permanganate oxidation. During *in situ* permanganate treatment, oxidation-reduction potential (ORP) values within the active treatment zone are typically greater than 200 millivolts (mV). Ion exchange is impacted with the addition of high K^+ or Na^+ concentrations. Cation exchange between Mg^{2+} and Na^+ are common. Similarly, anion desorption occurs as the MnO_4^- reacts, resulting in a net positive charge imbalance (Clayton *et al.*, 2000).

The permanganate ion oxidizes MTBE less rapidly (by two to three orders of magnitude) than other advanced oxidizing agents such as O_3, O_3/Ultrasonic Irradiation, Ultraviolet (UV)/H_2O_2, and Fenton's reagent because of the smaller oxidizing potential of the permanganate ion. In the subsurface, permanganate has a longer half-life compared to the more powerful oxidants such as O_3 or H_2O_2. The use of permanganate for MTBE as a passive risk management remediation strategy is possible as opposed to using permanganate as a rapid treatment strategy (Damm *et al.*, unpublished).

PERSULFATE

Description of Process. Sodium persulfate ($N_2S_2O_8$) is a radical-based ($SO_4^-\bullet$ and $OH\bullet$) oxidant that has been tested under laboratory conditions for the degradation of MTBE. For *in situ* applications, radical-based oxidation occurs under heat or metal catalyzed reactions. Formed $SO_4^-\bullet$ can convert to $OH\bullet$ in alkaline solutions (pH greater than 8.5). Similar to other radical reactions, MTBE is initially degraded indirectly via abstraction of the alpha-hydrogen by $SO_4^-\bullet$ and $OH\bullet$. A carbon-centered radical reacts with oxygen and forms a peroxyl radical. The peroxyl radical undergoes an acid-catalyzed hydrolysis reaction resulting in the formation of TBF (Huang *et al.*, unpublished).

Proven Effectiveness in Field or Laboratory.   There are no known field tests on the use of persulfate for MTBE oxidation. Shown in Table 11-5 are the pseudo-first order rate constants determined in laboratory studies for MTBE degradation by $N_2S_2O_8$ in a phosphate-buffered solution (Huang *et al.*, unpublished). In a persulfate solution at 40°C (104°F), four MTBE oxidation intermediate products, TBF, TBA, methyl acetate, and acetone, were readily degraded (Huang *et al.*, unpublished).

Parameters that increase the activation energy of the reaction (24.5 ± 1.6 kilocalories per mole) include temperature, persulfate concentration, pH, and ionic strength; however, it was determined that increasing pH over the range of 2.5 to 11 and ionic strength over the range of 0.11 to 0.53 moles per liter (M) decreased the reaction rate. In addition, persulfate oxidation at atmospheric pressure and ambient temperature (*i.e.*, uncatalyzed) and in the absence of transition metal ions has been found to be ineffective for the oxidation of MTBE (Huang *et al.*, unpublished).

Practical Design Considerations.   Bicarbonate in ground water may scavenge the radicals, decreasing the efficiency of the reaction. Otherwise, persulfate has high solubility and stability in normal subsurface conditions; therefore, it is a good candidate for MTBE oxidation under catalyzed conditions (Huang *et al.*, unpublished).

ULTRASOUND

Description of Process.   The chemistry of ultrasound is mostly attributed to cavitation. The three-step process of cavitation includes: nucleation, growth, and collapse of a gas or vapor filled bubble within the bulk solution (Chang

Table 11-5. **Pseudo-First-Order Rate Constants Determined by Huang et al. (unpublished) for MTBE Degradation by $N_2S_2O_8$ in a Phosphate-Buffered Solution**

Persulfate (31.5 mM ≅ 8 g/l) pH 7.0 Ionic Strength 0.11 M	
Temperature °C	k_{MTBE}/s
20	0.13×10^{-4}
30	0.48×10^{-4}
40	2.4×10^{-4}
50	5.8×10^{-4}

and Yen, 2000). Sonolytically induced microbubbles are enlarged with each ultrasonic frequency cycle. Once the microbubble reaches a critical resonance frequency size, the microbubble collapses violently (Kang and Hoffman, 1998). When the bubble implodes, intense heating of the bubble vapor occurs. Temperatures of approximately 5,000°C (9,000°F), pressures of approximately 500 atmospheres, and lifetimes of a few seconds can occur in localized hotspots (Chang and Yen, 2000). Under these conditions, water vapor undergoes thermal dissociation yielding H_2O_2 and the reactive radicals H•, OH•, and O•. The formation of these radicals and H_2O_2 allow for secondary reactions (oxidation) to take place between solute molecules and the reactive radicals (Kang and Hoffman, 1998).

Proven Effectiveness in Field or Laboratory.   A chemical-assisted ultrasound process has proven successful under laboratory conditions for the oxidation of MTBE. Ultrasound with Fenton's reagent plus Cu^{2+} and ultrasound combined with ozone are two examples of chemical-assisted ultrasound oxidation.

Ultrasound with Fenton's Reagent.   Both ultrasonic irradiation and Fenton's reagent are capable of generating OH• in solution. The OH• is a nonselective and powerful oxidant and ultrasound combined with Fenton's reagent has greater degradation strength when compared to other organic substrates. The list below is in order of decreasing degradation strength. The symbol (>>) indicates an oxidant that is much stronger than the oxidant listed below. The symbol (>) indicates an oxidant is stronger than the oxidant listed below.

- Ultrasound with Fenton's reagent plus Cu^{2+} >>
- Ultrasound with Fenton's reagent only>
- Fenton's reagent>>
- Ultrasound with H_2O_2 oxidation>
- H_2O_2 alone>
- Ultrasound alone>
- Ferrous iron alone

As shown above, ultrasound and H_2O_2 are a powerful oxidant combination. In studies without the addition of ultrasound, optimal H_2O_2 concentrations indicated a molar ratio of H_2O_2:organic substrates as (50 to 500):1. In combination with ultrasound, H_2O_2 is added to initiate the reaction only. The addition of copper (Cu^{2+}) enhances the MTBE degradation further by removing reaction byproducts including TBA, acetone, and acetic acid via ligand transfer or organocopper intermediate formation and precipitation (Chang and Yen, 2000). Laboratory results have shown 98% MTBE degradation in 60 minutes using the ultrasound with Fenton's reagent plus Cu^{2+} reaction (Chang and Yen, 2000).

To create the ultrasound/Fenton's reagent/Cu^{2+} reaction in the field, Fenton's reagent can be supplied by sonolysis of water and naturally occurring metal salts in the ground water. If not naturally occurring, Cu^{2+} can be added. A robotic self-powered mining head containing ultrasonic transducers that attach to the inner wall of a double-wall pipe is used to supply ultrasound. The drilling head moving through the subsurface impacted zone applies the chemical-assisted ultrasound treatment. The process combined with directional drilling technology can track the plume of a constituent. The chemicals used in the ultrasound/Fenton's reagent/Cu^{2+} process are environmentally benign (Chang and Yen, 2000).

Ultrasound with Ozone. Batch studies were performed to determine the kinetics and mechanisms of ultrasound and ultrasound enhanced by O_3 for the rapid degradation of MTBE in an aqueous solution. Initial batch studies were done using ultrasound only. The unbuffered batch studies had an initial pH of 6.6 to 6.8, the temperature was held constant at 20°C (68°F), and irradiation intensity was 200 watts per liter with a frequency of 205 kilohertz (kHz). Initial MTBE concentrations ranged from 0.01 to 1.0 millimoles/liter (mM). Time required for 90% oxidation of MTBE was 45 minutes (k = 8.5×10^{-4}/s) at 0.01 mM. Increased MTBE concentration (1.0 mM) required 93 minutes (k = 4.1×10^{-4}/s) for 90% oxidation. The reaction was limited by constituent contact with OH•.

When ultrasound and O_3 are combined, the rate of destruction of MTBE is accelerated by a factor of 1.5 to 3.9, depending on the initial concentration of MTBE (Kang and Hoffman, 1998). Table 11-6 summarizes the rate constants for the Kang and Hoffman (1998) chemical-assisted ultrasound oxidation batch experiments.

Practical Design Considerations. As previously discussed, AOPs, for example those that utilize ultrasound and O_3, can form bromate ion from bromide (Cater *et al.*, 2000). Carbonate and bicarbonate are also problematic in

Table 11-6. **Kang and Hoffman (1998) Chemically Assisted Ultrasound Oxidation Batch Study**

[MTBE] (mM)	[O_3] (mM)	k° (10^{-4}/s) Ultrasound	O_3 + ultrasound	E[a]
0.01	0.30	8.5	33.2	3.9
0.05	0.31	8.7	31.3	3.6
0.25	0.32	6.9	14.9	2.2
0.50	0.34	5.4	12.2	2.3
1.00	0.26	4.1	6.3	1.5

[a] Enhancement factor = k°(ozone + ultrasound)/k°(ultrasound alone)
k° = reaction rate constant

that they compete for OH•. Kang and Hoffman (1998) show the OH• and bicarbonate reaction as shown in Equation 11-5.

$$OH• + HCO_3^- \rightarrow H_2O + CO_3•^- \tag{11-5}$$

The second-order rate constant for the reaction shown above is 1.5×10^7/moles/liter per second (M-s). The second-order rate constant for MTBE and OH• is 1.57×10^9/M-s. Although the latter rate constant is two orders of magnitude larger, Kang and Hoffman (1998) found the effects of HCO_3^- on the rate of sonolytic degradation of MTBE to be negligible over the bicarbonate concentration range of 1 to 2 mM. Based on these findings, Kang and Hoffman (1998) suggest that the MTBE and OH• reaction is in the vapor phase of the cavitating bubbles and not in the aqueous solution.

Kang *et al.* (1999) continued the batch studies, varying frequencies to find the optimum frequency in order to obtain the most beneficial chemical effects, which include the production of H_2O_2. At a power density of 240 watts per liter and a frequency of 358 kHz, maximum H_2O_2 was produced, and the rate constant for MTBE degradation was maximized. At higher frequencies rarefaction cycles were too short to permit the microbubbles to grow sufficiently to cause optimal disruption of the liquid. Kang *et al.* (1999) also found that naturally occurring organic matter seemed to have negligible effect on the rate of MTBE degradation.

ISCO COSTS

Costs for each of the ISCO processes are discussed below where information is available.

Hydrogen Peroxide. For the Leetham (2001) field project (gasoline station tankhold with MTBE concentrations ranging from 411 to 475 mg/l), over a treatment duration of three to four months, the total cost of a chemical oxidation treatment utilizing 7,600 liters (2,000 gallons) of (33%) H_2O_2, 800 liters (220 gallons) of ferrous sulfate (12% concentration), and using lance injection techniques was $55,000 to $65,000. The total cost included remediation, sampling, management, and reporting.

Nyer and Vance (1999) describe an 1,800-cubic meter (64,000-cubic foot) test area at the Westinghouse Savannah River Site in Aiken, South Carolina. The site was impacted by dense non-aqueous phase liquid (DNAPL) consisting mainly of perchloroethene (PCE) and TCE. H_2O_2 and ferrous sulfate solution batches of 1,900 to 3,800 liters (500 to 1,000 gallons) were injected each day for six days. Total cost for the project was $511,000.

Ozone. The ozone panel used during the Long Island, New York microsparging pilot study previously described cost approximately $24,000 to buy and $2,000 per month to rent. Additional costs would be incurred for drilling, well installation, and monitoring (Nichols and Voci, 2001).

Permanganate. Xpert Design & Diagnostics (XDD) has done extensive ISCO applications using permanganate and has created a bar chart, depicted in Figure 11-2, illustrating the cost of a full-scale chemical oxidation application on a per cubic yard (0.76 cubic meter) basis. The cost per 1 cubic yard (0.76 cubic meters) includes procurement and delivery of the oxidant batching system, injection point installation, oxidant (assuming 5 g/kg SOD), and post injection monitoring. As shown on Figure 11-2, the costs per 1 cubic yard (0.76 cubic meters) decrease dramatically as the volume of soil treated increases.

Cost information for permanganate was presented by Brown *et al.* (2001) and includes $173 and $312 (cost per 1,000 equivalents) for potassium and sodium permanganate respectively. In oxidation/reduction reactions, an equivalent mass is calculated by the molar mass divided by the number of electrons consumed by the oxidation/reduction agent in the half-reaction. The equivalent weight in grams per equivalent (g/eq) is converted to pounds per equivalent (lb/eq) and multiplied by a price per pound ($/lb) to determine the cost per equivalent.

Persulfate. A price range for persulfate ($260 to $315) was presented by Brown *et al.* (2001) on the basis of dollars per 1,000 equivalents.

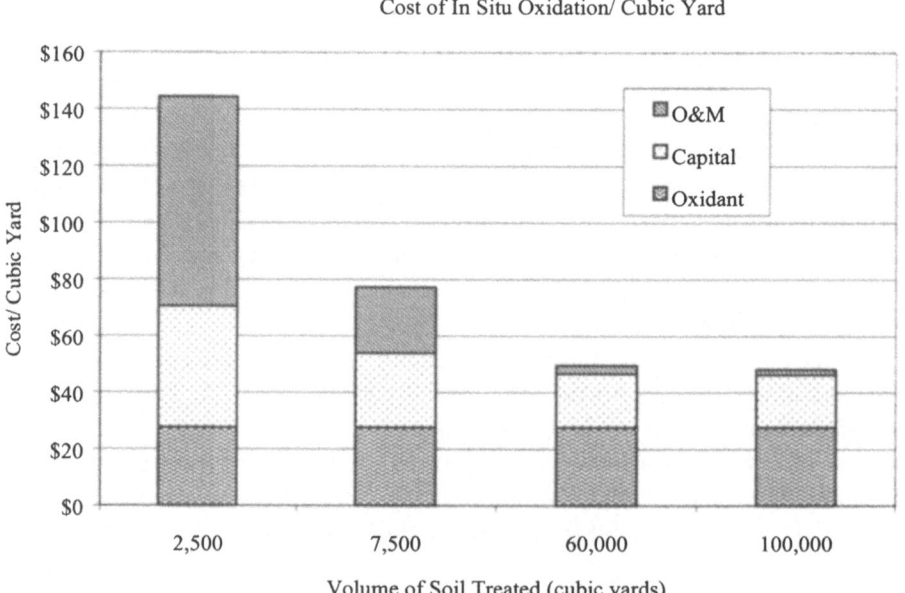

Figure 11-2. **Permanganate Full-Scale ISCO Costs by XDD**

Ultrasound.   At this stage, the technology development costs for fluid implementation of the ultrasound-oxidation approach are not available.

REFERENCES

Acero, J.L., Haderlein, S.B., Schmidt, T.C., Suter, M.J., and Gunten, U.V. 2001. MTBE oxidation by conventional ozonation and the combination ozone/hydrogen peroxide: Efficiency of the processes and bromate formation. *Environmental Science and Technology.* 35, 4252–4259.

Brown, R.A., Fiacco, J., Skladany, G., McTigue, J.W., and Robinson, D. 2001. Comparing permanganate and persulfate treatment effectiveness for various organic contaminants. In: *Proceedings of The First International Conference on Oxidation and Reduction Technologies for In-Situ Treatment of Soil and Groundwater,* Niagara Falls, Ontario, Canada. June 25–29, 2001. Anticipated publication in the summer of 2002.

Burbano, A.A., Dionysiou, D.D., Richardson, T.L., and Suidan, M.T. 2001. Remediation of MTBE-contaminated water: Studies on MTBE mineralization using the Fenton's reagent. In: *Proceedings of The Seventh International Conference on Advanced Oxidation Technologies for Water and Air Remediation,* Niagara Falls, Ontario, Canada. June 25–29, 2001. Anticipated publication in the summer of 2002.

Buxton, G.V. *et al.* 1988. Critical review of rate constants for hydrated electrons, hydrogen atoms, and hydroxyl radicals (OH/O^-) in aqueous solution. *Journal of Physical and Chemical Reference Data.* 17:2, 513.

Cater, S.R., Stefan, M.I., Bolton, J.R., and Safarzadeh-Amiri, A. 2000. UV/H_2O_2 treatment of methyl *tert*-butyl ether in contaminated waters. *Environmental Science and Technology.* 34, 659–662.

Chang, H.L. and Yen, T.F. 2000. An improved chemical-assisted ultrasound treatment for MTBE. In: *Proceedings of The Second International Conference on Remediation of Chlorinated and Recalcitrant Compounds, Chemical Oxidation and Reactive Barriers: Remediation of Chlorinated and Recalcitrant Compounds.* Volume C2-6, pp. 195–200. (Wickramanayake, G.B., Gavaskar, A.R., and Chen, A.S.C. Eds.). Columbus and Richland, Battelle Press.

Clayton, W.S., Marvin, B.K., Pac, T., and Mott-Smith, E. 2000. A multisite field performance evaluation of in-situ chemical oxidation using permanganate. In: *Proceedings of The Second International Conference on Remediation of Chlorinated and Recalcitrant Compounds, Chemical Oxidation and Reactive Barriers: Remediation of Chlorinated and Recalcitrant Compounds.* Volume C2-6, pp. 101–108. (Wickramanayake, G.B., Gavaskar, A.R., and Chen, A.S.C. Eds.). Columbus and Richland, Battelle Press.

Damm, J.H., Hardacre, C., Kalin, R.M., and Walsh, K.P. Unpublished. Oxidation of methyl tert-butyl ether by potassium permanganate. In: *Proceedings of The First International Conference on Oxidation and Reduction Tech-*

nologies for In-Situ Treatment of Soil and Groundwater, Niagara Falls, Ontario, Canada. June 25–29, 2001. Anticipated publication in 2002.

Deeb, R.A., Scow, K.M., and Alvarez-Cohen, L. 2000. Aerobic MTBE biodegradation: An examination of past studies, current challenges and future research directions. *Biodegradation.* 11, 171–186.

Huang, K., Couttenye, R.A., and Hoag, G.E. Unpublished. *Kinetics of Heat-Assisted Persulfate Oxidation of Methyl tert-Butyl Ether (MTBE).* Environmental Research Institute, Storrs, Connecticut.

Kakarla, P. 2002. Personal communication with Mr. Kakarla, Technical Manager for ISOTEC, West Windsor, New Jersey.

Kang, J.W. and Hoffmann, M.R. 1998. Kinetics and mechanism of the sonolytic destruction of MTBE by ultrasonic irradiation in the presence of ozone. *Environmental Science and Technology.* 32, 3194–3199.

Kang, J.W., Hung, H.M., Lin, A., and Hoffmann, M.R. 1999. Sonolytic destruction of MTBE by ultrasonic irradiation: The role of O_3, H_2O_2, frequency, and power density. *Environmental Science and Technology.* 33, 3199–3205.

Karpel Vel Leitner, N., Papailhou, A.L., Croué, J.P., Peyrot, J., and Doré, M. 1994. Oxidation of methyl *tert*-butyl ether (MTBE) and ethyl tert-butyl ether (ETBE) by ozone and combined ozone/hydrogen peroxide. *Ozone Science and Engineering.* 16, 41–53.

Kealy, J., Roth, R., and Pezzullo, J. 2001. In-situ remediation of an MTBE contaminated soil and groundwater using OxyVac® technology. In: *Proceedings of The First International Conference on Oxidation and Reduction Technologies for In-Situ Treatment of Soil and Groundwater,* Niagara Falls, Ontario, Canada. June 25–29, 2001. Anticipated publication in the summer of 2002.

Kerfoot, W.B. 2001a. Thin-layer Criegee-like oxidation for removal of PAHs and PCBs in sediments. Presented at *The First International Congress on Petroleum Contaminated Soils, Sediments, and Water,* Imperial College, London, U.K., August 14–17, 2001.

Kerfoot, W.B. 2001b. AOP/Bioremediation: A technology breakthrough with aquifer heart-lung machines. *In Situ Aeration and Aerobic Remediation.* Battelle Press, Columbus Ohio.

Leetham, J.T. 2001. In-situ chemical oxidation of MTBE and BTEX in soil and groundwater: A case study. *Contaminated Soil Sediment and Water.* Spring (Special Issue), 54–58.

Liang, S., Yates, R.S., Davis, D.V., Pastor, S.J., Palencia, L.S., and Bruno, J.M. 2001. Treatability of MTBE contaminated groundwater by ozone and peroxone. *Journal of the American Water Works Association.* June, 110–120.

Mitani, M.M., Keller, A.A., Bunton, C.A., Rinker, R.G., and Sandall, O.C. 2001. Kinetics and products of reactions of MTBE with ozone and ozone/hydrogen peroxide in water. *Journal of Hazardous Materials.* 89, 197–212.

Nichols, E.M. and Voci, C.J., 2001. Evaluation of an ozone-air sparging pilot test to remediate MTBE in groundwater on Long Island, NY. Presented at *The 17ᵗʰ Annual International Conference on Contaminated Soils, Sediments and Water,* University of Massachusetts, Amherst, Massachusetts. October 22–25, 2001.

Nyer, E.K. and Vance, D. 1999. Hydrogen peroxide treatment: The good, the bad, the ugly. *Ground Water Monitoring and Remediation.* Summer, 54–57.

Oberle, D.W. and Schroder, D.L. 2000. Design considerations for in-situ chemical oxidation. In: *Proceedings of The Second International Conference on Remediation of Chlorinated and Recalcitrant Compounds, Chemical Oxidation and Reactive Barriers: Remediation of Chlorinated and Recalcitrant Compounds.* Volume C2-6, pp. 91–99. (Wickramanayake, G.B., Gavaskar, A.R., and Chen, A.S.C. Eds.). Columbus and Richland, Battelle Press.

Shia, C., Templeton, J., Delp, D., Hoke, B., Kerfoot, W.B., Stone, P., and Hill, S. 2001. Pilot test results of ozone-enhanced microbubble air sparging technology for volatile organic compound removal. *Contaminated Soils.* 6, 315–343.

Suthersan, S.S. 1997. *Remediation Engineering Design Concepts,* pp. 222–224. Boca Raton, New York, London, and Tokyo, CRC-Lewis Publishers.

USEPA, 1998. *Stage 1 Disinfectants and Disinfection Byproducts Rule.* EPA 815-F-98-010. December, 1998. http://www.epa.gov/safewater/mdbp/dbp1.html. Accessed January 24, 2002.

Voci, C.J. and Nichols, E.M. 2001. Evaluation of an ozone-air sparging pilot test to remediate MTBE in groundwater on Long Island, NY. *Contaminated Soils.* 7.

Wagler, J.L. and Malley Jr., J.P. 1994. The removal of methyl tertiary-butyl ether from a model groundwater using UV/peroxide oxidation. *Journal — New England Water Works Association.* September, 236–260.

West Hertfordshire Health Authority. 2001. Health News Press Release. Investigation of Bromate Found in Groundwater in Hertfordshire. www.wherts-ha.nthames.nhs.uk/ha/archive/news/Jul00/25072000.html. Accessed April 17, 2002.

Wheeler, K.P. 2001. *In situ* ozone remediation of MTBE in groundwater. *Contaminated Soils.* 7.

Yeh, C.K. and Novak, J.T. 1995. The effect of hydrogen peroxide on the degradation of methyl and ethyl tert-butyl ether in soils. *Water Environment Research.* 67, 828–834.

CHAPTER 12

Aerobic *In Situ* Bioremediation

John T. Wilson, Ph.D., USEPA National Risk Management Laboratory, Office of Research and Development

As they exist naturally, almost all plumes of MTBE contain detectable concentrations of TBA. Depending on the geochemistry of the ground water containing the plume of MTBE contamination and on the presence or absence of natural anaerobic biodegradation, the MTBE plume will also contain greater or lesser amounts of TBA. Because TBA and MTBE exist together in ground water impacted by fuel spills, an aerobic biological treatment system for MTBE must also be effective against TBA.

Many plumes of MTBE contain concentrations of TBA that are in excess of the concentration of MTBE, while other plumes show little accumulation of TBA. Some of the TBA in ground water is TBA that was present in the technical grade of MTBE originally added to the gasoline. In some plumes, most of the TBA is probably produced by natural anaerobic biodegradation of MTBE. Little is known about natural anaerobic biodegradation of MTBE in aquifers. In some anaerobic ground waters there is almost a stoichiometric replacement of MTBE with TBA as water moves along the flow path. The TBA persists after MTBE is consumed. In other plumes, any TBA that is produced is consumed, and TBA does not accumulate.

Bradley *et al.* (2001) compared the extent of MTBE biodegradation under anaerobic conditions in the bed sediments of streams that received plumes of MTBE. At the three sites they studied, the sediments as they were originally collected were depleted of oxygen, nitrate, and sulfate, and there was little accumulation of Fe(II) or Mn(II); however, the sediments were actively methanogenic. In these sediments, MTBE was degraded to TBA, which was not further degraded. When they amended the sediments with oxygen or nitrate, MTBE was degraded to carbon dioxide, and TBA did not accumulate. When the sediments were amended with sulfate, MTBE was degraded primarily to carbon dioxide, and small amounts of TBA were produced. If these results are extrapolated from bed sediments to aquifers, an accumulation of TBA from MTBE would be expected in ground waters that are methanogenic, and where sulfate, nitrate, and oxygen are not available to support complete metabolism of MTBE to carbon dioxide.

MICROBIOLOGY AND BIOCHEMISTRY
OF AEROBIC MTBE BIODEGRADATION

Organisms that grow on MTBE are difficult to isolate and culture in the laboratory. Yeh and Novak (1995) found no evidence for aerobic biodegradation of MTBE in laboratory microcosms after 100 days of incubation. Jensen and Arvin (1990) found no degradation of MTBE after 60 days in samples of activated sludge, topsoil, or aquifer material. Salanitro *et al.* (1994) published the first report of the aerobic biodegradation of MTBE by a mixed culture of microorganisms; Mo *et al.* (1997) published the first report of aerobic biodegradation of MTBE by pure cultures of bacteria; and Borden *et al.* (1997) published the first report of aerobic MTBE biodegradation in microcosms. In the study of Borden *et al.* (1997), aerobic biodegradation of MTBE in the microcosms was incomplete. Degradation ceased after 100 days of incubation, leaving approximately 1.0 mg/l of MTBE in the pore water. The difficulty in isolating and culturing MTBE-degrading microorganisms has created the perception that MTBE-degrading organisms are rare in natural environments and that MTBE degradation by native microorganisms is slow or altogether absent. In their comprehensive review, Squillace *et al.* (1997) concluded "in general, most studies to date have indicated that MTBE is difficult to biodegrade, and some have classified MTBE as recalcitrant."

The initial difficulty in isolating and culturing MTBE-degrading bacteria may reflect the impatience of the scientists as much as the metabolic capability of the microbes. Microbiologists working in the laboratory base their experimental protocols on the behavior of organisms that grow rapidly. They rarely incubate their enrichment studies for more than a few months.

The difficulty in isolating MTBE-degrading microorganisms can be understood if we compare the growth rate of microorganisms that degrade MTBE (Table 12-1) to the growth rate of microorganisms that degrade ordinary petroleum hydrocarbons such a benzene, toluene, and xylenes (Table 12-2). Typical strains of bacteria growing aerobically on petroleum hydrocarbons can divide and double their numbers every few hours. As a consequence, laboratory enrichment cultures will grow up and remove the hydrocarbons in a few days. Cultures of bacteria using MTBE as a growth substrate require several days to several weeks to double their numbers. This 10- to 100-fold difference in the growth rate has an important effect on the time required for a culture to grow to densities that will entirely consume MTBE.

The metabolism of 1 mg/l of MTBE will produce a final density of bacteria of approximately 10^6 per milliliter. If the initial density of MTBE-degrading bacteria were one per milliliter and the doubling time were two days, it

Table 12-1. **Rate of Growth of Microorganisms that Aerobically Biodegrade MTBE**

Growth Substrate	Culture or Organism	Doubling Time (days)	Reference
MTBE	BC-1 culture	>14	Salanitro *et al.* (1998)
MTBE	Enrichment from Refinery Activated Sludge	2.4	Park and Cowan (1997)
MTBE	Enrichment from Biofilter	30	Fortin and Deshusses (1999)
MTBE	ENV735 *Hydrogenophaga flava*	1.7	Steffan *et al.* (2000b)
Pentane	*Pseudomonas aeruginosa*	0.15	Garnier *et al.* (1999)

Table 12-2. **Rate of Growth and Kinetics of Aerobic Biodegradation of Benzene, Toluene, and Xylenes by Microorganisms that Are Not Known to Degrade MTBE**
Data are from the review of Suarez and Rifai (1999). Tabulated are the median and the range of reported values.

Growth Substrate	Doubling Time		Maximum Specific Degradation Rate		Half-Saturation Constant for Growth Substrate K_s	
	Number of Studies	Time (days)	Number of Studies	Rate (gram per gram cells per day)	Number of Studies	Concentration (mg/l)
Benzene	10	0.086 0.78 to 0.042	10	8.3 25 to 0.78	11	6.6 20 to 0.3
Toluene	15	0.12 3.3 to 0.02	16	11 59 to 0.49	15	1.9 20 to 0.1
Xylenes	8	0.21 0.47 to 0.05	6	9.3 51 to 3.0	7	4.6 16 to 0.75

would require 40 days for the culture to grow up and degrade MTBE. If the doubling time were two weeks, it would require 280 days for the culture to grow up and degrade MTBE. Fortin and Deshusses (1999) note that most cultures of MTBE-degrading bacteria that are available were acquired from bioreactors or biofilters that had already acclimated to degrade MTBE.

Some organisms can use MTBE as their sole substrate for growth. Other organisms can biodegrade MTBE, but they cannot grow on MTBE alone. They require another substrate for growth. Biodegradation of MTBE under these circumstances is termed co-metabolism.

The organisms that grow on other substrates and co-metabolize MTBE can grow rapidly. Compare the growth rate of Pseudomonas aeruginosa when growing on pentane to the growth rate of microorganisms growing on MTBE (Table 12-1). At a doubling time of 0.15 days, this organism could grow up to densities that can degrade 1 mg/l of MTBE in only three days. Many of the natural hydrocarbons in gasoline can support the growth of organisms that will degrade MTBE. This is particularly true of the straight-chained alkanes and iso-alkanes (Hyman *et al.*, 2000). Because they grow more rapidly, adding oxygen to environmental samples that contain a mixture of petroleum hydrocarbons and MTBE will most likely enrich for organisms that co-metabolize MTBE. Unfortunately, the capacity to degrade MTBE is lost soon after the primary substrate is consumed (Garnier *et al.*, 1999; Steffan *et al.*, 2000a).

KINETICS OF METABOLISM

The rate of microbial metabolism of MTBE, petroleum hydrocarbons, or oxygen is a function of the concentration of the particular substrate in water. Across a wide range of concentrations, the biodegradation of MTBE or petroleum hydrocarbons or consumption of oxygen follows a hyperbolic rate law:

$$V = V_{max} [C/(K_s + C)] \qquad (12\text{-}1)$$

Where $V =$ the achieved rate of biodegradation or consumption (mg/l per day).

$V_{max} =$ the maximum possible rate of biodegradation or consumption at high concentrations (mg/l per day).

$C =$ the particular concentration of MTBE, hydrocarbon, or oxygen (mg/l).

$K_s =$ half-saturation constant, the concentration of MTBE, hydrocarbon, or oxygen that produces one-half of the maximum possible rate of biodegradation or consumption (mg/l).

The maximum possible rate of biodegradation or consumption (V_{max}) is a function of the specific activity of individual microbes and of the total density of active microbes. To correct for differences in the density of microbial biomass, it is often more convenient to compare the specific activity of different cultures or isolates, rather than the maximum rate achieved in cultures or microcosms. The maximum specific degradation rate as presented in the tables in this chapter (milligrams substrate consumed per milligram microbial biomass per day) is equivalent to the maximum possible rate (V_{max} in milligrams substrate consumed per liter per day) divided by the density of microbial biomass (milligrams biomass per liter on a dry weight basis).

Each microbe has a separate value of K_s for MTBE, for each separate petroleum hydrocarbon, and for oxygen. The values of K_s for MTBE, for each separate petroleum hydrocarbon, and for oxygen will vary from organism to organism.

When the concentration of a substrate (C) is significantly lower than the corresponding half-saturation constant (K_s), their sum ($K_s + C$) is approximately equivalent to K_s. Because V_{max} and K_s are constants, the rate of biodegradation or consumption (V) is proportional to the concentration of substrate (C). As the concentration of substrate goes down through biodegradation, the rate of biodegradation declines in proportion to the decline in concentration of substrate. The rate of biodegradation may then be approximated by a first-order rate law.

$$C = C_0 * e^{-kt} \qquad\qquad (12\text{-}2)$$

Where C =   the concentration of substrate at time t
       C_0 =   the original concentration
       e =   the base of the natural logarithms
       –k =   the first-order rate constant for the rate of change in concentration with time. The constant (k) is the rate of attenuation with time. A negative sign ($-k$) means that the concentrations are decreasing with time. The constant has units of reciprocal time.
       t =   the elapsed time.

When hydrocarbon concentrations are significantly higher than the half-saturation constant, the sum of ($K_s + C$) is approximately equivalent to C. As $C/(K_s + C)$ approaches 1.0, the rate of biodegradation (V) approaches the maximum rate (V_{max}). As a result, a constant amount of substrate is metabolized per unit time, regardless of the concentration of substrate. The rate of degradation of MTBE or hydrocarbon, or the consumption of oxygen, may then be approximated by a zero-order rate law.

$$C = C_0 - (k_z * t) \qquad\qquad (12\text{-}3)$$

Where C_0 and C are as defined above and
       k_z =   the zero-order rate constant (mg/l per day)

When C is more than 10 times K_s, the rate of biodegradation is more than 90% of the maximum rate. As a practical matter, a zero-order rate law is considered to be an acceptable approximation to the true rate of biodegradation whenever C is more than 10 times K, because the difference between the approximate rate and the true rate is less than the usual statistical variation in the measurements.

BIODEGRADATION OF MTBE, PETROLEUM HYDROCARBONS, AND CONSUMPTION OF OXYGEN

The capacity to degrade MTBE is found in a wide variety of microorganisms, including both gram positive and gram negative bacteria and one strain of fungus (Table 12-3). The maximum specific rate of MTBE degradation (gram MTBE per gram dry weight of cells per day) varies 20-fold among various MTBE-degrading bacteria.

Compare the maximum specific rate of MTBE degradation by MTBE-degrading organisms (Table 12-3) to the median rate and range of rates reported by Suarez and Rifai (1999) for degradation of benzene, toluene, and the xylenes (Table 12-2). The rates of degradation of MTBE and TBA are equivalent to the low end of the range of rates of degradation of benzene, toluene, and the xylenes. In general, the specific rate of MTBE degradation is lower in bacteria that co-metabolize MTBE, compared to bacteria that use it as the pri-

Table 12-3. **Rates of Aerobic Biodegradation of MTBE and TBA**

| Growth Substrate | Culture or Organism | Maximum Specific Degradation Rate (gram per gram cells per day) | | Reference |
		MTBE Degradation	TBA Degradation	
MTBE	BC-1 culture	0.82	0.34	Salanitro et al. (1994)
MTBE	Enrichment from Refinery Activated Sludge	0.36	Yes, K_s and max. spec. deg. rate similar to MTBE	Cowan and Park (1996)
MTBE	Enrichment from Biofilter	0.26 0.13		Fortin and Deshusses (1999)
MTBE	ENV735 Hydrogenophaga flava	3	≈ 0.6	Steffan et al. (2000b)
Butane	Graphium sp. (fungus)	0.022	Yes, but no rate data	Hardison et al. (1997)
Propane	ENV425	0.25	0.11	Steffan et al. (1997)
Pentane	Pseudomonas aeruginosa	0.25	No	Garnier et al. (1999)
Iso-butane	Five Strains Rhodococcus, Norcardia, Pseudomonas, Alcaligenes, Rhizobium.	0.13 to 0.89		Hyman et al. (2000)
Cyclohexane	Mixed Culture	0.15	0.024	Corcho et al. (2000)
TBA	Pseudomonas	Not degraded	0.43	Fayolle et al. (1999)

mary substrate. The specific rate of MTBE metabolism by the fungus was much lower than the rate by bacteria.

MTBE-degrading organisms require higher concentrations of oxygen than the usual aerobic bacteria. Maintaining adequate concentrations of oxygen is an important design consideration for aerobic biodegradation of MTBE. Ordinary bacteria use dissolved oxygen as the terminal electron acceptor for oxidative respiration. The effective half-saturation constant for oxygen respiration is usually less than 0.1 mg/l, and aerobic metabolism proceeds at the maximum rate at oxygen concentrations above 0.5 mg/l (Longmuir, 1954). Bacteria that metabolize MTBE have a second requirement for oxygen as a substrate for the mono-oxygenase enzymes that degrade MTBE. Koenigsberg *et al.* (1999) reported the relationship between the concentration of oxygen and the rate of MTBE metabolism by the mixed culture described by Fortin and Deshusses (1999). Their data are presented in Figure 12-1. The half-saturation constant for oxygen was near 3 mg/l. To allow active growth of their culture MV-100, Salanitro *et al.* (2000) maintained oxygen concentrations above 4 mg/l in their demonstration plots. Salanitro *et al.* (1998) reported that degradation of MTBE by their culture BC-1 was inhibited at concentrations less than 1 mg/l. Park and Cowan (1997) reported a value of K_s for dissolved oxygen for their mixed culture of 0.9 mg/l.

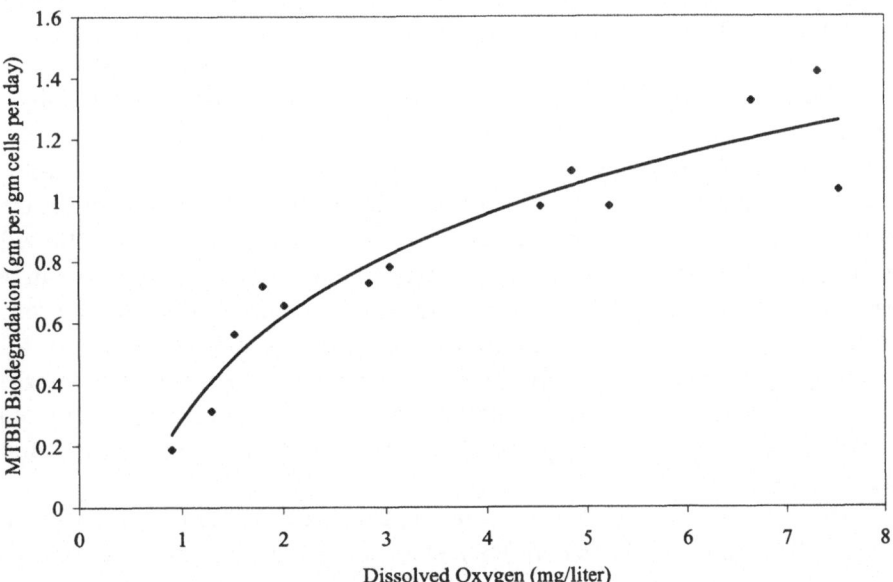

Figure 12-1.   **Effect of Concentration of Dissolved Oxygen on the Rate of Biodegradation of MTBE**
Data from Koenigsberg *et al.* (1999).

Table 12-4.   **Half-Saturation Constants for Aerobic Biodegradation of MTBE**

Growth Substrate	Culture or Organism	Half-Saturation Constant for MTBE K_s (mg/l)	Reference
MTBE	Enrichment from Refinery Activated Sludge	0.33	Park and Cowan (1997)
MTBE	Enrichment from Refinery Activated Sludge	4.8	Cowan and Park (1996)
MTBE	PM-1	≈ 50	Hanson et al. (1999)
Pentane	Pseudomonas aeruginosa	185	Garnier et al. (1999)
Iso-butane	Five Strains Rhodococcus, Norcardia, Pseudomonas, Alcaligenes, Rhizobium.	10 to 44	Hyman et al. (2000)

The half-saturation constant for MTBE for aerobic biodegradation varies over two orders of magnitude among different cultures (Table 12-4). The range in half-saturation constants for MTBE degradation (Table 12-4) is equivalent to the range of half-saturation constants for degradation of benzene, toluene, and xylenes compounds (Table 12-2). The range of half-saturation constants is far above the expected cleanup goals (0.005 to 0.02 mg/l). The range includes the starting concentration of MTBE at many sites. As a consequence, the rate law for biodegradation of MTBE can be expected to change as the concentration of MTBE is reduced from the initial concentration to the cleanup goal. At many sites, the initial concentration of MTBE will be high, and there will be an interval when the rate of biodegradation will be best described by the zero-order rate, followed by an interval where the complete hyperbolic equation is required to describe the rate remediation, followed by an interval where a first-order rate is the most appropriate description. At many sites, a simple calculation using just a first-order rate or just a zero-order rate will not be appropriate to forecast the time required to remediate a site.

Many of the organisms or cultures that degraded MTBE could also degrade TBA (Table 12-3). When TBA was degraded, the rate was usually slower than the rate of degradation of MTBE. As a consequence, TBA accumulated in the cultures while MTBE was being degraded, and then was depleted after the MTBE was exhausted. Both MTBE and TBA require the action of a mono-oxygenase enzyme for their metabolism. In two pure cultures that readily degrade MTBE, the mono-oxygenase for MTBE is a different enzyme than the mono-oxygenase for TBA. Hatzinger *et al.* (2001) identified two separate mono-oxygenase enzymes in Hydrogenophaga flava strain ENV735, and Deeb *et al.* (2000a) identified two mono-oxygenases in Rubrivivex species PM-1.

As a practical matter, the only intermediate of MTBE biodegradation that has been shown to accumulate in natural ground waters is TBA. Depending on the relative rates of biodegradation of TBA and MTBE, TBA may or may not accumulate. Because TBA is readily degradable in aerobic ground water, it accumulates for a brief period while MTBE is being degraded to produce TBA. Once MTBE is depleted, then the accumulated TBA is degraded.

Pruden *et al.* (2001) examined the effects of other oxygenates and natural petroleum hydrocarbons on the biodegradation of MTBE in aerobic bioreactors. One experimental reactor was fed a solution of MTBE, a second reactor received MTBE and diethyl ether, a third received MTBE and DIPE, a fourth received MTBE and ethanol, and a fifth received MTBE and BTEX compounds. The concentrations of MTBE and the other substrates supplied to the reactors were near 70 or 140 mg/l. After acclimation, the concentration of MTBE in the effluent of all the reactors varied from 0.00005 to 0.005 mg/l. The concentration of TBA was less than 0.01 mg/l.

Figure 12-2 presents a simplified and generalized pathway for complete aerobic metabolism of MTBE. The figure combines features in the pathways published by Steffan *et al.* (1997) and Deeb *et al.* (2000a). In the organisms studied, the first transformation is carried out by a mono-oxygenase enzyme. These enzymes insert one oxygen atom from molecular oxygen into the organic compound being metabolized. The other oxygen atom is reduced to form water. The first stable products are formaldehyde and TBA. The TBA can be transformed by a mono-oxygenase to form 2-methyl-2-hydroxy-1-propanol (MHP). MHP is oxidized to 2- hydroxyisobutyric acid (HIBA). Elimination of the carboxylic acid group from HIBA produces 2-propanol (IPA), which in turn can be oxidized to acetone. Acetone and 2-propanol are rapidly degraded in aerobic ground waters to carbon dioxide, water, and biomass.

PROSPECTS FOR BIODEGRADATION OF MTBE IN THE FIELD BY NATIVE MICROORGANISMS

In 1988, the University of Waterloo conducted a large controlled-release study of MTBE degradation at Canadian Forces Base Borden in Ontario (Hubbard *et al.*, 1994; Schirmer and Barker, 1998; Schirmer *et al.*, 1999). To simulate the effects of a gasoline spill on ground water, they injected ground water containing 19 mg/l BTEX and 269 mg/l MTBE into a shallow sandy water table aquifer and monitored the development of the plume. After 476

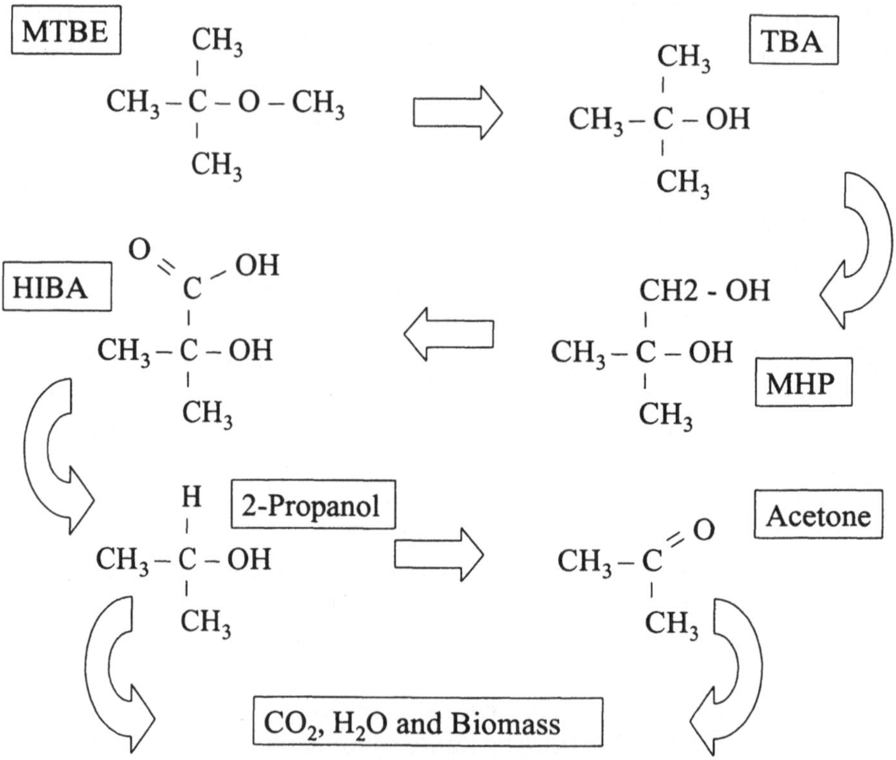

Figure 12-2.   **Generalized Pathway for Aerobic Biodegradation of MTBE MHP is 2-methyl-2-hydroxy-1-propanol, and HIBA is 2-hydroxyisobutyric acid.**

days, the BTEX compounds were gone; however, based on their mass balance calculations, there was no statistically significant evidence of biodegradation of MTBE. They reported the absence of biodegradation of MTBE (or rather the lack of evidence for biodegradation of MTBE) in 1994 (Hubbard *et al.*, 1994). This report supported and reinforced the conventional wisdom at the time that MTBE did not biodegrade in aquifers.

In 1995, the University of Waterloo sampled the MTBE plume again. The concentrations of MTBE were much lower than expected based on dilution and dispersion alone. In 1996, they sampled the plume using a fine grid to give themselves confidence in their estimate of the mass of MTBE remaining. After 3,000 days, only 3% of the MTBE injected into the aquifer remained in the aquifer (Schirmer and Barker, 1998; Schirmer *et al.*, 1999). They attributed the disappearance of MTBE in the plume to aerobic biodegradation. They were able to confirm the potential for biodegradation of MTBE in laboratory microcosm studies; however, acclimation was a rare event, only 3 of 40 microcosms acclimated after 20 months of incubation. Once acclimation occurred in laboratory microcosms, biodegradation was rapid and extensive. The prospects for aerobic biodegradation by native microorganisms may be related to the age of the spill and the time that has been available for acclimation of the native microorganisms to MTBE.

There seems to be a wide variation from one site to another in the distribution and activity of native microorganisms that can degrade MTBE. At a site in Port Hueneme, California, Salinitro *et al.* (2000) showed that natural bacteria in microcosm studies could consume 10 mg/l of MTBE in 10 days. In a field-scale demonstration, a plot containing only native bacteria consumed MTBE, but not as well as a plot that had been augmented with an MTBE-degrading culture. In contrast, at a site in New Jersey, an air sparging system had little effect on the concentration of MTBE (Steffan *et al.*, 2001). This implies that adding oxygen to the ground water was not sufficient to stimulate *in situ* biodegradation at this site. Sparging with propane and augmentation with the propane-oxidizing strain ENV425 resulted in reductions in the concentrations of MTBE of 93% within three months. Kane *et al.* (2001) constructed microcosms with material from four MTBE spills in California. In material from a site at Palo Alto, native organisms reduced 4.5 mg/l of MTBE to less than 0.1 mg/l in 15 days. In material from a site at Travis Air Force Base, native organisms reduced 4.2 mg/l to less than 0.2 mg/l in less than 20 days. However, material from a site in Sacramento showed no significant degradation in 75 days and material from a site in Sunnyvale showed no degradation in 23 days. Salinitro *et al.* (1998) surveyed sites for the presence of MTBE-degrading bacteria. They examined ground water and soil from 10 sites: 2 retail sites in California, refineries in Louisiana and Illinois, distribution terminals in Nevada and Ohio, a pipeline in Texas, and retail sites in Michigan, Texas, and New Jersey. They were able to isolate MTBE-degrading organisms from 2 of the 10 sites and demonstrate MTBE degradation in microcosms constructed with material from 2 sites.

REMEDIAL TECHNOLOGY FOR GROUND WATER

There are two engineering challenges associated with *in situ* aerobic bioremediation of MTBE. In general, ground water impacted by MTBE from fuel spills is devoid of dissolved oxygen. Dissolved oxygen must be added to the ground water to meet the biochemical oxygen demand of MTBE, the biochemical demand of other organic materials in the ground water or associated with the aquifer matrix, and non-biological oxygen demand associated with reduced minerals in the aquifer. As discussed above, the native microbes at some sites do not readily degrade MTBE at rates required for timely remediation; therefore, the second challenge at these sites is to introduce and distribute cultures of bacteria that are capable of degrading MTBE.

At some sites, adding dissolved oxygen to the ground water is the only action that is necessary to stimulate aerobic biodegradation of MTBE. At two sites, MTBE biodegradation was achieved by simply sparging with air or oxygen (Carter *et al.*, 1997; Javanmardian and Glasser, 1997). The concentration of MTBE in monitoring wells was reduced by an order of magnitude.

At other sites it was necessary or advantageous to augment the aquifer with MTBE-degrading microorganisms. Salanitro *et al.* (2000) added a culture of MTBE-degrading microorganisms to treat a plume at Port Hueneme, California. Oxygen and active organisms were introduced into the aquifer to create an *in situ* biobarrier. The aquifer containing the plume was naturally anoxic.

The biobarrier was constructed along a transect that was perpendicular to the direction of ground water flow and downgradient of the area with residual gasoline. The biobarrier is sustained by continued addition of oxygen over time. As the ground water moved past the transect, MTBE was degraded.

Sparge wells were installed at the bottom and in the middle of the aquifer being treated. Sparged gas can reduce the water-filled porosity, and thus the hydraulic conductivity of the aquifer being treated. To avoid this gas blockage, Salinitro *et al.* (2000) sparged with pure oxygen instead of air, and sparged on an intermittent cycle rather than continuously. The oxygen injection system was able to sustain dissolved oxygen concentrations in excess of 20 mg/l. The MTBE-degrading culture MC-100 was injected into the aquifer using Geoprobe® tools and a grout pump. After 261 days of operation, the concentration of MTBE in shallow monitoring wells ranged from 0.001 to 0.004 mg/l. Concentrations in deep monitoring wells ranged from 0.007 to 0.02 mg/l. In a second demonstration plot, oxygen was injected, but the culture MC-100 was not. The native aquifer bacteria established a biobarrier, but the reductions in concentration of MTBE were not as great as the reductions in the plot receiving the MTBE-degrading culture. (See Chapter 26 for further discussion of the Port Hueneme site.)

Spinnler *et al.* (2001a and 2001b) applied the same technology to a MTBE plume in Connecticut. Eighteen oxygen injection points were installed along a line that was downgradient of the spill and perpendicular to ground water flow. The oxygen injection wells were spaced approximately 1.5 meters (5 feet) apart. Oxygen was sparged at two depths at each location. The culture was injected in a vertical cross section downgradient of the source area of the plume. Individual injection points were on two-foot centers laterally and vertically. The initial concentration of MTBE in the most impacted well in the plume in Connecticut was greater than 100 mg/l. This is much higher than the concentration treated successfully at Port Hueneme, California. The reduction in concentration of MTBE in individual monitoring wells varied from one to four orders of magnitude (Table 12-5). (See Chapter 28 for additional information on this site.)

Table 12-5.   **Performance of Technology to Treat MTBE in Water**

Technology	Initial Concentration mg/l	Final Concentration mg/l	Time Required days	Reference
In situ treatment of ground water				
Bioremediation with culture MC-100	7	0.01 to 0.004	130 to 200	Salanitro *et al.* (2000)
Bioremediation with native organisms	7	0.07	130 to 200	Salanitro *et al.* (2000)
Bioremediation with culture MC	120	0.060	250	Spinnler *et al.* (2001a)
	13	1.9	250	
	2.0	0.003	250	
Bioremediation with propane-oxidizing organisms	320	70	90	Steffan *et al.* (2001)
	320	30	90	
	87	2.5	75	
Permeable In Situ Treatment Zone	1.5	0.01	4	Wilson *et al.* (2001)
Bioremediation with proprietary mixture of MgO$_2$	19.6	2.66	60	Landmeyer *et al.* (2001)
	29.8	6.42		
	2.0	0.2		
Air sparging/soil vapor extraction	190	30	300	Giattino *et al.* (2000)
	7	1.8		
	26	21		
	33	68		
Bioreactors for aboveground treatment				
Membrane bioreactor with ENV735	1000	<0.1	3	Steffan *et al.* (2000b)
	10	<0.1	0.125	
Fluidized bed reactor with ENV735	10	<0.005	0.0104	Steffan *et al.* (2001)
Fluidized bed reactor	2.7 to 9.8	0.067 to 0.013	0.3	Chapter 29

Steffan *et al.* (2001) used a propane-oxidizing organism to establish a bio-barrier in an MTBE plume in New Jersey. An air sparging system at the site successfully removed BTEX from the ground water, but not MTBE. Propane was added to the sparge air for 10 minutes every three hours at a concentration of 0.2% propane. After one month of the system's operation, a culture of propane-oxidizing bacteria (POB ENV425) was added directly to the sparge wells. The initial concentrations of MTBE ranged from approximately 90 mg/l to approximately 320 mg/l. Within three months the concentration of MTBE was reduced approximately 10-fold in the monitoring wells with elevated levels of dissolved oxygen (see Table 12-5). The maximum removal was 97%, from 87 mg/l to 2.5 mg/l. When the sparge system was turned off, the concentrations of MTBE rebounded in the monitoring wells. When the native microorganisms were not capable of degrading MTBE, it was possible to augment the microbes in the aquifer with organisms that can readily degrade MTBE. (See Chapter 27 for more information on this site.)

Wilson *et al.* (2001) evaluated MTBE degradation in a plume at Vandenberg Air Force Base, California. The aquifer containing the plume was naturally anaerobic. In addition to the oxygen demand associated with MTBE, there was appreciable demand associated with other soluble organic materials in the ground water and with the aquifer matrix itself. In aquifers of this type, it can be difficult to achieve high concentrations of dissolved oxygen by direct air sparging or by sparging pure oxygen. Wilson *et al.* (2001) diverted the flow of ground water into a treatment cell. To create the cell, they excavated the aquifer matrix and replaced it with pea gravel. Under the natural gradient, ground water entered the cell, flowed past a device that introduced oxygen into the water, then flowed through the pea gravel, and exited back to the aquifer at the other end of the treatment cell. The cell was essentially a treatment gate in a funnel-and-gate system. The flow path through the pea gravel was 4.5 meters (15 feet) long and the residence time of water in the treatment cell was five days. The concentration of MTBE at the entrance of the treatment cell was maintained at 1.5 mg/l. The concentration of oxygen near the device that introduced oxygen was near 15 mg/l. The rate of oxygen consumption as water moved through the treatment cell was approximately 4 mg/l per day. After full acclimation, the concentration of MTBE in the cell was below 0.01 mg/l in the downgradient portions of the treatment cell. The pilot-scale system was a facsimile of a funnel and treatment gate system. The treatment gate returned water to the aquifer with concentrations of MTBE that were less than 0.01 mg/l.

Landmeyer *et al.* (2001) injected a slurry of a proprietary mixture of MgO_2 into an anoxic plume of MTBE. The material slowly reacted with water to produce dissolved oxygen. A grout pump was used to inject the slurry using direct push tools. A slurry containing 30 kilograms of MgO_2 was injected into each of 18 locations. The slurry was injected along a single transect that was 29 meters (95 feet) long. The transect was oriented perpendicular to the direction of ground water flow, and was approximately 150 meters (492 feet) downgradient from the spill. The slurry was injected on 1.7-meter (5.6-foot) centers; at each location the slurry extended vertically from the bottom of the contaminated interval to the elevation of the expected seasonal high water table. The ground water seepage velocity was at least 33 meters (108 feet) per year. At a monitoring well that was 2 meters (7 feet) downgradient of the injected slurry, the concentration of dissolved oxygen reached 12 mg/l, and remained at or above 2 mg/l for one year. At a monitoring well that was 8 meters (26 feet) downgradient, the concentration of dissolved oxygen reached 8 mg/l, and remained at or above 2 mg/l for a year. Increased concentrations of dissolved oxygen did not reach wells that were 12 or 16 meters (39 or 52 feet) downgradient. In the region of the aquifer with elevated concentrations of dissolved oxygen, the concentration of MTBE was reduced from 4-fold to 10-fold (Table 12-5).

In general the published work on *in situ* bioremediation of MTBE is promising. Based on the work published to date, when the initial concentrations of MTBE were in the range of 1 to 10 mg/l, it was often possible to reduce the concentration of MTBE to concentrations below the USEPA health advisory of 0.02 to 0.04 mg/l. At concentrations of MTBE in excess of 100 mg/l, it was possible to achieve reductions in concentration of at least 10-fold. There is no theoretical reason why bacteria in an aquifer cannot degrade MTBE at concentrations in excess of 100 mg/l to concentrations below the USEPA health advisory, as these low concentrations are achieved in laboratory microcosms studies routinely. The failure to reach these low levels in the field may result from spatial heterogeneity in the pneumatic or hydraulic conductivity of the aquifer, which restricts the ability to supply oxygen to certain regions of the MTBE plume.

Aerobic bioremediation of MTBE may be a side benefit of technologies that are designed to extract contaminants from the aquifer. Giattino *et al.* (2000) reported reductions in concentrations of MTBE in ground water after air sparging and SVE (Table 12-5). These technologies remove MTBE directly from the subsurface as a vapor, but they also supply oxygen in air to support aerobic bioremediation. At the demonstration site, air sparging and SVE were implemented for 300 days. At two locations the concentration of MTBE in ground water were reduced by 85% and 82%, while the removal of benzene was 96% and 88%, and the removal of xylenes was 93% and 98%. The removal of MTBE was roughly equivalent to the removal of benzene and xylenes.

If removal was controlled by partitioning to air and removal as a vapor in air, the extent of removal should be related to the Henry's Law constant for partitioning between air and water. The dimensionless Henry's Law constant for MTBE is approximately 0.024 parts in air per part in water. The Henry's Law constants for benzene and xylenes are 0.22 and 0.29, which are 10 times larger. Benzene and xylenes should partition from ground water to air and be stripped approximately 10 times as rapidly as MTBE. Giattino *et al.* (2000) did not claim that biodegradation made a substantial contribution to removals; however, the fact that the removals of MTBE and of benzene and the xylenes were equivalent suggests that biodegradation by native microorganisms made a substantial contribution to these removals. At two other locations, there was not a significant reduction in the concentration of MTBE in ground water, probably because there was not good exchange of air to these portions of the contaminated aquifer (Table 12-5).

Successful implementation of aerobic bioremediation is controlled by site-specific factors. Spinnler *et al.* (2001a) identified suitable sites as those with plumes in unconsolidated sediments, where the direction and velocity of the plume is constant or near constant, where the fluctuation in the elevation of the water table is less than 3 meters (10 feet), and the depth to water is less than 7.6 meters (25 feet) below ground surface (bgs).

The direction of ground water flow in unconsolidated sediments can be predicted from the slope of the water table. In bedrock aquifers the flow of ground water follows fractures, and the direction of flow is much less predictable. When the direction of the plume is constant or near constant, it is possible to place a line of air injection wells or MgO_2 injection points that include and treat the entire plume as it moves away from the source. If the plume changes flow direction, it may move out of the influence of the air injection wells or MgO_2 injection. When the velocity of the plume is constant or near constant, it is possible to estimate the amount of air or MgO_2 to be injected to meet the oxygen demand of the plume. The radius of influence of an air sparge well changes as the depth of the water changes. When the water table elevation varies more than 3 meters (10 feet), it is difficult to design a sparge system that will operate efficiently under all conditions.

Another important feature is the spatial variation in hydraulic or pneumatic conductivity in the sediments containing the plume. In sites with great variation in conductivity, sparged air or water amended with oxygen will move in the most conductive regions. Fluids bringing oxygen may fail to thoroughly perfuse contaminants trapped in less conductive sediments, and treatment may not be complete. As a consequence, the plume may rebound when the sparge wells are turned off or when MgO_2 is exhausted.

When the flow of ground water through sediments is uniform and there are no confining layers to prevent the escape of air from the aquifer, direct sparging may be the most efficient way to deliver oxygen. If there are confining layers in the aquifer that prevent the escape of air, it may be necessary to sparge with pure oxygen on an intermittent schedule.

As outlined above, many site-specific factors influence the feasibility and selection of *in situ* oxygen amendment methods. These include ground water elevation, flow direction, and velocity (and variability in these parameters); stratigraphy; soil heterogeneity and potential preferential flow pathways; and the presence of nearby sensitive receptors (*e.g.*, basements of nearby homes subject to vapor intrusion if sparging were to be implemented at a site). Another important factor is the magnitude of the demand for oxygen. Oxygen is needed for biodegrading both the constituents of concern (*e.g.*, gasoline components) and non-target organics (*e.g.*, total organic carbon [TOC]) and inorganics (*e.g.*, ferrous iron). Ground water flow velocity will also influence the amount of oxygen required, sites with more rapid ground water flow requiring more oxygen. Table 12-6 summarizes available methods for amending the subsurface with oxygen and indicates generally the relative

Table 12-6.   **Available Methods for Amending the Subsurface with Oxygen**

Oxygen Amendment Method	Description and Example Vendors	Applicable Oxygen Demand
Hydrogen peroxide injection	Injection of H_2O_2 below the water table; H_2O_2 dissociates to oxygen and water	High
Air or oxygen sparging	Injection of air or oxygen below water table	Medium
Diffusive oxygen emitters	Diffusion of oxygen through hollow fiber membranes (tubing) into water flowing past the tubing (*e.g.*, iSOC®, University of Waterloo Diffusive Emitters)	Medium
In-well oxygenation	Aeration stone (similar to what is used in an aquarium) introducing oxygen at the bottom of a well via an air compressor (Raymond patent)	Medium
Electrolysis	Electrolytic cells dissociate water molecules into hydrogen and oxygen (*e.g.*, Iso-Gen®)	Low
Solid forms of oxygen	Magnesium/oxygen compounds (*e.g.*, Oxygen Release Compound [ORC®], Permeox®) slowly release dissolved oxygen to groundwater	Low

Notes: Oxygen demand is proportional to the concentrations of the constituents of concern (*e.g.*, MTBE); the concentrations of non-target constituents (*e.g.*, total organic carbon; ferrous iron); and groundwater velocity.

The table does not include chemical oxidation methods (which often add oxygen in addition to their direct chemical oxidation activity).

Mention of example vendors or products should not be construed as endorsement. (Lack of mention should not be construed as lack of endorsement.)

oxygen demand that each can satisfy. Oxygen demand and the other site-specific factors mentioned above must be considered in selecting an appropriate method.

If there is a great deal of spatial variation in hydraulic conductivity, it may be necessary to collect the plume into an artificial treatment cell or an aboveground bioreactor to ensure good mixing of oxygen into the impacted ground water. Solid experimental work has been done on laboratory bioreactors to treat MTBE in water or in air (Fortin and Deshusses, 1999; Steffan et al, 2000b; Steffan *et al.*, 2001). See Chapter 29 for a description of a full-scale bioreactor to treat MTBE. Based on the performance of the laboratory bioreactors and field-scale bioreactor, it should be possible to build bioreactors that can reduce the concentration of MTBE in ground water from initial concentrations of 1 to 10 mg/l to concentrations below the USEPA health advisory (see Table 12-5).

For additional information on aerobic biodegradation of MTBE, a number of excellent reviews are available (Deeb *et al.* 2000b; Fayolle *et al.*, 2001; Prince, 2000; Stocking *et al.*, 2000; Salinitro, 1995).

DISCLAIMER

The USEPA through its Office of Research and Development funded the research described here under in-house task 5857 (Natural Attenuation of MTBE) at the Ground Water and Ecosystems Restoration Division. It has not been subjected to Agency review and therefore does not necessarily reflect the views of the Agency. No official endorsement should be inferred.

REFERENCES

Borden, R.C., Daniel, R.A., LeBrun IV, L.E., and Davis, C.W. 1997. Intrinsic biodegradation of MTBE and BTEX in a gasoline-contaminated aquifer. *Water Resources Research.* 33(5), 1105–1115.

Bradley, P.M., Chapelle, F.H., and Landmeyer, J. 2001. Effect of redox conditions on MTBE biodegradation in surface water sediments. *Environmental Science and Technology.* 35(23), 4643–4648.

Carter, S.R., Bullock, J.M., and Morse, W.R. 1997. Enhanced biodegradation of MTBE and BTEX using pure oxygen injection. In: *In Situ and On-Site Bioremediation: Papers from the Fourth International In Situ and On-Site Bioremediation Symposium*, New Orleans, Louisiana. April 28–May 1, 1997. Volume 4, p. 147. (Alleman, B.C. and Leeson, A., Ed.). Columbus and Richland, Battelle Press.

Corcho, D., Watkinson, R.J., and Lerner, D.N. 2000. Cometabolic degradation of MTBE by a cyclohexane-oxidizing bacteria. Bioremediation and phytoremediation of chlorinated and eecalcitrant compounds. In: *The Second International Conference on Remediation of Chlorinated and Recalitrant Compounds*, Monterey, California. May 22–25, 2000. pp. 183–189. (Wickramanayake, G.B., Gasvaskar, A.R., Alleman, B.C., and Magar, V., Eds.). Columbus and Richland, Battelle Press.

Cowan, R.M. and Park, K. 1996. Biodegradation of the gasoline oxygenates MTBE, ETBE, TAME, TBA, and TAA by aerobic mixed cultures. In: *Proceedings of the 28th Mid-Atlantic Industrial and Hazardous Waste Conference*, Buffalo, New York. July 15–17, 1996. pp. 523–530.

Deeb, R.A., Nishino, S., Spain, J., Hu, H.-Y., Scow, K., and Alvarez-Cohen, L. 2000a. MTBE and benzene biodegradation by a bacterial isolate via two independent monooxygenase-initiated pathways. In: *Preprints of the Extended Abstracts*, Vol 40(1), pp. 280–282. Symposium paper presented before the Division of Environmental Chemistry, American Chemistry Society, San Francisco, California, March 26–30, 2000.

Deeb, R.A., Scow, K.M., and Alvarez-Cohen, L. 2000b. Aerobic MTBE biodegradation: An examination of past studies, current challenges and future research directions. *Biodegradation.* 11, 171–186.

Fayolle, F., Le Roux, F., Hernandez, G., and Vandecasteele, J.-P. 1999. Mineralization of ethyl t-butyl ether by defined mixed bacterial cultures. *In situ* bioremediation of petroleum hydrocarbon and other organic compounds. In: *The Fifth International In Situ and On-Site Bioremediation Symposium*, San Diego, California. April 19–22, 1999. pp. 25–30. (Alleman, B.C. and Leeson, A., Eds.). Columbus and Richland, Battelle Press.

Fayolle, F., Vandecasteele, J.-P., and Monot. F. 2001. Microbial degradation and the fate in the environment of metyl *tert*-butyl ether and related fuel oxygenates. *Applied Microbiology and Biotechnology*. 56, 339–346.

Fortin, N.Y. and Deshusses, M.A. 1999. Treatment of methyl tert-butyl ether vapors in biotrickling filters. 1. Reactor startup, steady state performance, and culture characteristics. *Environmental Science and Technology*. 33(17), 2980–2986.

Garnier, P., Auria, R., Magana, M. and Revah, S. 1999. Cometabolic biodegradation of methyl *t*-butyl ether by a soil consortium. In: *In Situ Bioremediation of Petroleum Hydrocarbon and Other Organic Compounds. The Fifth International In Situ and On-Site Bioremediation Symposium*, San Diego, California. April 19–22, 1999. pp. 31–33. (Alleman, B.C. and Leeson, A., Eds.). Columbus and Richland, Battelle Press.

Giattino, R., Gibbs, J.T., Desilva, S., Lingens, R., and Sullivan, M. 2000. Remediation of MTBE-contaminated groundwater using conventional air sparging and soil vapor extraction. Case studies in the remediation of chlorinated and recalcitrant compounds. In: *The Second International Conference on Remediation of Chlorinated and Recalcitrant Compounds*, Monterey, California. May 22–25, 2000. pp. 65–72. (Wickramanayake, G.B., Gavaskar, A.R., Gibbs, J.T., and Means, J.L., Eds.). Columbus and Richland, Battelle Press.

Hanson, J.R., Ackerman, C.E., and Scow, K.M. 1999. Biodegradation of methyl *tert*-butyl ether by a bacterial pure culture. *Applied and Environmental Microbiology*. 65(11), 4788–4792.

Hardison, L.K., Curry, S.S., Ciuffetti, L.M., and Hyman, M.R. 1997. Metabolism of diethyl ether and cometabolism of methyl *tert*-butyl ether by a filamentous fungus, a *Graphium* sp. *Applied and Environmental Microbiology*. 63(8), 3059–3067.

Hatzinger, P.B., McClay, K., Vainberg, S., Tugusheva, M., Condee, C.W., and Steffan, R.J. 2001. Biodegradation of methyl tert-butyl ether by a pure bacterial culture. *Applied and Environmental Microbiology*. 67(12), 5601–5607.

Hubbard, C.E., Barker, J.F., O' Hannesin, S.F., Vandegriendt, M., and Gillham, R.W. 1994. *Transport and Fate of Dissolved Methanol, Methyl-tertiary-Butyl Ether, and Monoaromatic Hydrocarbons in a Shallow Sand Aquifer*. Health & Environmental Sciences Department, Washington DC. American Petroleum Institute Publication 4601.

Hyman, M., Taylor, C., and O'Reilly, K. 2000. Cometabolic degradation of MTBE by *iso*-alkane-utilizing bacteria from gasoline-impacted soils.

Bioremediation and phytoremediation of chlorinated and recalcitrant compounds. In: *The Second International Conference on Remediation of Chlorinated and Recalitrant Compounds,* Monterey, California. May 22–25, 2000. pp. 149–55. (Wickramanayake, G.B., Gasvaskar, A.R., Alleman, B.C., and Magar V., Eds.). Columbus and Richland, Battelle Press.

Javanmardian, M. and Glasser, H.A. 1997. In-situ biodegradation of MTBE using biosparging. American Chemical Society, Division of Environmental Chemistry, *Preprints of Extended Abstracts.* 37(1), 424.

Jensen, H.M. and Arvin, E. 1990. Solubility and degradability of the gasoline additive MTBE, methyl *tert*-butyl ether and gasoline compounds in water. In: *Contaminated Soil '90,* pp. 445-448. (Arendt, F., Hinsenveld, M., and van Brink, W.J., Eds.). Dordrecht, the Netherlands, Kluwer.

Kane, S.R., Beller, H.R., Legler, T.C., Koester, C.J., Pinkart, H.C., Halden, R.U., and Happel, A.M. 2001. Aerobic biodegradation of methyl tert-butyl ether by aquifer bacteria from leaking underground storage tank sites. *Applied and Environmental Microbiology.* 67(12), 5824–5829.

Koenigsberg, S., Sandefur, C., Mahaffey, W., Deshusses, M. and Fortin, N. 1999. Peroxygen mediated bioremediation of MTBE. *In situ* bioremediation of petroleum hydrocarbon and other organic compounds. In: *The Fifth International In Situ and On-Site Bioremediation Symposium,* San Diego, California. April 19–22, 1999. pp. 13–18. (Alleman, B.C and Leeson, A., Eds.). Columbia and Richland, Battelle Press.

Landmeyer, J.E., Chapelle, F.H., Herlong, H.H., and Bradley, P.M. 2001. Methyl tert-butyl ether biodegradation by indigenous aquifer microorganisms under natural and artificial oxic conditions. *Environmental Science and Technology.* 35(6), 1118–1126.

Longmuir, I.S. 1954. Respiration rate of bacteria as a function of oxygen concentration. *Biochemistry.* 57, 81–87.

Mo, K., Lora, C.O., Wanken, A.E., Javanmardian, M., Yang, X., and Kulpa, C.F. 1997. Biodegradation of methyl t-butyl ether by pure bacterial cultures. *Applied Microbiology and Biotechnology.* 47, 69–72.

Park, K. and Cowan, R.M. 1997. Effects of oxygen and temperature on the biodegradation of MTBE. In: *Proceedings of the 213th ACS National Meeting, Division of Environmental Chemistry,* San Francisco, California. pp. 421–424.

Prince, R.C. 2000. Biodegradation of methyl *tertiary*-butyl ether (MTBE) and other fuel oxygenates. *Critical Reviews in Microbiology.* 26(3), 163–178.

Pruden, A., Suidan, M.T., Venosa, A.D., and Wilson, G.J. 2001. Biodegradation of methyl tert-butyl ether under various substrate conditions. *Environmental Science and Technology.* 35(21), 4235–4241.

Salinitro, J.P., Diaz, L.A., Williams, M.P., and Wisniewski, H.L. 1994. Isolation of a bacterial culture that degrades methyl t-butyl ether. *Applied and Environmental Microbiology.* 60, 2593–2596.

Salinitro, J.P. 1995. Understanding the limitations of microbial metabolism of ethers used as fuel octane enhancers. *Current Opinion in Biotechnology.* 6, 337–340.

Salanitro, J.P., Chou, C.-S., Wisniewski, H.L., and Vipond, T.E. 1998. Perspectives on MTBE biodegradation and the potential for *in situ* aquifer bioremediation. In: *Southwestern Regional Conference of the National Ground Water Association*, Anaheim, California. June 3–4, 1998. pp. 40–54.

Salanitro, J.P., Johnson, P.C., Spinnler, G.E., Maner, P.M., Wisniewski, H.L., and Bruce, C. 2000. Field-scale demonstration of enhanced MTBE bioremediation through aquifer bioaugmentation and oxygenation. *Environmental Science and Technology.* 34(19), 4152–4162.

Schirmer, M. and Barker, J.F. 1998. A study of long-term MTBE attenuation in the Borden Aquifer, Ontario, Canada. *Ground Water Monitoring and Remediation.* Spring 1998, 113–122.

Schirmer, M., Butler, B.J., Barker, J. F., Church, C.D., and Schirmer, K. 1999. Evaluation of biodegradation and dispersion as natural attenuation processes of MTBE and benzene at the Borden field site. *Physics and Chemistry of the Earth, Part B: Hydrology, Oceans and Atmosphere.* 24(6), 557–560.

Spinnler, G.E., Salanitro, J.P., Maner, P.M., and Lyons, K.A. 2001a. Enhanced bioremediation of MTBE (BioRemedy) at retail gas stations. *Contaminated Soil Sediment and Water.* Spring (Special Issue), 47–49.

Spinnler, G.E., Salanitro, J.P., Maner, P.M., and Johnson, P.C. 2001b. MTBE remediation at retail gas stations by bioaugmentation. In: *2001 Petroleum Hydrocarbons and Organic Chemicals in Ground Water: Prevention, Detection, and Remediation Conference and Exposition*, Houston, Texas. November 14–16, 2001. pp. 244–251.

Squillace, P.J., Pankow, J.F., Korte, N.E., and Zorgorski, J.S. 1997. Review of the environmental behavior and fate of methyl *tert*-buty ether. *Environmental Toxicology and Chemistry.* 16(9), 1836–1844.

Steffan, R.J., McClay, K., Vainberg, S., Condee, C.W., and Zhang, D. 1997. Biodegradation of the gasoline oxygenates methyl *tert*-butyl ether, ethyl *tert*-butyl ether, and *tert*-amyl ether by propane-oxidizing bacteria. *Applied and Environmental Microbiology.* 63(11), 4216–4222.

Steffan, R.J., Condee, C., Quinnan, J., Walsh, M., Abrams, S.H., and Flanders, J. 2000a. *in situ* application of propane sparging for MTBE bioremediation. Bioremediation and Phytoremediation of Chlorinated and Recalcitrant Compounds. In: *The Second International Conference on Remediation of Chlorinated and Recalitrant Compounds*, Monterey, California. May 22–25, 2000. pp. 157–164. (Wickramanayake, G.B., Gasvaskar, A.R., Alleman, B.C., and Magar, V., Eds.). Columbus and Richland, Battelle Press.

Steffan, R.J., Vainberg, S., Condee, C., McClay, K., and Hatzinger, P. 2000b. Biotreatment of MTBE with a new bacterial isolate. Bioremediation and phytoremediation of chlorinated and recalcitrant compounds. In: *The*

Second International Conference on Remediation of Chlorinated and Recalitrant Compounds, Monterey, California. May 22–25, 2000. pp. 165–173. (Wickramanayake, G.B., Gasvaskar, A.R., Alleman, B.C., and Magar, V., Eds.). Columbus and Richland, Battelle Press.

Steffan, R.J., Hatzinger, P.B., Farhan, Y., and Drew, S.R. 2001. *In situ* and *ex situ* biodegradation of MTBE and TBA in contaminated groundwater. In: *2001 Petroleum Hydrocarbons and Organic Chemicals in Ground Water: Prevention, Detection, and Remediation Conference and Exposition*, Houston, Texas. November 14–16, 2001. pp. 252–264.

Stocking, A.J., Deeb, R.A., Flores, A.E., Stringfellow, W., Talley, J., Brownell, R., and Kavanaugh, M.C. 2000. Bioremediation of MTBE: A review from a practical perspective. *Biodegradation*. 11, 187–201.

Suarez, M.P. and Rifai, H.S. 1999. Biodegradation rates for fuel hydrocarbons and chlorinated solvents in groundwater. *Bioremediation Journal*. 3(4), 337–362.

Wilson, R.D., Mackay, D.M., and Scow, K.M. 2001. *In situ* MTBE biodegradation supported by diffusive oxygen release. *Environmental Science and Technology*. 36(2), 190–199.

Yeh, C.K. and Novak, J.T. 1995. The effect of hydrogen peroxide on the degradation of methyl and ethyl tert-butyl ether in soils. *Water Environment Research*. 67, 826–834.

CHAPTER 13

Anaerobic *In Situ* Bioremediation

Kevin T. Finneran, Ph.D., GeoSyntec Incorporated
Derek R. Lovley, Ph.D., University of Massachusetts

INTRODUCTION

Many laboratories have been researching methods to accelerate *in situ* bioremediation of MTBE and TBA. Most of the focus has been on aerobic bioremediation because there are a number of aerobes known to utilize MTBE as a sole carbon and energy source; however, past research with the BTEX compounds indicates that anaerobic bioremediation can be just as effective as aerobic bioremediation (Lovley, 1997).

Early studies that focused on anaerobic MTBE bioremediation were inconclusive, but led to a general speculation that MTBE will not biodegrade to any significant extent in the absence of oxygen (Mormile *et al.*, 1994; Yeh and Novak, 1994; Salanitro, 1995). MTBE persisted for several hundred days in anaerobic laboratory incubations (Yeh and Novak, 1994). In the few reported instances when MTBE did biodegrade, TBA accumulated in the sediment as a metabolite in the degradation process (Yeh and Novak, 1994; Suflita and Mormile, 1993).

More recent research has indicated that MTBE and TBA biodegradation is possible. Evidence that MTBE degrades with each of the most dominant anaerobic electron acceptors — nitrate, Mn(IV), Fe(III), sulfate — as well as under methanogenic conditions has been provided in both laboratory and field studies, as will be described below. Finding that anaerobic degradation is possible was the first step in developing bioremediation strategies in which these processes will be accelerated to treat MTBE and TBA in the highly reduced source zone of petroleum-impacted aquifer or aquatic sediments. The following reviews the research to date with each of the anaerobic terminal electron accepting processes (TEAPs).

ANAEROBIC PROCESSES IN SUBSURFACE SEDIMENT

The source zone of petroleum-impacted subsurface environments is typically anaerobic. Aerobic microbial respiration rapidly depletes the oxygen with the influx of large concentrations of electron donor (Lovley, 1997). The anaerobic zone can become quite large in a short period of time, and depending on the factors controlling natural attenuation, constituents can spread downgradient from the source even after it is removed.

The anaerobic portion of the aquifer develops distinct areas dominated by specific TEAPs. The competing electron acceptors most prevalent in anaerobic aquifers are nitrate, Fe(III), and sulfate (Lovley, 1997) (Figure 13-1). Oxidation of BTEX compounds and MTBE is thermodynamically favorable with all of these electron acceptors (Figure 13-2); therefore, free energy values by themselves cannot explain why these distinct zones arise.

Competition amongst the microorganisms for the electron donors (*e.g.*, MTBE, BTEX) leads to this TEAP distribution. Nitrate yields the most energy of the anaerobic electron acceptors, and the denitrifying organisms can metabolize the electron donors at concentrations that are too low to sustain the Fe(III) reducers. When nitrate is depleted, the next most favorable acceptor is Fe(III). The Fe(III)-reducing microbes metabolize electron donors at concentrations too low to sustain the sulfate-reducers. In turn, sulfate-reducers can out-compete the methanogenic microorganisms. This is how the TEAP zones

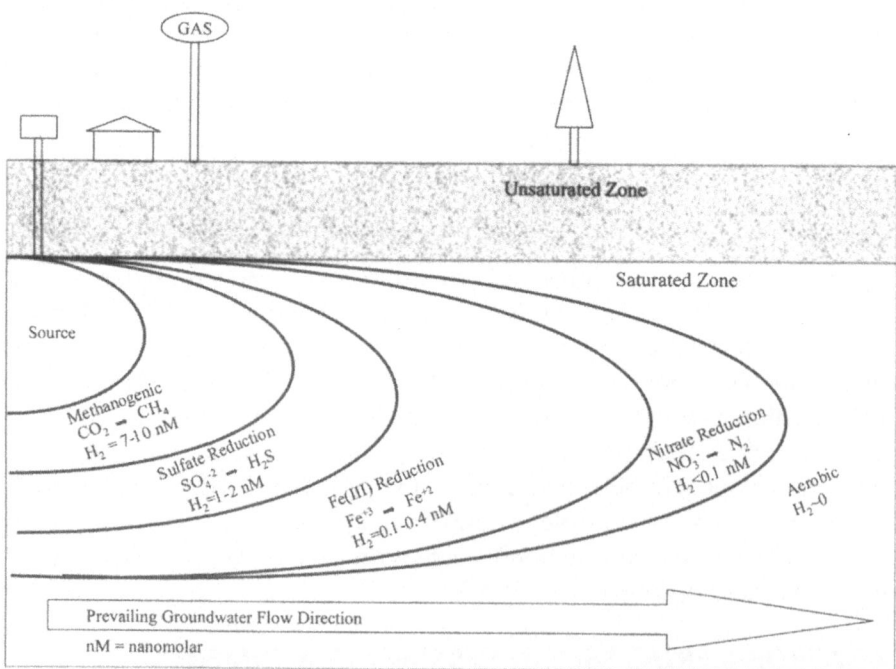

Figure 13-1.   **Distribution of Dominant Anaerobic Terminal Electron Accepting Processes in Gasoline Impacted Aquifer Sediment and Ground Water**

Reaction	$\Delta G^{\circ\prime}$ in kJ/Reaction
Aerobic Respiration $C_5H_{12}O + 7.5O_2 \rightarrow 5HCO_3^- + 5H^+ + H_2O$	-3245.8
Denitrification $C_5H_{12}O + 6NO_3^- + H^+ \rightarrow 5HCO_3^- + 3N_2 + 4H_2O$	-3054.8
Nitrate Reduction $C_5H_{12}O + 3.75NO_3^- + 2.5H^+ + 2.75H_2O \rightarrow 5HCO_3^- + 3.75NH_4^+$	-1951.2
Fe (III) reduction $C_5H_{12}O + 30Fe(OH)_3 + 55H^+ \rightarrow 5HCO_3^- + 30Fe^{2+} + 76H_2O$	-347.4
Sulfate Reduction $C_5H_{12}O + 3.75SO_4^{2-} + 2.5H^+ \rightarrow 5HCO_3^- + 3.75H_2S + H_2O$	-275.2
Methanogenesis $C_5H_{12}O + 2.75H_2O \rightarrow 3.75CH_4 + 1.25HCO_3^- + 1.25H^+$	-238.7

Figure 13-2. **Free Energy Yielded by the Theoretical Complete Oxidation of MTBE with Different Terminal Electron Acceptors** All free energies are calculated from the free energy of formation of each compound under biochemical steadystate conditions.
ΔG° is the free energy yield under biochemical standard state conditions (pH 7, 25°C) — free energy yielded per mole of MTBE oxidized to carbon dioxide.

arise; a hierarchy of processes is created that can be identified with existing monitoring wells using dissolved hydrogen, a key metabolic intermediate, as an indicator of anaerobic respiration (Lovley *et al.*, 1994a). Methanogenesis generally dominates in the area immediately adjacent to the source of petroleum contamination. Methanogenesis is the least energetically favorable anaerobic process and dominates only after nitrate, Fe(III), and sulfate are depleted. Within close proximity to the source, these electron acceptors are depleted almost immediately because of the high concentration of electron donors localized in the area.

ANAEROBIC BIOREMEDIATION STRATEGIES

Adding oxygen to the subsurface to stimulate the aerobic microbial community is a typical strategy for accelerating contaminant degradation; however, this is complicated in the source zone because of the highly reduced geochemical conditions and heavy biochemical and chemical oxygen demand (Hutchins, 1991). The difference between oxidized and reduced sediment is even visually apparent, with oxidized sediment being a reddish-brown color and reduced sediment being blackish-gray. Oxygen is not very soluble; adding it to ground water is an inefficient process within the source zone. In addition, oxygen that enters the source area can be depleted by chemically reacting with reduced compounds such as Fe(II). These newly formed Fe(III) oxides can plug oxygen injection wells or monitoring wells, further complicating the process. The electron acceptors used for anaerobic respiration are more soluble and less technically difficult to introduce into the subsurface.

Previous results with benzene indicate that anaerobic bioremediation may be just as effective as aerobic bioremediation. Benzene was previously thought to be recalcitrant under all but aerobic conditions, but it is now generally regarded as susceptible to anaerobic degradation. Different TEAPs support benzene degradation including Fe(III) reduction, sulfate reduction, and methanogenesis (Lovley *et al.*, 1994b; Anderson *et al.*, 1998; Weiner and Lovley, 1998b). Depending on the process, some of the acceptors have a higher electron accepting capacity than oxygen. For example, sulfate accepts eight electrons per mole reduced, as opposed to oxygen, which accepts two electrons per mole reduced — regardless of the electron donor. In the appropriate environment, this increased electron accepting capacity compensates for the lower free energy yield of the anaerobic electron acceptors.

Nitrate reduction is a high-energy yielding process that has been used to stimulate BTEX degradation (Hutchins *et al.*, 1991). In laboratory and field studies, benzene and other fuel components degraded quickly in the presence of added nitrate. Several studies indicate that benzene is degraded with the concomitant reduction of Fe(III) in impacted environments. At one petroleum-impacted site in Minnesota, benzene was degraded under *in situ* conditions within the Fe(III)-reducing zone. Upon further analysis, it was found that the bacteria responsible were members of the *Geobacteraceae*, known Fe(III) reducers (Anderson *et al.*, 1998). Several methods for stimulating Fe(III) reduction were successful in also stimulating benzene degradation, to be discussed in detail below.

Benzene degradation with sulfate as an electron acceptor is well characterized and has been successfully applied to a field site in Oklahoma for full-scale cleanup of benzene-impacted sediment (Anderson and Lovley, 2000). Finally, methanogenic aquatic sediment from the Potomac River has been used to develop a benzene-degrading enrichment culture (Weiner and Lovley, 1998b). The successes with benzene provide a foundation on which to develop MTBE-degradation strategies.

ANAEROBIC MTBE BIODEGRADATION WITH DIFFERENT TERMINAL ELECTRON ACCEPTORS

Nitrate Reduction.   Dissimilatory nitrate reduction and denitrification are two anaerobic processes that have been tested (See Figure 13-2). Yeh and

Novak (1994) reported anaerobic MTBE degradation under denitrifying conditions in soil that had been amended with starch and nutrients; however, it is uncertain if the conditions were strictly anaerobic. This study differentiated between "anoxic" and "anaerobic," leading to the conclusion that oxygen may have been present in the denitrifying incubations. Many of the nitrate-utilizing pathways are carried out by facultative anaerobes. Borden et al. (1997) attributed loss of MTBE at a field test site to anaerobic degradation, but the laboratory incubations did not confirm the results. Denitrification was the dominant anaerobic process in the laboratory incubations and the only anaerobic process in the aquifer; however, no direct evidence was presented confirming anaerobic MTBE degradation.

More recently, Bradley et al. (2001b) reported that [14C]-labeled MTBE is rapidly mineralized to $^{14}CO_2$ in denitrifying surface water sediment without a lag phase. Between 30% and 60% of the initial [14C]-MTBE was converted to $^{14}CO_2$ in 78 days. It was not reported that TBA accumulated during MTBE biodegradation. Denitrification was the dominant process (N^2 being the end product of this process; if NH^{4+} is the end product, the process is dissimilatory nitrate reduction), and other anaerobic processes such as sulfate reduction or methanogenesis were insignificant. These data were from two different streambed sites, indicating that the potential for nitrate-dependent MTBE oxidation may be more widespread than previously considered.

Fe(III) Reduction.   A wide variety of organic electron donors are oxidized by Fe(III)-reducing microorganisms. Since Fe(III) is often the most prevalent anaerobic electron acceptor in subsurface environments, Fe(III) reducers may be selected over time for MTBE biodegradation. In addition, there are several ways to increase the bioavailability of Fe(III) and stimulate Fe(III) reduction. Consequently, MTBE oxidation can be stimulated as well.

One of these techniques is adding humic acid substances to accelerate Fe(III) reduction (Figure 13-3). Humic substances are naturally-occurring compounds that result from the incomplete breakdown of complex organic matter. Fe(III)-reducing bacteria can directly reduce humic substances, and oxidized Fe(III) can abiotically accept electrons from reduced humic substances. The humic substances become re-oxidized and are again free to accept electrons in microbial metabolism. As such, the humic substances act as an electron shuttle between the microorganism and the Fe(III). This is significant because the prevalent form of iron in subsurface environments is Fe(III) oxides. It is generally accepted that bacterial cells must physically contact the insoluble Fe(III) oxides in order to reduce them; however, because they are not freely soluble, the Fe(III) oxides are occluded from rapid reduction. Soluble humic substances stimulate Fe(III) reduction because of this electron shuttling phenomenon (Lovley *et al.*, 1998), and increased Fe(III) reduction hastens the degradation of organic electron donors.

Synthetic electron shuttles have also been tested because of the recent

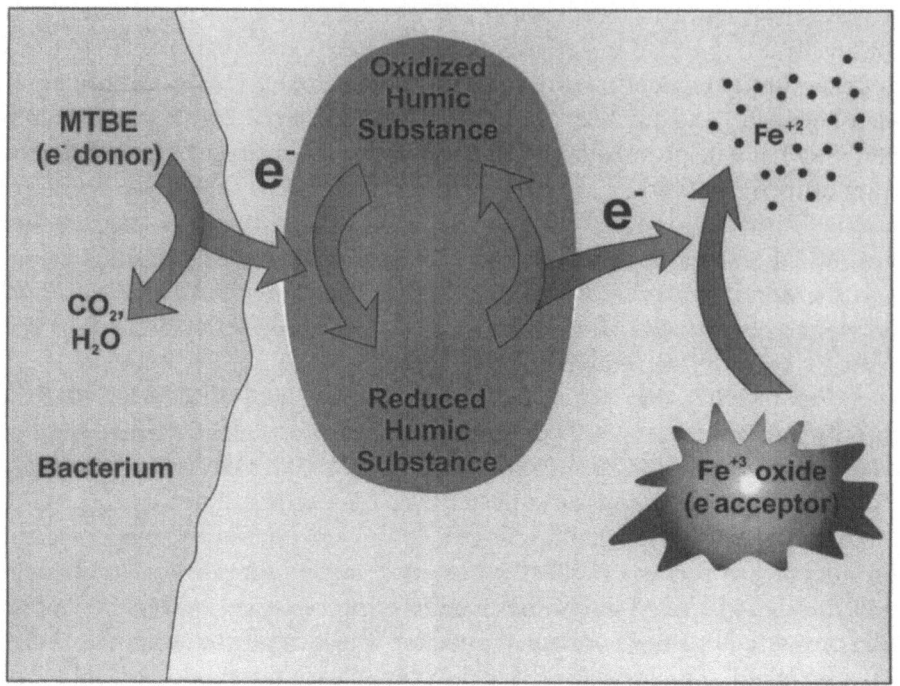

Figure 13-3.   **Electron Shuttling Via Humic Substances to Fe(III)**

discovery that quinone moieties on the humic substances are responsible for the electron accepting and, consequently, the electron shuttling capacity (Scott *et al.*, 1998). One such synthetic humic substance analog is anthra-quinone-2,6-disulfonate (AQDS). In laboratory incubations it was shown that AQDS stimulates Fe(III) reduction to the same extent as naturally occurring humic substances (Lovley et al., 1998). Because of its catalytic nature, very little AQDS (or humic substance) is needed to accelerate Fe(III) reduction. It is constantly recycled — alternating between variant oxidized and reduced states.

One gasoline-impacted site in coastal South Carolina was tested for its capacity to degrade MTBE. This site was characterized by the USGS, and the MTBE plume was found to be more extensive than the BTEX plume. Land-meyer et al. (1998) set up laboratory incubations with sediment from the Fe(III)-reducing zone of this aquifer. No exogenous Fe(III) or electron shut-tling compounds were added. In order to determine if indigenous microor-ganisms could degrade MTBE, [^{14}C]-MTBE was added. After seven months, 3% of the [^{14}C]-MTBE was mineralized to $^{14}CO_2$. This loss was attributed to Fe(III) reduction because oxygen and nitrate were depleted, and there was no significant Mn(IV) or sulfate. This was the first demonstration of the possi-bility of anaerobic mineralization of MTBE.

Further studies with the same sediment by Finneran and Lovley (2001) demonstrated that MTBE could be degraded to a much greater extent under Fe(III)-reducing conditions if humic substances were added as an electron shuttle. Although Fe(III) chelators have previously been shown to stimulate anaerobic benzene degradation in similar sediments, the addition of chelators did not stimulate MTBE degradation. When humic substances and additional Fe(III) oxide were added to the sediments, MTBE was degraded rapidly and repeatedly each time it was introduced. When Fe(III) was depleted, MTBE degradation ceased, but degradation ensued upon adding more Fe(III) oxide. Humic substances were never re-added to the sediment because they are constantly recycled. MTBE was also degraded when AQDS was added along with Fe(III) oxide.

Freshwater aquatic sediment from the Potomac River that had previously been used to develop a benzene-degrading enrichment (Weiner and Lovley, 1998b) was screened for its capacity to degrade MTBE. Uniformly labeled [^{14}C]-MTBE added to this sediment was oxidized to $^{14}CO_2$ and $^{14}CH_4$ if no exogenous electron acceptors were added. Three times as much $^{14}CO_2$ was produced as $^{14}CH_4$. This ratio of carbon dioxide to methane production suggested that there were probably several different microbial populations involved in MTBE degradation in this sediment. When Fe(III) alone or Fe(III) plus electron shuttling compounds was added to the sediment, no $^{14}CH_4$ was produced, only $^{14}CO_2$. Mineralization was rapid under these conditions, with almost one-third of the initial MTBE added converted to carbon dioxide in as little as 60 days (Finneran and Lovley, 2001).

MTBE has also been reported to biodegrade with Mn(IV) as the terminal electron acceptor (Bradley *et al.*, 2001a). Many well-characterized Fe(III) reducers can also reduce Mn(IV), and perhaps similar microbial communities are reducing Mn(IV) when it is available in the subsurface, as it yields slightly greater free energy than Fe(III) reduction. The reduction of Fe(III) would be obscured in these sediments because any Fe(II) produced would be abiotically oxidized by Mn(IV) (Lovley *et al.*, 1991).

Sulfate Reduction. Previous studies with aquifer sediment indicate that MTBE can be converted to TBA coupled to sulfate reduction (Suflita and Mormile, 1993). This activity was not continuous, as MTBE did not continue to be degraded in the sediment incubations. The studies mentioned above (Finneran and Lovley, 2001) provide initial evidence that MTBE is mineralized to CO_2 in the presence of sulfate reducers; however, the relative contribution of sulfate reduction versus Fe(III) reduction in the mixed sediment was not directly measured.

More recently, Somsamak et al. (2001) reported that MTBE was degraded in sulfate-reducing marine sediment that also degraded TAME. TBA accumulated as MTBE was removed from the sediment in agreement with the theoretical stoichiometry. In addition, as MTBE was degraded, sulfate was reduced to sulfide as an indicator of the dominant process.

Methanogenic Conditions. Recently, researchers with the USEPA have provided evidence for the degradation of MTBE under methanogenic conditions in the Elizabeth City aquifer in North Carolina (Wilson *et al.*, 2000). MTBE concentrations at selected monitoring wells were tested over the length of the methanogenic portion of a constituent plume. Over time the concentration of MTBE decreased, and TBA was detected in some wells as a metabolite of MTBE. Laboratory incubations support MTBE degradation within the methanogenic portion of the aquifer Wilson *et al.*, 2000).

ANAEROBIC TBA BIODEGRADATION

A major concern with MTBE biodegradation is the possible accumulation of the intermediate TBA. Several studies have suggested that, under anaerobic conditions, MTBE is degraded to TBA without further degradation. Previous research by Yeh and Novak (1994) indicated that TBA was degraded in soil under denitrifying conditions; however, as noted above, conditions may not have been strictly anaerobic. Studies with aquatic sediment (Potomac River) found that $[^{14}C]$-TBA was readily mineralized to both $^{14}CO_2$ and $^{14}CH_4$ without any lag. As much as 30% of the $[^{14}C]$-TBA was recovered as $^{14}CO_2$ and 9% as $^{14}CH_4$. When 900 micromoles per kilogram of unlabelled TBA was added to sediments, it was degraded in the sediment in less than 50 days. Degradation products included CO_2 and CH_4. The rate of TBA consumption was much greater than the rate of MTBE degradation. These results suggest that TBA is unlikely to accumulate under anaerobic conditions (Finneran and Lovley, 2001).

In subsequent experiments, sulfate was added to the Potomac River aquatic sediment. Although methanogenesis was inhibited, there was no significant increase in the rate and extent of $[^{14}C]$-TBA mineralization (Figure 13-4). When sodium molybdate, a specific inhibitor of sulfate reduction, was added, the production of $^{14}CO_2$ was decreased but did not completely cease (Figure 13-4). These results indicate that two processes are probably responsible for the $[^{14}C]$-TBA mineralization — sulfate reduction and Fe(III) reduction (because nitrate was not detected in these sediments).

More recent studies by Finneran and Lovley (unpublished) indicate that TBA is readily mineralized in aquifer sediment that had not been previously exposed to TBA (Figure 13-5). The sediment was impacted by petroleum from a pipeline spill and has been reported to rapidly degrade benzene (Anderson *et al.*, 1998). In a short period of time, $[^{14}C]$-TBA was readily mineralized to both $^{14}CO_2$ and $^{14}CH_4$. In highly methanogenic sediment that contained petroleum free product, both $^{14}CO_2$ and $^{14}CH_4$ were produced. Initial

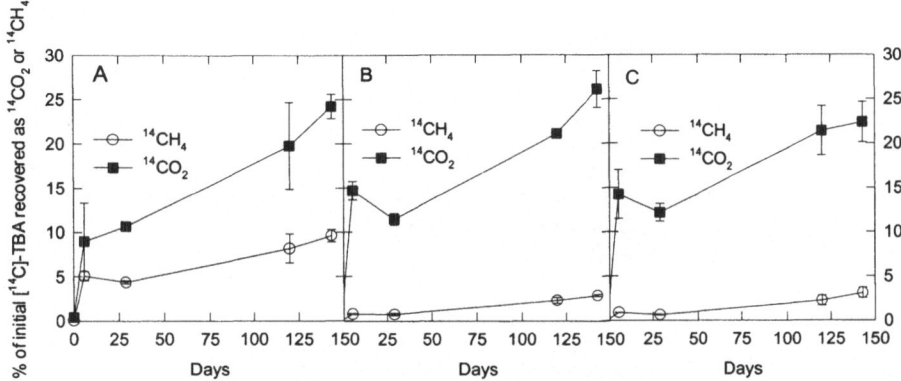

Figure 13-4.   **Anaerobic Production of $^{14}CO_2$ and $^{14}CH_4$ from [^{14}C]-TBA in Aquatic Sediment that Was (A) Unamended; (B) Amended with 10mM Sulfate without Molybdate Added at t = 123; (C) Amended with 10mM Sulfate and 10mM Molybdate Added at t = 123**
Data are the means of triplicate analyses. Bars designate one standard deviation.

characterization of this sediment with 2-[^{14}C]-acetate indicated that methanogenesis was the only TEAP; however, more $^{14}CO_2$ was produced than $^{14}CH_4$ when amended with [^{14}C]-TBA. When the results of highly methanogenic sediment are contrasted with Fe(III)-reducing sediment, of the [^{14}C]-TBA that was mineralized only $^{14}CO_2$ was produced. The Fe(III)-reducing sediment had the highest rate and extent of [^{14}C]-TBA mineralization. These results together indicate that methanogenesis alone most likely cannot be responsible for TBA degradation. In all cases there was production of $^{14}CO_2$, but in Fe(III)-bearing sediment, [^{14}C]-TBA was mineralized without methane being produced.

IMPLICATIONS FOR MTBE AND TBA BIOREMEDIATION

Anaerobic strategies are potential alternatives for source zone remediation when pump-and-treat technologies are too expensive and aerobic bioremediation may not be feasible or practical. If successful, these anaerobic strategies could remove MTBE before it spreads. In contrast, current aerobic bioremediation strategies may be most effective when they are employed farther downgradient from the source after MTBE has migrated.

All of the anaerobic strategies benefit from being relatively inexpensive and easy to implement. They require delivering the appropriate electron acceptor (and in the case of electron shuttling, the appropriate shuttling com-

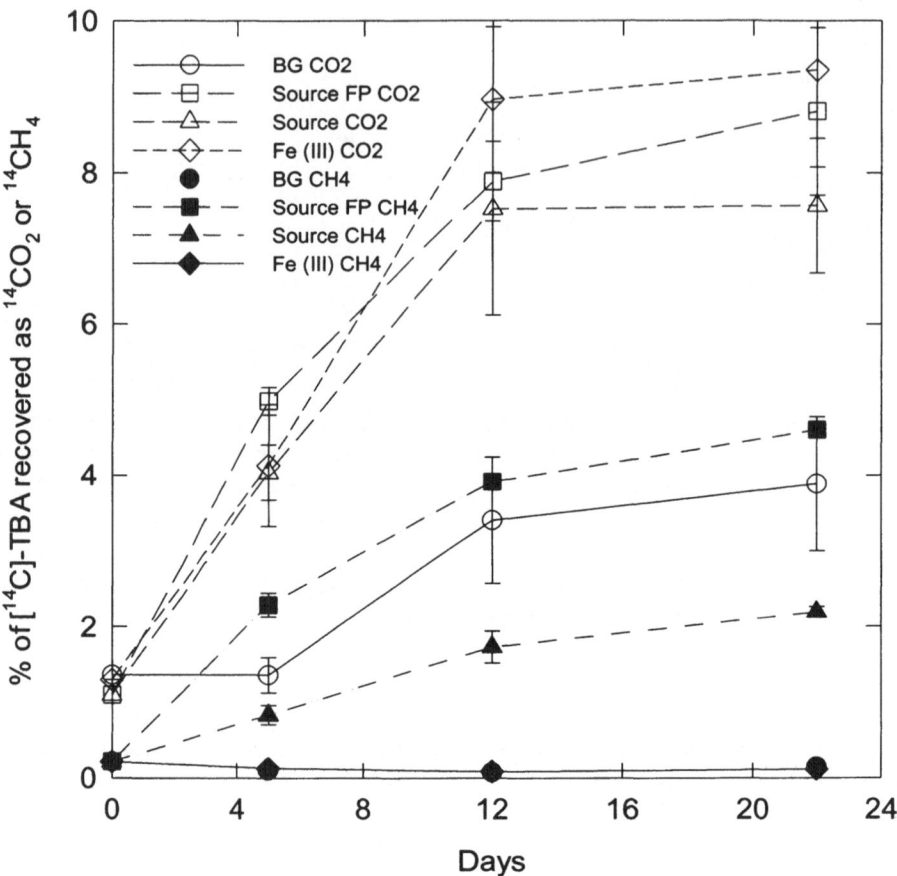

Figure 13-5.   **Anaerobic Production of $^{14}CO_2$ and $^{14}CH_4$ from [^{14}C]-TBA in Petroleum-Contaminated Aquifer Sediment Not Previously Exposed to TBA**
FP denotes free-product in the source area. Data are the means of triplicate analyses. Bars designate one standard deviation.

pound) to the subsurface, and monitoring ground water to document the loss of MTBE downgradient from the injection gallery (Figure 13-6). All of the electron acceptors that have been discussed above can be easily delivered to the subsurface via injection wells. Once the injection gallery is in place and operational, it would require infrequent monitoring.

Purified humic substances can be expensive; therefore, they are limited to laboratory incubations. For large-scale field applications, humic compounds should be taken from natural sources such as leaf litter or peat. Studies performed by Lovley and Nevin (unpublished) indicate that certain types of leaf compost leachate have electron shuttling capacities similar to purified

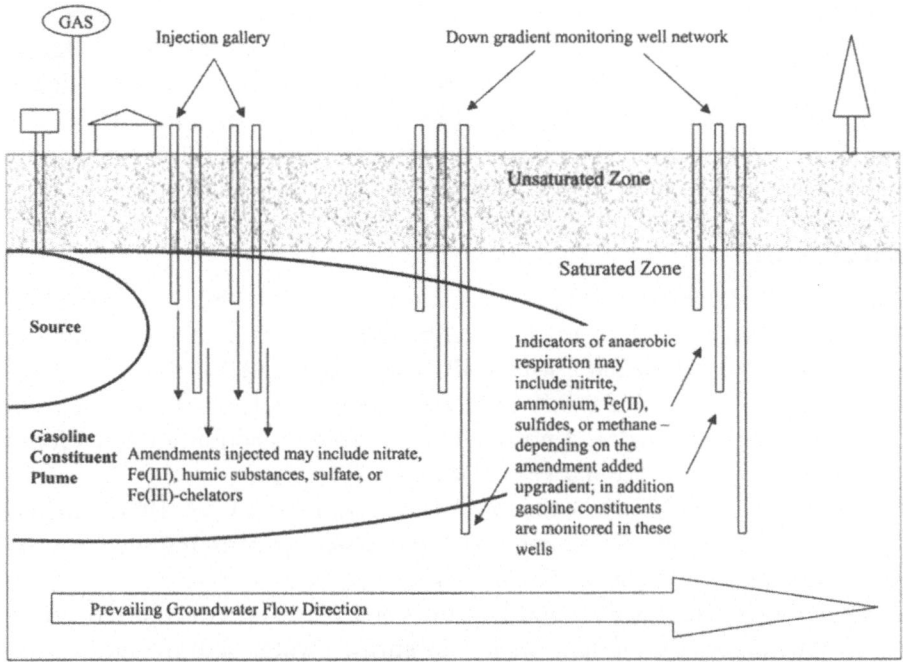

Figure 13-6.   **Diagram for Implementing the Proposed Anaerobic Bioremediation Strategies in an MTBE-Impacted Aquifer with Metabolites that Serve as an Indicator of Anaerobic Respiration**
HS denotes humic substances.

humic substances. Field applications of this technology should rely on these less expensive forms of humic substances.

In some instances, bioaugmentation with microorganisms capable of anaerobically degrading MTBE might accelerate the degradation process. For example, laboratory studies of anaerobic benzene oxidation coupled to sulfate reduction demonstrated that, in some instances, the addition of a benzene-oxidizing, sulfate-reducing consortia enhanced anaerobic benzene degradation better than the addition of sulfate alone (Weiner and Lovley, 1998a). Bioaugmentation might "prime the pump" to help anaerobic MTBE-degrading microbes gain a foothold in the source zone and increase initial degradation rates.

More research is necessary to optimize these anaerobic strategies. Compared to the work that has been done for anaerobic BTEX degradation, very little is actually known about anaerobic MTBE remediation. Nevertheless, the results to date are promising.

These enhanced anaerobic techniques may be the best option when the contamination is not immediately impinging on a drinking water source.

They may allow source zone remediation with minimal oversight. Few sites have been explored for potential anaerobic communities, which may degrade MTBE with little environmental manipulation. Preliminary studies suggest that microorganisms eventually adapt to anaerobically degrade MTBE. Perhaps techniques for stimulating this metabolism can be applied to hasten the process and add another remedial option to a growing list of solutions.

REFERENCES

Anderson, R., Rooney-Varga, J., Gaw, C., and Lovley, D. 1998. Anaerobic benzene oxidation in the Fe(III) reduction zone of petroleum-contaminated aquifers. *Environmental Science and Technology.* 32, 1222–1229.

Anderson, R.T. and Lovley, D.R. 2000. Anaerobic bioremediation of benzene under sulfate-reducing conditions in a petroleum-contaminated aquifer. *Environmental Science and Technology.* 34, 2261–2266.

Borden, R.C., Daniel, R.A., LeBrun IV, L.E., and Davis, C.W. 1997. Intrinsic biodegradation of MTBE and BTEX in a gasoline-contaminated aquifer. *Water Resources Research.* 33, 1105–1115.

Bradley, P.M., Chapelle, F.H., and Landmeyer, J.E. 2001a. Effect of redox conditions on MTBE biodegradation in surface water sediments. *Environmental Science and Technology.* 35, 4643–4647.

Bradley, P.M., Chapelle, F.H., and Landmeyer, J.E. 2001b. Methyl tert-butyl ether mineralization in surface-water sediment microcosms under denitrifying conditions. *Applied and Environmental Microbiology.* 67, 1975–1978.

Finneran, K.T. and Lovley, D.R. 2001. Anaerobic degradation of methyl tert-butyl ether (MTBE) and tert-butyl alcohol (TBA). *Environmental Science and Technology.* 35, 1785–1790.

Finneran, K.T. and Lovley, D.R. Unpublished data.

Hutchins, S.R. 1991. Optimizing BTEX biodegradation under denitrifying conditions. *Environmental Toxicology and Chemistry.* 10, 1437–1448.

Hutchins, S.R., Downs, W.C., Wilson, J.T., Smith, G.B., Kovacs, D.A., Fine, D.D., Douglass, R.H., and Hendrix, D.J. 1991. Effect of nitrate addition on biorestoration of fuel-contaminated aquifer: Field demonstration. *Ground Water.* 29, 571–580.

Landmeyer, J.E., Chapelle, F.H., Bradley, P.M., Pankow, J.F., Church, C.D., and Tratnek, P.G. 1998. Fate of MTBE relative to benzene in a gasoline-contaminated aquifer (1993–1998). *Ground Water Monitoring and Remediation.* Fall 1998, 93–102.

Lovley, D.R. 1997. Potential for anaerobic bioremediation of BTEX in petroleum-contaminated aquifers. *Journal of Industrial Microbiology and Biotechnology.* 18, 75–81.

Lovley, D.R., Phillips, E.J.P., and Lonergan, D.J. 1991. Enzymatic versus nonenzymatic mechanisms for Fe(III) reduction in aquatic sediments. *Environmental Science and Technology.* 25, 1062–1067.

Lovley, D.R., Chapelle, F.H., and Woodward, J.C. 1994a. Use of dissolved H_2 concentrations to determine distribution of microbially catalyzed redox reactions in anoxic groundwater. *Environmental Science and Technology.* 28, 1205–1210.

Lovley, D.R., Woodward, J.C., and Chapelle, F.H. 1994b. Stimulated anoxic biodegradation of aromatic hydrocarbons using Fe(III) ligands. *Nature.* 370, 128–131.

Lovley, D.R., Woodward, J.C., and Chapelle, F.H. 1996. Rapid anaerobic benzene oxidation with a variety of chelated Fe(III) forms. *Applied and Environmental Microbiology.* 62, 288–291.

Lovley, D.R., Fraga, J.L., Blunt-Harris, E.L., Hayes, L.A., Phillips, E.J.P., and Coates, J.D. 1998. Humic substances as a mediator for microbially catalyzed metal reduction. *Acta Hydrochimica et Hydrobiologica.* 26, 152–157.

Lovley, D.R. and Nevin. Unpublished data.

Mormile, M.M., Liu, S., and Suflita, J.M. 1994. Anaerobic biodegradation of gasoline oxygenates: Extrapolation of information to multiple sites and redox conditions. *Environmental Science and Technology.* 28, 1728–1732.

Salanitro, J.P. 1995. Understanding the limitations of microbial metabolism of ethers used as fuel octane enhancers. *Current Opinion in Biotechnology.* 6, 337–340.

Scott, D.T., McKnight, D.M., Blunt-Harris, E.L., Kolesar, S.E., and Lovley, D.R. 1998. Quinone moieties act as electron acceptors in the reduction of humic substances by humics-reducing microorganisms. *Environmental Science and Technology.* 32, 2984–2989.

Somsamak, P., Cowan, R.M., and Haggblom, M.M. 2001. Anaerobic biotransformation of fuel oxygenates under sulfate-reducing conditions. *FEMS Microbiolology Ecology.* 37, 259–264.

Suflita, J.M. and Mormile, M.R. 1993. Anaerobic degradation of known and potential gasoline oxygenates in the terrestrial subsurface. *Environmental Science and Technology.* 27, 976–978.

Weiner, J.M. and Lovley, D.R. 1998a. Anaerobic benzene degradation in petroleum-contaminated aquifer sediments after inoculation with a benzene-oxidizing enrichment. *Applied and Environmental Microbiology.* 64, 775–778.

Weiner, J.M. and Lovley, D.R. 1998b. Rapid benzene degradation in methanogenic sediments from a petroleum-contaminated aquifer. *Applied and Environmental Microbiology.* 64, 1937–1939.

Wilson, J.T., Cho, J.S., Wilson, B.H., and Vardy, J.A. 2000. *Natural Attenuation of MTBE in the Subsurface under Methanogenic Conditions.* USEPA, Ada, Oklahoma. EPA/600/R-00/006. www.epa.gov/ada/pubs/reports.html.

Yeh, C.K. and Novak, J.T. 1994. Anaerobic biodegradation of gasoline oxygenates in soils. *Water Environment Research.* 66, 744–752.

CHAPTER 14

Phytoremediation of MTBE—A Review of the State of the Technology

Lee A. Newman, Ph.D., University of South Carolina and Savannah River Ecology Laboratory

Charles W. Arnold, California State Water Resources Control Board

Phytoremediation is the use of plants to clean up environmental problems. While some have termed this technology a niche strategy, there are almost as many uses of phytoremediation as there are sites and problems. No plant will be able to survive in pure phase contamination, but in areas with low to moderate levels of contaminants, plants can perform a variety of functions. They can enhance the metabolism of microorganisms in the soil and thus the degradation of contaminants by the soil bacteria and fungi. Plants can be used in alternating methods to either immobilize soil metals, or to transport those metals out of the soil and into the aboveground portions of the plants for harvest and removal of the metals from the site. Plants have been used in aquatic situations, in wetlands, and in mountainous regions. But perhaps the most promising use of plants is in the control of ground water pollutants. Plants can be used to minimize the recharge of soil-based contaminants into the ground water, exert hydraulic control of a contaminant plume, or take up impacted water from the aquifer and degrade the contaminant within the plant tissues.

Much work has been done with ground water-based contaminants, including industrial solvents like TCE (Newman *et al.*, 1999), pesticides such as atrazine (Burken and Schnoor, 1997), or energetics such as trinitrotoluene (TNT) and cyclotrimethylenetrinitramine (RDX) (Thompson *et al.*, 1998). Early studies with soluble compounds often focused on the ability of a plant to take up the contaminant. There has been much said about the Briggs coefficient (Briggs *et al.*, 1982), where it was stated that a plant could only take up a compound with a mid-range octanol-water partition coefficient (K_{ow}), and that compounds that were strongly hydrophobic or hydrophilic would not be able to move into the plant root. This theory was debated within the phytoremediation community, as many believed that strongly hydrophilic compounds could not be taken up into a plant.

After information was obtained about plant uptake, researchers would try to discover the fate of the contaminant within the plant. Was the compound of interest merely passed through the plant and transpired to the atmosphere? Was it accumulated within the plant tissue, and if so, what were the toxic affects of this accumulation? Or was the compound metabolized within the plant, and if so, what were the metabolites, and what was their subsequent fate within the plant?

The physical properties of MTBE had some researchers skeptical about the ability of plants to take up the compound. MTBE is highly water soluble, and well outside the K_{ow} range that Briggs had hypothesized would be able to move into the plant. In spite of this, some groups did attempt to discover if plants could interact with MTBE.

CASE STUDIES

University of Washington.   One of the first groups to look at MTBE interactions with plants was at the University of Washington (Newman *et al.*, 1999). Their work involved three phases: using cell cultures to determine if plant cells could take up and metabolize MTBE, using whole plants to determine the uptake rates, and finally to examine plants growing in the field with their roots penetrating an existing MTBE plume.

Axenic immortalized cell cultures are sterile cultures of plant line that have been genetically altered to remain as a suspension of cells indefinitely and to not die or revert to a tissue type. Axenic immortalized cell cultures of a hybrid poplar line were grown for five days in a solution dosed to 10 mg/l MTBE spiked with [^{14}C]-uniformly labeled MTBE (1.95×10^6 disintegrations per minute [dpm]). After five days, the cells, headspace, and growth solution were analyzed to look for the presence of [^{14}C]-labeled products, such as CO_2, and soluble nonvolatile transformation products, such as chlorinated alcohols or acetic acids, in the culture to determine if plant cells were capable of metabolizing MTBE. As the cell suspensions were non-photosynthetic, any CO_2 generated from the mineralization of MTBE would be liberated rather than taken up into tissue production as in the case of whole plants. This study demonstrated that the poplar cells were able to oxidize 0.04% of the dosed MTBE to CO_2 while 0.05% was fixed in cell tissue over a three day testing period. During the same time period, only 0.01% was mineralized in pure medium or by heat-killed cells. While the numbers are low, they did show that uptake and mineralization was occurring.

Studies were done in mass balance chambers to determine the uptake rate for MTBE by whole plants. Hybrid poplars (*Populus trichocarpa* × *P. del-*

toides, clonal line H11-11) and eucalyptus were studied to determine their ability to take up [^{14}C]-labeled MTBE from the soil. Triplicate chambers were set up with each plant, as well as unplanted controls. Chambers had two sections, with a break between the soil section (2 liter volume) and the leaf section (6 liter volume) to prevent confusing leakage of MTBE with transpiration. The soils were dosed to 5 mg/kg MTBE spiked with 1×10^6 dpm [^{14}C]-uniformly labeled MTBE, and the systems were run for 14 days. Recoveries of radiolabel were 45% in the unplanted controls, 51% for the eucalyptus, and 60% for the hybrid poplars. Hybrid poplars incorporated 0.37% of the dosed MTBE into their tissues while transpiring 5.1%; eucalyptus incorporated 0.40% of the dosed MTBE while transpiring 16.52%. The higher transpiration rate by the eucalyptus trees can be explained by the greater leaf area of the plants; when transpiration amounts were normalized to leaf mass, both plants transpired approximately 0.7% of the dose per gram of leaf tissue (Table 14-1). In comparison, hybrid poplars are able to take up and transpire only 0.02% of dosed TCE per gram of leaf (Newman *et al.*, 1997).

Studies performed on field growing plants whose roots are believed to be in contact with an MTBE plume on the Port Hueneme Navy Base in California were inconclusive. The exposure concentration was an order of magnitude lower in the field than in the laboratory experiments; therefore, it was anticipated that detection would be difficult in the plant tissue. As expected, there was no detectable MTBE transpired by the plants. This finding could be due to a lack of sensitivity in the analytical methods, or due to the fact that the plant is metabolizing the MTBE as it moves through the plant. As this is the fate of TCE in field-grown plants, this hypothesis is not without some merit (Newman *et al.*, 1999). Soil and water samples taken in the field immediately upgradient and downgradient of the largest tree under study showed a 50% drop in concentrations of MTBE and other soluble gasoline additives in the downgradient samples. This result indicates plant interaction with the contaminant plume; however, there is no absolute proof of metabolism in the plants. Studies are still underway, with the majority of the work now being performed at the University of South Carolina.

Table 14-1.   **Volatilization of MTBE by Poplar and Eucalyptus Compared to Unplanted Controls and to Volatilization of TCE by Hybrid Poplars**

	No Plants	Eucalyptus	Hybrid poplar	Hybrid poplar
Dosed compound	MTBE	MTBE	MTBE	TCE
Grams of Volatiles in Headspace	0.12	16.52	5.11	0.11
Grams of Leaf Tissue	n/a	23.70	6.35	7.07
Volatiles per Gram Leaf Tissue (%)	n/a	0.70	0.80	0.02

Kansas State University. The research team at the Kansas State University (Zhang *et al.*, 2000) tried to determine the uptake/degradation rate of MTBE by plants as compared to microbial degradation rates in unplanted versus rhizosphere soils. Six self-contained channels (110 centimeters [43 inches] by 65 centimeters [26 inches] deep by 10 centimeters [4 inches] wide) were used in this experiment. Five of the channels were planted with alfalfa (*Medicago sativa*). Two of the five planted channels were seeded with the microorganism *Rhodococcus* (#33), and two channels were seeded with the microorganism *Arthrobacter* (#41). Both microorganisms have demonstrated a capacity to degrade MTBE (Mo *et al.*, 1997). For the pairs of inoculated channels, one of each was aerated. Ten days after inoculation of the microorganisms, a solution containing 0.84 millimolar (mM) of MTBE and was fed into each channel at the rate of one 1/day until a stable MTBE concentration was established (84 days) in the effluent water. After that, the feed solution was switched back to distilled water for 96 days.

Ground water effluents and soil gases were collected from the start of dosing until no MTBE was detected in either phase after the switch to distilled water. Plants were harvested monthly, and biomass production was recorded. Plant MTBE concentrations were determined by placing fresh plant parts into sealed jars and analyzing the headspace gases.

The unplanted channel had the greatest recovery of MTBE in the effluent water (91%), while the five planted channels had similar recovery rates (50 to 65%) (Table 14-2). This was probably due to the fact that there was no transpiration of the MTBE solution in the unplanted channel, and thus more water was recovered from that channel. The recovery of soil gas MTBE was

Table 14-2. **Volatilization of MTBE from Soil**

	1	2	3	4	5	6
Microorganisms added	yes	yes	no	no	yes	yes
Planted	yes	yes	yes	no	yes	yes
Sparged	yes	no	no	no	no	yes
MTBE Groundwater recovery (percent)	62	50	55	91	65	57
MTBE volatilized from soil (percent) I	**22**	**25**	**32**	*17*	**28**	**20**
MTBE volatilized from soil (percent) II	**22**	**25**	*32*	NC	**28**	**20**
MTBE volatilized from soil (percent) III	*22*	**25**	NC	NC	**28**	*20*

NC = not considered for this evaluation

Data I shows that the presence of plants increases the upward flow of MTBE and thus increases volatilization of MTBE. Data II shows that the presence of the innoculated bacteria increases degradation of MTBE, thus decreases volatilization of MTBE. Data III shows that the presence of aeration increases the oxygen in the soil, increase the MTBE degradation and thus decreases the volatilization of MTBE.

slightly higher in the planted, inoculated, non-aerated channels (25 to 32%) than in the aerated channels (20 to 22%), but the authors did not know if this was due to enhanced biodegradation of the MTBE by the microorganisms or a sampling artifact.

MTBE was found in the plant tissue, but at a lower concentration than in the feed solution (2.7 to 18% of influent concentrations). It was not known if this was due to degradation of MTBE in the plant tissue or if MTBE was released from the tissue after being taken up with the water.

The results showed that the presence of plants enhanced the removal of MTBE from a water stream, and that the presence of added microbial strains may also enhance the degradation of the MTBE. This group is also planning additional studies to obtain a better understanding of the biodegradation versus plant transformation rates.

University of Iowa.   The group at the University of Iowa (Hong *et al.*, 2001; Winnike-McMillan *et al.*, 2002) also started looking at MTBE/plant interactions. Their work was supported by Equilon Enterprises, who wanted to treat an MTBE plume using phytoremediation. Their work was done in phases: the first phase was laboratory studies that looked at plant uptake; the second phase was the design and installation of a field site to control the movement of the MTBE plume.

Poplar cuttings (*Populus deltoides* × *P. nigra*, clonal line DN-34) were grown in a hydroponic solution containing MTBE at a concentration of 11 mg/l spiked with 7.07 microcurie (μCi) of [^{14}C]-uniformly labeled MTBE for 10 days. Triplicate plants then had additional MTBE added to concentrations of 100, 1,000, or 10,000 mg/l to determine toxicity. All plants grew well except for the 10,000 mg/l plants, which died within a few days of dosing.

At the 11 mg/l MTBE level, systems included a completely sealed chamber, a glass rod as a control, and an excised cutting or a developed cutting. The 11 mg/l MTBE plants were further analyzed. Less than 1% of the MTBE was detected as ^{14}C-CO$_2$. Plants accumulated 2.21% of the label within their tissues, and different experimental runs produced data showing between 16.81 to 54.54% of the label was believed to be transpired by the plants. However, these data were calculated and not analyzed for directly, therefore the loss from the system due to leaks as opposed to plant transpiration was unknown.

Plants grown in soil were examined next. In this experiment, chambers were set up to directly collect MTBE that was either transpired from the leaves or off-gassed from the stem. Systems were set up with 500 grams of soil and approximately 180 milliliters of water. The systems were dosed with 3.95 mg of MTBE and 7.1 μCi of [^{14}C]-uniformly labeled MTBE. The headspace traps had an airflow of 1 to 3 liters per minute (l/min). The systems were allowed to run for 33 days. Total recovery of radiolabel was approximately 67%. Approximately 2.77% of the label was recovered in the plant tissue, and 55% was transpired by the plants.

The field site that was planted has an MTBE plume with a maximum concentration of 40 mg/l downgradient from the source area. The depth to ground water is 2.7 to 3 meters (9 to 10 feet), well within the rooting range for poplar trees. Extensive modeling using SWMS_3D, a finite element-based three-dimensional unsaturated/saturated flow/transport code and UNSAT-H, a finite difference-based one-dimensional code (Hong *et al.*, 2001) was done of the site to determine if trees would be able to have an appreciable impact on the aquifer. Once it was determined that phytoremediation was workable, the site was planted. Hybrid poplar DN-34, the same plant used in the laboratory studies, was used. The plants were placed in deep-drilled holes, with approximately 0.3 meters (1 foot) of the cutting aboveground. The hole was filled with a fertilizer amended mixture of sand and mulch. Field measurements are continuing, and include vadose zone moisture profiling, transpiration rates, leaf area indexing, and ground water elevations; however, it is still too early to determine the effectiveness of phytoremediation on this site.

University of Colorado. Researchers from the University of Colorado (Rubin and Ramaswami, 2001) have conducted work similar to that done at Washington and Iowa. Hybrid poplar cuttings (*Populus deltoides* × *P. nigra*, clonal line not given) were exposed to MTBE at either 300 or 1,600 µg/l. Cuttings were pushed through a septum and grown in soil until they reached 15 centimeters (6 inches) above the septum. They were then removed from the soil, the roots rinsed, and the plants placed in 250-milliliter jars with a hydroponic solution containing MTBE. The set-up was weighed daily to monitor water loss from transpiration, and at the end of one week the plants were analyzed. Controls included a jar with a solid septum, a jar with a copper tube through the septum to look at loss due to puncturing the septum, and a jar with a balsa wood rod through the septum that would represent passive transpiration through the porous wood. The solid septum and copper tube controls showed minimal water loss through the system; the balsa wood lost 12 milliliters, and the live plants averaged 24 milliliters. The sealed control and copper tube had a mass loss of MTBE of 1%; the balsa wood had a 15% reduction in mass of MTBE; and the planted system had a 30% loss of MTBE. Plant tissue concentration of MTBE was between 3 and 7% of the concentration in the solution.

A mass balance study was performed by this group, but only one plant was used in a closed chamber for this experiment. Therefore, it is difficult to determine the validity of those results.

State of California Water Resources Control Board. Researchers at the CASWRCB (Arnold and Parfitt, in preparation) reported the first functional phytoremediation extraction system of ground water impacted with high concentrations (greater than 100,000 µg/l) of MTBE. Five mature Monterey pines were identified adjacent to a UST that was shown to be leaking gasoline and MTBE into the aquifer. The trees are approximately 3.6 meters (12 feet) apart, and the row is roughly perpendicular to the flow of the plume. The growth of the trees predated the leaking of the MTBE from the tanks. Approximately 20 transpirate (condensate) samples were collected using a system of full branch enclosure from the five trees. Transpirate samples collected from each stand confirmed phytovolatilization of MTBE from the shallow ground water (Figure 14-1). Most interestingly, all of the condensate samples analyzed were positive not only for MTBE, but also for TBA. MTBE concentrations in the samples ranged from 2 to 460 µg/l and TBA ranged from 37 to 4,100 µg/l. Ground water concentrations showed the opposite trend, with MTBE levels far exceeding TBA levels.

Ground water monitoring was performed using three monitoring wells that run parallel with the plume and bisect the row of trees. MTBE concentrations declined by three orders of magnitude across the 6-meter (20-foot) distance spanning the up and downgradient wells. Concentrations at the

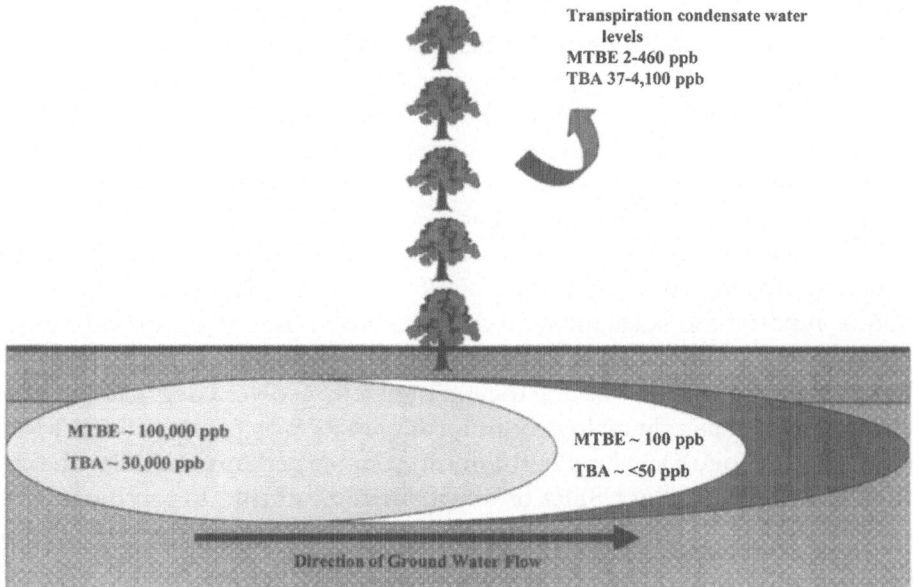

Figure 14-1.   **Effect of Trees on MTBE Concentration in Ground Water and Transpiration Concentrations**

source were in the range of 100,000 µg/l for MTBE and approximately 30,000 µg/l for TBA. Downgradient of the trees, MTBE concentrations dropped to 100 µg/l, and TBA was below 50 µg/l. Ground water elevation measurements indicate that the stand is also serving as a partial hydraulic barrier to flow of ground water and the MTBE plume. While the results of this study presented here are still preliminary, they are important as this is the first time that it has been shown under field conditions that trees are capable of taking up MTBE. Further, this study shows that the trees not only take up MTBE, but also have a substantial positive impact on the removal of MTBE from an aquifer. There is potential degradation of the MTBE within the plant system, assuming that the MTBE is not accumulating in the plant relative to the TBA, which could skew the transpiration data. While the exact mechanism and location of this metabolism is still under study, there can be little doubt as to the positive impact that the trees are having on the MTBE plume. These data are a strong argument for the use of plants to remediate MTBE-impacted shallow aquifers under conditions dominated by horizontal flow.

CONCLUSIONS AND FUTURE WORK

Phytoremediation has the potential to be a low-cost way to remediate large, dispersed plumes. All research to date indicates that plants are readily able to take up MTBE; however, the fate of MTBE is still unknown. Limitations on analytical methods may hinder studies trying to track MTBE degradation in plant tissues. The National Water Resource Council's MTBE panel recognized this as an area that needs further study and development.

It is also unknown how plants may influence the microbial degradation of MTBE. If there is a way to link the inoculation of MTBE degraders into a plume with plants that can supply necessary nutrients and aeration for the microorganism, it may be possible to remove MTBE from the aquifer faster than either system can remove it alone.

Another area that needs improvement is communication between research groups to avoid duplication of effort. While independent collaboration is important to substantiate results, having multiple groups repeating the same work will not advance the science as rapidly as it needs to be done. Just as there are groups working together to look at solvent degradation and enhancement of petroleum hydrocarbon degradation by plants (USEPA's Research Technology Development Forum groups), perhaps a similar group should be sponsored by USEPA or the petroleum industry to coordinate the research efforts of multiple teams looking at MTBE degradation within the plants.

In conclusion, phytoremediation of MTBE is a viable, cost-effective option for the cleanup of impacted aquifers. Additional work needs to be done on the scientific level before we can tell regulators and citizen groups the ultimate fate of MTBE in the plants. However, in areas where technologies such as air sparging and ground water pump-and-treat are already being used, phytoremediation can be utilized now as a low-cost alternative to these more expensive technologies.

REFERENCES

Briggs, G.G., Bromilow, R.H., and Evans, A.A. 1982. Relationships between llipophicity and root uptake and translocation on non-ionized chemicals by barley. *Pesticide Science*. 13, 495–504.

Burken, J.G. and Schnoor, J.L. 1997. Degradation of atrazine and metabolites by poplar trees. *Environmental Science and Technology*. 31(5), 1399–1406.

Hong, M.S., Farmayn, W.F., Dorth, I.J., Chiang, C.Y., McMillan, S.K., and Schnoor, J.L. 2001. Phytoremediation of MTBE from a groundwater plume. *Environmental Science and Technology*. 35, 1231–1239.

Mo, K., Lor, C.O., Wanken, A.E., and Javanmardian, J. 1997. Biodegradation of methyl-t-butyl ether by pure bacterial cultures. *Applied Microbial Biotechnology*. 47, 68–72.

Newman, L.A., Cortellucci, R., Crampton, R.S., Domroes, D., Duffy, J., Ekuan, G., Gordon, M.P., Hashmonay, R.A., Heilman, P., Karscig, G., Muiznieks. I.A., Newman, T., Ruszaj, M., Wang, X., Yost, M.G., and Strand, S.E. 1999. Remediation of trichloroethylene in an artificial aquifer with trees: A controlled field study. *Environmental Science and Technology*. 33, 2257–2265.

Rubin, E. and Ramaswami, A. 2001. The potential for phytoremediation of MTBE. *Water Resources*. 35, 1348–1354.

Thompson, P.L., Ramer, L.A., and Schnoor, J.L. 1998. Uptake and trans-formation of TNT by a poplar hybrid. *Environmental Science and Technology*. 32(7), 975–980.

Winnike-McMillan, S.K., Zhang, Q., Davis, L.C., Erickson, L.E., Newman, L., Schnoor, J.L. Phytoremediation of MTBE. In: *Phytoremediation*. (McCutcheon, S. and Schnoor, J., Eds.). In Press.

Zhang, Q., Davis, L.C., and Erickson, L.E. 2000. An experimental study of phytoremediation of methyl-tert-butyl ether (MTBE) in groundwater. *Journal of Hazardous Substances Research*. 2, 4.1–4.19.

CHAPTER 15

Ground Water Recovery and Treatment

Tie Li, Ph.D., URS Corporation

Raaj U. Patel, P.G., and David K. Ramsden, Ph.D.,
URS Corporation, Houston

Jonathan Greene, P.E., Malcolm Pirnie

PERSPECTIVE OF GROUND WATER RECOVERY AND TREATMENT

Until the environmental revolution, the only ground water that was routinely treated to remove contamination was the impacted ground water that was extracted for beneficial use. With the recognition that contamination could cumulatively impact drinking water wellfields or entire areas and basins, there was an incentive to remediate ground water not yet extracted for use. Although it is now recognized that impacted ground water can be treated *in situ* or *ex situ*, at the time it was logical to extract impacted water, using wells similar to potable water wells, and treat and/or dispose of that water into surface waters. The underlying strategy was based on the concept that by removing the impacted water, you would gradually remove the underground contamination. Consequently, the first ground water remediation systems were generally extraction and *ex situ* treatment or simple discharge to surface water. This process is called pump-and-treat (Liu, 1997). As experience with these pump-and-treat systems accumulated, it became apparent that for many applications these systems were going to operate for a very long time, possibly decades or more, to approach the ground water standards for drinking water in the Safe Drinking Water Act regulations. In addition, it soon became clear that the rules of solubility still applied, and contaminant mass removed per mass of water diminished over time due, ironically, to the success of the prior removal. This was due in large part to the physics and chemistry of contamination by pure products of low solubility but high toxicity at low concentrations. Extractions became prolonged, not necessarily because the total mass of contaminant to be removed was very high, but because the dissolved mass of contaminants at any time was low. Then as the contaminant mass is lowered by treatment, the dissolved concentrations diminish correspondingly per constant time unit due to the lower equilibrium gradient. In addition, many free-phase or high concentration contaminants had penetrated lower permeability materials driven by gravity and/or high concentrations. Their extraction without a similar extraction driving force, *i.e.*, overcoming the hysteresis effect, meant that after the cleanup goals were apparently reached and the system was shut down, sequestered contaminant concentrations would gradually equilibrate with the recoverable ground water, again increasing the contaminant concentrations (USEPA, 1996).

The next trend was *in situ* ground water treatments that would facilitate destruction or immobilization of the contaminants in place in the ground. This was initially achieved by enhancing the naturally occurring bioremediation, by aerating or oxygenating and adding nutrients to the extracted water that was then reinjected.

Despite problems with all of these pump-and-treat techniques, in many circumstances the pump-and-treat *ex situ* systems are the practical mechanism for ground water treatment (USEPA, 2001a). For MTBE, with its high solubility and low adsorptivity, *ex situ* processes such as pump-and-treat processes appear to offer a rapid mechanism for preferentially removing MTBE from ground water compared to BTEX with its lower solubility and higher adsorptivity.

RELATIONSHIP TO POTABLE WATER

As discussed in the previous section, ground water pump-and-treat systems are closely related to ground water-based drinking water production and treatment systems. Much of the potable water in the U.S. comes from ground water, particularly in small towns and rural areas. In many locations, the shallow or first ground water is the preferred or sole drinking water source. This is often because shallower ground water is economical to reach and recover and because vertical recharge usually provides a local sustainable, renewing mechanism.

Compared to ground water remedial extraction wells, potable water wells are often larger diameter, frequently constructed of more robust and durable materials such as steel, and often deeper in the aquifer to facilitate the common practice of screening across a large vertical interval including more than one transmissive zone. In general concept and construction principles, potable water and remedial extraction wells are much the same. Treatment systems for potable water are where the two systems currently diverge. Typical treatment trains for ground water that is to be used as potable water include very few unit treatment processes. Potable ground water may not be treated at all or may undergo treatment for particulates (filtration, etc.) and possibly microbes (usually chlorination). In some locations there is treatment for natural organics (GAC) and/or nuisance characteristics such as odor,

taste, and color (GAC, chlorination, ozonation). In general, the only treatment to reduce health risk is antimicrobial. Most treatments address aesthetic
issues with the water. Natural organics are sometimes removed because they
might be precursors of halomethanes from chlorination, not because they are
toxic contaminants themselves. Under circumstances with significant anthropogenic impacts, treatment can be more extensive including nitrate removal, metals removal (coagulation, precipitation), and more contaminant
specific variations of the above treatments.

GROUND WATER RECOVERY

General. Pump-and-treat has been one of the most widely used ground
water remediation technologies in the past two decades. Ground water
pump-and-treat is a group of technologies, including various ground water
extraction methods as well as many aboveground or end of pipe treatment
technologies. Figure 15-1 depicts a generalized ground water recovery and
treatment system. Pump-and-treat systems remain a necessary component of
many ground water remediation efforts, and are appropriate for both plume
remediation and containment. Pump-and-treat has been most successful
where it has been used to maintain a hydraulic control. In recent years, new
in situ remediation technologies have emerged and have become more successful with ground water remediation/restoration projects.

Extraction. In general, the methods for extraction of ground water, the
pump portion of the pump-and-treat technology, are not linked to or limited
by the *ex situ* treatment technologies. Any method that gets the water out can
supply the water to virtually any *ex situ* treatment system that can treat the
contaminant(s). Because water is the only *in situ* transport medium for
pump-and-treat, ground water extraction technologies are specific and opti-
mized for the recovery of only one compound, water. Treatment technologies

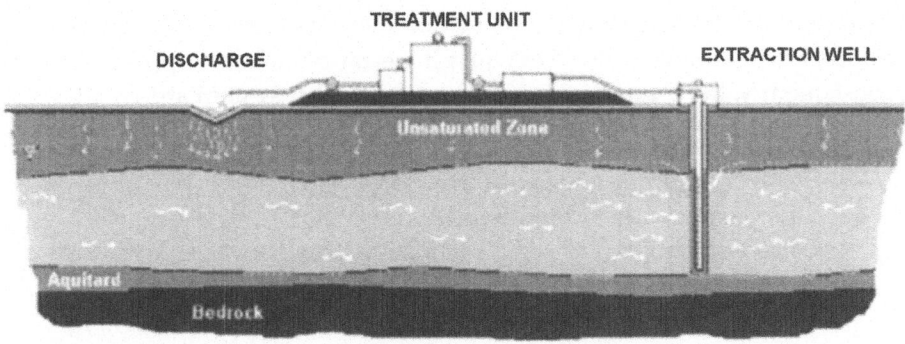

Figure 15-1. **Generalized Pump-and-Treat System**

are specific to the characteristics and concentrations of the contaminant or multiple contaminants of concern, and treatment design is often considered more challenging than design of extraction systems. While this may generally be true, one of the challenges of pump-and-treat is designing an efficient, effective extraction system at locations in intricate geological and hydrogeological locations such as karst or complexly faulted sites.

DESIGN

The real complexity of the pump-and-treat project is more the need to integrate the process from site characterization through design and operations to closure. To do this requires integration of the science (hydrogeology and geochemistry), regulatory analysis (cleanup goals, agreed endpoints), field services (construction and operations), and engineering design in one project team. Although ground water well systems are well understood, the design effort is different from the single drinking water well installation. The pump portion of pump-and-treat at most sites involves multiple wells and/or trenches for extraction and/or injection, often with differential control of extraction rates to optimize operations. Designing the extraction system to optimize contaminant mass or concentration recovery and/or containment can make for a very complex evaluation and design. Consequently, design of extraction still requires the full creative collaboration of the different skills: hydrogeologists, modelers, regulatory analysts, and civil, mechanical, and electrical design engineers.

Design Components. Most pump-and-treat systems involve extracting ground water via recovery wells constructed within the contaminant plume downgradient from the source. These are typically vertical wells, but in the last decade or so have included horizontal wells installed by horizontal boring or by trenching. In addition, simple trench drains with porous backfill media and pumps can be used for ground water recovery in areas where ground water is shallow enough to be economically reached by excavation. Even multi-phase and slurping extraction systems constitute ground water pump-and-treat operations. They just have more elaborate pump systems. Consequently, the components of a pump-and-treat system are:

- Extraction wells or trenches;
- Extraction pumps and controls;
- Collection piping and pumps;
- Collection tank(s);
- Treatment system with associated equipment;
- Post-treatment distribution piping and pumps, if needed;
- Injection/infiltration wells/trenches/fields, if needed; and/or
- Discharge point.

Well Array Design. The extraction wells/trenches, their locations, number, arrangement, and extraction rates, are designed and controlled so they act as a barrier to additional downgradient movement of the plume. Prior to installation of a pump-and-treat system, aquifer tests (slug tests, or preferably pumping tests) are conducted on enough wells or test trenches at the site to gather information on localized aquifer properties across the site. Through these tests, aquifer characteristics are determined and are used to model and/or design the extraction system configuration. Design factors include placement of recovery wells, number and size of wells, compatible well material (typically PVC or stainless steel), well screen placement, sustainable pumping rates, water transport systems, and well completion strategies.

Capture Zone Analysis. Appropriate well placement is designed using capture zone analysis, or fate and transport modeling, and data obtained from initial aquifer pumping tests. Numerous mathematical models are available to perform capture zone analysis and to evaluate recovery/injection system arrays. Capture zone analysis for MTBE is not different from plumes with other gasoline components since the contaminant type is not a factor in determining the well arrays capture zone(s).

Materials of Construction. Materials of construction for wells/trenches and the extraction pumps must be compatible with the contaminants of concerns. For MTBE at normal ground water concentrations, PVC well casings, high density polyethylene (HDPE) trench geotextiles, liners, and/or piping, and the usual gasket materials in ground water pumps are adequate. With MTBE, stainless steel is appropriate for ground water extraction pumps. Electrical and air driven submersible pumps are the two common types of pumps used in ground water recovery systems, and there is no particular reason to select one type over another with MTBE. More important is selection of pump properties as determined by the dimensions of the recovery well, required flow rate, depth to ground water, contaminants of concern, other groundwater chemistries, overall system configurations, and modes of managing and operating the system. For example, pneumatic pumps require careful adjustment initially, to avoid delivering water as well as air, and thus misleading the measuring system. Pneumatic pumps require a source of air that often requires a dedicated compressor station. Unless compressed air is available on site, electrical systems may be simpler at many sites, but may be less dependable than pneumatic pumps. Electrical pumps are inherently less safe and may require transformers to power multiple pumps located across large sites.

Typical Extraction Well Construction. Figure 15-2 depicts a typical recovery well design used in the industry today. The well can be completed below

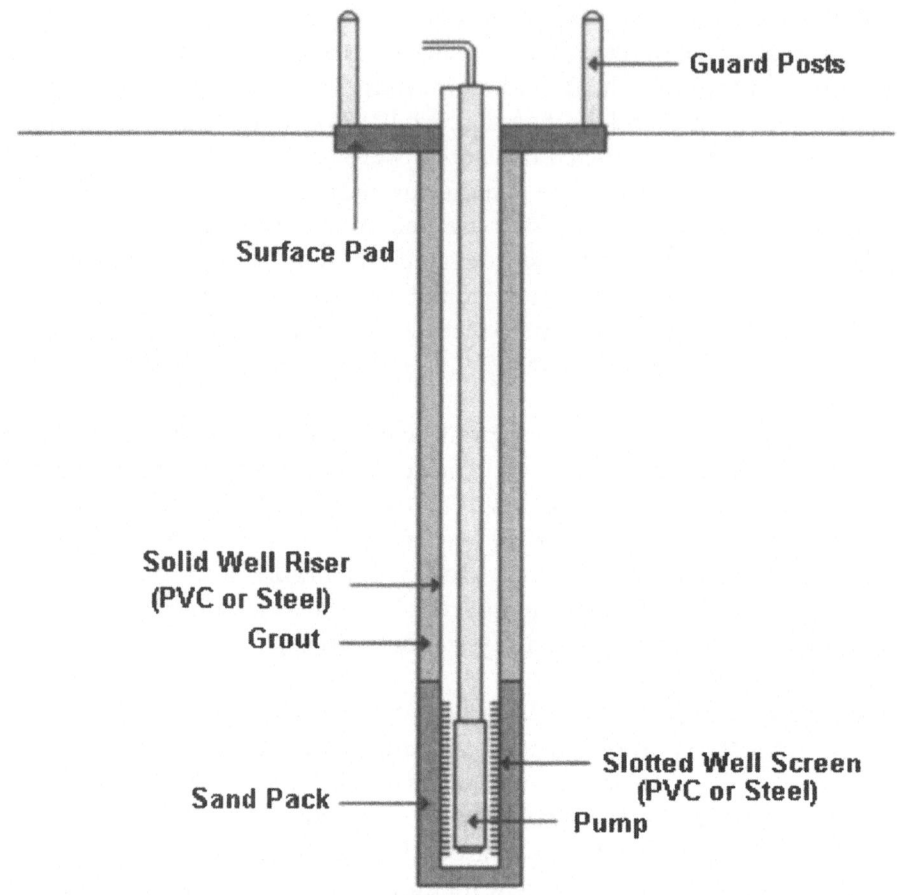

Figure 15-2.   **Typical Recovery Well Design**

the ground surface in a vault, with a manhole cover, that houses the pump cables and controllers, valves, flow meter, and related components. The vault provides protection against weather, restricts access, deters vandalism at unmanned locations, and protects the well from vehicular traffic.

Trench Construction.   Trenches can be thought of as wide long wells with a geotextile well screen and a porous internal sand/gravel pack. Biopolymer-stabilized trenching techniques are good for installations below the ground water table and particularly for deeper installations. The polymer can be digested (broken) after completion of excavation and installation of piping to restore the higher permeability of the trench base matrix. The MTBE-impacted ground water being transported from the well to the treatment system may be a hazardous waste if it contains other gasoline components. If the RCRA regulations apply to the site, this would require that underground piping be double-walled. Even if not required by regulation, double-walled piping is a good practice to minimize the possibility of a new leak and source. Figure 15-3 illustrates a typical trench extraction or re-infiltration system.

Optimization.   Optimizing the recovery of impacted ground water is a dynamic process that uses the response of the ground water remediation system to improve the extraction and thus overall remediation efficiency. Optimization can be achieved by phased construction, by adaptive management of the well pumping rates, by periodic modeling of the initial and supplemented well arrays, and by pulsed pumping. These optimization practices work for MTBE just as with any contaminant.

Phased construction of extraction wells allows data from the response of the aquifer to pumping operations to be used in determining the location or in-filling of subsequent wells to address areas of poor recovery, high contaminant concentrations, or hydrogeological heterogeneities.

Adaptive pumping involves designing the wellfield with sufficient flexibility so that extraction and injection can be varied to reduce zones of stagnation. Extraction wells can be periodically shut off, others turned on, and pumping rates varied to ensure that contaminant plumes are remediated at the fastest rate possible.

Pulsed pumping has the potential to increase the ratio of contaminant

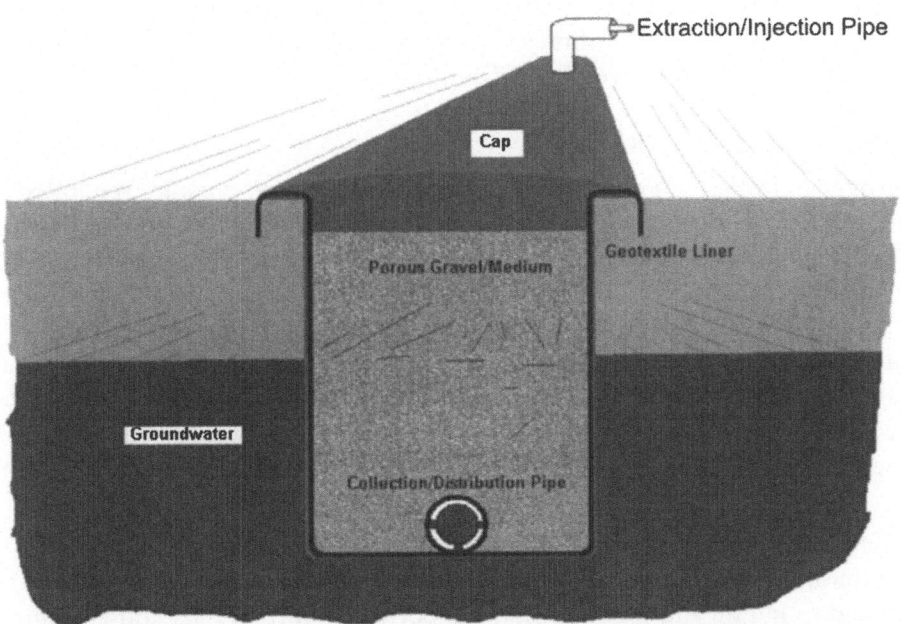

Figure 15-3.   **Simple Trench Extraction/Reinfiltration System**

mass removed to ground water volumes removed where mass transfer limitations restrict dissolved contaminant concentrations during continuous pumping. During the resting phase of pulsed pumping, contaminant concentrations increase toward equilibrium due to diffusion, desorption, and dissolution from soils or free-phase materials into the ground water. Once pumping is resumed, ground water with higher concentrations of contaminants is removed. There is a practical limitation to the benefits available through pulsed pumping. As the water approaches equilibrium in dissolution, the rate of further dissolution declines, producing a diminishing return in contaminant mass for time.

The phenomena of tailing and rebound are commonly observed during pump-and-treat. Tailing is the progressive decline in dissolved contaminant concentration with continued operation of a pump-and-treat system. Tailing is primarily a product of the contaminant equilibria between water and soil and/or phase-separated contaminants. As the remediation reduces the concentration of contaminants adsorbed on the soils or in the free-phase, there is a proportionally lower concentration of contaminants in the ground water. Rebound is an increase in contaminant concentration after pumping is temporarily or permanently discontinued. Rebound is primarily caused by the continued, equilibrium-driven, dissolution of contaminants into ground water after the cessation of the steady-state pumping. Rebound can be exploited as an indicator of when remediation goals will be achieved by evaluating the curve generated by the sequential rebound peaks of a number of pulsed operations of the recovery system. Without an ongoing source, rebound diminishes with continued treatment and eventually becomes small enough to not recover beyond the cleanup goals.

REINJECTION/INFILTRATION

Recovered, impacted ground water, with any free-phase organics separated, and treated by one or more of the extensive array of treatment technologies, can then be discharged to a publicly owned treatment works, surface water, or injected/infiltrated back to the ground water, usually upgradient from the plume. Consequently, the injection/infiltration system is the alter ego of the extraction system, capable of using the same basic components, wells, trenches, pumps, and control systems to put water back into the ground water. This can promote an enhanced flow downgradient to the original extraction system, potentially speeding the remediation process.

SPECIALIZED EXTRACTION SYSTEMS

Modifications to ground water extraction systems have been developed and applied to tailing and rebound issues associated with NAPLs and strongly adsorbed contaminants. Typical modification to pump-and-treat is MPE including slurping, and surfactant-enhanced recovery (SER). The detailed discussion on MPE is provided in Chapter 9.

MPE is the simultaneous recovery of two or three phases; ground water, LNAPL, and/or soil vapor. MPE is applied with the well screen extending from below the water table to above the water table in the vadose zone. Ground water is pumped from the deeper screened interval (below the water table). This enhances LNAPL recovery and allows the designer to select the optimal ground water drawdown. An added benefit to three-phase recovery is that the lift of the recovered material is dependent on the mass being lifted, and the net lightening effect of the air mixed with the water and LNAPL product enhances the lift. MPE is applicable to MTBE particularly where both an LNAPL and dissolved plume exist in an area of the site. Since MTBE's solubility and high concentration in many gasolines are keys to its ability to form a plume, rapid removal of LNAPL concomitant with plume recovery is a plus.

SER uses injection of surfactant(s) and usually a cosolvent (such as ethanol or isopropanol) to mobilize contaminants adsorbed to the soils or present as a dense or light NAPL into the readily recoverable ground water flow. The cosolvent enhances contaminant solubilization. More importantly, it is a viscosity modifier for the usually viscous surfactant, allowing easier injection of the aqueous surfactant solution. SER enhances the recovery and shortens the time to recover contaminant mass by the formation of ground water-mobile micelles with contaminants sequestered within. This also increases the solubility of contaminants in water, and thus altering the surface activity properties of the water with soils and NAPL. This enhances the movement of water into sequestered areas, lessens the capillary interactions, and generally increases the ability to flow and recover dissolved contaminants. The surfactant and contaminants are subsequently extracted through pumping wells. Aboveground processes are used to treat the impacted ground water and recycle the surfactant if practical. SER is probably not useful for MTBE ground water recovery and treatment since MTBE adsorbs poorly and thus is readily mobilized in ground water.

MTBE SPECIFIC ISSUES

High solubility and low adsorption make pump-and-treat an effective approach for the remediation of MTBE impacted ground water. Tailing and rebound effects occur with MTBE, particularly when LNAPL is present, but are believed to be less significant for ground water dissolved plumes.

The most important consideration in designing an extraction system for MTBE or other contaminants is placement of wells/trenches and vertical screen or trench depth intervals. The extraction system is placed within the contaminant plume so it effectively captures the contaminants. Placement of the system wells and or trenches can incorporate two approaches. The design can incorporate a risk-based capture zone that allows areas of lower contaminant concentration, and lower risk, to escape the extraction system and naturally attenuate. With MTBE this needs to be considered carefully since its low taste and odor threshold make it liable to be a nuisance issue at concentrations that would probably not be of risk to humans or the environment. Alternatively, the extraction system can capture virtually all the ground water with detectable contaminants. This approach is used for sites where the contaminant is very high risk at low concentrations (acutely hazardous) or where escape of contaminants off-site would pose high liability, such as lawsuits. Although MTBE would not qualify as an acutely hazardous contaminant, its easy detectability by odor and taste in water could trigger lawsuits regardless of merit.

Well screens or trenches should be installed so they extend above the annual high water table to effectively capture LNAPL and accommodate seasonal ground water level fluctuations. They also must extend deep enough within the aquifer to capture the full, anticipated depth of the contaminant plume around or moving to the extraction point(s). MTBE, based on its specific gravity, does not produce a DNAPL itself and does not produce sinking plumes. Plumes of any dissolved contaminant sink only if there is significant downward movement of the ground water (along with all the constituents dissolved in it). A good example is drawdown by a heavily pumped well. However, MTBE, as with all dissolved contaminants, will tend to disperse and diffuse vertically and horizontally to produce a gradually deeper plume with time and distance. Extraction systems should be designed to capture the needed vertical and horizontal span of the plume.

Equally important is the design of the reinjection system if the treated water is to be reinjected back to the aquifer. Ideally, the reinjection wells/trenches are screened for the entire depth of the aquifer. Adequate reinjection capacity must be planned to accommodate this activity. The water management design must include plans for managing this excess water, such as discharge to a surface receiving water.

GROUND WATER TREATMENT

Although critical to the success of a remediation project, the extraction component is usually considered the simple part of pump-and-treat. This is because the extraction technologies are old and familiar, the variables are considered well-understood and definable, and the systems are considered to be forgiving, so that a few changes and additions can address operational issues. The challenges for extraction and reinjection are more likely to come in the hydrogeology and regulatory portions of the project.

The treat aspect of pump-and-treat has been the area of broadening technical possibilities in the last few decades. There are many new, potential, and refined treatment technologies applicable to extracted, impacted water. In this section, treatment technologies for MTBE in extracted ground water are discussed.

Granular Activated Carbon (Liquid Phase).   GAC is prepared from organic materials such as coal, lignite, wood, and coconut shell. The preparation involves controlled heating of the material to produce a porous high-carbon matrix that is activated by further treatment with heat and steam. Activation means that the carbon is altered by the treatment to maximize its potential to adsorb certain organic and inorganic compounds. The product of the preparation process can be ground and sieved to produce GAC or powdered activated carbon (PAC) of different sizes. The starting material and treatment processes produce GACs with varied properties, including porosity and pore size. There are experimental activated carbons impregnated with iron so that adsorbed materials can then be oxidized by peroxide to regenerate some of the adsorptive capacity. Despite relatively uniform production processes and starting materials, GAC purported to be of the same type can vary by manufacturer and by batch from a particular manufacturer. Usually these variations are not significant in changing the function of the GAC for actual field applications, but this should always be considered if new GAC batches do not behave as expected.

Adsorption of a substance is an interaction and accumulation of that substance at the surface of one phase, the solid phase, of a two-phase system interface, *i.e.*, liquid:solid phases or vapor:solid phases. The production of activated carbon results in a highly porous material with total surface areas reported up to 3,000 square meters per gram (m^2/g) (32,000 square feet per gram [ft^2/g]), but more typically 600 to 1,300 m^2/g (6,500 to 14,000 ft^2/g). The material is full of pores ranging from 10 to 1,000 angstroms that attract and adsorb organic molecules. Debate remains about the mechanisms of carbon adsorption, between a straightforward interaction with a surface and physical trapping of molecules in micropore spaces, or a combination of these mechanisms with others. It is critical to understand that some types and brands of GAC may perform better for adsorption of particular contaminants or contaminants from particular ground water sources. This is probably due to a variety of properties of the particular GAC and to water chemistry such as pH, natural occurring organics, mineralization, and similar characteristics. The current perception is that coconut shell GAC removes MTBE better than other GAC varieties (California MTBE Research Partnership, 1999).

TBA can be a degradative intermediate and co-contaminant of MTBE. Due to its higher solubility and lower Henry's Law constant than MTBE, TBA may be more difficult to remove by GAC than MTBE. Resin adsorption is demonstrated to better absorb TBA than GAC and thus may be a better option for sites with high concentrations of TBA.

GAC adsorption is not a destructive technology. Consequently, the GAC must be handled in a way that prevents release of the adsorbed materials to the environment again or must be regenerated for reuse in a way that removes but captures or destroys the released contaminants. Used GAC can be regenerated with heat and steam treatment, which releases and/or destroys the contaminants.

The efficiency of the GAC is defined as the mass of adsorbed material per mass of activated carbon. Generally, it declines with each subsequent regeneration and use, so that eventually it becomes cost-effective to dispose of the spent GAC. Spent GAC can be incinerated, burned as an alternative fuel, or, if free of contaminants, can be landfilled. In most applications for drinking water treatment, regenerated GAC cannot be used by regulation; only virgin GAC can be used. This is a precaution to be sure that the GAC does not itself become a source of contamination to the drinking water from residual contaminants from a previous use. If the water being treated for MTBE is to be used for drinking water, regeneration will probably not be a cost-cutting approach available to the operator.

Carbon adsorption is relatively nonspecific and has been shown to be effective for removing halogenated and nonhalogenated VOCs and SVOCs, and polychlorinated biphenyl (PCBs) compounds from liquid streams. From the perspective of MTBE treatment, this means that many other organic compounds, which may be co-contaminants, will also be consuming the GAC adsorptive capacity. Since BTEX and smaller alkanes and alkenes commonly accompany MTBE from gasoline and other fuels, sometimes this competition can be significant. In addition, free-phase hydrocarbons in the ground water can physically blind the GAC by accumulating as a layer on the surface of the GAC, limiting water and its contaminants access to the GAC porosity and thus limiting or preventing removal of dissolved components by the GAC.

Figure 15-4 is a simple diagram of a series of GAC beds in a simple treatment system such as might be used in a temporary ground water pump-and-treat remedial project. GAC beds can be operated either in parallel or in series. The usual design reason for using multiple bed approaches is to limit the frequency of the GAC changes or regenerations. For contaminants with lower affinities for GAC and with a breakthrough point at a fraction of GAC saturation, such as with MTBE, GAC reactors in series are appropriate. The series arrangement allows monitoring for breakthrough at the first carbon bed or canister while still protecting from total system effluent breakthrough with the second, third, or additional GAC beds or canisters.

The performance of a GAC system is affected by the properties of the GAC, chemistry of influent streams, and design parameters. The properties

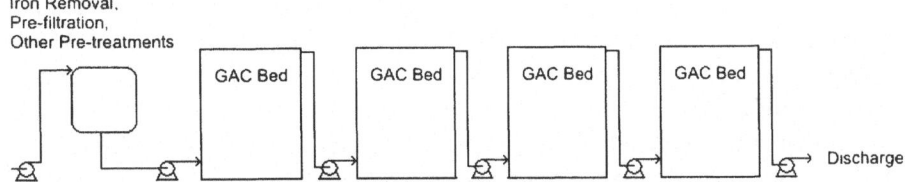

Figure 15-4.   **Granular Activated Carbon Treatment Beds in Series**

of GAC include particle size, surface area, pore structures, surface chemistry, and also the type of carbon used. The adsorption capacity is dependent on the surface area accessible to organic molecules within the pore structures. The pH of the influent stream affects the affinity of weak organic acids or bases for activated carbon but probably is insignificant for MTBE.

Adsorption of anything, including MTBE, is a dynamic equilibrium phenomenon. In a static equilibrium (not flow through) with only one adsorbable constituent, the adsorption system comes to an equilibrium with some fraction of the constituent on the adsorbent and some fraction remaining in solution. At any given concentration, the equilibrium ratio, the fraction adsorbed and fraction not adsorbed, will be the same as long as the adsorption capacity of the adsorbent is not exceeded. Regardless of high or low total concentrations, the concentrations on the GAC and in the water will be proportional to the total available contaminant in the stream as long as the GAC adsorbent capacity is not exceeded. However, even in a static mode, individual molecules desorb randomly per unit time and others adsorb to maintain the equilibrium, basically an off and on phenomenon. In a mixed contaminant situation where the contaminants have different adsorption characteristics for GAC, this allows stronger adsorbers, such as benzene, to displace weaker adsorbers, such as MTBE. When an MTBE molecule desorbs, its place can be taken by a benzene molecule that is much less likely to desorb due to its equilibrium gradually and progressively shifting the equilibrium towards dissolution of the MTBE within the carbon bed. The equilibrium constant of a contaminant depends on its relative affinities for water and GAC but results in a constant proportional ratio in each phase, water or GAC. Thus, if the concentration in the influent drops significantly, MTBE, and other contaminants, now present on the GAC at considerably above the equilibrium ratio can desorb, increasing the water concentration until the equilibrium is reestablished. MTBE adsorbs to GAC and has a very early breakthrough relative to its saturation on GAC. The consequences of this are that MTBE is readily displaced by stronger adsorbing compounds and breaks through so early that GAC, if replaced at breakthrough, is prohibitively costly in most applications. Both problems can be resolved by designing longer or guarded systems. Since many of the compounds competing for MTBE adsorption locations are stronger adsorbers, placing multiple GAC beds ahead of the actual MTBE adsorption beds allows those beds to remove the displacing contaminants and allows MTBE to then adsorb on the later beds. The first beds must be monitored for breakthrough of the other contaminants and replaced when saturated to continue to guard the MTBE beds. Likewise, the guarded MTBE beds must be monitored for breakthrough of MTBE. By using a sufficient number of MTBE beds and enough guard beds, the first MTBE bed in each series can be operated after breakthrough until nearer saturation, maximizing use and minimizing cost per mass of MTBE removed per GAC mass.

Empty bed contact time (EBCT), bed depth, and hydraulic loading rate are design parameters for GAC adsorbers. EBCT is the time for the volume of water in a carbon bed to flow through the bed. Thus for a given flow, to increase the EBCT, the size of the bed and its associated water volume must be increased. Typically, an increase in EBCT will increase the bed life of the GAC for a given flow and contaminant concentration. However, as EBCT increases, at a constant hydraulic application rate, the capital cost increases for the larger tank and associated components of the system. Yet operating costs can decrease because of lower frequency GAC replacement and associated costs or downtime.

MTBE does not adsorb to activated carbon as well as hydrocarbons such as BTEX. Gasoline components including BTEX are usually present with MTBE and compete for adsorption locations. Natural organic matter in ground water will compete with MTBE for adsorption locations. Consequently, single component isotherms are not applicable for such mixtures. Bench jar or beaker tests can be conducted to estimate carbon usage, but dynamic column testing with natural site water is recommended for design of MTBE treatment systems. Poorer adsorption and co-contaminant competition causes a long mass transfer zone and early breakthrough of MTBE. Systems with three or more GAC adsorbers in series should be considered to enhance the total percent saturation of the first adsorber prior to its replacement. This decreases carbon usage rates and thus carbon replacement frequency, and prevents shutdowns due to potential breakthrough at discharge. Some regulatory agencies require shutdown when the next-to-last GAC bed in the series breaks through.

Coconut-based GAC is thought to have smaller pores that are more effective at adsorbing MTBE. Laboratory tests have demonstrated coconut shell GAC adsorbs more MTBE than coal-based GAC. However, variations in raw materials and manufacturing processes make it difficult to predict GAC performance without testing. Virgin carbon is preferred for the treatment of MTBE due to its higher adsorption capacity than regenerated carbon.

GAC is a well-developed and proven technology with a long history as a treatment for municipal, industrial, and hazardous waste streams. Modular

GAC column or bed reactors are readily available from a variety of vendors. The following factors may limit the applicability and effectiveness of the process:

- Spent carbon regeneration or disposal;
- Multiple contaminant competition;
- Fouling by suspended solids or solids formed by precipitation of reduced iron or manganese species;
- Free-phase hydrocarbon; and
- Poor adsorption for water-soluble compounds and small molecules.

For these reasons GAC is generally selected for lower concentrations of MTBE in water. A rule of thumb is 500 μg/l or less, and GAC may only be economical at half that loading.

INTERFERENCES

As with GAC applications for target contaminants other than MTBE, a variety of materials also present in the water to be treated may interfere with adsorption of MTBE.

Iron. Soluble reduced ferrous iron (Fe^{+2}) dissolved in the extracted water may oxidize to less soluble ferric iron (Fe^{+3}) forming iron hydroxide ($Fe(OH)_3$) during extraction and pumping by contact with oxygen introduced from the air. The $Fe(OH)_3$ can precipitate, both in piping and in or within the GAC bed. This can result in piping obstruction and/or progressive blinding of the flow channels for water and pore structure of the GAC, resulting in increased backpressure in the bed, poorer flow and distribution of flow, and inaccessible adsorption sites in the pore structure. Regeneration of the GAC may also be less effective because iron may remain trapped in the pore structures. During regeneration, such minerals, possibly acting as catalysts, in the GAC contribute to a degradation of the GAC carbon structure causing an accelerated and progressive loss of adsorptive capacity with each reuse, beyond the normal loss from regeneration. This type of deterioration can be determined by iodine or molasses tests, which essentially determine the small and large pore capacity of the carbon respectively.

If iron in the recovered water is expected to significantly interfere with the adsorption of MTBE, consideration should be given to using an alternate treatment technology or a pretreatment system to remove the iron. Iron concentrations of several mg/l in extracted water can usually be tolerated, but concentrations of tens of mg/l or higher will usually cause problems. Commonly used pretreatment systems can consist of aeration or chemical oxidation, followed by separation of the precipitated iron by filtering. In this application, removing MTBE by aeration or chemicals would not be the focus of the pretreatments since one aeration or stripper unit would generally not be adequate. One mechanical hydraulic stripper or aeration unit will probably be adequate for iron precipitation and could be designed to provide recirculation adequate to precipitate most of the iron. Pretreatment chemical oxidation with hydrogen peroxide or other oxidizers can be used to precipitate iron and has proven successful in this application for MTBE remediation. Following oxidation, the precipitate can be removed by a separator, clarifier, filtration, or other technologies. Coagulant technologies for iron should be carefully considered with respect to the impact of residual coagulant and its potential to also interfere with the GAC.

Manganese. Like iron, dissolved manganese can precipitate and blind or plug flow paths and pores in the GAC. The same pretreatment technologies as used for iron generally will work for manganese. Both iron and manganese should be evaluated when one or the other is suspected to be in the water as often the same geochemical conditions and parent soils or rock can produce both soluble iron and manganese.

Total Organic Carbon. TOC or dissolved organic carbon includes the contaminants and naturally occurring organic materials. Naturally occurring organic materials can include tannins, a range of humic materials, particularly the more soluble humic and fulvic acid ranges of compounds, and in some areas naturally occurring concentrations of hydrocarbons from native shales, sands, limestones, and sandstones. These smaller humic and other materials are usually GAC adsorbable and thus consume adsorption capacity on the GAC. In addition, if their affinity for GAC is strong, they may significantly contribute to displacement of MTBE from the GAC. Co-contaminant gasoline range hydrocarbons may also contribute to the TOC and displace or replace MTBE on the GAC. Some fraction of the TOC may not be adsorbable on GAC and will emerge as breakthrough. Consequently, TOC as a measured parameter should generally be avoided as an indicator of breakthrough and particularly in areas with natural TOC or non-target contaminants in the ground water.

Pretreatment can help prevent MTBE from simply being displaced or non-contaminant TOC consuming the GAC capacity intended for MTBE. TOC can be removed by using a guard GAC bed that pretreats, leaving the subsequent beds for adsorption of MTBE.

Mineralization. In addition to iron and manganese, hard water compounds such as calcium carbonate and similar compounds may precipitate or scale in the equipment due to oxygenation and pH changes. Treatments for excessive mineralization include standard pretreatment with ion exchange systems and pH control to minimize precipitation.

Coagulants and Additives. Coagulants and additives, used to pre-treat water for removal of metals or turbidity, could interfere themselves with the GAC adsorption. These coagulants, just like turbidity, can adsorb to the GAC, blinding pores and blocking adsorption sites. Discussions with and between vendors of GAC systems and pretreatment systems should be expedited to select and match the appropriate systems to minimize the chances of interference caused by pretreatment for another interference.

Turbidity. Turbidity can include natural organic or inorganic turbidity components such as fine clay particles, particulate plant decay debris, silts, and even fine sands. However, it can also include free-phase organic material dispersed as droplets, micelles, or fine emulsions in the recovered water. Fine soil particles can physically bind the GAC pores, thus limiting the water flow channels, and generally disrupting good water distribution throughout the GAC bed. Free-phase organic material can smear over the surface of the bed face, coat GAC particles, and generally blind pores and cover adsorption sites. In addition, the organics may contain compounds that specifically adsorb, competing with MTBE.

Turbidity can be pretreated using sand, diatomaceous earth, or similar filter systems and/or coagulation systems.

Co-contaminants. Related to the free-phase organics discussed above are the dissolved organics, particularly those considered co-contaminants. It is appropriate to use the GAC system to remove co-contaminants along with MTBE, but the behavior and impact of the expected mix and individual components targeted for removal must be assessed. Generally, the design of the treatment system for mixed contaminants will be driven by the treatment fate of the contaminant most likely to breakthrough. This will usually be MTBE. Its breakthrough during interactions with the other contaminants will probably dictate the design (*e.g.*, the number of beds needed to maximize carbon usage while protecting from an unguarded breakthrough) and the need to use alternative technologies. It may make sense to design a GAC system knowing that the first bed will be used to remove BTEX, for example, and only subsequent beds will be expected to adsorb MTBE at significant concentrations, as discussed above. The first bed would then be monitored for BTEX breakthrough, and the bed changed when that happens. In essence such a system uses the first bed to guard the other beds from MTBE co-contaminants, freeing up the remaining beds for MTBE with minimal adsorption competition.

Biological Growth. Adsorbed contaminants, renewing oxygenated flows of water, and high surface area GAC combine to provide very good locations for biological growth. This is exploited in fluidized bed reactors (FBRs) with

GAC but can become a problem or a blessing in a simple GAC adsorption system. To some extent this can be a problem related to the lifetime of the first bed and the biodegradable nature of one or more of the contaminants. If the first bed is in place for weeks or months, this allows time for selection, adherence, and growth of a fixed biofilm on the GAC with the resulting blinding of pores and some water flow. Growth can occur across the interface of the GAC bed in extreme cases, suggesting that perhaps the wrong treatment system was selected. However, if the growth does not significantly impact the operation of the GAC, it can extend the life of the carbon by degrading adsorbed contaminants, allowing for additional adsorption at freed up locations. In addition to growth within the GAC, growth can occur within the recovery wells, piping, and tanks, becoming a source of sloughed biomass and inoculum to the GAC with similar effects as discussed for growth on the GAC. Simple filtration systems can remove biomass sloughed from upstream sources. These filters can be sand or mixed media but will need to be backwashed periodically. Such filters are often a standard component of GAC systems and can usually be supplied by the vendor of the GAC beds and equipment.

Costs.   The costs of GAC treatment are generally well-understood for simple adsorption of the contaminants, even for MTBE with its recent emergence as a contaminant of concern. The California MTBE Research Partnership assembled cost numbers for GAC treatment.

Estimated costs for MTBE treatment with GAC vary from $0.41/3,800 liters (1,000 gallons) to $2.80/3,800 liters (1,000 gallons) treated. By comparison to the estimated costs of other technologies, GAC should be considered for sites at which influent MTBE concentrations are 200 µg/l or less. Other organic matter concentrations in ground water should be taken into account during the design. System design life will also affect the technology selection. GAC systems are more likely to be cost-effective for sites requiring short duration treatment compared to other systems, such as advanced oxidation, which require higher capital cost.

RESIN ADSORPTION

Resin adsorption is an adsorptive technology similar to GAC adsorption. Resin adsorption uses synthetic adsorptive resins to adsorb contaminants from ground water. Figure 15-5 illustrates the basics of such a system that typically uses at least two beds, so one is adsorbing while the other is being regenerated. The resins are typically spherical, microporous beads or prills, produced with a high surface area and a high adsorption capacity for organic molecules. The resins are polymer matrices with side functional groups that bind the contaminants, adsorbing them from the water. They adsorb selec-

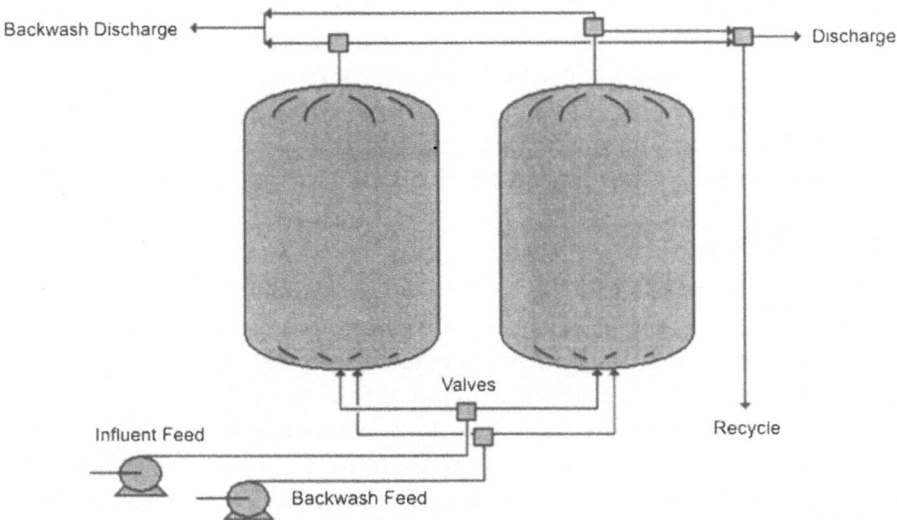

Figure 15-5. **Resin Adsorption System**

tively based on the chemistry, structure, and steric position of these side functional groups and the contaminant.

Resin adsorption is applicable for removing halogenated and nonhalogenated VOCs and SVOCs, PCBs, and explosive compounds from water. The synthetic resins typically provide 5 to 10 times higher mass loading of contaminants than does GAC at low contaminant concentrations. The affinity of the resins is higher for the contaminant molecule compared to GAC. Some resins have a higher capacity for MTBE compared to GAC. In addition, resin adsorbents are less likely than GAC to adsorb competing inorganic matter. Resins can be regenerated on-site through microwave irradiation or steam stripping. However, data are limited on the effectiveness and efficiency of MTBE treatment with synthetic resin adsorbents, suggesting that treatability studies need to be done if it is to be used for a specific application. Ground water impacted by MTBE is likely to contain TBA as well. For sites impacted by both MTBE and TBA, synthetic resin adsorption should be considered as a possible treat component of pump-and-treat.

There are some factors limiting the applicability and effectiveness of the process. Unit costs of resin are higher than GAC, typically requiring resin regeneration. Preliminary experience indicates that resins can be regenerated more times than GAC and with less loss of adsorption capacity than GAC. The economic tradeoff of resin's higher capacity, more regeneration cycles, and sustained adsorption capacity to the resin's higher cost has not been fully assessed. Suspended solids or solids formed by precipitation of reduced iron and manganese species in solution can physically foul a resin bed, requiring pretreatment to remove the fouling agents. Finally, there is the issue of treatment or disposal of the concentrated waste stream from the regenerated resin. Suggested options, but not demonstrated in the field, have included biological treatment of the waste stream from steam regeneration and thermal oxidation after microwave regeneration/volatilization.

AIR STRIPPING

Air Stripping is a process that exploits the phase equilibrium of contaminants between water and air. The transfer of contaminants from water to air is a surface area mass transport process. Impacted water is contacted by air, typically a larger volume of air, in a manner that greatly increases the two phases' interface surface area. This greatly increases the mass transport of the contaminants from water to air in response to the equilibrium gradient. The treatability of a contaminant by air stripping is indicated by its Henry's Law constant. The higher the Henry's Law constant, the easier the compound is to air strip. The dimensionless Henry's Law constants for MTBE and TBA are approximately 0.026 and 0.0009 respectively. Stripping is best for compounds with dimensionless Henry's Law constants of greater than 0.05. Nevertheless, MTBE can be air stripped and the CDHS has approved air stripping as their best available technology for MTBE. Recycling of water may be necessary to strip the MTBE down to the needed concentration for discharge. A typical volumetric ratio of air to water for effective MTBE stripping is greater than or equal to 200:1.

Air strippers of several types are commercially available. Types of air stripping methods include packed tower, tray stripper, diffused aeration, low profile aeration, and mechanical stripping.

STRIPPING TECHNOLOGIES

Packed Tower Stripper.   A PTS usually consists of a cylindrical tower containing a perforated plate, near the bottom, to support the packing material. Packing materials can be various ceramic, stainless steel, or plastic shapes that promote turbulent flow of the water and air and maximize air:water interface surface area. Typical packed towers includes water distribution mechanisms at the top to distribute impacted water over the packing in the column. Water descends as air is blown up through the column, stripping off the volatile compounds. A sump at the bottom of the tower collects decontaminated water for recycling and/or discharge. PTSs can be permanent, or skid or trailer mounted. Most states require capture and treatment of the PTS off-gas prior to discharge to the atmosphere. Typical off-gas treatment technologies applicable to MTBE are GAC, thermal oxidation, catalytic oxidation, and biofilter treatment. Figure 15-6 illustrates a PTS.

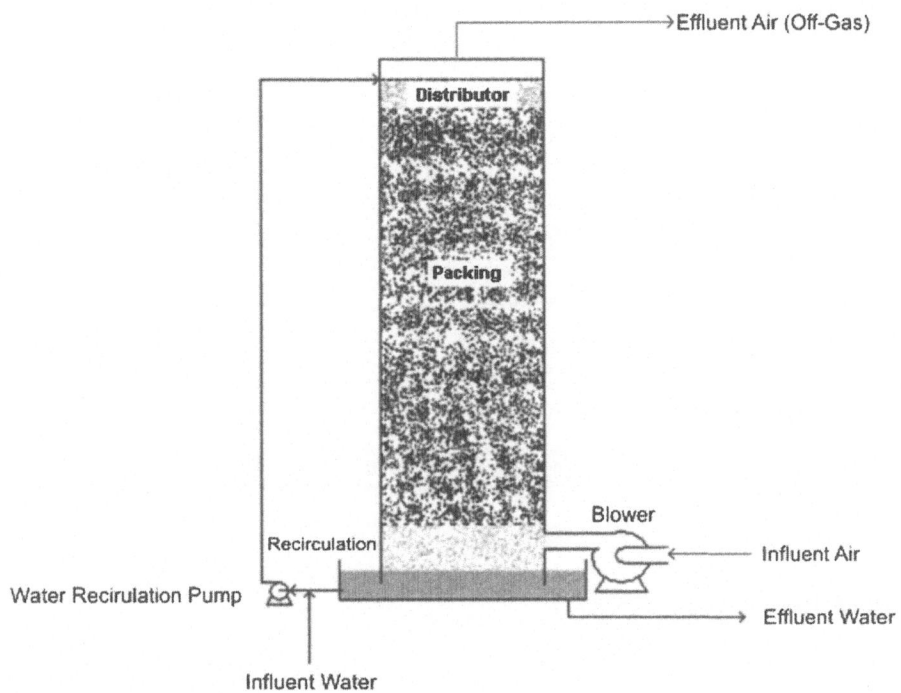

Figure 15-6. **Packed Tower Air Stripper**

Low-Profile Air Stripper.   Low-profile air strippers (LPSs) operate on the same principles as PTSs. LPSs flow impacted water across stacks of trays that are perforated with small aeration holes. Water flows over a weir on each tray to the next lower tray until the water exits the bottom of the stripper. Air is bubbled through holes in the trays to facilitate stripping. The trays are packed in a very small chamber to maximize air-water contact while minimizing space. Because of the significant vertical and horizontal space savings, these units are increasingly being used for ground water treatment. LPSs may also be a good choice regarding concerns for visual impact at sites located in residential areas. These systems can also use heated trays to increase the stripping efficiency. Figure 15-7 illustrates an LPS.

Diffused Aeration Stripper.   Diffused aeration strippers (DASs) are tanks or reservoirs with fine bubble or other air diffusers placed near the bottom. Air bubbles move up through the liquid, stripping contaminants, are captured by the tank roof, and then exit at the top of the vessel to off-gas treatment. Stripping efficiency is increased with deeper tanks or reservoirs (longer contact time) and finer bubbler or air diffuser systems (higher contact surface area). DASs are generally unimpacted by suspended solids. How-

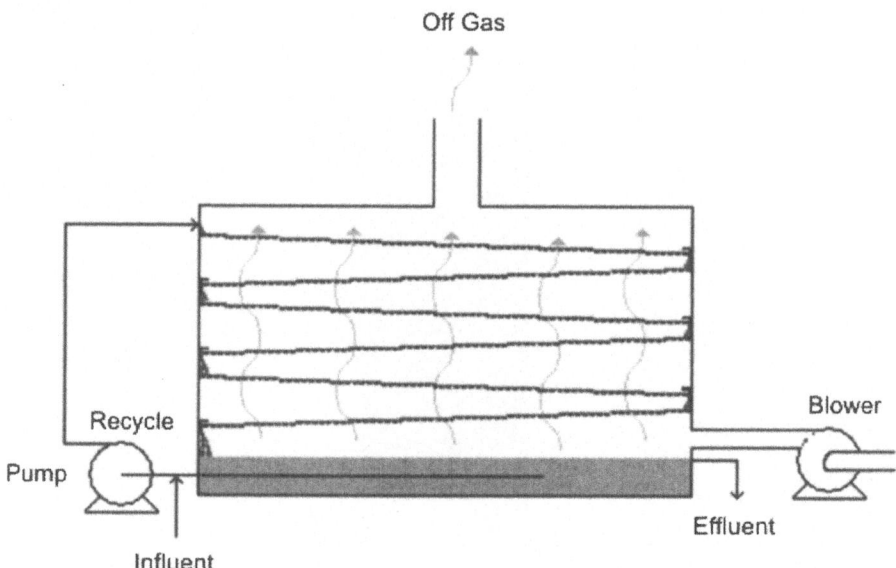

Figure 15-7. **Tray Stripper**

ever, for stripping MTBE, which needs a higher ratio of air to water, DASs are generally not effective unless the starting MTBE concentrations are very low, near the target discharge or cleanup concentration. As with any stripper system, poorer stripping performance can be overcome by larger, deeper tanks and extended air to water contact and residence time. Figure 15-8 illustrates a simple tank DAS.

Mechanical Stripper.   Mechanical strippers are devices that exploit thin film or venturi effects to transfer contaminants from water to air. An example of a mechanical stripper is the Hazleton Maxi-strip® System. The venturi devices are mounted on the tops of tanks. The impacted water is jetted down the inside of a venturi by ring nozzles, producing highly turbulent jets of water that shear in the open venturi bore. The turbulence achieved in the venturi creates the large surface area needed to enhance mass transfer. The water jetting also induces airflow down the venturi, eliminating the need for blowers. The water can be recycled from the attached tanks to the venturies as many times as needed to achieve the treatment concentrations. Units can be assembled in series to meet the needed water capacity. The system has series of air intake hoods to allow outside air to be drawn directly to the strippers. In a stripper series, off-gas from the least contaminated module can be recycled counter to water flow as stripper gas. This minimizes the amount of total off-gas flow and increases the concentration of the contaminants in the eventually discharged off-gas, usually making for more efficient off-gas treat-

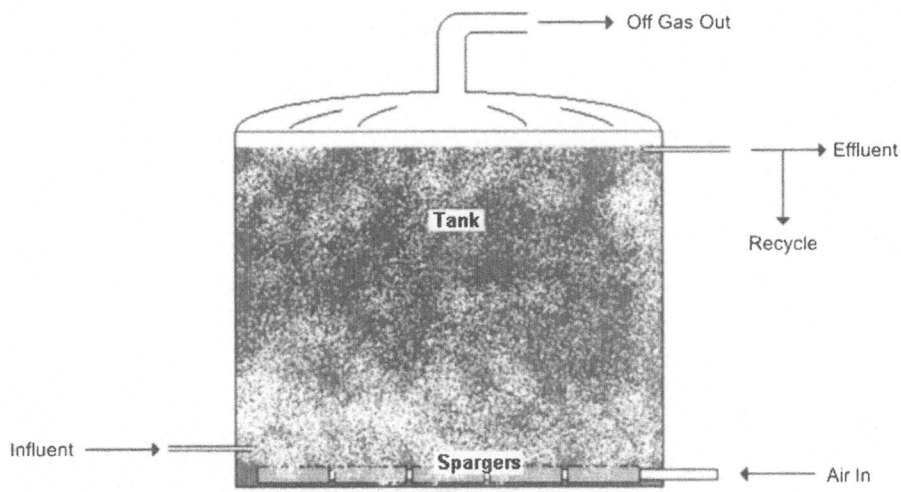

Figure 15-8. **Diffused Aeration Stripper**

ment. The off-gas can be treated as needed by GAC or other technologies. The mechanical stripper system is flexible for different influent concentrations or flows. Additional stripper modules can be added and modules can be turned off as required. Figure 15-9 illustrates the principles of a mechanical venturi air stripper system.

OFF-GAS TREATMENT

The air stream leaving the stripping system may require off-gas treatment if contaminant levels exceed regulatory emissions standards. Several treatment processes, such as GAC, catalytic and thermal oxidation, and biofilters can be cost-effective for treating the off-gas emissions.

Thermal and Catalytic Thermal Oxidation.   Both thermal oxidation and catalytic thermal oxidation can be used to destroy organic compounds. In a thermal oxidizer, VOCs are oxidized in a flame at high temperature. In catalytic oxidation, VOCs in the emission stream are oxidized at lower temperatures with the aid of a catalyst. Catalysts typically employ platinum, palladium, or other metal oxides. The performance of a catalytic oxidizer is affected by several factors including: operating temperature, space, velocity, VOC type and concentration, catalyst properties, and presence of catalyst inhibitors or "poisons" in the emission stream. Typical poisons are metals and other inorganics that can deposit on the catalyst and reduce its reactivity with the hot organics. Figure 15-10 presents a simple illustration of a catalytic (thermal) off-gas treatment system. Thermal only systems are similar but without a catalyst to lower the temperature needed for destruction. Thermal

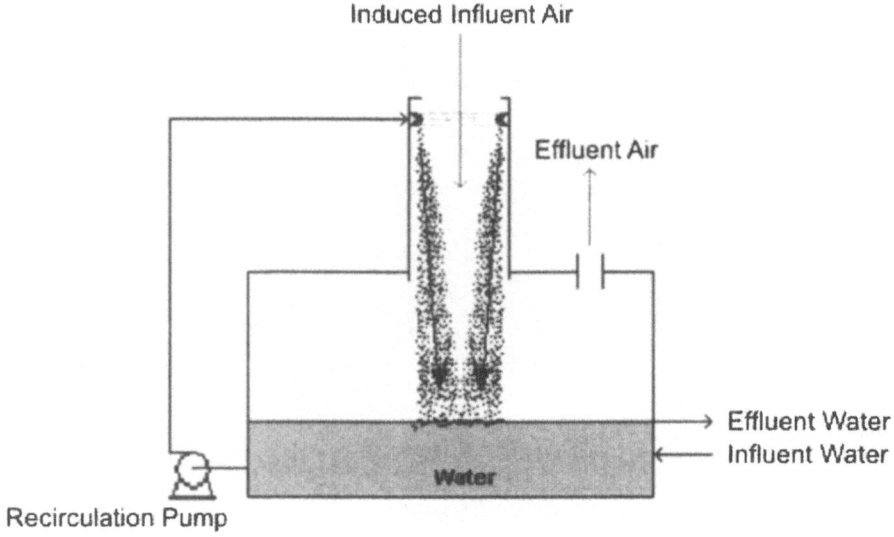

Figure 15-9.   **Mechanical Venturi Air Stripper**

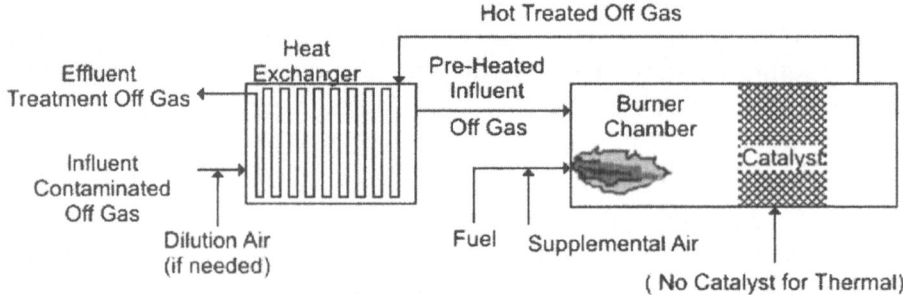

Figure 15-10.   **Catalytic and Thermal Off-Gas Treatment System**

systems often use specially designed burners to assure destruction of the gas stream.

Granular Activated Carbon.   Vapor phase GAC adsorbers can be used to treat off-gas streams. The activated carbon removes contaminants from the gas stream by adsorption, just as from water, until available active sites are occupied. Commercial grades of activated carbon are available for specific use in vapor-phase applications. As with GAC use in water systems, MTBE can be removed from off-gas by GAC. Many of the same principles apply for GAC adsorption from air streams as for water steams. Figure 15-4 illustrates a water based adsorption system. An air adsorption system would be similar except pretreatment would include a knockout pot or other demisting/de-humidifier system to remove moisture in the off-gas stream. This is particularly critical when off-gas is originating from a stripper system.

Biofilters. Biofilters are high surface area beds with wet biofilms that degrade the organic contaminants in air streams as they pass through the filters. The bed and biofilm must be moist so that organics will partition from the air stream to the thin water layer over the surface of the microbial biofilm where the microbes can rapidly access and degrade them. Uptake and degradation by the biofilm is a sink for the contaminant, effectively maintaining a steep mass transfer gradient from the air to the water to the biofilm. A blower and an air dispersion system are used to move the air through the biofilter and evenly distribute the flow. Water is misted or sprayed over the beds and/or the off-gas is humidified to increase transfer to the microbial biofilm. The biofilm is attached to the bed materials, which can be plastic, wood chips, or similar inexpensive and durable materials with high surface areas. Figure 15-11 illustrates an engineered tank system and a pile system for treatment of off-gas.

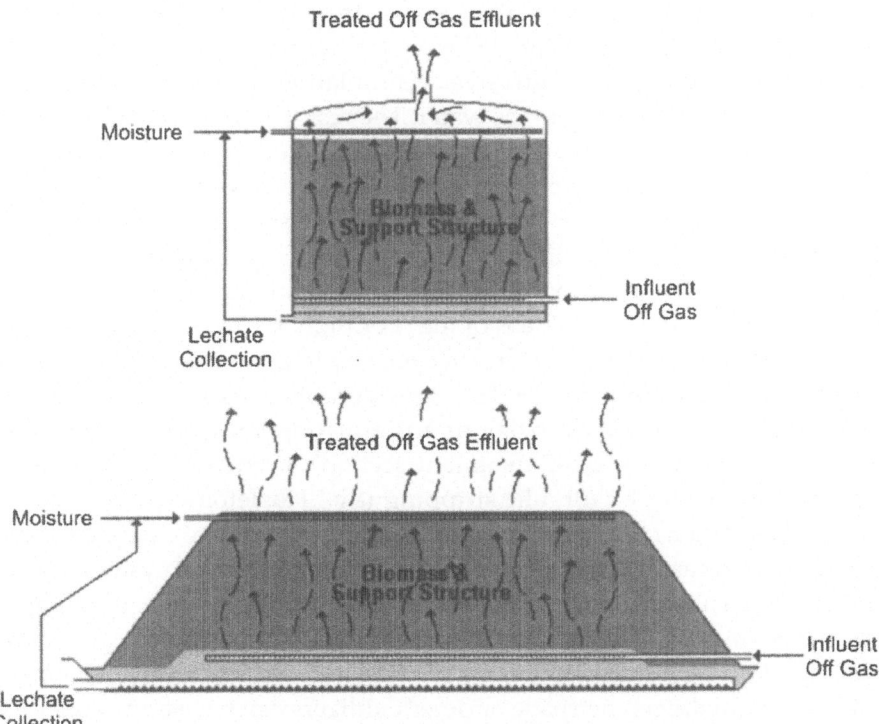

Figure 15-11. **Off-Gas Treatment with Biofilters**

Off-Gas Treatment Costs. The startup costs are comparable to conventional technologies such as GAC and thermal, but operations and maintenance costs are often substantially lower. There are no external fuel costs such as those associated with thermal oxidation technologies. Biofiltration becomes more economical than GAC or oxidation when airflow rates are high and pollutant concentrations are low. Biofilters can treat MTBE but may require larger filters for increased residence time to degrade MTBE and prevent accumulation in the water phase. At least one of the consortia and one of the microbes capable of degrading MTBE in a pure culture were isolated from a large biofilter application where the organisms were presumably degrading MTBE in the off-gas from a wastewater treatment plant. With appropriate conditions, trickling biofilters can be used directly to treat MTBE impacted water, eliminating the need to strip MTBE.

INTERFERENCES FOR STRIPPING

Interferences for effective stripping are water conditions or constituents that impact effective stripping of the target contaminant(s).

Iron. Ferrous iron dissolved in the extracted water may oxidize to produce iron hydroxide during extraction and pumping and will almost certainly be oxidized by contact with oxygen introduced from the air stripping processes. The iron hydroxide can precipitate before reaching the air stripper or within the stripper (resulting in progressive accumulations or fouling on the packing or trays), or as turbidity in the water. The result is decreased stripping efficiency as surface area is reduced and air and water flow and distribution are reduced. Since effective stripping of MTBE requires a very high air to water ratio, MTBE stripper systems are potentially more sensitive to losses of stripper effectiveness due to fouling compared with BTEX stripper systems.

Iron concentrations in extracted water of several mg/l can usually be tolerated, but concentrations of tens of mg/l or higher will usually cause problems. If iron in the recovered water is expected to generate significant solids and interfere with stripping of MTBE, consideration should be given to using an alternate treatment technology or a pretreatment system to remove the iron (see discussion in GAC treatment section). Pretreatment systems can consist of mechanical hydraulic stripping (*e.g.*, Hazleton-type systems), or chemical oxidation/precipitation, followed by separation of the precipitated iron. With a mechanical hydraulic stripper as a pretreatment system, the separation can be integral to the tank beneath the stripper module. In this application, one unit is usually adequate for iron precipitation and can be set up to provide adequate precipitation of most of the iron. Pretreatment chemical oxidation with hydrogen peroxide or other oxidizers can be used to precipitate iron. Following oxidation, the precipitate can be separated by corrugated plate interceptor or other physical separation technologies.

Manganese.   Like iron, dissolved manganese precipitate can accumulate on packing and piping in strippers. The same pretreatment technologies used for iron generally will work for manganese. Both iron and manganese should be evaluated when one or the other is suspected. Often the same geochemical conditions and parent soils or rock could produce both soluble iron and manganese.

Mineralization.   In addition to iron and manganese, hard water compounds, such as calcium carbonate, may precipitate or scale in the equipment due to oxygenation and pH changes. Treatments for excessive mineralization include pretreatment with ion exchange systems and pH control to minimize precipitation.

Temperature.   Low temperature decreases the stripping efficiency and the removal of contaminants from the stream. Winter operations may require pre-heating the air and/or pre-heating the water prior to the treatment. Heating can also enhance the overall stripping efficiency year-round.

MTBE APPLICATIONS

Air stripping is a proven technology, successfully used to remove MTBE from drinking water. PTSs appear to be the most cost-effective for MTBE removal, in part because of their simplicity and robustness and in part because they are well understood. At low flows and removal efficiencies, prefabricated LPSs are competitive with PTSs, despite their slightly higher treatment costs. This is because of their availability as premanufactured package units, simple installation, ease of operation, and small space requirements.

Selection of a treatment system for off-gas containing MTBE is dependent on the off-gas MTBE concentration and flow rate. At low (less than 0.5 ppm) off-gas MTBE concentrations, vapor phase adsorption is the most cost effective off-gas control technology because of very low initial capital cost. For higher off-gas flow (greater than 300 cubic meters per minute [cmm] [10,000 cubic feet per minute {cfm}]) and higher MTBE concentrations (greater than 5 ppm), thermal oxidation becomes the most cost-effective treatment technology.

BIOREACTORS

Biological reactors have been widely used in the treatment of domestic and industrial wastewater. Organic pollutants in wastewater can be converted to biomass and mineral end-products, such as carbon dioxide and water, through microbial metabolism. Common types of wastewater bioreactors include activated-sludge, fluidized bed, and fixed-film reactors (rotating disk or trickling filter) systems.

Activated Sludge. An activated sludge process treats wastewater in an aeration basin, tank, or pond with an active mass of microorganisms (the activated sludge or mixed liquor volatile suspended solids [MLVSS]) capable of aerobically degrading organic matter into carbon dioxide, water, new cells, and other end products. Diffused or mechanical aeration maintains the aerobic environment in the basin and keeps reactor contents (mixed liquor) mixed. After a specific residence time (average time of water in the basin), the mixed liquor passes into a gravity clarifier of some type, where the activated sludge settles under relatively quiescent conditions and a clarified effluent is produced for discharge or tertiary treatment. The process recycles a portion of settled activated sludge back to the aeration basin to maintain the required activated sludge (MLVSS) concentration and discharges or wastes the remainder for disposal. Figure 15-12 illustrates a simple activated sludge system. These can usually be purchased as packaged or modular systems intended for small municipal utility districts but suitable for remediation sites.

MTBE can be treated with an activated sludge process. Most facilities that manufacture MTBE are successfully treating wastewater with significant MTBE and TBA concentrations using the activated sludge process. With MTBE, the residence time is usually extended to assure adequate time for treatment. In some activated sludge applications for TBA and MTBE, zoned aeration and zoned recycled activated sludge appears to enhance treatment efficiencies and end concentrations. A benefit of biological systems over other non-biological systems is that they typically improve their treatment capacity with time due to selection and adaptation. This is particularly true with activated sludge systems where a portion of the microbial biomass is returned to the treatment to maintain a designed MLVSS, allowing continuing adaptation and selection of the returned biomass. A second benefit of biological systems is that intermediates produced are usually rapidly degraded as well, since their generation also selects for an activated sludge population capable of degrading them.

The activated sludge process is a well-understood, conventional treatment process. The major design parameters for an activated sludge process include:

- Organic loading (food) (usually milligrams biological oxygen demand [BOD] per liter);

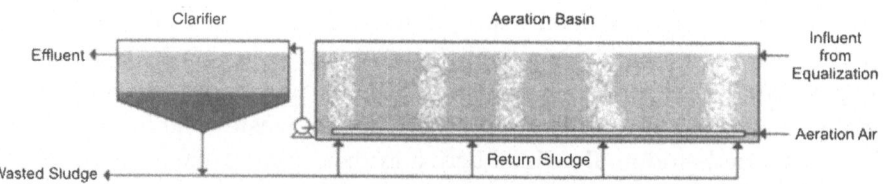

Figure 15-12.   Simple Aerated Activated Sludge Biotreatment System

- Microbial biomass (MLVSS or mixed liquor suspended solids) concentration (mg/l);
- Food to (bio)mass ratio (mg BOD/mg MLVSS), also called sludge loading rate (SLR);
- Hydraulic (contact) retention time (HRT) or hydraulic loading rate;
- Mean cell retention time (MCRT) or sludge age (ratio of total MLVSS/ wasted MLVSS per day; and
- Rate of aeration (total pounds per hour) or as pounds per hour of oxygen transferred to water.

Simplistically, what these parameters mean, in sequence, are how much food is entering the system, how much biomass is there and is that enough active biomass (some of the biomass is dead at all times) to treat that food, is the food and biomass together in the reactor long enough to treat the food, is enough of the biomass active (not to old) to treat the food, and is there enough oxygen to support the biomass in oxidizing the food.

For MTBE, because of its apparent slower degradation rate, the design and operational issues that may greatly impact an activated sludge system are HRT or hydraulic loading rate, MTBE and other food concentration(s), effluent treatment goal, and use of a sludge recycling system for maximum adaptation. These issues will determine whether a temporary system or a large, complex, and more permanent system is needed. The latter can greatly increase costs, particularly if the treated water is not reused or recycled for later use. One could expect that, in a mixed contaminant food stream, MTBE will be the target compound that determines the HRT or hydraulic loading rate (*i.e.*, the critical target compound) since it will probably be the slowest to degrade of the target organics. However, this should be confirmed with treatability studies when treating a complex mixed organic wastewater stream containing MTBE. Activated sludge systems are not applicable to the low concentrations of MTBE often seen at field sites unless additional degradable organic carbon is available in the wastewater or can be added to supplement the feed. To maintain an activated sludge system, typically a minimal mass of food (*e.g.*, 100 mg/l of BOD) is needed to sustain a viable biomass. As the BOD concentration goes down, the biomass concentration declines and becomes unhealthy. The use of fluidized bed GAC systems can typically spare this effect so that lower concentrations can sustain the system, but again this is more typically in the 10 to 100 mg/l range, still requiring additional contaminants in the mix or supplementation of the food.

Fixed-Film Reactors. Fixed-film reactors are those that biologically treat wastewater using a thin biological film fixed (self adhered) to a solid surface. Microbes prefer to adhere to solid surfaces and readily do so if the surface has fine structure that allows attachment. The more surface area the more biomass that can attach. Consequently, a supporting solid surface with the maximum surface area in contact with water is the optimum for these systems. Such systems occur naturally in ground water and streams where microbes adhere to soil particles and rocks in the flow of the water. The two most common engineered fixed-film reactors are packed trickling filter systems (trickling filters) and rotating disk systems.

Trickling filters range from older, low concrete tanks with crushed, sized rock packing to modern tank systems with tortuous, complex, high surface area plastic packing over which contaminated water flows. A rotary or stationary mechanism distributes wastewater across the top of the filter, from which it percolates, generally as a thin film, through the interstices of the film-covered medium. Trickling filters can be designed with vertical plug flow, but this makes aeration difficult and control of water feed more complex. Most trickling filters operate with thin film flows over most of the packing. As the wastewater moves through the filter, the organic matter diffuses into the film where it is available to the microbes and is degraded by those microbes. The oxygen can be supplied in the wastewater or by air forced up through the filter, but most typically is supplied by induced natural eduction of air into the interstices of the packing as water drains down over the packing, leaving voids between the packing. An underdrain system collects treated effluent for recycling for additional treatment and/or discharge. Biomass sloughs from the surfaces over time and must be separated from the discharge. Sloughing can generally be controlled by hydraulic loading to the filter but is desirable at some rate, since it removes aged and dead biomass just as wasting does in activated sludge systems. Figure 15-13 illustrates a

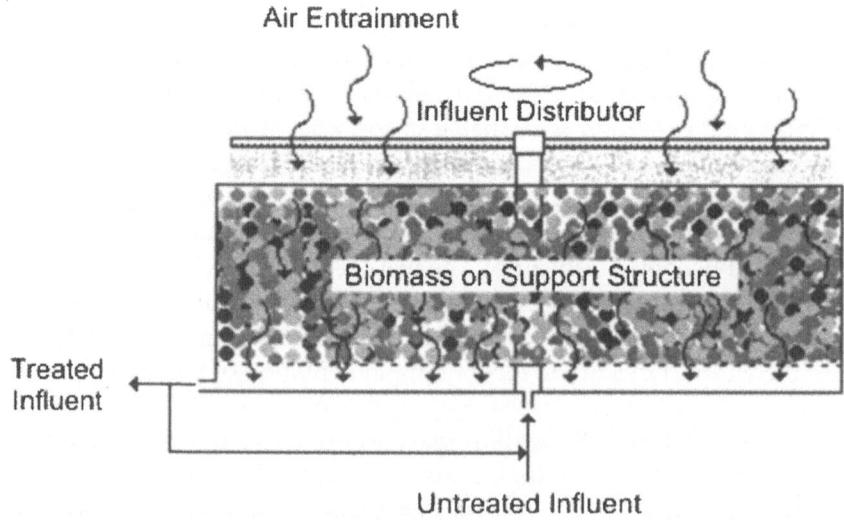

Figure 15-13.   **Biological Trickling Filter**

simple biological trickling filter with a rotating overhead distribution of untreated influent water.

Rotating disk systems usually consist of closely spaced disks or perforated plates that rotate through a tank or trough carrying the impacted water. Microbial biofilm grows adhered to the disks or plates as they are repeatedly immersed in water. Aeration occurs as the disks are above water and entrain air into the water. Some systems are supplemented with aeration in the water flow and some are enclosed overhead in sheaths in which air or oxygen can be added. As with other fixed-film systems, the film biomass sloughs as it grows and must be managed prior to discharge of the treated water. Figure 15-14 illustrates the principle of rotating disk operation.

Fixed-film systems should be capable of treating MTBE, but the long residence times expected for MTBE and TBA treatment opposed to the typically low contact times in fixed-film systems indicate that such systems will require high recycle ratios and larger reactors than needed for rapidly degraded contaminants. However, these systems may be applicable to wastewaters with low concentrations of MTBE. Rotating disk systems are probably not good candidates for MTBE or TBA treatment particularly at MTBE concentrations in the mg/l range because of the slow degradation rate and low cleanup goals. The problem associated with any fixed-film system is the mass transfer limitation of oxygen into the inner layer of the attached biofilms. This can be critical for MTBE where high oxygen concentrations appear to enhance the degradation rate.

Fluidized Bed Bioreactor. FBRs use the water to be treated to suspend (fluidize) a bed of fine-grained material such as sand or GAC, using an upward water velocity. Chapter 29 describes a case study that employs an FBR. As

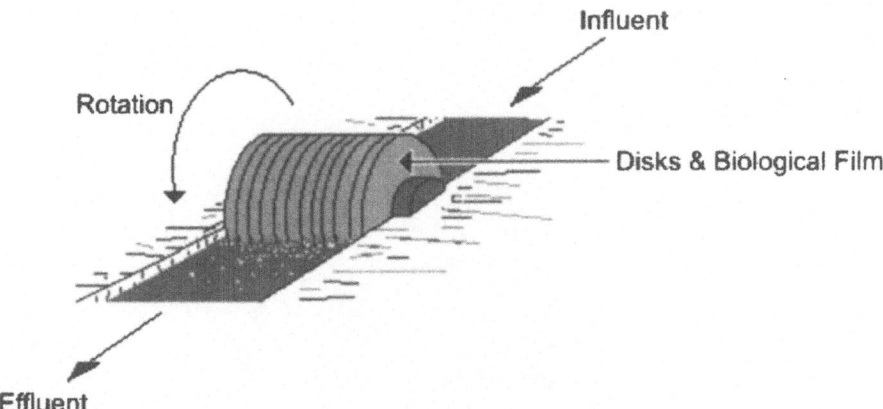

Figure 15-14. **Rotating Biological Disk Contactor**

with other fixed-film processes, the particles develop a microbial film. Fluidization significantly increases the specific surface area available for biomass and thus degradation of contaminants. Use of GAC as the fluidizing bed medium also adds to the specific surface area available for microbial colonization. FBRs avoid the bed plugging problems associated with a fixed bed bioreactor and trade it for a narrow range of operational flows due to the need to fluidize but not wash out the bed. Water being treated is usually recycled repeatedly to the fluidizing flow for extended treatment. GAC-FBRs have the added aspect of adsorbing the contaminants to potentially extend the contaminants' residence time in the reactor and possibly enhance the contaminants availability to the biofilm. The fluidization tends to abrade the biofilm, sloughing excess biomass that passes out of the reactor and which typically must be managed and disposed of. This sloughing is necessary to allow new growth and remove aged biomass. Figure 15-15 illustrates the operation of an FBR.

FBRs typically require a higher degree of operator maintenance and process control than other readily available treatment processes. Additionally, this process is sensitive to high variations in influent flow and contaminant concentrations and thus a substantial equalization capacity may be required.

FBRs using GAC potentially combine two technologies demonstrated to treat MTBE, biodegradation and GAC adsorption. FBRs have been used to treat an MTBE impacted ground water stream in Sparks, Nevada. GAC-FBRs have also been shown to be capable of degrading MTBE to below detection limit from high concentrations in bench studies.

MEMBRANE SEPARATION (REVERSE OSMOSIS)

A membrane, as discussed here, is a filter material with very small pores. It acts as a selective barrier that permits the diffusive separation of solutes in a fluid by mechanical size-sieving and/or charge effect mechanisms. Mem-

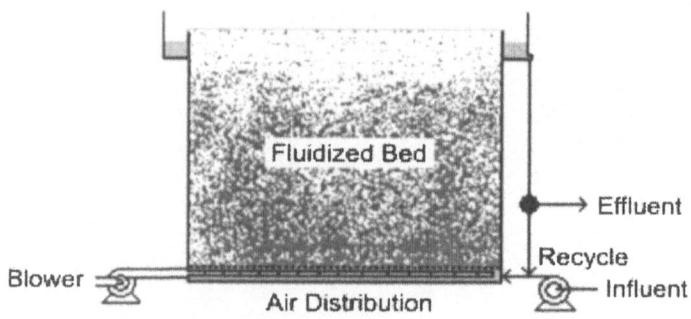

Figure 15-15. Fluidized Bed Bioreactor

branes can be selected to separate components over a wide range of particle sizes, molecular weights, and charges. Membranes are available in several different configurations — tubular, hollow-fiber, plate-and-frame, and spiral-wound.

RO is actually an ultrafiltration. Osmosis is the movement of some component (a solvent such as water) of a solution through a membrane from the more dilute solution to the more concentrated solution with a net long-term effect of equalizing the concentrations on both sides of the membrane. RO moves water or other solvent against the normal direction of flow by applying high pressures on the concentrated side of the membrane to overcome the high osmotic pressures across the membrane. At the same time it excludes the passage of most or all of the solute molecules by pore size and/or by charge effects. The solution passed through the membrane is called permeate or filtrate. The material retained is often called rejectate or retentate. Figure 15-16 illustrates the membrane mechanism for possible separation of MTBE from a water stream.

RO membranes can reject organic molecules greater than or equal to 150 molecular weight (MW) and a percentage of those between 25 and 150 MW. RO is a possible technology for the removal of MTBE (nominal MW equals 88) from ground water, but its practicality has not yet been demonstrated. RO membranes are subject to fouling by particulates, colloids, chemical scaling, and biological growth. Treatment may be required to remove the fouling agents. The concentrates of an RO process generally require post-treatment before disposal. In the case of MTBE, the higher concentration rejectate would require management by some other treatment.

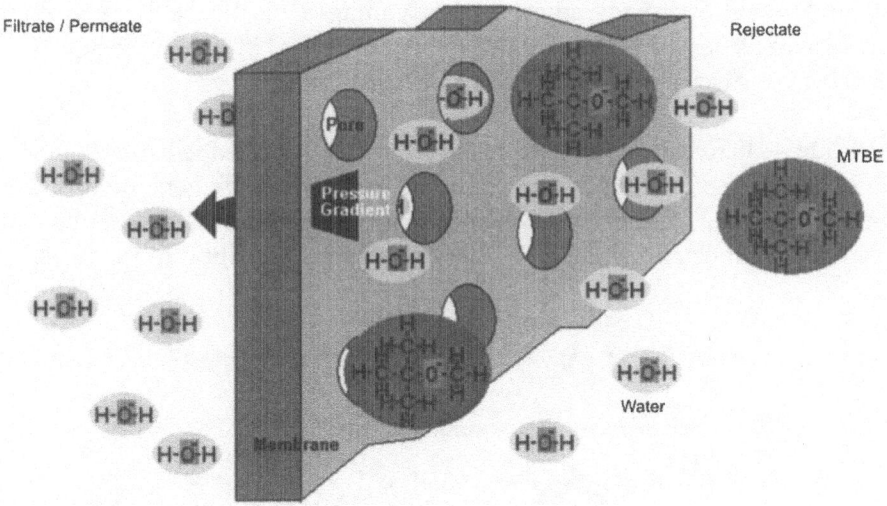

Figure 15-16.   **Membrane Filter Separation**

ADVANCED OXIDATION PROCESSES

AOPs use oxidants (chemical or chemical/physical) to treat organic and oxidizable inorganic components in water. Although other oxidants can oxidize some organics, virtually all AOPs rely on hydroxyl radicals (OH•) to react with pollutants. The AOPs are capable of completely oxidizing organic materials to carbon dioxide and water, if treated long enough with enough oxidant. The common oxidants used as hydroxyl radical sources are hydrogen peroxide and ozone. Hydroxyl radical generation is stimulated by UV light, ferrous iron, cavitation, and sonication as well. A wide variety of AOPs are available to generate hydroxyl radicals for the oxidation of contaminants including: UV/hydrogen peroxide, UV/ozone, Fenton's reagent (peroxide and ferrous iron), and combined ozone and peroxide. The hydroxyl radicals react with organic contaminants, which by definition are more reduced than carbon dioxide, to oxidize them to carbon dioxide, thereby destroying them. If these organics contain elements such as chlorine, these become inorganic end products such as halide salts.

TYPES OF AOPS

Fenton's Reagent. Figure 15-17 shows the cyclical chemistry of Fenton's reagent generation of hydroxyl radicals. This process is among the oldest and simplest to implement. This is because hydrogen peroxide is widely available in bulk at various strengths and can be mixed easily *in situ* or in a reactor system at the process site with ferrous salts to generate the active hydroxyl radicals. As shown in Figure 15-17, an advantage of this system is the cyclical regeneration of the ferrous iron by a reaction of the ferric iron with another hydrogen peroxide molecule, regenerating the ferrous iron to continue to produce more hydroxyl radicals. Disadvantages of the Fenton's reagent process are its sensitivity to pH and the issue of iron *in situ* in the aquifer if reinjected or discharged in the surface effluent stream.

Peroxide — Ozone. Mixing ozone and hydrogen peroxide can also generate hydroxyl radicals. However, ozone must typically be generated on site with an electrical ozone generator, and ozone is an air emission health problem in operations and a water discharge problem, resulting in the need for

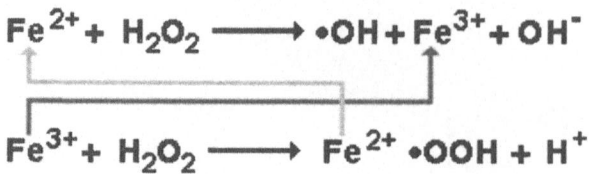

$$Fe^{2+} + H_2O_2 \longrightarrow \bullet OH + Fe^{3+} + OH^-$$

$$Fe^{3+} + H_2O_2 \longrightarrow Fe^{2+} \bullet OOH + H^+$$

Figure 15-17. Generalized Fenton's Reagent Chemistry

treatment to remove ozone from the process effluent. This type of AOP has been successfully applied in the removal of MTBE from drinking water and extracted ground water (Liang *et al.*, 1999; Hydroxyl Systems and URS Corporation, 1999). The peroxone process is one that has been developed based on the use of ozone in conjunction with hydrogen peroxide to provide effective, economical treatment of MTBE. The peroxone process technology involves two essential steps, ozone dissolution and hydrogen peroxide addition in the water stream.

Cavitation/Sonication. AOPs can also be enhanced by cavitation and sonication. Both systems operate on a similar physical-chemical basis. By applying mechanical cavitation or high-energy ultrasonic vibrations to a water phase, microbubbles form, grow to a critical size (on the order of a few angstroms), and then implode. When these bubbles collapse, temperatures of approximately 3,000 to 5,000 Kelvin are generated, and products similar to those found in combustion systems are formed. Water vapor in the bubble is dissociated into hydroxyl radicals ($OH\bullet$) and hydrogen atoms (H^+). The destruction of organic contaminants occurs within the bubble via thermolysis and in the bubble interface and bulk solution through reaction with hydroxyl radicals. These systems are more effective if combined with other AOPs to enhance the generation of hydroxyl radicals.

UV Driven Systems. Another mechanism for generating hydroxyl radicals is excitation of oxidants such as hydrogen peroxide with UV radiation. These systems mix the oxidant into the water to be treated and pass the water through reactors around powerful UV lamps whose radiation produces hydroxyl radicals that oxidize the contaminants. These systems have been used for years for chlorinated contaminants such as pentachlorophenol and chlorinated solvents. These systems also work on MTBE because the key is generation of the hydroxyl radical.

UV driven systems suffer from all the other hydroxyl radical issues, such as oxidation of non-target compounds and quenching reactions. In addition, turbidity and color can act as barriers to the transmission of light in the impacted water inhibiting the efficiency of oxidation. Finally, UV systems are capital intensive and expensive to operate, requiring substantial maintenance. For small footprint systems, treating low flows and low concentrations ($\mu g/l$) they can be an appropriate selection for MTBE or chlorinated contaminants. For high flows and concentrations, other technologies seem to have cost advantages.

Electron Beams. High-energy electron beams (E-beams) have also been utilized at pilot scale as an AOP. The radiolysis (radiation driven splitting) of water by an E-beam forms oxidizing hydroxyl radicals ($OH\bullet$), reducing hydrated electrons (e_{aq^-}), and hydrogen atoms (H$^+$), also hydrogen radicals (H•), hydrogen gas, and H$_2$O$_2$. The high-energy E-beam irradiation has been demonstrated to simultaneously destroy an array of organic compounds due to the oxidation power of hydroxyl radicals and the reducing capacity of the various hydrogen forms generated. This technology is still in the research phase; reaction chemistry must be sufficiently understood prior to its availability for full-scale applications. This technology is very unlikely to ever be used specifically for MTBE or at a remedial field pump-and-treat site. First the electrical energy requirements are enormous, the system requires extensive radiation shielding (particularly massive in the target area), the technology control is a specialized technical area, and the size and cost of the systems are prohibitive for most applications. Their best applications are in mixed contaminants in sole source or high value drinking water where the extracted water can be expected to remain contaminated for decades.

LIMITATIONS OF AOPS

Chemical oxidations and the particular type to be used for a particular treatment can be limited or inhibited in their cost-effectiveness by various site and process characteristics including:

- *Contaminants* — Some contaminants, such as carboxylic acids and dicarboxylic acids, may be slower to oxidize, requiring longer residence (oxidant contact) times while others will oxidize very rapidly;
- *Intermediates* — Residence (contact) times or polishing treatments must take into account the time needed to oxidize potentially harmful intermediates. For example MTBE chemical oxidation may produce TBF, acetone, or TBA, all of which could be harmful in not sequentially destroyed as well;
- *TOC* — AOPs are indiscriminate oxidizers, leading to oxidation of non-target contaminants or natural organics (humic substances and natural hydrocarbons) along with the constituent of concern. As indicated by the bullet above, other compounds (*e.g.*, phenolics) oxidize rapidly and preferentially, rapidly consuming the oxidant;
- *Turbidity/Color* — Turbidity and color are indicators of other materials, possibly organic, in the stream that will consume the oxidants. In addition, for UV driven processes, turbidity and color can interfere with or quench the effective UV reaching the oxidant and contaminants, thereby reducing the efficiency of the oxidant system;
- *Free Radical Scavengers* — Free radicals are scavenged by specific chemicals such as the dissolved inorganics discussed below. However, excess chemical oxidant can also be a free radical scavenger, and the design dosing of the process should take this into account and change as needed as the treatment continues; and

• *Dissolved Inorganics* — Carbonates, bromine, and iron can quench oxidant reactions and alter the kinetics of oxidation of target compounds.

ADVANTAGES OF AOPS

The AOPs offer several advantages over biological or physical processes, including; destruction of contaminants, absence of secondary wastes, killing of bacteria and microorganisms, and process operability. However, AOPs are not cost-effective for high contaminant concentrations because of the large amount of oxidizing agent required. Free radical scavengers such as carbonate can inhibit contaminant destruction efficiency. The consumption of oxidants by natural organic matters in the water will also increase treatment costs and create the potential for forming undesirable byproducts. Suspended solids or solids formed by precipitation can interfere with UV transmission. There is also a concern of producing intermediates by incomplete oxidation of MTBE.

OTHER AOPS

Permanganate.   Other AOPs are available, such as permanganate systems. However, preliminary testing of permanganate indicates that it is ineffective at MTBE treatment. This is not surprising, because permanganate oxidation is most effective when used to oxidize organics with double bonds (vinyl or alkene bonds).

COSTING PUMP-AND-TREAT SYSTEMS

The costs of ground water recovery are generally well-understood for the various recovery approaches (USEPA, 2001b). Variability in recovery costs is generally due to contaminant properties, localized geology, and water chemistry. These factors mostly impact the number of wells, depth of wells, simplicity/complexity of the pumping system, operation and maintenance of the recovery system, remediation time (time to capture a given volume of water and contaminant), and pre- or post-treatment for iron, suspended solids, and similar characteristics. Costs can be considerably higher for locations where the geology, ground water chemistry, or treated ground water disposal limitations require more elaborate recovery systems, pre- or post-treatment(s), or discharge fees, pipelines, extensive testing, and similar regulatory requirements.

Table 15-1 provides typical ranges of capital and operation and maintenance costs for the ground water recovery and treatment systems. The costs provided on Table 15-1 apply to approximately 70% of the pump-and-treat systems with varying treatment technologies discussed in this section and assume simple or no pretreatment or specialized post-treatment. However, the

Table 15-1. Costs for Typical Pump-and-Treat System

Recovery System	Capital Costs[1]	*Recovery Wells:* • $2,000 to $3,000/well (fixed costs) PLUS • $100 to $150/foot/well (variable costs)	Fixed costs include: System design, mob/demob, setup for drilling, well completion and development. Variable costs include: Labor for field oversight, well drilling and sampling, well casing and screen, pump, controllers, well piping, and disposal of waste generated during field activities.
		Transfer Piping: • $15 /linear foot (aboveground piping) OR • $60 /linear foot (below-ground piping)	Aboveground piping: 4-inch High Density Polyethylene pipe placed on ground or support. Belowground piping: Saw cut 6-inch concrete slab, trench, install double-wall (2½ inch, 4 inch) PVC piping, and backfill.
	Operation and Maintenance Costs[2] (per month)	• $130/well/month (smaller system) to • $100/well/month (larger system)	Operation and Maintenance costs include: Labor and equipment for inspection, routine mechanical maintenance, rehabilitating wells (50% each year), sampling and analysis for MTBE (semi-annual sampling), and utilities.
Treatment System	Low Costs[3]	$0.37 /1,000 gallons	Simple system (600 gallons per minute [gpm]): High flow, no pretreatment, low influent concentrations, relaxed treatment goals.
	High Costs[4]	$2.00 /1,000 gallons	Complex System (600 gpm): Low flow, pre/post-treatment required, high influent concentrations, and low cleanup goals.

Notes: [1] RS Means(r) Remediation Cost Data Assemblies Cost Book — 8th Edition: Recovery system cost is based on one (1) well drilled with hollow stem auger up to a depth of 100 feet bgs and installed with a 6-inch diameter recovery well with manhole cover as well completion.

[2] RS Means(r) Remediation Cost Data Assemblies Cost Book — 8th Edition: Operation and Maintenance costs are provided for a smaller system for retail stations (5 recovery wells, each pumping at 10 gpm, with total of 50 gpm) and larger system for industrial facilities (25 recovery wells, each also pumping at 10 gpm, with total of 500 gpm).

[3] From Table 6.1 California MTBE Research Partnership (2001). Assumptions: 30-year operation and maintenance life, 600-gpm flow rate, very clean effluent, no pre-treatment (See Appendix A of the report for details).

[4] From Table 6.1 California MTBE Research Partnership (2001), with 60% added for pretreatment and interference. Assumptions: 30-year operation and maintenance life. Smaller complex system (about 6 gpm) could easily be an order of magnitude more expensive/1,000 gallons.

cost drastically goes up with low flow, high concentration, and high interference to the treatment system.

REFERENCES

California MTBE Research Partnership. 1999. *Evaluation of the Applicability of Synthetic Resin Sorbents for MTBE Removal from Water.* The National Water Research Institute, Fountain Valley, California. December 1999. (714-378-3278).

California MTBE Research Partnership. 2001. *Treating MTBE — Impacted Drinking Water Using Granular Activated Carbon.* The National Water Research Institute, Found Valley, California. December 2001. (714-378-3278).

Hydroxyl Systems & URS Corporation. 1999. *Hydroxyl Case Study Treatment of Groundwater with High Concentrations of MtBE: JFK Airport.* Sidney, British Columbia, Canada, Hydroxyl Systems.

Liang, S., Yates, R.S., Palencia, L.S., and Bruno, J.M. 1999. Oxidation of methyl tertiary-butyl ether (MTBE) by ozone and peroxone and identification of by-products, [CD ROM]. In: *Proceedings of American Water Works Association Annual Conference,* Chicago, Illinois. June 20–24, 1999. American Water Works Association.

Liu, D.H.F. (Ed.). 1997. *Environmental Engineers' Handbook, Second Edition.* New York, Lewis Publishers.

RS Means®. 2002. Environmental cost handling options and solutions. In: *Remediation Cost Data Assemblies Cost Book,* 8th Edition. ISBN 0-87629-6509. Kingston, Massachusetts, RS Means®.

USEPA (U.S. Environmental Protection Agency). 1996. *Pump-and-Treat Ground-Water Remediation, A Guide for Decision Markers and Practitioners.* EPA/625/R-95/005. July 1996.

USEPA (U.S. Environmental Protection Agency). 2001a. *Groundwater and Remediation: Public Policy, Technology, Risk Reduction, and Cost.* Kevin Garon, DuPont. June 27, 2001. www.epa.gov/reg3wcmd/ca/pdf/gwremediation.pdf.

USEPA (U.S. Environmental Protection Agency). 2001b. *Groundwater Pump and Treat Systems: Summary of Selected Cost and Performance Information at Superfund-Financed Sites.* U.S. EPA 542-R-01-021b. December 2001.

CHAPTER 16

Monitored Natural Attenuation of MTBE

Bruce E. Rittmann, Ph.D., Northwestern University

BACKGROUND ON MONITORED NATURAL ATTENUATION

Since the mid-1990s, managers and regulators of sites contaminated with petroleum hydrocarbons have increasingly turned to a remedial strategy called MNA. This strategy relies on naturally occurring processes — without human intervention or enhancement — to reduce concentrations and exposure risks from ground water contaminants. Although the name MNA is commonly used, at least two other names are associated with the same strategy (NRC, 2000): natural attenuation and intrinsic remediation. This chapter routinely uses the shorter name, natural attenuation, but the three terms can be taken as synonyms.

According to the USEPA's policy directive on MNA (USEPA, 1999), this strategy is the "reliance on natural attenuation processes (within the context of a carefully controlled and monitored site cleanup approach) to achieve site-specific remediation objectives within a time frame that is reasonable compared to that offered by other more active methods." The natural attenuation processes work without human intervention and, according to USEPA (1999), include biodegradation, dispersion, dilution, adsorption, volatilization, radioactive decay, and chemical or biological reactions that stabilize, transform, or destroy the contaminants.

The rapid increase in sites where natural attenuation was proposed or accepted provoked careful scrutiny of the strategy by scientists, engineers, policy makers, and the public. The "all inclusive" list of possible natural attenuation processes allowed by the USEPA (1999) directive raised the ire of environmentalists and community activists; it also fomented considerable debate within the technical community. Towards the end of the 1990s, interest in and concern about natural attenuation led to an NRC study sponsored by 10 government and private institutions and carried out by a committee of 15 experts in all aspects of the science, technology, and policy of natural attenuation (NRC, 2000; Rittmann and MacDonald, 2000). The author of this chapter was chairman of the committee.

In August 2000, the National Academy Press released *Natural Attenuation for Groundwater Remediation* (NRC, 2000), which provides expert guidance on when and how natural attenuation should be applied to remediate ground water and soil contamination. The approach recommended by the NRC (2000) was later endorsed by a committee of the USEPA's Science Advisory Board (SAB) in the spring of 2001 (SAB, 2001). NRC (2000) set the standard by which natural attenuation is to be evaluated for any type of contaminant, including MTBE.

Although NRC (2000) affirmed the principle that natural attenuation should not involve human intervention, the report indicated that only a subset of all natural processes should be relied upon when evaluating a natural attenuation remedy. In particular, the NRC (2000) recommended that natural attenuation be accepted only when contaminants are destroyed or strongly immobilized by a natural process. In most cases, this limits acceptable natural attenuation processes to biodegradation, precipitation, strong adsorption, and radioactive decay.

The NRC committee chose to limit the acceptable natural attenuation processes for two complementary reasons. The first reason is technical. Because human intervention is not used to remove or control contaminants, the "technology" of natural attenuation is the knowledge that a natural process is actively preventing the exposure of humans or sensitive environments to contaminants. Destruction and strong immobilization can be positively documented now and in the future. They are active means by which an exposure route to humans or a sensitive environment is prevented. Other mechanisms — particularly dispersion, dilution, and volatilization — may create new exposure routes that are not easily detected. The second reason is that members of communities affected by contaminated sites normally will accept natural attenuation only when they are given strong proof that the contaminants are destroyed or permanently sequestered. The communities believe that their risks are too high unless they are certain that an active destruction or immobilization process is at work. Because local communities are de facto asked to bear much of the risk during natural attenuation, their agreement is essential. The community's desire to have proof of destruction or strong immobilization is consistent with sound technical reasoning.

THE NRC STRATEGY FOR EVALUATING NATURAL ATTENUATION

NRC (2000) outlines a strategy for evaluating whether or not natural attenuation protects humans and the environment from unacceptable risks at a contaminated site. The two keys to the strategy are a site conceptual model and measurements of footprints of the controlling process.

A site conceptual model is a picture of the flow and reaction processes at a site. It includes the direction and velocity of ground water flow, as well as the location of the source and possible receptors. It also includes the destruction or immobilization processes that are postulated to prevent exposures.

Footprints occur because the processes that destroy or immobilize contaminants also consume or produce other materials, many of which can be measured. Quantitatively coupling the loss of a contaminant with the observation of several footprints provides strong evidence of cause and effect. In other words, footprints can actively demonstrate the destruction or immobilization reactions postulated in the site conceptual model.

NRC (2000) gives many examples of footprints and illustrates them in case studies. Commonly useful footprints for fuel-based organic constituents include loss of electron acceptors (*e.g.*, O_2 for aerobic respiration, NO_3^- for denitrification, and SO_4^{2-} for sulfate reduction); formation of reduced products (*e.g.*, CH_4 for methanogenesis and Fe^{2+} for Fe(III) reduction); increases in inorganic carbon; characteristic changes in alkalinity; and formation of characteristic intermediates. Two examples for toluene, one of the most common fuel components, show how the different types of footprints are identified.

The first example is aerobic biodegradation of toluene (C_7H_8). Equation 16-1 for aerobic oxidation of toluene shows the stoichiometry that leads to footprints.

$$C_7H_8 + 3H_2O + 9O_2 = 7H_2CO_3 \qquad (16\text{-}1)$$

The loss of toluene leads to the stoichiometric loss of dissolved oxygen, the production of inorganic carbon (H_2CO_3), and no change in alkalinity.

A second example is for anaerobic biodegradation of toluene via sulfate reduction.

$$C_7H_8 + 4.5SO_4^{2-} + 3H_2O + 9H^+ = 7H_2CO_3 + 4.5H_2S \qquad (16\text{-}2)$$

The footprints are loss of sulfate (SO_4^{2-}), production of inorganic carbon (H_2CO_3), and production of alkalinity (shown as a loss of acidic hydrogen, H^+). Production of sulfide (*e.g.*, H_2S) also could be a footprint, although sulfide precipitation may make this footprint difficult to observe.

MTBE AND THE NRC REPORT

NRC (2000) addressed the scientific basis for natural attenuation of MTBE and the likelihood that natural attenuation will succeed as a remedy. About the scientific basis, NRC (2000) stated that MTBE is "generally resistant to biodegradation," but that microorganisms having certain oxygenase enzymes can "fortuitously insert oxygen into the MTBE molecule" rendering the products susceptible to further breakdown. Although field and laboratory studies showed the possibility of MTBE biodegradation, NRC (2000) cautioned that partial breakdown to TBA is likely. In summary, the NRC report advised that the findings on MTBE biodegradation are preliminary and the "natural attenuation potential is unclear at this time."

NRC (2000) also offered general guidance about the likelihood of success for common ground water contaminants. Likelihood of success depends on the state of scientific knowledge about processes that can destroy or immobilize contaminants and on the probability that the conditions allowing those processes occur at contaminated sites in general. A high likelihood of success in NRC (2000) does not guarantee that natural attenuation will succeed at a given site; site-specific monitoring and footprints are needed for all sites. A low likelihood of success does not mean that natural attenuation cannot succeed at a specific site. A low likelihood of success means that the level of effort to gather evidence that supports or refutes the success of natural attenuation may be very high (NRC, 2000).

Concerning MTBE, NRC (2000) stated that the level of understanding in early 2000 was moderate, while the likelihood of success (given the then-current level of understanding) was low. The low rating stems from a lack of firm understanding and also from a perceived need for special conditions to allow aerobic biodegradation. The most important special conditions are significant dissolved oxygen and the presence of microorganisms containing the key oxygenase enzymes. These issues are discussed in detail in the following sections.

NRC (2000) did not provide guidance on footprints for the biodegradation of MTBE.

RECENT FINDINGS ON MTBE AND NATURAL ATTENUATION

Interest in the fate of MTBE is a growing phenomenon worldwide, and new research is being carried out and reported regularly. Two excellent review articles (Deeb *et al.*, 2000; Stocking *et al.*, 2000) recently appeared in a special issue of *Biodegradation* (Volume 11, Number 2–3, 2000) devoted to the biodegradation of fuel components. Together, these two reviews cover many recent advances in laboratory, pilot, field, and technology studies on MTBE biodegradation. In addition, a committee of the USEPA's SAB issued a report (SAB, 2001) on research needs for natural attenuation, including for MTBE. This chapter highlights the key findings from these reviews and also reviews very recent information since these reviews were completed.

Aerobic Biodegradation.   Although evidence is now emerging that MTBE can be biodegraded under anaerobic conditions, the bulk of evidence supports that aerobic biodegradation is the pathway that holds the greatest promise for rapid biodegradation. Therefore, this section focuses on aerobic metabolism, although anaerobic biodegradation is reviewed later.

A substantial body of research (reviewed in detail by Deeb *et al.*, 2000; Stocking *et al.*, 2000; with more recent finding by Deeb *et al.*, 2001; Fortin *et al.*, 2001; Liu *et al.*, 2001; Bradley *et al.*, 1999, 2001; Garnier *et al.*, 2000; Chapter 12) indicates that monooxygenase enzymes catalyze the first two steps of MTBE metabolism. Figure 16-1 shows those first two steps, which produce TBA and 2-methyl-1,2-propanediol (MHP), intermediates that have been detected in

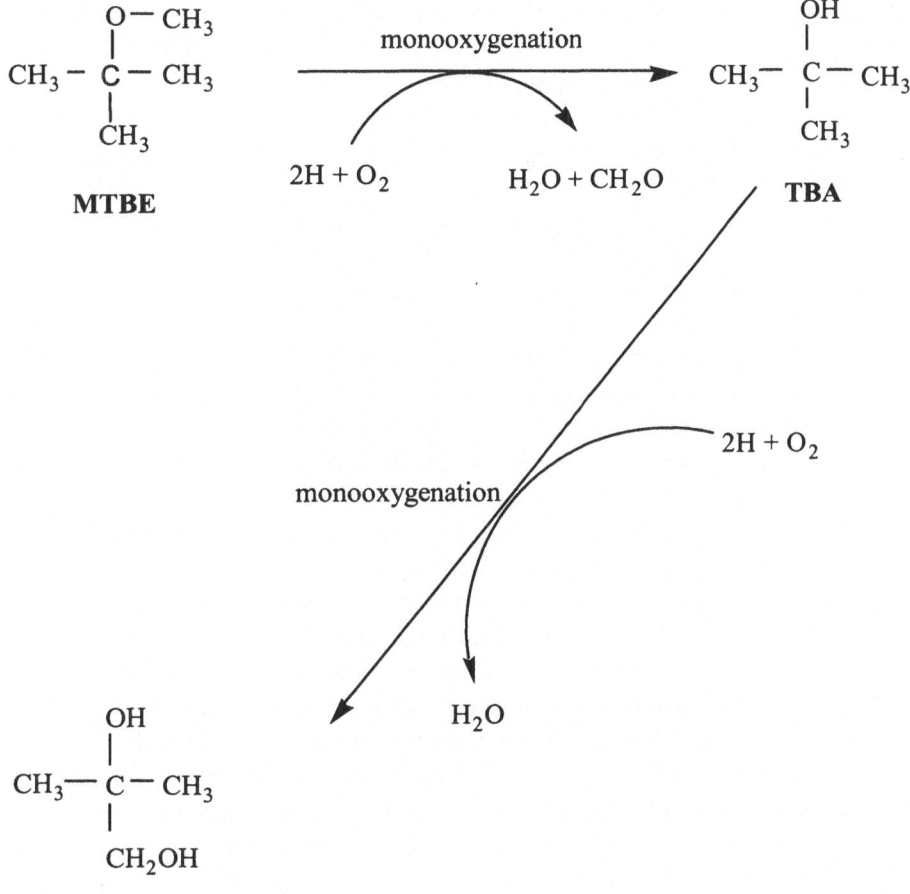

Figure 16-1.   **Initial Monooxygenation Reactions from MTBE to TBA to MHP**
MHP = 2-methyl-1,2-propanediol, CH_2O = formaldehyde, 2H = $2H^+ + 2e^-$.

various studies. These early steps have several features that are critical for the success of MTBE biodegradation. These features are discussed in the next four paragraphs.

The first feature is that monooxygenase enzymes catalyze both reactions. This means that microorganisms containing these enzymes must be present with sufficient biomass. At this time, it is not clear whether or not the presence of MTBE by itself is always sufficient to support the growth of the key microorganisms and to induce the monooxygenase enzymes. Co-substrates may be needed in some cases, and these may include propane, butane, pentane, diethyl ether, and benzene. A co-substrate is needed if TBA or MHP is the terminal product of metabolism; mixed cultures can degrade MTBE completely without addition of co-substrates.

The second feature is that molecular oxygen (O_2) is required as a co-reactant in both monooxygenase reactions. In general, monooxygenases require relatively high concentrations of dissolved oxygen, compared to respiratory use of O_2, to avoid severe oxygen limitation (Malmstead *et al.*, 1995; Fortin *et al.*, 2001). When MTBE contaminates ground water, it normally occurs with BTEX, which are more readily biodegraded. If dissolved oxygen is present, aerobic biodegradation of BTEX can consume all of the dissolved oxygen, making the initial steps of MTBE biodegradation impossible within the plume. Therefore, aerobic MTBE biodegradation seems likely to occur only on the fringes of the plume, where dissolved oxygen has not been depleted.

The third key feature of the monooxygenase reactions is that they require an internal source of two electrons, represented as 2H in Figure 16-1. The two electrons are required to reduce both oxygen molecules of O_2. The oxygen that is substituted into the organic molecule as an -OH group obtains the electrons from MTBE (or TBA), but the other oxygen requires another source of electrons. If TBA or MHP is not further oxidized, those electrons must come from another source, perhaps oxidation of a co-substrate. The first monooxygenation releases formaldehyde (CH_2O), which is a potential source of four electrons if it is oxidized by normal dehydrogenase and hydroxylation reactions (Rittmann and McCarty, 2001). Thus, oxidation of formaldehyde can supply exactly enough electrons to "fuel" the two monooxygenations, as long as O_2 is present (Woo and Rittmann, 2000). This is shown by an overall reaction from MTBE to MHP:

$$C_5H_{12}O + 2O_2 = C_4H_{10}O_2 + H_2CO_3 \tag{16-3}$$

A fourth key feature of the monooxygenase reactions is what is not released: electrons and organic carbon immediately available for synthesis. This is shown most succinctly in Equation 16-3, shown directly above this paragraph. The reaction in Equation 16-3 does not show electrons, which would be represented as H, because reduction of O_2 consumes all the electrons liberated by oxidation of formaldehyde. Likewise, MHP is the only form of organic carbon on the right side, and it is far from being a common metabolite that can be used for biomass synthesis. Although MTBE has 30 electron equivalents and 5 carbon equivalents, while MHP has 22 electron equivalents and 4 carbon equivalents, the microorganisms gain no useful electrons or carbon from the transformation. This means that the first two monooxygenation steps must be viewed as "activation" steps that cannot support biomass growth (Woo and Rittmann, 2000).

Figure 16-2 shows likely metabolic steps from MHP to acetone. It shows 2-hydroxyisobutyric acid (HIBA), 2-propanol, and acetone, which have been

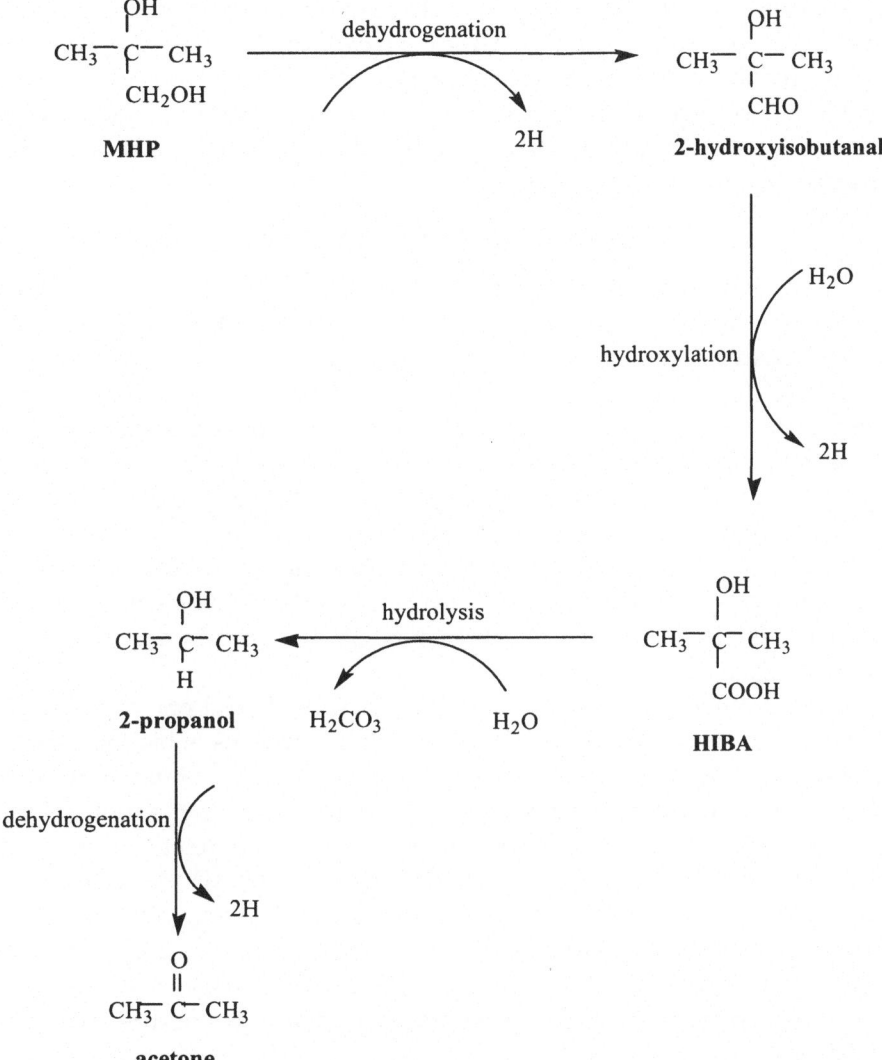

Figure 16-2.   **Reactions Leading from MHP to Acetone**
MHP = 2-methyl-1,2-propanediol, HIBA = 2-hydroxyisobu-
tyric acid.

detected as intermediates of MTBE biodegradation (Deeb *et al.*, 2000). Figure 16-2 also shows other necessary intermediates and the types of reactions that must take place. Unlike the reactions shown in Figure 16-1, the reactions in Figure 16-2 release electrons and available carbon. Two dehydrogenations and one hydroxylation release 2H each. Thus, the microorganisms capture all six of the electrons released from MHP (22 electron equivalents) to acetone

(16 electron equivalents). If they send those electrons to a terminal electron acceptor, such as O_2, in respiration, they can capture the energy for synthesis and maintenance (Rittmann and McCarty, 2001). The electrons can be used for the monooxygenation reactions in Figure 16-1. The following reaction shows the complete oxidation of acetone, which occurs via a series of hydroxylation and dehydrogenation steps.

$$C_3H_6O + 8H_2O = 3H_2CO_3 + 16H \qquad (16\text{-}4)$$

Furthermore, the organic product, acetone, is easily available for complete oxidation or conversion to common organic metabolites that can be used for biomass synthesis.

Field Experience.   Field experience with MTBE biodegradation suggests that engineered bioremediation, or bio-enhancement, is usually needed to see significant biodegradation (Stocking *et al.*, 2000). The addition of oxygen via air and oxygen sparging, oxygen-release compounds, or hydrogen peroxide decay has been the most successful strategy so far. Addition of co-substrates also has had some benefit. These bio-enhancement effects are consistent with the concept that MTBE is biodegraded aerobically via initial monooxygenation reactions.

Field evidence for natural attenuation is limited, and one of the major shortcomings of studies reported to date is that footprints of biodegradation reactions are seldom available. This means that natural attenuation must be inferred from information that is incomplete, such as whether or not a plume is advancing or receding and its rate of movement compared to another constituent, such as benzene (Mace and Choi, 1998; Chapter 31). One of the most important messages of NRC (2000) is that natural attenuation cannot be demonstrated by measurements of the constituent alone. Sampling problems, transients in hydrodynamics, and dilution are causes for decreases in constituent concentration, even when biodegradation is not destroying the constituent.

Many field studies of possible MTBE natural attenuation through 1999 (reviewed by Stocking *et al.*, 2000) suggest that natural biodegradation is not stopping the movement of MTBE plumes, although some studies show reductions in MTBE concentration over time. More recent studies (Chapter 31) show a mixture of outcomes for MTBE, and the authors suggest that more information must be collected on footprints to bolster the evidence for natural attenuation. One case showing apparent biodegradation is a study of 149 BP gasoline stations in Florida (Reid *et al.*, 1999), where the MTBE plumes behaved similarly to benzene plumes. Stocking *et al.* (2000) interpret these findings as being "consistent with environmental factors unique to Florida." In particular, they cite low ground water flow rates due to flat hydraulic gradients and high ground water oxygenation rates due to shallow water tables and high rates of recharge. These findings of apparent natural attenuation of

MTBE are consistent with the above interpretation of the bio-enhancement results: that special conditions are needed to allow significant natural attenuation.

SAB Report. When offering advice to the USEPA on critical research needs in natural attenuation, the committee of the SAB (2001) emphasized MTBE as key for evaluating sites with LUSTs. In particular, the committee stressed that basic information about the biodegradability of MTBE is needed for the range of conditions likely to occur in the field. This includes the effects of different electron acceptors and other organic constituents, the accumulation of hazardous intermediate products, and the development of biogeochemical footprints. The committee indicated that fundamental laboratory and field studies would be needed to fill in the major gaps in understanding of MTBE.

Evidence on Anaerobic Biodegradation of MTBE. Evidence for the possible biodegradation of MTBE under anaerobic conditions was available in the early to mid 1990s (Yeh and Novak, 1995; Mormile *et al.*, 1994). Much more recent findings confirm that MTBE can be mineralized under methanogenic conditions (Wilson *et al.*, 1999, 2000; Hurt *et al.*, 1998) and coupled to the reduction of ferric iron (Finneran and Lovley, 2001; Chapter 13). Decoupling MTBE biodegradation from the obligate need for molecular oxygen would greatly expand the conditions under which MTBE could be biodegraded in a plume. Information on the kinetics of anaerobic MTBE biodegradation and on the occurrence of anaerobic MTBE degraders is in the infant stage. This makes the likelihood of anaerobic biodegradation low for a natural attenuation process.

An interesting corollary to anaerobic biodegradation of MTBE is hybrid biodegradation, in which molecular oxygen is used for the first monooxygenation steps, but an alternate electron acceptor is used for respiration of the electrons generated by oxidations beyond MHP. If MTBE-degrading bacteria were able to scavenge scarce O_2 molecules selectively for the monooxygenation steps, they could carry out a very efficient hybrid biodegradation. This hybrid strategy would be greatly enhanced if the same bacteria carried out the monooxygenation reactions and the subsequent anaerobic respiration reactions. The apparently low affinity that the monooxygenases have for O_2 (Fortin *et al.*, 2001; Malmstead *et al.*, 1995) works against this hybrid scenario.

UPDATING THE NRC GUIDANCE FOR NATURAL ATTENUATION OF MTBE

NRC (2000) indicated that the state of understanding of biodegradation processes for MTBE is moderate, while the likelihood of success for natural attenuation is low. The report did not explicitly discuss footprints that would be most valuable for assessing natural attenuation of MTBE. Based on new studies, should the NRC guidance on understanding or likelihood change or remain the same? Can guidance be given on workable footprints?

Scientific Understanding. The new information on laboratory and field studies generally reinforces the guidance given by NRC (2000). While more studies (Deeb *et al.*, 2001; Fortin *et al.*, 2001; Liu *et al.*, 2001; Bradley *et al.*, 2001; Garnier *et al.*, 2000; Finneran and Lovley, 2001; Wilson *et al.*, 2000; Chapter 13) have verified that aerobic biodegradation occurs frequently, that monooxygenations are critical first steps for aerobic biodegradation, and that anaerobic biodegradation might be possible, fundamental questions remain:

- Why do TBA and MHP accumulate in some situations, but not others?
- When are co-substrates required? Which co-substrates work best?
- Can MTBE be totally biodegraded under anaerobic conditions?
- Can anaerobic respiration be used for the oxidation of intermediate products after MHP?
- When and how do other organic contaminants (particularly BTEX) enhance or inhibit the biodegradation of MTBE?

In sum, confidence that MTBE is biodegradable has improved, but the understanding of what controls success is still at the moderate level.

Likelihood of Success. Laboratory and field studies suggest that somewhat specialized conditions are needed to obtain significant MTBE biodegradation. Most important is the presence of significant dissolved oxygen, which is required for the monooxygenase reactions. In addition, it may be necessary to have special co-substrates in order to grow the biomass and induce the monooxygenase enzymes (Garnier *et al.*, 2000; Liu *et al.*, 2001; Deeb *et al.*, 2001; Fortin *et al.*, 2001). Further complicating matters is that negative interactions between MTBE and the BTEX compounds have been observed (Deeb *et al.*, 2001) and that the MTBE degraders seem to be slow growers (Fortin *et al.*, 2001).

Except for one study in Florida (Reid *et al.*, 1999), field evidence suggests that these special conditions are not achieved. The Florida sites appear to have unique features that may enhance the likelihood of success: slow ground water flow velocities, shallow water tables, and high recharge rates. Although promising, the Florida study may be "the exception that proves the rule" that special conditions are necessary for good MTBE biodegradation. The likelihood of success in a general sense still seems to be in the low category.

Footprints. Even if the likelihood of success remains low, it may be worthwhile to evaluate natural attenuation at sites that appear to have favorable conditions. Therefore, having a good set of footprints is very valuable. This section first reviews footprints for aerobic biodegradation, then it provides likely footprints for anaerobic and hybrid biodegradations.

Footprints of Aerobic Biodegradation. The aerobic oxidation of MTBE to H_2CO_3 is given by the following reaction when the electrons needed to fuel the monooxygenations come from an MTBE intermediate (such as formaldehyde) and O_2 is used as the terminal electron acceptor:

$$C_5H_{12}O + 7.5O_2 = 5H_2CO_3 + H_2O \qquad (16\text{-}5)$$

This overall reaction identifies the most valuable footprints of complete MTBE biodegradation:

- A substantial consumption of dissolved oxygen: 7.5 moles O_2/mole MTBE, or 2.73 grams O_2/gram MTBE;
- A large release of inorganic carbon: 5 moles C/mole MTBE, or 0.68 grams C/gram MTBE; and
- No change in alkalinity, since acidic hydrogen (H^+) is neither consumed nor produced.

Small concentrations of intermediate products, most likely TBA, may be detected and can add further support for the biodegradation mechanism. Likewise, detecting MTBE-degrading bacteria can be a useful type of supporting evidence.

Should MTBE be only partially oxidized to TBA or MHP, the footprints would be different. They can be illustrated from Equation 16-3 for conversion of MTBE to MHP:

- A smaller consumption of dissolved oxygen: 2 moles O_2/mole MTBE, or 0.73 grams O_2/gram MTBE;
- A smaller release of inorganic carbon: 1 mole C/mole MTBE, or 0.14 grams C/gram MTBE;
- No change in alkalinity; and
- Accumulation of MHP (or TBA, if the reaction were to stop there).

Having the biotransformation of MTBE stop at TBA or MHP probably would not be considered a successful outcome, as these compounds have health and aesthetic concerns.

Table 16-1 summarizes footprints that make a direct connection between the loss of MTBE and its aerobic biodegradation. The list is divided into two categories (Rittmann *et al.*, 1994; Rittmann and McCarty, 2001): principal evidence and confirmatory evidence. Principal evidence is equally capable of documenting success or failure of the natural attenuation process. Confirmatory evidence usually can only be used to support success, but its absence does not prove failure. Confirmatory evidence falls into this one-sided category because it is difficult to detect (*e.g.*, specific microorganisms), can be con-

Table 16-1. Footprints for Aerobic Biodegradation of MTBE

Footprint	Principal or Confirmatory?	Comments
Consumption of oxygen	Principal	At least 0.73 g O_2/gram MTBE for monooxygenation steps. About 2.73 g O_2/gram MTBE for full aerobic oxidation. Oxygen supply rates from advection and gas transfer are important to estimate.
Release of inorganic carbon	Principal	Approximately 0.68 gram C/gram MTBE.
Alkalinity change	Principal when coupled to other footprints	No change for aerobic respiration.
TBA or other intermediates	Confirmatory	Intermediates are degraded and may not accumulate to measurable concentrations.
MTBE-degrading microorganisms	Confirmatory	Most microorganisms are attached to the aquifer or soil solids and are difficult to sample.

For more details on stoichiometry and the distinction between principal and confirmatory evidence, see NRC (2000), Rittmann and McCarty (2001), and Rittmann *et al.* (1994).

sumed itself (*e.g.*, intermediates), or can be affected by processes other than biodegradation (*e.g.*, by adsorption, volatilization, or precipitation/dissolution).

Footprints of Anaerobic Biodegradation.   The documented situations of anaerobic biodegradation of MTBE are methanogenesis and ferric-iron reduction. For methanogenesis of MTBE, the stoichiometry is:

$$C_5H_{12}O + 2.75\ H_2O = 1.25\ H_2CO_3 + 3.75\ CH_4 \tag{16-6}$$

The footprints of methanogenesis of MTBE are:

- Release of a relatively low amount of inorganic carbon: 1.25 moles C/mole MTBE, or 0.17 grams C/gram MTBE;
- Release of significant methane gas: 3.75 moles CH_4/mole MTBE, or 0.68 grams CH_4/gram MTBE; and
- No change in alkalinity.

If MTBE were totally mineralized with ferric iron (originating as $Fe(OH)_{3(s)}$) accepting all the electrons, the reaction would be:

$$C_5H_{12}O + 30\ Fe(OH)_{3(s)} + 60\ H^+ = 5\ H_2CO_3 + 30\ Fe^{2+} + 76\ H_2O \tag{16-7}$$

The footprints are:

- Full release of inorganic carbon: 5 moles C/mole MTBE, or 0.68 grams C/gram MTBE;

- A large release of ferrous iron: 30 moles Fe^{2+}/mole MTBE, or 19.1 grams Fe^{2+}/gram MTBE; and
- A large increase in alkalinity: 60 eqH^+/mole MTBE, or 34.1 grams alkalinity as $CaCO_3$/gram MTBE.

It is possible that the Fe^{2+} would precipitate, which would alter the observed stoichiometry.

Footprints of Hybrid Biodegradation. It may be possible that the total oxidation of MTBE could be supported by a combination of O_2 and another electron acceptor (or several other acceptors). In this case, enough dissolved oxygen must be present to drive at least the two monooxygenation steps. The 22 electron equivalents in MHP might be accepted by NO_3^-, SO_4^{2-}, H_2CO_3, or Fe^{3+} (NRC, 2000), or by O_2.

As an example of what might be possible, Equation 16-8 shows the stoichiometry when NO_3^- reduction is used to accept the 22 electron equivalents in MHP:

$$C_5H_{12}O + 2O_2 + 4.4NO_3^- + 4.4H^+ = 5H_2CO_3 + 2.2N_2 + 3.2H_2O \quad (16\text{-}8)$$

The footprints of this hybrid reaction with denitrification are:

- A smaller consumption of dissolved oxygen: 2 moles O_2/mole MTBE, or 0.73 grams O_2/gram MTBE;
- A full release of inorganic carbon: 5 moles C/mole MTBE, or 0.68 grams C/gram MTBE;
- A large consumption of NO_3^--N: 4.4 moles N/mole MTBE, or 0.7 grams N/gram MTBE; and
- A significant increase in alkalinity, represented by the consumption of H^+: 4.4 eqH^+/mole MTBE, or 2.5 grams alkalinity as $CaCO_3$/gram MTBE.

It is also possible that the ground water is supersaturated with N_2 gas, which may evolve upon sampling.

A second example couples reduction of ferric iron (originating as $Fe(OH)_{3(s)}$) to the release of the 22 electron equivalents from MHP.

$$C_5H_{12}O + 2\,O_2 + 22\,Fe(OH)_{3(s)} + 44\,H^+ = 5\,H_2CO_3 + 22\,Fe^{2+} + 56\,H_2O \quad (16\text{-}9)$$

The footprints of this hybrid reaction with ferric-iron reduction are:

- The smaller consumption of dissolved oxygen: 2 moles O_2/mole MTBE, or 0.73 grams O_2/gram MTBE;
- Full release of inorganic carbon: 5 moles C/mole MTBE, or 0.68 grams C/gram MTBE;
- A large release of ferrous (Fe^{2+}) iron: 22 moles Fe^{2+}/mole MTBE, or 14 grams Fe^{2+}/gram MTBE; and

- A large increase in alkalinity: 44 eqH$^+$/mole MTBE, or 25 grams alkalinity as $CaCO_3$/gram MTBE.

It is possible that the Fe^{2+} might precipitate, which would alter the observed stoichiometry.

Table 16-2 summarizes the potential types of footprints for anaerobic and hybrid biodegradation of MTBE. For two reasons, these footprints likely can be distinguished from footprints of BTEX biodegradation. First, MTBE biodegradation often takes place after, or downstream, of BTEX biodegradation. Second, the high solubility of MTBE means that its footprints should be at relatively high concentrations relative to the footprints of BTEX biodegradation.

CONCLUSIONS

Knowledge about the biodegradation of MTBE is growing rapidly and is likely to continue to grow. Based on laboratory and field experience so far, the most promising type of MTBE biodegradation is aerobic and involves initial monooxygenase reactions. This situation requires that the supply of dissolved oxygen be significant and that MTBE degraders expressing the monooxygenase enzymes be present in significant numbers. Having both requirements present is not a high probability circumstance for natural attenuation; therefore, the original NRC (2000) judgment that the likelihood of success is low for natural attenuation of MTBE seems to remain valid. This judgment is analogous to the one made for natural attenuation of chlorinated

Table 16-2.   **Footprints for Anaerobic or Hybrid Biodegradation of MTBE**

Footprint	Principal or Confirmatory?	Comments
Consumption of oxygen	Principal	At least 0.73 grams O_2/gram MTBE for monooxygenation steps when hybrid.
Release of inorganic carbon	Principal	Approximately 0.68 grams C/gram MTBE, except for methanogenesis, in which it is about 0.17 grams C/gram MTBE.
Release of methane gas	Principal	For methanogenesis, about 0.68 grams CH_4/gram MTBE is released, but may not be measured, due to its off-gassing.
Alkalinity change	Principal when coupled to other footprints	No change for methanogenesis, but large increases for other respiratory reactions: e.g., 34 grams as $CaCO_3$/gram MTBE for reduction of $Fe(OH)_{3(s)}$.
TBA or other intermediates	Confirmatory	Intermediates are likely with hybrid processes, but subsequent steps also degrade them.
MTBE-degrading microorganisms	Confirmatory	Most microorganisms are attached to the aquifer or soil solids and are difficult to sample.

solvents, such as TCE, by reductive dechlorination (NRC, 2000). A substantial source of electron donor must be present to drive reductive dechlorination of TCE, and this is not a routine occurrence.

If natural attenuation is proposed for an MTBE-contaminated site, footprints of the biodegradation reaction must be measured. This chapter extends what is in NRC (2000) by providing the most important footprints for aerobic and anaerobic biodegradation reactions. The collection of good footprint data is the essential step for estimating whether or not natural attenuation of MTBE is likely in a generic sense, as well as for assessing the role of natural attenuation at a particular site. In fact, the measurement of footprints takes on elevated importance for low-likelihood constituents because of the higher level of effort needed to document biodegradation (NRC, 2000).

REFERENCES

Bradley, P.M., Landmeyer, J.E., and Chappelle, F.H. 1999. Aerobic mineralization of MTBE and *tert*-butyl alcohol by stream-bed sediment microorganisms. *Environmental Science and Technology*. 33, 1877–1879.

Bradley, P.M., Landmeyer, J.E., and Chappelle, F.H. 2001. Widespread potential for microbial MTBE degradation in surface-water sediments. *Environmental Science and Technology*. 35, 658–662.

Deeb, R.A., Scow, K.M., and Alvarez-Cohen, L. 2000. Aerobic MTBE biodegradation: an examination of past studies, current challenges and future research directions. *Biodegradation*. 11, 171–186.

Deeb, R.A., Hu, H.Y., Hanson, J.R., Scow, K.M., and Alvarez-Cohen, L. 2001. Substrate interactions on BTEX and MTBE mixtures by an MTBE-degrading isolate. *Environmental Science and Technology*. 35, 312–317.

Finneran, K.T. and Lovley, D.R. 2001. Anaerobic degradation of methyl *tert*-butyl ether (MTBE) and *tert*-butyl alcohol (TBA). *Environmental Science and Technology*. 35, 1785–1790.

Fortin, N.Y., Nakagawa, Y., Focht, D.D., and Deshusses, M.A. 2001. Methyl-*tert*-butyl ether (MTBE) degradation by a microbial consortium. *Environmental Microbiology*. 3, 307–316.

Garnier, P.M., Auria, R., Augur, C., and Revah, S. 2000. Cometabolic biodegradation of methyl *tert*-butyl ether by a soil consortium: Effect of components present in gasoline. *Journal of General and Applied Microbiology*. 46, 79–84.

Hurt, K.L., Kwon, P., Williamson, K., and O'Reilly, K. 1998. Anaerobic biodegradation of MTBE in a contaminated aquifer. In: *Proceedings of the Fifth International In Situ and On Site Bioremediation Symposium* 5 (1), pp. 103–108. (Alleman, B. C. and Leeson, A., Eds.). Columbus, Ohio, Battelle Press.

Liu, C.Y., Speitel, G.E., and Georgiou, G. 2001. Kinetics of methyl *tert*-butyl ether cometabolism at low concentrations by pure cultures of butane-degrading bacteria. *Applied and Environmental Microbiology.* 67, 2197–2201.

Mace, R.E. and Choi, W.-J. 1998. The size and behavior of MTBE plumes in Texas. In: *Proceedings American Petroleum Institute/National Ground Water Association Conference on Petroleum Hydrocarbons and Organic Chemicals in Ground Water*, Houston, Texas. November, 1998, pp. 1–11.

Malmstead, M.J., Brockman, F., Valocchi, A.J., and Rittmann, B.E. 1995. Modeling biofilm biodegradation requiring cosubstrates: The quinoline example. *Water Science and Technology.* 31(1), 71–84.

Mormile, M.R., Liu, S., and Suflita, J.M. 1994. Anaerobic biodegradation of gasoline oxygenates: extrapolation of information to multiple sites and redox conditions. *Environmental Science and Technology.* 28, 1727–1732.

NRC (National Research Council). 2000. *Natural Attenuation for Groundwater Remediation.* Rittmann, B. E. (chairman). Washington, DC, National Academy Press.

Reid, J.B., Reisinger, H.J., Bartholomae, P.G., Gray, J.C., and Huilnian, A.S. 1999. A comparative assessment of the long-term behavior of MTBE and benzene plumes in Florida, USA. In: *Proceedings of the Fifth International In Situ and On Site Bioremediation Symposiu.* 5 (1), pp. 97–102. (Alleman, B. C. and Lesson, A., Eds.). Columbus, Ohio, Battelle Press.

Rittmann, B.E. and MacDonald, J.A. 2000. National Research Council guidance on natural attenuation. In: *Natural Attenuation Considerations and Case Studies: Remediation of Chlorinated and Recalcitrant Compounds*, 1–8. (Wickrananayake, G. B., Gavaskar, A. R., and Kelley, M. E., Eds.). Columbus, Ohio, Battelle Press.

Rittmann, B.E. and McCarty, P.L. 2001. *Environmental Biotechnology: Principles and Applications.* New York, McGraw-Hill Book Co.

Rittmann, B.E., Seagren, E., Wrenn, B.A., Valocchi, A.J., Ray, C., and Raskin, L. 1994. *In Situ Bioremediation*, second edition. Park Ridge, New Jersey, Noyes Publishers, Inc.

SAB (Science Advisory Board). 2001. *Monitored Natural Attenuation: USEPA Research Program—An EPA Science Advisory Board Review.* Science Advisory Board, Washington, DC. June, 2001.

Stocking, A.J., Deeb, R.A., Flores, A.E., Stringfellow, W., Talley, J., Brownell, R., and Kavanaugh, M.C. 2000. Bioremediation of MTBE: a review from a practical perspective. *Biodegradation.* 11, 187–201.

USEPA (United States Environmental Protection Agency). 1999. *Use of Monitored Natural Attenuation at Superfund, RCRA Corrective Action, and Underground Storage Tanks Sites.* Environmental Protection Agency Office of Solid Waste and Emergency Response, Washington, DC. Directive number 9200.17P.

Wilson, J.T., Cho, J.S., Wilson, B.H., and Vardy, J.A. 2000. *Natural Attenuation of MTBE in the Subsurface under Methanogenic Conditions.* National Risk

Management Research Laboratory, Office of Research and Development, U. S. Environmental Protection Agency, Cincinnati, Ohio. EPA/600/R-00/006.

Wilson, R.D., Schirmer, M., Naas, C.N., Smith, A., Smith, C., Scow, K.M., Hyman, M.R., and Mackay, D.M. 1999. Natural attenuation of MTBE in ground water: What do we need to know? In: *Proceedings 2000 Petroleum Hydrocarbons and Organic Chemicals in Groundwater: Prevention, Detection, and Remediation.* pp. 167–175. National Ground Water Association, Westerfield, Ohio.

Woo, S.H. and Rittmann, B.E. 2000. Microbial energetics and stoichiometry for biodegradation of aromatic compounds involving oxygenation reactions. *Biodegradation.* 11, 213–227.

Yeh, C.K. and Novak, J.T. 1995. Anaerobic biodegradation of gasoline oxygenates in soils. *Water Environment Research.* 67, 828–834.

Section III

Remediation Case Studies

CHAPTER 17

Remedial Costs for MTBE in Soil and Ground Water

Barbara H. Wilson, Dynamac Corporation
John T. Wilson, Ph.D., USEPA National Risk Management Laboratory, Office of Research and Development

INTRODUCTION

The contamination of MTBE in ground water has introduced concerns about the increased cost of remediating MTBE/BTEX releases compared to remediating sites with BTEX only contamination. In an attempt to evaluate these costs, cost information for 311 MTBE-blended gasoline release sites was furnished by the USEPA Office of Underground Storage Tanks (OUST) (USEPA, 2000), several states, BP/Amoco, Creek and Davidson (1998), and other sources. The majority of the sites were from South Carolina (183), Kansas (53), and New York (32), with information from sites in Maine, Texas, California, and Illinois also included. The reported information were project costs, actual costs, estimated project costs, or estimated project costs to date and included site assessment costs, capital expenditures, and operation and maintenance expenses. Site-specific cost data, average operation and maintenance expenses, and total project costs for several sites, primarily in California and on the East Coast, were provided by BP/Amoco (Kolhatkar, 2000).

Remedial decisions for USTs are frequently based on the location of a release in relation to a receptor. Most gasoline sites that do not impact drinking water wells are actively remediated by some combination of air sparging, SVE, and soil excavation. Bioremediation and natural attenuation are also widely used as remedies. Most gasoline sites impacting drinking water wells are remediated using multiple technologies, with pump-and-treat/air stripping/carbon treatment of the water prior to distribution. Increased costs are frequently due to the need to provide an alternate drinking water supply or home (point of entry) treatment for one or more homes.

The majority of the gasoline sites are remediated under the various state UST programs, and the individual states may vary in their responses to gasoline releases. These programs provide funds for UST cleanups that must meet specific cost requirements. The types of programs include reimbursable, pre-approval, and pay-for-performance. These cost guidelines include rates for labor, drilling, analytical costs, and any other potential costs associated with the UST cleanup. Most of the states have a deductible that must be met prior to the state issuing any funds. Basically in many states, the cost of MTBE cleanup will depend on the allowable cost expenditures for the individual states.

COST OF CLEANUP

To determine the effectiveness and cost of remedial actions currently used at LUST sites, data from actual sites were compiled and analyzed. The 311 sites included 183 sites from South Carolina that provided costs for remediation and operation and maintenance, but no site assessment costs; however, costs for the other states did include site assessment in addition to remediation and operation and maintenance costs. Cost averages for both the 183 South Carolina sites and the 128 remaining sites were included in Table 17-1. The South Carolina sites were classified as service station/petroleum sites because no indication of impact to a drinking water well was found. The average cleanup cost for South Carolina was $146,132, compared to $279,022 for the remaining 128 sites. The average cleanup costs for the 93 service station/petroleum sites not in South Carolina were $231,270. This suggests that average site assessment costs may range from approximately $85,000 to $140,000.

Table 17-1.   **Total Project Cost by Type of Site**

Type of Site	No. of sites	Mean cost	Median cost	Std. Dev.	Minimum cost	Maximum cost
Total Sites: 311						
Drinking water supply	32	$414,273	$262,550	$379,490	$31,004	$1,203,168
Service station/ Petroleum	276	$174,820	$139,790	$139,890	$21,000	$1,147,000
Hazardous waste	3	$316,667	$400,000	$144,337	$150,000	$400,000
Total sites	311	$200,827	$150,000	$193,210	$21,000	$1,203,168
Sites Outside South Carolina: 128						
Drinking water supply	32	$414,273	$262,550	$379,490	$31,004	$1,203,168
Service station/ Petroleum	93	$231,270	$195,158	$173,506	$21,369	$1,147,000
Hazardous waste	3	$316,667	$400,000	$144,337	$150,000	$400,000
Total non-SC sites	128	$279,022	$207,000	$252,192	$21,369	$1,203,168
South Carolina Sites*: 183						
Service station/ Petroleum	183	$146,132	$117,000	$109,066	$21,000	$454,000

*South Carolina cost data are for remediation and operation and maintenance; no costs for site assessment were provided.
Cost data for states other than South Carolina included site assessment, remediation, and operation and maintenance costs.

The provided cost data varied among the individual states that partici- pated in the survey. New York provided cost data for 52 UST sites impacted with MTBE that had been closed between July 1998 and July 2000. These costs ranged from a minimum of $99 to $567,136, with 18 of the 52 sites less than $20,000. No explanation was provided on how the costs were incurred. In- cluding these low costs in their data resulted in the average cost of $97,399 to treat a MTBE/BTEX site in New York compared to $200,827 for all sites in this report. New York may have included costs for sites that were evaluated as no-risk, whereas Kansas, South Carolina, and other states did not include no- risk costs. Cost data for New York less than $20,000 were arbitrarily excluded from calculated averages presented in Table 17-1. A frequency distribution of cleanup costs for the 311 sites is presented in Figure 17-1.

COST COMPARISONS FOR MTBE AND BTEX REMEDIATIONS

A comparison of the mean cleanup costs calculated in this study with mean costs for the technologies of MNA, pump-and-treat, AS/SVE, SVE/pump- and-treat, and AS/SVE/pump-and-treat used in the States of Tennessee and Texas is presented in Table 17-2. The individual technology cost values were obtained from a nationwide site survey and from state UST programs as part of a critical review of the cost effectiveness of MNA in the risk management of petroleum hydrocarbon plumes (Chen and Fishman, 1999). In general, the cost of remediating a UST site with both MTBE and BTEX contamination ranges from less than $100,000 to $300,000. However, if a private or munici-

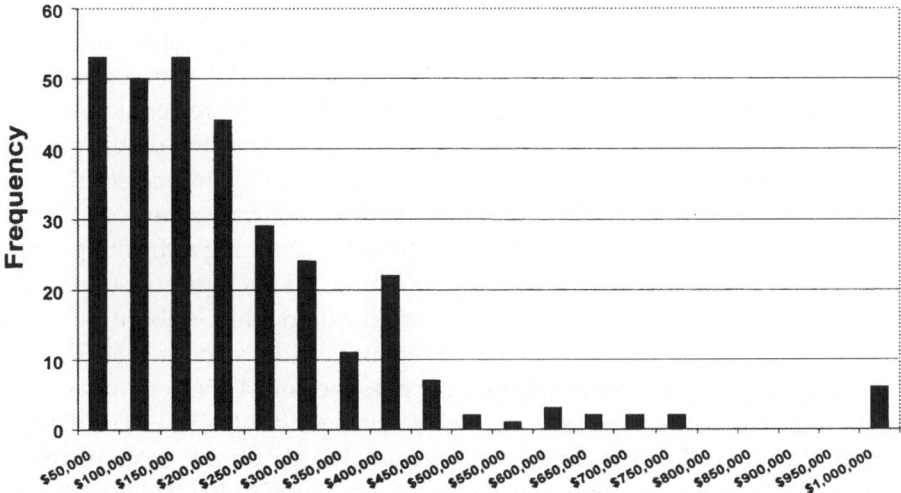

Figure 17-1. **MTBE/BTEX Cleanup Costs for 311 Sites**

Table 17-2.   Cost Comparison of Several Remedial Technologies

Source	MNA	P&T	AS/SVE	SVE/P&T	AS/SVE/ P&T	Mean Project Cost
Tennessee UST*	$74,000	$122,209		$226,366	$328,017	
Texas UST*	$113,574	$364,700	$179,429	$265,948	$316,343	
This study: South Carolina			$176,556		$270,141	$146,132
Creek and David- son, 1998						$321,590
This study: 275 USTs						$171,284
This study: Water supply						$414,273
This study: All sites						$200,827

MNA = Monitored Natural Attenuation; P&T = Pump-and-Treat; SVE = Soil Vapor Extraction;
AS = Air Sparging
*Chen and Fishman, 1999.

pal well is impacted or may potentially be impacted, the average cost increases to approximately $415,000.

The USEPA (2000) conducted a survey of state regulators and regulators in the USEPA regions to gather information on the relative costs of cleanup of MTBE and BTEX in their states. The USEPA OUST and the States of New York, Maine, Illinois, Ohio, Florida, Kansas, and South Carolina provided general information about the cost of UST site cleanups both with and without MTBE and about actions taken at UST release sites. The following summarizes their responses. The average costs to clean up sites varied greatly from $750,000 in Illinois to $42,000 in Maine. In the State of Illinois, the average cost to successfully remediate sites with MTBE was estimated to be $750,000; the average cost to remediate sites without MTBE was estimated to be $95,000. Illinois expects 90% of the MTBE ground water contamination sites to significantly add costs to the total cleanup, depending on the ground water receptor impacts, because most of these sites have impacted high capacity drinking water wells. The State of Ohio estimates remediation of MTBE/BTEX sites to be approximately $200,000, assuming a drinking water receptor is or may become impacted. The estimated average cost to remediate a BTEX-only ground water contamination site in Ohio is about $130,000. The historical average to clean up a UST site in Florida is about $250,000. It is estimated that in the State of Florida, the presence of MTBE will cause a 1% to 2% increase in cleanup cost compared to sites without MTBE. The State of South Carolina did not find a difference in treatment costs between MTBE/BTEX and BTEX-only sites, with an average of $140,000 for each type of site. The average cost in the State of Maine to remediate UST sites both with and without MTBE is approximately $42,000, although costs associated with MTBE/BTEX impact of drinking water supplies ranged from $200,000 to greater than $1,000,000. New York also observed a cost difference with an average of $97,400 for MTBE/BTEX sites and $56,393 for BTEX-only sites.

To determine if any relationship could be identified between cleanup costs and site or plume characteristics, the MTBE/BTEX cleanup costs for the South Carolina sites were plotted against BTEX soil concentrations, BTEX ground water concentrations, MTBE ground water concentrations, BTEX plume length, MTBE plume length, BTEX plume volume, and MTBE plume volume (data not shown). No clear correlation between cost of cleanup and any plume characteristic could be identified. For the South Carolina data, the cost associated with cleanup of MTBE sites could not be predicted based on size of release, length of plume, concentration of MTBE or BTEX, or any other measurable plume characteristic.

SOUTH CAROLINA COST DATA

South Carolina provided an extensive data set that included the technologies selected as remedies, plume length and width, saturated thickness of the impacted aquifer, and concentration information on MTBE and BTEX. Information on both active corrective action (ACA) and MNA sites were provided for a total of 215 sites. Cost information was provided for 183 of the ACA sites.

The cost data were sorted to determine if any association between cost and technology or between cost and soil type could be identified. The soil types were categorized into three major types: clay (clay, clayey sand, clayey silt), sand (sand, gravel, sand/clay, sand/rip rap, sand/silt, sandy clay), and silt (silt, silty clay, silty sand). The technologies were sorted among air sparging, AS/SVE, AS/SVE/ground water pumping, AS/SVE/soil excavation, bioremediation, free product recovery, and other combinations of technologies. Not all of the sites had information on both cost and soil type or cost and chosen remedial technology. The information is presented in Tables 17-3 and 17-4.

The mean cost for the various soil types were $120,648 for clay, $150,295 for sand, and $180,478 for silt. The higher costs associated with silt may reflect the difficulty of source removal in finely textured soils. The average cleanup costs associated with technologies were $127,585 for air sparging, $176,556 for AS/SVE, $270,141 for AS/SVE/ground water pumping, $308,000 for AS/SVE/soil excavation, $118,170 for bioremediation, $98,408 for free product recovery plus other technologies, and $80,396 for other technology combinations. For these data, increased costs were observed for soil excavation and ground water pumping, technologies requiring physical removal of soil or ground water with subsequent disposal or treatment. No information was provided for the treatment of the pumped ground water.

Table 17-3.   South Carolina Cleanup Costs Sorted by Soil Type

Soil Type	Number of Sites	Mean Cost
Clay	21	$120,648
Sand	66	$150,295
Silt	44	$180,478
All South Carolina cost sites	183	$146,132

Table 17-4.   South Carolina Cleanup Costs Sorted by
Remedial Technology

Technology	Number of Sites	Mean Cost
AS	28	$127,585
AS/SVE	57	$176,556
AS/SVE/GW Pumping	10	$270,141
AS/SVE/Soil Excavation	3	$308,000
Bioremediation	57	$118,170
Free Product Recovery + other technologies	12	$98,408
Other technology combinations	13	$80,396
All South Carolina cost sites	183	$146,132

REMEDIAL TECHNOLOGIES USED AT USTS IN NEW YORK STATE

The data provided by New York indicate that the use of a single technology for remediation of USTs is not common: normally two or three major technologies are implemented at each site. Table 17-5 compares the frequency of application of remedial technologies for cleanup of 1,563 UST releases in the State of New York. No cost information was provided for the majority of these sites. Most frequently, a suite of technologies was used at a particular site; the most common approaches were combinations of soil excavation, SVE, air sparging, and ground water pumping. The lowermost cells in the rows indicate the number of times a particular technology was used for cleanup at the 1,563 UST sites. The values in the other columns designate the additional technologies used. For example, out of 1,563 sites, soil excavation was used at 601 sites. Out of these 601 sites, 206 sites also used SVE, 182 used pump-and-treat, 79 used air sparging (ground water), 106 used natural attenuation (ground water), 30 used other ground water treatments, 24 used dual-phase extraction (DPE), 28 used O_2 injection (ground water), 28 used soil bioremediation, and 22 used other non-specified treatments. In the State of South Carolina, natural attenuation, air sparging, and SVE were the most ordinarily used technologies for MTBE sites (data not shown). However, in the States of Texas and Tennessee, natural attenuation was the remedy chosen most frequently for USTs, followed by pump-and-treat and SVE (data not shown).

Table 17-5.   **Cleanup Technologies Used for 1,563 USTs in the State of New York**

	Soil Excav.	SVE	GW P&T	GW AS	GW NA	GW Other Treat.	GW DPE	GW O$_2$ Inject.	Soil Biorem.	Others
Soil Excav.	601	206	182	79	106	30	24	28	28	22
SVE		525	234	176	48	26	33	33	22	27
GW P&T			393	67	30	17	19	15	18	18
GW AS				186	10	4	7	10	8	10
GW NA					181	6	6	9	18	15
GW Other Treat						82	5	0	12	9
GW DPE							76	3	2	3
GW O$_2$ Inject.								75	0	15
Soil Biorem.									60	13
Others										127

EFFICIENCY OF REMEDIAL TECHNOLOGIES

Data from 1,563 UST sites in the State of New York were previously sorted to identify key technologies used for UST remediation (Wilson *et al.*, 2001). Typically, two or three major technologies were implemented at each site. The most common remedial approaches were combinations of soil excavation, SVE, air sparging, and pump-and-treat. In order to evaluate the efficiency of cleanup technologies, the maximum MTBE concentration was divided by the current MTBE concentration, and the resulting ratios were compared to individual technologies. All sites with the maximum concentration equal to the current concentration were discarded because of an inability to distinguish between sites with ineffective technologies and those where cleanup had not yet begun.

A total of 1,012 UST sites remained for evaluation, 463 sites with aquifer impacts and 549 sites without aquifer impacts. The "no aquifer impact" designation may refer to LNAPL present only in the vadose zone, although ground water contamination could occur due to diffusion gradients or a fluctuating water table. Each group was divided into subgroups with concentration reduction of 1,000x or greater, 100 to 999x, 10 to 99x, and 1.1 to 9.9x. The technologies used at these sites were SVE, soil excavation, soil bioremediation, pump-and-treat, air sparging, ground water bioremediation, DPE, other treatment for soil and ground water, and no action. The results are shown in Table 17-6 and Table 17-7. Of the 463 sites with aquifer impacts, the observed ratio of maximum to current concentration was 10% for 1,000x+, 16% for 100x to 999x, 37% for 10x to 99x, and 37% for 1.1x to 9.9x. The most frequently used technologies for cleanup were SVE (43%), soil excavation (38%), pump-and-treat (34%), air sparging (21%), and bioremediation (18%).

Table 17-6. Comparison of Treatment Efficiencies of Various Technologies for Aquifer Impacted UST Sites

Ratio of Maximum to Current Concentration	Total Sites[a]	Soil Treatments				Ground Water Treatments					
		Soil Vapor Extraction[b]	Excavation[b]	Bioreme-diation[b]	Other Treatment[b]	Pump-and-Treat[b]	Air Sparging[b]	Bioreme-diation[b]	Dual Phase Extraction[b]	Other Treatment[b]	No Action[b]
1,000x+	10% (46)	61% (28)	41% (19)	7% (3)	2% (1)	48% (22)	33% (15)	17% (8)	4% (2)	7% (3)	0% (0)
100x–999x	16% (76)	37% (28)	41% (31)	3% (2)	1% (1)	41% (31)	16% (12)	22% (17)	7% (5)	11% (8)	4% (3)
10x–99x	37% (169)	44% (75)	35% (59)	6% (10)	1% (2)	38% (65)	20% (34)	13% (22)	5% (9)	7% (11)	1% (20)
1.1x–9.9x	37% (172)	40% (69)	38% (65)	1% (2)	1% (2)	22% (38)	22% (38)	22% (38)	6% (11)	5% (8)	3% (6)
Total	100% (463)	43% (200)	38% (175)	4% (17)	1% (6)	34% (156)	21% (99)	18% (85)	6% (27)	6% (30)	2% (11)

[a]Percentage (number) of sites in ratio out of 463.
[b]Percentage (number) of sites using a specific technology in a particular ratio.

Table 17-7. Comparison of Treatment Efficiencies of Various Technologies for UST Sites With No Aquifer Impacts

Ratio of Maximum to Current Concentration	Total Sites[a]	Soil Treatments				Ground Water Treatments					
		Soil Vapor Extraction[b]	Excavation[b]	Bioreme-diation[b]	Other Treatment[b]	Pump-and-Treat[b]	Air Sparging[b]	Bioreme-diation[b]	Dual Phase Extraction[b]	Other Treatment[b]	No Action[b]
1,000x+	6% (34)	56% (19)	41% (14)	9% (3)	6% (2)	38% (13)	15% (5)	26% (9)	0% (0)	3% (1)	3% (1)
100x–999x	12% (65)	38% (25)	48% (31)	18% (2)	3% (2)	26% (17)	8% (5)	26% (17)	5% (3)	61% (3)	0% (0)
10x–99x	30% (164)	36% (59)	41% (68)	9% (10)	1% (1)	30% (50)	9% (14)	20% (32)	5% (9)	4% (7)	1% (2)
1.1x–9.9x	52% (286)	34% (96)	48% (136)	5% (2)	2% (6)	29% (82)	8% (23)	23% (66)	5% (15)	5% (15)	3% (4)
Total	100% (549)	36% (199)	45% (247)	8% (42)	2% (11)	30% (162)	9% (47)	23% (124)	5% (27)	5% (27)	2% (7)

[a]Percentage (number) of sites in ratio out of 549.
[b]Percentage (number) of sites using a specific technology in a particular ratio.

For the sites with greatest maximum/current concentration ratio (1,000x+), the most commonly used technologies were SVE (61%), soil excavation (41%), pump-and-treat (48%), and air sparging (33%). For the sites with the least maximum/current concentration ratio (1.1x to 9.9x), the most frequently used technologies were the same, but the percentages differed with SVE used at 40% of sites, soil excavation used at 38%, and pump-and-treat and air sparging each used at 22%.

Slightly different results were observed for the 549 UST sites with no aquifer impacts. The most consistently used technologies were SVE (36%), soil excavation (45%), pump-and-treat (30%), and ground water bioremediation (23%). Although the sites were identified as no aquifer impacts, ground water treatment technologies were still utilized as cleanup remedies. The technologies used for the greatest maximum/current concentration ratio (1,000x+) were SVE (56%), soil excavation (41%), pump-and-treat (38%), and ground water bioremediation (26%). The technologies used for the least maximum/current concentration ratio (1.1x to 9.9x) were also SVE (34%), soil excavation (48%), pump-and-treat (29%), and ground water bioremediation (23%).

SUMMARY

The average cleanup cost of the 311 sites was $200,827 with a range of $21,000 (a UST in South Carolina) to $1,203,168 (for cleanup of a PWS in Kansas), a standard deviation of $193,210, and a median of $150,000. The majority of the costs ranged from less than $100,000 to $300,000 per site. The 311 sites included 183 sites from South Carolina that provided costs for remediation and operation and maintenance, but no site assessment costs. The average cleanup cost for South Carolina was $146,132 compared to $279,022 for the remaining 128 sites.

Out of the 311 MTBE/BTEX sites, 276 were service stations or other petroleum-related facilities, 32 of the sites were impacted drinking water wells/supplies, and 3 were hazardous waste sites. Overall, cleanup costs for MTBE contamination of drinking water wells were found to be higher than cleanup costs for USTs. The average cleanup cost of sites impacting drinking water supplies was $414,273 while the average cleanup cost for 276 service station/petroleum sites was $174,820. Excluding one major pipeline spill, the average cleanup costs for 275 service stations was $171,284. For many states, the presence of MTBE did not affect the cost of UST remediation or the technologies used for cleanup. The cost of cleanup of MTBE sites appeared to be controlled primarily by two factors: (1) MTBE contamination is the responsibility of most states' UST trust funds/programs, and (2) MTBE contamination impacting private or municipal drinking water wells substantially increases cleanup costs compared to cleanup costs for USTs that do not.

There are several generalizations that may be made from the data presented in this chapter. The average cost to remediate an MTBE/BTEX site with drinking water impacts is approximately twice the cost of cleanup of sites where no drinking water impacts occur ($414, 273 compared to $174,820). Estimation of remedial cost based on quantifiable characteristics of the release, the resulting plume(s), the contaminant type, or the contaminant concentration for the data in this study has not been possible. The cleanup technologies most often utilized in the states surveyed were soil excavation, SVE, pump-and-treat, and air sparging, with multiple technologies most frequently applied for site remediation, and of these, the more efficient technologies appear to be SVE, pump-and-treat, and air sparging, based on the ratios of maximum to current concentrations. MTBE treatment at most sites utilizes technology that is also intended, and originally developed, to treat BTEX contamination, because often the MTBE plume is congruent to or is contained within the BTEX plume.

Although states respond with individual approaches to MTBE contamination, remediation decisions are frequently based on the location of a release in relation to a receptor. The approach that many states, other than California, exercise for the cleanup of USTs may be reflected in the following statement from Greg Hattan (2000), Kansas Department of Health and Environment (KDHE), "The philosophy of our remedial effort at all sites is an aggressive source removal with treatment to the receptor, if it becomes impacted. We rarely address the plume other than air sparging in the hottest areas. We have found this to be efficient and cost-effective. We have and do use risk-based decision making (based on proximity to receptor) in determining which sites go into remediation."

DISCLAIMER

The views expressed in this Chapter are those of the individual authors and do not necessarily reflect the views and policies of the USEPA. Scientists in the USEPA's Office of Research and Development have prepared the USEPA sections, and these sections have been reviewed in accordance with USEPA's peer review and administrative review policies and approved for presentation and publication.

ACKNOWLEDGMENT

This paper would not have been possible without the assistance and information provided by the following agencies and individuals: Doug Maddox and Steve McNeely, USEPA OUST; Art Shrader, South Carolina Department of Health and Environmental Control (SCDHEC); Bill Reetz and Greg Hattan, KDHE; Tom Conrardy, Florida Department of Environmental Protection (FLDEP); Bruce Hunter, Maine Department of Environmental Protection

(MEDEP); D. Darmer, NYSDEC; Gilberto Alvarez, USEPA Region 5 UST Program; Linda Fiedler, USEPA, DC; and Tetra Tech EM Inc., Reston Virginia.

REFERENCES

Chen, J.S. and Fishman, M. 1999. *Critical Review: Cost-Effectiveness Analysis of Natural Attenuation in the Risk Management of Petroleum Hydrocarbon Plumes.* Prepared by Dynamac Corporation for USEPA Subsurface Protection and Remediation Division, National Risk Management Research Center, Ada, Oklahoma.

Creek, D.N, and Davidson, J.M. 1998. The performance and cost of MTBE remediation technologies. In: *Proceedings of the 1998 Petroleum Hydrocarbons and Organic Chemicals in Groundwater Conference.* Houston, Texas. November 11–13, 1998.

Hattan, G. 2000. Kansas Department of Health and Environment. Personal communication.

Kolhatkar, R. 2000. BP/Amoco. Personal communication.

USEPA (U.S. Environmental Protection Agency). 2000. Office of Underground Storage Tanks. Washington, DC. Doug Maddox and Steve McNeely. *OUST State Survey*: State of South Carolina: Art Shrader, DHEC. State of Kansas: Bill Reetz and Greg Hattan, KDHE. State of Florida: Tom Conrardy, DEP. State of Maine: Bruce Hunter, Maine DEP. State of New York: D. Darmer, NY DEC. Information concerning Ohio and Illinois was provided by Gilberto Alvarez, U.S. EPA, Region 5 UST Program. Some information was provided to OUST by Linda Fiedler, U.S. EPA, DC through Tetra Tech EM Inc., Reston VA.

Wilson, B.H., Shen, H., Pope, D., and Schemelling, S. 2001. Cost of MTBE remediation. In: *Bioremediation of MTBE, Alcohols, and Ethers, The Sixth International In Situ and On-Site Bioremediation Symposium,* San Diego, California. June 4–7, 2001. (Magar, V.S., Gibbs, J.T., O'Reilly, K.T., Hyman, M.R., and Leeson, A., Eds.). Columbus, Ohio, Battelle Press.

CHAPTER 18

Remediation Experiences in Finland

Martti R. Suominen, Neste Marketing Ltd.
Nancy E. Milkey, P.G., Tighe & Bond, Inc.

This chapter discusses remediation practices for addressing gasoline-impacted soil and ground water at several hundred Neste Marketing Limited (Neste) retail sites in Finland. The first systematic investigation and remediation program was initiated in 1994 at approximately100 stations in environmentally sensitive areas.

BACKGROUND

Legislation for Soil and Ground Water Protection in Finland.   Basic legislation for soil and ground water protection has been in place for decades. Important refinements have been introduced since the 1960s, and a new Environmental Protection Act was passed in 2000 and refined in 2001.

Current regulations prohibit releases of gasoline into soil and are very protective of ground water quality. Law enforcement has not been very proactive in the past; however, recent legislative developments and a change in the public attitude regarding environmental protection have brought considerable improvements. The current legislation requires anyone with knowledge of a release or of soil or ground water contamination to report it to the authorities. Noncompliance with environmental laws may lead to criminal charges; however, the number of actual court cases has been small.

The Finnish Oil and Gas Federation has been an active and effective partner with the legislative bodies in preparing legislation and discussing the interpretation of the regulations and practices with regard to retail site related matters. In general, the relationship between industry and the authorities is good. Consequently, a common sense approach to retail site remediation can be used in most cases.

Geology.   In Finland, the geology and landscape were reshaped by the latest glacial period, resulting in wide and fairly shallow sand and till eskers and ridge systems, a large number of separated aquifers, 56,000 lakes, and a number of rivers. The receding ice effectively peeled off soft soil layers and smoothed the bedrock hilltops leaving the bedrock fairly close to the ground surface. The bedrock formations in some areas are fractured and faulted.

The eskers often have a coarse sandy inner core, which is an excellent high flow water conduit. At the foot of esker ridges water discharges occur transversally, making these areas preferred locations for drinking water wells.

Human populations, roads, and service stations are primarily located in river valleys and especially on eskers and other elevated areas where the land is dry and stable enough for construction of roads and buildings. Consequently, there is considerable potential for ground water and surface water contamination of high yield formations.

Aquifers and Water Service in Finland.   The annual average rainfall is about 600 millimeters (24 inches), which recharges Finland's 7,141 designated aquifers (2,226 of which are in the "important" category). Twenty-five percent of important aquifers are currently in use, while the utilization rate of all the aquifers is 12%. Fifty-seven percent of Finland's drinking water comes from ground water. Municipalities typically supply drinking water service in urban areas, whereas private wells or surface water supplies are commonly used in rural areas and by some industries.

Gasoline Usage.   In 2000, 2.4 billion liters (0.63 billion gallons) of gasoline were sold in Finland, including 0.26 billion liters (69 million gallons) of MTBE/TAME. Since 1991, nearly all gasoline is blended with 11 to 13% oxygenates — the highest level in Europe. MTBE alone was used between 1991 and 1995; a combination of TAME and MTBE has been in use since 1995.

Gasoline consumption grew rapidly between 1960 and the 1980s and is presently growing at a rate of 1 to 2% per year. Gasoline distribution is provided by nine retail chains operating approximately 1,850 retail stations.

Retail sites in Finland are typically small in area, 1,500 to 4,000 square meters (16,000 to 43,000 square feet); self-service outlets are even smaller. The sites are densely developed with one or more buildings that often have basements and underground utility conduits. The subsurface is often filled with blocks of rock, crushed rock, gravel, and moraine. Screened sand is used beneath the pavement and in the tankholds.

FUEL HANDLING AT RETAIL STATIONS — TECHNOLOGY AND PRACTICES

Technology and fuel handling practices originated from the large international oil companies that began operating in Finland in the late 1920s. The Finnish retail chains began to emerge in the 1940s.

The current technology standard is consistent with the best international practices and includes the use of double containment and other best available technology elements, where applicable. Within important aquifers, standard equipment includes double-walled USTs equipped with jacket leak indicators and computerized tank level control devices, which often include a level trend indicator program capable of identifying even minor leaks. The tanks are provided with a durable corrosion-resistant coating both on the outside and the inside. To further reduce the risk of releases, oil companies voluntarily launched a tank inspection program in the early 1980s, which later (in 1983) was adopted into law.

In addition to the equipment requirements, the pavement must be watertight and fuel-proof, and the forecourt (the area where the pumps are located) and the tank refilling area must drain via oil/water separators that are equipped with electronic oil level alarms. Water tightness is often provided by a heavy-duty plastic liner installed beneath the forecourt and refilling areas. Recently, the authorities have required the installation of plastic liners beneath the entire tankhold at several facilities in very sensitive areas.

In 30 cases, SVE piping has been installed when major forecourt and tank renovations have been performed. Such a low cost arrangement allows nearly immediate response in case of a release and provides an excellent opportunity to monitor soil gas at any time and location along the pipes.

PRACTICES IN SOIL AND GROUND WATER INVESTIGATION AND RISK ASSESSMENT AT NESTE SITES

Site investigations are performed when a major release has occurred or is suspected to have occurred; when there is a major renovation planned for a site; or when the site will be closed, sold, acquired, or leased.

Generally, the retail site investigation approach used in Finland is simple and follows the same principles and practices as elsewhere. It focuses on known or suspected hot spots and ground water downgradient of potential releases, especially if there are sensitive receptors. The law does not specify the scope of the investigation; however, the mandatory remediation permitting process, which includes submitting a proper investigation and risk assessment report, gives the authorities the leverage they need to indirectly influence the approach and content of the investigation and risk assessment.

The site investigation typically includes soil and ground water sampling at or near tankholds, pump islands, oil/water separators, waste oil tanks, tank refill areas, vent pipes, and oil and solvent storage areas. Also, if the station is located within an important aquifer, at least one downgradient ground water sample must be collected.

The majority of retail sites are quite old and have been rebuilt several times. Prior to conducting a subsurface investigation, a historical review is necessary as it is not uncommon to find sites with more than one UST still in place, often containing fuel.

During the initial sampling round, soil samples are collected at four to six locations, from varying depths. In addition, soil gas samples are collected from four to eight locations, again from varying depths of typically 1 to 5 meters (3 to 16 feet) bgs. The soil samples are screened on site with field equipment and, based on the results, a few are selected for laboratory analysis. The soil gas and ground water samples are analyzed in the laboratory by GC applying USEPA methods or modifications thereof.

Contracted consultants carry out the work according to investigation programs developed jointly with Neste. The investigations and remediation are carried out by three to four key consultants selected in a periodic bidding process to work in different areas of the country.

Close contact between Neste and the consultants is maintained at all times, and subcontractors' services are coordinated either by the consultant or Neste, on a case-by-case basis. Neste approves all contacts with the authorities and other information releases by the consultants.

The results of the investigation are summarized in a report, which includes information on site location and operational history, hydrogeology, sensitive receptors, history of known fuel incidents, sample locations, field and analytical laboratory results, boring logs, maps of the surrounding area, and a preliminary risk assessment. Recommendations for additional actions such as remediation are also provided in the report.

In the event that the investigation identifies contamination levels in excess of the "limit values" (see Table 18-1), the case must be reported to the governing authorities and remediation discussed. If the remediation will include excavation as the first step and the contamination is not extensive, then additional delineation is not often conducted. The extent of the contamination is then determined during soil removal.

The cost of a basic retail site investigation is typically around $4,000 to $6,000 (U.S. dollars, used throughout this chapter). In serious cases where an important aquifer is impacted, the costs can increase significantly to several tens of thousands of dollars.

PRACTICES IN SOIL AND GROUND WATER REMEDIATION AT NESTE SITES

Remediation is generally initiated on an as-needed basis. The key deciding factors are the risks to ground water that may be a potential drinking water source, surface water, adjacent lots (including streets and parks), or basements at the site or adjacent lots. Releases close to streets may present a problem due to the preferential migration of gasoline as LNAPL or even as a

Table 18-1. **Soil Contamination Guideline Values**

	Current Values		Proposed Values, Fall 2001	
Constituent	Lower Limit (mg/kg)	Upper Limit (mg/kg)	Lower Limit (mg/kg)	Upper Limit (mg/kg)
Gasoline (C_4–C_{10})	100	500	100	500
Middle Distillates (C_{10}–C_{23})	300	1,000	200	1,000
Heavier Oil Products (C_{20}–C_{35})	500	2,000	600	2,000
Other Gasoline Components				
Benzene	0.5	25	0.06	2.5
Ethylbenzene	5	50	5	50
Xylenes	0.5	25	0.5	25
Toluene	2	120	2	120
MTBE	none	none	5	100

Notes: No numerical limit values exist for gasoline components in drinking water or ground water.

The regulations state that drinking water cannot have any off taste or odor.

The limit values are based on various studies (e.g., of toxicity) and on limit values in other countries.

The lowere limit is a target value in remediation cases; the upper limit has been used as an action limit.

The current values were never included in regulations; however, the authorities have applied them in remediation goal setting since 1992.

vapor in the soil pores along the uniform sand-filled utility excavations. This creates numerous problems, including potential hazards to municipal utility service and maintenance crews and the contamination of drinking water supply lines since gasoline is capable of penetrating certain plastic pipes.

Remediation is frequently carried out due to the exceedance of contamination limit values in connection with a site being closed, sold, or leased. The intention is to protect a new site owner, leaseholder, or operator. At times, remediation is required due to the exceedance even if there is no foreseeable risk involved.

Neste's reason for remediating a site is straightforward — to eliminate real risk to receptors. At times, discussion with the remediation permitting authorities is difficult, as there are differing philosophies, values, and interests involved. The remediation methods are selected based on cost efficiency; exotic or untested methods are generally not used. The selected remedial measure is often excavation, after which other methods may be applied, if necessary. It is important to keep in mind that the cost difference between real risk elimination and complete remediation is often substantial.

The following methods are routinely used to remediate impacted sites in Finland.

- Excavation is by far the most common remedial option. Practically every remediation includes excavation and subsequent disposal or on- or off-site treatment.
- SVE is the second most preferred method, at times enhanced with soil heating via re-injected clean off-gas. Others in Finland have also used steam heating. In most cases the off-gas must be treated, usually with GAC, but occasionally by catalytic oxidation.
- Hydraulic control of a hot spot or a contaminant plume is frequently used in difficult cases, often including GAC-treatment of the extracted water.
- Air sparging and air stripping are only used in very basic applications.
- Biotreatment has been applied in impacted soil composting and in several long-term passive bioventing cases. There has been little activity using bioreactors to treat extracted ground water. One bioreactor is currently being used at a large tank yard with positive results. The use of *ex situ* and *in situ* bioremediation is expected to increase in the near future.
- MNA is often used as the last step in the remediation process to monitor residual concentrations until levels acceptable to the authorities are achieved.
- Chemical oxidation has not been implemented for gasoline releases; however, other companies have used chemical oxidation on soils impacted by heavier oils.
- Two experiments with electro-osmosis/electrical soil heating have been carried out with some success; however, considerable application development is needed if the technology is to become cost-competitive.

The presence of ether oxygenates has not had much influence on the selection of remediation methods since the methods that work with traditional gasoline components usually work with ether oxygenates; however, when ether oxygenates are involved in gasoline releases, a greater emphasis is placed on the determination of the ground water status in critical areas and acute releases are intercepted immediately.

Cost of Remediation of Retail Sites in Finland. Neste's average remediation cost, based on experience at several hundred sites, is about $40,000, with a range between $10,000 and $150,000. This figure excludes a few $150,000 to $500,000 cases where large soil volumes and serious ground water impact — and in some cases an inappropriate initial investigation or remediation approach — resulted in long-term and intensive remediation. MTBE has been the driver in some of these cases; however, if these releases had been detected and intercepted soon after the release, the costs would have been dramatically lower.

This information is generally consistent with cost data compiled by SOILI, the Finnish oil industry branch's mutual remediation organization for closed sites. SOILI's average cost for the remediation of 87 older (30 to 50 years old) retail sites is about $60,000.

Table 18-2 gives examples of typical unit costs for investigation/remediation work in Finland.

Table 18-2. Typical Retail Station Remediation Cost in Finland—2001

Element	Unit	Range (FIM)	Range ($)
Remediation Permit		0–4,000	0–615
Typical Retail Station Investigation		20,000–40,000	3,076–6,152
Environmental Consulting	Hour	190–400	29–62
Technical Assistance	Hour	170–300	29–46
Travel	Kilometer	2	0.3
Daily Allowance	Day	160	25
Travel Time	Hour	100–250	15–39
50 mm Sample Holes (5–6 m)	12 holes	4,000–6,000	615–922
SVE Well	Meter	800–1,000	123–153
Ring Well	Meter	1,000–1,5000/m	153–230
Soil Venting Compressor	Month	1,500–4,000	230–615
Equipment Installation	Typical case	4,000–8,000	615–1,230
GAC Cartridge	Month	1,500	230
GAC	Kilogram	10–30	1.5–5.0
Regeneration of GAC	Kilogram	6–10	1–1.5
Catlytic Burner	Month	15,000–20,000	2,300–3,076
Magnetic Ground Survey (Radar)	Typical case	5,000	770
Soil or Water Analysis	TVOC	400–600	61–92
	MTBE/BTEX	500–800	77–123
	Oil	400–600	61–92
	Package of above	700–1,000	107–153
SV Off-Gas Analysis	TVOC	400–600	61-92
	MTBE/BTEX	400–700	61–107
Field Screening PID/Petroflag	Sample	150–250	23–38
Excavator and Operator	Hour	200–400	31–62
Excavation Work	Ton	30–50	5–8
Contaminated Soil Transport	Ton × kilometer	1.2–2.4	0.18–3.6
Clean Fill Delivered and Compacted	Ton	20–70	3–11
Disposal of Mildly Contaminated Soil	Ton	10–150	1.5–23
Composting at Dump	Ton	120–700	18–108
Thermal/Incineration Treatment	Ton	250–800	38–123
Contaminated Water into Sewer (Mild Contamination)	m^3	0–7.5	0–1.2
Tank Lifting, Dismantling and Reinstallation (Tanks & Piping Reinstalled)	Typical 4 tank case	50–100,000	770–1,530
Lifting an Old Tank	1 tank	1,500–3,000	230–460
Pavement (Tarmac)	m^2	50–100	8–16

FIM = Finnish Markka; FIM/$ = 6.5; PID = Photoionization detector; TVOC = Total volatile organic compounds; SV = Soil vapor; GAC = Granular activated carbon

CASE STUDIES

Case 1 — Traditional Practices, High Hopes, and Not Enough Information.
In May 1991, about 8,000 liters (2,000 gallons) of gasoline were released at a
service station, which had been in business since 1970 in southern Finland.
The tank was punctured during repeated striking of the unprotected metal
bottom of the tank with a steel-tipped gauging stick.

Site Hydrogeology.   The site is located on a very important large esker/ridge
area, consisting primarily of various grades of sand, including very coarse
sand. Ground water is present at depths between 3 and 9 meters (10 and 30
feet) bgs. The ground water gradient is relatively flat at the site, but increases
downgradient of the site. A ridge runs perpendicular to the main esker and
extends from the site to a municipal well (Figure 18-1). Water conductivity in
the ridge is approximately 1.2×10^{-3} cm/s (3.3 ft/day).

Nature and Extent of Contamination.   A site investigation was conducted in
late 1994, following initial remediation efforts in 1991. Contaminants were
detected in site ground water at concentrations of up to 3.3 mg/l of MTBE
and 9.7 mg/l of total VOCs. At a monitoring well located 150 meters (490 feet)
downgradient of the site, MTBE was detected at a concentration of 890 µg/l,
with BTEX compounds below detection limits; at a monitoring well located
600 meters (2,000 feet) downgradient of the site, MTBE was detected at
19 µg/l; and at a municipal well 800 meters (2,600 feet) downgradient of the
site, MTBE was identified at a concentration of 70 µg/l, again with no de-

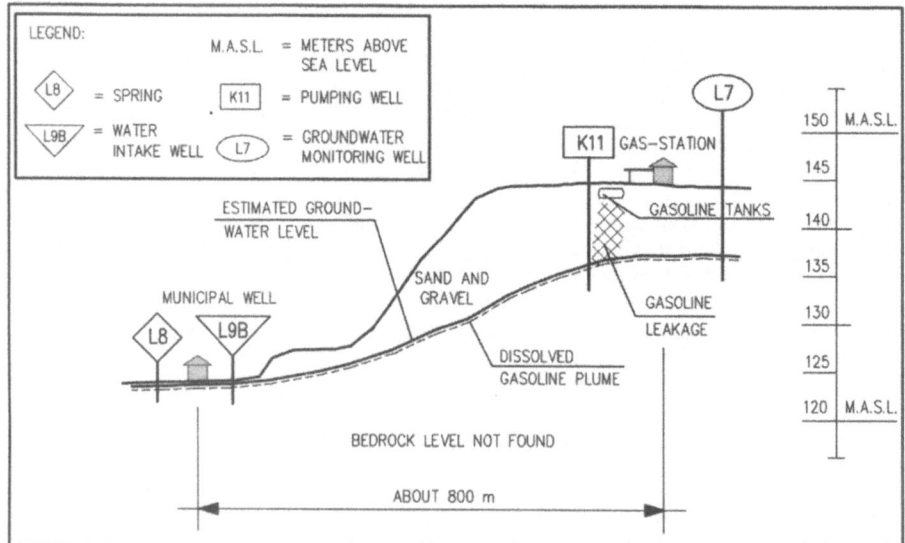

Figure 18-1.   Case 1 Schematic Cross-Section of Site and Receptor

tected BTEX compounds. Soil samples collected close to the source area, at depths of 1 to 4 meters (3 to 13 feet), had very low concentrations of gasoline constituents. Only one sample showed a trace (less than 1 mg/kg of total VOCs) of toluene and xylene. Soil gas samples collected at depths of 1 to 4 meters (3 to 13 feet) had very low concentrations, whereas soil gas samples collected close to the tanks at depths of 4 to 6 meters (13 to 20 feet) had concentrations of MTBE up to 1,400 mg/m^3 and BTEX at 44 mg/m^3.

Receptors.   In the fall of 1993, MTBE was detected at a concentration of 70 µg/l in a municipal drinking water supply well located 800 meters (2,600 feet) from the release site. The well was immediately decommissioned and arrangements were made for an alternate water supply. Neste partially financed the installation of a new well.

Remediation.   The leak was detected during a routine inventory control check several days after the release. The initial remedial actions included the removal of approximately 100 cubic meters (130 cubic yards) of impacted sand from around and under the tank. The soil was disposed of at a municipal landfill. Limited laboratory analytical data were collected during the initial response actions; however, a sheen was observed in a ground water sample, and analysis detected MTBE and BTEX at a total concentration of 200 mg/l.

A recovery well was constructed for the removal of impacted ground water. The water was pumped through an oil/water separator, and the effluent was directed to a nearby ditch running into a small wetland area. The pumping provided hydraulic control to reduce contaminant migration. This was later verified by analytical data from sampling at adjacent monitoring wells. No additional actions were taken to complete the hot spot removal.

The recovery well produced approximately 100 m^3/day (26,400 gpd) through March 1995, at which time the analyses indicated that the contaminant concentrations in the well were no longer decreasing.

In 1996, the authorities agreed to shut down the pump-and-treat system and initiate MNA. No additional remedial activities were required. The status of the plume was monitored from several locations in progressively lengthening intervals. Fluctuations in ground water elevations resulted in changes to contaminant concentrations. The reduction in MTBE concentrations was very slow; the original concentration of 70 µg/l identified at the municipal well was reduced to approximately 20 µg/l in eight years.

Required Cleanup Levels.   Forecasting based on concentration trends indicates that the contamination will eventually decrease to a safe level over an extended period of time (greater than five years) unless additional soil hot spot removal is completed (See Figures 18-2 and 18-3).

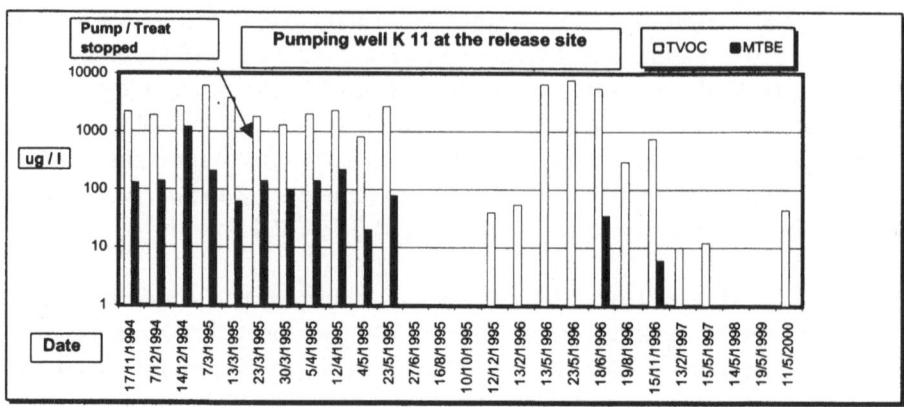

Figure 18-2.   **Case 1 Source Area Contamination Level Versus Time**

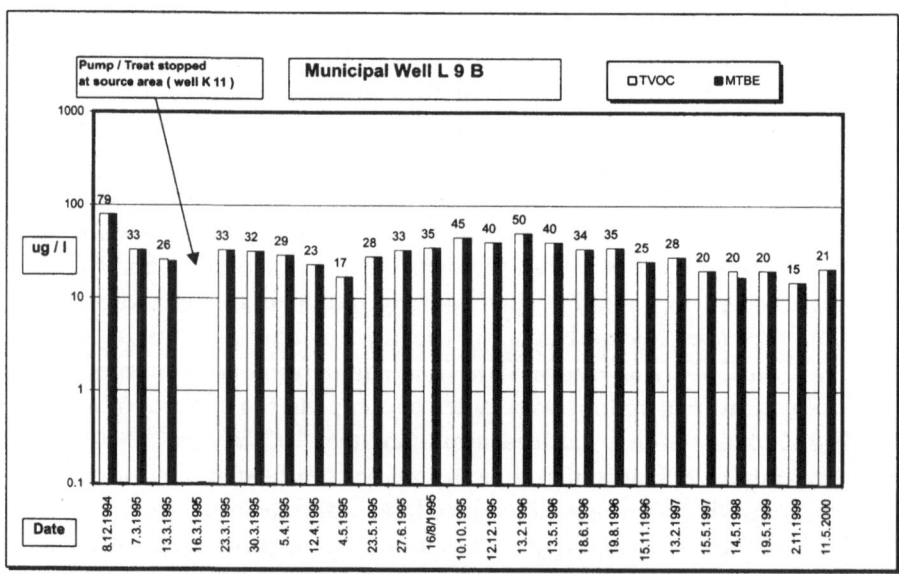

Figure 18-3.   **Case 1 Municipal Well Contamination Over Time**

With the MTBE concentrations in the municipal well remaining at approximately 20 µg/l, the municipality has been considering use of the well for a small amount of drinking water. The water from the well would be blended with water from other sources. There is a risk that activating the well will change the ground water flow pattern and create a potential for new impacts.

Table 18-3. **Case 1 Cost Summary**

Remediation Task	Cost
Immediate remediation actions	$67,692
Site investigation/risk assessment	46,154
Design	21,538
Installations	53,846
Well replacement	167,692
Operation and maintenance	84,615
Cost of supplying alternative drinking water	96,463
Total cost to-date	$538,000

Costs. To date, the cost of the remediation has been approximately $540,000 as shown in Table 18-3. This case highlights the necessity of immediate aggressive interception and careful hot spot removal.

Case 2 — Traditional Approach and Methods Applied Successfully to Remediate a Service Station Site and Natural Spring. The site was constructed in the 1940s as a service station and has been owned and operated by several oil companies. In 1995, gasoline-impacted soil was encountered during a routine UST replacement.

Site Hydrogeology. The site is situated on a major sand and gravel esker within the area of a Category III aquifer (no immediate need for use as a drinking water supply). The ground water level is 5 to 7 meters (16 to 23 feet) bgs, and ground water flow is towards a lake and a natural spring with a discharge rate of 100 m³/day (26,400 gpd) (Figure 18-4). Although it is not an official drinking water source, the spring is used by local people.

Nature and Extent of Contamination. Analysis of soil samples from the immediate area of the release detected MTBE at concentrations of a few tens of mg/kg. Ground water contamination was detected beneath the tank yard at concentrations up to: 270 mg/l MTBE, 14 mg/l benzene, 45 mg/l toluene, 23 mg/l xylenes, and 0.6 mg/l ethylbenzene. Initially, SVE using a 2.2 kilowatts (kw) compressor with an output of 260 m³/hour was applied to two wells with screens located at 3.7 to 4.7 meters (12.1 to 15.4 feet) bgs. To provide replacement airflow, three additional vent wells were installed around the area, with screens at 4- to 5-meter (13- to 16-foot) depths. The SVE system ran for 57 days and produced 330 kilograms (730 pounds) of total VOCs (as determined by GC analysis of off-gas). The system was turned off when concentrations of total VOCs in the effluent were reduced to 211 mg/m³.

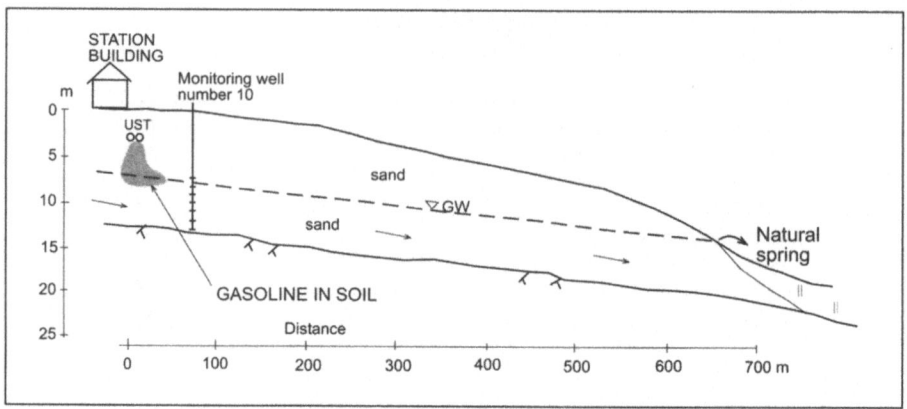

Figure 18-4.   Case 2 Schematic Cross-Section of Site and Receptor

Receptors.   In October 1995, the authorities agreed that there was no immediate need for further remediation and that additional remediation could
wait until a major upgrade of the site was scheduled. This decision was based
in part on that fact that there were no important sensitive receptors immediately downgradient of the site. Nevertheless, a ground water monitoring
program was instituted.

Additional Remediation.   In September 1998, MTBE was detected in the natural spring at a concentration of 87 µg/l. An additional sampling round indicated that remediation of the site was needed, particularly to address
residual gasoline found in another area of the site at a depth of 6 to 7 meters
(20 to 23 feet). In March 1998, MTBE concentrations in the natural spring had
decreased slightly to 85 µg/l. Ethylbenzene was also detected in the spring at
a trace amount of 4 µg/l.

   An SVE system was installed and minor excavation performed at the end
of 1999, at the same time as a major renovation occurred at the site. After the
remediation, MTBE concentrations in the spring steadily declined until the
concentration was under 10 µg/l in June 2001. In September 1999, MTBE was
unexpectedly identified at 300 µg/l in a monitoring well located 60 meters
(200 feet) downgradient of the tank yard. By June 2001, the MTBE concentration in that monitoring well had decreased to 60 µg/l.

   There is an understanding with the authorities that the remaining risk
will be addressed by natural attenuation. Removal of hot spots has allowed
MNA to do its job. The site and the spring are still monitored.

Costs.   To date, the cost of the remediation has been approximately $54,000,
as shown in Table 18-4.

Table 18-4.   Case 2 Cost Summary

Remediation Cost	Cost
Site investigation/risk assessment	$10,000
Design	2,308
Excavations, installations	34,923
Operation and maintenance	6,154
Total cost	$53,385

Case 3 — Emergency Remediation Operation.   On February 18, 1998, a gasoline release of 880 liters (230 gallons) occurred during the refilling of a UST at a service station in Central Finland. The truck driver recovered 256 liters (67 gallons); however, the remainder infiltrated the sand-filled tank-hold. The oil company's emergency response personnel were immediately activated, and a consultant, the insurance company, and the authorities were contacted.

Site Hydrogeology.   The site is located in a Class I aquifer area with sandy soil and a low hydraulic gradient. The distance to the closest municipal supply well is 1.4 kilometers (km) (0.9 miles). Residences in the vicinity of the site are supplied with municipal water; no private wells are located near the site. Ground water is located at a depth of 5 meters (16 feet) bgs and the initial assessment indicated that a thin LNAPL layer was present on the ground water surface.

Nature and Extent of Contamination.   Due to obvious conditions indicating a release had occurred, no soil or ground water samples were collected initially.

Remediation.   The authorities immediately gave permission, via telephone, to initiate SVE. Two vertical suction tubes were installed to the depth of the ground water table, a 2.2 kw/250 m^3/hour (150 cfm) compressor was installed, and the system was started within two days of the release.

SVE continued until April 8, 1998, at which point 491 kilograms (1,082 pounds) of volatile gasoline components had been removed (as calculated from analysis of the off-gas). Soil and ground water sampling conducted on both sides of the tank and a few meters away indicated that the SVE was effectively remediating the soil and ground water. The highest total VOC concentrations in soil and ground water were approximately 100 mg/kg and less than 20 µg/l, respectively.

Costs.   The cost was fully covered by the trucking company's insurance policy. The total cost for the remediation was approximately $12,000, as shown in Table 18-5.

Table 18-5. **Case 3 Cost Summary**

Remediation Cost	Cost
Site investigation/risk assessment	$1,333
Planning & design	833
SVE installation/equipment rental	6,300
Operation and maintenance	333
Authorities, analytical, reporting	2,833
Total cost	$11,632

FORENSIC FINDINGS — THE REASONS FOR THE RELEASES

Unexpectedly, most releases have occurred due to operational failures. The strong predominance of failures occurred during tank refilling and included improperly emptying the hose before or after filling, overfilling due to misinformation on the fuel level in the tank, and defective or missing automatic shut-offs leading to vent pipe overflow. A significant number of spill catch basins around filling connections were inadequate or missing. Filling connection camlock and gasket faults have also been common in the past. Another issue contributing to releases was discontinuous bitumen corrosion protection layer on the sides of older tanks; this has been caused by gasoline contacting the bitumen layer as a result of carelessly performed refills, leading to spills or overfills.

Numerous releases have resulted from missing or improper spill and leak containment provisions in combination with leaking fixtures (*e.g.*, pumps, piping, joints, concrete ring oil separators with bitumen gaskets), cracked concrete tank bunkers, or cracked tank vent pipes. Recurring small drips and spills not properly intercepted have created cumulative and costly problems. Fortunately, increased awareness and enhanced operational practices and protection systems have improved the situation considerably.

Although a convenient and simple name to the problem, LUST is not descriptive enough and certainly not the primary reason for contamination at retail sites in Finland. There have been only two LUST cases, out of about 1,800 USTs currently active at Neste retail sites, that have had any environmental impact in the 1990s.

In 1991, a LUST was caused by a level-gauging puncture (Case Study 1). In 1993, a second LUST was the result of corrosion and a level-gauging puncture. The site was remediated by excavation, soil venting, and air sparging from 1994 through 1998 with no impact to the public drinking water well located 1,000 meters (3,300 feet) downgradient of the release.

Tank inspection programs have proven to be an effective means of protecting the environment. Since 1983, about 200 suspect tanks have been replaced with new ones as a result of timely inspections.

LESSONS LEARNED

Neste has learned numerous lessons, sometimes the hard way, with regard to soil and ground water remediation. The lessons confirm common knowledge:

- Incomplete, low-budget site investigations have in some cases become expensive as they did not produce a satisfactory hydrogeological picture of the release. Based on this incomplete information, inappropriate approaches were initially attempted.
- Detailed risk assessment saves money by allowing remediation to focus on key risks.
- Aggressive removal of hot spots is necessary to shorten ground water remediation time.
- Immediate action in simple acute cases yields tangible benefits. By initiating remediation quickly, there is little time for gasoline to migrate, resulting in smaller volumes of impacted soil and ground water and less damage compensation to negotiate from insurance policies. Quick response usually results in better publicity by the news media.
- Doing it right the first time means that one does not need to do it again, facing high costs and potentially unpleasant publicity due to initial substandard work.

Risk reduction of soil and ground water impact at retail stations is simple and based on common sense:

- The site operator should keep him/herself fully aware of the daily operations and incidents at the site, especially at sites located in environmentally sensitive areas (*e.g.*, active or potential aquifers, sensitive urban areas, etc.).
- On-site fuel distribution equipment must be suitable for the designed purpose and properly installed (including environmental protection elements), maintained, and periodically leak-tested.
- To become free of the burden of past releases, sites in sensitive areas need to be investigated, risk assessed, and, where necessary, remediated.
- Personnel must be trained and motivated in proper procedures. When gasoline releases occur, time is a critical factor. SVE works quickly and efficiently and should be installed as quickly as possible. If ground water is impacted, remediation is not likely to be quick and inexpensive, unless the release can be immediately intercepted. The importance of the previous statement cannot be overestimated. Even new remediation technologies still require rapid release detection and response to produce good results.
- The authorities should have the legal means and resources to exercise tight guidance and control.

CHAPTER 19

USEPA Case Studies Database for MTBE Remediation

David K. Ramsden, Ph.D. and Tie Li, Ph.D., URS Corporation, Houston

PURPOSE OF DATABASE

As of January 2002, the USEPA has established a database of 244 MTBE case studies on their website (USEPA, 2002). MTBE remediation case studies spanning the last 10 years are tabulated in the database. The purpose of the database is to provide information and guidance to project managers trying to determine the best approach to managing their MTBE release. The database can be accessed and searched on-line.

Table 19-1 presents a compacted version of the recoverable information provided by searching the database on the USEPA's website. Because of the size of the database, information summarized in Table 19-1 uses abbreviations. A key to the abbreviations is provided at the bottom of the table.

Table 19-2 presents a summary of the statistics for the case studies and the case studies' associated matrices, technologies, contaminants, and types of studies. Files with more detailed information are available on the website for 21 of the case studies. These are readily accessible from icons on the tables generated from the site by the search function. These expanded case studies share a common format (Figure 19-1) for each document and diverge in style and detail, as necessary, to describe the project and site. For 223 cases studies, only the tabulated information in the website table was available at the time of the review, essentially the same as the formatted data found at the beginning of the 21 case studies.

The case studies, as a whole, carry a strong message that MTBE can be managed and remediated, and by a variety of technologies. While there are some negative or failure studies, these are a small fraction of the total. As Table 19-2 summarizes, searching the database for all technologies shows the range of technologies that have or are being employed to remediate MTBE, particularly in ground water. Trends in the application of particular technologies are identifiable in the database (see Table 19-2), but whether these are reflective of national trends is not known.

Table 19-1. List of USEPA Case Studies (Adapted from USEPA Database)

Site No.	ID	Site Location	Contaminants	Matrix	Technology Description	Scale	MTBE Concentrations (µg/l or µg/kg)			Operation Status
							Goal	Initial	Last Reported	
1	1	Brooksville, FL	MTBE	GW	ISB — Bioaugmentation	FS	35	440	5	Completed
2	2	Houston, TX	MTBE, TBA, B, EB, Sty	S/GW	ESB — Slurry Phase/MNA	FS	NP	150	10	Ongoing
3	3	Lake Tahoe, CA	MTBE, BTEX, TPHg	S/GW	DPE/ASp/GAC/ISOX	FS	0.5	790	1,500	Ongoing
4	5	Lake Tahoe, CA	MTBE, BTEX, TPHg	S/GW	DPE	FS	0.5	32,000	2,300	Ongoing
5	6	Lacrosse, KS	MTBE, BTEX, TPH	S/DW	AStrp(Tower)	FS	10	340	< 0.2	Ongoing
6	7	Livermore, CA	MTBE	S/GW	ESB	BS	NP	420	<5	Completed
7	8	Park City, KS	MTBE, BTEX, TBA, 1, 2-DCA	S/GW	ASp/SVE/GAC	FS	NP	285,000	72,100	Ongoing
8	9	Riverhead, NY	MTBE	GW	PT/AStrp(Tower)	FS	50	1,800	23	Completed
9	10	Rockaway, NJ	MTBE, TCE, DIPE	DW	AStrp/GAC	FS	NP	35	1	Ongoing
10	11	Rush Center, KS	MTBE, BTEX, TBA, 1, 2-DCA	S/GW	ASp/SVE/ISB — ORC®	FS	NP	191,000 (GW)	47,000	Ongoing
11	12	Santa Monica, CA	MTBE, BTEX, TBH	DW/GW	PT/Thermal-[SAVRS]/(PARS)	PS	NP	33,000	14.5	Ongoing
12	13	Florence, SC	MTBE, BTEX, Naphthalene	GW	PT/ASp	FS	2,495	87,000	7,410	Ongoing
13	14	Service Station, MA	MTBE, BTEX	GW	ASp	FS	NP	215	115	Completed
14	15	CA	MTBE, BTEX	GW	ISB — ORC®	FS	NP	2,200	2,300	Ongoing
15	16	Sebato, ME	MTBE, BTEX	GW	ASp	FS	NP	62,000	980	Completed
16	17	Lake Geneva, WI	MTBE, BTEX	GW	ISB — ORC®	FS	NP	1,800	<2	Ongoing
17	18	Cheshire, CT	MTBE, BTEX, B, T, TPH	GW	ISB — EnzCat(DO-IT) — O_2	FS	70	6,000	1,500	Completed
18	19	Pensacola, FL	MTBE, BTEX	GW	ASp	FS	NP	230	<5	Completed
19	20	South Lake Tahoe, CA	MTBE, BTEX, TPHd,g, TCA, Acetone	S/GW	ISOX/PT	FS	0.5	9	22	Ongoing
20	21	Hilton Head, SC	MTBE, BTEX, Naphthalene	GW	ISB — Bioaugmentation	FS	40	3,140	41	Ongoing
21	22	SC	MTBE, BTEX	S/GW	ISB — Bioaugmentation	FS	NP	229	3	Completed
22	23	Sparks, NV	MTBE, PCE, TCE, TPH	GW	PT/ESB — FBRGAC	FS	NP	2,400	55	Ongoing
23	24	Lake Tahoe, CA	MTBE, BTEX, TBA, EtOH, TPHg	GW	PT/ASp/SVE/GAC	FS	0.5	31,000	270	Ongoing
24	25	Oxnard, CA	MTBE, BTEX, TBA	GB	ISB — O_2	PS	5	1,000	<1	Ongoing
25	26	Lompoc, CA	MTBE	GW	ISB — O_2	PS	NP	420	2	Ongoing
26	27	Lake Tahoe, CA	MTBE, BTEX, TPHd,g	S/GW	DPE/ASp/ISOX	FS	0.5	260	27	Ongoing
27	29	Bayport, TX	MTBE, TBA, IsoBE	S	EXCV/ESB/LF	FS	NP	1,200	NP	Ongoing

28	30	Lake Tahoe, CA	MTBE, BTEX, TPHg, TAME	GW	PT/BV	FS	0.5	1,400	50	Ongoing
29	31	Laboratory, MA	MTBE	S/GW	ISB — Butane	BS	NP	10,000	<1	Completed
30	32	Chattanooga, TN	MTBE, BTEX, TPH	S/GW	ISB — EnzCatalysis	FS	NP	5,000	90	Ongoing
31	33	Channelview, TX	MTBE, TBA, B, EB, Sty	GW	ISB — O_2	FS	NP	10,000	NP	Ongoing
32	34	Santa Monica, CA	MTBE, TBA	DW	AOP/GAC/Resin	PS	13	1,000	3	Ongoing
33	35	Pasadena, TX	MTBE, BTEX, TBA	S/GW	ISB/MNA	FS	NP	60,000	NP	Ongoing
34	36	N. Windham, ME	MTBE	GW	PT/AStrp(Tray)/GAC	FS	500	6,000	2	Completed
35	37	Ravenel, SC	MTBE, BTEX, Naphthalene	S/GW	EXCV/ISB — ORC®	FS	NP	239	BDL	Ongoing
36	38	Hemingway, SC	MTBE, BTEX, Naphthalene	S/GW	ASp/SVE	FS	80	5,110,000	1,400	Ongoing
37	39	MD	MTBE, BTEX	S/GW	SVE/PT	FS	NP	6,139	791	Completed
38	40	Albuquerque, NM	MTBE, BTEX	GW	ASp	FS	NP	1,600	27	Completed
39	42	Smithtown, NY	MTBE, BTEX	S/GW	PT/ASp/SVE	FS	NP	15,000	NP	Ongoing
40	43	Lindenhurst, NY	MTBE, BTEX	S/GW	PT/SVE/CatOX	FS	NP	>10,000	NP	Ongoing
41	44	Baldwin, NY	MTBE, BTEX	S/GW	PT/ASp/SVE/CatOX	FS	NP	NP	NP	Ongoing
42	45	Bayshore, NY	MTBE, BTEX	S/GW	ASp/SVE	FS	NP	110	NP	Ongoing
43	46	Wilmington, NC	MTBE, BTEX	S/GW	ASp/SVE	FS	NP	2,483	2	Ongoing
44	47	Grinnell, KS	MTBE, BTEX	S	SVE	FS	NP	175	21.5	Ongoing
45	48	Cimarron, KS	MTBE, BTEX	S/GW	ASp/SVE	FS	NP	92,000	2,700	Ongoing
46	49	Cimarron, KS	MTBE, BTEX	S/GW	ASp/SVE	FS	NP	280	46	Ongoing
47	50	Greenfield, KS	MTBE, BTEX	S/GW	ASp/SVE	FS	NP	255	<250	Ongoing
48	51	Great Bend, KS	MTBE, BTEX	S/GW	ASp/SVE/PT	FS	NP	390	4.8	Completed
49	52	Hays, KS	MTBE, BTEX	S/GW	ASp/SVE/EXCV	FS	NP	3,350	140	Ongoing
50	54	Wichita, KS	MTBE, BTEX	S/GW	ASp/SVE	FS	NP	8,850	6	Ongoing
51	55	Wichita, KS	MTBE, BTEX	S/GW	ASp/SVE/EXCV	FS	NP	1,560	106	Ongoing
52	56	Peabody, KS	MTBE, BTEX	S/GW	ASp/SVE/EXCV	FS	NP	4,840	4,840	Ongoing
53	57	Hutchinson, KS	MTBE, BTEX	S/GW	ASp/SVE	FS	NP	3,580	74	Ongoing
54	58	Sylvia, KS	MTBE, BTEX	GW	ASp/SVE	FS	NP	NP	NP	Ongoing
55	59	Peabody, KS	MTBE, BTEX	GW	Product Recovery	FS	NP	NP	NP	Ongoing
56	61	Selden, KS	MTBE, BTEX	S/GW	ASp/SVE	FS	NP	5.32	NP	Ongoing
57	62	Ness City, KS	MTBE, BTEX	S/GW	ASp/SVE	FS	NP	<250	NP	Ongoing
58	63	Andale, KS	MTBE, BTEX	S	SVE	FS	NP	850	65	Ongoing
59	64	Andale, KS	MTBE, BTEX	S/GW	ASp/SVE	FS	NP	128	NP	Ongoing
60	65	Ellinwood, KS	MTBE, BTEX	S/GW	ASp/SVE	FS	NP	232	26	Ongoing

Table 19-1. Continued

Site No.	ID	Site Location	Contaminants	Matrix	Technology Description	Scale	MTBE Concentrations (µg/l or µg/kg) Goal	Initial	Last Reported	Operation Status
61	66	Ellinwood, KS	MTBE, BTEX	S/GW	ASp/SVE	FS	NP	232	26	Ongoing
62	67	Ellinwood, KS	MTBE, BTEX	S/GW	ASp/SVE	FS	NP	27.7	NP	Ongoing
63	68	Ellinwood, KS	MTBE, BTEX	S/GW	ASp/SVE	FS	NP	190	190	Ongoing
64	69	Ellinwood, KS	MTBE, BTEX	S/GW	ASp/SVE	FS	NP	194	252	Ongoing
65	70	Ellis, KS	MTBE, BTEX	S/GW	ASp/SVE	FS	NP	8,140	6,430	Ongoing
66	71	Ellis, KS	MTBE, BTEX	GW	ISB/EXCV	FS	NP	1,400	980	Ongoing
67	72	Ellis, KS	MTBE, BTEX	S/GW	ASp/SVE	FS	NP	46.3	47.7	Ongoing
68	73	Oberlin, KS	MTBE, BTEX	S/GW	ASp/SVE	FS	NP	127	189	Ongoing
69	74	Ness City, KS	MTBE, BTEX	S/GW	ASp/SVE	FS	NP	97	45.3	Ongoing
70	75	Ness City, KS	MTBE, BTEX	S	SVE	FS	NP	280	868	Ongoing
71	76	Alden, KS	MTBE, BTEX	S/GW	ASp/SVE	FS	NP	46	47	Ongoing
72	78	Pratt, KS	MTBE, BTEX	S	SVE	FS	NP	49	15	Ongoing
73	79	Goodland, KS	MTBE, BTEX	S	SVE	FS	NP	270	1.92	Ongoing
74	80	Quinter, KS	MTBE, BTEX	S/GW	ASp/SVE	FS	NP	280	15.1	Ongoing
75	81	Great Bend, KS	MTBE, BTEX	S/GW	SVE/PT	FS	NP	79	NP	Ongoing
76	82	Hanston, KS	MTBE, BTEX	S	SVE	FS	NP	6,900	3,200	Completed
77	85	Dodge City, KS	MTBE, BTEX	S/GW	ASp/SVE	FS	NP	4,340	18,400	Ongoing
78	86	Junction City, KS	MTBE, BTEX	S/GW	ASp/SVE	FS	NP	<500	120	Ongoing
79	87	Witchita, KS	MTBE, BTEX	S/GW	ASp/SVE	FS	NP	72.5	NP	Ongoing
80	88	Atwood, KS	MTBE, BTEX	S	SVE	FS	NP	480	93	Ongoing
81	90	Colby, KS	MTBE, BTEX	S/GW	ASp/SVE	FS	NP	426	236	Ongoing
82	91	Hugoton, KS	MTBE, BTEX	S/GW	ASp/SVE	FS	NP	510	3	Ongoing
83	92	Hugoton, KS	MTBE, BTEX	S/GW	ASp/SVE	FS	NP	21,000	NP	Ongoing
84	93	Luka, KS	MTBE, BTEX	S/GW	ASp/SVE	FS	NP	NP	115	Ongoing
85	94	Macksville, KS	MTBE, BTEX	S/GW	ASp/SVE	FS	NP	<100	108	Ongoing
86	95	Marienthal, KS	MTBE, BTEX	S/GW	ASp/SVE	FS	NP	22,300	1,850	Ongoing
87	96	Oakley, KS	MTBE, BTEX	S/GW	ASp/SVE	FS	NP	666	NP	Ongoing

88	97	Pratt, KS	MTBE, BTEX	S/GW	ASp/SVE	FS	NP	105	NP	Ongoing
89	98	Wichita, KS	MTBE, BTEX	S/GW	ASp/SVE	FS	NP	2,770	46	Ongoing
90	99	Rush Center, KS	MTBE, BTEX	S	SVE	FS	NP	191,000	99,800	Ongoing
91	100	Park City, KS	MTBE, BTEX	S/GW	ASp/SVE	FS	NP	NP	1,060	Ongoing
92	101	Wichita, KS	MTBE, BTEX	S/GW	ASp/SVE	FS	NP	1,250	98	Ongoing
93	103	Oakley, KS	MTBE, BTEX	S/GW	ASp/SVE	FS	NP	9,158	55.6	Ongoing
94	104	Junction City, KS	MTBE, BTEX	S/GW	ASp/SVE	FS	NP	3,871	NP	Ongoing
95	105	Dodge City, KS	MTBE, BTEX	S	SVE	FS	NP	383	74	Ongoing
96	106	Dodge City, KS	MTBE, BTEX	S	SVE	FS	NP	22	NP	Ongoing
97	107	Dodge City, KS	MTBE, BTEX	S/GW	SVE/PT	FS	NP	NP	NP	Completed
98	108	Dodge City, KS	MTBE, BTEX	S/GW	ASp/SVE/PT	FS	NP	15	NP	Ongoing
99	109	Stuttgart, KS	MTBE, BTEX	S/GW	ASp/SVE/EXCV	FS	NP	6,401	NP	Ongoing
100	110	Santanta, KS	MTBE, BTEX	S	SVE	FS	NP	460	NP	Ongoing
101	111	Clay Center, KS	MTBE, BTEX	S/GW	ASp/SVE	FS	NP	NP	25	Ongoing
102	112	Clay Center, KS	MTBE, BTEX	S	SVE	FS	NP	11	NP	Ongoing
103	113	Castleton, KS	MTBE, BTEX	S/GW	ASp/SVE/EXCV	FS	NP	20.3	108	Ongoing
104	114	Summerfield, KS	MTBE, BTEX	S	SVE/EXCV	FS	NP	NP	450	Ongoing
105	115	Offerley, KS	MTBE, BTEX	S/GW	ASp/SVE/EXCV	FS	NP	755	542	Ongoing
106	116	Phillipsburg, KS	MTBE, BTEX	GW	Product Recovery	FS	NP	NP	NP	Completed
107	117	Wichita, KS	MTBE, BTEX	S/GW	ASp/SVE	FS	NP	355	NP	Ongoing
108	118	Dodge City, KS	MTBE, BTEX	S	SVE	FS	NP	1,400	12	Ongoing
109	119	Hays, KS	MTBE, BTEX	S/GW	SVE/PT	FS	NP	<2,500	1,942	Ongoing
110	120	Grantville, KS	MTBE, BTEX	S/GW	ASp/SVE	FS	NP	337	NP	Completed
111	121	Grantville, KS	MTBE, BTEX	S/GW	ASp/SVE	FS	NP	95	NP	Ongoing
112	122	Scott City, KS	MTBE, BTEX	S	SVE	FS	NP	63	4	Ongoing
113	123	Dodge City, KS	MTBE, BTEX	S/GW	ASp/SVE	FS	NP	360	<100	Ongoing
114	124	Scott City, KS	MTBE, BTEX	S/GW	ASp/SVE	FS	NP	69,100	NP	Ongoing
115	125	Scott City, KS	MTBE, BTEX	S/GW	ASp/SVE	FS	NP	21,600	NP	Ongoing
116	126	Ronan, MT	MTBE, TBA	GW	ISB- Bioaugmentation	PS	NP	NP	NP	Completed
117	127	Oxnard, CA	MTBE	GW	ISB — Bioaugmentation — O_2	PS	NP	2,000	1-25	Completed
118	128	Oxnard, CA	MTBE	GW	ISB — Bioaugmentation /ASp	PS	NP	6,000	<10-200	Ongoing
119	129	Oxnard, CA	MTBE	GW	ESB-O_2 Butane, Propane	BS	NP	10,000	NP	Ongoing
120	130	Bucks County, PA	MTBE, BTEX, Naphthalene	S/GW	DPE/ISOX-Ozone	FS	2,900	17,000	31	Completed

Table 19-1. Continued

No.	ID	Site Location	Contaminants	Matrix	Technology Description	Scale	Goal	Initial	Last Reported	Operation Status
121	131	TX	MTBE	S	PT/ISOX	FS	NP	475,000	68,400	Completed
122	132	Houston, TX	MTBE	GW	PT/Phytoremediation	PS	NP	46.7	NP	Ongoing
123	133	NV	MTBE	GW	PT/ESB/AOP	FS	100	4,000	30	Ongoing
124	134	Jamaica City, NY	MTBE, BTEX	GW	DPE/PT	FS	50	100,000	1,100	Ongoing
125	135	NY	MTBE, BTEX	S/GW	ISB — O_2	FS	10	870,000	42,000	Ongoing
126	136	Medina, NY	MTBE, BTEX	GW	ISB — O_2	FS	50	1,800	40	Completed
127	137	Clifton Park, NY	MTBE, BTEX	S/GW	ISB — O_2	FS	50	2,800	30	Completed
128	138	CA	MTBE	GW	ISB — Bioaugmentation, O_2	FS	NP	>20,000	NP	Ongoing
129	139	CT	MTBE	GW	ISB — Bioaugmentation, O_2	FS	NP	100,000	1,100	Completed
130	140	Elizabeth City, NC	MTBE, BTEX	S/GW	ISB — ORC®	FS	NP	NP	NP	Completed
131	141	Lake Tahoe, CA	MTBE	DW	PT	FS	NP	NP	NP	Ongoing
132	142	NJ	MTBE, BTEX	GW	ISOX	FS	70	6,400	<70	Completed
133	143	NJ	MTBE, BTEX, TBA	GW	ISB — ORC®	FS	NP	140	19	Completed
134	144	NJ	MTBE, BTEX	GW	ISB — ORC®	FS	NP	1,570	1,160	Completed
135	145	Oakland, CA	MTBE, BTEX	GW	ISB — ORC®	FS	NP	48,000	33,000	Completed
136	146	PA	MTBE, BTEX	GW	ISB — ORC®	FS	NP	15,000	12,500	Completed
137	147	Long Island, NY	MTBE, BTEX	GW	ISOX — Ozone	PS	NP	6,300	79	Completed
138	148	CA	MTBE, TPH	GW	PT/ FBR	PS	NP	NP	NP	Completed
139	149	NJ	MTBE	S/GW	ISB- Bioaugmentation, Propane	PS	NP	NP	NP	Ongoing
140	150	PA	MTBE	GW	ISB- Bioaugmentation	PS	NP	NP	NP	Ongoing
141	151	Allenton, WI	MTBE, BTEX	GW	EXCV/ISB — ORC®	FS	NP	58,000	7,000	Completed
142	152	FL	MTBE, BTEX, Naphthalene	GW	ISB — ORC®	FS	NP	15	0	Completed
143	153	Lompoc, CA	MTBE, BTEX	S/GW	EXCV/MNA/Phytoremediation	FS	NP	NP	NP	Ongoing
144	154	CA	MTBE, TBA	GW	Phytoremediation	PS	NP	NP	NP	Completed
145	155	Beaufort, SC	MTBE, BTEX	GW	Phytoremediation	PS	NP	NP	NP	Ongoing
146	156	Novato, CA	MTBE, BTEX	S/GW	ISB	BS	NP	10,000	NP	Completed
147	157	NY	MTBE, BTEX, Naphthalene	S/GW	ISOX-Ozone/DPE	FS	NP	160	40	Ongoing

MTBE Concentrations (µg/l or µg/kg) — columns Goal, Initial, Last Reported

148	158	Saratoga Springs, NY	MTBE, BTEX	GW	PT/ESB	PS	NP	NP	NP	Ongoing
149	159	Newport Beach, CA	MTBE, BTEX	GW	PT/AOP	FS	NP	300	NP	Completed
150	160	Jerico Springs, MO	MTBE, BTEX, TPH	S/GW	EXCV/PT/SVE	FS	NP	7,762	354	Ongoing
151	161	MA	MTBE, BTEX	S/GW	EXCV/ISB	FS	NP	370	12	Completed
152	162	Lake Tahoe, CA	MTBE, BTEX, TPHd,g	GW	PT/ASp/SVE	FS	NP	52,100	760,000	Ongoing
153	163	Lake Tahoe, CA	MTBE, BTEX, TPHg, EtOH, TCE	GW	PT	FS	NP	65,500	21,600	Ongoing
154	164	Lake Tahoe, CA	MTBE, BTEX, TPHg	S/GW	ASp/SVE	FS	NP	296	240	Ongoing
155	165	Lake Tahoe, CA	MTBE, BTEX, EtOH, TPHg	S	SVE	FS	NP	NP	1,800	Completed
156	166	Lake Tahoe, CA	MTBE, BTEX, TPHd,g	GW	EXCV/ASp/SVE/PT/GAC	FS	NP	5	2	Completed
157	167	Lake Tahoe, CA	MTBE, BTEX, TPHd,g	S/GW	ASp/SVE	FS	NP	5	437	Ongoing
158	168	Lake Tahoe, CA	MTBE, TCE, PCE, TPH	S/GW	EXCV/PT	FS	NP	NP	1	Ongoing
159	169	Lake Tahoe, CA	MTBE, BTEX, TPHd,g	GW	EXCV/ISB — ORC®	FS	NP	<5	8	Completed
160	170	Lake Tahoe, CA	MTBE, BTEX, TPHd,g	S/GW	ASp/SVE/GAC	FS	NP	58	13	Ongoing
161	171	Lake Tahoe, CA	MTBE, BTEX, TPHd,g, TBA	GW	PT/GAC	FS	NP	460	20	Ongoing
162	172	Lake Tahoe, CA	MTBE, BTEX, TPHd, TPHmo	GW	PT/GAC	FS	NP	2,300	2.9	Ongoing
163	173	Lake Tahoe, CA	MTBE, BTEX, TAME, PCE, Naphthalene	GW	PT/GAC	FS	NP	110	35	Ongoing
164	174	Lake Tahoe, CA	MTBE, BTEX, TPHg	GW	PT/GAC/ASp	FS	NP	8,700	8,700	Ongoing
165	175	Lake Tahoe, CA	MTBE, BTEX, TPHg, TAME	S/GW	EXCV/PT/ASp	FS	NP	11,000	2,070	Completed
166	176	Tahoe City, CA	MTBE, BTEX, TPHd,g	GW	ISB/PT	FS	NP	22,300	8,500	Ongoing
167	177	Lake Tahoe, CA	MTBE, BTEX, TPHg	S	SVE	FS	NP	48	NP	Completed
168	178	Lake Tahoe, CA	MTBE, BTEX, TPHg	GW	DPE/GAC/ASp	FS	NP	55	79	Completed
169	179	Lake Tahoe, CA	MTBE, BTEX, TPHg	GW	PT/GAC	FS	NP	167	26	Ongoing
170	180	Lake Tahoe, CA	MTBE, BTEX, TPHd,g	S	EXCV/SVE	FS	NP	40	4.5	Ongoing
171	181	Lake Tahoe, CA	MTBE, BTEX, TPHg	GW	PT	FS	NP	53	75	Ongoing
172	182	Lake Tahoe, CA	MTBE, BTEX, TPHd,g	GW	PT/GAC/Resin	FS	NP	70,000	27,000	Ongoing
173	183	Lake Tahoe, CA	MTBE, BTEX, TPHg	GW	PT/GAC/ASp	FS	NP	1,200	18	Ongoing
174	184	Lake Tahoe, CA	MTBE	GW	ISOX/ASp	FS	NP	6	0	Ongoing
175	185	Beaufort, SC	MTBE, BTEX	GW	ISB — ORC®	PS	NP	30,000	5,000	Ongoing
176	200	Island Lake, IL	MTBE	GW	PT/AStrp	FS	NP	100	NP	Ongoing
177	201	East Alton, IL	MTBE	GW	PT/AStrp	FS	NP	100	NP	Ongoing
178	202	Turtle Bayou, TX	MTBE, TBA, Naphthalene, 1,2-DCA, B	S/GW	ISB/SVE/ESB/EXCV	FS	NP	NP	NP	Ongoing
179	203	Liberty, TX	MTBE, TBA, 1,2-DCA, B	S/GW	EXCV/ESB/ISB/Soil Washing	FS	NP	3,000	NP	Completed

Table 19-1. Continued

Site No.	ID	Site Location	Contaminants	Matrix	Technology Description	Scale	MTBE Concentrations (µg/l or µg/kg) Goal	Initial	Last Reported	Operation Status
180	204	Boston, MA	MTBE, BTEX, TBA	S/GW	EXCV/ISB/MNA	BS	NP	2,500	NP	Completed
181	205	Cape Cod, MA	MTBE, BTEX, TPH, B	S/GW	PT/ASp/SVE	FS	70	2,600	3	Completed
182	206	NH	MTBE, BTEX	S/GW	EXCV/SVE/PT	FS	NP	1,000,000	200	Completed
183	207	NJ	MTBE	GW	PT/GAC/AStrp	FS	NP	96	NP	Completed
184	208	FL	MTBE	GW	PT/AStrp	FS	NP	580	NP	Ongoing
185	209	FL	MTBE	GW	PT/AStrp	FS	NP	130	NP	Ongoing
186	210	NJ	MTBE	GW	PT/GAC	FS	NP	1,200	NP	Ongoing
187	211	NJ	MTBE	GW	PT/GAC/AOP	FS	NP	1,610	NP	Completed
188	212	CA	MTBE, TPHg	S	SVE	FS	NP	8,900	21	Completed
189	213	VT	MTBE	S/GW	PT/SVE	FS	NP	2,720	NP	Ongoing
190	214	MA	MTBE	S/GW	PT/SVE/GAC	FS	NP	5,000	NP	Ongoing
191	215	MA	MTBE	S/GW	PT/SVE/GAC	FS	NP	34,440	NP	Ongoing
192	216	NJ	MTBE	S/GW	PT/SVE/AStrp/GAC	FS	NP	56,000	NP	Ongoing
193	217	MA	MTBE	S/GW	PT/SVE/ASp/GAC	FS	NP	2,290	NP	Ongoing
194	218	CO	MTBE	S	ASp/SVE	FS	NP	NP	NP	Ongoing
195	219	WI	MTBE	S	ASp	FS	NP	NP	NP	Ongoing
196	220	Windham, ME	MTBE	DW	PT	FS	NP	100	NP	Ongoing
197	221	Standish, ME	MTBE	S/DW	EXCV/PT	FS	NP	6,500	25	Completed
198	222	ME	MTBE	S/GW	EXCV/PT	FS	NP	NP	NP	Ongoing
199	223	Limington, ME	MTBE	S/GW	PT	FS	NP	NP	NP	Ongoing
200	224	Fryeburg, ME	MTBE	S/GW	EXCV/PT	FS	NP	NP	NP	Ongoing
201	225	Barnwell, SC	MTBE	GW	ISB	FS	NP	21	NP	Ongoing
202	226	Yemassee, SC	MTBE	GW	Product Recovery	FS	NP	NP	NP	Ongoing
203	227	Bluffton, SC	MTBE	S/GW	PT/ASp/SVE	FS	NP	33	NP	Ongoing
204	228	North Charleston, SC	MTBE	GW	ASp	FS	NP	5,111	NP	Ongoing
205	229	North Charleston, SC	MTBE	GW	ISB	FS	NP	10	NP	Ongoing
206	230	Charleston, SC	MTBE	S/GW	ASp/SVE	FS	NP	808	NP	Ongoing

207	231	Charleston, SC	MTBE	GW	Product Recovery	FS	NP	NP	NP	Ongoing
208	232	North Charleston, SC	MTBE	S/GW	ASp/SVE/PT	FS	NP	20	NP	Ongoing
209	233	Charleston, SC	MTBE	S/GW	ASp/SVE	FS	NP	21	NP	Ongoing
210	234	Blacksburg, SC	MTBE	GW	ASp	FS	NP	135	NP	Ongoing
211	235	Chester, SC	MTBE	GW	ISB	FS	NP	10,221	NP	Ongoing
212	236	Cheraw, SC	MTBE	S/GW	ASp/SVE/PT	FS	NP	5,111	NP	Ongoing
213	237	Hartsville, SC	MTBE	S/GW	ASp/SVE/PT	FS	NP	270	NP	Ongoing
214	238	Dillion, SC	MTBE	S	SVE	FS	NP	NP	NP	Ongoing
215	239	Winnsboro, SC	MTBE	GW	ISB	FS	NP	283	NP	Ongoing
216	240	Johnsonville, SC	MTBE	GW	ASp/ISB	FS	NP	NP	NP	Ongoing
217	241	Georgetown, SC	MTBE	GW	ASp	FS	NP	49	NP	Ongoing
218	242	Greenville, SC	MTBE	S/GW	ASp/SVE	FS	NP	49	NP	Ongoing
219	243	Travelers Rest, SC	MTBE	S/GW	ASp/SVE	FS	NP	NP	NP	Ongoing
220	244	Fountain Inn, SC	MTBE	GW	ASp	FS	NP	10,227	NP	Ongoing
221	245	Myrtle Beach, SC	MTBE	S/GW	ASp/SVE	FS	NP	3,204	NP	Ongoing
222	246	Conway, SC	MTBE	GW	ISB	FS	NP	10	NP	Ongoing
223	247	Green Sea, SC	MTBE	S/GW	ASp/SVE/PT	FS	NP	15	NP	Ongoing
224	248	West Columbia, SC	MTBE	GW	ASp	FS	NP	999	NP	Ongoing
225	249	West Columbia, SC	MTBE	GW	ISB	FS	NP	5,113	NP	Ongoing
226	250	Bennetsville, SC	MTBE	S/GW	ASp/SVE	FS	NP	12	NP	Ongoing
227	251	Westminster, SC	MTBE	GW	ISB	FS	NP	10,442	NP	Ongoing
228	252	Orangeburg, SC	MTBE	GW	ISB	FS	NP	21	NP	Ongoing
229	253	Columbia, SC	MTBE	GW	ISB	FS	NP	41	NP	Ongoing
230	254	Spartanburg, SC	MTBE	S/GW	ASp/SVE	FS	NP	5,132	NP	Ongoing
231	255	Rock Hill, SC	MTBE	S/GW	ASp/SVE	FS	NP	16	NP	Ongoing
232	256	Clover, SC	MTBE	S/GW	ASp/SVE	FS	NP	35,773	NP	Ongoing
233	257	Pageland, SC	MTBE	GW	ASp/Product Recovery	FS	NP	10,242	NP	Ongoing
234	258	Newberry, SC	MTBE	S/GW	ASp/SVE	FS	NP	12	NP	Ongoing
235	259	Spartanburg, SC	MTBE	S/GW	ASp/SVE/PT	FS	NP	36	NP	Ongoing
236	260	Conway, SC	MTBE	GW	ISB	FS	NP	10,223	NP	Ongoing
237	261	Latta, SC	MTBE	GW	Product Recovery	FS	NP	NP	NP	Ongoing
238	262	Dillon, SC	MTBE	S/GW	ASp/SVE	FS	NP	NP	NP	Ongoing
239	263	Conway, SC	MTBE	GW	ISB	FS	NP	20,449	NP	Ongoing

Table 19-1. Continued

Site No.	ID	Site Location	Contaminants	Matrix	Technology Description	Scale	MTBE Concentrations (µg/l or µg/kg)			Operation Status
							Goal	Initial	Last Reported	
240	264	Olanta, SC	MTBE	GW	ISB	FS	NP	23,999	NP	Ongoing
241	265	Abbeville, SC	MTBE	S/GW	ASp/SVE	FS	NP	20,553	NP	Ongoing
242	266	Orangeburg, SC	MTBE	S/GW	ASp/SVE	FS	NP	36	NP	Ongoing
243	267	Chester, SC	MTBE	G	ISB	FS	NP	15,335	NP	Ongoing
244	268	Branchville, SC	MTBE	S	SVE	FS	NP	NP	NP	Ongoing

CONTAMINANTS: 1,2-DCA = 1,2-dichloroethene ("EDC"); B = Benzene; BTEX = Benzene, Toluene, Ethylbenzene, Xylenes; DIPE = Diisopropyl Ether; EB = Ethylbenzene; EtOH = Ethanol; IsoBE = Isobutylene; MTBE = Methyl tertiary Butyl Ether; PCE = 1,1,2,2-Tetrachloroethene (Perchloroethene); Sty = Styrene; T = Toluene; TAME = Tertiary Amyl Methyl Ether; TBA = Tertiary Butyl Alcohol; TCA = Trichloroethane; TCE = 1,1,2-Trichloroethene); TPH = Total Petroleum Hydrocarbons; TPHd = TPH diesel range; TPHg =TPH gasoline range; TPH$_{mo}$ = TPH motor oil range

TECHNOLOGIES: AOP = Advanced Oxidation Processes; ASp = Air Sparging; AStrp = Air Stripping; BV = Bioventing; CatOX = Catalytic Oxidation; DPE = Dual Phase Extraction; ESB = Ex Situ Bioremediation; EXCV = Excavation; FBRGAC = GAC Fluidized Bed Reactor; GAC = Granular Activated Carbon; ISB = In Situ Bioremediation; ISOX = In Situ Chemical Oxidation; LF = Landfilling; MNA = Monitored Natural Attenuation; ORC® = Oxygen Release Compound; PT = Pump-and-Treat; Resin = Resin Adsorption/Desorption; SVE = Soil Vapor Extraction

MATRIX: DW = Drinking Water; GW = Groundwater; S = Soil

SCALE: BS = Bench-Scale; FS = Full-Scale; PS = Pilot-Scale

MISCELLANEOUS: BDL = Below Detection Limits; Gray Shading = Case Studies with PDF Document; O&M = Operations and Maintenance; NP = Not Provided; SCF = State Compensation Fund

Table 19-2. **Study Characteristics Summary Statistics**

Property

	Matrix[1]	Study Type	Technology[2]	*In Situ*	*Ex Situ*	Contaminants[3]
GW	213					
Soil	149					
DW	7					
Combo	124					
Bench		4				
Pilot		15				
Full		225				
Bioremediation			57	53	4	
Chemical Oxidation Processes			13	10	3	
Air Stripping			11		11	
Air Sparging			108	108		
Soil Vapor Extraction			118	118		
Dual-Phase Extraction			8	8		
Pump-and-Treat			60	60	60	
Granular Activated Carbon			23		23	
Resin Adsorption			2		2	
Other			5	5		
Methyl tertiary Butyl Ether (MTBE)						244
Benzene, Toluene, Ethylbenzene, Xylenes (BTEX) (as group)						151
Tertiary Butyl Alcohol (TBA)						17
Total Petroleum Hydrocarbons (TPH)						46
Total Petroleum Hydrocarbons — gasoline range (TPHg)						26
Total Petroleum Hydrocarbons — diesel range (TPHd)						12
Total Petroleum Hydrocarbons — motor oil range (TPHmo)						1
Benzene						5
Ethylbenzene						2
Acetone						1
Isobutylene						1
Ethanol						3
Styrene						2
Diisopropyl Ether (DIPE)						1
Tertiary Amyl Methyl Ether (TAME)						3
Perchloroethene (PCE) (Tetrachloroethene)						3
Trichloroethene (TCE)						4
Trichloroethane (TCA)						1
Dichloroethane (1,2 DCA) ("EDC")						4
Naphthalene						38

[1] Most projects were ground water with about 50% soil as well.

[2] Technologies were almost all in combinations due to multiple matrices, or complementary or sequential treatments.

[3] All sites had MTBE but many case studies had multiple contaminants.

```
┌─────────────────────────────────────────────────────────────────────┐
│ ┌─────────────────────────────────────────────────────────────────┐ │
│ │                        MTBE Case Study                          │ │
│ │                  Air Sparging/Soil Vapor Extraction             │ │
│ │                       No Where, Some State                      │ │
│ │                                                                 │ │
│ │   Site Name:                                                    │ │
│ │   Site Location:                                                │ │
│ │   Contaminants:                                                 │ │
│ │   Media:                                                        │ │
│ │   Technology:                                                   │ │
│ │ . Technology Scale:                                             │ │
│ │   Type of Cleanup:                                              │ │
│ │   Period of Operation:                                          │ │
│ │   State Contact:                                                │ │
│ │   Contractor:                                                   │ │
│ │   Site History:                                                 │ │
│ │   Technology Description:                                       │ │
│ │   Technology Performance:                                       │ │
│ │   Technology Cost:                                              │ │
│ │   Observations and Lessons Learned:                             │ │
│ │   References:                                                   │ │
│ └─────────────────────────────────────────────────────────────────┘ │
└─────────────────────────────────────────────────────────────────────┘
```

Figure 19-1.   **Format for Case Study Write-Ups**

SITE SELECTION

No particular selection basis for the case studies' associated sites is provided. A review of the sites suggests that they are:

(a) High-visibility sites;
(b) Remediation studies demonstrating a variety of proprietary or pseudo-proprietary technologies;
(c) Pioneering (early) MTBE remediation locations; and
(d) Many cookie-cutter UST locations.

High visibility case studies include Port Hueneme, Vandenberg Air Force Base, Santa Monica, and Lake Tahoe, all in California. Pioneering sites include LaCrosse, Kansas and Rockaway, New Jersey, early sites where MTBE remediation occurred in drinking water. It is likely that sites included are those whose responsible parties were willing to make data available or for which the USEPA already had data.

These case study sites range east to west in the U.S. Figure 19-2 indicates those states with at least one case study in the USEPA database. However, most of the studies are clustered in a limited number of states. Figure 19-3 graphically illustrates the number of case studies in each of those states. The majority of the case studies, 173 of 244 (approximately 71%), are from Kansas, South Carolina, and California, providing an east, central, and west balance to the bulk of the case studies.

Figure 19-2. **States with Case Studies Featured in the USEPA MTBE Website**

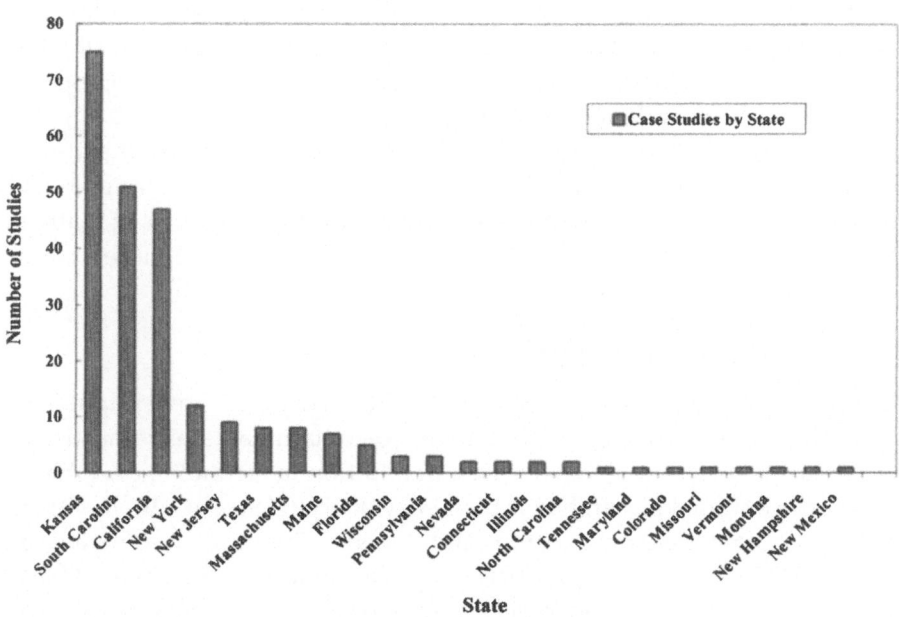

Figure 19-3. **Number of Case Studies by State**

SITE CHARACTERISTICS

Matrices treated in the case studies include soil, ground water, and drinking water. Although California drinking water issues triggered the MTBE furor, only seven of the case studies are listed as addressing impacted drinking water. Among those are the pioneering LaCrosse, Kansas and Rockaway, New Jersey sites, sites where remediations were started years before the California alarm.

Of the 244 case studies, about 84 of them are clearly remediations associated with UST leaks. Many of these UST case studies are actually multiple UST locations, essentially collations of remediations at multiple, related sites. Other case studies may also involve UST releases but may not be clear in the database.

TECHNOLOGY VARIETY

Technologies include chemical, physical, and biological treatments and *in situ* and *ex situ* applications. SVE and air sparging are the two most common remedial technologies (Table 19-2). Sometimes there is no indication of whether SVE or other off-gas management technologies were used with the air sparging. Pump-and-treat and bioremediation technologies are the next most common. Almost all of the bioremediation case studies are *in situ*. Less common are case studies using GAC, chemical oxidation, air stripping, and DPE. Table 19-1 includes the technologies used for each site, including sites where multiple technologies were used. Table 19-2 summarizes the number of times technologies were used in the 244 case studies. Since some sites used more than one technology, either in concert or for different areas of matrices, the numbers of applications of all the technologies totals are more than the number of sites in the database.

CO-CONTAMINANT VARIETY

Among the case studies, in addition to MTBE, many sites have multiple co-contaminants. In many case studies the co-contaminants are not just the BTEX expected from gasoline releases. In addition to technologies, Table 19-2 also lists contaminants and their frequency of occurrence at sites in the case studies. Co-contaminants include naphthalene, chlorinated solvents, and possible MTBE breakdown products such as acetone and TBA. These USEPA case study statistics probably reflect conditions at many of the sites of potential database users, and thus is a valuable component of the database.

TRENDS

The database studies, in aggregate, reflect some trends seen previously in the treatment of MTBE. This is very reassuring for the environmental industry, to find that technologies already well-understood can be effective at remediating MTBE. Users of this database, inexperienced with MTBE remediation, should be encouraged that they will be able to remediate their site, and probably with well-known and understood technologies proven to work for MTBE. Several of these trends in MTBE remediation are discussed below.

- *Treatability of MTBE* — The case studies reflect that a variety of long-used, well-understood technologies are effective at treating releases of gasoline with MTBE.
- *SVE and Air Sparging* — SVE and air sparging are the two most common cited treatment technologies among the 244 case studies. SVE is used for *in situ* removal of MTBE and other contaminants from the vadose zone in 118 case studies. Air sparging is used for *in situ* removal of MTBE and other contaminants from the saturated zone in 108 case studies with or without SVE or other mechanisms for capturing the off-gas.
- *Bioremediation* — Bioremediation is cited as a treatment technology in 57 of the 244 case studies. While MTBE was held by some to be biorecalcitrant, many of these case studies (as well as other studies) show that MTBE is biodegradable.
- *Bioaugmentation* — Bioaugmentation was used in at least 16 of the USEPA case studies and possibly in several others. Table 19-3 summarizes the database information concerning these bioaugmentation case studies. Products used range from axenic cultures to microbial consortia, and even enzyme preparations. The bioaugmentation trend is periodically repeated when a compound emerges as recalcitrant contaminant. Degraders of the recalcitrant compound are discovered and attempts to bioaugment follow. Bioaugmentation with MTBE degraders can be an attempt to overcome the anticipated slow degradation of MTBE with biomass additions or can be a search for a better mousetrap.

SUMMARY

The USEPA database is a useful starting point for a remediation project manager trying to develop an approach to remediating MTBE as part of a mixed plume. The user needs to assess what probably made the technology and approach successful at the case study site and determine if those same factors apply to the user's site. The user should not expect the database to have details for designing the remediation. The USEPA database provides contacts for acquiring more detailed information on the specific projects. Generally, the database should be used as a guide to screening which technologies and techniques need to be investigated in more detail for application to the user's site.

Table 19-3. Bioaugmentation in USEPA Case Studies

	Site	Scale	Status	Bioaugmentation Type	Placement	Outcome
1	Brooksville, FL	Full	C	Proprietary Microbial Mixture	Deep Injection (method not specified)	Achieved 5 µg/l goal
7	Lawrence Livermore Laboratory, CA	Bench	C	Various Cultures including PEL-Pg, PEL-CON, M. vaccae JOB5, BioPetro, and Whittier Joint Water Pollution Control Biofilter Culture	Laboratory Inoculations	Only M. vaccae JOB5 showed ability to rapidly degrade MTBE. Biofilter culture was slow degrader.
21	South Beach Marina Hilton Head, SC	Full	O	Liquid microbial suspension reportedly consisting of Pseudomonas, Bacillus, and Corynebacterium	1 Round of Geoprobe® and 4 Rounds of Well Injections	MTBE, BTE and naphthalene not at goal, xylene met goal
22	Unnamed Site, SC	Full	C	Slurry Mix of Facultative Microbes	"Pressure Injection"	Achieved 3 µg/l goal
23	Sparks Solvent/ Fuel Superfund Site, NV	Full	C	FBR Inoculated from South Carolina FBR, (probably mixed inoculum)	20 Gallon Addition to Reactor	Inoculated reactor developed greater % MTBE removal earlier but uninoculated reactor developed as well with approximately 1-month lag
25	Port Hueneme, CA	Field Pilot	C	MC-100 (old BC-4)	Injection	Moderate rate of degradation from 1 mg/l to <1 µg/l
26	Vandenberg AFB, CA	Field Pilot	P	Single Organism, Culture PM-1	Well Injection	No results available in case study report
31	Laboratory, MA	Bench	C	Butane Oxidizing Organism	Laboratory Inoculations	<1 µg/l MTBE achieved
32	Chattanooga, TN	Full	O	Enzyme-Catalyzed In Situ Dissolved Oxygen Treatment (DO-IT) Proprietary Multi-Enzyme Complexes and Consortium of Bacteria	Vertical and Horizontal Injection Wells	Claim of >70% reduction of MTBE, BTEX and TPH but not at goals
126	Ronan, MT	Full	O	Aerobic PM-1 Bacteria	Horizontal Subsurface Fractures	No results available in case study report
127	Oxnard, CA	Field Pilot	C	MC-100 or SC-100	Barrier	Reduce MTBE to 1-25 µg/l in ground water
128	Oxnard, CA	Field Pilot	O	Strain PM-1	Not Available	<10-200 µg/l MTBE

138	Service Station, CA	Full	O	Mixed Culture of MTBE-Degrading Microbes	"Biobarrier"	No results available in case study report
139	Service Station, CT	Full	C	Mixed culture of MTBE-Degrading Microbes	"Biobarrier"	Reduce MTBE from 100,000 µg/l to 1,100 µg/l
149	Service Station, NJ	Field Pilot	O	Strain ENV425	Injection through Air Sparging Wells	No results available in case study report
150	Site PA-B, PA	Field Pilot	O	Hydrogenophaga flava (ENV735)	Injection Using a Modified Push-Pull Approach	No results available in case study report

Status is for time of Case Study Report; this status may have changed in the intervening time.
P = proposed; C = completed; O = ongoing

REFERENCES

USEPA (U.S. Environmental Protection Agency). 2002. *MTBE Treatment Case Study website.* http://www.ttclients.com/mtbe/summary_table.htm. Last accessed January 2002.

CHAPTER 20

Remediation of Releases Containing MTBE at Gasoline Station Sites—ENSR International's Experience

Robert M. Cataldo, P.G., LSP, ENSR International

This chapter summarizes ENSR's national and international experience remediating MTBE and other gasoline constituents in soil and ground water at retail gasoline station sites. ENSR has extensive experience in this area, having conducted work in over 22 states and several foreign countries as a result of numerous long-standing contracts with petroleum distribution companies and convenience store owners.

When remediation began at sites impacted by gasoline containing MTBE, there was the general impression that the recovery and treatment of MTBE was either not possible, or would require specialized technologies and strategies that were different than those used for other gasoline constituents such as BTEX. In addition, it was also the prevailing belief that the presence of MTBE significantly influenced the final remediation strategies, and the specialized technologies would add substantial costs to the overall cleanup budget and significant time to the ultimate remediation schedule. In fact, it has been ENSR's experience that, although MTBE might alter some of the above parameters, recovery and treatment options, costs, and timetables are more typically a function of site-specific conditions. Over the past 10 years, ENSR has routinely addressed gasoline releases using long-proven conventional technologies and the same array of cleanup strategies, whether or not MTBE is present in the gasoline released.

This chapter describes the recovery and removal technologies currently used by ENSR and summarizes ENSR's experience applying these technologies at sites impacted by MTBE-blended gasoline releases. Cost information is summarized, and lessons learned are shared. Finally, some thoughts as to the future of gasoline station remediation are presented. In addition, Chapters 21 through 23 describe three case studies in detail.

WHY MTBE MAKES A DIFFERENCE AND HOW DO WE EXPLOIT ITS PROPERTIES FOR REMEDIATION?

MTBE and other gasoline constituents are relatively soluble in ground water, and therefore, mobile in the environment. For this reason, prompt release detection and source control are essential in reducing both the time of, and ultimate costs for, remediation. Although MTBE is readily soluble in water, it has a low Henry's Law constant, which allows MTBE that is dissolved in water to remain in the aqueous phase. MTBE also has a high vapor pressure, and prior to its dissolving in water, it volatilizes readily from the liquid phase to the vapor phase.

MTBE has a lower tendency than BTEX to adsorb to soil particles; therefore, it moves more readily through the subsurface. ENSR has observed that the greater the percentage of silts and clays present in the soil matrix, the greater the separation or difference in retardation factor between MTBE and the other constituents. Therefore, when gasoline is released, the soil can act as a natural strainer, causing an increase in the relative ratio of MTBE to BTEX in downgradient ground water with increasing distance from the source.

The same physical properties that make MTBE more mobile in the subsurface can also make it more easily recovered, compared to other gasoline constituents, if early detection and remediation occur.

REMEDIATION TECHNOLOGIES

Remediation technologies for the treatment of gasoline with MTBE can generally be divided into two phases, the recovery phase and the treatment phase. In the recovery phase, the physical characteristics of MTBE can most effectively be used to advantage in its removal from the environment.

Recovery of MTBE in Soil.   When dealing with a gasoline release in soil, MTBE's relatively higher vapor pressure causes it to be more responsive to air-based removal technologies, such as SVE, and therefore, more readily recoverable before the released gasoline has had an opportunity to enter the ground water. With the advent of smaller and less complex recovery equipment that can be mounted on trailers and mobilized to a site within hours of release detection, soil vapor points (either drilled or driven into the ground) can be quickly manifolded aboveground and the collection of vapors initiated. In addition to allowing for quick and easy access to the released gasoline vapors, the early usage of negative pressure with SVE also works to slow the downward migration of gasoline as an LNAPL through the soil. If the LNAPL does encounter ground water due to a high water table, the negative pressure recovery technology can also provide some volatilization or stripping of the gasoline constituents, including MTBE, from the upper portions of the saturated zone and capillary fringe.

Recovery of MTBE in Ground Water. As discussed above, MTBE has greater propensity to enter the ground water than other gasoline constituents. Once in ground water, gasoline constituents including MTBE will tend to migrate at or near the velocity of the natural ground water movement, although there will be some natural retardation imposed by adsorption to soil. In areas where the sedimentary conditions are defined by finer grained silts and clays, ground water movement can be as slow as 0.3 meters (1 foot) per year. However, in areas of well-sorted sands or shallow, fractured bedrock, the relatively high ground water velocity punctuates the need to address releases in a timely manner.

The high solubility and low adsorption of MTBE make ground water extraction a very successful technology for the efficient removal of large quantities of MTBE from the environment. Ground water extraction is often conducted by using existing tank pit observation wells or on-site drinking water wells already impacted by MTBE. Mobile technology similar to that used for SVE can also be employed for quick installation of additional ground water withdrawal points within the gasoline plume.

TREATMENT OF MTBE

Treating MTBE, once it is recovered from the subsurface can be accomplished with a number of proven technologies. Its low tendency to adsorb and low Henry's Law constant make MTBE more difficult to treat using GAC adsorption or air stripping, relative to the other gasoline constituents, although these technologies are still applicable under the right circumstances and widely used by ENSR at gasoline station sites. Use of GAC is limited to situations with relatively low MTBE concentrations or where MTBE is the predominant contaminant of concern. Air stripping of MTBE typically requires higher air to water ratios than other gasoline constituents. Other methods, such as thermal or catalytic oxidation, are equally effective for MTBE and other gasoline constituents. ENSR has also had success with oxygen release compound (ORC®) and bioaugmentation.

DRIVING FORCES TO SITE REMEDIATION

The technology chosen to deal with the release of gasoline containing MTBE is based largely on site-specific conditions that often change over the lifespan of a project. Receptor protection and source control are the first tasks once a release of gasoline is detected; however, the technologies that are successful in quickly addressing these needs may not be the most efficient or cost-effective methods as time progresses and site conditions change.

ENSR has found that one of the more rapid and successful methods in the isolation of the contaminant source is the excavation and removal of heavily impacted soil in tandem with ground water extraction from an open trench or pit. The removal of material in the area of the release prevents the soil from acting as a continuing source of gasoline constituents to ground water. Once the immediate objectives of receptor protection and source control have been achieved, further assessment is usually needed in order to determine if all the residual gasoline has been removed from the environment, or if additional remediation and a change of technologies will be required.

TECHNOLOGY SEQUENCING

Based on the particular site conditions (*e.g.*, geology, depth to ground water, contaminant concentrations, and proximity to receptors), site remediation can consist of several different recovery and treatment technologies carried out in sequence over the duration of the project. Major factors that determine if and when sequencing is necessary are the efficiency with which the contaminants are being removed and treated, and the cost-effectiveness of continuing to utilize an existing technology.

ENSR often changes technologies for the treatment of residual contamination as concentrations decrease over time. The presence of high concentrations of gasoline constituents in either the soil or ground water may require very aggressive removal methods utilizing high volume vapor and ground water removal rates accompanied by complex treatment technologies (such as catalytic/thermal oxidation) that can efficiently and effectively deal with the high concentrations of recovered contaminants. However, as the areal extent of the gasoline plume decreases and the residual concentrations decline, high operation costs combined with the high maintenance requirements of complex equipment call for the reevaluation of the remediation strategy. Sequencing after catalytic/thermal oxidation may include the use of GAC or air stripping, biological/oxygen enhanced treatment, and eventually, MNA. Regardless of the remediation strategy, the important point is to monitor the effectiveness of the recovery and treatment technology and have milestones where current activities are reviewed and progress is evaluated.

ENSR'S EXPERIENCE REMEDIATING MTBE

ENSR's project experience includes over 1,000 gasoline/convenience store sites, with tasks ranging from preliminary site assessment through full remediation and closure. ENSR has conducted remediation activities at many sites where gasoline containing MTBE has been released. Information on 70 of these sites is summarized in Table 20-1. The majority of these sites have received official regulatory closure; the remainder are anticipated to be closed in the near future after several additional rounds of post-remediation monitoring or other relatively minor pre-closure activities are completed. Our experience with current ongoing remediation projects at gasoline station sites is consistent with that presented in Table 20-1.

Remediation typically involved some combination of immediate source control, such as soil excavation and/or separate-phase product removal (38 sites), and (if necessary) follow-up remediation of residual product and dissolved contamination in the subsurface by ground water pump-and-treat (31 sites), SVE (26 sites), and/or air sparging (5 sites). It should be noted that source removal activities may have occurred at some sites prior to ENSR starting work on these sites; information pertaining to prior consultant activities was not always available. Recovered water and vapors were treated most often by GAC adsorption (32 sites). Air stripping was used to treat recovered water at a total of 8 sites, 3 of which were in conjunction with GAC adsorption. Catalytic/thermal oxidation was used at 7 sites to treat recovered vapors. MNA was used as a final remediation step at 20 sites and was the only technology necessary to accomplish closure at 7 of the sites.

Soil disposal from the majority of the 34 sites where excavation was conducted was accomplished by recycling the material at an asphalt batching plant. Disposal of soil via this method is considered to be highly desirable over other soil disposal options (*e.g.*, landfill cover, untreated re-use at lower risk sites). This is because the owner or site operator obtains a destruction certificate (informally called a "death certificate") from the batching facility that typically precludes the possibility of the soil becoming considered part of another impacted site. Typically when soil was not sent to an asphalt batching plant, the reasons were either the presence of clay and other fine-textured materials or cold weather, as plants typically close during the winter months in the northern states.

SITE-SPECIFIC CONDITIONS

MTBE, the most important factors in the selection of the treatment technology for recovered ground water or vapors are site-specific conditions and the need to protect human or environmental receptors. Some of the parameters that are critical in the design and implementation of a successful remediation technology are as follows:

- Nature and extent of contamination in all media;
- Ground water depth and velocity or flow rate;
- Soil permeability and heterogeneity;
- Location of known or potential drinking water sources (either surface or subsurface);
- Location of structures, buildings, and utilities;
- Ongoing activities (environmental, operational) at the site;
- Location of surface water bodies; and
- Regulatory closure requirements.

Sometimes these site-specific factors become more critical than the actual

Table 20-1.　ENSR's Experience Remediating Gasoline Containing MTBE

Location		Recovery Methods					Treatment Methods					
Town	State or Country	Groundwater Recovery	Soil Vapor Extraction	Air Sparging	Soil Removal	Product Removal	Catalytic Oxidation	Granular Activated Carbon Adsorption	Air Stripping	Soil Recycling (Asphalt Batching)	In Situ Oxygen Amendment	Monitored Natural Attenuation
Anchorage	AK				•							
Anchorage	AK				•							
Eagle River	AK			•								
Watsonville	CA	•										•
Enfield	CT	•	•		•		•	•				
Christiana	DE	•	•		•		•	•		•		
Newport	DE							•				•
Oakforest	IL				•							
Griffiths	IN	•	•	•			•	•				
Confidential	MA	•	•					•				
Confidiential	MA				•			•		•		
Amherst	MA				•					•		
Avon	MA	•	•					•				
Beverly	MA			•	•		•	•		•		
Bridgewater	MA	•	•									
Brighton	MA				•			•		•		
Carlisle	MA	•			•	•	•	•	•	•		
Chatham	MA	•			•			•		•		
Clinton	MA	•			•					•	•	
Dudley—1	MA				•					•		•
Dudley—2	MA				•					•		•

Holbrook	MA
Hudson	MA
Medford	MA
Needham	MA
New Seabury	MA
North Andover	MA
Plymouth	MA
Raynham	MA
Reading	MA
Sandwich	MA
Swampscott	MA
Wakefield	MA
Worcester	MA
Yarmouth	MA
Baltimore	MD
Liberty Town	MD
Severna Park	MD
Windham	ME
Confidential	NH
Exeter	NH
Hampton	NH
Nashua	NH
Hackensack	NJ
Edison	NJ
Caldwell	NJ
Binghamton	NY
Buffalo	NY
Poughkeepsie	NY
Wappingers Falls—1	NY
Wappingers Falls—2	NY
Allentown	PA
Drexel Hill	PA
Hummelstown	PA

Table 20-1. Continued

| Location | | Recovery Methods | | | | | Treatment Methods | | | | | |
Town	State or Country	Groundwater Recovery	Soil Vapor Extraction	Air Sparging	Soil Removal	Product Removal	Catalytic Oxidation	Granular Activated Carbon Adsorption	Air Stripping	Soil Recycling (Asphalt Batching)	In Situ Oxygen Amendment	Monitored Natural Attenuation
Newport	RI	•						•	•			
South Kingston	RI				•			•		•	•	
Woonsocket	RI		•				•	•				
Houston	TX				•							
Kenosha	WI											
Milwaukee	WI											
Marlowe	WV	•	•					•				
Shepardstown	WV	•	•					•			•	
Bypass	England											
Tile Cross	England				•							
Anagni	Italy		•									
Chianciano	Italy		•								•	
Cinisello	Italy	•	•									
Solanas	Italy	•	•									
Villadossola	Italy	•	•									
SS Jaya	Malaysia				•							
Totals	70	31	26	5	34	4	7	32	8	19	6	20

volume of gasoline released, as they affect the determination of whether an immediate response is necessary or a more long-term solution is warranted. An example would be the release of a large volume of gasoline from a failed line or UST, where the leak is limited to the area within the tank grave. If the tank grave were situated in a dense silt or clay, product migration would be minimal. In this situation, product removal could be highly effective using existing tank pit observation wells. However, in a similar scenario where the tank grave has been blasted into fractured bedrock with a high water table, rapid migration is typical and immediate actions may be required, such as extraction from off-site, deep recovery wells.

REMEDIATION SELECTION FACTORS

After the site-specific considerations have been evaluated, the next most important driving force is the volume or concentration of the gasoline present within the environment. At sites where initial total VOC concentrations exceed approximately 10,000 µg/l, ENSR would consider employing ground water extraction and air stripping followed by catalytic oxidation of the stripped vapor-phase VOCs. VOC concentrations of this magnitude contribute enough British Thermal Units (BTUs) to the catalytic oxidation process to preclude the need for large amounts of fuel, making this technology economical. In the 1,000 to 10,000 µg/l range of total VOCs, air stripping followed by GAC adsorption is often the best approach. For VOC concentrations less than about 1,000 µg/l, GAC adsorption or MNA *in situ* are appropriate. In cases where MTBE comprises greater than 50% of the total VOC concentration, the above guideline numbers would be reduced by roughly 25%.

For vapor-phase treatment, catalytic/thermal oxidation is often favored for total VOC concentrations greater than 75 ppm and GAC adsorption is often applied for lower concentrations. Again, in cases where MTBE is the major constituent being treated, the above number can be reduced by roughly 25%.

ENSR has recently been using more biologically-based technologies, such as *in situ* oxygen addition by means of solid- or liquid-phase oxygen-containing compounds (at six sites in Table 20-1). This technology has worked well at some sites and can offer less intrusive methods to meet cleanup objectives. However, it should be noted that these technologies may not be suitable for some sites, such as when a shallow water table is being used for drinking water. Other cases where ENSR has found these technologies less effective is when there is LNAPL, very high VOC concentrations, or high concentrations of non-target compounds (*e.g.*, iron or TOC). In addition, we have had limited success at sites where there has been incomplete source control (unidentified continued gasoline source) as well as at sites with low soil permeability.

REMEDIATION COSTS

Costs to conduct a successful remediation strategy are greatly influenced by site-specific conditions and receptor protection. Also, the final cost-to-closure numbers generally increase the more time it takes to detect and respond to the release. At some sites, MTBE drives the remediation strategy due to high concentrations; however, at many of the sites, benzene is the chief driving force for the selected type and duration of remediation activities due to its higher toxicity and more stringent cleanup requirements.

The impact of MTBE on remediation cost is likewise site-specific. In many cases, there is little or no impact to costs. Generally, releases that are limited to impacted soil cost on the order of $100,000 to remediate (on average). By comparison, releases that have impacted both soil and ground water may cost on the order of $250,000 to remediate (on average). The most expensive gasoline station remediation to our knowledge was approximately $4 million (MTBE was not the primary constituent driving the cleanup). At sites where the MTBE plume is larger than the BTEX plume, remediation costs can be higher due to the presence of MTBE. ENSR has found that such sites are often in areas of relatively shallow bedrock or where soil permeability is low and the silt content is high.

FUTURE TRENDS IN REMEDIATION

While ENSR continues to evaluate other technologies for remediating releases of gasoline containing MTBE, we will continue to use proven physical treatment technologies such as ground water pump-and-treat, SVE, and air sparging. ENSR has also had success using *in situ* oxygen amendment and anticipates using this and other biological methods more extensively in the future. In addition, ENSR is increasingly using MNA as a final polishing step.

At the end of 2000/beginning of 2001, ENSR started using bioaugmentation at three sites with MTBE. Although early indications appear positive at two of the sites, it is too early to speculate on its long-term effectiveness. At the third site, a gasoline station/convenience store in Connecticut, BTEX and MTBE concentrations had remained fairly stable throughout seven years of assessment and pump-and-treat/SVE remediation, and MNA was not considered a viable option. ENSR began injections of cultured aerobic microorganisms in late 2000 and observed BTEX concentrations fall from 10,000 to 12,000 µg/l to less than 100 µg/l, and MTBE concentrations fall from 1,000 to 2,000 µg/l to less than 500 µg/l over a two month period. The most recent sampling event conducted in October 2001 shows a continuation of these lower concentrations.

COMPLIANCE, EARLY DETECTION, AND QUICK RESPONSE

Although not a remediation technology, sometimes the best offense in dealing with the complexities and costs of gasoline releases is a good defense.

This would include spill prevention and early detection of releases. Over the past several years, ENSR has become more involved in the compliance aspects of our clients' operations of gasoline stations, and in addition to conducting environmental assessments, we have also performed cursory compliance inspections of sites. At a gasoline site, these compliance inspections can be as simple as opening up tank manholes, fill ports, and dispenser housings, or simply asking the store operator if they have had any inventory fluctuations or problems with equipment. Because of the demand on operators to deal with in-store customers and the advent of self-service pump and pay-at-the-pump dispensers, store personnel typically do not continually monitor the operation of the UST system.

In addition to helping site operators stay in compliance, these cursory site inspections have been instrumental at many sites in the early detection of small leaks and spills. Although the time of the release may not be known, the duration of the release in many cases can be timed to the prior inspection. The inspections not only provide early detection, but limit the duration of the release. In addition, they allow for quick response to initiate source control and begin remediation. At several recent sites situated on highly permeable, sandy aquifers, early detection and quick response have been responsible for addressing the release of gasoline containing MTBE to weeks, instead of months or years. This approach has prevented the release from extending off-site and has limited (or will limit) the costs for remediation to probably less than $100,000, compared to $250,000 or more for a more typical case.

CONCLUSIONS

The main lesson from ENSR's remediation experience with gasoline stations is the importance of release prevention, early detection, and prompt source identification and control in order to minimize total site remediation costs. In our opinion, gasoline station owners are increasingly realizing that it is in their best interest to focus more on these activities and less on post-release alternatives. Smaller plumes of gasoline constituents result from early detection and prompt control. In addition, *in situ* bioremediation may come to be more widely used at gasoline station sites because of its relatively quick application and non-intrusive nature.

ENSR has found that during ongoing remediation, changing or sequencing technologies becomes more efficient and cost-effective for the treatment of residual and dissolved contamination as concentrations decrease with time. For example, it is common to start with catalytic/thermal oxidation and then switch to GAC adsorption once the total VOC concentrations become low enough to make GAC economical.

As states become more aware of the costs of remediation and the actual risks (or lack thereof) posed by the presence of gasoline, risk-based corrective actions (RBCA) have become a driving mechanism in the need for and duration of remediation technologies. For these reasons, MNA has become a more acceptable application for the final "polishing" step and in some cases, the only remediation technology needed to achieve closure at a site

Bioremediation of MTBE is becoming a more widely accepted technology. Like chlorinated solvents, MTBE was initially considered to be recalcitrant to biological treatment; however, research and field data continue to demonstrate that MTBE is biodegradable under a variety of conditions, both *in situ* and *ex situ*. Bioremediation technology is evolving rapidly, and ENSR anticipates that bioremediation will become more prevalent in the future to treat gasoline releases containing MTBE.

CHAPTER 21

Source Control and Point of Entry Treatment
at a Massachusetts Site

Christopher G. Mariano, P.G., LSP, ENSR International

INTRODUCTION

A release of gasoline to the environment occurred as a result of historical operation of a gasoline dispensing facility. The release was discovered upon removal of the UST system, when the owner of the USTs ceased operation of the facility. The site is located in an interim wellhead protection area for a PWS well and in close proximity to numerous private water supply wells. Assessment activities identified soil impacts proximate to the gasoline UST system and ground water impacts in both the overburden and fractured bedrock. Remediation consisted of source removal and point of entry treatment to eliminate human exposure through contaminated water supplies. In some cases where bedrock ground water is impacted, the best approach is to actively address the source and the receptors, relying on natural attenuation, if possible, to address the bedrock ground water located between the source and the receptors.

SITE DESCRIPTION

The facility consists of a 0.37-hectare (0.92-acre) lot, located in a commercial and residential area in Massachusetts. Between 1964 and May 1996, the facility had intermittently been used as both a gasoline dispensing and automobile repair facility. In 1998, the facility was redeveloped and has since been used for gasoline dispensing only. In May 1996, three 15,000-liter (4,000-gallon) steel gasoline USTs, one 23,000-liter (6,000-gallon) steel gasoline UST, and associated product piping and dispensers were removed from the facility.

Topography at the site is generally flat with the exception of a steep incline along the southern portion of the property and a gentle decline that leads to the drainage swale along the western property line.

All of the properties surrounding the site are residential with the exception of a retail property located approximately 90 meters (300 feet) west of the site. The site and surrounding residences are serviced by private drinking water wells and septic systems. Several of the private water supply wells are located less than 150 meters (500 feet) from the site. The site is also located approximately 75 meters (250 feet) from a PWS well that serves a neighboring commercial property (Figure 21-1). The majority of the area water supply wells, including the on-site well and the nearby public supply well, are bedrock wells; however, some of the nearby residential wells are overburden wells.

RELEASE HISTORY

In May 1996, the four gasoline USTs and associated piping and dispensers were excavated and removed from the site. Soil samples collected from the UST area were screened for volatile organic vapors with a portable PID and exhibited jar headspace readings of more than 1,000 ppm, indicating that a gasoline release to the environment had occurred. The likely potential sources of gasoline impacts identified at the site include the former gasoline UST system as well as historical use of the facility for dispensing gasoline.

Petroleum-impacted soil was excavated and removed from the UST and dispenser island areas at the time of the UST system removal. Follow-up response actions included installation of several monitoring wells both on and downgradient of the facility, soil and ground water contamination characterization and delineation, determination of hydrogeologic parameters to evaluate contaminant fate and transport, and periodic sampling of several area water supply wells. In February 1998, a new gasoline dispensing facility was constructed and additional petroleum-impacted soil was excavated from the area of the former dispenser.

SITE HYDROGEOLOGY

Information regarding the site hydrogeology was obtained based on subsurface investigations completed at the site and published information as referenced herein.

Surficial Geology. The naturally occurring soils observed beneath the facility consisted of tan, fine- to coarse-grained sand with some fine- to coarse-grained gravel, indicative of glaciofluvial and/or glacial outwash deposits. A silty fine sand present from depths of 0.3 to 1.5 meters (1 to 5 feet) was likely artificial fill.

Bedrock Geology. According to the Bedrock Geologic Map of Massachusetts, the bedrock underlying the site is part of the Nashoba Formation (Ordovician or Proterozoic Age), which consists of sillimanite schist and gneiss;

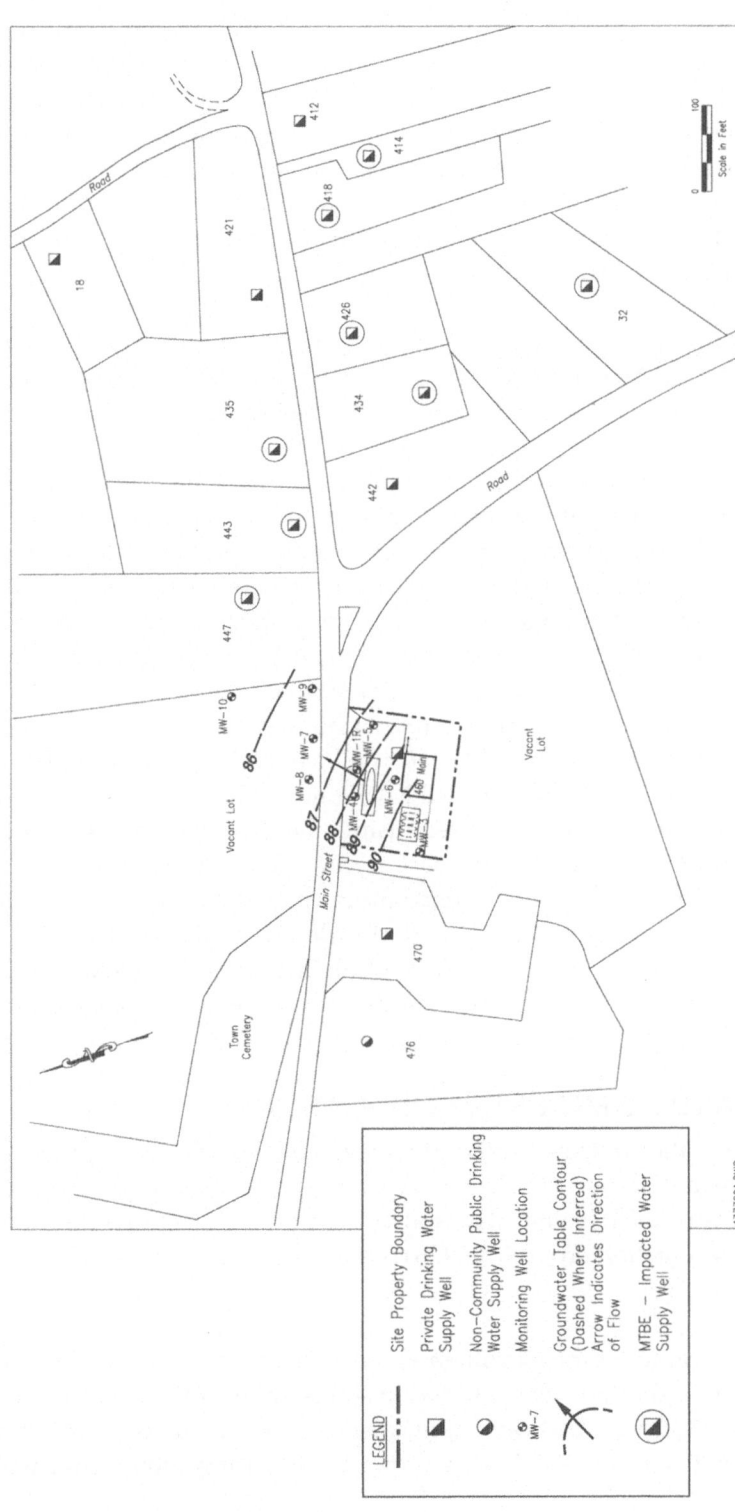

Figure 21-1.   Site Plan and Area Water Supply Wells

partly sulfidic, amphibolic, biotite gneiss; calc-silicate gneiss; and marble
(Zen, 1983). Bedrock outcrops are located near the southwestern site border.
As observed during subsurface investigations, the depth to bedrock generally declines across the site in an easterly direction, ranging from approximately 1.5 meters (5 feet) bgs adjacent to the southwestern side of the site
building to greater than 5 meters (16 feet) bgs along the northeastern site border. Bedrock, as observed in excavations and borings, appears competent and
exhibits minimal surface fractures.

Hydrogeological Parameters.   Depth to ground water in the site monitoring wells ranged from approximately 2.5 meters (8 feet) bgs (west of the UST
pads) to 5 meters (16 feet) bgs (north of the facility, in the vacant lot across
Main Street).

Ground water flow direction in the overburden aquifer was determined
using a modified straight-line interpolation method. Generally, ground water
flow in the overburden is to the northeast with an approximate horizontal hydraulic gradient of 0.033 in the southern side of Main Street and a shallower
horizontal hydraulic gradient of approximately 0.011 on the northern side of
Main Street (Figure 21-1). Although a vertical hydraulic gradient was not
measured, the petroleum impacts to area bedrock water supply wells indicates a downward vertical gradient (either natural or pumping-induced).

Hydraulic conductivities of 2×10^{-2} cm/s (60 ft/day) for well MW-7 and
2.6×10^{-3} cm/s (7.5 ft/day) for well MW-6 were obtained from rising head
slug tests conducted and analyzed using the Bouwer and Rice method
(Bouwer and Rice, 1976). Using these hydraulic conductivities and an assumed porosity of 25% based on field observation, the estimated average linear velocity for ground water in the overburden was calculated at 282 meters
(925 feet) per year for the north side of Main Street and 110 meters (360 feet)
per year for the south side of Main Street. Hydrogeologic characteristics of
the bedrock aquifer were not evaluated.

NATURE AND EXTENT OF CONTAMINATION

The contamination detected at the site consisted of gasoline constituents, and
the source was determined to be the historical spillage/leakage related to the
former gasoline UST system. The source was determined based on the observed contamination around the UST system during its excavation and removal.

Soil.   Concentrations of gasoline-related constituents (*i.e.*, BTEX, MTBE,
naphthalene, and TPH) were detected in soil samples collected from the area
of the former gasoline USTs and associated dispenser island. Based on samples collected from the soil borings completed during monitoring well in-

stallation, the most highly impacted soil (9.7 mg/kg BTEX; 5.4 mg/kg MTBE; 4.5 mg/kg naphthalene; 84 mg/kg TPH) was limited to the areas immediately around and beneath the USTs and dispensers.

Ground Water. At various times throughout the monitoring period, concentrations of one or more gasoline-related constituents were detected in ground water from all site monitoring wells (all overburden wells), except well MW-9. The highest concentrations of BTEX and MTBE were detected in wells MW-4, MW-7, and MW-8, located hydraulically downgradient of the former UST system. (Refer to Figure 21-1 for monitoring well locations.) Dissolved-phase gasoline impacts extend downgradient of the former UST system across Main Street and under the vacant lot, as well as northeast of the site.

MTBE has been consistently detected in 8 of the 14 water supply wells sampled from 1995 through 2001. The highest MTBE concentrations have historically been detected in the water supply wells at 435 Main Street (up to 43 μg/l), 443 Main Street (up to 38.7 μg/l), and 447 Main Street (up to 16 μg/l). MTBE was never detected in the on-site drinking water supply well. In addition, concentrations of toluene and xylenes have been detected infrequently in samples from three residential supply wells. Ranges of BTEX and MTBE concentrations detected in site monitoring wells and area water supply wells are presented in Table 21-1.

The horizontal extent of the dissolved-phase ground water contaminant plume has been defined to below the applicable drinking water standards to the south by monitoring well MW-3; to the east by monitoring well MW-5; the north/northwest by MW-9, MW-8, and MW-10; and to the west by monitoring well MW-4. The vertical extent of the dissolved-phase ground water contaminant plume was considered to be defined to below drinking water standards based on the historical concentrations detected in several of the area bedrock water supply wells sampled as part of the investigation. With the exception of the wells at 435 Main Street and 443 Main Street, all of the water supply wells sampled as part of this investigation are bedrock wells. The depths of most of the wells are unknown.

FATE AND TRANSPORT

As previously mentioned, the source of the contamination identified on-site appeared to be the former gasoline UST system. The migration pathway of release(s) from the source area was via ground water, toward the adjacent vacant lot and residences. In addition, the contamination migrated from the overburden downward to the underlying fractured bedrock and was transported through the bedrock fracture network.

Low concentrations of dissolved petroleum-related compounds were de-

Table 21-1.   **Ground Water Concentration Ranges and**
Applicable Cleanup Standards, Former Gasoline
Dispensing Facility, Massachusetts

Sample Location	Benzene (µg/l)	Toluene (µg/l)	Ethyl-benzene (µg/l)	Xylenes (µg/l)	MTBE (µg/l)
MW-1/MW-1R	<0.5 to 1.8	<0.5	<0.5 to 2.9	<1.0 to 5.9	<2.0 to 102
MW-2	<0.5	<0.5	<0.5	<20.	<5.0 to 7.6
MW-3	<0.5	<0.5	<0.5	<1.0	<1.0 to 11.4
MW-4	<1.0 to 5.8	<1.0 to 2.0	<1.0 to 9.4	<4.0 to 31	<1.0 to 910
MW-5	<1.0	<1.0	<1.0	<1.0	<1.0 to 31
MW-6	<2.0 to 12.0	<2.0 to 10.0	<2.0 to 41	<4.0 to 168	<1.0 to 29
MW-7	<1.0 to 13.6	<1.0 to 3.1	<0.5	<1.0 to 3.0	<1.0 to 627
MW-8	<1.0	<1.0	<1.0	<1.0	<1.0 to 234
MW-10	<2.0	<2.0	<2.0	<4.0	<2.0 to 12.0
414 Main Street	<0.5	<0.5	<0.5	<1.0	<0.5 to 5.8
418 Main Street	<0.5	<0.5	<0.5	<1.0	<0.6 to 8.5
426 Main Street	<0.5	<0.5	<0.5 to 65	<0.5	<0.5 to 7.7
434 Main Street	<0.5	<0.5	<0.5	<1.0	<0.5 to 2.2
435 Main Street	<0.5	<0.5	<0.5	<1.0 to 1.2	1.2 to 43
443 Main Street	<0.5	<0.5	<0.5	<1.0	0.8 to 38.7
447 Main Street	<0.5	<0.5	<0.5	<1.0	<0.5 to 16.0
32 Road	<0.5	<0.5	<0.5	<1.0	<0.5 to 2.8
GW-1	5	1,000	700	10,000	70
GW-2	2,000	6,000	30,000	6,000	50,000
GW-3	7,000	50,000	4,000	50,000	50,000

GW-1 = Massachusettes category GW-1 groundwater standards in µg/l.
GW-2 = Massachusettes category GW-2 groundwater standards in µg/l.
GW-3 = Massachusettes category GW-3 groundwater standards in µg/l.

tected in the ground water downgradient of the former gasoline USTs and dispensing island. Ground water in this area occurs at a depth of approximately 2.5 to 5 meters (8 to 16 feet) bgs. The private drinking water well for the facility is located cross- to downgradient of the former USTs and upgradient of the detected impacts. Low concentrations of MTBE (below the Massachusetts drinking water guideline of 70 µg/l) have been detected in eight drinking water supply wells located downgradient of the former gasoline UST system. In addition, low levels of toluene and xylenes (less than the Massachusetts drinking water standards of 1,000 µg/l and 10,000 µg/l, respectively) have been detected in samples from the water supply wells at 426 Main Street and 435 Main Street, respectively.

The well for 447 Main Street is 50 meters (165 feet) deep (depth to bedrock is approximately 18 meters [60 feet] at that location) and the wells for 435 and 443 Main Street are overburden wells of unknown depth. Well completion details for the other area supply wells were not available. Based on the presence of MTBE in the bedrock well at 447 Main Street, dissolved-phase MTBE at concentrations below drinking water standards has migrated vertically into the bedrock fractures in addition to horizontally north/northwest of the source area. No underground utilities are present in this area; therefore, underground utilities are not considered potential migration pathways for the dissolved-phase impacts detected on-site. Vertical migration of vapor-phase compounds in the area of detected impacts was determined to be insignificant due to the fact that VOCs were detected at very low concentrations in the ground water and no impacted soil was detected in proximity to occupied buildings.

RECEPTORS

The determination of ecological and human receptors was based on the site setting and contaminant concentrations, distribution, and potential migration pathways identified in the assessment.

Ecological.   No surface water bodies or wetland areas were identified at or in close proximity of the site. The vacant woodland on the north side of Main Street may be a wildlife habitat; therefore, wildlife in this area may be potential receptors.

Human.   Potential human receptors include the following:

- Construction workers performing excavation in the vicinity of the current/former gasoline UST system;
- Consumers of drinking water from the on-site and surrounding drinking water supply wells; and
- Occupants of the gasoline dispensing facility building and nearby residences exposed to potential petroleum vapor accumulations in buildings.

EXPOSURE POTENTIAL

The determination of the potential for exposure to contamination from the site was based on the results of the site assessment activities including the current and foreseeable future use of the site, the location and concentration of detected impacts, the identified potential migration pathways, and the exposure points identified at the site.

Ecological.   No environmental exposure potential was identified since there were no wetlands or surface water receptors identified in proximity to the site. Although the vacant lot on the north side of Main Street was identified as potential wildlife habitat, the extent of contamination in this area is limited to ground water at a depth of approximately 4.9 meters (16 feet) bgs. No surficial soil impacts were identified in this area; therefore, there was no ecological exposure potential associated with the site.

Human. Potential human exposures include inhalation or dermal contact by construction workers performing excavations at the site and ingestion via impacted residential water supply wells. The occupants of the site building and area residences were considered potential receptors of vapor accumulations from volatilization of the ground water contaminant plume. The petroleum concentrations in the ground water are well below vapor-related and drinking water risk levels; therefore, exposure potential is considered insignificant.

REQUIRED CLEANUP LEVELS AND TIMEFRAMES

The Massachusetts Contingency Plan (MCP) (MADEP, 1999) addresses site risk characterization similar to the American Society for Testing and Materials (ASTM) RBCA process, whereby appropriate cleanup levels are developed using a three-tiered approach. Method 1 involves comparison of concentrations in site soil and ground water to generic risk-based screening levels. Method 2 allows the determination of site-specific target levels. Method 3 is a more complex risk evaluation, which may involve additional site assessment, probabilistic evaluations, and more sophisticated fate and transport models than Method 2. Based on site conditions and proximity of potable water supply wells, the Method 1 approach was employed for the site.

Soil. Under the MCP Method 1 risk characterization, contaminated soils are divided into three categories (S-1, S-2, and S-3) based on frequency and intensity of potential exposures to children and adults. The S-1 category assumes maximum exposure potential; therefore, it has the lowest cleanup levels. In general, soils that exceed the S-1 cleanup levels require a property deed restriction limiting future property use in order to maintain a level of acceptable risk. Therefore, although the site soils were categorized as S-2 and S-3, the S-1 cleanup levels applicable to drinking water areas (10 mg/kg benzene; 90 mg/kg toluene; 80 mg/kg ethylbenzene; 500 mg/kg xylenes; 0.3 mg/kg MTBE; 200 mg/kg TPH) were used in order to prevent the need for a deed restriction to prevent unacceptable exposure and to remove an ongoing source of contaminants to ground water.

Ground Water. The MCP Method 1 risk standards for ground water include three categories (GW-1, GW-2, and GW-3) based on receptor and exposure route. The GW-1 and GW-2 standards consider human exposures through ingestion and vapor inhalation, respectively, and the GW-3 standards address environmental risk to surface water. Based on the site setting and potential receptors, the cleanup standards associated with all three ground water categories, as shown in Table 21-1, apply to the site.

Cleanup Timeframe. Pursuant to the MCP, compliance with applicable cleanup goals is required within six years of notifying the regulatory agency of the release. However, where it is recognized that this timeframe is not always feasible, the MCP does allow an open-ended timeframe, provided the remedial alternative selected to achieve the cleanup goals is implemented within four years of release notification.

REMEDIAL ACTIONS

Remedial actions conducted at the site consisted initially of source removal, followed by long-term point of entry treatment of several of the impacted water supply wells.

Source Removal. As stated previously, the site contamination was first detected upon removal of the gasoline UST system, which was determined to be the source of the soil and ground water contamination. As such, the removal of the UST system constituted primary source removal.

Concurrent with the UST system removal, a total of 480 tons of impacted soil from around the USTs and beneath the dispensers was excavated and transported off-site where the soil was recycled in an emulsified cold mix asphalt manufacturing process. Soil samples collected from the limits of the excavation around the USTs contained no BTEX, MTBE, or TPH concentrations above laboratory detection limits, while soil samples from the limits of the dispenser island excavation contained concentrations of xylenes (572 mg/kg), MTBE (5.4 mg/kg), and TPH (2,920 mg/kg) above residential (S-1/GW-1) standards.

An additional 35 tons of petroleum-impacted soil were excavated and removed from beneath the former gasoline dispensers in February 1998 during the redevelopment of the site. Laboratory analysis of soil samples collected from the limits of the excavation indicated minor concentrations of toluene, xylenes, and naphthalene, well below residential (S-1/GW-1) cleanup standards.

Point of Entry Treatment. As mentioned previously, 8 of the 14 drinking water supply wells sampled contained MTBE concentrations above laboratory detection limits, but consistently below the Massachusetts drinking water standard of 70 µg/l. All of the impacted wells were residential supply wells. As a proactive measure and at the request of the water supply well owners, the responsible party offered to install and maintain treatment systems at each of the impacted water supplies in order to eliminate exposure.

From early 1999 to 2000, water treatment systems were installed on seven of the eight impacted residential water supplies. Each system consisted of two 0.06-cubic meter (2-cubic foot) virgin GAC vessels connected in series located in the basements of the residences. The systems were monitored monthly for the first three months of operation, and quarterly thereafter. Monitoring consisted of collecting samples from the influent to the treatment system and between carbon vessels (midfluent) for laboratory analysis of VOCs by EPA Method 524.2. Midfluent samples were collected instead of effluent in order to detect and respond to contaminant breakthrough of the first GAC vessel by replacing the used vessel while the second vessel continued to treat the impacted water prior to consumption. On average, each treatment system is replaced with virgin GAC on a yearly basis, and the used carbon is regenerated off-site.

COSTS

Project costs are subdivided into five categories, as identified below. The operation and maintenance costs reflect the costs to operate and maintain the residential water supply treatment systems through the end of 2001. Note that, due to steady MTBE concentrations in the residential wells, the operation and maintenance of the treatment system is anticipated to continue for several years.

Site Assessment/Risk Characterization	$103,000
Source Removal (soil excavation, disposal, and backfill)	$49,000
Treatment System Design and Installation	$17,500
Operation and Maintenance	$52,000
Site Closure	$2,500
TOTAL PROJECT COST:	**$224,000**

TIMELINE

The project timeline spanned from May 1996 through December 2001. The timeframes associated with various major components of the project are indicated below. Pursuant to the random site audit program of the Massachusetts Department of Environmental Protection (MADEP), the site was audited in December 2001. The MADEP found the response actions, risk characterization, and site closure in compliance with the MCP and no further action was required at the site.

Release Detection	May 1996
Assessment Activities	May 1996–March 2001
Source Removal	May 1996; February 1998
Point of Entry Treatment	May 1999–Ongoing
Regulatory Closure	April 2001

REFERENCES

Bouwer, H. and Rice, R.C. 1976. A slug test for determining hydraulic conductivity of unconfined aquifers with completely or partially penetrating wells. *Water Resources Research.* 12(3).

MADEP (Massachusetts Department of Environmental Protection). 1999. *Massachusetts Contingency Plan 310 CMR 40.0000.* October 1999.

Zen, E. 1983. *Bedrock Geologic Map of Massachusetts.* United States Geological Survey prepared in cooperation with the Commonwealth of Massachusetts Department of Public Works, and Joseph A. Sinnot, State Geologist.

CHAPTER 22

Physical Treatment at a New Hampshire Site

David L. Espy, ENSR International

INTRODUCTION

In response to a sudden release of RFG from a UST in late September 1996, emergency response actions and assessment activities were completed. Ultimately, a combined ground water recovery and SVE system was installed and operated for a period of four years. As a result of source removal and physical treatment of soil and ground water, an estimated total of 5,118 kilograms (11,282 pounds) or 6,833 liters (1,805 gallons) of petroleum hydrocarbons were recovered from beneath the site. The site is currently in a natural attenuation and limited enhanced bioremediation program consisting of triannual sampling and periodic microbe and nutrient addition. This site demonstrates that conventional technologies and strategies are effective for remediating gasoline constituents, including MTBE.

SITE DESCRIPTION

The facility consists of a 0.18-hectare (0.44-acre) lot, located in a residential section of New Hampshire. The site has been used for the dispensing of gasoline since the 1970s. Prior to 1970, the site was vacant. The facility is occupied by a gasoline retail/convenience store, a gasoline UST system, and paved and landscaped areas. The UST system present at the time of the release consisted of three 30,000-liter (8,000-gallon) single-walled steel gasoline USTs and associated steel product piping.

Surficial topography at the site slopes in a southeasterly direction with a steep decline between the southwestern and southeastern borders of the facility property. All of the properties surrounding the site are occupied by residential structures with basements constructed of concrete block and/or stonewall foundations. Surficial topography further south of the facility is relatively flat with a slight slope towards a wetland area to the southeast. The site and surrounding residences are serviced by municipal water and sewer systems. No public or private water supply wells are located within a 1.6-km (1-mile) radius of the site.

RELEASE HISTORY

In September 1996, the manager of the facility detected an inventory discrepancy in the 30,000-liter (8,000-gallon) super unleaded gasoline UST. Due to the inventory discrepancy, the tank was tightness tested and it failed the test. Based on a review of the inventory records, it was estimated that a total of approximately 7,950 liters (2,100 gallons) of RFG was released from the UST in a 48-hour time period. Prior to the September 1996 release, no other releases of oil and/or hazardous materials had been documented on-site or in the site vicinity, and ground water monitoring data at the site indicated no significant prior releases.

In response to the sudden release, several investigation and remedial response actions were implemented. Response actions included the removal of the gasoline UST system and associated impacted soil; the installation of 25 monitoring wells and 20 soil borings; manual and automated removal of separate phase hydrocarbons (SPH); ground water recovery and soil venting pilot testing; the installation and operation of soil and ground water remediation systems; indoor air sampling; and venting of the basement of an adjacent residence.

SITE HYDROGEOLOGY

Information regarding the site hydrogeology was obtained based on subsurface investigations completed at the site and published information as referenced herein.

Surficial Geology. Stratigraphy at the site generally consists of sandy fill material overlying loose glacial till. The fill material consists of poorly sorted brown fine to coarse sand and gravel with some silt and cobbles to depths of approximately 1.2 to 2.4 meters (4 to 8 feet) bgs. At these depths, the sandy fill changes to a glacial till, which consists of brown fine to medium sand and some fine to medium gravel and cobbles with traces of silt to a depth of approximately 3 to 4.6 meters (10 to 15 feet) bgs. Bedrock underlies the loose glacial till.

Bedrock Geology. According to the Bedrock Geologic Map of New Hampshire (Lyons *et al.*, 1997), bedrock underlying the site consists of early to late Devonian rocks of the New Hampshire Plutonic Suite, specifically, two-mica granite of northern and southeastern New Hampshire, similar to Concord Granite. During the UST removal and drilling activities, a 0.3- to 0.6-meter (1- to 2-foot) layer of weathered bedrock was observed overlying a competent bedrock layer. Based on information obtained from the series of borings advanced across the site, it appears that bedrock generally slopes downward to the south-southeast with surface topography.

Hydrogeological Parameters. Depth to ground water in the site monitoring wells ranged from approximately 1 meter (3 feet) bgs (south of the facility, beneath the residential properties) to 3 meters (9 feet) bgs (adjacent to the UST area). In the drier summer months, the water table drops up to 1.2 meters (4 feet) and appears to approach or intersect the top of the weathered bedrock.

Ground water flow in the overburden is to the south-southeast with an approximate horizontal hydraulic gradient of 0.03. A hydraulic conductivity of 2×10^{-2} cm/s (58 ft/day) was measured from a single well obtained from rising head slug tests conducted and analyzed using the Bouwer and Rice method (Bouwer and Rice, 1976). Using an assumed porosity of 30% based on field observation, the estimated average linear velocity for ground water in the overburden was calculated at 2×10^{-3} cm/s (6 ft/day) or 600 meters/year (2,000 feet/year).

NATURE AND EXTENT OF CONTAMINATION

The contamination detected at the site consisted of gasoline constituents including BTEX, MTBE, naphthalene, TPH, and alkyl benzenes. Although the site had been used as a gasoline dispensing facility since the 1970s, the main source of contamination was determined to be a sudden release from a super unleaded gasoline UST. This was based on the fact that ground water samples collected downgradient of the UST system immediately after the release contained low concentrations of total gasoline-related VOCs, less than 500 µg/l. Within two months of the release, migration from the UST area resulted in the same monitoring point containing gasoline-related VOCs up to 10,000 µg/l.

Soil. Based on soil samples collected during the UST removal and during the numerous subsurface investigations completed at the site, the extent of vadose zone soil contamination was determined to be limited to the immediate vicinity of the former UST area. Due to seasonal fluctuations in the ground water table, unsaturated soils located in areas of historical SPH were also impacted.

Ground Water. Concentrations of one or more gasoline-related constituents were detected in ground water from all site monitoring wells at various times throughout the monitoring period with the exception of one upgradient monitoring well. The highest concentrations of BTEX (up to 120,000 µg/l) and MTBE (up to 145,000 µg/l) were detected in wells located hydraulically downgradient of the former UST system. Pre-remediation concentrations of dissolved-phase gasoline impacts extended downgradient of the former UST system across and beneath five occupied residential structures. Refer to Figures 22-1 and 22-2 for pre-remediation MTBE and BTEX contaminant plume maps.

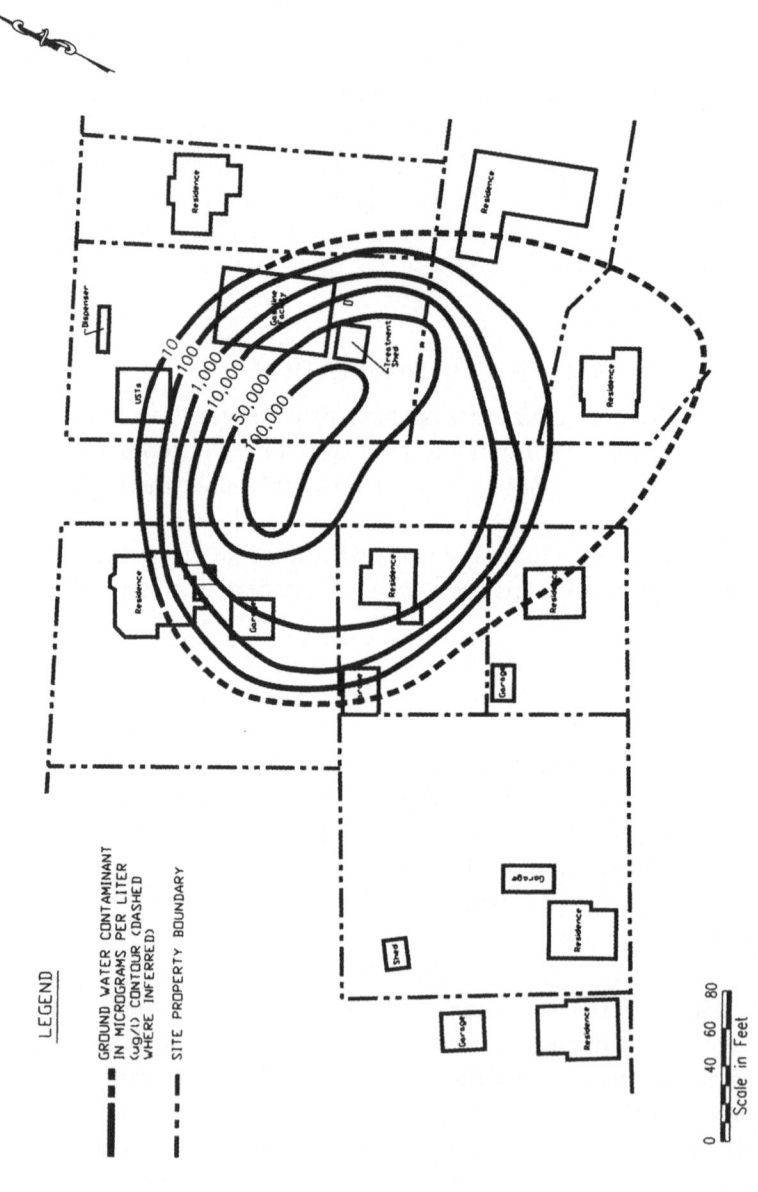

LEGEND

▬▬▬ GROUND WATER CONTAMINANT
     IN MICROGRAMS PER LITER
     (ug/l) CONTOUR (DASHED
     WHERE INFERRED)

—··—·· SITE PROPERTY BOUNDARY

0 40 60 80
Scale in Feet

Figure 22-1.   Pre-Remediation MTBE Contaminant Plume

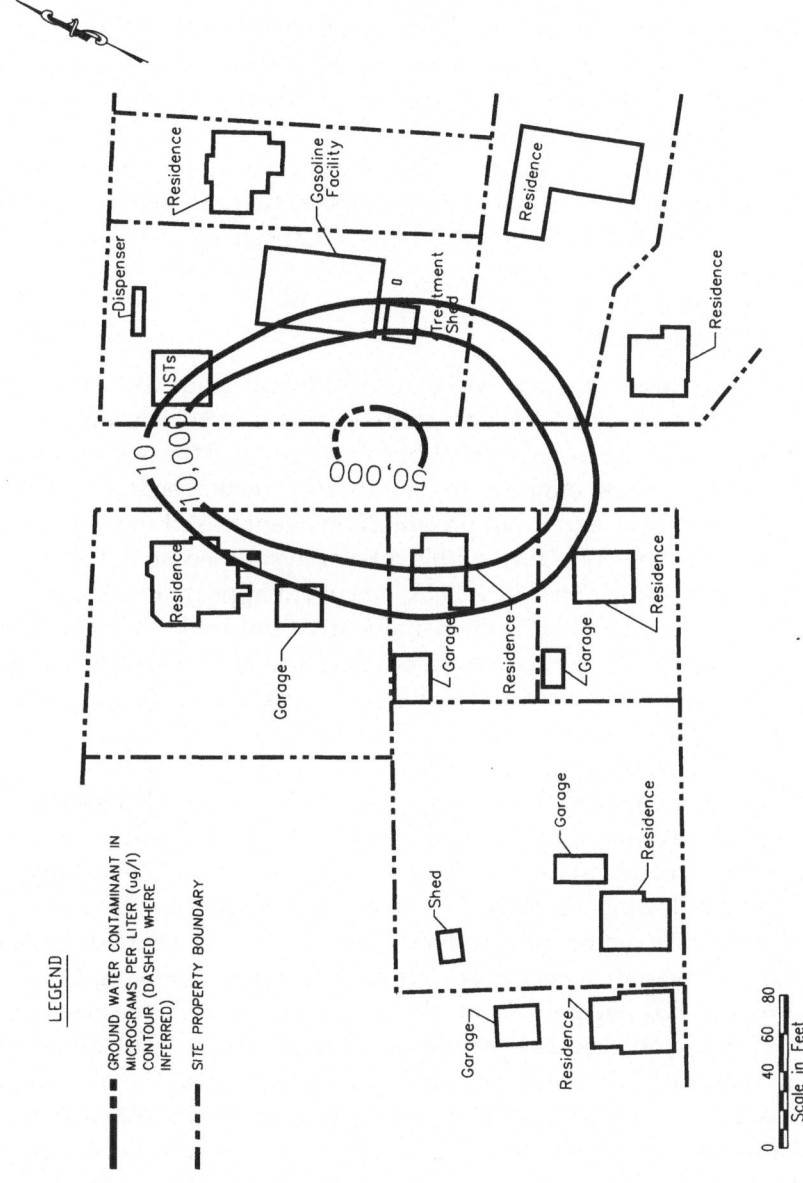

Figure 22-2. Pre-Remediation BTEX Contaminant Plume

Contaminant concentrations in ground water have significantly decreased over time. Current (April 2002) post-remediation BTEX concentrations range from below detection limits to 12,700 µg/l and MTBE concentrations range from below detection limits to 9,500 µg/l. The horizontal extent of dissolved BTEX impact has been delineated to concentrations below New Hampshire Ambient Ground Water Quality Standards (NHAGQS). The extent of dissolved MTBE impact has been defined to below NHAGQS in all but the downgradient direction. In this area, MTBE has been approaching NHAGQS in the last five sampling rounds. ENSR's recommendation to evaluate the vertical extent of impact in the bedrock was not approved by the state for this UST trust fund reimbursement site. Refer to Figures 22-3 and 22-4 for the most recent post-remediation MTBE and BTEX contaminant plume maps.

FATE AND TRANSPORT

Following the release, the gasoline spread outward from the UST to the north, west, and south as evidenced by the occurrence of SPH in several wells downgradient of the USTs. The SPH plume then followed the general ground water flow direction and migrated in a southerly direction across the site.

In October 1996, a significant precipitation event raised the water table approximately 1.2 meters (4 feet) within a few days. Based on data obtained from multiple investigations since 1996, this event appears to have had the following effects on the SPH: 1) the rising water "smeared" the SPH within the soil normally located above the water table; and 2) the rising water table influenced ground water flow to the west, resulting in the transport of SPH to the west of the UST area (see Figures 22-1 and 22-2). The SPH that is bound in the soil is a continuing source of contamination to ground water. After the storm event, only minor SPH thickness (less than 0.02 meter [0.05 feet]) was periodically observed in wells located downgradient of the UST system.

At the request of the state, indoor air sampling was conducted at five residences located downgradient of the site on a monthly basis between November 1996 and December 1998 and on a periodic basis in 1999 and 2000. Indoor air samples were collected on carbon trap tubes following Modified USEPA Method TO1 and New Hampshire Department of Public Health Services (NHDPHS) protocol. The samples were analyzed for BTEX and MTBE using an instrument detection limit of 0.002 micrograms. The air samples were collected from the basement and first floor over an eight-hour time period. In addition, an outside ambient air sample was also collected during each sampling event. Approximately 24 hours prior to each sampling event, any potential petroleum containing compounds were removed from the residence, and the area was vented for 15 to 30 minutes.

Shortly after the release, indoor air concentrations of BTEX (up to 855

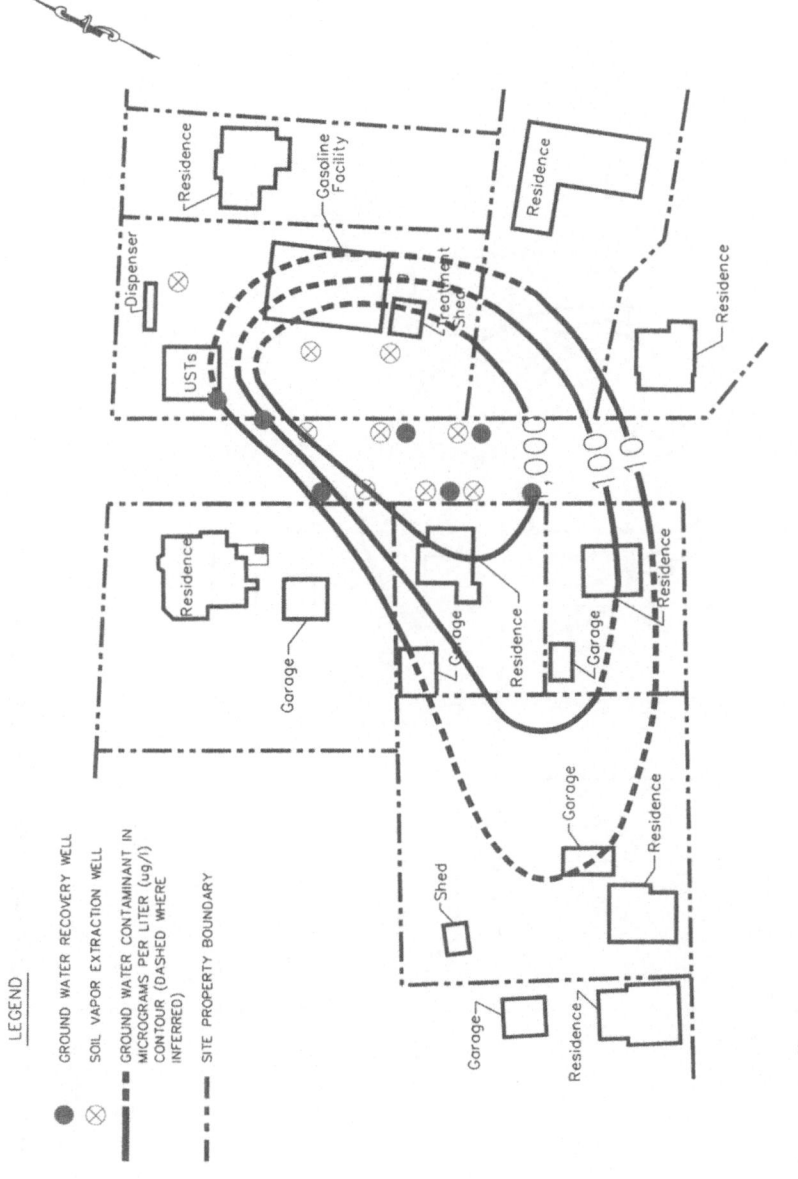

LEGEND

● GROUND WATER RECOVERY WELL

⊗ SOIL VAPOR EXTRACTION WELL

▬▬ GROUND WATER CONTAMINANT IN MICROGRAMS PER LITER (ug/l) CONTOUR (DASHED WHERE INFERRED)

—··— SITE PROPERTY BOUNDARY

0 40 60 80
Scale in Feet

Figure 22-3.   Post-Remediation MTBE Contaminant Plume

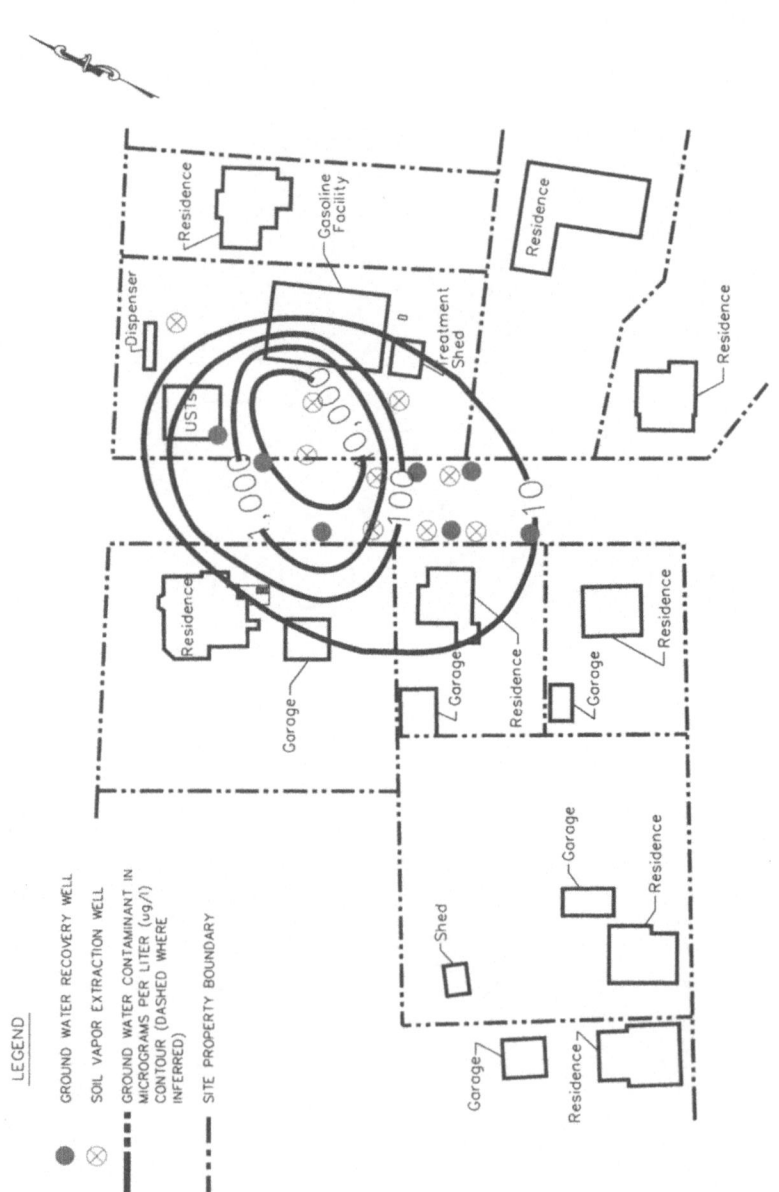

Figure 22-4.   Post-Remediation BTEX Contaminant Plume

$\mu g/m^3$) and MTBE (up to 1,200 $\mu g/m^3$) were detected in the basements of five residences located downgradient of the site. In response, ventilation blowers were installed and operated periodically in the basement of the two most downgradient residences until 1999. The 1996 to 2000 average indoor air BTEX concentrations ranged from 23 to 218 $\mu g/m^3$, and the average indoor air MTBE concentrations ranged from 19 to 335 $\mu g/m^3$. Based on the evaluation of the cumulative air data, the state recommended that no restrictions were necessary regarding the occupancy of the homes. Discussion of the state indoor air standards is provided below.

From November 1996 until April 2000, an active ground water recovery system was operated at the site. Based on an evaluation of historical ground water data and ground water elevation data collected during pumping conditions, the ground water recovery system adequately intercepted the contamination migrating from the release area. The BTEX and MTBE concentrations observed in monitoring wells south of the recovery wells stabilized or decreased, indicating that the southerly edge of the plume was "retreating," presumably due to the ground water recovery system. The presence of minor amounts of SPH detected in monitoring wells located beyond the recovery well network indicated that some SPH had spread beyond (west of) the recovery wells, most likely due to the initial flush from the aforementioned rainfall event before the wells were installed and operating. In addition, the plume of BTEX and MTBE has been observed west of the recovery system. In April 2000, the state gave approval to shut down the remediation system based on the reduction in ground water concentrations across the site. Since April 2000, ground water concentrations have fluctuated, but overall concentrations have continued to decrease.

RECEPTORS

Based on the site setting, the only potential environmental receptors identified were flora and fauna in a wetland area located 150 meters (500 feet) downgradient of the former USTs. Dissolved-phase concentrations of MTBE (ranging from less than 5 to 367 $\mu g/l$) have been detected in the wetland; however, MTBE was not detected in a monitoring well located between the downgradient edge of the hydrocarbon plume and the wetland. These results indicate that concentrations of MTBE detected in the wetland may not be a result of impacted ground water from the site. Surface water runoff was identified as a suspected source of the MTBE.

Potential human receptors include the following:

- Workers performing excavation in the vicinity of the current/former gasoline UST system, at the gas station facility, at catch basins, or along downgradient subsurface utilities; and

- As discussed above, human exposure from the migration of vapors into the basements of the residential structures was evaluated as part of the investigation activities completed at the site. In addition to the ventilation blowers, additional receptor protection was considered including pressurization of the basements, subslab ventilation, and refinishing and sealing the basements. No additional measures were completed based on a risk-based evaluation of the indoor air data. Ultimately, the petroleum company responsible for the release purchased four of the homes to settle litigation regarding impact to the residential properties.

REQUIRED CLEANUP LEVELS AND TIMEFRAMES

Ground Water.   New Hampshire has established a risk-based approach to characterize risks to human health and the environment posed by the release of contaminants at sites within the state (NHDES, 1998). Two categories of ground water contamination have been established to characterize the risks associated with impacted ground water. The ground water categories are described as GW-1 (applicable use as drinking water) and GW-2 (as a potential source to indoor air contamination). Ground water category GW-1 standards are the same as NHAGQS (5 µg/l benzene; 1,000 µg/l toluene; 700 µg/l ethylbenzene; 10,000 µg/l xylenes; 13 µg/l MTBE). All sites in New Hampshire are required to meet GW-1 standards. Ground water category GW-2 standards are applicable to ground water within 9 meters (30 feet) of an existing building and where average depth to water is less than 5 meters (15 feet). The GW-2 standards are intended to provide guidelines on when it may be appropriate to examine the indoor air exposure pathway.

Soil.   Like Massachusetts, the state of New Hampshire divides soils into three categories (S-1, S-2, and S-3) based on frequency and intensity of potential exposures to children and adults, with the S-1 category assuming maximum exposure potential and the lowest cleanup levels. In general, soils that exceed the S-1 cleanup levels require a property deed restriction limiting future property use in order to maintain a level of acceptable risk. Therefore, although the site soils were categorized as S-2 and S-3, the S-1 cleanup levels (0.3 mg/kg benzene; 100 mg/kg toluene; 140 mg/kg ethylbenzene; 500 mg/kg xylenes; and 2 mg/kg MTBE) were used in order to prevent the need for a deed restriction.

Indoor Air.   Indoor air sampling results were evaluated using a method developed by the NHDPHS. Carcinogenic and non-carcinogenic effects of subchronic (less than seven years) and chronic exposures were considered. In order to characterize risk, HIs and a Risk Index (RI) were calculated and compared to regulatory risk limits. An RI or HI of one or less indicates that adverse health effects from the chemicals of concern are not anticipated.

The HI was calculated by taking the ratio of the average daily exposure and a concentration considered to be without adverse non-carcinogenic effects. As provided by NHDPHS, the accepted daily exposure concentrations in air considered to be without adverse non-carcinogenic effects are 200 $\mu g/m^3$ for toluene, 500 $\mu g/m^3$ for ethylbenzene, 150 $\mu g/m^3$ for xylenes, and 1,500 $\mu g/m^3$ MTBE.

The RI was calculated by taking a ratio of the concentration an individual is exposed to over time and a concentration considered to be below the concentrations that would increase lifetime cancer risk by 1/100,000. An RI was not calculated for toluene, ethylbenzene, and xylenes since these compounds are not classified as carcinogens. Although USEPA has not classified MTBE as a carcinogen, an RI was calculated for MTBE since there is a provisional USEPA CSF for MTBE. As approved by NHDPHS, the accepted daily exposure concentrations in air associated with a cancer risk of 1/100,000 are 15 $\mu g/m^3$ for benzene and 500 $\mu g/m^3$ for MTBE.

Cleanup Timeframe.   There has been no overall cleanup timeframe established by the state for this site; however, there are deadlines established during the approval of each work scope submitted to the state for each phase of the project. In addition, the site is currently being monitored under a Ground Water Management Permit (GMP) issued by the state. The permit is in effect through May 2004, at which time the site conditions, the status of response actions, and remedial action goals would need to be reevaluated as part of the reapplication process.

REMEDIAL ACTIONS

Remedial actions conducted at the site consisted of initial immediate response actions, source removal, and physical treatment of impacted ground water and soil, followed by a combination of MNA and limited enhanced bioremediation. The remedial response actions implemented at the site resulted in the removal of over 5,118 kilograms (11,282 pounds) or 6,833 liters (1,805 gallons) of petroleum hydrocarbons, which is equivalent to approximately 85% of the original release of 7,950 liters (2,100 gallons) (Table 22-1).

Immediate Response Actions.   In addition to the residential vapor abatement activities discussed above, additional immediate response actions were implemented including manual and automated SPH recovery. Within several days of release, an automated pneumatic SPH recovery system was installed in three on-site monitoring wells. This installation, in addition to manual

Table 22-1.   **Petroleum Hydrocarbons Recovered (Pounds) 1996 to 2000**

Remedial Action	Volume of Material Recovered	Pounds of Petroleum Hydrocarbon Recovered
Immediate Response Actions		
SPH Removal	120 gallons	750[1]
Source Removal		
Groundwater	27,000 gallons	158[2]
Soil	860 tons	2,574[2]
SVE System Operation	—	3,500[3]
Ground Water Recovery	3.5 million gallons	4,300[4]
TOTAL PETROLEUM HYDROCARBONS RECOVERED		11,282 pounds or 1,805 gallons

Quantity recovered is equivalent to approximately 85% of the original release of 2,100 gallons.

[1] Immediate Response Actions: 1 gallon of SPH = 6.25 pounds.
[2] Source Removal: *Pounds of Hydrocarbons Recovered From Soil* = TPH soil concentration (mg/kg) * (soil volume (tons) * 907 (kg/ton)) / 454,000 (mg/lb). *Pounds of Hydrocarbons Recovered From Groundwater* = Total VOC groundwater concentration (μg/l) * (groundwater recovered (gal) * 3.78 (l/gal)) / 454,000,000 (μg/lb).
[3] SVE System: *Pounds of Hydrocarbons Recovered (lbs/day)* = 1,440 (minutes/day) * Flow Rate (cfm) * Influent (ppm) * Molecular Weight BTEX (mg/m3) / 35.31 (ft3/m3) * 454,000 (mg/lb).
[4] Hydraulic Control: *Pounds of Hydrocarbons Recovered (lbs/flow period)* = (Influent (μg/l) — Effluent (μg/l)) * (Flow for Period (gal) * 3.78 (l/gal)) / 454,000,000 (μg/lb).

SPH bailing, recovered a total of approximately 450 liters (120 gallons) or 340 kilograms (750 pounds) of SPH.

Source Removal.   The primary source removal was considered to be the excavation of the entire UST system and associated impacted soil, which occurred in October 1996. During the UST removal, approximately 860 tons of petroleum-impacted soil was removed from the site. In addition, approximately 102,000 liters (27,000 gallons) of ground water were pumped from the excavation and removed from the facility. Based on analytical data and the volume of soil and ground water removed during the UST removal, it was estimated that approximately 1,239 kilograms (2,732 pounds) or 1,654 liters (437 gallons) of petroleum hydrocarbons were recovered during the source removal. Soil samples collected from the base of the UST excavation contained concentrations of MTBE and naphthalene above S-1 standards.

Physical Treatment.   Between October 1996 and December 1996, remediation system design pilot testing was conducted, and the ground water recovery and AS/SVE remediation system was installed. The permanent system operated from November 1996 until April 2000.

SVE System. Soil vapors were extracted through nine 10-centimeter (4-inch) diameter vertical vapor extraction wells; two 10-centimeter (4-inch) diameter horizontal SVE wells installed in the former UST excavation; and two 10-centimeter (4-inch) diameter horizontal wells installed adjacent to residential structures. Vapors were extracted from the subsurface under negative pressure using a skid-mounted SVE system utilizing a 15-horsepower positive displacement blower. The blower typically operated at a flow rate of 20 cmm (750 cfm), and at a vacuum between 89 and 127 mm Hg (3.5 and 5 in Hg). Pilot test results indicated that the system was capable of extracting vapors from a 9- to 15-meter (30- to 50-foot) radius around the vertical wells and a 3- to 6-meter (10- to 20-foot) radius around the horizontal wells.

Total VOC concentrations in air extracted by the SVE system, as measured by a PID, ranged from less than 1 ppm to 105 ppm. It is estimated that approximately 1,600 kilograms (3,500 pounds) or 2,120 liters (560 gallons) of petroleum hydrocarbons were removed from the subsurface through the operation of the SVE system. Prior to the system installation, the state made a determination that an operating permit was not required for the system, thus no effluent vapor treatment was installed.

Ground Water Recovery. Impacted ground water was recovered from seven 10-centimeter (4-inch) diameter recovery wells at a total flow rate of approximately 23 to 26 liters per minute (l/min) (6 to 7 gallons per minute [gpm]). The recovery wells are screened to the bedrock surface, with a 1.5-meter (5-foot) long, lined sump installed into the bedrock to accommodate the body of the pump. Ground water was pumped via bottom loading pneumatic pumps to the treatment system enclosure. Once in the enclosure, the ground water was pumped into an oil/water separator, then gravity fed into a 760-liter (200-gallon) equalization tank where it was then pumped through a particulate filter and into the top tray of an LPS. The ground water cascades through five stainless steel stripper trays and accumulates in a 190-liter (50-gallon) sump at the bottom of the stripper. The airflow rate through the stripper was typically 70 to 85 cmm (2,500 to 3,000 cfm).

Effluent water exiting the air stripper was pumped into the municipal sewer system under an industrial user wastewater permit. The discharge limits under the wastewater permit were 5,000 µg/l total VOCs and 25 mg/l TPH. In 1999, after a drop in influent concentrations, discharge was switched to the Somersworth storm drain system under a National Pollutant Discharge Elimination System (NPDES) permit exclusion. The discharge limits under the NPDES exclusion were 5 µg/l benzene, 100 µg/l total BTEX, 70 µg/l MTBE, 100 µg/l naphthalene, and 5 mg/l TPH.

A total of 13.4 million liters (3.5 million gallons) of impacted ground water was treated and discharged during system operation. MTBE influent concentrations ranged from 200 µg/l to 1,670,000 µg/l and BTEX concentrations ranged from 250 µg/l to 438,000 µg/l. On average, the air stripper was successful in removing over 90% of the petroleum hydrocarbons prior to discharge to municipal sewer. It is estimated that approximately 1,950 kilograms (4,300 pounds) or 2,604 liters (688 gallons) of SPH were removed from the subsurface through the operation of the pump-and-treat system.

Air Sparging System.   As part of the remediation system installation, nine vertical air sparging points were installed, manifolded into two separate legs, and piped back to the remediation system enclosure. Due to concerns over potential vapor migration into adjacent occupied structures, the state never gave approval to activate the air sparging system.

Monitoring and Enhanced Bioremediation.   A GMP was approved by the state in May 1999. The state approved tri-annual ground water sampling of 13 monitoring wells as part of the GMP. In April 2000, the state approved the shutdown of the remediation system due to decreasing ground water concentrations. Although overall ground water concentrations have either stabilized or continued to decrease, an increase in MTBE and BTEX concentrations adjacent to the former source area was observed in December 2000 and April 2001. Additional active remediation was deemed necessary to expedite site closure and to afford protection of the adjacent residents. Several options were evaluated including reactivating the existing system (ground water recovery, air sparging, and SVE), implementing the new *in situ* submerged oxygen curtain technology, and enhanced bioremediation (using microbial injections). Microbial injections were selected based on cost, the lack of disruption to the neighborhood, proven effectiveness at similar sites, and the limiting subsurface conditions.

The first bimonthly microbial injection (using existing wells) was completed in October 2001. Microbial injections were temporarily suspended in winter 2001 and restarted in spring 2002 when conditions were more favorable for bioremediation. Approximately 110 liters (30 gallons) of microbial solution consisting of microbes, nutrients (20-10-10 fertilizer), and oxygen is periodically injected into four existing recovery wells, three existing SVE wells, and one monitoring well. Since the initiation of the microbial injections, there has not been a significant change in dissolved-phase contaminant concentrations observed. However, there have only been two sampling events since the injections were started. The effectiveness of this remedial alternative will continue to be evaluated in the third and fourth quarter of 2002.

COSTS

Project costs are subdivided into the following six categories. Please note that the ongoing monitoring and enhanced bioremediation costs have been included in the operation and maintenance category.

Site Assessment/Risk Characterization	$480,000
Source Removal (soil excavation, disposal, and backfill)	$50,000
Treatment System Design	$50,000
Treatment System Installation	$240,000
Operation and Maintenance	$250,000
Site Closure (Estimated Future Costs)	$150,000
TOTAL PROJECT COST:	**$1,220,000**

This site is eligible for reimbursement under the New Hampshire Oil Discharge Cleanup Fund. To date a total of $839,000 of investigation and remediation costs have been reimbursed by the state. The cap on reimbursement in the State of New Hampshire is $1.5 million dollars per release.

TIMELINE

The timeframes associated with various major components of the project are indicated below.

Release Detection	September 1996
Assessment Activities	September 1996–September 1997
Source Removal	May 1996; February 1998
Physical Treatment	December 1996–April 2000
Monitoring Only Program	April 2000–ongoing
Enhanced Bioremediation	December 2001–ongoing

REFERENCES

Bouwer, H. and Rice, R.C. 1976. A slug test for determining hydraulic conductivity of unconfined aquifers with completely or partially penetrating wells. *Water Resources Research.* 12, No. 3, 424–428.

Lyons, B., Bothnor, W., Moench, R., Thompson, J., Lincoln, P., and Stewart, G. 1997. *Bedrock Geologic Map of New Hampshire.* Prepared in cooperation with the U.S. Department of Energy and the State of New Hampshire.

NHDES (New Hampshire Department of Environmental Services). 1998. *Risk Characterization and Management Policy.* January 1998.

CHAPTER 23

Physical Treatment at a Massachusetts Site

Christopher G. Mariano, P.G., LSP, ENSR International

INTRODUCTION

A release of gasoline to the environment occurred as a result of historical operation of a gasoline dispensing facility. The release was discovered during a real estate acquisition site assessment, which was followed by the removal of the gasoline UST system. The site is located in a sole source aquifer area; however, it is not located within the zone of contribution for the municipal water supply, and a portion of the site is classified as a non-potential drinking water source area (NPDWSA). Assessment activities identified ground water impacts in the overburden beneath the site. Remediation consisted of source removal and physical treatment to achieve regulatory site closure. The site is of special interest because the MTBE concentrations were generally lower than the BTEX concentrations and, as such, MTBE was not a driving factor in the remediation efforts.

SITE DESCRIPTION

The facility consists of a 0.22-hectare (0.55-acre) parcel of land, located in a commercial and residential section of Massachusetts. The site was used as a service station and gasoline dispensing facility from the 1930s to 1981 when it was converted into a gasoline dispensing facility and convenience store. In 1966, a total of six USTs (three gasoline, two waste oil, and one kerosene) were removed during site facility renovations. Three 19,000-liter (5,000-gallon) gasoline USTs were installed, and the existing pump islands were replaced with larger islands. In 1986, gasoline-dispensing activities ceased, and the three 19,000-liter (5,000-gallon) gasoline USTs and associated dispensing islands and piping were removed (Figure 23-1).

Topography at the site is generally flat with a gradual downward slope to the south. The site is surrounded by commercial properties, with some residential properties located nearby. The site and surrounding properties are serviced by municipal drinking water. There are no private water supply

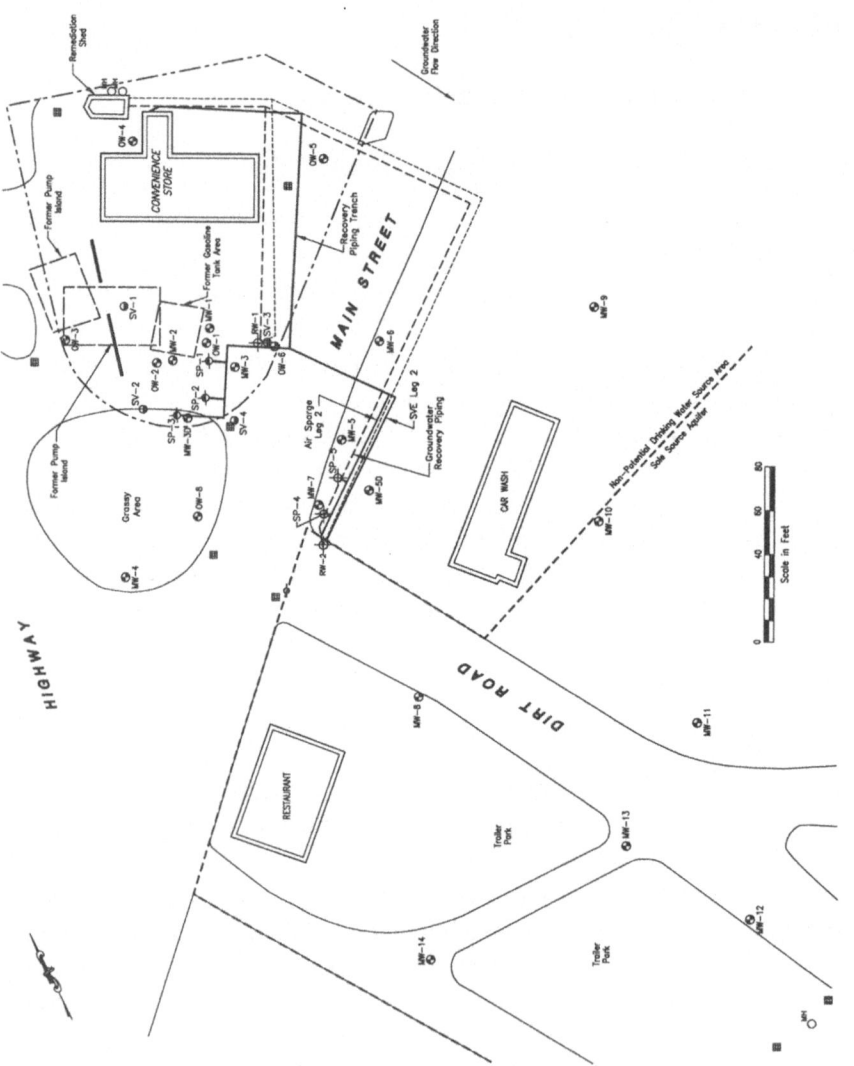

Figure 23-1. Site Plan

wells located within 400 meters (1,200 feet) of the site; however, there are seven private water supply wells located to the south and east of the site between 400 meters (1,200 feet) and 800 meters (2,650 feet) of the site. The nearest PWS well is located 900 meters (3,000 feet) north and upgradient of the site, and the site is not located within the zone of contribution for these wells.

RELEASE HISTORY

In 1985, during a site investigation, petroleum hydrocarbon impacts were detected in soil and ground water samples collected from the site. Subsequent tank tightness testing performed in January 1985 indicated no leaks in the tanks or associated piping. In December 1986, the three gasoline USTs and associated dispensers were excavated and removed from the site and were observed to be in good condition with no evidence of stained soils. During a hydrogeologic investigation performed in January 1988, dissolved-phase gasoline constituents were detected in ground water beneath the site, indicating that a gasoline release to the environment had occurred. The likely potential sources of gasoline impacts identified at the site included the original USTs and piping (removed in 1966) and the historical use of the facility for dispensing gasoline.

Follow-up response actions included installation of several monitoring wells both on and downgradient of the facility, hydraulic containment, the installation and operation of a ground water extraction/treatment system, and the installation and operation of an SVE system. In 1994, during the addition of an air sparging system to the existing remediation system, approximately 180 tons of impacted soil were removed for off-site recycling.

SITE HYDROGEOLOGY

Information regarding the site hydrogeology was obtained based on subsurface investigations completed at the site and published information as referenced herein.

Surficial Geology. A silty-sand and gravel was present from depths of 0.3 to 1.5 meters (1 to 5 feet) bgs, which is likely artificial fill. Below the fill is approximately 1.5 meters (5 feet) of tan, fine micaceous sand, indicative of glacial outwash deposits, beneath which is a layer of coarse sand that extends to the deepest boring depth of 19 meters (62 feet) bgs.

Bedrock Geology. According to information provided by the USGS (Odale, 1974), bedrock is located beneath the site at a depth exceeding 100 meters (400 feet) below mean sea level. Borings advanced as part of investigative activities at the site reached a maximum depth of 19 meters (62 feet) bgs, and bedrock was not encountered.

Hydrogeological Parameters.   Depth to ground water in the site monitoring wells ranged from approximately 1.4 meters (4.5 feet) bgs (MW-3D), to 1.5 meters (5.0 feet) bgs (OW-4). Ground water flow direction in the overburden aquifer was determined using a modified straight-line interpolation method. Generally, ground water flow in the overburden is to the south-southeast with an approximate horizontal hydraulic gradient of 0.001 (Figure 23-1). Although a vertical hydraulic gradient was not calculated for the site, a minor downward vertical gradient is suggested based on the occasional detection of gasoline constituents in deeper wells MW-3D and MW-5D.

Based on literature values (Freeze and Cherry, 1979) for fine sand, the hydraulic conductivity in the upper portion of the saturated zone was estimated at 10^{-3} cm/s (3 ft/day). With an assumed porosity of 30% (Driscoll, 1986), the estimated average linear velocity for ground water in the overburden was calculated at 2.9×10^{-6} cm/s (8.2×10^{-3} ft/day).

NATURE AND EXTENT OF CONTAMINATION

The contamination detected at the site consisted of gasoline constituents, and the source was determined to be the historical spillage/leakage related to the former gasoline UST system (removed in 1966) and minor release(s) associated with the subsequent UST system.

Soil.   Concentrations of gasoline-related constituents were detected in soil samples collected from the area of the former gasoline USTs and associated dispenser island. Based on laboratory analysis of soil samples collected during monitoring well installation, the soils with the highest concentrations were limited to the vicinity of the USTs and dispensers (PID screening readings up to 220 ppm; TPH up to 2,700 mg/kg).

Ground Water.   Concentrations of one or more gasoline-related constituents were detected in ground water from on- and off-site monitoring wells at various times throughout the monitoring period. Dissolved-phase gasoline impacts extended from the former UST system in a south-southeasterly direction as far as monitoring well MW-11, located approximately 85 meters (280 feet) downgradient of the former UST system. (Figure 23-1) The highest concentrations were detected in wells MW-1, MW-2, MW-3, and OW-2, all of which are located in close proximity to the former UST system.

The horizontal extent of the dissolved-phase ground water contaminant plume was defined to below the applicable regulatory standards by wells OW-4, MW-4, MW-9, and MW-12. Prior to conducting remediation at the site, BTEX and MTBE concentrations in the area of the former UST system were detected as high as 28.5 mg/l and 4.1 mg/l, respectively, while BTEX and MTBE on the south side of Main Street were detected at concentrations of

7.4 mg/l and 0.8 mg/l, respectively. Ranges of BTEX and MTBE concentrations detected in site monitoring wells are presented in Table 23-1.

FATE AND TRANSPORT

The migration pathway of release(s) from the source area was with ground water, toward a river, located approximately 700 meters (2,300 feet) south-southeast of the site. Based on the downward vertical hydraulic gradient measured at the site, vertical migration of BTEX and MTBE has occurred; however, as indicated by the sample results from deep monitoring wells MW-3D and MW-5D, the vertical migration of the dissolved gasoline constituents has been relatively minor.

No underground utilities are present in this area; therefore, underground utilities were not considered potential migration pathways. Vertical migra-

Table 23-1. **Ground Water Concentration Ranges and Applicable Cleanup Standards, Former Gasoline Dispensing Facility, Massachusetts**

Sample Location	Benzene (μg/l)	Toluene (μg/l)	Ethylbenzene (μg/l)	Xylenes (μg/l)	MTBE (μg/l)
MW-1	ND to 150	ND to 2,400	ND to 600	ND to 4,400	ND to 2,600
MW-2	ND to 1,000	ND to 3,700	ND to 1,500	ND to 9,300	ND to 4,100
MW-3	ND to 2,600	ND to 9,100	ND to 2,900	ND to 14,00	ND to 850
MW-3D	ND	ND	ND	ND	ND to 110
MW-5	ND to 1,900	ND to 1,300	ND to 920	ND to 2,600	ND to 650
MW-5D	ND to 31	ND to 3	ND to 23	ND to 9	ND to 52
MW-6	ND	ND to 1.7	ND	ND	ND to 6
MW-7	ND to 1,100	ND to 1,100	ND to 2.100	ND to 5,170	ND to 96
MW-8	ND to 6.2	ND to 8.7	ND to 21	ND to 28	ND to 86
MW-10	ND	ND	ND	ND	ND to 18
MW-11	ND to 4	ND to 44	ND to 1.1	ND to 4.9	ND to 800
OW-1	1,400	9,500	2,600	8,900	NA
OW-2	660 to 2,300	2,200 to 11,000	290 to 9,600	3,000 to 5,600	NA
OW-3	ND to 1.2	ND to 3.2	ND to 24	ND to 98	ND
OW-4	ND to 0.2	ND to 0.5	ND	ND	ND to 97
OW-5	ND to 0.6	ND to 1.7	ND to 2.8	ND to 15	ND
OW-6	470	440	390	380	NA
OW-8	ND to 880	ND to 2,400	ND to 610	ND to 2,800	ND to 20
GW-1	5	1,000	700	10,000	70
GW-2	2,000	6,000	30,000	6,000	50,000
GW-3	7,000	50,000	4,000	50,000	50,000

GW-1 = Massachusettes category GW-1 ground water standards in μg/l.
GW-2 = Massachusettes category GW-2 ground water standards in μg/l.
GW-3 = Massachusettes category GW-3 ground water standards in μg/l.
ND = Not Detected above laboratory minimum detection limit.
NA = Not Analyzed.

tion of vapor-phase compounds to the on-site convenience store was initially considered a potential migration pathway; however, following active reme-diation, the residual BTEX and MTBE concentrations were well below the volatilization cleanup standards. In addition, there were no vadose-zone im-pacted soils detected in proximity to occupied buildings.

RECEPTORS

The determination of ecological and human receptors was based on the site setting and contaminant concentrations, distribution, and potential migra-tion pathways identified in the assessment.

Ecological.   No natural resource areas, surface water bodies, or wetland areas were identified at or in close proximity of the site. The nearest surface water body is a river, located approximately 700 meters (2,300 feet) down-gradient of the site. Based on visual assessment, no portion of the site is con-sidered a potential wildlife habitat.

Human.   Potential receptors to soil contamination included construction workers who may perform excavation at the site. Post remediation condi-tions indicated that the site soils had been remediated to below laboratory de-tection limits; therefore, construction workers were no longer considered po-tential receptors.

The site is partially located in a USEPA-designated sole source aquifer area. However, as shown on Figure 23-1, the majority of the site, including the former UST system, is located in an area that is designated as a NPDWSA. In 1996, the MADEP designated portions of potentially productive aquifers as NPDWSAs based on population density and zoning. The site is not located within the zone of contribution of any of the nearby municipal pumping wells; the two nearest municipal wells are located 900 meters (3,000 feet) north and upgradient of the site. Seven private drinking water wells are lo-cated south and east of the site; the closest private well is 400 meters (1,200 feet) from the site. Based on the observed and calculated contaminant migra-tion rate and the distance to the drinking water supply wells, petroleum con-stituents in ground water from the site would be expected to attenuate before reaching any of the water supply wells. Thus, the exposure potential from pe-troleum-impacted ground water at the site is considered negligible. Remedi-ation of the ground water impact was warranted since the contaminants were migrating onto the portion of the sole source aquifer that is designated as a potential drinking water source area.

In addition, occupants of the on-site convenience store and downgradi-ent buildings were considered potential receptors of petroleum vapor accu-mulations in the buildings.

EXPOSURE POTENTIAL

The determination of the potential for exposure to contamination from the site was based on the results of the site assessment activities, including the current and foreseeable future use of the site, the location and concentration of detected impacts, the identified potential migration pathways, and the exposure points identified at the site.

Ecological.   No wetlands or surface water receptors were identified in proximity to the site, and no surficial soil impacts were identified at the site. Therefore, no environmental exposure potential was identified.

Human.   Prior to conducting active remediation at the site, potential human exposures included inhalation or dermal contact by construction workers performing excavations at the site, and potential future exposure through ingestion if the potential drinking water source area across Main Street was developed into a water supply. Following remediation, the potential construction worker exposure via inhalation or dermal contact was eliminated. The occupants of the site building and downgradient buildings were considered potential receptors of vapor accumulations from volatilization of the ground water contaminant plume. Because the petroleum concentrations in the ground water are well below risk levels, the exposure potential is considered insignificant.

REQUIRED CLEANUP LEVELS AND TIMEFRAMES

The MCP (MADEP, 1999) addresses site risk characterization similar to the ASTM RBCA process, whereby appropriate cleanup levels are developed using a three-tiered approach. Method 1 involves comparison of concentrations in site soil and ground water to generic risk-based screening levels. Method 2 allows the determination of site-specific target levels, and Method 3 is a more complex risk evaluation, which may involve additional site assessment, probabilistic evaluations, and more sophisticated fate and transport models than Method 2. Based on site conditions, the Method 2 approach was employed for the site.

Soil.   Under the MCP Method 1 risk characterization, impacted soils are divided into three categories (S-1, S-2, and S-3) based on frequency and intensity of potential exposures to children and adults, with the S-1 category assuming maximum exposure potential and the lowest cleanup levels. In general, soils that exceed the S-1 cleanup levels require a property deed restriction limiting future property use in order to maintain a level of acceptable risk. Therefore, although the site soils were categorized as S-3, the S-1 cleanup levels in a non-drinking water area (benzene-40 mg/kg; toluene-500 mg/kg;

ethylbenzene-500 mg/kg; xylenes-500 mg/kg; MTBE-100 mg/kg; naphthalene-100 mg/kg) were used in order to prevent the need for a deed restriction.

Ground Water.   The MCP Method 1 risk standards for ground water include three categories (GW-1, GW-2, and GW-3), based on receptor and exposure route. The GW-1 and GW-2 standards consider human exposures through ingestion and vapor inhalation, respectively, and the GW-3 standards address environmental risk to surface water. Based on the site setting and potential receptors, the cleanup standards associated with all three ground water categories, as shown in Table 23-1, apply to the site. Note that the GW-1 standards (exposure through ingestion) only apply to the downgradient part of the site located within the potential drinking water source area of the sole source aquifer.

Cleanup Timeframe.   Pursuant to the MCP, compliance with applicable cleanup goals is required within six years of notifying the MADEP of the release. Where it is recognized that this timeframe is not always feasible, the MCP does allow an open-ended timeframe, provided the remedial alternative selected to achieve the cleanup goals is implemented within four years of release notification.

REMEDIAL ACTIONS

Remedial actions conducted at the site consisted initially of source removal, followed by physical treatment of soil and ground water beneath the site.

Source Removal.   As stated previously, the former gasoline UST system was determined to be a likely source of the soil and ground water contamination. As such, the removal of the UST system constituted primary source removal. Additionally, approximately 180 tons of impacted soil were removed from the site during remediation system upgrade activities.

Physical Treatment.   In 1990, ground water and soil treatment systems were installed at the site. The ground water treatment system involved the extraction of impacted ground water through one ground water recovery well (RW-1) and treatment with an air stripper prior to discharge to two on-site recharge galleries. The soil treatment consisted of an SVE system, which was comprised of four on-site vertical SVE wells, one off-site horizontal SVE well, and two 90-kilogram (200-pound) carbon adsorption vessels connected in series. Following carbon treatment, extracted air was discharged to the atmosphere.

In November 1991, the carbon treatment system was expanded in order to more effectively treat the MTBE. The expanded system consisted of the addition of two 900-kilogram (2,000-pound) carbon adsorption vessels to treat ground water pumped from the on-site recovery well. Following treatment, the ground water was discharged into a catch basin located on Main Street. Ground water samples were collected on a quarterly basis to monitor decreases in petroleum hydrocarbon concentrations in the ground water beneath the site. At least 28 ground water monitoring events were conducted at the site.

In 1994, an air sparging pilot test was performed to evaluate the effectiveness of the addition of an air sparging system to the existing treatment system at the site. The conditions were favorable, and in 1995, five sparge points and one additional ground water recovery well (RW-2) were added. The treatment systems remained in operation until October 1996 when they were discontinued due to decreasing concentrations. Post-remediation soil sampling and ground water monitoring confirmed that cleanup objectives had been achieved, and the site was closed in August 1998.

COSTS

Project costs are subdivided into the following five categories.

Site Assessment/Risk Characterization	$150,000
Source Removal (soil excavation, disposal, and backfill)	$10,000
Treatment System Design and Installation	$463,500
Operation and Maintenance	$180,000
Site Closure	$2,500
TOTAL PROJECT COST:	**$806,000**

TIMELINE

The project timeline spanned from January 1985 through August 1998. The timeframes associated with various major components of the project are indicated below. Note that the site achieved regulatory closure in August 1998. Pursuant to the MADEP's random site audit program, the site was audited in 1999. The DEP found the response actions, risk characterization, and site closure in compliance with the MCP and no further action was required at the site.

Release Detection	January 1985
Assessment Activities	January 1985–August 1998
Source Removal	December 1986, 1995
Physical Treatment	August 1990–October 1996
Regulatory Closure	August 1998

REFERENCES

Driscoll, F.G. 1986. *Applied Hydrogeology*, Second Edition. Columbus, Ohio, Merrill Publishing Company.

Freeze, R.A. and Cherry, J.A. 1979. *Groundwater*. Englewood Cliffs, New Jersey, Prentice-Hall, Inc.

MADEP (Massachusetts Department of Environmental Protection). 1999. *Massachusetts Contingency Plan 310 CMR 40.0000*. October 1999.

Odale, R. 1974. *Geologic Map of the Dennis Quadrangle, Barnstable County, Cape Cod, Massachusetts*. USGS Map GQ-1114.

CHAPTER 24

Strategic Pumping to Divert an MTBE/BTEX Plume from Municipal Water Supply Wells

Evan T. Johnson, P.E., LSP and Tracy J. Adamski, AICP, Tighe & Bond, Inc.
Michael Scherer, Massachusetts Department of Environmental Protection

INTRODUCTION

Three USTs were installed as part of the opening of a new convenience store in Palmer, Massachusetts in 1989. Shortly after the tanks were filled with gasoline, one of the tanks indicated low product volume. The storeowner ordered a second delivery of fuel to refill the tank. Again, the tank indicated low product volume, the result of a significant release of fuel from the tank. However, the store owner did not initially notify MADEP of the release, stating later that he thought someone was stealing his gasoline. Upon removal, it was determined that the tank had failed, and premium gasoline containing a high percentage of MTBE had been released into the subsurface, impacting soil and ground water.

Initial data from an MPE system, which included ground water/product recovery wells, installed at the convenience store indicated that, although LNAPL was present on the water table around the facility, the total LNAPL area was decreasing and the gasoline plume had been contained on the site. However, MADEP required that additional monitoring wells be installed and sampled, and that samples be collected from the new and existing monitoring wells located downgradient of the site. Samples from these downgradient monitoring wells revealed that an MTBE/BTEX plume had migrated at depth beyond the original release site and was migrating at a rapid rate deep within the aquifer. Further, the plume was approaching the active drinking water supply wells for the municipal Water District, located approximately 600 meters (2,000 feet) immediately downgradient of the gasoline release.

Given the concern for the municipal supply wells, two additional recovery wells (RW-5 and RW-6) were installed between the release point and the municipal wells to capture the plume and prevent it from impacting the wellfield. With MADEP and municipal approval, these mid-plume recovery wells were allowed to discharge to the municipal wastewater treatment facility.

Despite success with mid-plume recovery wells, MADEP required installation of an additional recovery well (RW-7) as monitoring data indicated that the leading edge of the plume had passed beyond the effective recovery zone of RW-5 and RW-6. Recovery well RW-7 was installed just upgradient of the main water supply well and, with MADEP and USEPA approval, was allowed to discharge to the Quaboag River.

An additional recovery well (RW-8) was later installed as a backup to RW-7 and connected to a header for river discharge. Recovery wells continue to operate, and monitoring data collected by both the state and the Water District indicate that the plume has been effectively captured and the municipal wells have not been impacted by MTBE or BTEX compounds. The recovery well system installed as an Immediate Response Action will continue to operate as part of any long-term remedial plan at the site.

SITE DESCRIPTION

The release site is a 1.1-acre parcel of land located at the intersection of two major roads in western Massachusetts. The site includes a convenience store/gas station, a warehouse, and one 10,000-gallon and two 8,000-gallon gasoline USTs associated with one fuel dispensing pump island (Figure 24-1). The surrounding area consists primarily of industrial, commercial, and transportation uses. A railroad abuts the site on the south and a warehouse is adjacent to the west. Industrial properties are located across the roadways to the north and east of the site and across the railroad to the south and west. Graves Brook and its associated wetland are located approximately 60 meters (200 feet) to the southwest, and the Quaboag River is located approximately 600 meters (2,000 feet) downgradient. The municipal Water District's wellfield is also located approximately 600 meters (2,000 feet) to the southwest (and downgradient) of the release site.

RELEASE HISTORY

Approximately 12,000 gallons of gasoline were released into the ground in 1989 when a UST failed. When the tank was removed, a weld failure was identified as the cause of the leak. The tank manufacturer was eventually identified as the responsible party for remediation of the release.

The owner of the property initially failed to notify the MADEP of the spill, and no remedial actions were undertaken for two months. Initial response actions included the installation of three recovery wells at the release site to capture the plume and to remove the floating product. Shallow monitoring wells were also installed to determine the extent of the plume. According to the consultant at the time, the information collected from these initial response actions and monitoring indicated that the contamination was contained within the release site.

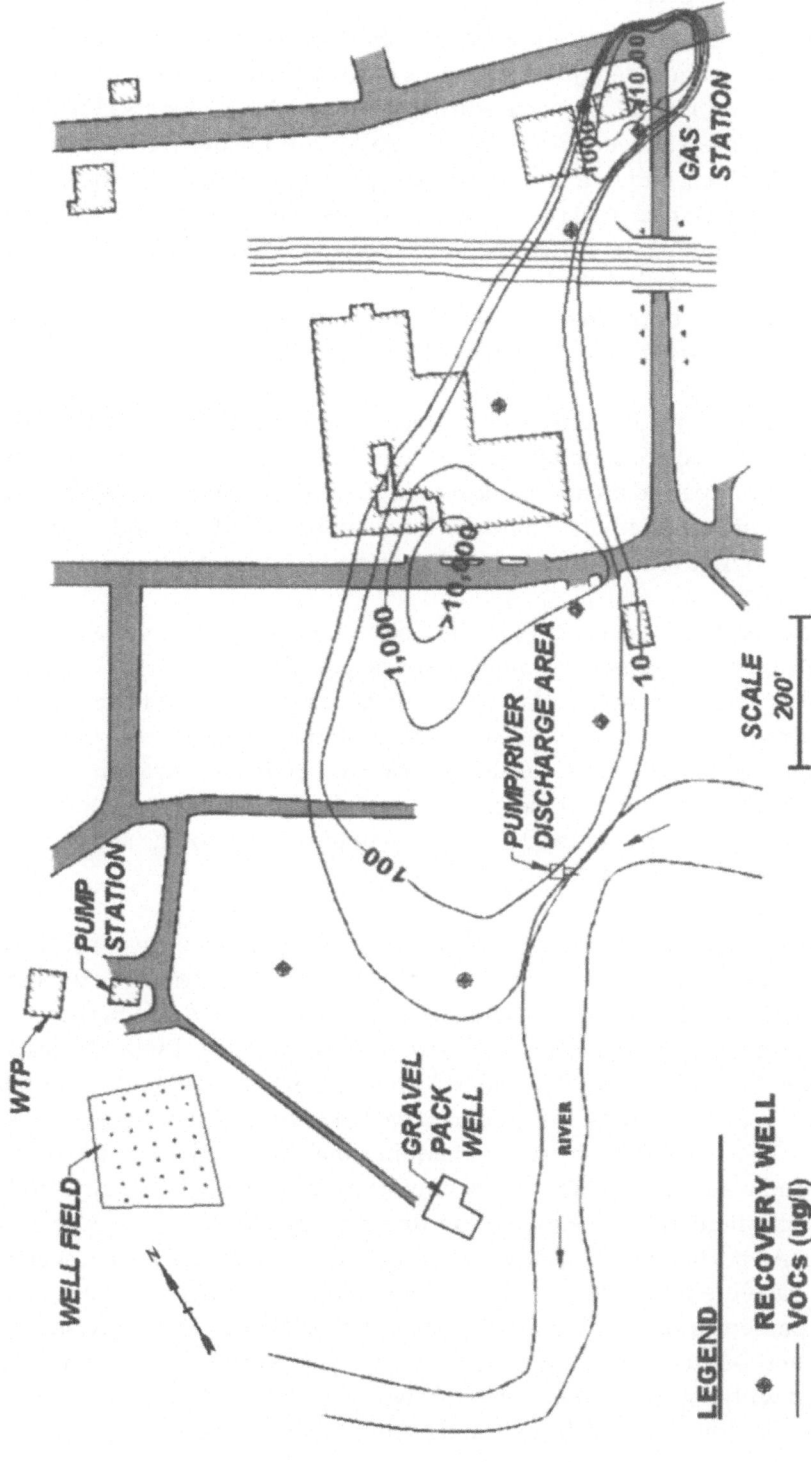

Figure 24-1.   Site Location and Total VOCs under Pumping Conditions (1999)

SITE HYDROGEOLOGY

Several borings and monitoring wells were installed to evaluate hydrogeo-logic conditions and determine the limits of the gasoline release. The data indicated that an LNAPL layer measuring 40 meters by 30 meters (120 feet by 90 feet) was present beneath the facility, that ground water was at a depth of 6 meters (20 feet), and that ground water elevations fluctuated by as much as 3 meters (10 feet) due to seasonal variations (Sweitzer, 1999). Due to the fluctuation in ground water levels and the discrete sudden nature of the release, an approximate vertical thickness as large as 5 meters (16 feet) of soils in the source area has been impacted by the separate-phase gasoline plume at depths ranging from 5.5 to 10 meters (18 to 32 feet) bgs.

Surficial Geology. Data from subsurface investigations indicated that soils at the release site are sand and gravel over fine sand and silt. The regional aquifer and impacted area were located in the deeper fine sand and silt. Typical hydraulic conductivity values for fine sands and silts are in the range of 10^{-2} to 10^{-3} cm/s (10^1 to 10^0 ft/day) (Freeze and Cherry, 1979).

Bedrock Geology. According to the Bedrock Geologic Map of Massachusetts (Zen, 1983), bedrock beneath the site is identified as Monson Gneiss described as a massive biotite, plagioclase gneiss interspersed with amphibolite and microline augen gneiss. Seismic refraction data were used to determine the elevation of the bedrock at the release site. Bedrock elevation was estimated at approximately 30 meters (100 feet) bgs at the release site and between 24 and 29 meters (80 and 94 feet) bgs in the vicinity of the adjacent wetland system (ECS, 1998).

Hydrogeological Parameters. Ground water flow is to the west/southwest from the release site, towards the Quaboag River and the District's wellfield. Initially, Graves Brook and its associated wetland located downgradient of the release site were considered to be a hydrogeological divide. Instead, it was later determined that the wetland system created a perched water table that may have influenced the ground water in the vicinity of the release site (Figures 24-2 and 24-3). The perched aquifer may act as a barrier to horizontal ground water flow in the shallow aquifer by increasing the downward hydraulic gradient, thereby submerging the contaminant plume deeper in the aquifer. An additional likely factor contributing to the downward movement of the gasoline plume within the aquifer is the continual pumping of the downgradient public water supply wells. As such, the release constituents MTBE and benzene have migrated to the deeper portion of the aquifer and traveled with regional ground water flow toward the wellfield.

The shallow and mid-level downgradient wells were not detecting con-

Figure 24-2.   Graves Brook and Wetlands: Perched Water Table Between Source and Wellfield

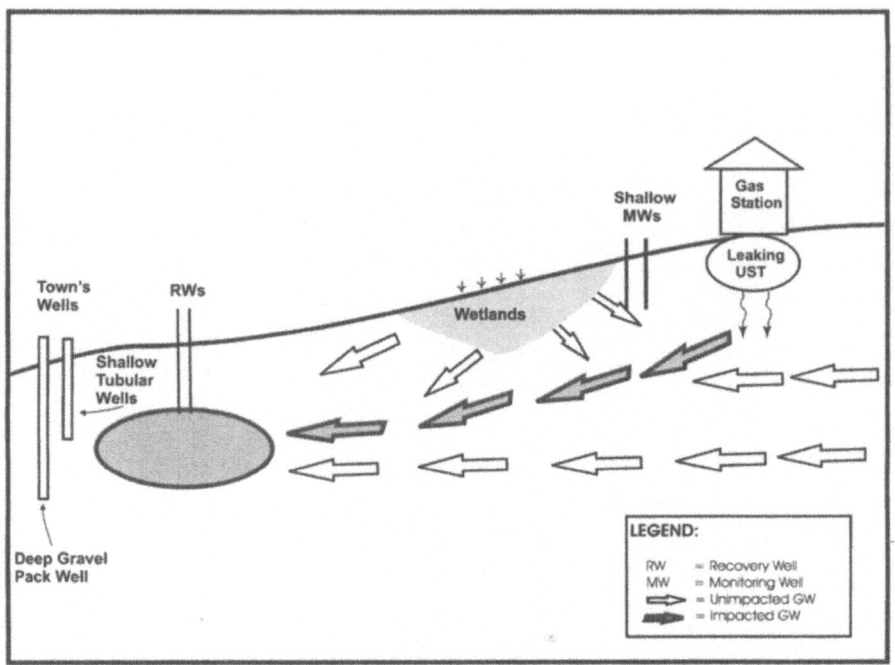

Figure 24-3.   **Cross-Section Depicting Impact of Perched Ground Water Table**

tamination because the plume was located as deep as 32 meters (105 feet) below grade, which was approximately 17 to 19 meters (50 to 60 feet) below the ground water surface. MADEP required that additional investigations be conducted downgradient of the release site. In 1993, elevated levels of MTBE (11,200 µg/l) were detected in a deep monitoring well, screened at 25 to 30 meters (90 to 100 feet), located approximately 300 meters (1,000 feet) downgradient of the site. Elevated benzene levels (460 µg/l) were detected there two years later. In total, four additional recovery wells were installed at downgradient properties at depths of 20 to 30 meters (70 to 100 feet) bgs.

NATURE AND EXTENT OF CONTAMINATION

Soil and ground water were impacted by the release. LNAPL was smeared over a 5-meter (16-foot) vertical area in the immediate release area due to the sudden, discrete subsurface releases and large seasonal changes in ground water depths. Ground water was identified in the immediate response area at approximate depths ranging from 6 meters (20 feet) bgs up to a depth of 8.5 meters (28 feet) bgs. As noted above, the contaminant plume migrated into deeper ground water zones, up to 20 meters (60 feet) below the ground water table and approximately 30 meters (100 feet) bgs. The vertical thickness of the MTBE plume was approximated at 3 to 4 meters (9 to 12 feet).

RECEPTORS

Potential receptors included the community serviced by the District's ground water supply wells, workers at the convenience store and affected properties from possible exposure to vapors entering basements of the buildings, and flora and fauna related to the nearby Graves Brook, its associated wetland, and the Quaboag River.

MPE for product and ground water recovery was installed on the release site to recover product and protect workers at the release site. Three MPE wells were installed on other affected properties to recover contaminated ground water and prevent downgradient migration. The initial recovery wells (RW-1, RW-2, RW-3, and RW-4) were in use from four to nine years and were permitted to discharge to the sanitary sewer system after pretreatment with liquid phase carbon. Recovery wells RW-5 and RW-6 continue to operate and discharge to the sanitary sewer.

With the support of the MADEP, USEPA issued a NPDES permit for discharging recovery well flows from RW-7 and RW-8 to the adjacent Quaboag River. The USEPA approval was predicated on the fact that the concentrations of total VOCs remain below surface water standards (100 µg/l) and that the benefits to the municipal wellfield outweighed any minor impacts to surface water. Further, the minor discharge volumes are easily diluted by the river, even under low flow conditions.

The Water District's wellfield downgradient of the release site provides up to 80% of the potable water supply for the District's customers. If the water supply was to be impacted, numerous customers would be affected. The wells draw from a large regional aquifer, which includes the release site. The aquifer supports approximately 800 gpm of withdrawal from the District's wells, which consists of a 26-meter (85-foot) deep gravel pack well and a tubular wellfield consisting of thirty-five 11-meter (35-foot) deep wells.

The contamination had migrated to a deeper portion of the aquifer, apparently influenced by a number of factors including a perched water table associated with the wetland system adjacent to the release site resulting in a slight downward gradient; and possible preferential pathways developed under years of constant supply well withdrawal conditions. By the time this was discovered, the contaminant plume had traveled approximately 300 meters (1,000 feet) downgradient from the release area. Based on these findings, MADEP determined that an imminent hazard to public health and safety existed due to the immediate threat to the public water supply wells.

The District had previously installed a GAC treatment facility to remediate VOC contamination already present in the ground water supply from an upgradient chlorinated solvent release. However, the use of GAC for treating minor concentrations of VOCs was expensive for the District to maintain. Treating the water supply using the existing GAC system for MTBE in addition to VOCs would require design changes and more frequent GAC recharge, resulting in additional cost to the financially strapped District.

REMEDIAL ACTIONS

In the source area, the three original recovery wells, installed to remove LNAPL and control dissolved petroleum hydrocarbons, have been replaced with an SVE system. Additional assessment actions including monitoring well installation and sampling to evaluate remedial measures to address the residual source area are underway. While separate phase has not been detected for seven years, recent sampling results (Kiernan and Gailor, 2001; Kiernan, 2002) indicate that high levels of dissolved gasoline (greater than 20,000 µg/l VOCs) continue to exist in the soils and ground water in the residual source area. Liquid phase carbon was used for the removal of VOCs, including BTEX and MTBE from the MPE system effluent.

Downgradient from the release location, off-site remedial actions include the installation of five ground water recovery wells. These wells were installed to hydraulically control the contaminant plume with the goal of recovering MTBE/BTEX to prevent impacts to the drinking water supply wells. Two of the five recovery wells (RW-7 and RW-8) are located immediately upgradient of the District's wellfield. RW-8 was installed in 1999 as a redundant backup well system to RW-7 and to provide a wider capture zone at the leading edge of the plume. Both of these wells are on automatic alarm and telephone dialer systems to notify the MADEP and municipal officials in the event that either well stops pumping. In addition to the selected remedial measures, a natural attenuation program has been established to document the natural remediation in the affected areas, extending from the original release site to the wellfield.

As of January 2000, remedial actions had resulted in the recovery of approximately 23,000 liters (6,000 gallons) of gasoline. The off-site recovery wells RW-5, RW-6, RW-7, and RW-8 remain in use to recover contaminants and maintain hydraulic control of ground water contaminant migration. The hydraulic head is reduced at the recovery well, thereby inducing a hydraulic gradient towards the well (Sweitzer, 1999). As part of continuing response actions, selected monitoring wells will be sampled four times per year for VOCs, the SVE system will continue to operate, and selected wells will be sampled to monitor natural attenuation.

CLEANUP LEVELS

Site remediation has been performed in accordance with the MCP. Under the MCP, the cleanup standards are determined at the site by the potential exposures to soil and ground water receptors and the exposure routes and exposure point concentrations for contaminants of concern. The soils at the site are classified under the MCP as S-3 (least restrictive soil category) due to the developed nature of the site and depth to the release area; however, the soils in the immediate release area could be exposed should the property be developed for a new use. Therefore, the most restrictive soil category (S-1) is also used for evaluation in the immediate release area at this site. Ground water is included under the most restrictive MCP category (GW-1) due to the site's location within the recharge area (Zone II) of a drinking water supply. S-1 soil cleanup levels applicable to drinking water areas (GW-1) are 0.3 mg/kg MTBE; 10 mg/kg benzene; 90 mg/kg toluene; 80 mg/kg ethylbenzene; and 500 mg/kg xylenes. The applicable ground water cleanup levels (GW-1, drinking water areas) are 70 µg/l MTBE; 5 µg/l benzene; 1,000 µg/l toluene; 700 µg/l ethylbenzene; 10,000 µg/l xylenes; 400 µg/l C_9 to C_{10} aromatic hydrocarbons; 400 µg/l C_5 to C_8 aliphatic hydrocarbons; and 4,000 µg/l C_9 to C_{12} aliphatic hydrocarbons (MADEP, 2001). Within the release area and 300 meters (1,000 feet) downgradient, concentrations in soils and ground water remain above cleanup levels (Figure 24-1).

COSTS

To date, the costs of the site assessment and remediation work total $1.5 million (Scherer, 2002). The tank manufacturer covered these costs until it went bankrupt in 2000. The total cost of the remediation is anticipated to reach $2.0 million. Operation and maintenance costs for the four most downgradient recovery wells and the SVE system are approximately $25,000 to $30,000 per year.

TIMELINE

Site remediation and oversight was taken over in 2000 by the MADEP with assistance from a state contractor. To date, the recovery wells have successfully prevented impacted ground water from reaching the District's drinking water supply. The existing system will continue to operate as part of a long-term remedial plan at the site.

REFERENCES

ECS (Environmental Compliance Services, Inc.). 1998. *Phase II Comprehensive Site Assessment Winton's Food and Fuel Palmer, Massachusetts.* File No. J10933.21, Document No. 14386, March 1998.

Freeze, R.A. and Cherry, J.A. 1979. *Groundwater.* p.29. Englewood Cliffs, New Jersey, Prentice-Hall, Inc.

Kiernan, R.J. and Gailor, D.G. Clean Harbors Environmental Services Inc. 2001. *Revised Investigative Summary at Winton's Food and Fuel.* June 14, 2001.

Kiernan, R.J. Clean Harbors Environmental Services Inc. 2002. *Sampling Summary Report Winton's Food and Fuel.* January 25, 2002.

MADEP (Massachusetts Department of Environmental Protection). 2001. *Massachusetts Contingency Plan 310 CMR 40.0000.*

Sweitzer, F.J. Environmental Compliance Services, Inc. 1999. *Phase III Remedial Action Plan Winton's Food and Fuel Park Street Palmer, Massachusetts RTN 1-0716.* File No. J10933.23, October 1999.

Zen, E. 1983. *Bedrock Geologic Map of Massachusetts.* United States Geological Survey prepared in cooperation with the Commonwealth of Massachusetts Department of Public Works, and Joseph A. Sinnot, State Geologist.

CHAPTER 25

Ozone Microbubble Sparging at a California Site

William B. Kerfoot, Ph.D, K-V Associates, Inc.
Paul LeCheminant, Cirrus Engineering Services

Engineered microbubble systems with ozone have proven to be a powerful means of targeting and effectively eliminating petroleum spill products from ground water, particularly the gasoline additive MTBE (Kerfoot, 2000; Kerfoot and McGrath, 2001). The primary reaction involves a low molar ratio of ozone to MTBE to decompose the contaminant molecule (Karpel vel Leitner *et al.*, 1994). For field applications, a site in Lincoln, California, is presented with mass oxidation demand and equations for attenuation of MTBE, ETBE, BTEX, and naphthalene. Under field conditions of ozone injection, TBA and TAME also exhibit rapid degradation.

Ozone treatment of ground water and soil has become one of three generally recognized oxidation systems for VOC treatment (Siegrist and Watts, 2001). A clear benefit of treatment with ozone has become apparent. Ozone decomposes the MTBE and releases oxygen in the process. Numerous breakdown products (*e.g.*, TBA, TBF, acetate) are readily degraded by bacteria in oxygenated conditions (Miller *et al.*, 2001). Ozone is 12 times more soluble than oxygen (Weast, 1972). An increase in dissolved oxygen occurs when ozone is supplied in quantities greater than the decomposition reaction, particularly during treatment of MTBE where oxygen or hydrogen peroxide is an end product.

This chapter will briefly review the state of science in ozone microbubble oxidation technology for MTBE removal and provide guidance to engineers for application. A case study is presented with stoichiometry and simple approaches to answering the critical questions of how much and for how long. A brief cost comparison is also presented.

TREATMENT TECHNOLOGY OVERVIEW — OZONE OXIDATION AND MICROBUBBLE TREATMENT

The KVA C-Sparge(tm) process combines the unit operations of air stripping and oxidative decomposition into a single process. This patented process involves injecting air and ozone directly into the ground water and soil column (U.S. Patents: #5,855,775; #6,083,407; #6,284,143; #6,306,296; #6,312,605).

When air is bubbled through ground water in soil pores, dissolved VOCs transfer from the liquid to gas phase in accordance with Henry's Law. The spargers used in the pilot test process (C-Sparger®) produce extremely small "microbubbles" (0.3 to 200 micron) with a very high surface area-to-volume ratio. This high surface area-to-volume ratio maximizes VOC transfer from the liquid phase to the gas phase. If the air bubbles contain sufficient ozone for decomposition, the VOCs react with the ozone and are destroyed while still in the water column.

With MTBE decomposition, hydrogen peroxide is released, creating an OH• radical coating that also speeds reaction. This *in situ* combined VOC recovery and destruction not only obviates the need for an additional process step, but also enhances the physical and chemical kinetics of the process (Figure 25-1).

For field applications, the process engineer defines the oxidant requirement for the site based on: stoichiometric oxidation requirement for the chemicals of concern; SOD; aqueous oxidant needs for metals, carbonates, and sulfides; and the *in situ* decomposition rate of the ozone. The theory section defines a practical stoichiometric oxidant demand. This oxidant demand computation procedure is then contrasted with the time to completion of treatment at a spill site.

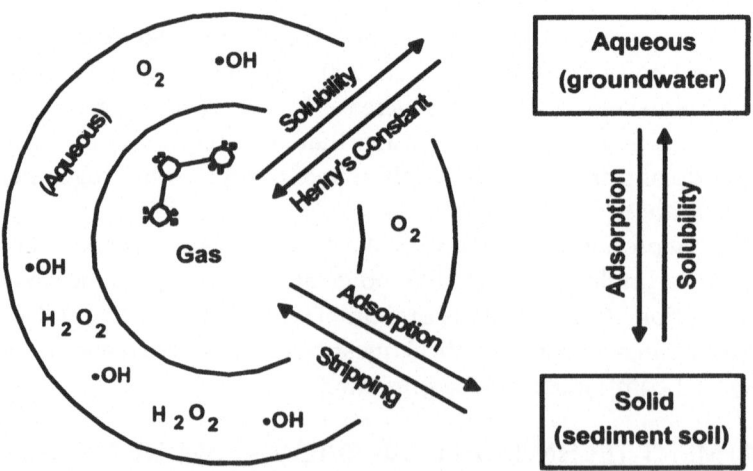

Figure 25-1.   **Microbubble Organic Oxidation Reactions and Partitioning Environment for Ozone Reactions**

Theory. When ozone is bubbled into an aqueous solution containing dissolved VOCs, ozonation may occur in either the aqueous phase or the gas phase. Whether the VOC transfers into the ozone-containing bubble and is destroyed in the gas phase, or the ozone dissolves in the water around the skim surface of the bubble and destroys the VOC in the aqueous phase is primarily dependent upon the rate of reaction of each VOC with ozone. Table 25-1 shows the oxidation capacity of different ozone states compared with those of other common oxidants.

In accordance with Henry's Law, the dissolved VOCs will be driven into the gas phase, and the gaseous ozone will be driven into the aqueous phase. This will result in the various reactions occurring at the bubble-liquid interface, whether in the gas-film or liquid-film of the bubble (Figure 25-1). Whether the primary decomposition reaction is occurring in the gaseous or liquid phase, the oxygenates are driven by partitioning into the bubble environment. The smaller the bubble, the greater the surface-to-volume ratio and ability to "strip" volatile organics (Kerfoot *et al.*, 1996). The thin film theory of Henley and Seader (1981), summarized by Kerfoot (2002), describes the mass transfer of a reactant across a liquid and a gas film before it contacts the other reactant.

MTBE has a Henry's Law constant of 6.9×10^{-04} atmosphere cubic meters per mole (atm m^3/mol), about one-eighth that of BTEX compounds; however, the high surface to volume ratio of micron-sized bubbles enhances the *in situ* stripping capacity (partitioning from aqueous to gaseous phase) to allow effective extraction.

Oxidation Chemical Mechanisms. Karpel vel Leitner *et al.* (1994) elucidated the reaction pathways of ozone and MTBE in dilute aqueous solution using controlled experimental conditions. The primary reaction (90% of the consumed mass of MTBE) results in the formation of TBF and hydrogen peroxide. A second parallel reaction (less than 10% of the consumed mass of MTBE) generates formaldehyde, TBA, and oxygen (Figure 25-2). Initially, the peroxide will be concentrated around the shell of the bubble. Since peroxide will later decompose to oxygen and water, the surrounding ground water

Table 25-1. **Oxidation Potential**

Oxidant	E_0
Catalyzed Ozone (OH•)	2.80V
Fenton's Reagent	2.76V
Ozone (Gas)	2.42V
Ozone (Molecular)	2.07V
Permanganate	1.67V
Hydrogen Peroxide	1.50V

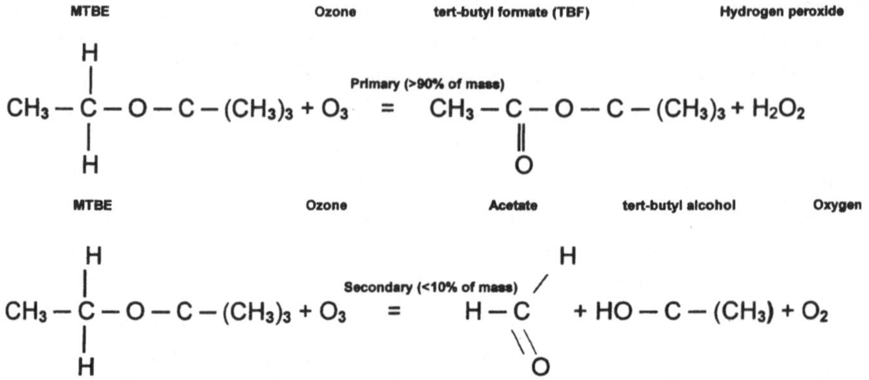

Figure 25-2. **Primary and Secondary Reaction of MTBE with Ozone (Karpel vel Leitner et al., 1994)**

will become highly oxygenated. Oxygen contents in excess of 10 mg/l are common (Figure 25-2).

Oxidant Application and Spread. Clayton (1998) developed a simple model of ozone transport for a chemical subject to first-order degradation. Subsurface ozone transport is limited by ozone reaction as it moves through the soil. The importance of ozone reaction rates on ozone transport is illustrated by considering simplified radial transport of ozone from an injection well.

Clayton's equation, combining the well drawdown equation with the standard first-order decay equation yields:

$$C = C_0 e^{-kt} = C_0 e^{-[k\pi H n S_g R^2/Q]} \tag{25-1}$$

This equation is an analytical solution for steady-state radial gas transport subject to first-order decay where C is ozone concentrations; C_0 is the initial ozone concentration; k is the degradation constant of 0.693/half-life; t is time; H is the height of flow zone; n is the soil porosity; S_g is the gas saturation (the fraction of ozone gas per void volume); R is the radial distance from the spargewell; and Q is the injection rate (Figure 25-3).

SITE DESCRIPTION AND RELEASE HISTORY

The case study site is an active bulk fuel and cardlock facility located at 210 G Street (State Highway 65) in Lincoln, California (Figure 25-4). The current facilities consist of a building housing a convenience store, office, and warehouse on a 0.2-hectare (0.5-acre) parcel owned and operated by Ramos Oil Company. The station previously included seven USTs, which were removed

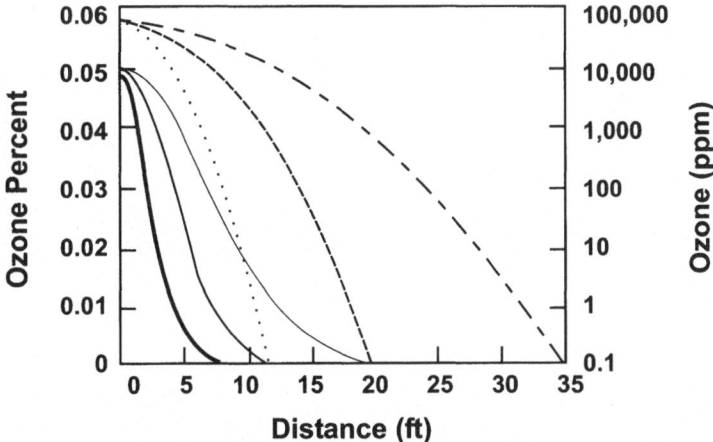

Figure 25-3.   **Steady-State Radial Ozone Transport at Ozone Half-Lives of 5, 15, and 45 Minutes (Coarse to Fine Lines) Ozone concentrations are shown in percent of saturation and ppm (dashed) (Reprinted from Clayton, 1998)**

from the site in 1998 (GeoTrans, 2001). After removal of the USTs, the service station was remodeled to include retail islands and a convenience store. Fuel is supplied to the retail island dispensers from aboveground tanks. Currently there are no USTs on the site. The topography in the vicinity is flat, lying at an approximate elevation of 50 meters (165 feet) above mean sea level. The site is bordered to the north, east, and south by commercial properties and to the west by Union Pacific Railroad tracks. Direction of ground water flow is to the south-southwest.

Previous Environmental Work. Petroleum hydrocarbon impacts were noted in the gasoline and diesel UST pit sidewalls, piping trenches, and dispenser locations. Impacted soils were removed based on visual observation and transported to and disposed of at a licensed facility. Confirmatory soil sampling indicated a small area of impacted soil that could not be removed safely because of underground utilities. This soil was left in place.

Ground water was found in the UST pits at approximately 2 meters (7 feet) bgs. Grab ground water sample results confirmed the presence of petroleum hydrocarbons and fuel oxygenates in the ground water. Water was removed from the UST pits on several occasions and transported off-site for disposal by Ramos Oil Company. The UST pits were backfilled with pea gravel and capped with asphalt.

Three ground water monitoring wells were installed on the site in 1993. During March 2000, Ramos Environmental Engineering completed 18 push-

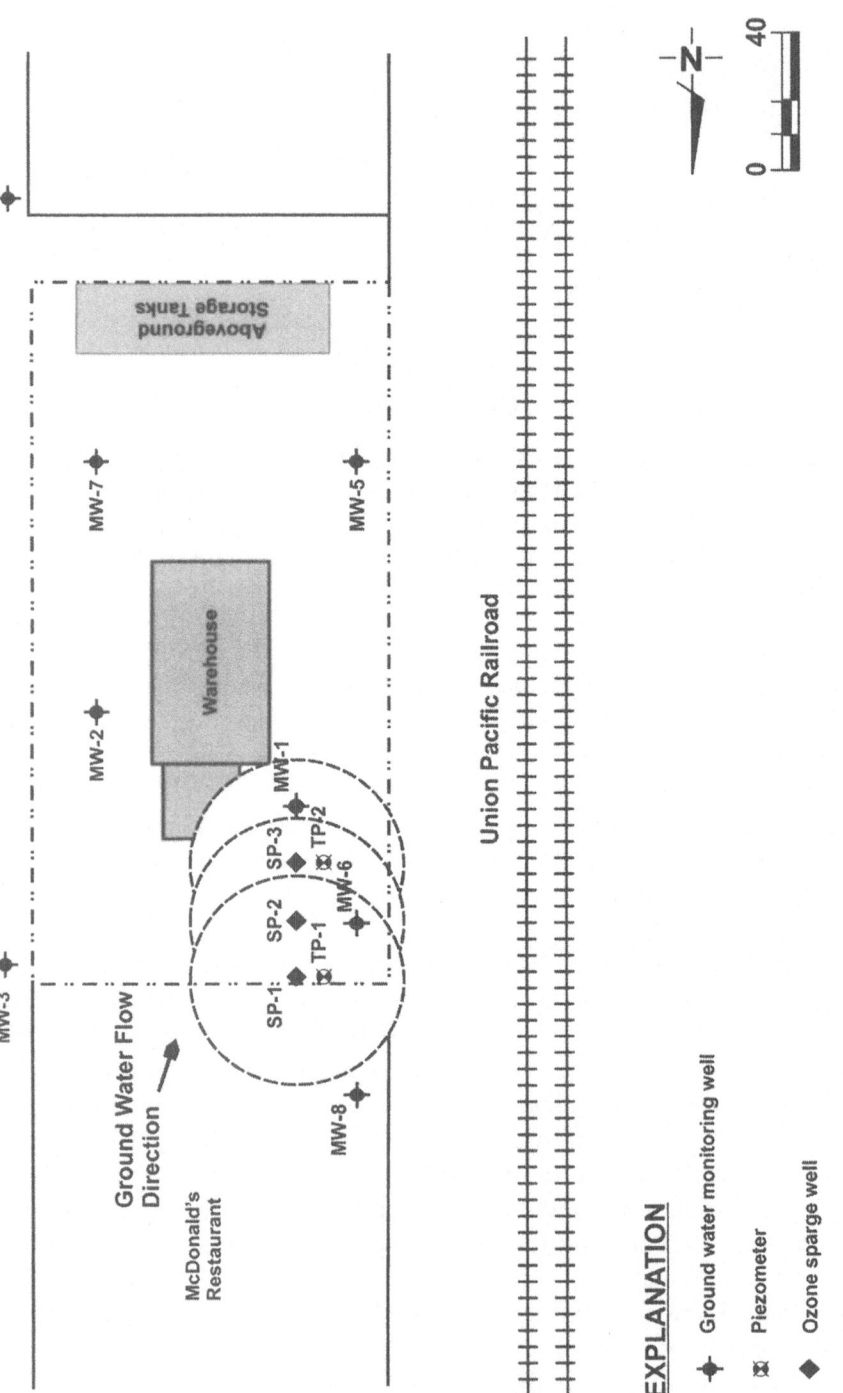

Figure 25-4.   Treatment Site Plan with Projected Radius of Influence (Modified from Gettler-Ryan, Inc., 2001)

probe borings to evaluate the distribution of petroleum hydrocarbons in the soil and ground water on and in the vicinity of the site. Petroleum hydrocarbons were reportedly detected in ground water on the eastern, southeastern, northern, and western portions of the site. Six additional monitoring wells, two piezometers, and three Spargepoints® were installed in June 2000.

During June 2001, Gettler-Ryan supervised the installation of three additional monitoring wells to further evaluate the downgradient extent of petroleum hydrocarbons in the ground water. Figure 25-5 shows an example log of installation of SP-2.

The most recent quarterly ground water monitoring event was performed by JJW Geosciences on October 2, 2001 (Gettler-Ryan, 2001). Prior to ozone injection, the highest reported contaminant levels were 270 µg/l TPH and 7,600 µg/l MTBE in MW-1 and 12,000 µg/l MTBE and 870 µg/l TBA in

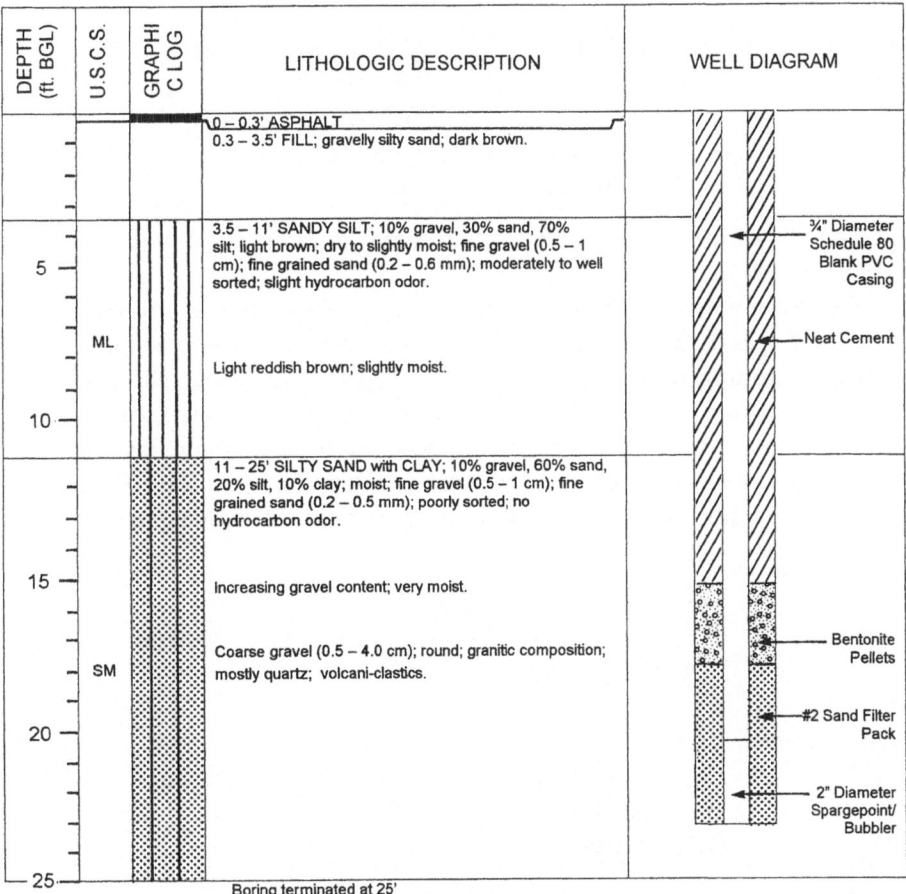

Figure 25-5. **Example Well Construction Log Showing Microporous Spargepoint® Installation (Modified from Gettler-Ryan, Inc., 2001)**

TP-2, with a mean value for MTBE of 6,400 µg/l. No benzene was reported above laboratory detection limits (1 µg/l).

SITE CONDITIONS

For determining the expected time course of removal and daily dosage of ozone, a mass balance is normally conducted in advance of treatment. A gasoline spill with predominantly MTBE contamination is used as an example. The following site information was obtained:

Soil type:	silty sand
Depth to water table:	1.8 meters (6 feet)
Vertical depth of contamination:	6.4 meters (21 feet)
Aqueous mean concentration:	MTBE 6,000 µg/l
	TBA 550 µg/l
	TAME 10 µg/l
Soil mean concentration:	MTBE 500 mg/kg
	TBA 500 mg/kg
	TAME 10 mg/kg
Saturated treatment zone depth:	7.9 meters (26 feet)
Ground water flow direction:	south-southwest (SSW)

A schematic of the ozone well installation and treatment volume are shown in Figure 25-6.

For this site the ratio of MTBE in aqueous phase compared to solid phase (adsorbed) was assumed to be 1:1. A liter of sand from the site weighed 1,600 grams, containing 1,300 grams solid and 300 grams of water (300 cubic centimeters or 30% volume). A circular area, 6.6 meters (25 feet) in radius and 6.4 meters (21 feet) deep was expected to be treated by a single Spargepoint® placed at 7.9 meters (26 feet) below water table depth (Figure 25-6). For a given cylinder, the liquid volume and solid volume are computed separately:

Liquid volume = $V_L = \pi r^2 h n$
$$= (3.14)(6.6)^2(6.4)(0.30)$$
$$= 263 \text{ cubic meters (344 cubic yards) or 263,000 liters}$$
$$(70,000 \text{ gallons}) \tag{25-2}$$

Solid volume = $V_S = \pi r^2 h (1 - n)$
$$= (3.14)(6.6)^2(6.4)(0.70) = 613 \text{ cubic meters (802 cubic yards)}$$
$$\text{or } 613,000 \text{ liters (162,000 gallons) or 796.9 kilograms}$$
$$(1,757 \text{ pounds}) \tag{25-3}$$

Where: V_L = liquid volume (liters)
V_S = solid volume (kilograms)
r = radius of influence (meters)
h = height of cylinder (meters)

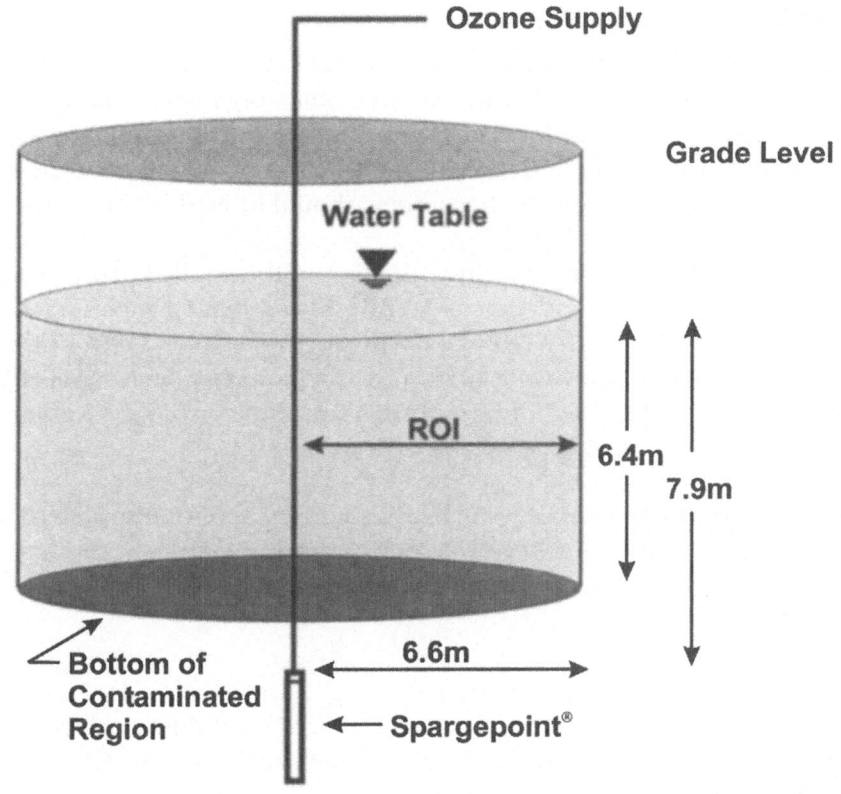

Figure 25-6.   **Treatment Zone for an Example Microporous Sparge-point®**

To convert to mass, the mean concentrations of MTBE are multiplied times the respective water volume and solid volume.

Liquid content = $V_L \times C_{aq}$
$$= 263,000 \text{ liters} \times 6,000 \text{ µg/l} = 1,578 \text{ grams MTBE} \qquad (25\text{-}4)$$

Solid content = $V_S \times C_S$
$$= 796.9 \text{ kilograms} \times 500 \text{ mg/kg} = 398 \text{ grams MTBE} \qquad (25\text{-}5)$$

Total content: 1,976 grams MTBE

Equivalent mean concentrations for TBA and TAME were similarly computed yielding:

	MTBE	TBA	TAME
Liquid content	1,578 grams	144 grams	2.6 grams
Solid content	398 grams	398 grams	7.9 grams
Total content	1,976 grams	542 grams	10.5 grams

EXPECTED OXIDANT DEMAND

The expected oxidant demand is a sum of stoichiometric VOC demand, oxidizable metals demand, soil demand, and other organic demand. Since BTEX compounds compete for ozone, and the oxidation demand can be estimated at over three mole equivalents for the benzene ring, other volatile organics, particularly aromatic hydrocarbons, should be analyzed for.

Stoichiometric VOC Demand. Investigation of the applied ozone dosage (mole/mole) for 80% degradation of MTBE showed that 5.5 moles of ozone would degrade 1.0 mole of MTBE at pH 8 (Karpel vel Leitner, 1994). This corresponds to 3.0-gram equivalents of ozone to degrade 1.0 gram of MTBE. For TBA, 3.9 grams of O_3 react with 1.0 gram of TBA; and for TAME, 3.0 grams of O_3 react with 1.0 gram of TAME (Table 25-2).

Oxidizable Metals Demand. With oxidizable metals, the ferrous iron (Fe^{+2}) demand can be estimated from the dissolved iron content in acidic soils. A mole of ozone can be consumed by two moles of Fe^{+2}.

$$O_3 + 2Fe^{+2} \; Fe_2O_3 \tag{25-6}$$

If the mean dissolved iron content is 0.1 mg/l, then the oxidant demand would be expected to be 0.05 mg/l or 0.05 mg/l $\times$ 263,000 liters or 13.1 grams total for the treatment region. Samples from the site showed no ferrous iron present during the treatment period.

Soil Demand. For sandy soils with aerobic conditions and acidic pH, the assumed oxidation demand of 30% of residual total organic carbon (non-VOC) has been used frequently. The gas reacts less with the soil substrate than aqueous oxidants (such as peroxide and permanganate). Bench-scale testing can be conducted on drained, uncontaminated soil samples to estimate the soil demand.

Table 25-2. **Oxidative Demand and Gram Equivalents of Ozone for Selected VOC Contaminants**

Compound	Ozone Molar Equivalents (moles per mole of contaminant)	Gram Equivalents (gram ozone per gram contaminant)
MTBE	5.5	3.0
ETBE	3.0	1.6
Benzene (other BTEX)	6.0	3.4
Naphthalene	8.0	3.0
TBA	6.0	3.9
TAME	5.0	3.0

Other Organics.   Early ground water analysis at the site showed toluene, ethylbenzene, and xylenes were present at a combined concentration of 100 µg/l or less in the aqueous phase. Toluene will require three moles of ozone to decompose the benzene ring and six mols to completely convert carbon to CO_2.

$$C_7H_8 + 6O_3 \, 7CO_2 + 4H_2O \qquad\qquad (25\text{-}7)$$

The weight ratio is 3.4 grams of ozone to decompose 1 gram of toluene. For the volume of water, 263,000 liters times 0.1 mg/l yields 26 grams of toluene (C_7H_8). This requires 27 grams of ozone (26 grams of toluene times 1.04 grams ozone per gram of toluene equals 27 grams of ozone).

Total Ozone Demand.   The portable (Model 3600) C-Sparger® system normally generates from 5 to 15 grams per hour (g/hr) or 120 to 360 grams per day (g/day), depending upon the amount of oxygen supplied to it. In the site example given here, three spargewells were operating, so each was set to receive one-third of the total loading or 1.7 g/hr to 5 g/hr.

PROJECTED TIME TO TREAT (DURATION) COMPUTATION — MASS BASIS

The supplied ozone is divided into the total ozone demand (Table 25-3) to yield treatment time. If ozone is supplied at 5 g/hr to three spargewells, the total ozone demand can be divided by the spargewell oxidant supply to yield an approximation of the time to treat.

TOD/S = 8,281 grams/5.0 grams/hour = 1,656 hours or 69 days   (25-8)

> Where   TOD = total ozone demand
> S = spargewell oxidant supply.

MONITORING THE VOC DECAY

The aqueous phase was sampled (ground water sampling) from monitoring wells situated at varying distances from the injection spargewells. Measurements of MTBE, TBA, and TAME concentrations were made at time intervals at three monitoring wells from the beginning of the injection of 5 g/hr. The distances of TP-2 and MW-6 from the nearest spargewell were 2.4 meters (8 feet) and 4.9 meters (16 feet) respectively.

FIELD RESULTS

Monitoring wells MW-1 and MW-6 and piezometer TP-2 were the most significantly affected by the pilot test. These wells had significant concentrations of MTBE prior to the test and were within the zone of influence of the ozone sparge system. MTBE concentrations prior to the test were much lower in

Table 25-3.   **Total Ozone Demand for Oxygenates**

	MTBE	TBA	TAME	Sum
Mean Total Mass	1,976 grams	542 grams	11 grams	2,529 grams contaminant
Stoichiometric VOC Demand	5,928 grams	2,114 grams	33 grams	8,075 grams O_3
Oxidizable Metals Demand				179 grams
Soil Oxidant Demand*				0 grams
Other Organics Demand				27 grams
Total Ozone Demand =				8,281 grams

* Bench-scale Soil Oxidant Demand testing not performed.

TP-1 and upgradient well MW-8. Sampling results are presented in Table 25-4. MTBE concentrations from MW-1 as measured over the course of the pilot test are depicted in Figure 25-7.

MTBE concentrations decreased significantly in all of the wells sampled as part of the field test. In MW-1 the MTBE concentration decreased 71% from 5,800 µg/l on July 24, 2001 to 1,700 µg/l on October 2, 2001. In this same time period, MTBE concentrations in MW-6 decreased from 170 µg/l to less than 0.5 µg/l, and in TP-2 the decrease was from 12,000 µg/l to 50 µg/l; these are decreases of more than 99.8% and 92%, respectively.

During the test period, TPH concentrations were seen to increase slightly in wells MW-1 and MW-6. Concentrations in both wells were below detection limits prior to the test. The highest concentration reported during the test was 270 µg/l.

Dissolved oxygen levels in MW-1, MW-6, TP-1, and TP-2 were elevated throughout the course of the test. The levels were between 4.9 and 9.0 mg/l in MW-1; 8.5 and 9.3 mg/l in MW-6; 8.3 and 9.6 mg/l in TP-1; and 8.4 and 10.1 mg/l in TP-2. In MW-8, which is outside of the direct area of influence of the ozone Spargepoints®, the levels were between 3.2 and 3.5 mg/l.

Ferrous iron levels decreased below detection limits in all samples collected during the test. The addition of ozone causes ferrous iron (Fe^{+2}) to convert to ferric iron (Fe^{+3}) which will precipitate out on soil. Dissolved iron normally will go to nondectable levels (less than 0.05 mg/l).

The rate of decay of MTBE is presented as k values for different distances from the spargewell (Table 25-5). In Figure 25-8, the change in both MTBE and TBA is graphed for TP-2. The rate of removal was found by solving a regression line on time versus concentration for the period of July 24 through October 2 and substituting into equation 25-1. For MTBE removal, C equals

Table 25-4. Groundwater Sampling Results Ramos Oil, Lincoln, California

Time	Dissolved O$_2$ (mg/l)					TPH(D) (µg/l)					MTBE (µg/l)					TBA (µg/l)					TAME (µg/l)				
	MW-1	TP-1	TP-2	MW-8	MW-6	MW-1	TP-1	TP-2	MW-8	MW-6	MW-1	TP-1	TP-2	MW-8	MW-6	MW-1	TP-1	TP-2	MW-8	MW-6	MW-1	TP-1	TP-2	MW-8	MW-6
09/28/00	—	—	—	—	—	ND	—	—	—	ND	6,000	—	—	160	1,000	800	—	—	ND	15	9.6	0.77	17	280	1.3
07/24/01	8.1	9.6	8.9	3.2	8.9	270	170	150	100	240	5,800	27	12,000	2.3	170	550	ND	870	ND	ND	ND	ND	17	1.1	1.0
08/07/01	9.0	9.2	9.1	3.3	8.5	210	ND	ND	ND	100	6,000	22	2,500	5.2	120	340	ND	ND	ND	ND	ND	ND	ND	2.8	ND
08/21/01	6.5	8.3	10.1	3.5	9.3	260	97	ND	NS	100	4,700	0.95	1,000	3.9	0.92	470	ND	30	ND	ND	ND	ND	ND	3.9	ND
09/04/01	4.9	9.0	8.4	NS	8.8	240	76	110	ND	140	5,700	0.64	710	NS	ND	140	ND	120	ND	ND	6.2	ND	ND	NS	ND
09/18/01	NS	NS	NS	NS	NS	ND	ND	ND	ND	ND	3,200	1.1	150	7.4	22	ND	ND	ND	ND	ND	ND	ND	ND	ND	ND
10/02/01	5.2	9.1	9.5	3.4	8.5	210	62	72	ND	72	1,700	ND	50	21	ND	420	ND	ND	ND	ND	1.8	ND	ND	7.5	ND

NS = not sampled
ND = nondetectable
Note: O$_3$ injection began on 7/10/01.

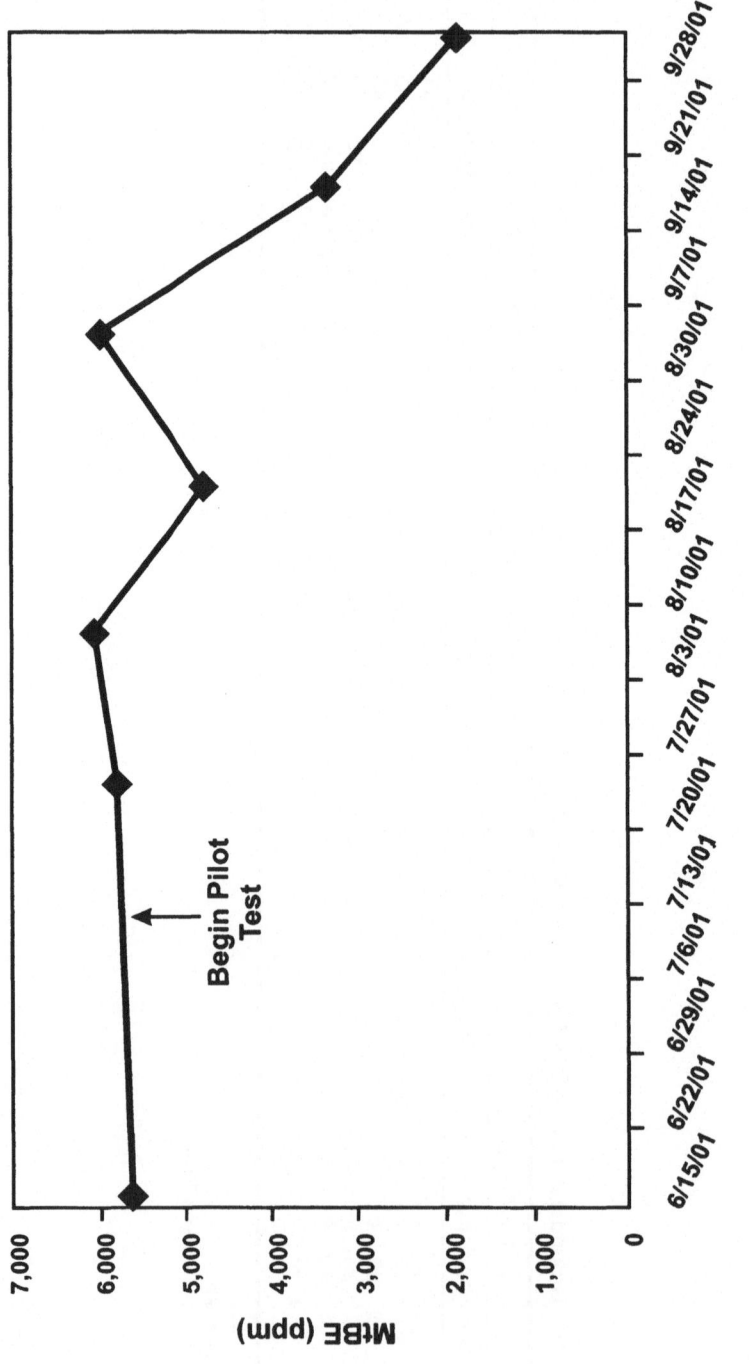

Figure 25-7.   Removal of MTBE at MW-1

Table 25-5.   **Rates of Decay of MTBE and TBA from Ground Water Samples Obtained at Different Distances from Spargepoints®**

	Distance from Spargepoint®	
	8 Feet (TP-2)	16 Feet (MW-1)
MTBE	−0.077	−0.057
TBA	−0.075	ND*

*ND = non-detectable

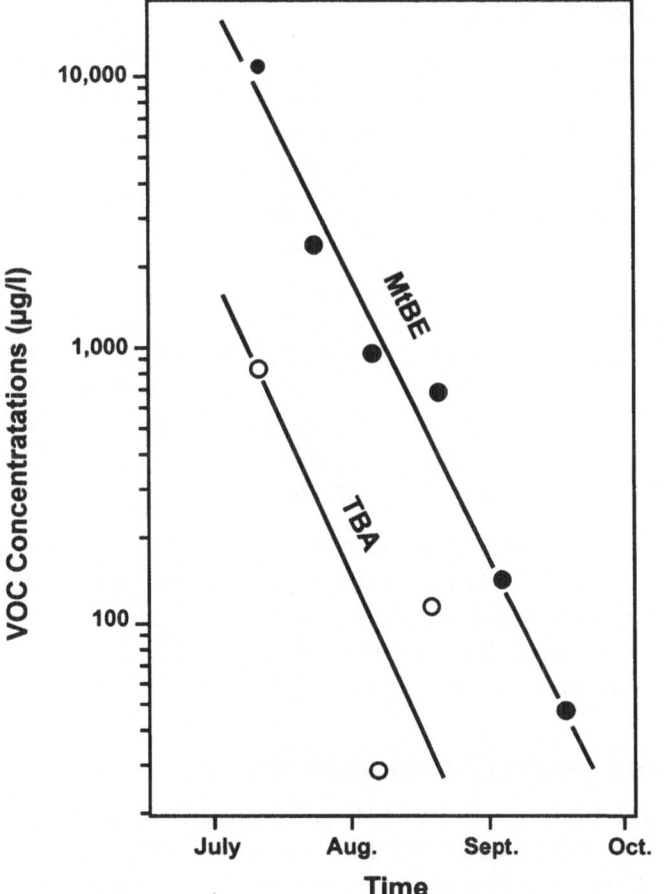

Figure 25-8.   **Projection of Time-Course Kinetics of Attenuation of VOCs Observed at Well TP-2**

$C_o e^{-0.077t}$ was obtained for the 2.4-meter (8-foot) distance (well TP-2), and C equals $C_o e^{-0.057t}$ for the rate of decay at the 4.8-meter (16-foot) distance (well MW-6). This corresponds to time to half concentration (half-lives) of 9 days and 12 days, respectively. An expected projection outwards would yield half-lives of 15 days and 18 days at distances of 7.5 meters (24.6 feet) and 10 meters (33 feet) from a source spargewell. Doubling the ozone mass rate would result in doubling of the rate of decay or halving of the half-lives.

Note that the rate of decay of TBA followed that of MTBE at well TP-2. Although TBA is considered to be a daughter product of oxidation of MTBE, it too is oxidized by ozone, resulting in a similar removal rate (Figure 25-8).

SITE COST COMPARISON

Ozone oxidation treatment has been compared to air sparging and use of re-circulation wells (Dreiling *et al.*, 1998). Since VOCs are decomposed before reaching the surface, the main cost savings come from low volume compressor use and limited or no activated carbon use. The costs from this site were:

Site/Risk Assessment:	$10,000
Design:	$ 3,500
Capital Equipment:	$25,000
Installation:	$15,000
Operation and Maintenance: (18 months)	$30–$50/month (electrical for compressor and ozone generator) $600/month (labor)
Closure:	$ 5,000
TOTAL PROJECT COST:	**$70,200**

A comparison between oxidant treatment and biological treatment leans toward oxidant treatment of sites. Oxidants more quickly target high health-risk compounds such as BTEX, naphthalene, oxygenates, and chlorinated solvents when comingled. If the volume of the spill is large, or extensive off-site migration has occurred, costs will be higher than those indicated above.

CONCLUSIONS AND RECOMMENDATIONS

The ozone microsparging system has been successful in significantly reducing MTBE concentrations in the ground water at this site. Reductions of 71% to over 99% were observed over the test period of approximately three months. Treatment goals for MTBE, TBA, and other oxygenates were satisfactorily achieved.

All oxygenates exhibited substantial decay. TBA showed a similar rate of decrease with no apparent increase in TBA concentrations observed during the most rapid MTBE attenuation.

Dissolved oxygen levels increased significantly in the test area. This will increase the natural biodegradation process in the treatment area.

Computations of the oxidation demand and comparison with the delivery rate provided an approximation of time to treat assuming logarithmic first-order decay. The radius of influence was found to be about 10.7 meters (35 feet). The observed field removal rate from well TP-2 exhibited a decay rate corresponding to that expected from oxidant demand (a time of 72 days).

REFERENCES

Clayton, W.S. 1998. Ozone and contaminant transport during in-situ ozonation. In: *Physical, Chemical, and Thermal Technologies, Remediation of Chlorinated and Recalcitrant Compounds*, pp. 389–395. (Wickramanayake, G.B. and Hinchee, R.E., Eds.). Columbus, Ohio, Battelle Press.

Dreiling, D.N., Henning, L.G., Jurgens, R.D., and Ballard, D.L. 1998. *Multi-Site Comparison of Chlorinated Solvent Remediation Using Innovative Technology*, pp. 247–253. (Wickramanayake, G.B. and Hinchee, R.E., Eds.). Columbus, Ohio, Battelle Press.

GeoTrans, Inc. 2001. *Monitor Well Installation Report, Ramos Cardlock and Bulk Facility, 210 "G" Street, Lincoln, Placer County, California*. February 23, 2001.

Gettler-Ryan, Inc. *Monitoring Well Installation Report, Ramos Cardlock and Bulk Facility, 210 "G" Street, Lincoln, California*. July 30, 2001.

Henley, E.J. and Seader, J.D. 1981. Equilibrium-state separation operations. In: *Chemical Engineering*, Chapter 16. New York, New York, John Wiley and Sons.

Karpel vel Leitner, R.N., Papailhous, A.L, Crove, J.P., Payrot, J., and Dore, M. 1994. Oxidation of methy tert-butyl ether (MTBE) and ethyl tert-butyl ether (ETBE) by ozone and combined ozone/hydrogen peroxide. *Ozone Science and Engineering*. 16, 41–54.

Kerfoot, W.B., McCulloch, W., and Connors, J. 1996. The use of micropores for fine bubble creation in the removal of PCE in groundwater. In: *Proceedings of the Sixth West Coast Conference on Contaminated Soils and Groundwater*, Newport Beach, California. March 13, 1996. (Calabrese, E.J. and Kostecki, P.T., Eds.). Amherst, Massachusetts, AEHS.

Kerfoot, W.B. 2000. Ozone microsparging for rapid MTBE removal. In: *Chemical Oxidation and Reactive Barriers, Remediation of Chlorinated and Recalcitrant Compounds*, pp. 187–194. (Wickramanayake, G.B., Gavaskar, A.R., and Chen, A.S.C., Eds.). Columbus, Ohio, Battelle Press.

Kerfoot, W.B. and McGrath, A. 2001. Microbubble oxidation smokes MTBE and BTEX. *Contaminated Soil Sediment and Water*. Spring (Special Issue), 77–78.

Kerfoot, W.B. 2002. Microbubble ozone sparging for chlorinated ethene spill remediation. In: *Innovative Strategies for the Remediation of Chlorinated Sol-*

vents and DNAPLs in the Subsurface. Washington, D.C., American Chemical Society, Division of Environmental Chemistry.

Miller, K.D., Heath, J.C., and Johnson, P.C. 2001. MTBE biobarrier demonstration at Port Hueneme. *Contaminated Soil Sediment and Water.* Special Oxygenated Fuels Issue. 6–9.

Siegrist, R.L. and Watts, A.J. 2001. Chemical processes for the in-situ oxidation of contaminants in soil and groundwater. In: *The First Annual Conference on Oxidation and Reduction Technologies for In-Situ Treatment of Soil and Groundwater*, Niagara Falls, Ontario, Canada. June 25–19, 2001. (Brown, R. and Al-Ekabi, H., Eds.). London, Ontario, Canada, Redox Technologies.

Weast, R.C., Ed. 1972. *Handbook of Chemistry and Physics*, pp. D111. Cleveland, Ohio, The Chemical Rubber Co.

U.S. Patents 5,855,775; 6,083,407; 1999, 2000; 6,284,143; 6,306,296; 6,312,605; 2001.

CHAPTER 26

MTBE Cleanup Technology Evaluations at the Port Hueneme NETTS

Ernest E. Lory, Naval Facilities Engineering Service Center

The U.S. Department of Defense (USDOD) Strategic Environmental Research and Development Program (SERDP) established four National Environmental Technology Test Sites (NETTS). The goal of the SERDP-funded test site program is to provide accessible, well-supported field locations for project proof-of-principle tests, applied research, and comparative demonstration, as well as to facilitate transfer of innovative environmental technologies from research to full-scale use. Since selection in 1994 as a NETTS, the Port Hueneme site on the Navy Base Ventura County (NBVC), Port Hueneme, California has supported demonstrations of over 15 innovative MTBE cleanup technologies. The technologies are designed to find, monitor, and remediate MTBE. Past and future demonstrations will benefit both the public and private sector, as well as have significant cleanup benefit for the plume at NBVC, Port Hueneme.

From late 1984 to early 1985, approximately 40,900 liters (10,800 gallons) of gasoline leaked from piping under the Naval Exchange (NEX) gas station at Port Hueneme. Since 1985, the Navy has taken actions to prevent any damage to the environment from the leaks.

As a result of the testing, the gasoline additive MTBE was detected in the ground water in December 1992. No specific cleanup actions or plans concerning the MTBE were made at that time because the USEPA and other scientific institutions did not recognize it as a risk to human health or the environment. Results indicate that the MTBE plume is located in areas underlying the industrial activities of the Base. The MTBE plume at NBVC begins at the NEX gas station (Figure 26-1) and extends about 1,500 meters (4,800 feet) to the control and containment system, with a maximum width of about 150 meters (500 feet). The MTBE plume is confined to a semi-perched aquifer located at a depth from 2 to 7 meters (6 to 25 feet) bgs. A clay layer, which ranges from approximately 6 to 15 meters (20 to 50 feet) in thickness in the vicinity of the

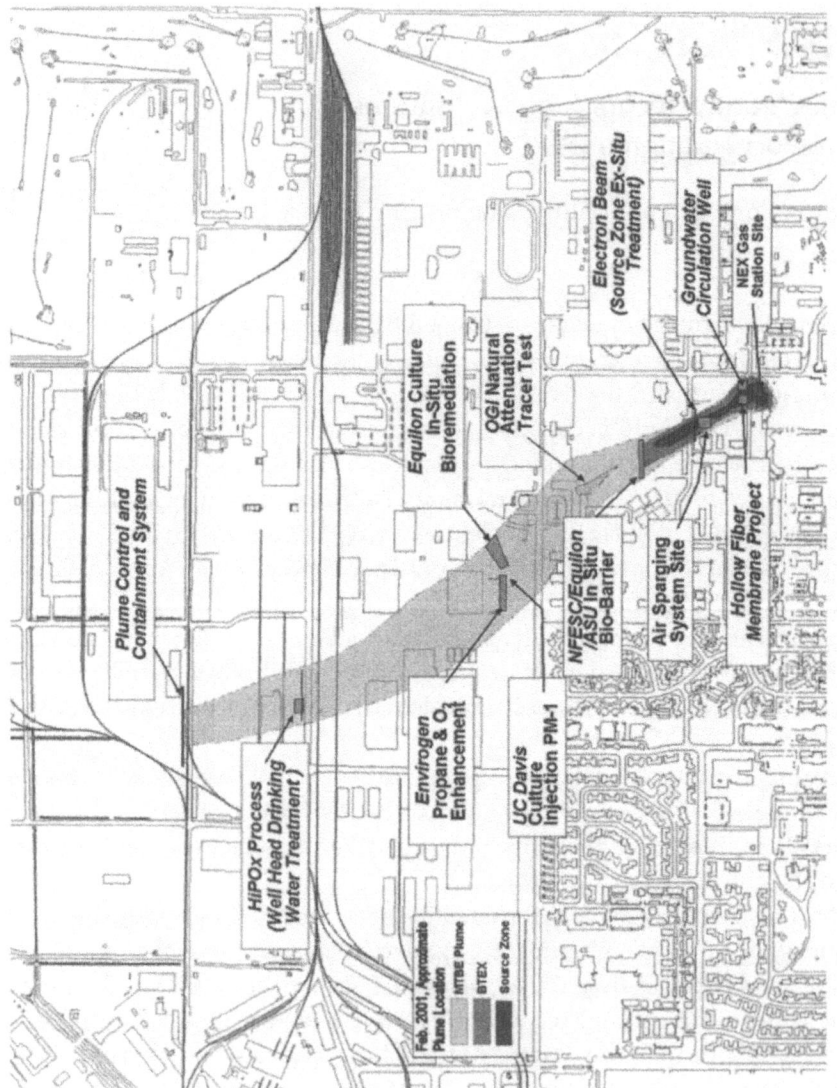

Figure 26-1.   Gasoline and MTBE Plume at NBVC Port Hueneme Site with Demonstration Locations

plume, isolates the impacted water from underlying aquifers. The MTBE plume is totally confined within the limits of the NBVC boundaries.

Drinking water production wells throughout the county pump water from an entirely different aquifer system, called the "lower aquifer system," which ranges from 200 to 300 meters (800 to 1,000 feet) deep.

Innovative MTBE remediation technologies demonstrated at the Port Hueneme NETTS include the following categories: (1) physical and chemical, (2) biological, and (3) natural attenuation assessments. The following sections describe a wide range of innovative technologies and some conclusions drawn from field tests and evaluations.

Additional information about the Port Hueneme NETTS, including various projects and technical reports can be found at: enviro.nfesc.navy.mil/erb/support/netts/main.htm.

GROUND WATER CIRCULATION WELL ENVIRONMENTAL CLEANUP SYSTEMS

Conducted by:
Barry Spargo, Naval Research Laboratory, Washington, D.C.
SBP Technologies, Inc., Gulf Breeze, Florida

The ground water circulation well technology is designed to stimulate microorganisms in ground water, thus optimizing environmental conditions in the capillary fringe, surrounding aquifer, and vadose zone for degrading petroleum constituents. The purpose of the ground water circulation well system at the NEX plume was to provide a viable, cost-effective method to remove the gasoline contaminant from the shallow water aquifer.

Remedial efficacy of a ground water circulation well is achieved through the combined use of physical and biological processes. The technology consists of a specially adapted ground water well, which serves as a negative pressure stripping reactor, an aboveground mounted blower, and an off-gas treatment system (Figure 26-2). As the water level rises in the well due to the negative pressure generated by the blower, fresh air is drawn into the system through a pipe leading to the stripping reactor and bubbles up through the raised water level. An uplift effect is created, pulling ground water up from the deep section of the well and circulating it out at a shallow point in the well, creating a circulation cell in the aquifer. Volatile contaminants from the ground water are released to the circulated air as a result of the system operation. The air used for circulation is pulled out of the well and passed through a carbon filter so that only clean air is released to the environment.

The ground water circulation also moves air through the soil, exposing indigenous soil bacteria to more contaminants and oxygen; therefore, more contaminant is degraded than would be in a static system. The bubbling effect aerates the water, supplying necessary oxygen to the bacteria.

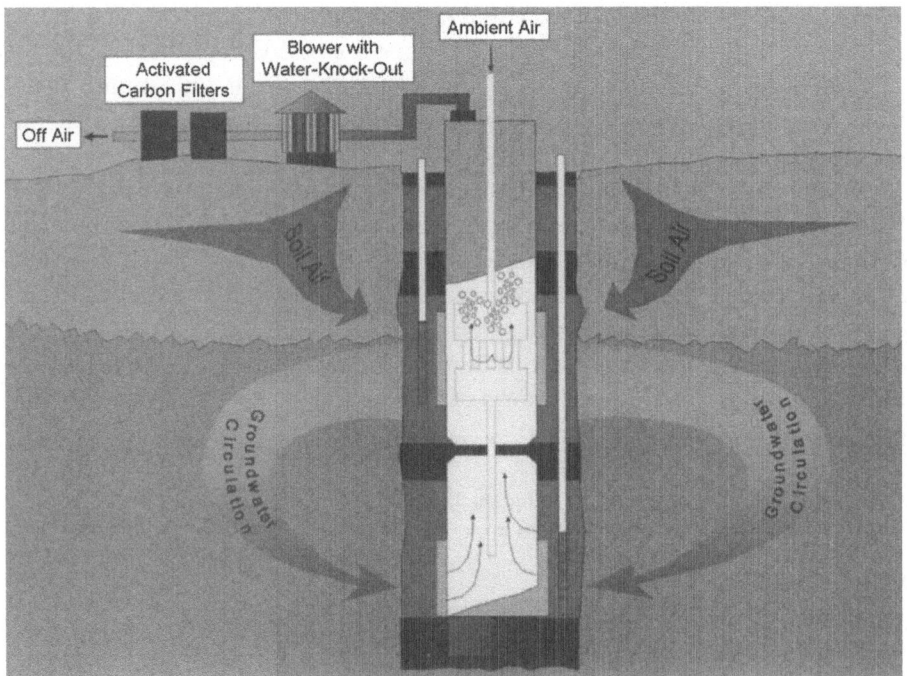

Figure 26-2.   **Simplified Cross-Section of Ground Water Circulation Well Treatment Process**

During a short field test, samples were analyzed for MTBE. Analyses indicated a minimal reduction in MTBE concentration downgradient of the ground water circulation well. Due to source area fluctuation in MTBE concentration and periodic downtime, this technology could not be fully evaluated for its ability to reduce MTBE concentrations.

For additional information:

Allmon, W.E., Everett, L.G., Lightner, A.T., Alleman, B., Boyd, T.J., and Spargo, B.J. 1999. *Groundwater Circulating Well Technology Assessment*. Naval Research Laboratory, NRL/PU/6115-99-384

Spargo, B. 1996. *In Situ Bioremedation and Efficacy Monitoring*. (Spargo, B. Ed.). Naval Research Laboratory, NRL/PU/6115-96-317.

IN SITU AIR SPARGING SYSTEM

Conducted by:

Catherine Vogel, Air Force Research Laboratory,
  Tyndall Air Force Base, Florida
Andrea Leeson, Battelle Columbus Division, Columbus, Ohio
Rick Johnson, Oregon Graduate Institute, Beaverton, Oregon
Paul Johnson, Arizona State University, Tempe, Arizona

Air sparging is the process of injecting clean air directly into an aquifer for remediation of impacted ground water (Figure 26-3). The objective of air sparging is to force air through the impacted aquifer materials and vadose zone to provide oxygen for bioremediation and to strip contaminants out of the aquifer. Bioremediation refers to enhancing the growth of naturally occurring microorganisms that use contaminants, such as petroleum products, as a food source, or as a cometabolite. In so doing, impacted areas can be remediated naturally, with contaminants detoxified. Volatilization can carry contaminants from the ground water to the unsaturated zone, where they can either be biodegraded or removed by an SVE system.

This demonstration validated an air sparging design paradigm. It was created during the scale-up of the air sparging system in an earlier project — Aerobic Biodegradation/Air Sparging. The design paradigm is a manual, intended for the USDOD environmental engineering community, which aims

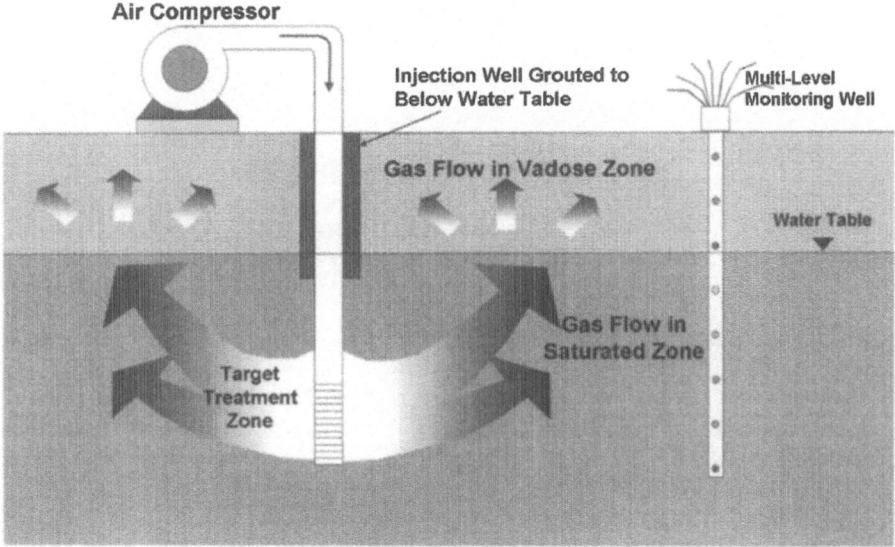

Figure 26-3. **Simplified Cross-Section of Air Sparging Treatment Process and Multi-level Monitoring Well Configuration**

to reduce the "trial and error" nature of the air sparging practice by providing a fundamental understanding of the air sparging process, basics design techniques, and cost-effective monitoring methods.

The air sparging system and monitoring devices installed included 18 sparge wells, 12 multi-level subsurface samplers, 12 neutron probe access tubes (to measure the volumetric water content of the vadose soil in order to detect air channels), 6 ground water monitoring wells, and 4 SVE wells (Figure 26-4). Monitoring techniques include SVE off-gas testing, ground water analysis, and testing of fuel hydrocarbon vapor off-gassing from soil. Tracer gasses used were sulfur hexafluoride (SF_6) for ground water and vapor samples, and helium for vadose zone sampling. Monitoring air sparging operations provided information to assess contaminant removal rates given different and changing environmental parameters.

Physical model studies have demonstrated that there appears to be an optimal air injection rate. Injection rates above this optimal rate appear to cause decreases in permeability. Studies also show that pulsing the airflow has little effect on oxygen transfer, but does improve volatilization. During evaluation of this technology, impacted ground water originally at concen-

Figure 26-4.   **Air Sparging System and Associated Monitoring Well Configurations**

trations of 30 to 40 mg/l MTBE was remediated to non-detect levels (detection limit of 5 µg/l).

For additional information:
Leeson, A., Johnson, P.C., Johnson, R.L., Hinchee, R.E., and McWhorter, D.B. 1999. *Air Sparging Design Paradigm*, Battelle Memorial Institute. www. estcp.org/documents/techdocs/Air_Sparging.pdf.

EXTRACTION OF MTBE BY A HOLLOW FIBER MEMBRANE

Conducted by:
Mark Kram and Arturo Keller, UC Santa Barbara, California

This report summarizes the results of a demonstration of selected technologies used to treat ground water impacted by dissolved MTBE and other VOCs. The specific technologies include spray aeration vacuum extraction (SAVE), developed by Remediation Service International, and hollow fiber membrane (HFM) degasification, developed by UC Santa Barbara.

The SAVE system is a mobile remediation system which uses the principles of air stripping and combustion to remove and treat hydrocarbon contaminants in ground water. The system incorporates a large air-water interface in combination with operating temperatures of 33 to 50°C (92 to 122°F) and a vacuum of −193 kilopascal (kPa) (−28 pounds per square inch [psi]). The stripping element of the process is enhanced using a spray nozzle to produce a fine water vapor that maximizes the surface area between the water and air. Added vacuum and a high system temperature lead to the partitioning of highly soluble VOCs, such as MTBE, into the vapor phase.

The structure of the HFM provides for a large contact area between vapor and water phases, allowing efficient mass transfer to occur with relatively smaller volumes of air than is typically required of air strippers. In addition, there is no further contact between the VOCs and water, minimizing partitioning of VOCs back into the liquid phase. Therefore, the removal efficiency is much higher than in conventional air-stripping. Due to its small size, it becomes practical to combine the HFM module with other treatment processes, such as the SAVE system, to achieve desired remediation goals.

Water containing VOCs and SVOCs was passed through the inside of the HFM, while a vacuum was applied to the outside of the fibers (Figure 26-5). The organic compounds were transferred to the gas phase outside of the fiber and then were destroyed using the internal combustion unit (ICU) component of the SAVE system. The hollow fibers are made of materials that retain the flowing water, yet allow for gaseous exchange. The fibers are bundled in a container that includes separate inlet and outlet ports for water and organic vapors. Impacted ground water enters the module and comes in contact with clean air. Treated water and organic laden vapor exit the system as separate streams.

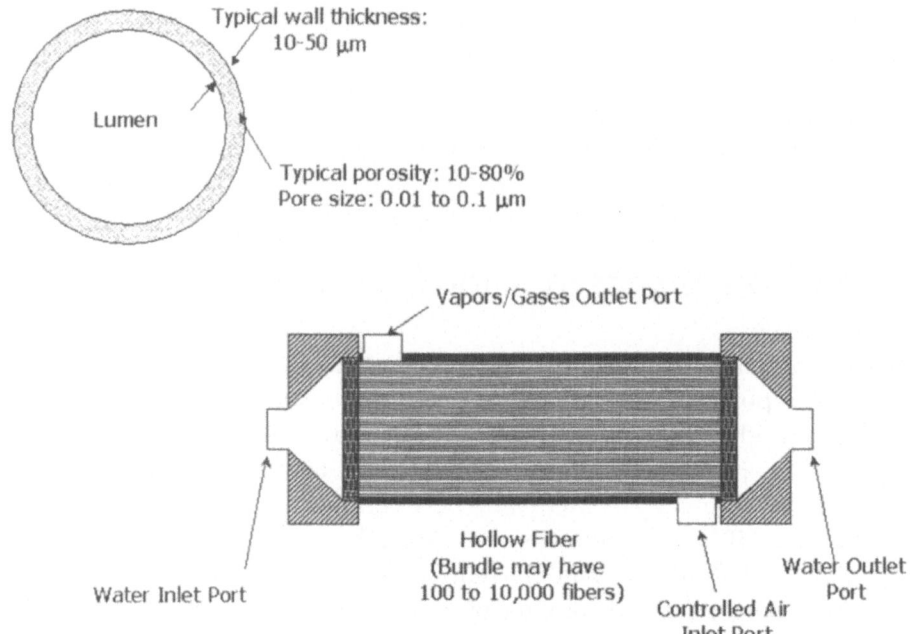

Figure 26-5.   **Cross-Section of Hollow Fiber Membrane Unit**

The actual field set up, from the beginning to the end of the treatment process, included water softener units, storage tanks, particulate bag filters, a SAVE system with an ICU (for off-gas treatment and supplemental power), one HFM unit, and two GAC units (Figure 26-6). The system was configured specifically for remediation of dissolved MTBE and other volatile VPH constituents extracted from a monitoring well. During the evaluation, the operating system's two main components were set in three configurations: (1) SAVE without the HFM, (2) HFM without the SAVE, and (3) both the SAVE and HFM operating. The removal efficiencies were evaluated for several system components at water flow rates between 4 and 40 l/min (1 and 10 gpm). The water temperature in the system was maintained at temperatures lower than 54°C (130°F).

Removal efficiency strongly depends on the vacuum applied to the exterior of the hollow fibers, temperatures, and residence time at a specific temperature range. Typically, higher vacuums result in higher removal efficiencies. Similarly, higher water temperature results in more volatilization of the organic compounds, thereby increasing the removal efficiency. Water flow rate is also a critical factor. Low flow rates (*e.g.*, 4 to 15 l/min [1 to 4 gpm]) result in relatively higher removal efficiencies since residence time is increased, thereby allowing for the water temperature to rise before entering the HFM.

Figure 26-6.   **Hollow Fiber Membrane Unit in Conjunction with SAVE System Operation**

Designed to serve as a self-contained unit, the ICU component of the SAVE system was used to augment heat and power requirements.

For the combined SAVE-HFM system, the MTBE removal efficiency ranged from 81% for the highest flow rates (30 l/mim [8 gpm]) to greater than 99% for the lowest flow rates (4 l/min [1 gpm]) for temperatures ranging from 40 to 43°C (105 to 110°F). The combined SAVE-HFM system removal efficiency for benzene was difficult to quantify due to the low influent concentrations and non-detectable benzene concentrations in the effluent.

The complete system configuration used during the demonstration (including SAVE, HFM, water softeners, and GAC units) consistently achieved removal efficiencies of MTBE, benzene, and toluene greater than 99.9% for flow rates between 4 and 34 l/min (1 and 9 gpm) and temperatures ranging from 32 to 56°C (89 to 133°F). No breakthrough of MTBE, benzene, or toluene occurred in either of the GAC units. During the five-day demonstration effort, approximately 34,000 liters (9,000 gallons) of impacted ground water was treated while continuously meeting the MTBE cleanup goal of 5 μg/l.

For additional information:

Kram, M., Sirivithayapakorn, S., Joy, M., Lory, E., and Keller, A. 2000. *MTBE Remediation Using Hollow Fiber Membrane and Spray Aeration Vacuum Extraction Technologies*. NFESC, CR 00-004-ENV.

HIGH ENERGY ELECTRON INJECTION

Conducted by:

Paul Tornatore, Haley & Aldrich Inc., Rochester, New York

William Cooper, University of North Carolina, Wilmington, North Carolina and High Voltage Environmental Applications, Miami, Florida

In Cooperation with: USEPA National Risk Management Research Laboratory, Cincinnati, Ohio

The E-beam technology is a unique AOP capable of treating hydrocarbon and solvent impacted water sources to drinking water standards. Substantial technology development related to drinking water and other environmental applications was developed by Dr. William Cooper at the University of North Carolina, Wilmington and Drs. Chuck Kurucz and Tom Waite at the University of Miami, Florida. Haley and Aldrich, Inc., assisted in developing the test plan and operated the E-beam for the demonstration project. Project oversight, sampling, laboratory analyses, and data quality assurance are provided by the USEPA.

The E-beam technology is an AOP based on the destruction of organic compounds in impacted water through irradiation with a beam of high-energy electrons. The actual process is characterized as indirect radiolysis where the high-energy electrons impinge on water molecules to create approximately equal amounts of oxidizing and reducing chemistries used to affect ultimate organic destruction. During irradiation, three primary transient reactive species are formed: aqueous electrons (e^-_{aq}) and hydrogen radicals (H•), both strong reducing species; and hydroxyl radicals (OH•), a strong oxidizing species. When high-energy electrons impact flowing water, the electrons slow down, lose energy, and react with water to produce the three reactive species responsible for organic compound destruction, as well as hydrogen (H_2), hydrogen peroxide (H_2O_2), and hydronium ions (H_3O^+). The combination of oxidizing and reducing species from a single technology is unique and differentiates the E-beam process from other AOPs. These reactive species stimulate sequential reactions during treatment, with the ability to fully mineralize the parent compound. Based on concentration, flow rate, and energy/dose, the process can destroy organic compounds initially present in water at ppm concentrations and achieve non-detectable concentrations. The entire sequence of reactions between organic compounds and reactive species occurs in the area where the E-beam penetrates the water and is completed in milliseconds.

The E-beam is produced using an electron accelerator. Within the electron accelerator, a stream of electrons is emitted when an electric current (beam current) is passed through a tungsten wire filament. The electron stream is accelerated by applying an electric field and is focused into a beam using collimating plates. The applied voltage determines the speed of the accelerated electrons, which affects the depth to which the E-beam penetrates the water or other media being treated.

The E-beam treatment technology does not generate residue, sludge, or spent media that require further processing, handling, or disposal. Target organic compounds are either mineralized or broken down into lower MW compounds. For certain recalcitrant compounds or complex molecular structures, higher E-beam doses are sometimes used to increase the concentration of reactive species. The reactions continue to produce intermediate chemical species that are ultimately oxidized to carbon dioxide, water, and salts. The process is computer controlled, allowing delivery of predictable E-beam doses at controlled energy as needed to prevent incomplete oxidation/treatment. In cases of incomplete treatment of gasoline with MTBE, it may be possible to have residual levels of unwanted chemical byproducts such as low MW aldehydes, organic acids, and SVOCs.

In a typical treatment configuration, impacted water is first passed through a coarse particle filter and then pumped at low pressure to a feed delivery system where high-energy electrons are injected. The feed delivery system controls the depth of the water, ensuring full penetration of the high-energy electrons, and thus completely treating the impacted water stream. Since the reactions occur in milliseconds, decontamination is accomplished before the water leaves the treatment unit.

The system is housed in a 2-meter by 15-meter (8-foot by 48-foot) trailer and is rated for a maximum flow rate of 190 l/min (50 gpm) (Figure 26-7). The trailer is divided into three separate compartments: the pump room, process room, and control room. The pump room contains all ancillary equipment for the E-beam unit for both water and air handling. The process room contains the E-beam generator (electron accelerator), which is shielded to prevent radiation exposure from X-rays generated by contact of the E-beam with the stainless steel surfaces of the contact chamber. The control room contains the process control console where the system operating conditions are monitored and adjusted.

This field demonstration will address the following primary objectives:

Objective No. 1: Will the E-beam technology reduce the concentration of MTBE to below 5 µg/l? Preliminary test results indicate MTBE concentrations have been reduced to below 5 µg/l, with the ability to meet any prescribed intermediate treatment goal through power adjustment.

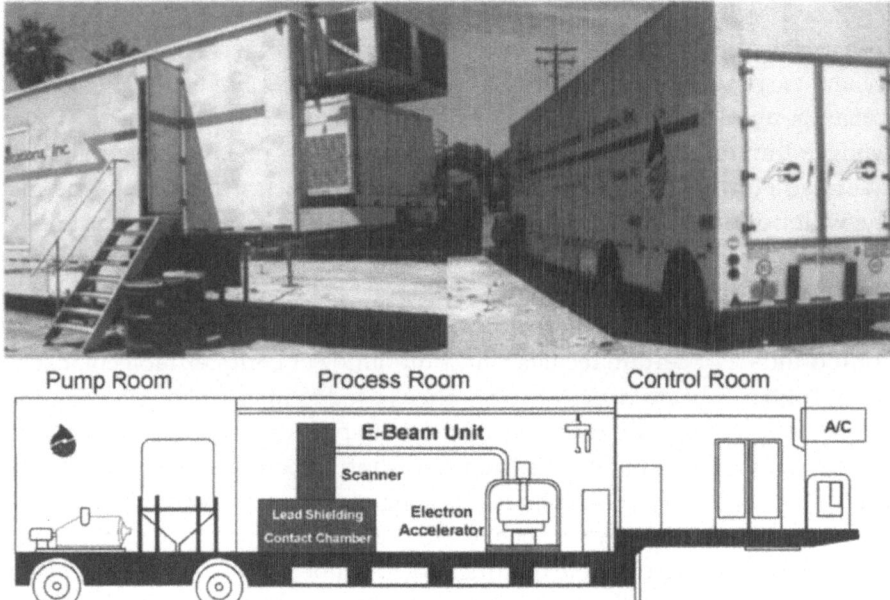

Figure 26-7.   **E-beam Trailer with Interior Cross-Section**

Objective No. 2: Will the E-beam technology produce an effluent that meets or exceeds all relevant California MCLs for applicable contaminants? Variable power testing has confirmed that the technology is capable of meeting any prescribed treatment objective for the contaminants of concern.

Objective No. 3: When subjected to uniform formation conditions, does the effluent meet total trihalomethanes and haloacetic acid requirements of the Stage 2 disinfection byproduct rule? Previous E-beam technology demonstrations have shown that the technology met these requirements. Interferences from bromate and possible impacts of dissolved organic carbon sources (from sediments not filtered in the original demonstration) are still under evaluation and are expected to yield a favorable outcome.

Objective No. 4: Will the removal efficiency (percent removal) change over time for this technology? Test data shows excellent repeatability and consistent treatment performance over time.

The preliminary results of the demonstration have shown that the E-beam is capable of meeting any prescribed treatment objective, including the production of water meeting California MCLs (*e.g.*, 13 µg/l for MTBE) from a source area where concentrations would preclude water use for any reason. During two different field tests, influent MTBE concentrations were between 1,400 and 1,640 µg/l, and the treated water MTBE concentrations were between 1 to 1.6 µg/l. The energy consumption for source area treatment suggests that earlier projections of $0.30/3,800 liters (1,000 gallons) for actual drinking water applications (lower concentrations and higher flow rates) are valid.

For additional information:

Cooper, W.J., Nickelsen, M.G., Tobien, T., and Mincher, B.J., 2001. The electron beam process for waste treatment. In: *Hazardous and Radioactive Waste Treatment Technologies Handbook*, pp. 5.5-1–15. (Oh, C., Ed.). Boca Raton, Florida, CRC Press LLC.

Mincher, B.J. and Cooper, W.J., 2001. High energy photons for waste treatment. In: *Hazardous and Radioactive Waste Treatment Technologies Handbook*, pp. 5.5-17–25. (Oh, C., Ed.). Boca Raton, Florida, CRC Press LLC.

Venosa, A.D. 2002. *High Energy Electron Injection (E-Beam) Technology for the Ex-Situ Treatment of MTBE-Contaminated Groundwater.* EPA/600/R-02/066. *Innovative Technology Evaluation Report.* September 2002.

HiPOx ADVANCED OXIDATION FOR THE REMEDIATION OF MTBE

Conducted by:

Reid Bowman, Applied Process Technology, Inc., San Francisco, California
In Cooperation with: USEPA National Risk Management Research
    Laboratory, Cincinnati, Ohio

The HiPOx technology, which can treat hydrocarbon impacted water to drinking water standards, was developed by and is being operated by Applied Process Technology, Inc. Demonstration oversight, sampling, laboratory analyses, and data quality assurance are provided by the USEPA.

The HiPOx process (Figure 26-8) uses ozone (O_3) and hydrogen peroxide (H_2O_2) to destroy organic compounds in a specially designed oxidation reactor. The reactants are injected directly into the water stream in precisely controlled ratios. Ozone dissociates and reacts with hydrogen peroxide to produce an intermediate hydroxyl radical (OH•). The OH• react very rapidly to oxidize organic contaminants into non-hazardous compounds. The OH• attack the covalent bonds in the MTBE molecules, progressively reducing these compounds and any resulting intermediate byproducts. The outcome of the breakdown process is benign end products of carbon dioxide, water, and salts (*e.g.*, NaCl).

The HiPOx system is composed of a series of injection modules (Figure 26-9). Influent MTBE impacted water is pumped out of the ground at a rate of 19 l/min (5 gpm). Ozone and hydrogen peroxide are immediately injected into the water stream at a pressure of 241 to 310 kPa (35 to 45 psi gas [psig]). Next, the water flows through a mixing section followed by a reaction zone,

Figure 26-8.   **Truck Mounted HiPOx Advanced Oxidation Unit**

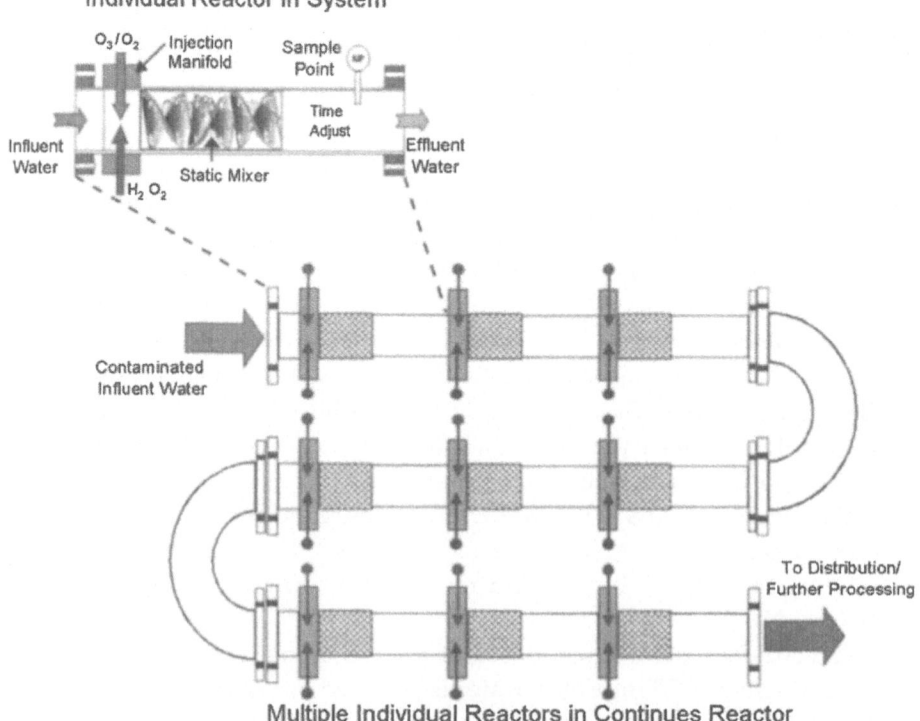

Figure 26-9.   **Cross-Section and Process Flow for HiPOx Advanced Oxidation Unit**

specifically designed for the required residence time. As soon as the water exits the reaction zone, it is injected with another dose of ozone and hydrogen peroxide. Again, it flows through a mixing section followed by a reaction zone. The process continues in this manner through 18 reactor sections. The HiPOx technology enhances and maximizes mass transfer of ozone into the water. Improved mass transfer is accomplished by using higher ozone concentrations of 8 to 10% by weight (as compared with other technologies that use 2 to 3% by weight), higher operating pressures of 241 to 310 kPa (35 to 45 psig) (as compared with other technologies that use ambient pressure) and highly effective mixing. During this demonstration, 18 reactors were used with hydrogen peroxide and ozone injected at 140 to 410 kPa (20 to 60 psig).

The traditional ozone/hydrogen peroxide AOP systems had limited drinking water applications due to the production of bromate (a human carcinogen). During this demonstration, HiPOx technology system will be in a batch type configuration where the impacted water will pass through the treatment system several times. The production of bromate is being evaluated.

Data quality objectives for this field demonstration evaluation are:

Objective No. 1: Will the HiPOx system reduce the concentration of MTBE to below 5 µg/l?

Objective No. 2: Will the HiPOx system produce an effluent that meets or exceeds California MCLs for applicable contaminants?

For additional information:
Speth, T.F. and Swanson, G. 2001. *Demonstration of the HiPOx Advanced Oxidation Technology for the Treatment of MTBE-Contaminated Groundwater.* USEPA Environmental Research Brief.

IN SITU BIOREMEDIATION OF MTBE

Conducted by:
Joseph Salanitro, Equilon Enterprises Ltd. (Shell/Texaco), Houston, Texas
Paul Johnson, Arizona State University, Tempe, Arizona

Equilon's Westhollow Technology Center teamed with Arizona State University to evaluate an *in situ* process for the enhanced bioremediation of MTBE.

Until recently it was commonly believed that MTBE was resistant to bioremediation. Over the past 10 years, a researcher from Equilon (formerly the Shell Development Company), Joseph Salanitro, has been working with a consortium of bacterium that showed potential MTBE degrading properties. Laboratory experiments and limited field studies have demonstrated that MTBE can be aerobically degraded by a mixed bacterial culture, MC-100. MC-100 was subcultured from activated sludge at industrial wastewater treatment plants. The MC-100 culture is a mixture of bacteria such as coryneforms, pseudomonads, and achromobacter species that had adjusted to the presence of MTBE for an extended period of time. Using conventional enrichment practices, MC-100 was acclimated to using MTBE as its only source of carbon.

The field trial focused on the use of MC-100 as an *in situ* biobarrier to downgradient MTBE migration. Before the field trial, MTBE had been detected at concentrations of 2,000 to 10,000 µg/l in the test area. The project design called for the injection of a narrow band of MC-100 slurry into the aquifer, followed by injection of oxygen.

The field-test layout includes four 6- by 12-meter (20- by 40-foot) test cells and two equally sized control plots (one with no injection and the other with oxygen-only injection) each aligned with the direction of ground water flow (Figure 26-10). Each test cell contains at least 15 paired monitoring wells (Figure 26-11). These wells are used to track MTBE concentrations upgradient and downgradient of the barrier and within the treatment zone. The test cells were injected with MC-100, or SC-100 (single culture of an MTBE degrader), and either oxygen or air to determine the culture's effectiveness under different conditions.

Figure 26-10.   **Equilon *In Situ* Bioremediation Site, with Original Installation in Foreground**

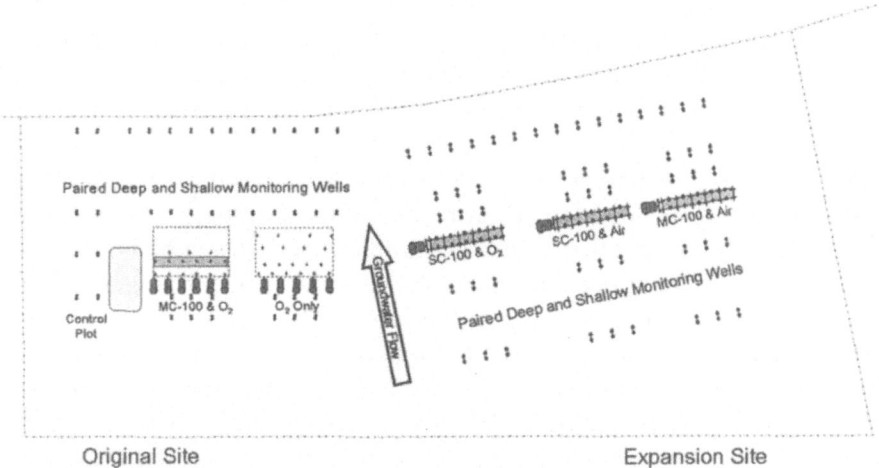

Figure 26-11.   **Equilon *In Situ* Bioremediation Site Layout with Injection and Monitoring Well Configuration**

Initial testing of the MC-100 in an MTBE only area of a naturally oxygen deficient aquifer at the NETTS Port Hueneme site has been successful in that the MTBE was degraded to or below 5 µg/l.

This project is in progress at the present time.

For additional information:

Salanitro, J. P., Spinnler, G. E., Neaville, C. C., Maner, P. M., Strearns, S. M., Johnson, P. C., and Bruce, C. Demonstration of the enhanced MTBE bioremediation (EMB) *in situ* process. Presented at the *Battelle, In Situ and On-Site Bioremediation, 5ᵗʰ International Symposium,* San Diego, California. April 19–22, 1999.

DIRECT INJECTION OF A BACTERIAL CULTURE TO BIODEGRADE MTBE-IMPACTED GROUND WATER

Conducted by:

Kate Scow, UC, Davis, California

Doug Mackay, University of Waterloo, Ontario, Canada

A field study was conducted at the NETTS Port Hueneme site to investigate the feasibility of using *in situ* bioaugmentation with a bacterial culture to reduce concentrations of MTBE in impacted ground water. The bacterial culture used in this project, designated PM-1, was isolated by UC Davis from a field-operated biofilter at the Joint Water Pollution Control Plant of the Los Angeles County Sanitation District. PM-1 utilizes MTBE as its sole carbon and energy source.

The PM-1 injection site was approximately 600 meters (2,000 feet) down-gradient from the gasoline release site. Three test plots were installed at the PM-1 injection site including: (1) Test Plot A, oxygen sparging only; (2) Test Plot B, PM-1 culture injection with oxygen sparging; and (3) Test Plot C, air sparging only.

Each plot was 3 meters (9 feet) wide perpendicular to the ground water flow, by 1.4 meters (4.5 feet) long (Figure 26-12). In addition, 36 monitoring wells were installed upgradient and downgradient of each test plot to collect samples for determining background water chemistry and changes to the water once it passed through each plot (Figure 26-13).

The primary objective of this field study was to determine if the PM-1 culture injected into ground water could effectively degrade MTBE under field conditions. A second objective was to compare differences between an injected MTBE biodegrading culture to indigenous microorganisms that are given either oxygen or air biostimulation. A third objective was to track the survival and movement of an inoculated bacterial strain using polymerace chain reaction (PCR) techniques.

Oxygen delivery to Plots A and B began in late October 1999. Intensive sampling of dissolved oxygen in ground water was conducted to determine

Figure 26-12.   **University of California Davis Site, Direct Injection of PM-1 Bacteria to Biodegrade MTBE-Impacted Ground Water**

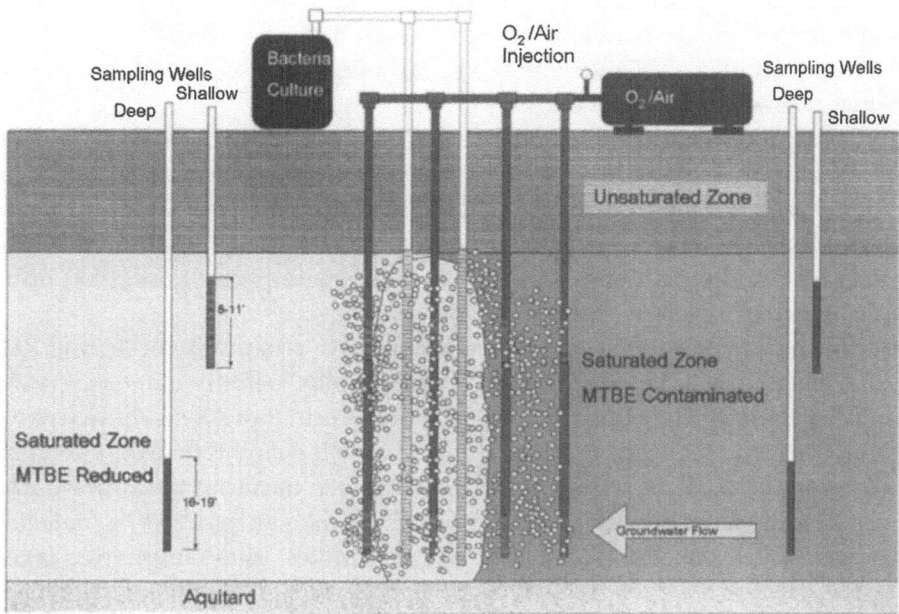

Figure 26-13.   **University of California Davis Site, Simple Cross-Section Showing Process and Monitoring Well Configuration**

when sufficient oxygen was present to support the activity of PM-1. Modifications had to be made to the original design of the wells to increase oxygen delivery to locations where high concentrations of MTBE were present. By early November, high concentrations of dissolved oxygen were measurable in almost all of the shallow wells and in some, but not all, of the deeper wells at the site. A Geoprobe® was used to inject the strain PM-1 into Plot B.

In the shallow zone, initial concentrations of MTBE ranged from 2.5 to 3.5 mg/l in Plot B and were much lower in Plot A (below 0.14 mg/l). During the project, MTBE concentrations in both plots decreased substantially in downgradient wells and in wells immediately upstream near the oxygen release wells. After six months of treatment, MTBE concentrations declined to 0.008 mg/l or non-detect (less than 0.005 mg/l) in Plot A and to 0.09 mg/l or non-detect in Plot B.

In the deeper zone, initial MTBE concentrations ranged from 5 mg/l upstream to less than 1 mg/l downstream in Plot A and from 5.7 to 9.3 mg/l in Plot B. Downstream MTBE concentrations decreased substantially in Plot A to below 0.11 mg/l, but only slightly in Plot B. Difficulties in delivery of oxygen to the deep zone in Plot B, as evidenced by the low dissolved oxygen concentrations present, was likely responsible for low rates of MTBE removal.

Well pump tests indicated that ground water flow was substantially slower in the shallow zones than in the deep zones and slower in Plot B than in Plot A. Degradation was also seen in Plot C (air injection only), but not to the extent as was seen in Plots A and B.

Using TaqMan PCR, the densities of PM-1 cells in samples (shallow and deep) from Plots B and A were quantified. Twenty-four days after injection, PM-1 was detectable at densities ranging from 10^2 to 10^5 per mililiter ground water. Detected densities of PM-1 were higher in the deep depths of both test plots than in the shallow depths. PM-1 cell densities were quantified up to one log order higher in Plot B than Plot A.

In conclusion, results suggest the presence of a naturally occurring PM-1-like bacterium at the NETTS Port Hueneme site. Definitive determination could not be made whether the presence of PM-1 in Plot A was due to movement from the inoculated plot (by sparging) or if it was native to the site. Tests were conducted to determine whether PM-1-like bacterium was present in locations far removed from the field site. PM-1-like bacterium was detected by two PCR techniques in only one of the samples. This sample was taken from a location 500 meters (1,500 feet) downgradient from the injection bed.

The overall significance of this study is that bioremediation, both through inoculation and by stimulating native organisms, shows promise as a technology for cleaning up MTBE-impacted ground water. Results indicate that both inoculated and uninoculated plots showed similar levels of MTBE removal. Strain PM-1 or a naturally occurring PM-1-like bacterium may be responsible for some (or all) of the biological removal of MTBE in the plots as evidenced by molecular detection of strain PM-1 sequences in all samples. Results of controlled microcosm studies provided further evidence that oxygen additions to uninoculated Port Hueneme aquifer sediments stimulate native MTBE-degrading organisms already present in these samples. Once the native community is adapted to MTBE, additional inputs of MTBE are degraded at rates similar to those measured for strain PM-1.

For additional information:
Hanson, J.R., Ackerman, C.E., and Scow, K.M. 1999. Biodegradation of methyl tert-butyl ether by a bacterial pure culture. *Applied and Environmental Microbiology.* 65, 4788–4792.

LARGE-SCALE BIOBARRIER DEMONSTRATION

Conducted by:
Karen Miller, Naval Facilities Engineering Service Center, Port Hueneme, California
Paul Johnson, Arizona State University, Tempe, Arizona

For the past three years, the USDOD's Environmental Security Technology Certification Program (ESTCP) has been sponsoring a large-scale MTBE plume biobarrier containment demonstration at the NEX gas station plume,

Port Hueneme site. This passive flow-through biobarrier system has been designed to degrade MTBE and other dissolved hydrocarbons to innocuous carbon dioxide and water. The design of the large-scale biobarrier system was based on pre-design characterization results as well as lessons learned from pilot-scale studies at the Port Hueneme site sponsored by Equilon Enterprises. In the anaerobic impacted aquifer, several different combinations of bioaugmentation and air and oxygen biostimulation are being evaluated in this large-scale demonstration, shown in Figure 26-14. In this biobarrier design, oxygen delivery to the aquifer is done by short duration intermittent gas injection bursts. Gas trapped in the aquifer between injections continues to deliver oxygen by gas dissolution for days.

The large-scale biobarrier system consists of an automated aquifer oxygenation system, both bioaugmented (seeded) and biostimulated (unseeded) sections, and an extensive monitoring well network. The system components include: 252 gas injection wells (on 0.6-meter [2-foot] centers), 174 monitoring wells, 25 satellite gas storage tanks, 154 solenoid valves, a 1 cmm (40 cfm) oxygen generator, automated timer circuits, and associated piping and electrical lines (Figure 26-15).

The oxygen gas and bioaugmented sections were placed in the central core of the dissolved plume where MTBE concentrations are as high as 1,000

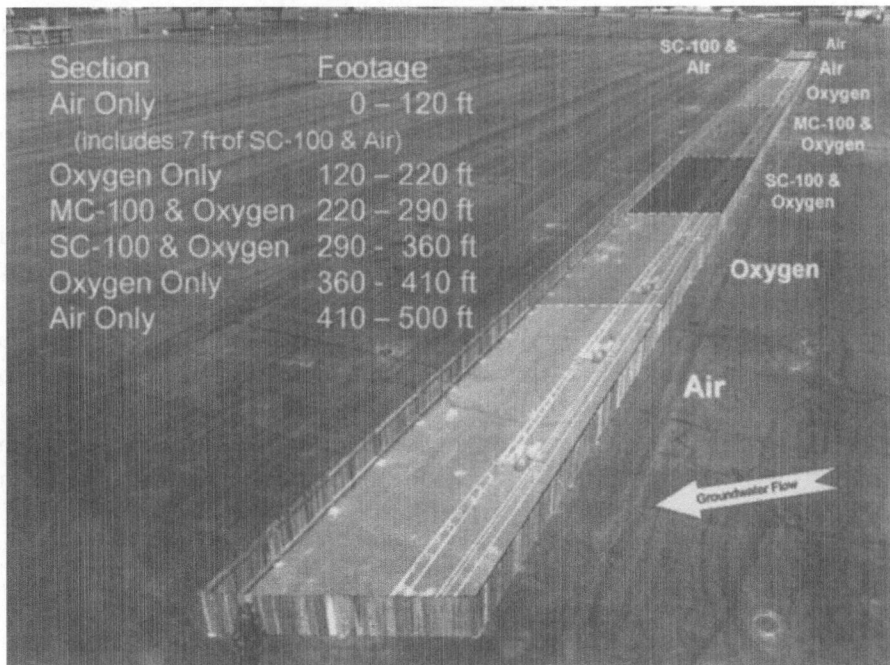

Figure 26-14. **Large Biobarrier Demonstration with Air/Oxygen and Microbial Regimes**

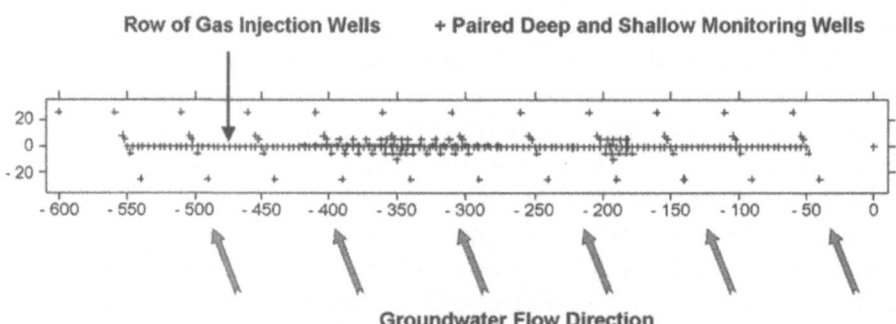

Figure 26-15.   **Large Biobarrier Demonstration Layout with Injection and Monitoring Well Configuration**

μg/l. Sections targeted for biostimulation by oxygen gas injection roughly correspond to MTBE-only regions of the plume where MTBE concentrations are approximately 100 to 1,000 μg/l. Air-only regions were reserved for the edges of the plume where only MTBE is present at concentrations less than 100 μg/l.

To collect detailed performance data, the monitoring well network at this site is fairly extensive. Wells were also placed at either end of the biobarrier to determine if operation of the gas injection system causes ground water and the MTBE plume to flow around, rather than through, the biobarrier.

Installation of gas injection and monitoring wells was done using innovative direct push well installation techniques. The oxygenation system was operated for approximately two months to ensure the desired stability of an aerobic zone. The bioaugmentation portions of the biobarrier were then inoculated with the Equilon Enterprises' mixed and single degrader cultures (MC-100 and SC-100, respectively). Inoculation was done by high-pressure injection of the cultures through direct push rods. The rods were driven down to 6 meters (20 feet) bgs and then pressurized to allow solution to flow out into the formation.

The treatment performance of the biobarrier system has been exceptional for all sections of the barrier. Ground water treatment efficiencies in excess of 99.9% have been achieved and sustained during the first 12 months of operation. The MTBE concentrations downgradient of the biobarrier are less than 5 μg/l. This demonstration will validate the biobarrier design and provide information on installation and operation costs.

This project is in progress at the present time.

For additional information:

Miller, K.D., Johnson, P.C., and Bruce, C.L. 2001. *Full-Scale In-Situ Biobarrier Demonstration for Containment and Treatment of MTBE Remediation,* Volume 12, No 1, pp. 25–36. Wiley Publishers.

Miller, K.D., Heath, J.C., and Johnson, P.C. 2001. MTBE biobarrier demonstration at Port Hueneme. *Soil, Sediment and Water.* Spring 2001, 6–9.

Bruce, C.L., Miller, K.D., and Johnson, P.C. 2002. Full scale bio-barrier demonstration for treatment of a mixed MTBE/TBA/BTEX plume at Port Hueneme. *Soil, Sediment and Water.* July/August 2002, 80–84.

IN SITU REMEDIATION OF MTBE IMPACTED AQUIFER USING PROPANE BIOSTIMULATION

Conducted by:

Yassar Farhan and Robert Steffan, Envirogen, Inc., Lawrenceville, New Jersey
In Cooperation with: USEPA National Risk Management Research
Laboratory, Ada, Oklahoma

The Envirogen propane biostimulation demonstration is providing an evaluation of an *in situ* treatment of hydrocarbon-impacted ground water in a plume that contains MTBE. The Envirogen technology is being demonstrated to evaluate its ability to reduce MTBE concentrations in water to 5 µg/l. This technology is being evaluated under the USEPA's MTBE Treatment Technology Evaluation Program. Project oversight, sampling, laboratory analyses, and data quality assurance are being provided by the USEPA.

The Envirogen technology involves injecting propane and oxygen into the MTBE-impacted aquifer to promote the growth and degradative activity of injected propane-oxidizing bacteria. The addition of these substrates promotes the growth of propane-oxidizing bacteria and the production of the enzyme propane mono-oxygenase that catalyzes the complete destruction of MTBE and its primary daughter product, TBA. To seed the aquifer to ensure activity or speed initiation of the treatment process, the exogenous propane-oxidizing bacteria strain ENV425 was injected across the width and depth of the test plot. This bacterium was isolated by Envirogen from an industrial site in New Jersey.

The efficiency of the treatment process in reducing contaminant concentrations will be determined by comparing treatment levels achieved in a test plot where propane, bacteria, and oxygen are being injected verses treatment levels achieved in a control plot where only oxygen is injected (Figure 26-16).

The demonstration system consists of a network of oxygen and propane injection points, pressurized oxygen and propane gas delivery and control systems, and a ground water and soil-gas monitoring network. The control plot is similar in configuration to the test plot, except that no propane injection points and fewer monitoring points were installed. In addition to Envirogen's monitoring network, the USEPA installed a series of multi-level ground water monitoring points, soil-gas monitoring points, and tracer injection points to collect performance-monitoring data (Figure 26-17).

Early data are suggesting there is an increased biodegradation rate of MTBE due to the introduction of propane-oxidizing bacteria in the test plot

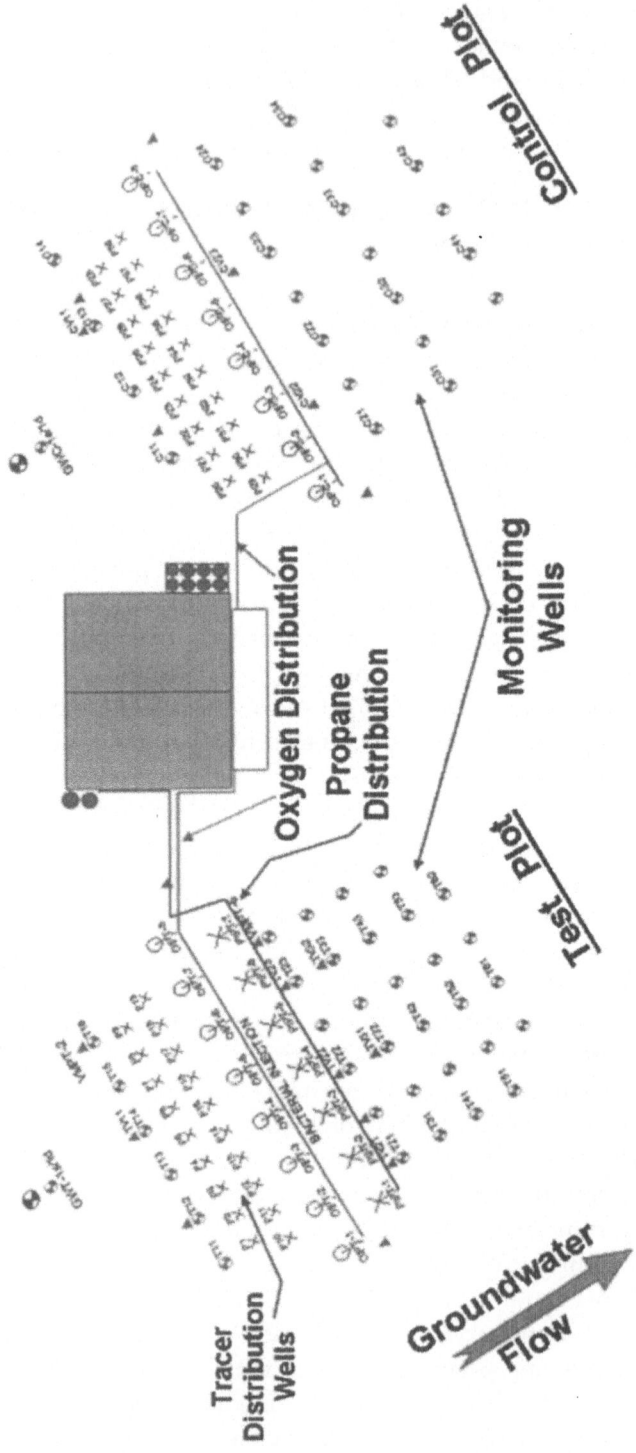

Figure 26-16.   Envirogen Propane Biostimulation Demonstration Site Layout with Injection and Monitoring Well Configurations

Figure 26-17.   **Ground Water Sampling at Envirogen Propane Biostimulation Demonstration Site**

(oxygen and propane injection) as compared with the control plot (oxygen only injection) and the indigenous bacterial population. Changes in geochemical parameters as well as metabolites (*e.g.*, dissolved oxygen, formaldehyde, acetone, carbon dioxide, TBA) provide additional evidence that biodegradation is occurring. The dissolved oxygen concentration in both plots has increased due to sparging. Slight increases in MTBE daughter products, such as TBA, as well as decreased levels of propane in ground water are indicators that biodegradation is occurring. This demonstration is assessing propane biostimulation barrier design and operational cost data.

For additional information:

Keeley, A. 2002. *Envirogen Propane Biostimulation Technology for the In-Situ Treatment of MTBE-Contaminated Groundwater.* EPA/600/R-02/092. *Innovative Technology Evaluation Report.* November 2002.

NATURAL ATTENUATION OF MTBE
IN AN ANAEROBIC GROUND WATER PLUME

Conducted by:

Illa Amerson and Rick Johnson, Oregon Graduate Institute, Beaverton, Oregon

In order to better understand the potential impact of MTBE on the subsurface environment, a natural attenuation study was conducted at the NETTS Port Hueneme site. The objective of the study was to gather information on the behavior and degradation of MTBE in an oxygen-depleted aquifer.

The study used a solution of perdeuterated MTBE ($[^2H_{12}]$-MTBE) as a tracer to measure the natural biodegradation of MTBE. In order to minimize any environmental impact, the tracer test was conducted in the middle of an existing MTBE plume. Since deuterated molecules have a slightly higher MW than the commonly occurring ($[^1H_{12}]$-MTBE) molecules, the $[^2H_{12}]$-MTBE tracer and its degradation products can be distinguished from existing MTBE using GC with a mass selective detector. The $[^2H_{12}]$-MTBE was mixed with shallow ground water using wells in a 9-meter by 9-meter (30-foot by 30-foot) grid. Immediately following their injection, the distribution and mass of $[^2H_{12}]$-MTBE in the subsurface was determined (Figure 26-18). The tracer was allowed to move with the natural ground water flow for one year.

Figure 26-18. **Injection of Perdeuterated MTBE Tracer for Natural Attenuation of MTBE Demonstration**

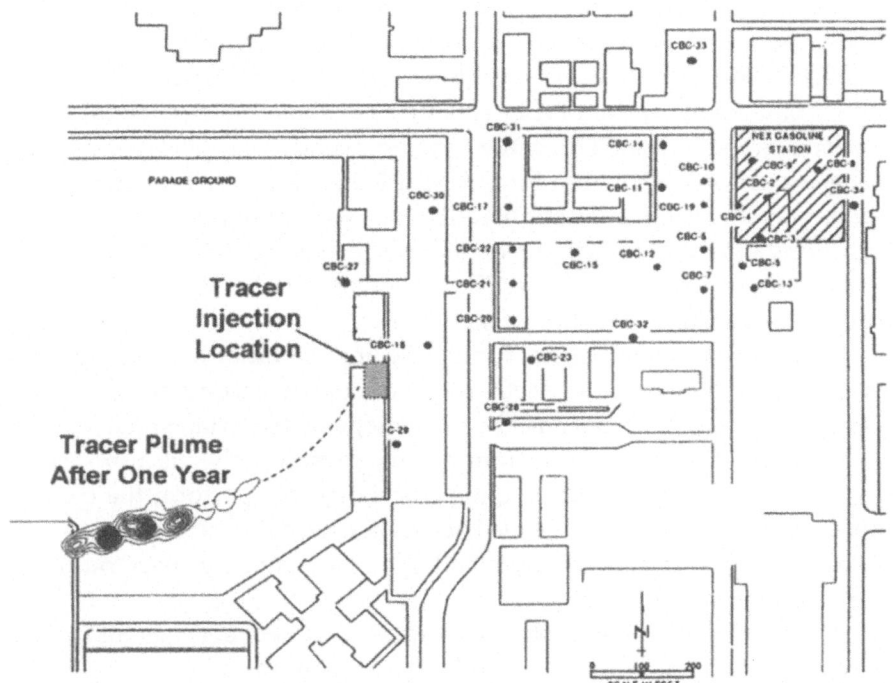

Figure 26-19.   **Perdeuterated MTBE Tracer Plume One Year After Injection**

After one year of plume migration, the mass of $[^2H_{12}]$-MTBE remaining in the plume was determined by integrating the ground water concentration within the mapped plume (Figure 26-19). This value was compared to the initial mass at the start of the test. While the observed concentrations of $[^2H_{12}]$-MTBE in the ground water were considerably reduced after one year due to dispersion, the mass balance data indicate that little if any MTBE degradation occurred during the year of transport. These data suggest that, in this aquifer with low oxygen and high sulfate concentrations, the natural degradation rate of the originally released MTBE is quite slow.

For additional information:

Amerson, I.L. and Johnson, R.L. 2002. A natural gradient tracer test to evaluate natural attenuation of MTBE under anaerobic conditions. Submitted to *Ground Water Monitoring and Remediation*.

NATURAL ATTENUATION OF MTBE IN GROUND WATER UNDER METHANOGENIC CONDITIONS

Conducted by:

John Wilson, USEPA National Risk Management Research Laboratory, Ada, Oklahoma

Studies at other sites have shown that anaerobic biodegradation of MTBE is often associated with methane production in the ground water. Studies at the NETTS Port Hueneme site have shown that adding oxygen will stimulate aerobic biodegradation of MTBE. This study evaluated the potential for natural biodegradation of MTBE under anaerobic conditions in the plume at Port Hueneme. A flow path in the plume of MTBE that emanates from the NEX gas station was characterized using 10 wells, each with 0.6-meter (2-foot) screens, to determine the true availability of dissolved oxygen and the true accumulation and attenuation of methane. The rate of attenuation of MTBE was compared to the rate of attenuation of methane.

Methane formation occurs during decomposition of diverse substrates found in fuel hydrocarbons. Methane production is carried out by a highly specialized group of organisms, the methanogenic bacteria, which are obligate anaerobic. Methanogenesis from organic carbon is always a process carried out by a mixture of bacteria, none of which can perform the complete process alone. Non-methanogenic bacteria ferment the substances to either acetate or hydrogen and carbon dioxide; the methanogenic bacteria can use these waste products.

A truck-mounted Geoprobe® was used to collect soil gas and aquifer samples to determine methane and oxygen concentrations from various depths at locations throughout the NEX plume area and upgradient of the release site (Figure 26-20).

The ground water in the aquifer containing the plume is naturally anaerobic, even in areas where the plume is absent. When careful precautions were taken to sample ground water and measure the content of dissolved oxygen, the concentration was always below the analytical detection limit of 0.1 mg/l. Oxygen is not available to contribute to the natural attenuation of MTBE in this plume. Methane is only produced in the portion of the plume that contains detectable concentrations of BTEX compounds (Figure 26-1). Due to natural anaerobic biodegradation, the region containing BTEX compounds was less than 300 meters (900 feet) long. In the downgradient portion of the plume where concentrations of BTEX compounds are below non-detect (5 µg/l), the concentrations of methane were attenuated by dilution and dispersion alone. In this portion of the plume, the bulk rate of attenuation of methane was 0.55 ± 0.15 year. The bulk rate of attenuation of MTBE was 0.37 ± 0.21 per year. The attenuation of MTBE was due entirely to dilution and dispersion. There is no evidence of an additional contribution due to natural biodegradation under anaerobic conditions.

For additional information:
Kolhatkar, R., Wilson, J.T., and Dunlap, L.E. 2000. Evaluating natural biodegradation of MTBE at multiple UST sites. In: *Proceedings of the Petroleum Hydrocarbons and Organic Chemicals in Ground Water: Prevention,*

Figure 26-20.   **Soil Gas and Ground Water Sampling to Determine Methane and Oxygen Concentrations**

Detection, and Remediation Conference, American Petroleum Institute, National Ground Water Association, STEP Conference and Exposition, Anaheim, California. November 15–17, 2000. pp. 32–49.

Wilson, J.T. and Kolhatkar, R. The role of natural attenuation in the life cycle of MTBE plumes. In review: *The Journal of Environmental Engineering,* ASCE.

CHAPTER 27

Bioremediation at a New Jersey Site Using Propane-Oxidizing Bacteria

Robert J. Steffan, Ph.D., Yassar H. Farhan, Ph.D., Charles W. Condee, and Scott Drew, Envirogen, Inc.

INTRODUCTION

In our early work we observed that propane-oxidizing microorganisms could mineralize MTBE to CO_2 and H_2O after growth on propane (Steffan *et al.*, 1997). Other hydrocarbon gases, such as methane and butane, have been used to stimulate cometabolic biodegradation processes *in situ*. In the most publicized application of this "biostimulation" approach, methane and oxygen were injected into a TCE-contaminated aquifer at the USDOE's Savannah River site (Lombard *et al.*, 1994). This procedure successfully stimulated *in situ* biodegradation of the chlorinated solvent. Therefore, it is likely that a similar application of biostimulation, whereby propane and oxygen are injected to stimulate MTBE degradation by indigenous organisms or seed cultures, is feasible (US Patent # 5,814,514, Sept. 29, 1998).

There are several potential advantages to using a biostimulation approach for degrading MTBE *in situ*. Biostimulation uncouples biodegradation of the contaminant from growth of the organisms. That is, the microbes can be supplied sufficient co-substrate (*e.g.*, propane) to support growth, so they do not have to rely on the utilization of low levels of contaminants to maintain their survival. Also, the technology can be applied in a number of configurations depending on site characteristics and treatment needs.

Possible application scenarios include: 1) reengineered or modified multi-point AS/SVE systems that deliver propane and air throughout an impacted site (suitable for use with existing AS/SVE systems or specially designed systems); 2) a series of air/propane delivery points arranged to form a permeable treatment wall to prevent off-site migration of MTBE and other gasoline constituents; 3) permeable treatment trenches fitted with air and propane injection systems; 4) *in situ* recirculating treatment cells that rely on pumping and reinjection to capture and treat a migrating contaminant plume; and 5) propane and oxygen injection through bubble-free gas injection devices to minimize off-gas release and contaminant stripping. Furthermore, propane is widely available, transportable even to remote sites, already present at many gasoline stations, and relatively inexpensive. Thus, propane biostimulation has the potential to be an attractive remediation option at a wide variety of MTBE-impacted sites.

We applied and evaluated propane biostimulation for MTBE remediation at a gasoline station in southern New Jersey. The primary purposes of this field demonstration included:

- Evaluating the effectiveness of propane biostimulation for MTBE remediation;
- Optimizing sparging and SVE flow rates and injection/extraction cycles;
- Quantitatively assessing the impact of propane sparging on soil gas and ambient air quality;
- Delineating the zone of influence of the treatment; and
- Assessing the potential for subsurface gas migration and fugitive emissions.

Microcosm testing with samples from the site revealed that the resident ground water had low indigenous microbial activity, presumably due to the aquifer's low pH of approximately 3.8 to 4.5 standard units. Consequently, the demonstration required the addition of a seed culture of propane-oxidizing bacteria to initiate biodegradation. Results of the demonstration suggest that propane biostimulation can be a suitable and low cost remedial technology for MTBE-impacted aquifers.

METHODOLOGY

Site Characterization.   The field demonstration site, located in Camden County, New Jersey, was actively used as a gas station and repair shop both before and during the field demonstration. A site investigation was initiated at the site after one of the site's gasoline USTs failed a tightness test in July 1988. The site has since undergone a range of remedial actions including soil excavation and air sparging. Six on-site ground water monitoring wells (MW5 to MW10) and two off-site wells (MW11 and MW12) were installed to monitor BTEX and MTBE. These wells are currently being monitored on a quarterly basis. Figure 27-1 illustrates the configuration of the site monitoring well network in the treatment area.

Site Hydrogeology.   The site is located in the western edge of the Coastal Plain Physiographic Province and is underlain by the Kirkwood-Cohansey Formation. The underlying soil consists of unconsolidated orange, medium- to fine-

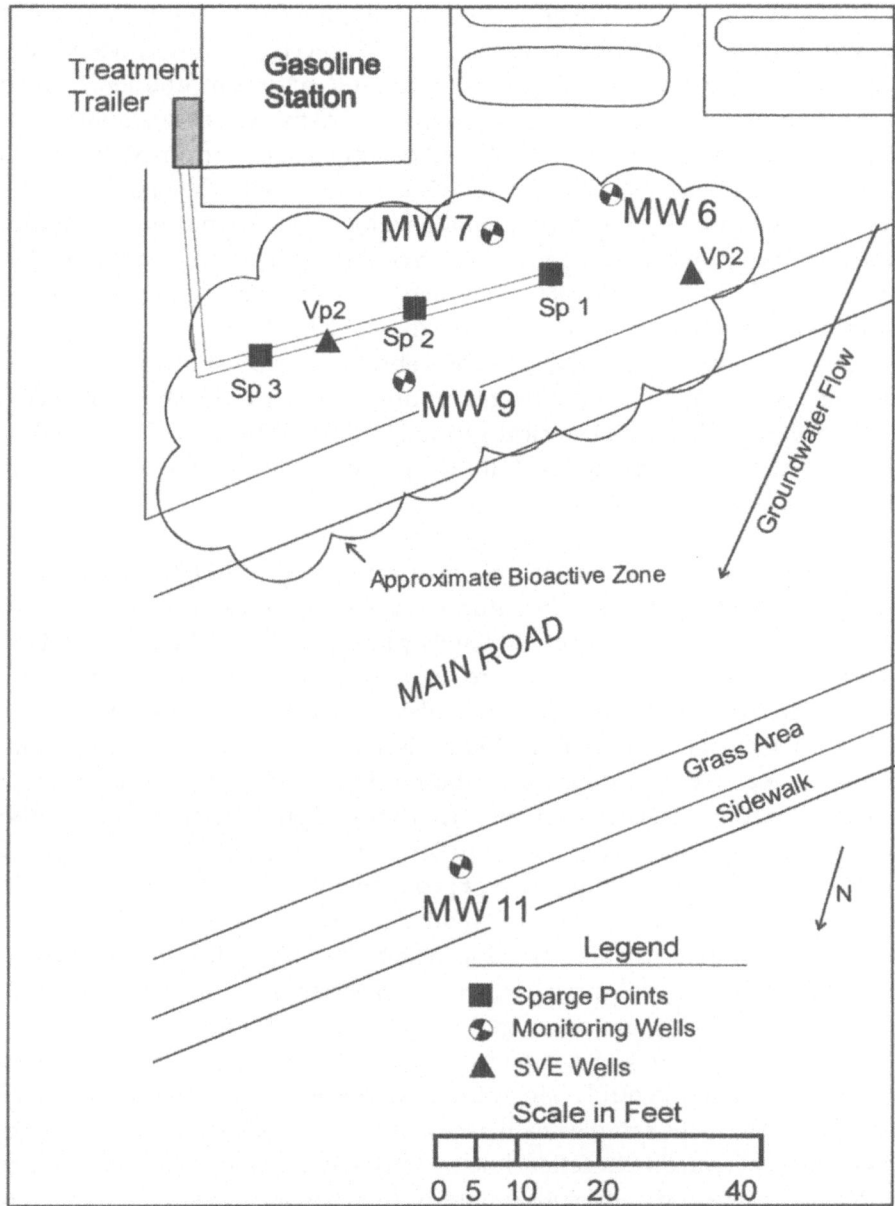

Figure 27-1.   **Field Demonstration Site and System Layout**
Propane and air were injected into three existing air sparging points (Sp1, Sp2, and Sp3), and MTBE concentrations were measured in MW6, MW7, MW9, and MW11.

grained soil and silt with some well-rounded gravel. The depth to water at the site varies between 9 to 11 meters (30 to 35 feet) bgs. Ground water flow is typically to the east-northeast under a variable hydraulic gradient, which averages approximately 0.01 but ranges from 0.003 to 0.02. The estimated calculated hydraulic conductivity (K) based on slug test data from MW8 is 1.80 $\times 10^{-4}$ cm/s (0.51 ft/day). This K value is consistent with the geology of the site and observations made during well development and purging activities. Based on the site hydraulic conductivity, average hydraulic gradient, and assuming an effective porosity of 0.3, average linear ground water velocities are estimated to be on the order of 7×10^{-5} cm/s (0.2 ft/day).

Ground water samples collected on February 9, 1999 showed site MTBE concentrations ranging from 170 µg/l (at upgradient monitoring well MW8) to 270,000 µg/l (MW6). Historical ground water MTBE data from 1990 to 1999 indicate increasing concentrations at monitoring wells MW6, MW7, MW9, and MW11.

Microcosm Testing. To perform microcosm testing of MTBE biodegradation, aquifer sediment (from the saturated zone) and ground water samples were collected from the site, immediately placed on ice, and transported directly to Envirogen. Microcosms were set up within 24 hours of receiving the samples. To construct the microcosms, subsamples of the sediment (50 grams [2 ounces]) were placed into sterile 150-milliliter serum vials, and 60 milliliters of ground water was added to each bottle. Four different treatments were evaluated as follows: 1) no additions; 2) addition of propane and oxygen only; 3) addition of propane, oxygen, and nutrients (nitrogen and phosphorus); and 4) pH adjustment to 7, and addition of propane, oxygen, nutrients, and strain ENV425 (approximately 1×10^6 cells/ml of slurry; Steffan *et al.*, 1997). A second microcosm test was performed in which the pH of all samples was adjusted to 7 and triplicate microcosms received either no amendments, oxygen only, oxygen and propane, or oxygen, propane, and ENV425 (approximately 1×10^6 cells/ml of slurry). Each microcosm test also included $HgCl_2$-poisoned microcosms ("killed controls") to evaluate abiotic losses. Propane (4-milliliter gas) was added to the sample headspace just prior to sealing the vials. The microcosms were sealed with Teflon-lined crimp seals and incubated on a shaker at 15°C (59°F). Periodically, subsamples of the headspace gas were removed and analyzed for MTBE on a GC equipped with an FID. Alternatively, subsamples of the slurry were removed and analyzed by purge-and-trap GC/MS by USEPA Method 8260. Method 8260 allowed for an MTBE detection limit of 5 µg/l, whereas the detection limit of the GC/FID method was approximately 300 µg/l. TBA was measured using USEPA Method 8015 with a heated purge and trap system. This method had a detection limit of approximately 5 µg/l. Additional oxygen was added to microcosms as needed to ensure aerobic conditions. Strain ENV425 was grown in basal salts medium (Hareland *et al.*,

1975) with propane as a carbon source (Steffan *et al.*, 1997), washed, and added to the microcosms to a final concentration of approximately 1×10^6 cells/ml.

Field-Scale System Implementation and Operation. The biosparging and propane injection system was designed to allow flexible and safe implementation of the field demonstration. The system consisted of injection and SVE components, and utilized existing sparge wells (SP-1, SP-2, and SP-3) and SVE wells (VP-1, VP-2, and MW10) at the site. The injection system consisted of two separate components: an air compressor and a propane supply system that was connected to the existing sparging distribution lines via a common manifold. An in-line filter was installed on the injection line to remove moisture and/or oil escaping the air compressor. The SVE system consisted of a vacuum blower that was connected to the existing SVE distribution lines and a carbon canister for treatment of the off-gas. Operation of the system was controlled using a common control panel with redundant control switches to ensure safe operations. An interlock device was used to prevent propane injection unless the SVE system was operational.

Because the existing air sparging wells were not designed and constructed for pulsed operation, operation of the wells in a pulsed mode resulted in an accumulation of silt in the wells and reduced airflow. Consequently, the sparging system was operated with a continuous low airflow of 0.4 scmm (13 scfm). A 4.5-kilogram (10-pound) propane gas cylinder (*e.g.*, similar in size to home barbecue propane tanks) was used as the propane supply. The discharge from the propane cylinder was controlled by a flow valve and pressure indicator mounted on the cylinder. A pressure control valve set at 276 kPa (40 psi) was utilized to monitor and control the propane pressure in the line. An in-line propane lower explosive limit (LEL) detector was installed to continuously monitor the LEL level and ensure safe operation of the system. Dedicated flow meters were installed on each line to control the flow to each sparge well. Propane was added to the air stream for ten minutes every three hours at a rate that ensured that the propane concentration did not exceed 0.2% propane in air (10% of the propane LEL). Approximately 0.2 kilograms (0.5 pounds) of propane and 143 kilograms (315 pounds) of oxygen were added to the site each day.

The system was initially operated for approximately one month without the additional MTBE degrading microorganisms. A total of 17 liters (4 gallons) of culture of strain ENV425 (approximately 1×10^{11} cells/ml) was then added to the three sparge points. Bacterial injection was followed by several cycles of air sparging to help distribute the microbes into the treatment zone and two days of continuous propane and air sparging to aid in establishing an active MTBE degrading microbial population. The system was operated for an additional five months before a scheduled shutdown. MTBE and BTEX concentrations in the ground water were measured using USEPA Method 8260.

Because the pH measured in ground water at the site was low (approximately pH 3) and unfavorable for biological activity, the ground water needed to be buffered to raise the pH to a range more favorable to MTBE biodegradation. A buffer solution of sodium bicarbonate was added to the sparge point periodically during the demonstration to achieve this goal. During each buffering event, a total of 450 liters (120 gallons) of a sodium bicarbonate solution was added to the sparge points, followed immediately by air sparging to disperse the buffer into the formation.

RESULTS

Microcosm Studies. Plate count analysis of the site samples indicated that microbial populations in the aquifer were low. Total heterotroph numbers were 1.7×10^5 and 4×10^3 colony forming unit (CFU) per milliliter in ground water from MW6 and MW7, respectively, but were less than 3×10^2 CFU/ml in ground water from MW9 and MW11. Propanotroph numbers were 3.6×10^4 CFU/ml in ground water from MW6, but less than 3×10^2 CFU/ml in ground water from the other monitoring wells. Therefore, microcosm studies were performed to assess our ability to stimulate MTBE biodegradation by indigenous microorganisms at the site and to evaluate our ability to seed the aquifer to enhance MTBE biodegradation. Biostimulation experiments were performed by adding oxygen, oxygen and propane or oxygen, propane, and nutrients (nitrogen and phosphorus) to the microcosms and incubating them at *in situ* temperatures. None of the biostimulation treatments resulted in significant MTBE degradation in the site samples during the greater than 30 days of incubation (Figure 27-2a). Adjusting the pH of the samples prior to the biostimulation did not improve degradation during the biostimulation treatments (Figure 27-2b). MTBE biodegradation occurred only in pH-adjusted microcosms that were augmented with approximately 1×10^6 cells/ml of strain ENV425 (Figures 27-2a and 27-2b). MTBE biodegradation continued in these microcosms for more than 30 days, and further additions of MTBE were rapidly degraded in the microcosms (Figure 27-2a).

Field Evaluation. Ground water monitoring during the project was performed in monitoring wells MW6, MW7, MW9, and MW11 (Figure 27-1). MW6 is located just upgradient of the treatment zone, but it was slightly influenced by the treatment as indicated by increased dissolved oxygen in the ground water during the treatment system operation. MW7 also was upgradient of the treatment wells, but clearly within the zone of influence of the propane and oxygen injection system. MW9 was immediately downgradient of the sparging points, and MW11 was far downgradient of the treatment system.

Dissolved oxygen concentrations in the treatment area increased slightly

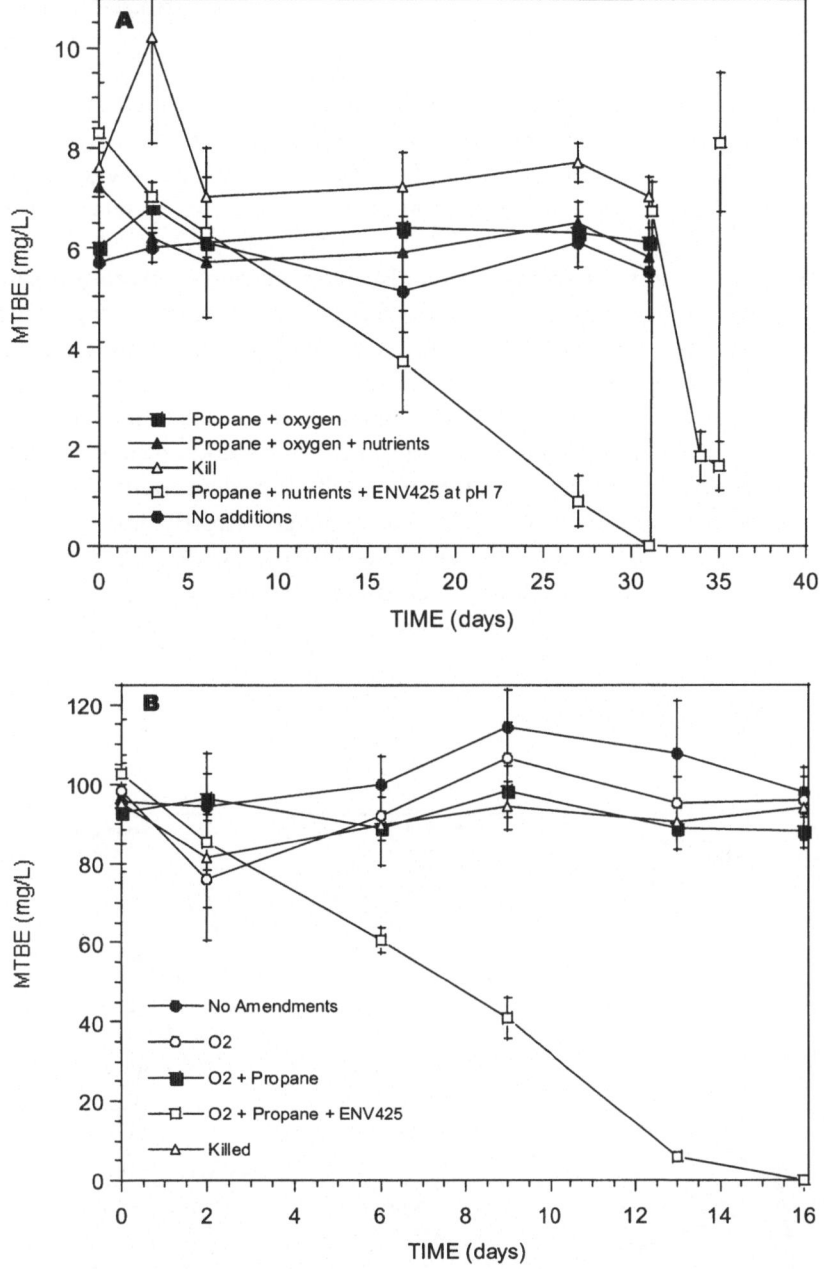

Figure 27-2.   **Results of Microcosm Testing**
Panel A represents microcosms incubated at ambient pH (approximately pH 4), with the exception of bioaugmented microcosms that were adjusted to pH 7. Panel B represents microcosms that were adjusted to pH 7. Symbols represent treatments as shown on the graphs. Values are mean (n=3) and error bars are one SD. ENV725 was added to a final concentration of 10^6 CFU/ml.

during operation of the treatment system, but they remained less than 11 mg/l throughout the treatment period (Figure 27-3). When the system was shut down after nearly five months of operation, dissolved oxygen concentrations throughout the treatment zone rapidly decreased to near historical background levels.

The pH within the treatment area also was relatively constant throughout the treatment period, despite repeated additions of buffering solution (Figure 27-4). The pH levels of ground water at the wells with the greatest MTBE degradation (MW7 and MW9; see below) were consistently lower than the pH at wells MW6 and MW11. This decrease in pH in the biologically active wells could be related to increased CO_2 production or the formation of acidic products during biodegradation of MTBE and propane. Late in the demonstration, sodium bicarbonate was injected directly into MW7 and MW11 to better evaluate the effect of pH adjustment on MTBE biodegradation. The direct injection of sodium bicarbonate resulted in high pH levels in these wells, and it also appeared to increase the pH in MW6 that was upgradient of MW7.

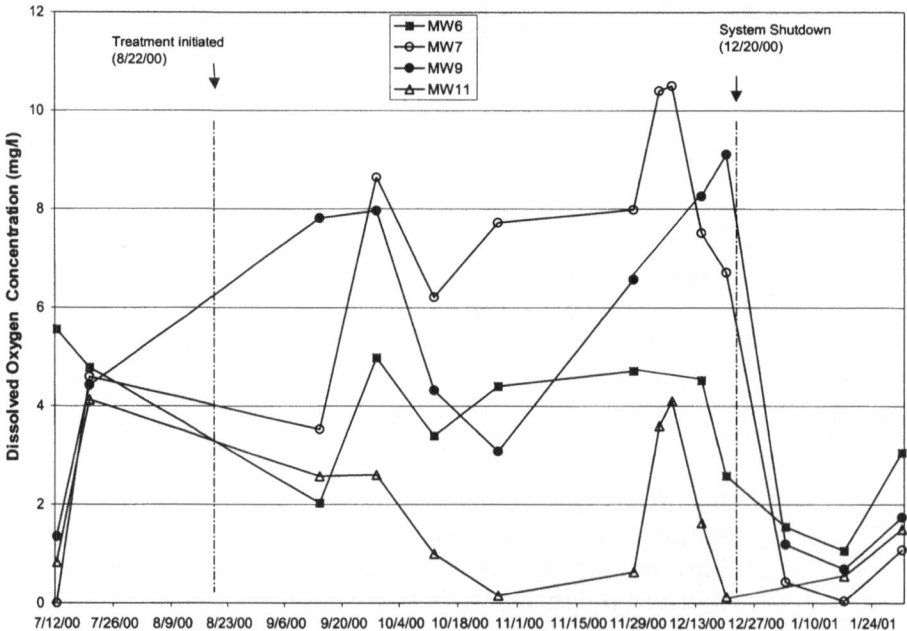

Figure 27-3.   **Dissolved Oxygen Concentration in Ground Water from the Demonstration Site Monitoring Wells**
Symbols are as follows: MW6, closed squares; MW7, open circles; MW9, closed circles; MW11, open triangles.

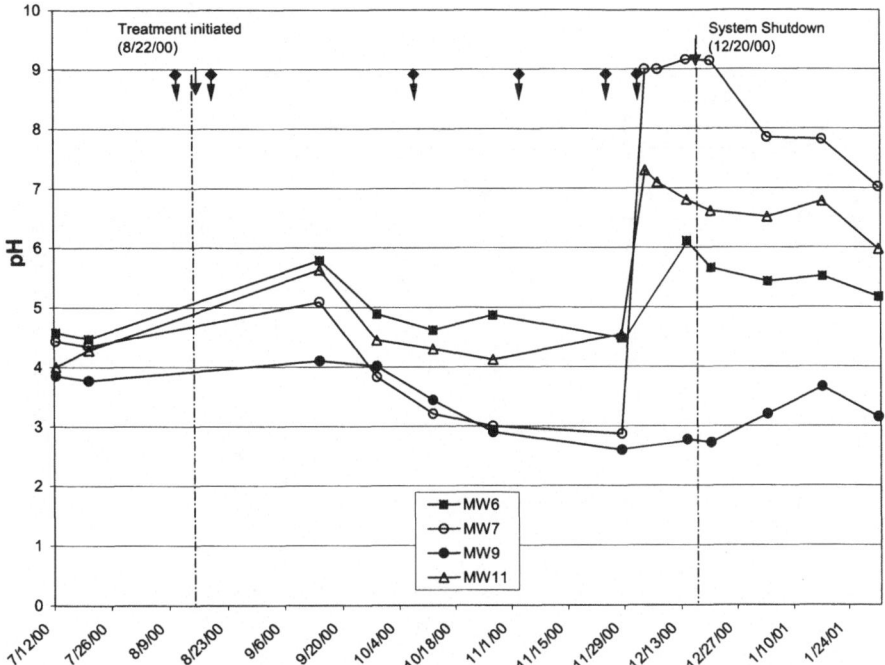

Figure 27-4. **Ground Water pH at the Demonstration Site Monitoring Wells**
Symbols are as follows: MW6, closed squares; MW7, open circles; MW9, closed circles; MW11, open triangles. Diamonds with arrows identify buffering events.

MTBE concentrations in MW6 were reduced by approximately 40% during the four-month treatment period (Figure 27-5a). Likewise, MTBE concentrations in MW7 were reduced by as much as 76% during biostimulation treatment (Figure 27-5b). MTBE concentrations in MW9 were reduced by as much as 98%, from 88 mg/l to 1.7 mg/l, during the treatment period (Figure 27-5c). MTBE concentrations in MW11 were relatively constant during the five-month demonstration (Figure 27-5d), presumably because it was too far downgradient for treated water to reach it during the demonstration period. First-order rate constants for MW6, MW7, and MW9 were calculated to be 0.0084, 0.0288, and 0.0027/day, respectively. This corresponded to MTBE half-lives of 82, 24, and 30 days, respectively. After nearly five months of operation the treatment system was shut down. In each of the treatment zone monitoring wells the MTBE concentration rebounded to near pretreatment levels (see Figures 27-5a-c). The rebound effect was attributed to a continuing source of MTBE contamination at the site. Ongoing work at the site has led to

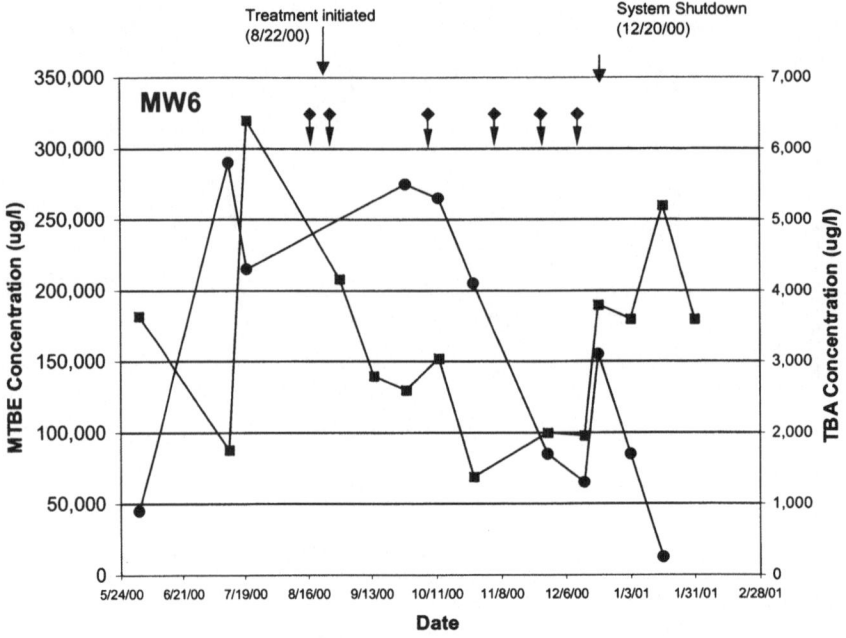

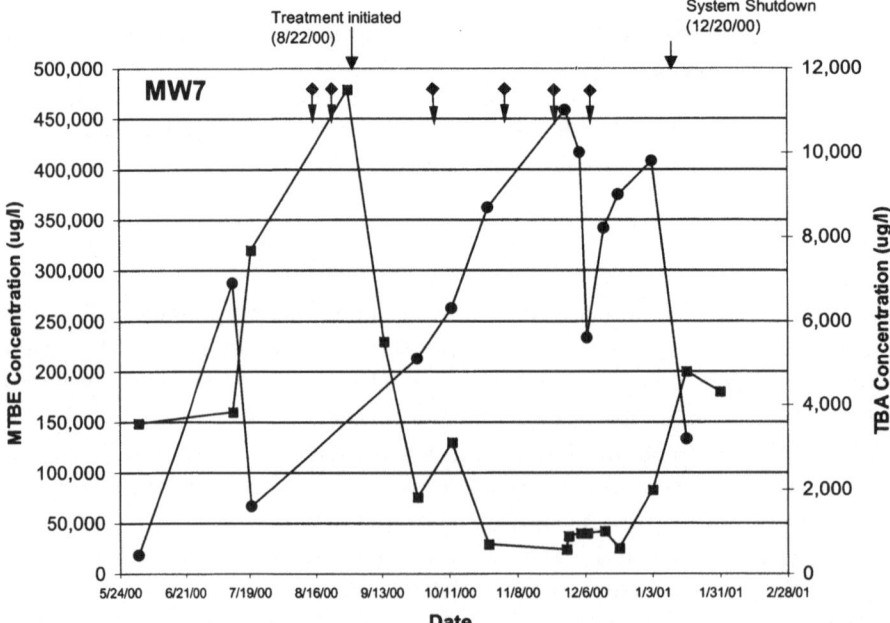

Figure 27-5.   **MTBE and TBA Concentrations in Ground Water from the Demonstration Site Monitoring Wells**
Symbols are as follows: MTBE, closed squares; TBA, closed circles. Diamonds with arrows identify buffering events.

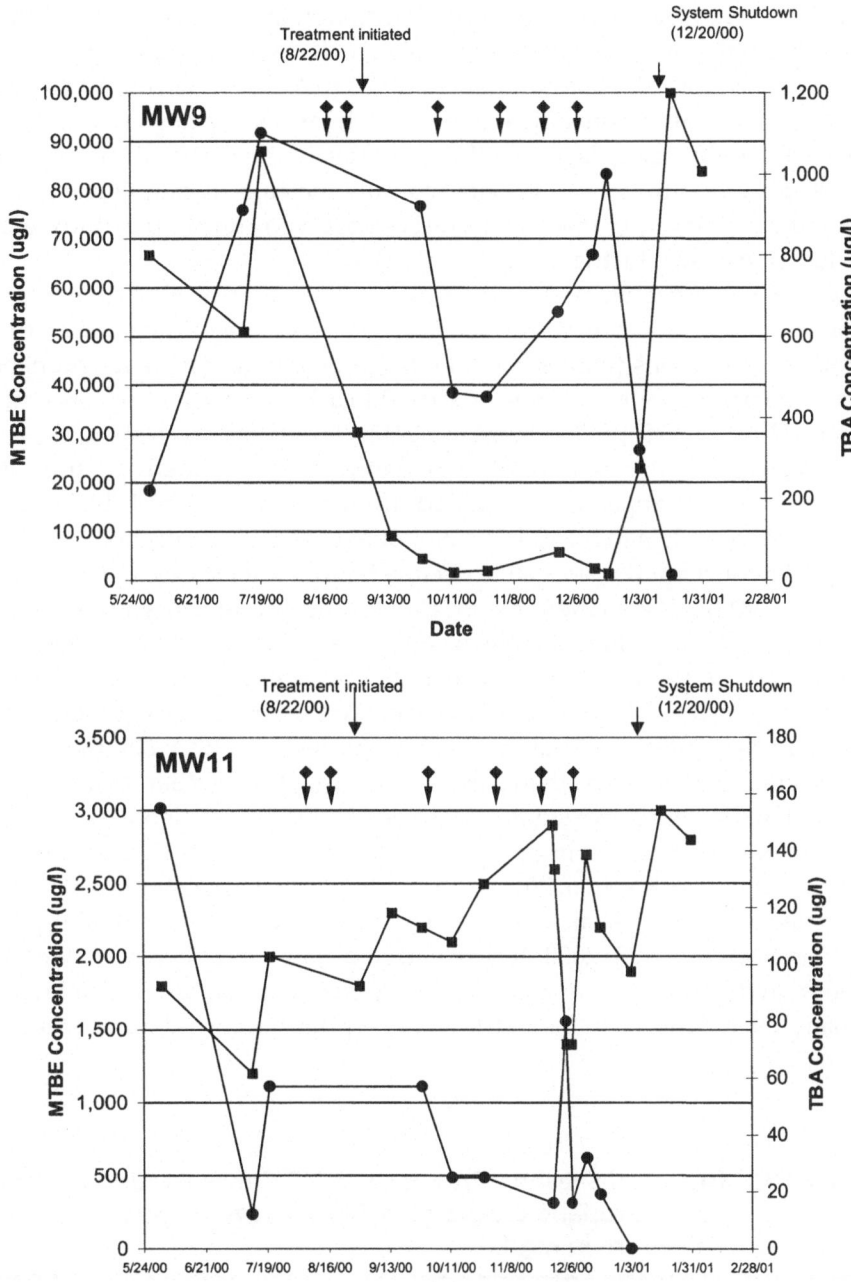

Figure 27-5. **Continued**

a repair of the leakage source and implementation of an expanded treatment system for full-scale remediation of the site, including the source area.

To confirm that the observed fluctuations in MTBE concentrations were not due to normal seasonal fluctuations, the treatment data were compared to historical monitoring data from the site (Table 27-1). This analysis revealed that MTBE concentrations at the site typically increase during the late summer and fall, indicating that the decreases observed during the study were not due to seasonal fluctuations.

TBA concentrations in the site ground water increased during MTBE biodegradation, but they were typically orders of magnitude lower than MTBE concentrations (Figure 27-5). During our initial work with propane-oxidizing bacteria, pure cultures produced nearly stoichiometric concentrations of TBA from MTBE (Steffan *et al.*, 1997). TBA concentrations in the cultures decreased only after MTBE was completely degraded. At this site, however, TBA was apparently degraded simultaneously with MTBE because it did not accumulate to levels near the initial MTBE concentration. Furthermore, TBA concentrations began to decline rapidly after the system was shut down and MTBE degradation ceased. The decline in TBA concentrations was accompanied by a decline in oxygen concentration. These data suggest that the propane oxidizers continued to degrade TBA after propane was no longer available to induce MTBE degradation, or that other TBA degraders were present in the system. During microcosm studies with ENV425, the organisms degraded TBA to less than 5 µg/l, indicating that similar levels will be achieved in the field provided the treatment period is sufficient.

With only two exceptions, propane concentrations in the treatment zone ground water were less than 20 µg/l throughout the demonstration. Propane was detected at 30 µg/l at MW7 on September 14, 2000, and at 40 µg/l in MW9 on September 28, 2000. It was detected only sporadically in samples taken from the SVE system operated during the demonstration. The highest propane levels detected in the SVE system, 59 ppmv, occurred during the first

Table 27-1.   **Historical Seasonal Trends in MTBE Concentration at the In Situ Propane Biosparging Demonstration Site Values are in mg/l.**

Monitoring well	July 1997	October 1997	June/July 1998	October 1998
MW6	9.3	150	110	240
MW7	32	41	34.5	120
MW9	4	10	22	47
MW11	BDL*	0.8	1	2

* = below detection limit

week of system operation. MTBE was detected only once in the SVE gas, and it was present at concentrations near the analytical detection limit for MTBE.

IN SITU BIOTREATMENT SUMMARY

The results of this demonstration show that MTBE-impacted ground water can be biologically remediated using propane-oxidizing bacteria and propane biosparging. This site presented a number of unique challenges to this technology, including low pH, high MTBE concentrations, and a continuing source of MTBE. Nonetheless, a significant mass of MTBE was removed during this demonstration, and MTBE reductions of greater than 90% were achieved in a relatively short time. The results suggest that this treatment approach also supports the degradation of TBA.

Propane has proven to be an excellent substrate for biostimulation applications; it is widely available, transportable even to remote sites, and relatively inexpensive. Application of propane injection in the field, however, may raise concerns about creating explosive mixtures of propane and air *in situ*. To address these concerns we injected propane in pulses and did not exceed 10% of the LEL of propane in the injection gas. We also used SVE to prevent *in situ* accumulation of propane; however, the results of our monitoring suggest that propane stripping is minimal, and SVE is likely unnecessary at most sites.

TECHNOLOGY COSTS

The costs of this field demonstration of the propane biotechnology system for remediation of MTBE and TBA are not indicative of the cost of implementation of the technology at other sites. First, as a pilot-scale demonstration, the system was operated for only six months, as allowed by the Permit-By-Rule issued by the NJDEP. Second, it was applied using an air sparging system and monitoring wells that were previously installed by others, and not optimum for this application. Finally, a great deal more analytical testing was performed for the project than would be required during a more routine application.

Estimates of the cost of implementation are similar to the costs of applying conventional air sparging/biosparging at a service station site. During the demonstration, propane costs were only $240 for the entire six months. The primary equipment cost for the application is a biosparging system that safely blends low levels of propane with sparging air. A typical system, fully engineered, constructed, and mounted in a trailer, is expected to cost $35,000 to $40,000. While stationary systems can probably be installed at a lower cost, the mobile system is suitable for repeated use at multiple sites, or it could be returned to a site to remediate future MTBE releases. Future applications of the technology probably will not require the use of SVE during biosparging, saving both the equipment and discharge permit costs. It also is recommended that pre-design treatability studies be performed with site ground water and soil. These tests are expected to cost approximately $4,000. The technology also could be applied in a number of alternative configurations — some employing existing systems — depending on site characteristics and treatment needs (see introduction). Thus, the complexity of the site and the selection of an application design will ultimately determine the total cost of the system.

REFERENCES

Hareland, W., Crawford, R.L., Chapman, P.J., and Dagley, S. 1975. Metabolic function and properties of 4-hydroxyphenylacetic acid 1-hydroxylase from *Pseudomonas acidovorans*. *Journal of Bacteriology*. 121, 272–285.

Lombard, K.H., Borthen, J. W., and Hazen, T.C. 1994. The design and management of system components for *in situ* methanotrophic bioremediation of chlorinated hydrocarbons at the Savannah River Site. In: *Air Sparging for Site Remediation*, pp. 81–96. (Hinchee, R.E., Ed.). Boca Raton, Florida, Lewis Publishers.

Steffan, R. J., McClay, K., Vainberg, S., Condee, C. W., and Zhang, D. 1997. Biodegradation of the gasoline oxygenates methyl *tert*-butyl ether, ethyl *tert*-butyl ether, and *tert*-amyl ether by propane-oxidizing bacteria. *Applied and Environmental Microbiology*. 63, 4216–4222.

U.S. Patent #5,814,514. September 29, 1998.

CHAPTER 28

Application of an *In Situ* Bioremedy Biobarrier at a Retail Gas Station

Gerard E. Spinnler, Ph.D., Paul M. Maner, Jeffrey D. Stevenson, Shell Global Solutions (US) Inc.

Joseph P. Salanitro, Ph.D., University of St. Thomas

Jennifer Bothwell, Equiva Services LLC and

John Hickey, Geologic Services Corporation

SITE LOCATION AND GEOLOGY/HYDROGEOLOGY

Located in western Connecticut, this site has been in use as a gas station since 1950. It is situated in a mixed commercial and residential area and is now operating as a retail gas station and convenience store. The site rests on fairly flat ground with a slight northwest topographic grade toward the Pomperaug River. Bedrock underlies the site approximately 12 meters (40 feet) bgs and drops to more than 60 meters (200 feet) bgs to the northwest. The unconsolidated material above the bedrock is stratified drift consisting of fine to coarse sand with silts, gravels, and cobbles. A dense cobble-rich layer was encountered at approximately 4 meters (14 feet) bgs beneath the site.

The depth-to-water on site ranges from approximately 3 to 5 meters (10 to 15 feet) bgs and varies seasonally approximately 1.5 meters (5 feet), with lower levels occurring in the winter months. The water table slopes to the northwest with a hydraulic gradient of approximately 0.01. The aquifer near the site is approximately 10 meters (35 feet) thick and thickens to almost 45 meters (150 feet) to the northwest. Based on pumping test data, the hydraulic conductivity ranges between 8.8×10^{-4} to 1.8×10^{-3} cm/s (2.5 to 5 ft/day). Regional data indicate hydraulic conductivity values ranging from 1.8×10^{-3} to 5.3×10^{-2} cm/s (5 to 150 ft/day).

NATURE AND EXTENT OF CONTAMINATION AND POTENTIAL RECEPTORS

For approximately two years prior to the biobarrier installation, the average MTBE ground water concentrations ranged from 7,000 to 40,000 µg/l. Lower concentrations of BTEX compounds were also observed, with average total concentrations less than 1,000 µg/l.

MTBE was detected throughout the vertical extent of the aquifer using depth discriminated ground water samples. Dissolved oxygen concentrations in the plume were less than 1 mg/l.

A municipal supply well and domestic water wells are within 800 meters (2,500 feet) of the gas station. MTBE was the constituent of concern, with a target cleanup level of 70 µg/l.

REMEDIATION

An *in situ* biobarrier system (BioRemedy®) was chosen to control the migration of the MTBE plume from this site. Previous attempts to remediate and control the MTBE plume used in-well stripping, pump-and-treat, and SVE. The success of these efforts was limited in remediating and controlling the plume. MTBE plume control using *in situ* biobarrier technology has been successfully demonstrated at Port Hueneme, California (Salanitro *et al.*, 2000, 2001).

Biobarrier.   The BioRemedy® biobarrier consists of specialized MTBE-degrading microorganisms (MC) and an oxygen gas injection network placed to intercept the plume. As the plume flows through the biologically active zone, MTBE is transformed to CO_2 and H_2O. Because the MC microorganisms are aerobic, oxygen is necessary for degradation.

Components of Biobarrier System.   Biobarrier system components (Figure 28-1) consist of MTBE-degrading microbes, an oxygen generation and distribution system, and monitoring wells. Ideally, MTBE-degrading microbes are introduced into the subsurface at or near the leading edge of the dissolved MTBE plume. Oxygen is distributed by pulsed-injection of oxygen gas through wells located within or near the microbe-rich zone. Monitoring wells are located upgradient and downgradient of the biobarrier to assess concentrations flowing into and out of the biobarrier. Figure 28-2 is a map view of the system illustrating the relationship of the MTBE plume and the biobarrier.

Microbes.   The MTBE-degrading culture used in this biobarrier is a proprietary aerobic mixed bacterial culture, designated MC (Salanitro *et al.*, 1994). The culture degrades MTBE, TBA, and other common fuel oxygenates. MC is grown in commercial quantities at Shell Global Solution's Westhollow Technology Center in Houston, Texas. The culture was placed into the saturated zone using direct push techniques and a specially designed injection pump.

Oxygen.   Since the transformation of MTBE by MC is aerobic, oxygen is necessary as an electron acceptor. Dissolved oxygen is supplied to the reaction

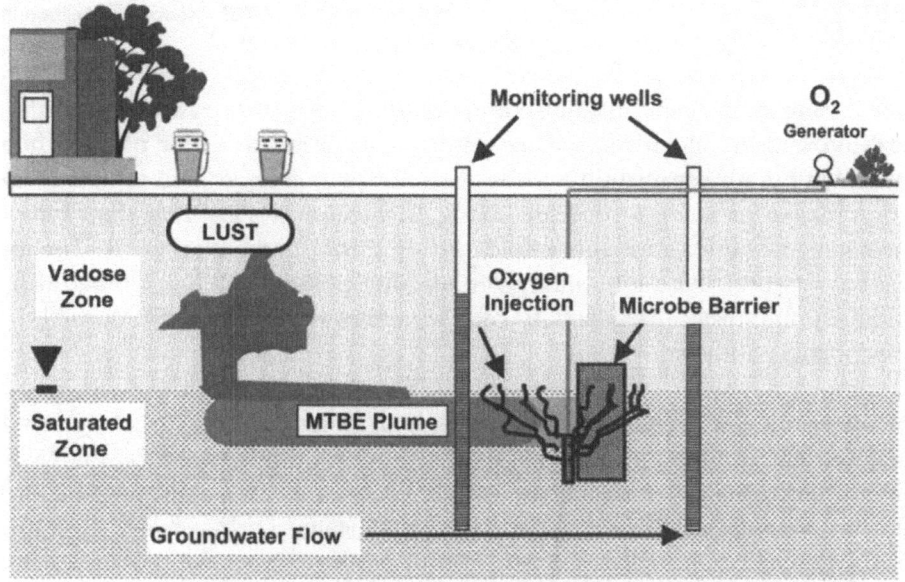

Figure 28-1.　**Idealized Cross-Section of an MTBE Biobarrier System**
MTBE-degrading microbes (MC) are located at the down-
gradient edge of the plume along with oxygen injection
wells. The MTBE plume flows through the microbe-rich re-
gion and degrades.

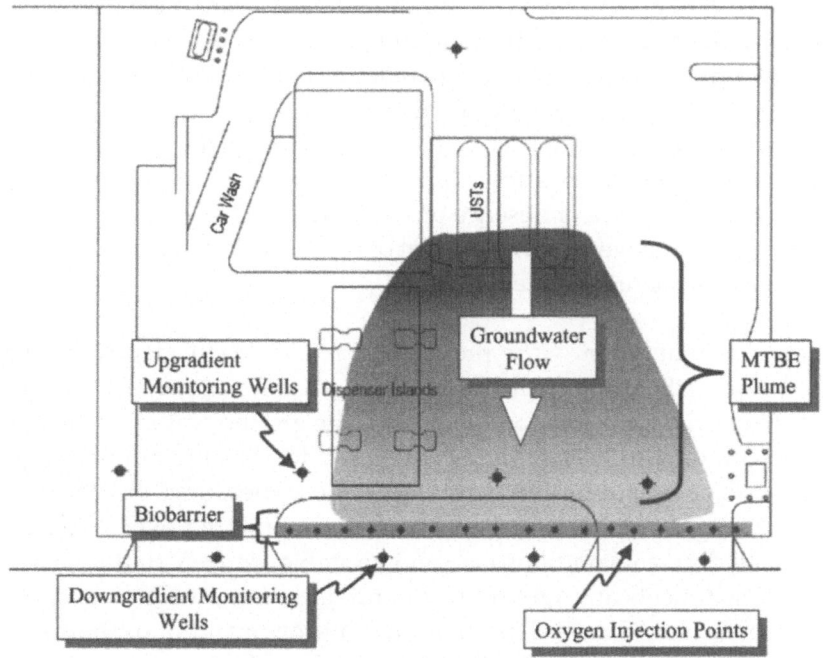

Figure 28-2.　**Map View of an Idealized MTBE Biobarrier System**
The biobarrier is located at the leading edge of the plume.

zone using an oxygen pulse-injection system. This system optimizes gas distribution in the biobarrier and oxygen transfer into the water phase while maintaining the permeability of the zone. Oxygen is generated on-site from compressed air using a pressure swing adsorption system and distributed using a solenoid-operated manifold system through injection wells. The location, spacing, and depth of oxygen injection points were determined during field pilot tests. Gas injection wells were installed using conventional direct push techniques.

Monitoring Well System.   A measurement system of strategically placed appropriate sampling points and analytical techniques is necessary to evaluate the performance of the biobarrier. MTBE concentrations entering the biobarrier are measured from monitoring wells located hydraulically upgradient, whereas concentrations from ground water exiting the biobarrier are measured from monitoring wells located hydraulically downgradient. Ground water samples from within the biobarrier can also be obtained from the gas injection wells. GC/MS techniques were used for detection of MTBE and BTEX compounds.

SITE APPLICATION

Field Pilot/Evaluation Test.   Implementation of the biobarrier technology requires an understanding of site conditions, contaminant distribution, oxygen distribution, and the activity of indigenous MTBE-degrading microorganisms. Prior to full-scale implementation, field investigations were conducted to determine the vertical distribution of dissolved MTBE in the proposed biobarrier location as well as distribution characteristics of injected gas. Subsurface sediment cores collected during the investigation were used to test for the presence or absence of indigenous MTBE-degrading microorganisms. The effectiveness and amount of MC to be added to the site was also determined using the sediments.

Microcosm Evaluation.   The presence of native MTBE degraders was assessed using laboratory microcosm assays. Aquifer material from the proposed biobarrier zone was combined with site ground water in sealed serum bottles with oxygen headspace. MTBE was added to the sediment/ground water system at concentrations similar to those measured on site (approximately 25 mg/l); MTBE concentrations were monitored over 90 days. Data from three different conditions are shown in Figure 28-3. In the presence of only site sediments and ground water, no decrease in MTBE concentration was observed for 90 days. The addition of nitrogen and phosphorous nutrients to enhance (stimulate) MTBE biodegradation had no effect since no decrease in MTBE concentrations were observed. Control samples containing

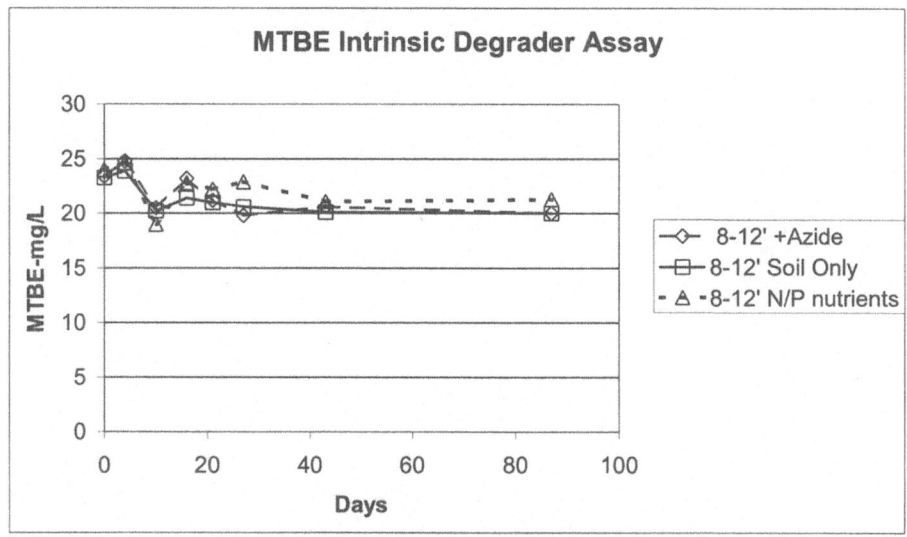

Figure 28-3.   **Microcosm Results from: Site Soil and Ground Water Only; Nutrients; and Sodium Azide (Control Sample) In all cases, MTBE was not observed to degrade in 90 days.**

2% sodium azide as a respiration inhibitor also showed no decrease in MTBE concentrations, indicating no biodegradation occurred. These data indicated there was no inherent MTBE-degrading activity in the sediments at this site. Microcosms were also conducted with MTBE concentrations of approximately 10 μg/l. Results from these assays demonstrated no inherent MTBE degrading activity, even with lower MTBE concentrations.

In order for the biobarrier to be effective, MC must be added in sufficient quantity to degrade MTBE within the contact time of the plume with the barrier. The contact time depends on ground water velocity and the width of the biobarrier and is typically a few days. Laboratory experiments estimated the amount of MC necessary to degrade various concentrations of MTBE within a few days of contact time. MTBE degradation curves were experimentally determined as a function of MC loading in milligrams (dry weight) of MC per kilogram of soil for two initial MTBE concentrations (Figure 28-4). Minimum MC loading was estimated by considering the contact time of the ground water and MTBE concentration. In this case, an MC loading of 400 mg/kg was sufficient to degrade MTBE concentrations of 10,000 μg/l within 4 days and 100,000 μg/l within five days.

Oxygen Delivery.   The oxygen generation system consists of a compressor and an oxygen generator designed by Matrix Technologies. It is housed on-

MC Soil Loading

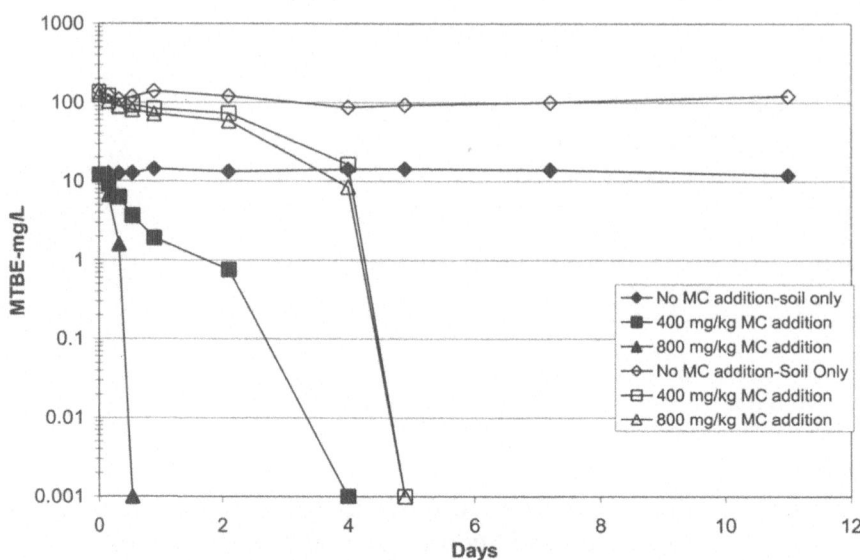

Figure 28-4.   **Microcosm Results for Two MTBE Starting Concentrations with Site Soil and Ground Water and Adding 400 mg/kg MC, 800 mg/kg MC**
In the case of soil-only, MTBE did not degrade. For both concentration ranges, MTBE degraded within five days.

site in an enclosed trailer and uses 220V power. Connections to the distribution wells were made using underground tubing. Each well was directly connected to a solenoid, ensuring uniform oxygen distribution across the site. The system is controlled by a sequence timer and equipped with a remote alarm with fax notification in the event of system malfunction.

Oxygen wells were located across the downgradient edge of the site as indicated in Figure 28-5. Nested oxygen wells, designated OP-X on Figure 28-5, were installed to approximately 5.5 meters (18 feet) bgs and 8.5 meters (28 feet) bgs. They were constructed with 1.3-centimeter (0.5-inch) polyvinyl chloride riser with 0.6-meter (2-foot) well screens. Wells were installed using conventional direct push techniques. Special care was taken to prevent short-circuiting.

The oxygen system operated for approximately two weeks prior to MC injection to ensure adequate dissolved oxygen distribution and system integrity.

MC Delivery.   MC was concentrated to a paste and shipped in refrigerated containers from our facility in Houston, Texas. Prior to injection, the MC

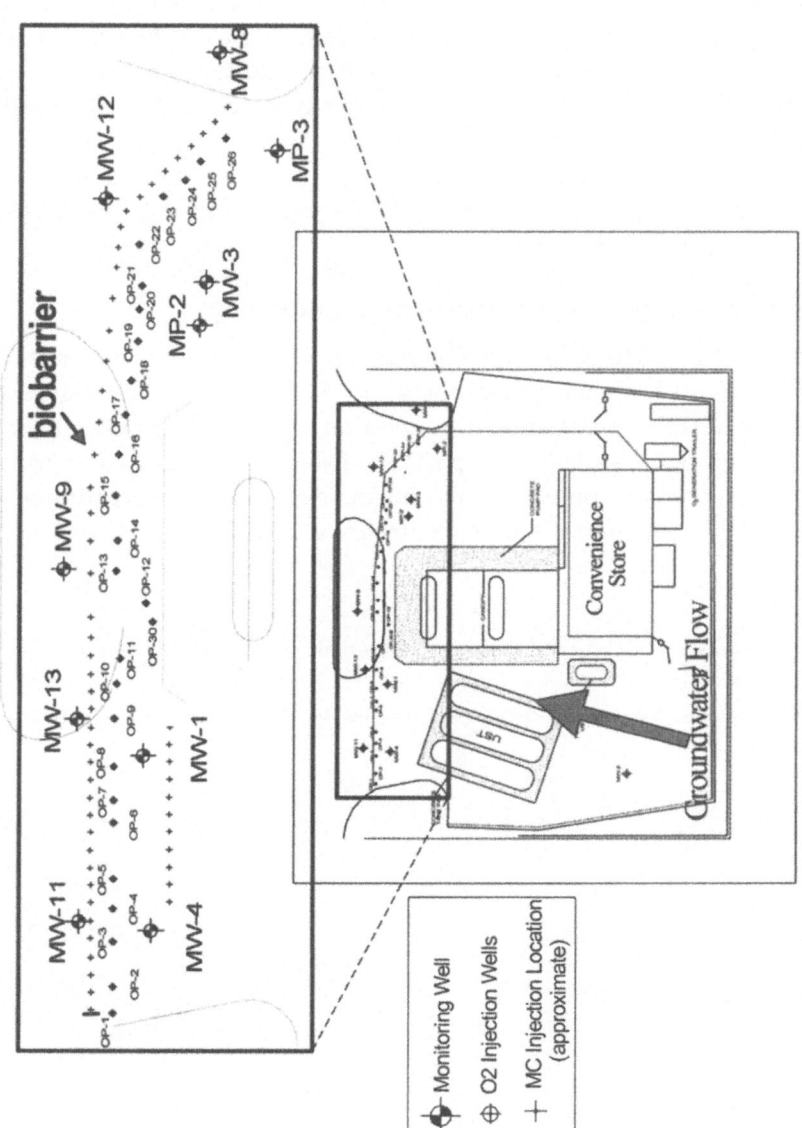

Figure 28-5.   **Site Map with Enlargement of Biobarrier Area**

paste was reconstituted with non-impacted ground water and aerated using compressed air. MC was injected in a linear array downgradient of the oxygen points across the site's downgradient edge. Biomass was injected as a suspension through an open drilling rod with either an expendable tip or a specially designed one-way valve that prevented backflow. Injections were spaced approximately 0.6 meters (2 feet) laterally and over the vertical range of approximately 4 to 9 meters (12 to 30 feet) bgs. Additional MC was injected upgradient of the oxygen points on the south side of the site (between MW-4 and MW-1, Figure 28-5) to provide additional biomass for higher MTBE concentrations. Approximately 380 liters (100 gallons) of suspension was injected in each boring. The total mass of MC injected at the site was 175 kilograms (382 pounds) (dry weight).

PERFORMANCE OF THE BIOREMEDY BIOBARRIER

To evaluate the effectiveness of the biobarrier, the average MTBE concentrations were plotted versus system operation time in Figure 28-6. The mean (geometric) MTBE concentration from monitoring wells in the vicinity of the biobarrier before the system was operational was 12,000 µg/l. After approximately 16 months of operation, the mean (geometric) concentration was 8 µg/l, a 99.9% reduction. Mean concentrations of "upgradient" wells not af-

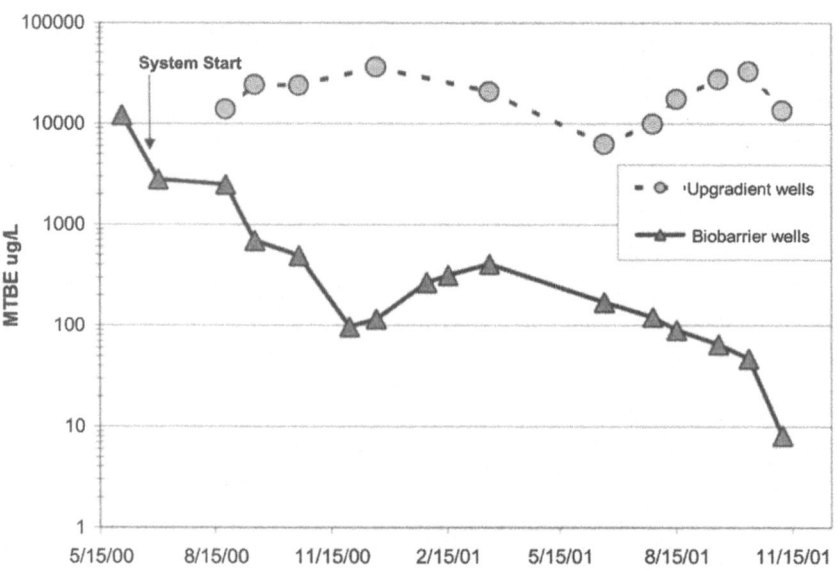

Figure 28-6.   **Mean of MTBE Concentration in Biobarrier Wells and Upgradient Wells since the Start of the Biobarrier System**

fected by the activity of the biobarrier were plotted to evaluate whether the decrease in concentration was due to the biobarrier or to an overall decrease in MTBE. The mean MTBE concentrations in upgradient wells remained relatively constant over the observation period and had a mean concentration of 11,000 µg/l. The concentration of MTBE in the upgradient wells represents the concentration of MTBE flowing into the biobarrier.

The amount of biomass at each injection point along the biobarrier was plotted in Figure 28-7. The southern end of the biobarrier received a higher MC mass than the other part of the barrier since the MTBE concentrations were higher in this area. The plot also indicates injection spacing near locations 34 through 52 were as much as 2 meters (6 feet) apart rather than 0.6 meters (2 feet) apart as in other parts of the barrier. Large spacings were due to refusal of the injection rod because of a cobble-rich zone at approximately 4 meters (14 feet) bgs. HSA methods were used in this area to drill through the problem layer and provided pilot holes for delivery of MC.

Individual monitoring well performance was assessed as percent MTBE reduction and plotted on Figure 28-7 along with their approximate (lateral) location. With the exception of wells MW-9, OP-13, and OP-21, all monitoring points showed MTBE reduction of 88 to 99.6%. MTBE was reduced in the three remaining wells (MW-9, OP-13, and OP-21) by 50 to 60%. These wells are located in areas where MC injections were greater than 0.6 meters (2 feet) and MC was not well distributed due to difficult drilling conditions. The amount and distribution of MC appears to be an important factor in the performance of the biobarrier.

In summary, the data indicate the BioRemedy® biobarrier has been effective in reducing MTBE leaving this site. A one-time injection of the specialized MTBE-degrading microorganisms, MC, together with ongoing pulsed oxygen injection has proved effective in maintaining the MTBE-degrading capacity of the system. The average concentration of the biobarrier is below state standards; however, some individual wells are not. Consideration has been given to inject additional microbes to areas where the MC was not evenly distributed. This is the first demonstration of an *in situ* biobarrier limiting MTBE plume migration with ground water concentrations of 10 to 100 mg/l.

SYSTEM COSTS

The costs of the system are listed in Table 28-1. Costs include MC, shipping, injection of biomass, oxygen well installation, oxygen system, installation, and construction costs. A pilot test was conducted to assess injected gas distribution for system design.

System installation took approximately three weeks although normal installation of this size biobarrier is about two weeks. This site took longer (and cost more) because of drilling difficulties. The installation of the system was conducted with the station in operation.

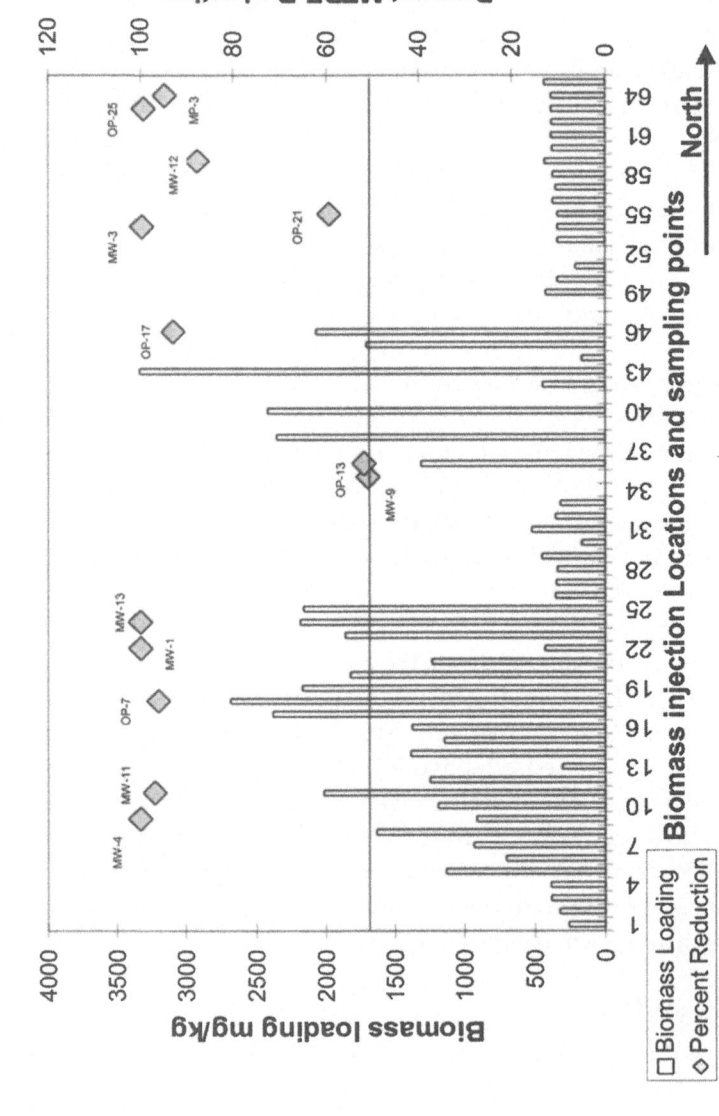

Figure 28-7. **Biomass Loading at Each Injection Point across the Site**
The percent MTBE degraded over the operational period of the biobarrier is also plotted for each well.

Table 28-1.   **Costs for Connecticut Site BioRemedy BioBarrier**

Site Assessment	$20,000
Design	$20,000
Installation	$203,000
Operation and Maintenance 1½ years of Operation	$30,000
Total	$273,000

CT-Site Timeline

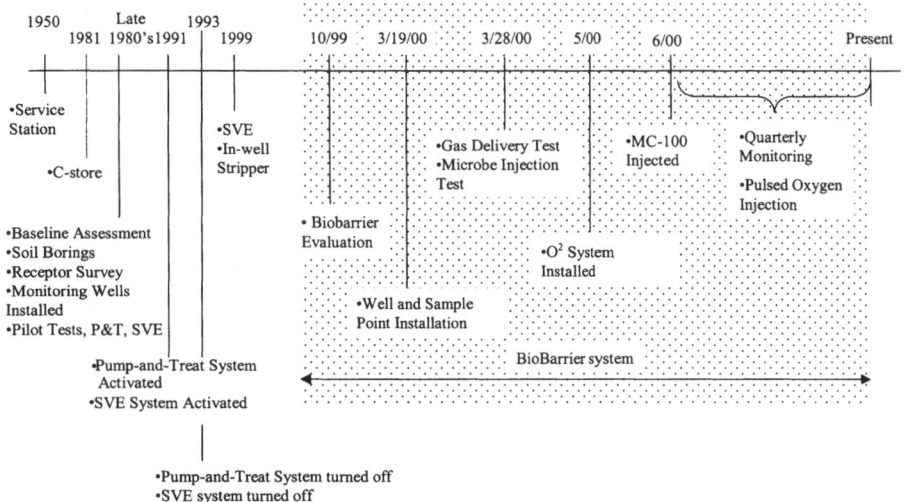

Figure 28-8.   **Timeline of Site Activities**

TIMELINE

Figure 28-8 illustrates a timeline for the site.

REFERENCES

Salanitro, J.P., Diaz, L.A., Williams, M.P. and Wisniewski, H.L. 1994. Isolation of a bacterial culture that degrades methyl-t-butyl ether. *Applied and Environmental Microbiology*. 60, 2593–2596.

Salanitro, J.P., Johnson, P.C., Spinnler, G.E., Maner, P.M., Wisniewski, H.L., and Bruce, C. 2000. Field-scale demonstration of enhanced MTBE bioremediation through aquifer bioaugmentation and oxygenation. *Environmental Science and Technology*. 34, 4152–4162.

Salanitro, J.P., Spinnler, G.E., Maner, P.M., Tharpe, D.L., Pickle, D.W., Wisneiwski, H.L., Johnson, P.C. and Bruce, C. 2001. *In situ* bio-remediation of MTBE using biobarriers of single or mixed cultures. In: *In situ and On-Site Bioremediation, Sixth International Symposium Proceedings*. June 4–7, 2001. Battelle Press.

CHAPTER 29

Ground Water Recovery and Bioreactor Treatment at a California Site

Joseph E. O'Connell, Sc.D., P.E. and Steve M. Zigan, R.G., C.H.G., REA,
Environmental Resolutions, Inc.

SUMMARY

The site is an active service station located in southern California. Currently, there are two 38,000-liter (10,000-gallon) gasoline USTs, one 45,000-liter (12,000-gallon) gasoline UST, and one 2,100-liter (550-gallon) waste oil UST located at the site. The current USTs were installed at the site in 1987 to replace older USTs. Although BTEX concentrations at the site were quite low (*e.g.*, benzene less than 50 µg/l), the site exhibited increasing concentrations of MTBE and TBA in ground water immediately below the dispensers and in the vicinity of the tank pit. Cluster wells screened over narrow ranges indicated a downward hydraulic gradient and migration of soluble fuel oxygenates into the lower aquifers. A combination pump-and-treat and SVE (DPE) system was installed and is beginning to localize the plume. Activated carbon is being used for remediation of the water; however, carbon is not effective for removal of TBA, and the site has low discharge limits for both MTBE and TBA. Therefore, a bioreactor, capable of consuming both TBA and MTBE, was used ahead of the carbon. DPE began continuous operation in September 2001. Initially the vapor flow was about 3 cmm (100 scfm), and the ground water flow was 8 l/min (2 gpm). The water flow was gradually increased to 15 l/min (4 gpm). Flows much greater than 15 l/min (4 gpm) tend to run the well system dry.

Site History.   The site was originally operated by a major oil company, but was closed in 1997 and sold to an independent dealer one year later. A fuel release had been reported in 1992, and subsequent water sampling indicated the presence of liquid product and BTEX, but not MTBE. TBA was not a constituent of concern at that time. The impacted wells were overpurged by pumping and bailing liquid hydrocarbons and water. Overpurging continued until the station was temporarily closed in 1997. BTEX has been present since water sampling began in 1992, but concentrations of each component are now less than 100 µg/l. MTBE began to appear in the ground water analytical results for the wells adjacent to the tank cavity about six months after the independent dealer took over the station in 1998. MTBE concentrations reached a high of 97,000 µg/l. The local oversight agency has indicated that the original owner is the responsible party and is being held accountable for the cleanup.

Hydrology.   The site is located on the western portion of the coastal plain of Orange County within the Orange County Ground Water Basin. The ground water basin lies within a synclinal, northwest-trending trough that extends from the Irvine area and deepens as it continues beyond the Los Angeles-Orange County line. The uplifted margins of the basin are formed by the Coyote Hills-Santa Ana Mountains to the north and east and the coastal mesas along the Newport-Inglewood Fault Zone to the south. The basin contains a total thickness of over 6,000 meters (20,000 feet) of Miocene-age sedimentary rocks, of which only the upper 1,000 meters (4,000 feet) contain fresh water.

Based upon the results of previous subsurface investigations at the site, the shallow sediments beneath the site consist of predominantly poorly graded sand with interbeds of silt and clay to a depth of approximately 14 meters (45 feet) bgs, the maximum depth drilled. In August 2000, continuous coring (sampling) to a depth of 14 meters (45 feet) bgs showed the following subsurface profile:

- 0 to 2 meters (0 to 6 feet) bgs: silt (ML);
- 2 to 4 meters (6 to 12 feet) bgs: silty sand (SM);
- 4 to 10 meters (12 to 34 feet) bgs: interbedded silty sand and sandy silt (SM and ML);
- 10 to 12.5 meters (34 to 41 feet) bgs: interbedded sand, silty sand, silt, clay, and peat (SP, SM, ML, CL, and PT); and
- 12.5 to 14 meters (41 to 45 feet) bgs: silty sand (SM).

The site is located within the "Pressure Area" hydrologic subunit of the Orange County Ground Water Basin. The Pressure Area is characterized by shallow, semi-perched, water-bearing zones that are separated from the deeper production aquifers in the basin by low permeability clays and silts. These low-permeability, near-surface sediments are known collectively as the Semiperched Aquifer. The sediments underlie most of the central and coastal portions of the basin and represent deposition in lagoonal and freshwater swamp environments.

A perched ground water aquifer occurs beneath the site at a depth of approximately 2 to 3 meters (7 to 9 feet) bgs and is separated by a thick unit of silt and clay from the first producible aquifer, the Talbert Aquifer. The Talbert

Aquifer is a relatively thin sand and gravel unit deposited by ancestral streams. In the region of the site, the Talbert Aquifer occurs at a depth of approximately 20 to 30 meters (60 to 90 feet) bgs. According to the Orange County Water District's November 1997 Ground Water Contour Map for the Coastal Area, the regional ground water flow direction in the vicinity of the site is southwest toward the Pacific Ocean.

Cluster wells have been installed to evaluate the presence of a vertical gradient. Well CW2D is screened from 13 to 13.5 meters (42 to 44 feet) bgs. Well CW2M is screened from 9 to 10 meters (30 to 32 feet) bgs. CW2S is screened from 1.5 to 6 meters (5 to 20 feet) bgs. A downward vertical gradient (difference in static ground water elevation) of about 1.5 meters (5 feet) was observed in the static water levels between the shallow and the deep zone.

There are 11 known active production wells within a one-mile radius of the site property. The closest, and of greatest concern, is located approximately 300 meters (900 feet) west-southwest (downgradient) of the site. It is an active, large system water supply well set at a depth of approximately 233 meters (764 feet) bgs.

Slug tests were performed and analyzed using the Bouwer and Rice method for unconfined aquifers. An average hydraulic conductivity of approximately 2.9×10^{-3} cm/s (8.1 ft/day) was calculated, which falls within the range of silty sand. A vapor extraction test (VET) was performed using an induced vacuum of approximately 140 to 229 centimeters (55 to 90 inches) of water. A radius of influence ranging from 10 to 12 meters (34 to 39 feet) was calculated. Influent vapor concentrations of TPH-modified for gasoline (TPHg) during the VET ranged from 888 to 4,320 ppmv. An aquifer pumping test was performed using well MW16 as the pumping well and wells MW4 and MW17 as observation wells. A hydraulic conductivity of between 1.3×10^{-2} and 1.7×10^{-2} cm/s (36 and 49 ft/day) and a transmissivity of 50 m³/ m/day (550 ft³/ft/day) were calculated. On September 4, 1998, an eight-hour high-vacuum DPE test was performed. Approximately 39 kilograms (85 pounds) of hydrocarbon vapors and 9,500 liters (2,500 gallons) of hydrocarbon-impacted ground water were extracted during the event. Vapor extraction flow rates ranged from approximately 1 to 1.5 cmm (47 to 52 cfm) at vacuums ranging from approximately 840 to 910 centimeters (330 to 360 inches) of water. Vapor TPHg concentrations ranged from 1,500 to 4,600 ppmv.

REMEDIAL ACTIVITIES

Remedial activities were implemented in stages, beginning with excavation; continuing with portable pumps and bailers (overpurging) and vacuum enhanced extraction; and culminating in the installation of a permanent system to treat both ground water and soil vapors. These activities are described in detail below.

Soil Excavation. In October 1987, impacted soil was excavated to a depth of approximately 4 to 5 meters (13 to 15 feet) bgs from the gasoline UST cavity and to a depth of approximately 4 meters (12 feet) from the used-oil UST cavity. The excavations were extended 1 meter (3 feet) below the water table in an effort to remove all the TPHg and BTEX impacted soil.

Overpurging. As an interim remedial measure, overpurging began in February 2000 and continued monthly until January 2001. Approximately 1,100 liters (300 gallons) were removed per event. Merely pumping water once a month had little effect on the ground water flow direction or the concentrations of fuel oxygenates in the ground water as determined from the quarterly ground water monitoring.

Interim Enhanced Vacuum Extraction. To address both soil vapors and hydrocarbons trapped below the ground water table, 72-hour high vacuum DPE events were undertaken once a month until a permanent system could be installed. A 25-horsepower oil-sealed liquid ring pump was used capable of producing a vacuum of 640 mm Hg (25 in Hg) and extracting up to 6 cmm (200 cfm) of soil vapors. As the ground water was removed, the water table beneath the site was lowered, and hydrocarbon-impacted soil in the capillary fringe and uppermost portion of the saturated zone was exposed. The vaporized hydrocarbons were incinerated, and the ground water was treated with carbon or taken to an appropriate disposal facility. These monthly events provided data to indicate that about 2 cmm (60 scfm) of soil vapor and 8 l/min (2 gpm) of ground water could be produced from each remedial well. Vapor concentrations of 1,500 ppmv TPHg could be sustained; ground water concentrations were about 15,000 µg/l of TBA, 10,000 µg/l of MTBE, and less than 100 µg/l BTEX.

Remedial Design. A fixed system, capable of handling 4 scmm (150 scfm) soil vapor and 30 l/min (8 gpm) of ground water, was installed. The soil vapors are extracted with a positive displacement blower and were treated initially with a catalytic oxidizer and eventually with activated carbon. The ground water is extracted with submersible electric pumps and treated with a bioreactor followed by activated carbon in order to meet the discharge requirements, which are 13 µg/l for MTBE and 12 µg/l for TBA. A footprint of the remediation layout is shown to scale in Figure 29-1. A diagram of the bioreactor used is shown in Figure 29-2. For this site, the MTBE/TBA loading was low enough that there was no need for an oxygen booster. Vapor and ground water are extracted from wells MW1, MW3, MW5, and CW3 as shown in Figure 29-3.

The bioreactor was sized to treat 30 l/min (8 gpm) of ground water containing 15,000 µg/l TBA and 10,000 µg/l MTBE. The bioreactor had to re-

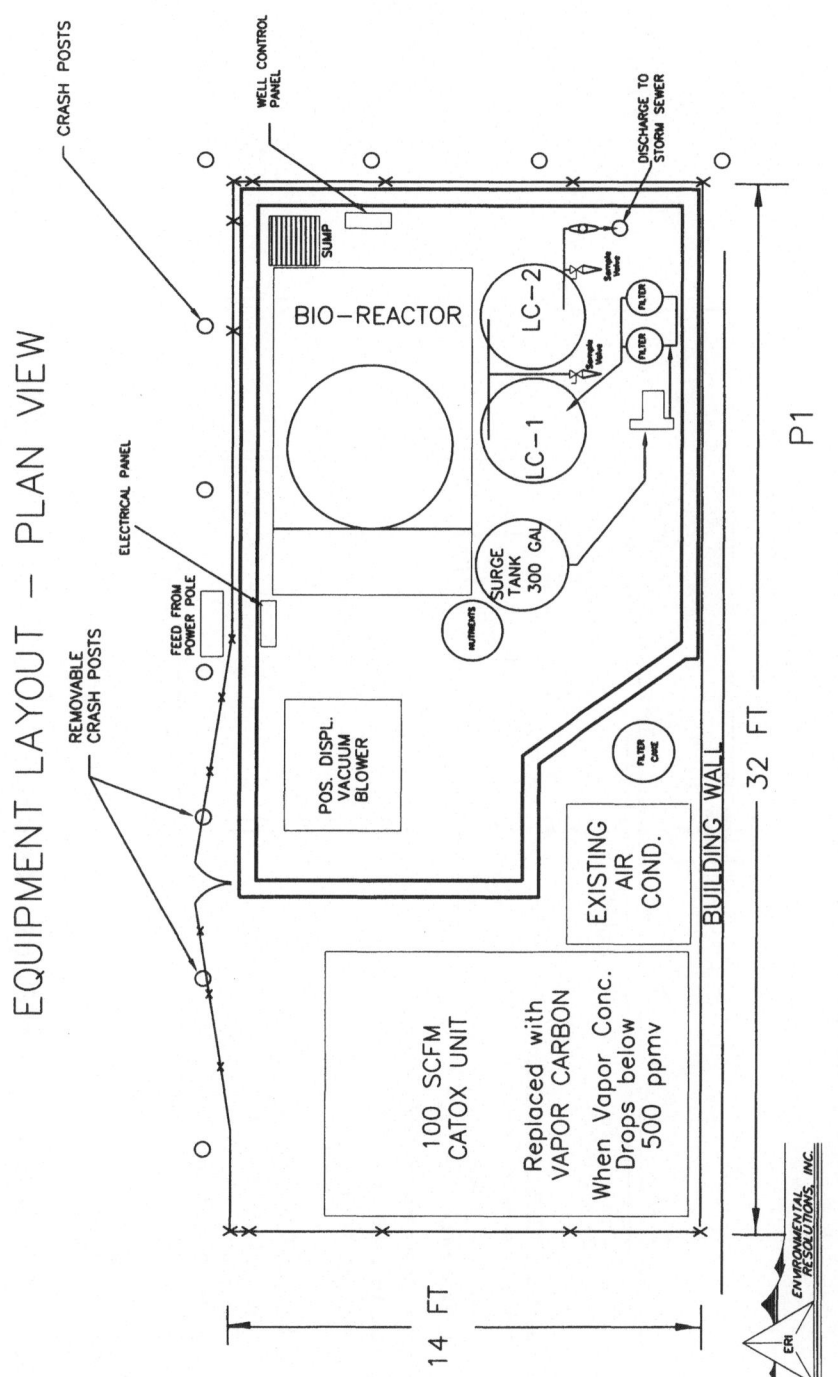

EQUIPMENT LAYOUT – PLAN VIEW

Figure 29-1.   Equipment Layout— Plan View

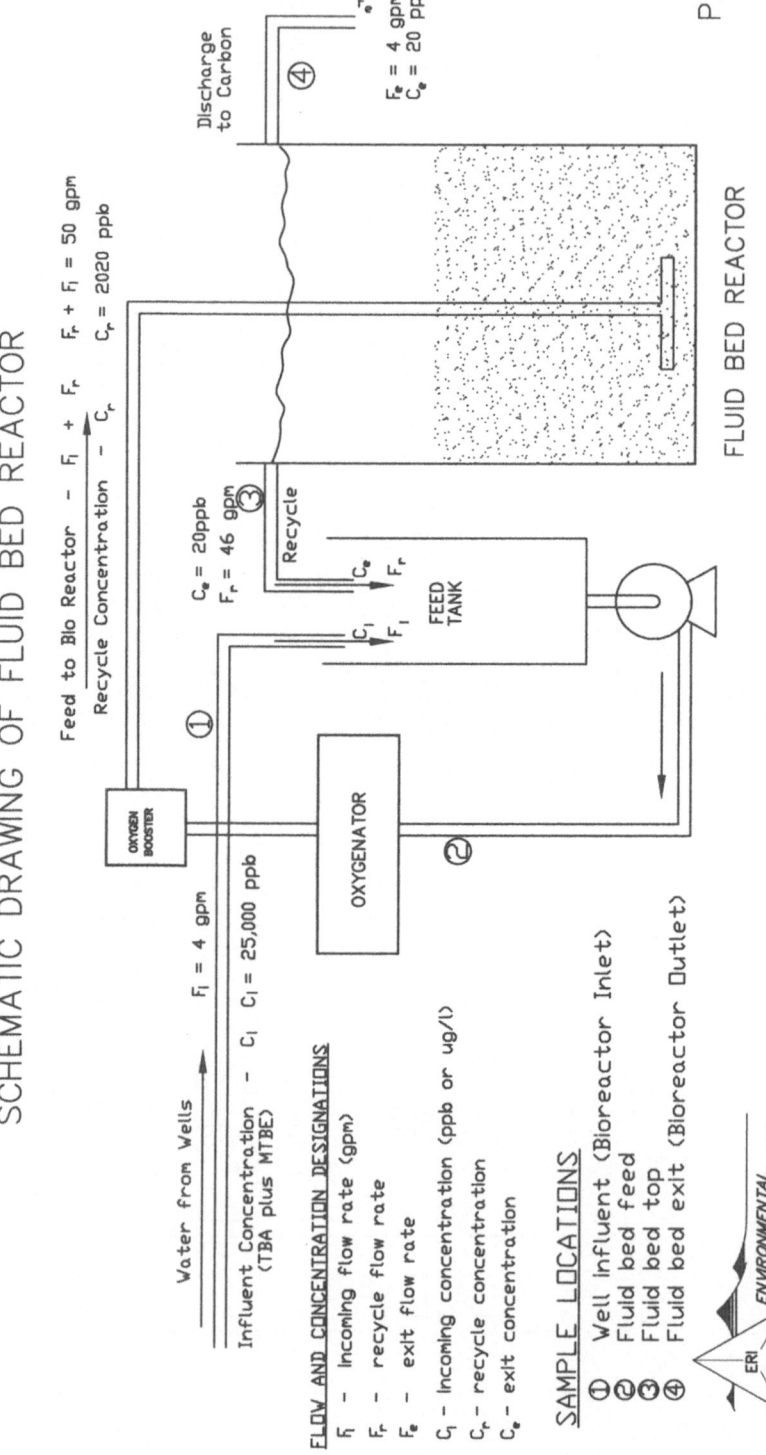

Figure 29-2.   **Schematic Drawing of Fluid Bed Reactor**

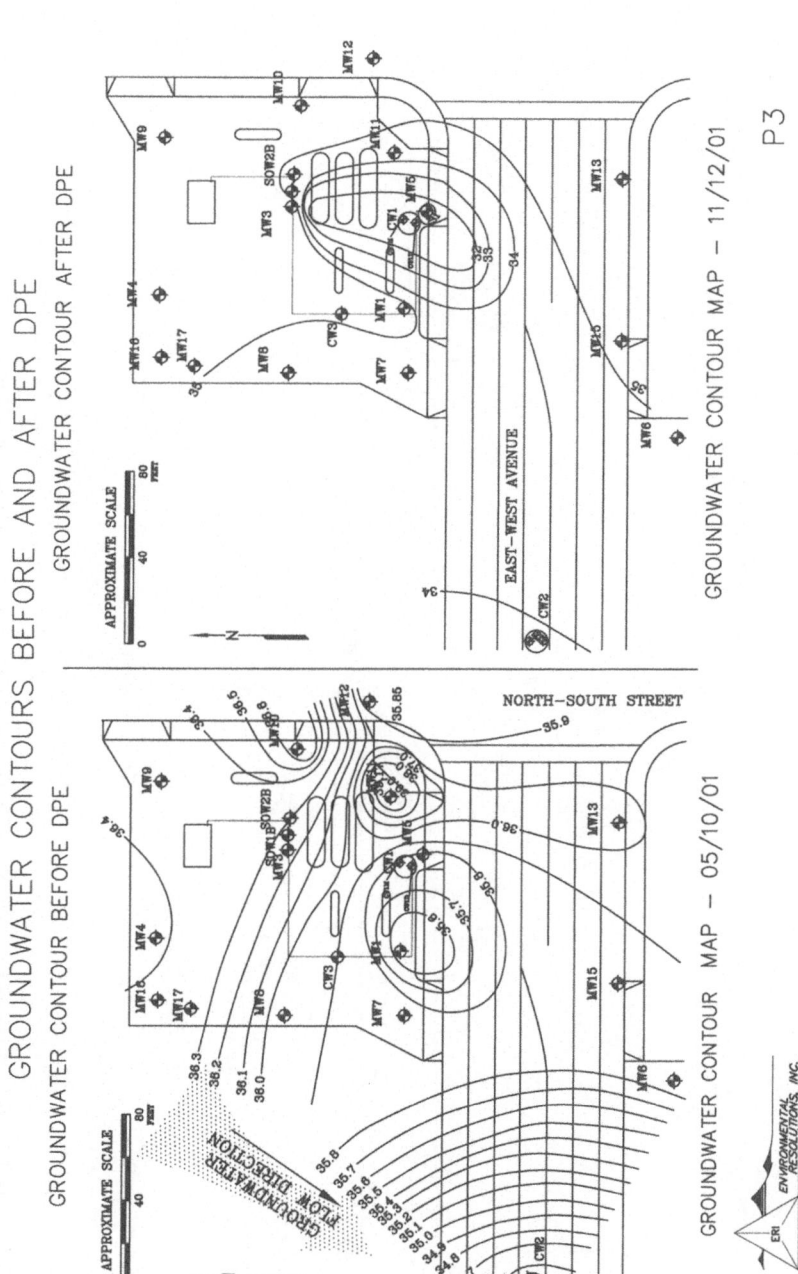

Figure 29-3. Groundwater Contours Before and After DPE

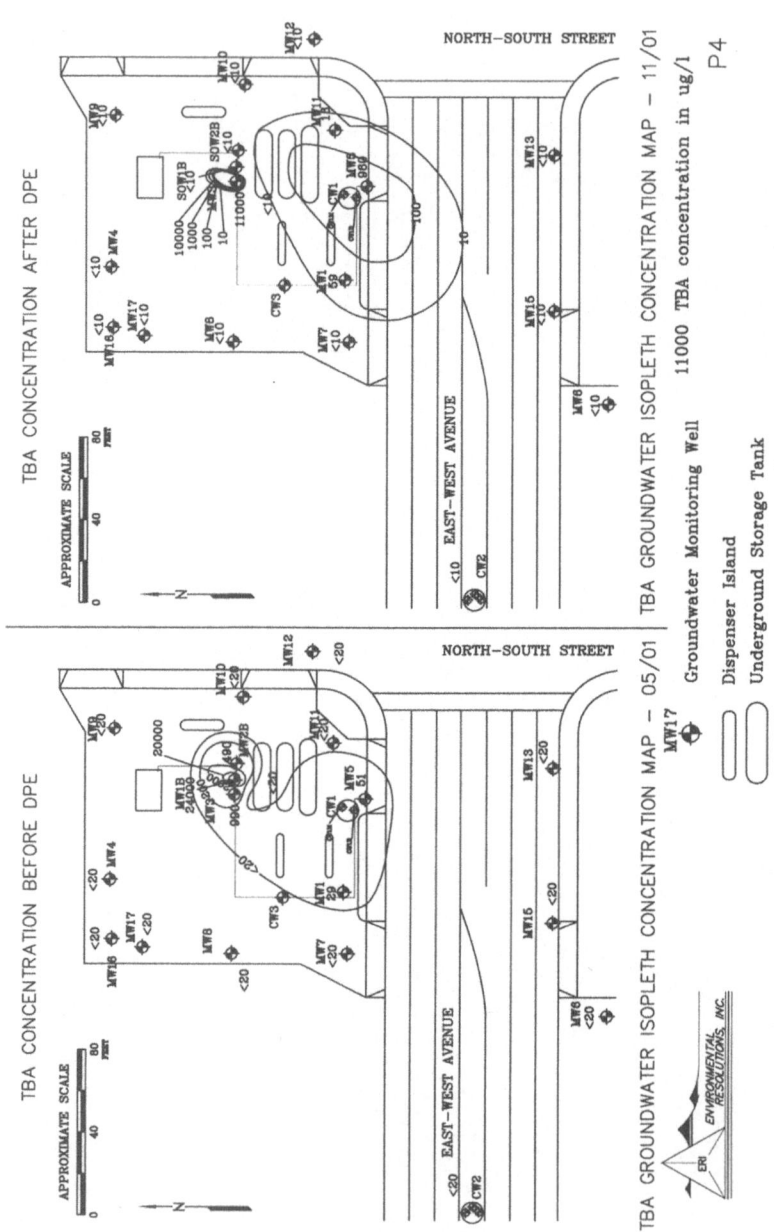

Figure 29-4. **TBA Groundwater ISO Concentration Map — Before and After DPE**

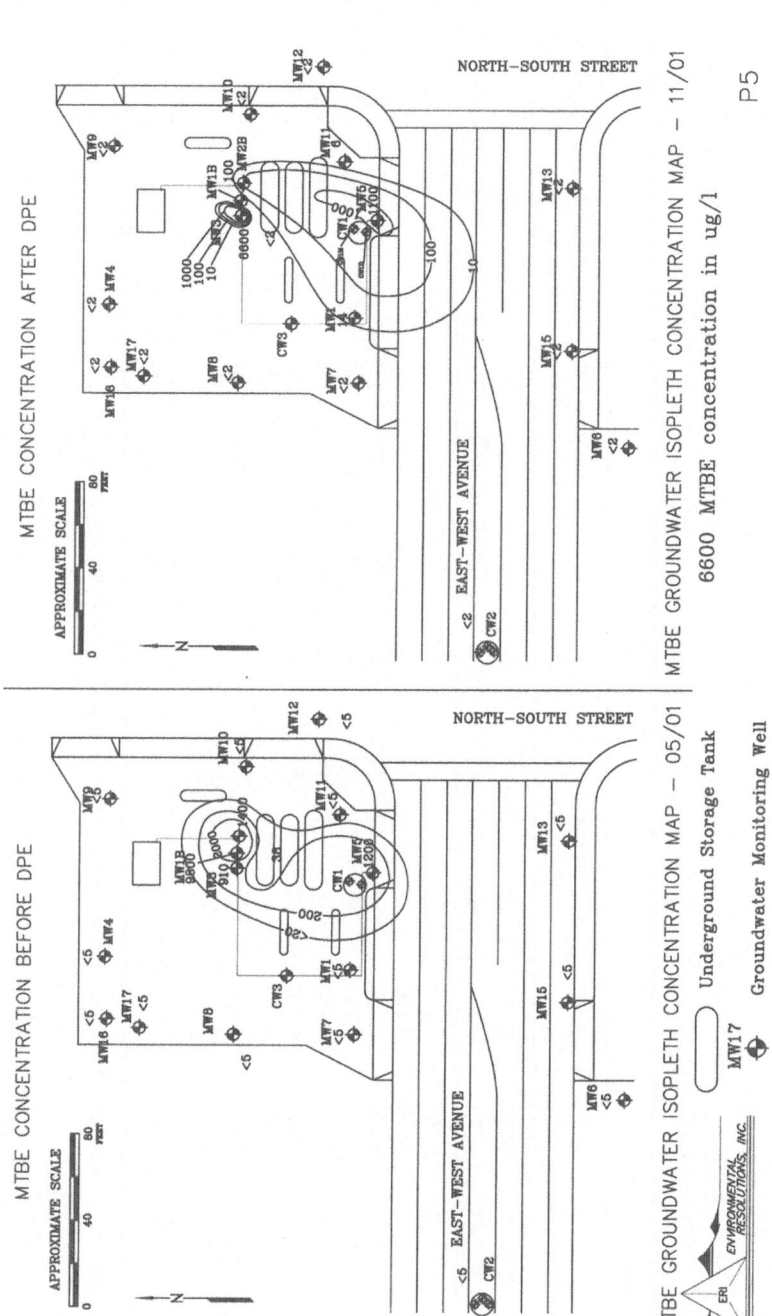

Figure 29-5. MTBE Groundwater Concentration Map — Before and After DPE

Table 29-1. Operating Data

Bioreactor Sampling Results

Date Sampled	Flow to unit-gpm	Sample #1* Well Influent						Sample #2* Fluid Bed Feed		Sample #3* Fluid Bed Top		Sample #4* Bioreactor Exit	
		MTBE (µg/l)	TBA (µg/l)	Benzene (µg/l)	Toluene (µg/l)	Et. Ben. (µg/l)	Xylene (µg/l)	MTBE (µg/l)	TBA (µg/l)	MTBE (µg/l)	TBA (µg/l)	MTBE (µg/l)	TBA (µg/l)
9/6/01	1.5	5,800	15,000	55	28	54	28					120	<50
9/20/01	1.2	2,700	4,200	36	19	120	12					99	<50
10/12/01	1.3	2,400	8,500	54	36	52	ND<100	110	160	67	<50	67	<50
10/19/01	1.3	3,300	7,200	49	32	33	ND<50	74	130	28	<50	25	<50
11/2/01	2	6,000	7,500	28	14	12	ND<20	230	260	23	86	25	<50
11/13/01	1.6	5,900	8,300	38	<50	<50	<100	180	280	23	<50	19	<50
11/20/01	2	6,900	14,000	45	34	31	<50	140	270	13	<50	11	<50
11/28/01	1.6	9,800	11,000	56	26	32	13	290	290	29	<50	31	<50
12/6/01	1.6	9,400	13,000	61	<100	<100	<100	360	260	14	<50	15	<50
Discharge Limit 1996–2001—Unofficial												<35	100
Proposed Discharge Limit 2002–2007												13	12

* Sample locations are shown on the Schematic Drawing of the Fluid Bed Reactor—Figure 29-2.

move most of the MTBE, to keep carbon usage low, and virtually all of the TBA, since carbon is very ineffective for TBA removal. The bioreactor with its pumps and feeders requires only about 3 kw of power.

RESULTS

Ground water flow elevation contours before (May 2001) and after (November 2001) initiation of DPE are shown on Figure 29-3. Concentration isopleths for May 2001 (before remediation began) and November 2001 (after remediation began) are shown for TBA on Figure 29-4 and for MTBE on Figure 29-5.

Ground water samples were taken from the bioreactor inlet, the bioreactor outlet, and from the discharge from the carbon units to the storm sewer discharge to track the progress of remediation. The results are presented in Table 29-1. The exit from the bioreactor meets the current unofficial discharge requirements (35 µg/l MTBE and 100 µg/l TBA), resulting in a very long life for the activated carbon canisters. Previously the precise limits were not specified, although some facilities were shut down for discharging more than 100 µg/l TBA. The discharge requirements (13 µg/l for MTBE and 12 µg/l for TBA) are being tightened considerably with the renewal of the general order issued in early 2002. Additional analytical work and evaluation will be required to determine if the bioreactor/activated carbon system will meet these new limits.

Based on the performance of other similar bioreactors, one of which has been operating for over a year, this system will continue to operate with infrequent carbon changeouts. The problems that have occurred are water-related. Precipitation of carbonates has required cleanout of valves, pump impellers, and carbon headers. The pH must be maintained between 6.5 and 8.5. Incoming ground water can vary from pH 5.5 to 9.5. Extensive temperature studies have not been performed, but the biomass loses its effectiveness below 16°C (60°F) and above 32°C (90°F). The bioreactor at this site cost about $65,000 to install. The carbon savings on MTBE remediation alone are over $2,000 per month. Without the bioreactor, it would not be possible to meet the TBA discharge requirements.

CHAPTER 30

Natural Attenuation of Tert Butyl Alcohol at a Texas Chemical Plant

Michael J. Day and Terry Gulliver, Applied Hydrology Associates, Inc.

INTRODUCTION

TBA is often detected in ground water affected by spills or leaks of fuels oxygenated with MTBE. TBA in ground water may originate from three separate sources. First, in some areas of the U.S., TBA has been directly added to fuels as an oxygenate or octane booster. Second, commercial MTBE may contain a small percentage of TBA. Third, TBA has been documented as an intermediate or transformation product of MTBE biodegradation. (In addition, as discussed in Chapter 6, TBA may be an artifact of sample preservation and analytical techniques.) TBA has been found to accumulate in some MTBE ground water plumes as a result of the natural degradation of MTBE. Consequently, there is considerable interest in the natural attenuation of TBA.

Field studies of TBA fate and transport are limited because the chemical usually occurs in association with MTBE, and the production of TBA from MTBE degradation tends to confound the study of TBA attenuation. This chapter presents results from a decade of ground water monitoring at a chemical plant in Pasadena, Texas that provides a unique opportunity to investigate the fate and transport of TBA in ground water where no significant MTBE is present. The plant produces TBA as a byproduct of propylene oxide manufacturing, and several areas of the plant have shallow ground water that has been affected by historic leaks and spills of TBA. Monitoring data from one ground water plume, in which there has been no remedial intervention to date, indicates generally declining concentrations and a shrinking plume area, suggesting that natural attenuation mechanisms are limiting the advective transport of TBA.

PREVIOUS WORK ON TBA DEGRADATION

Several studies have shown that TBA readily degrades under aerobic conditions. At Vandenberg Air Force Base, California, MTBE and TBA rapidly degraded when oxygen was added to a ground water plume (Mackay *et al.*, 2001). At Port Hueneme, California, TBA was degraded in demonstration plots under aerobic conditions, with and without bioaugmentation (Salanitro *et al.*, 2000). The bioaugmented plot showed a higher rate and greater extent of TBA degradation. Laboratory microcosm studies using propane-oxidizing bacteria under aerobic conditions have shown transformation of MTBE to TBA followed by further biodegradation, and in some cases complete mineralization, of TBA (Steffan *et al.*, 1997). One of several possible biodegradation pathways for MTBE and TBA developed from the microcosm study is illustrated in Chapter 12 (Figure 12-2). This study demonstrated that the degradation of TBA by propanotrophs was significantly slower than that of the parent MTBE, and that TBA degradation did not occur until nearly complete removal of MTBE. In this situation, TBA degradation may be the rate-limiting step in the total mineralization of MTBE (Steffan *et al.*, 1997). Other studies suggest that MTBE biodegradation can occur without significant TBA accumulation (Fortin and Deshusses, 1999).

TBA has also been observed to degrade under methanogenic conditions (Wilson *et al.*, 2000). Kolhatkar *et al.* (2000) examined the distribution of TBA at 74 gasoline stations in the eastern U.S. and were able to extract field-scale anaerobic biodegradation rate constants for TBA that were statistically significant at three sites where methanogenic conditions prevailed. First-order degradation rate constants varied from 0.61 to 13.9 per year, corresponding to half-lives of 1.1 years to 2.4 weeks.

Most work on the biodegradation of TBA under anaerobic conditions has been done using laboratory microcosms. J.T. Novak and his associates at Virginia Polytechnic Institute and State University have demonstrated anaerobic biodegradation of TBA at a variety of locations (Novak *et al.*, 1985; Hickman and Novak, 1989; Hickman *et al.*, 1989; Yeh and Novak, 1994). TBA degradation was generally rapid, with a median first-order degradation rate constant of 2.0 per year, corresponding to a half-life of four months. However, some sites showed slow degradation with rate constants as low as 0.16 per year, corresponding to a half-life of over four years (Hickman and Novak, 1989). In microcosm studies performed at the University of Oklahoma, TBA did not degrade in aquifer sediments under anaerobic conditions (Suflita and Mormile, 1993; Mormile *et al.*, 1994). These studies did demonstrate that TBA was produced as an intermediate product of MTBE microbial degradation.

INFLUENCE OF TBA PROPERTIES ON NATURAL ATTENUATION

TBA and MTBE both contain the tert butyl group of three methyl groups attached to a tertiary carbon $[(CH_3)_3-C-]$. TBA has an OH group, and MTBE an $O-CH_3$ group, attached to the tert butyl group. Both are highly soluble compounds (TBA is completely miscible with water) that weakly adsorb to aquifer matrix solids. Although TBA degradation has been demonstrated under aerobic and methanogenic conditions, the tert butyl structure tends to make the TBA molecule more resistant to biological degradation than less complex gasoline components such as benzene.

As described in more detail in Chapter 16, the term "natural attenuation" refers to any naturally occurring physical, chemical, or biological process that reduces the mass, toxicity, mobility, volume, or concentration of contaminants in soil or ground water (USEPA, 1999). *In situ* natural attenuation processes include adsorption, volatilization, dispersion, dilution, diffusion, biodegradation, and chemical reactions. The natural attenuation of TBA has not been widely investigated. The physical properties of TBA result in generally insignificant volatilization and adsorption, while the influence of dispersion, dilution, and diffusional mechanisms are largely dictated by site-specific hydrogeologic conditions. Similarly, the significance of biodegradation and chemical reactions are primarily influenced by site-specific geochemical and biological conditions.

SITE DESCRIPTION

Detailed environmental investigations and monitoring performed at the Pasadena chemical plant site since 1991 have characterized the hydrostratigraphy and the nature, extent, and migration pathways of chemical constituents in the shallow ground water. Quarterly ground water monitoring of an extensive network of wells (there are now over 200 monitoring and remedial wells in the facility) over the past decade has yielded a relatively long and detailed history of ground water conditions at the site.

A simplified geologic cross-section below the facility is shown in Figure 30-1. The facility is underlain by Quaternary sediments of the Beaumont Formation that are comprised primarily of low permeability clays and silts interspersed with moderately permeable fine-grained sands. The surficial sediments are clays (C1 unit in Figure 30-1) that are partially penetrated by plant excavations and buried utility lines. The C1 clay unit restricts significant infiltration of rainfall; however, historic operational spills and leaks have resulted in chemical migration to shallow ground water. Migration pathways are most likely through localized artificial penetrations of the C1 clay in production areas such as sumps, underdrains, and utility trenches, and through clay joints.

The shallowest permeable ground water zone below the site, designated the S1 zone (Figure 30-1), consists of 0.6 to 1.5 meters (2 to 5 feet) of silt and very fine sand with low organic content, from 5.5 to 7 meters (18 to 23 feet) bgs. Most ground water monitoring wells are screened within this sand zone. The S1 sand is confined by the overlying surficial C1 clay unit, with the potentiometric surface from 1.8 to 4.3 meters (6 to 14 feet) above the top of the

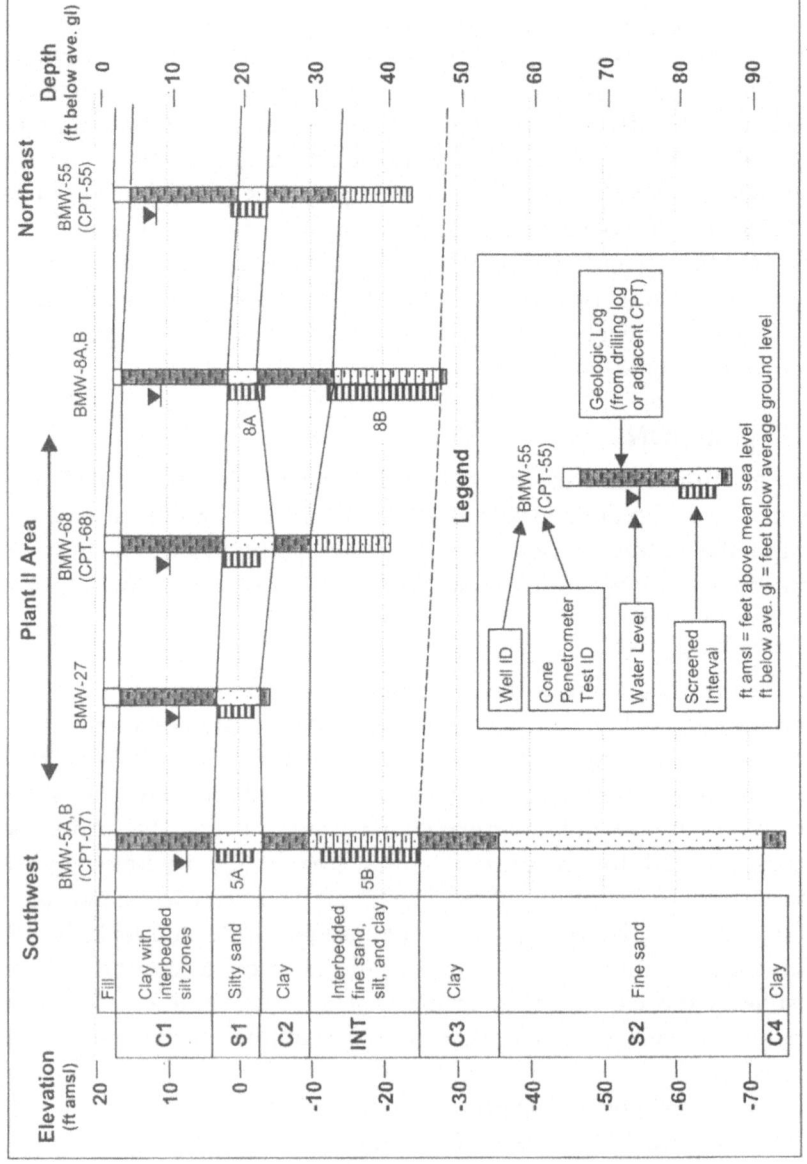

Figure 30-1.   **Schematic of Shallow Hydrostratigraphy**

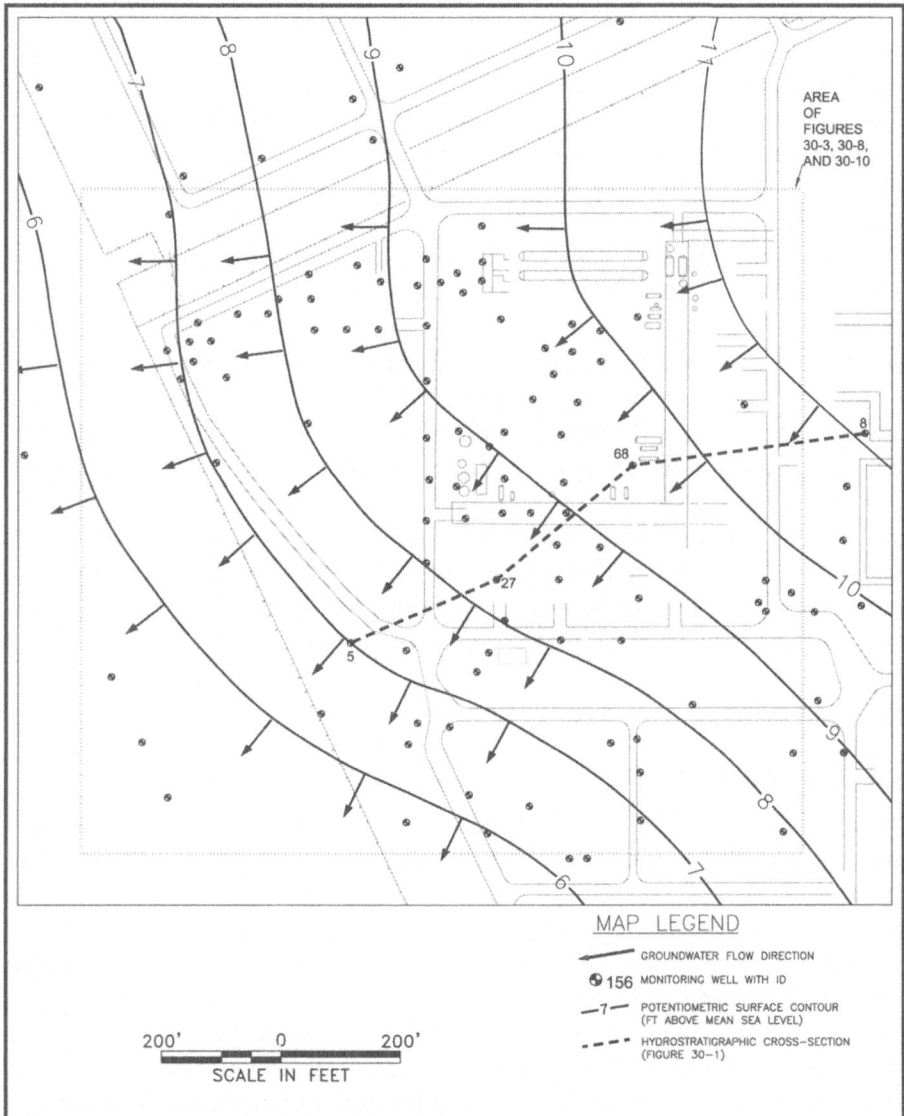

MAP LEGEND

⟶ GROUNDWATER FLOW DIRECTION

⊕ 156 MONITORING WELL WITH ID

—7— POTENTIOMETRIC SURFACE CONTOUR
        (FT ABOVE MEAN SEA LEVEL)

- - - - HYDROSTRATIGRAPHIC CROSS−SECTION
          (FIGURE 30−1)

200'   0   200'

SCALE IN FEET

Figure 30-2. **Potentiometric Surface of S1 Sand Unit in Plant II Area**

sand unit. A typical potentiometric surface for the S1 zone is illustrated in
Figure 30-2. Ground water flow is generally to the southwest under a lateral
hydraulic gradient of approximately 0.005. Quarterly monitoring of ground
water levels over the past decade indicates that the potentiometric surface for
the S1 zone fluctuates seasonally over a range of about 0.6 meters (2 feet), but
that the direction of ground water flow does not vary significantly, except
where remedial systems have been installed. The limited recharge of rainfall
results in minimal dilution of dissolved chemicals in the S1 ground water.

The hydraulic conductivity of the S1 sand has been estimated from slug and bailed-down recovery tests to be approximately 7×10^{-4} to 2×10^{-3} cm/s (2 to 5 ft/day). The average hydraulic gradient of 0.005 and an estimated effective porosity of 0.25 give an approximate lateral ground water flow velocity of between 4.6 and 11 meters per year (15 and 37 feet per year), or 130 to 310 meters (420 to 1,040 feet) in the 28 years that the chemical facility has been in operation.

A second permeable ground water interval, designated as the INT zone, occurs at a depth of about 9 meters (30 feet) and consists of interbedded clays and silty fine sands (Figure 30-1). The INT zone is separated from the S1 zone by 2.4 to 3 meters (8 to 10 feet) of clay (C2 unit). Several locations have paired monitoring wells screened separately within the S1 and INT units. Monitoring of these wells indicates minimal chemical impacts to the INT unit. Potentiometric head differences between the S1 zone and the INT zone are typically less than 0.06 meters (0.2 feet), resulting in low vertical hydraulic gradients less than 0.022. A third deeper sandy interval, designated the S2, occurs at a depth of about 18 meters (60 feet). Monitoring data from wells screened in the S2 unit show evidence of minimal chemical impacts. The potentiometric head in the S2 zone is typically within a few inches of the head in the S1 zone. There is thus no significant vertical leakage sink for chemicals detected in the S1 zone ground water.

PLANT II TBA PLUME

The Plant II Area of the chemical facility was constructed in 1974. Shallow ground water below the Plant II Area contains TBA that presumably originated from operational spills and leaks over the past 28 years. The TBA distribution in S1 ground water for the third quarter of 2001 is shown in Figure 30-3. Also shown in Figure 30-3 are historical trends of TBA concentrations in selected monitoring wells, some of which have been monitored since 1991.

Five suspected source areas for this plume have been identified along the production corridors and at an oxidizer clean-out area within the Plant II Area shown as boxes in Figure 30-3. The location of source areas, combined with the ground water flow pattern shown in Figure 30-2, results in a bifurcated plume. TBA is the only significant organic constituent in the southern lobe of the plume. The northern lobe of the plume contains TBA and chlorinated aliphatic constituents that originated from solvents formerly used in oxidizer clean-out activities. Active remedial measures have recently (late 2000) been initiated in the northern lobe.

Figure 30-4 illustrates the TBA concentration profile along three ground water flow paths shown in Figure 30-3. The stacked bars in Figure 30-4 indicate the lowest and highest concentrations reported for each well during the period from August 2000 to September 2001. During this time period, between three and six samples were taken from each well, at intervals ranging

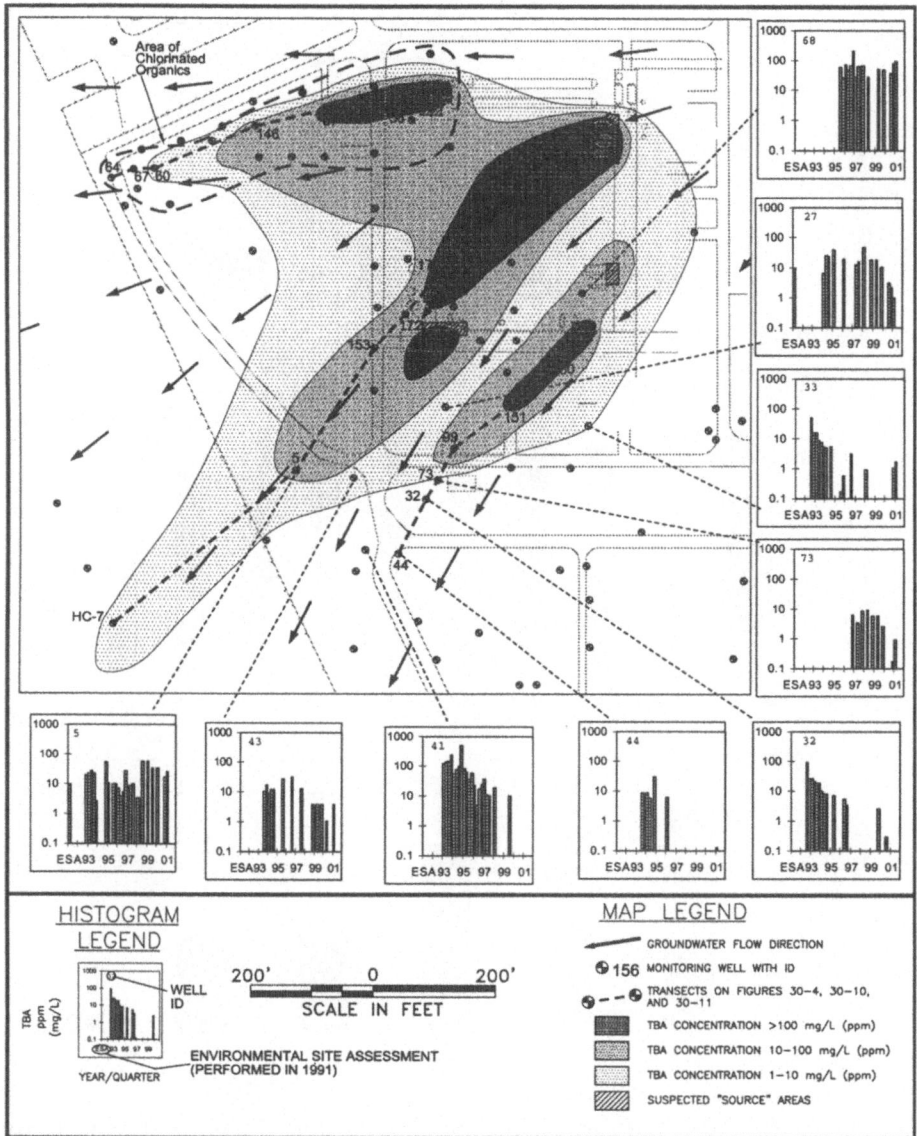

Figure 30-3. **TBA Concentration Distribution in Shallow Ground Water**

from three to six months. For the northern lobe, only monitoring data outside the area influenced by current remedial activities are included in Figure 30-4.

NATURAL ATTENUATION OF TBA IN THE PLANT II AREA PLUME

Due to the high solubility and low adsorption characteristics of TBA, it would be expected that the TBA plume should move with ground water with

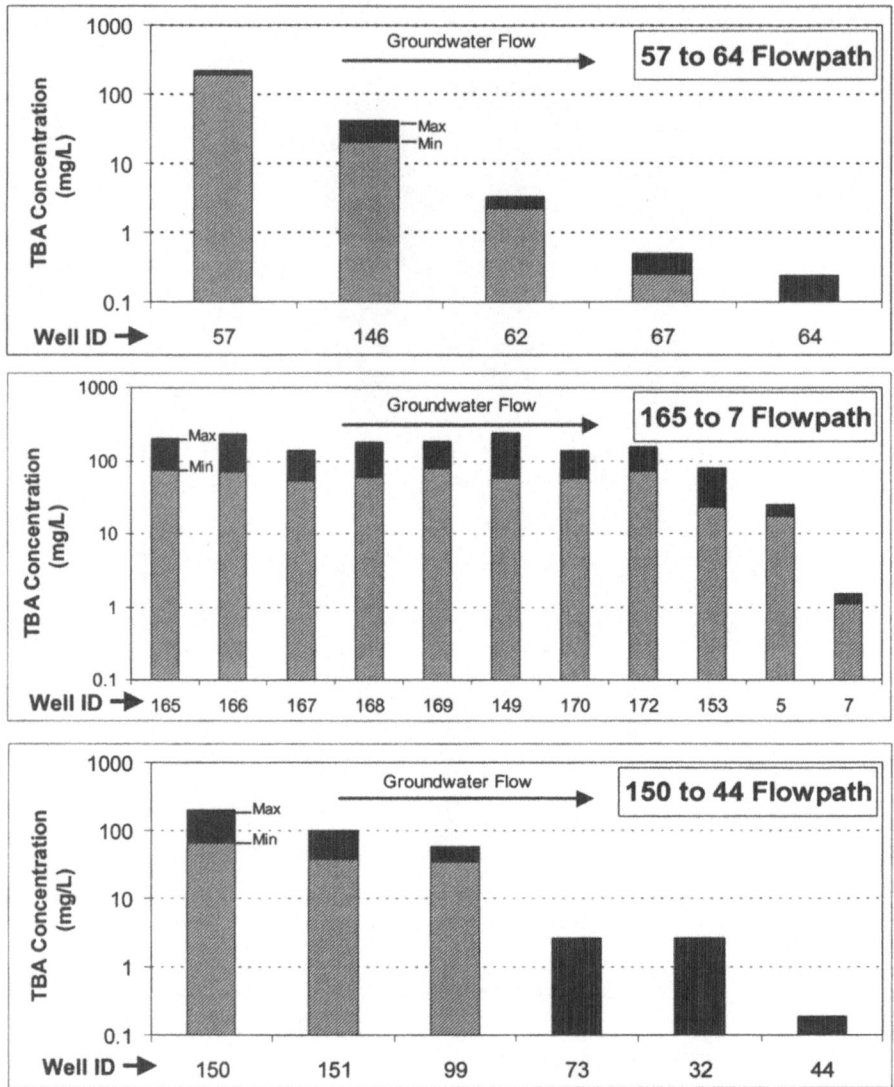

Figure 30-4. **TBA Concentration Profiles along Ground Water Flow Paths**

little retardation. If TBA was conserved in this plume, and the source concentrations were relatively constant, the plume should be expanding at the leading edge through advection and on the fringes by dispersion. Monitoring wells at these locations should show increasing concentrations over time. The chemical gradient spans roughly one order of magnitude in 50 meters (150 feet) on the plume front of Figure 30-3. In a decade (the length of the histogram date axis), ground water (and TBA) should have migrated on average between 50 and 110 meters (150 and 370 feet), and the rising concentration gradient should have increased the TBA concentrations in monitoring wells MW-5, MW-32, MW-41, MW-43, MW-44, and MW-73 by an order of magnitude. However, histograms of TBA over the last decade (Figure 30-3) show that these wells and other downgradient and fringe monitoring wells have seen steady or decreasing TBA concentrations. This trend suggests that natural attenuation of TBA is occurring.

Role of Diffusion in Plant II Area Plume Natural Attenuation. Given the relatively thin (0.6 to 1.5 meters [2 to 5 feet]) nature of the S1 sand unit, the length of the plume (up to 300 meters [1,000 feet]), and the potentially long residence time (up to 28 years), it might be expected that diffusion of TBA into the clay units overlying and underlying the S1 may be an important attenuation mechanism. The sequestering of TBA in the clay zones along the ground water flow path by this mechanism was estimated in two ways: first, by comparison with the pre-remediation transport of chlorinated hydrocarbons in the northern lobe of the plume, and second, by theoretical calculations using a numerical mass transport model.

Examination of TBA and chlorinated organics transport in the northern lobe of the Plant II plume allows an evaluation of relative diffusional mass loss along a flow path. Figure 30-3 shows the area of the TBA plume that contains chlorinated organics, and the wells along the ground water flow path that are used in this analysis. This northern lobe of the plume contains TBA, 1,1-DCE, and 1,1-dichloroethane (DCA). TBA and the solvent 1,1,1-trichloroethane (TCA) were spilled at the west end of two oxidizers and migrated in a narrow plume slightly divergent from the main Plant II TBA plume. DCA and DCE are daughter products of the original TCA, which is not detected in the ground water. The area east of well MW-58 is under remediation and not utilized in the following analysis.

Concentration profiles over the distal 120 meters (400 feet) of the northern flow path are shown for TBA, DCE, and DCA in Figure 30-5. The ratios of the diffusion coefficients for DCA and DCE to that of TBA, calculated from the Wilke-Chang equation based on atomic radii (Wilke and Chang, 1955), are 1.05 and 1.13 respectively. According to these ratios, the loss of TBA along the flow path due to dispersion and diffusion should be less than that for DCE and DCA. In addition, DCE and DCA would adsorb more readily than TBA. However, Figure 30-5 indicates that TBA is attenuating at a log-linear rate more rapidly than DCA and DCE, suggesting that diffusion and adsorption are not the major mechanisms causing concentrations to decrease along the flow path.

Modeling of diffusion effects was accomplished using the numerical mass transport code MT3D (Zheng, 1990) coupled with the flow code MOD-

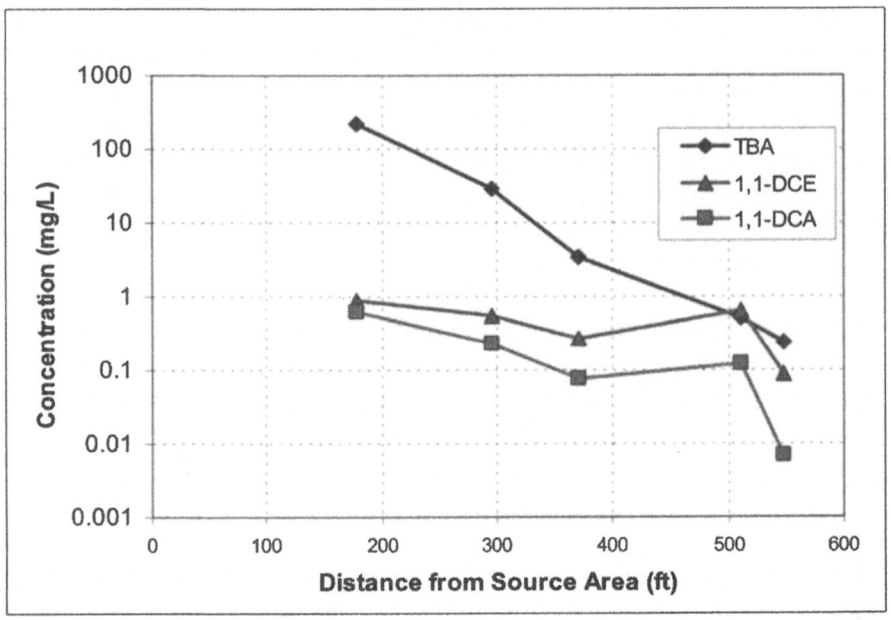

Figure 30-5.   **TBA, DCE, and DCA Concentration Profiles along Ground
Water Flow Path in Northern Lobe of Plant II TBA Plume**

FLOW (McDonald and Harbaugh, 1988). Transport of TBA was modeled under a range of hydrogeologic conditions observed in the Plant II Area. Major variables for these calculations were the saturated thickness (0.6 to 1.5 meters [2 to 5 feet]) and the hydraulic conductivity (7×10^{-4} to 2×10^{-3} cm/s [2 to 5 ft/day]) of the S1 zone. A source area recharging the S1 sand at a constant concentration of 300 mg/l for 20 years was assumed. Transport was considered for two cases: (a) with diffusion and (b) without diffusion into the C1 and C2 clay units bounding the S1 sand unit. A diffusion coefficient of 1.0×10^{-6} cm^2/s (9.7×10^{-5} ft^2/day) was assumed for TBA in porous media.

The results of the modeling indicated that diffusional effects are greater where the S1 sand thickness is less and also where the ground water flow velocity is less. The modeled TBA concentration profile along the flow path at various times for the case of a 0.6-meter (2-foot) thick S1 sand having a hydraulic conductivity of 7×10^{-4} cm/s (2 ft/day) is shown in Figure 30-6. Diffusion into the clays is most significant near the source area where concentration gradients between the S1 sand and the bounding clay units are highest. At the leading edges of the plume, diffusion is relatively insignificant. After 20 years, the source (in terms of continued leaks to the subsurface) is assumed to stop, simulating improvements in plant operational procedures that have occurred over the past few years. The modeled 25-year pro-

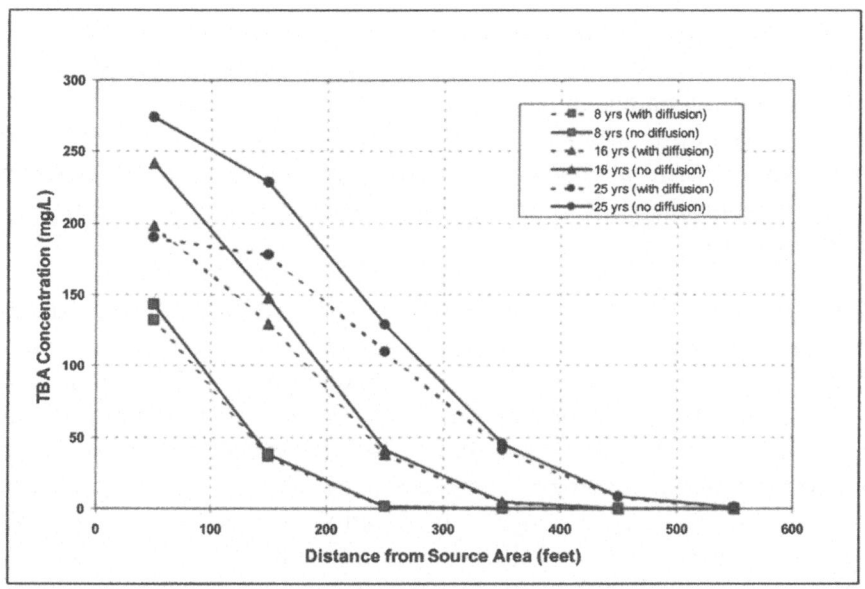

Figure 30-6.   **Predicted Transport of TBA along Ground Water Flow Path**

file shows that diffusional effects become more pronounced when the source ceases to contribute mass to the plume, particularly near the source area.

Use of Carbon Isotopes to Document TBA Biodegradation.   In order to provide direct evidence of TBA biodegradation, Day *et al.* (2002) examined the carbon isotope composition ($^{13}C/^{12}C$) of TBA within the Plant II plume. Hunkeler *et al.* (2001) observed that TBA biodegradation was accompanied by a noticeable carbon isotope fractionation. Biodegradation should lead to an enrichment of ^{13}C in TBA because the biodegradation rate for ^{12}C is slightly faster, owing to the weaker bonds between light isotopes. The carbon isotope ratios were analyzed in the Environmental Isotope Laboratory of the University of Waterloo, Canada, using a GC coupled with an isotope-ratio mass spectrometer. Results are reported as $\delta^{13}C$ values in units of per mill (‰):

$$\delta^{13}C = (R_s/R_r - 1) \times 1,000 \qquad (30\text{-}1)$$

where: $R_s = {}^{13}C/{}^{12}C$ ratio of the sample
$\quad\quad\quad R_r = {}^{13}C/{}^{12}C$ ratio of an international standard, Vienna Peedee Belemnite (VPDB)

Because fossil hydrocarbons (including the raw material for TBA) are depleted in ^{13}C relative to the present atmospheric carbon ($\delta^{13}C = -7$ ‰) the $\delta^{13}C$ in the original TBA product are negative (about –28.5 to –29 ‰ in our study). Therefore, a ^{13}C enrichment corresponds to a trend toward less negative $\delta^{13}C$ values.

The $\delta^{13}C$ values of TBA are illustrated as a function of the TBA concentration in Figure 30-7. Samples with low TBA concentrations near the fringe of the plume are characterized by more positive $\delta^{13}C$ values, ranging as high as –22 ‰. The areas where $\delta^{13}C$ are closer to –28 ‰ corresponded to areas where TBA concentrations were measured greater than 10 mg/l (Figure 30-8). These results indicate that substantial biodegradation occurs in the plume margins.

Mechanisms of TBA Biodegradation. Selected dissolved gases, electron acceptors, and some other microbial process indicators in ground water in the TBA plume were analyzed during the first quarter of 2002. A summary of the results for the 2002 sampling is included in Table 30-1. Preliminary findings indicate that the plume is mainly anaerobic with dissolved oxygen depleted in the plume (0.83 mg/l) compared to a background of approximately 2 mg/l. Dissolved hydrogen gas concentrations suggest that redox conditions in the plume cores are Fe(III)-reducing to sulfate-reducing. Iron (III) is available in the brownish sands, but Fe(II) is not elevated in the ground water.

Values for manganese, sulfate, and total inorganic carbon (TIC) show some differences within the TBA plume compared to background values (Figure 30-9). Manganese is slightly elevated in the plume core, averaging 0.3 mg/l compared to 0.1 mg/l in background wells. Sulfate concentration in the plume core averages 55 mg/l, while background concentrations are variable with an average of 128 mg/l. Sulfide is below the detection limit (less

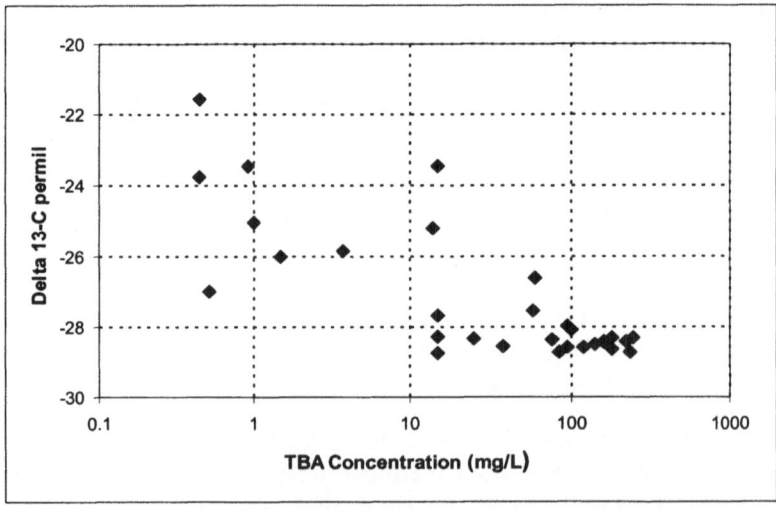

Figure 30-7.   **TBA Concentration Versus ^{13}C Values in TBA Samples**

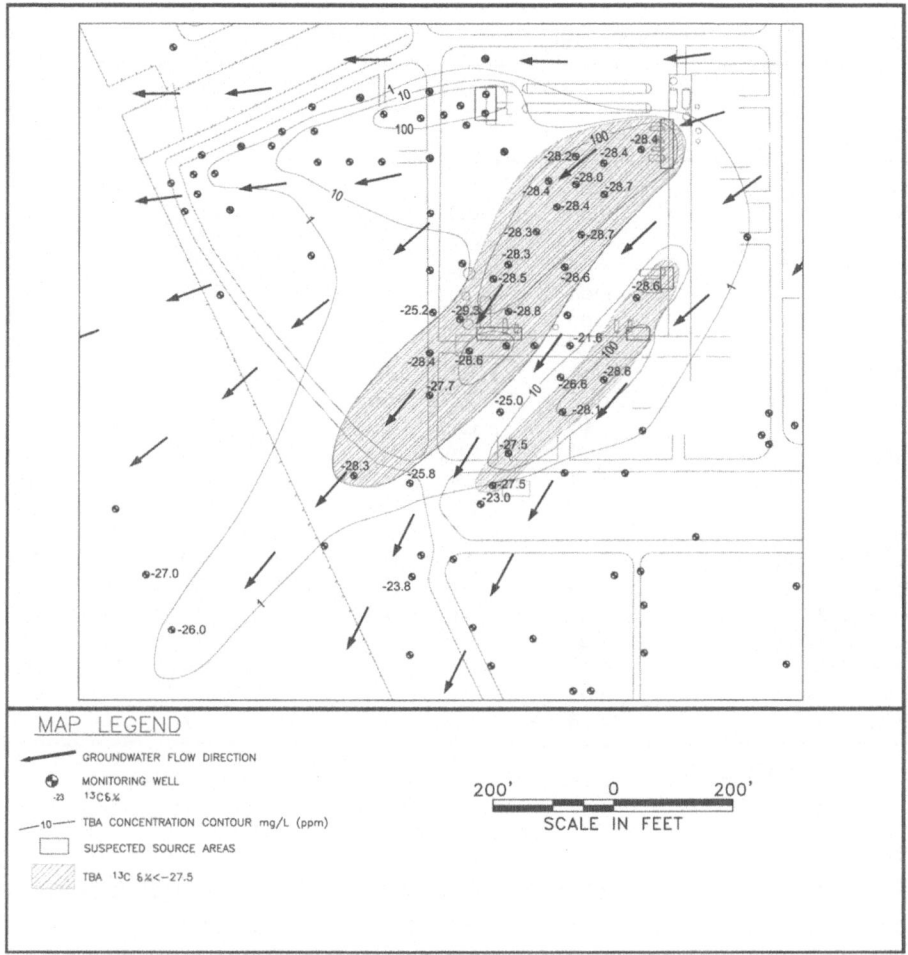

Figure 30-8. ^{13}C Ratios in Plant II TBA Plume

than 2 mg/l). High dissolved TIC concentrations (sum of dissolved carbon dioxide and bicarbonate) with values up to 2,100 mg/l and averaging 1,750 mg/l are observed in the plume. Background dissolved TIC concentrations average 1,300 mg/l. In addition, CO_2 concentrations in the plume are much higher than background concentrations (66 mg/l versus 32 mg/l). The increased TIC and CO_2 concentrations within the TBA plume may be indicative of TBA biodegradation. The TIC concentration differences between the plume and background are sufficient to account for all TBA degradation, but no other organic byproducts, such as light gases (methane, ethane, ethene, etc.), volatile acids, or acetone, are found in significant quantities.

Values of some of these parameters for monitoring wells along the center

Table 30-1.   **Summary of Results of 2002 Sampling**

Constituent	Units	In Plume (10 wells)			Background (9 wells)		
		Max	Min	Avg	Max	Min	Avg
Hydrogen	nM	2.7	0.56	0.91	1.5	0.65	1.22
Fe^{2+}	mg/l	0.8	0	0.16	0.11	0	0.22
Sulfate	mg/l	200	21	55	420	31	128
Sulfide	mg/l	<2	<2	<2	<2	<2	<2
Nitrate	mg/l	9.9	0.26	4.77	7.3	1.6	3.57
Manganese	mg/l	0.63	0.064	0.302	0.19	0.011	0.100
Oxygen	mg/l	1.5	0.4	0.83	3.6	0.9	2.0
Methane	ug/l	69	0.02	11.22	18	0.18	5.73
Ethane	ng/l	53	6	24	53	9	26
Ethene	ng/l	120	12	41	51	10	31
Propane	ng/l	130	71	101	150	<25	150
n-Butane	ng/l	59	<25	59	<25	<25	<25
CO_2	mg/l	93	46	66	62	8	32.2
Nitrogen Gas	mg/l	16	14	14.8	16	14	15.1
Total Inorganic Carbon	mg/l (CaCO3)	2100	1500	1750	1700	620	1322
TBA	mg/l	160	0.58	63.3	0.79	0.1	0.28
Acetic Acid	mg/l	<1	<1	<1	<1	<1	<1
Butyric Acid	mg/l	<1	<1	<1	<1	<1	<1
Lactic Acid	mg/l	<25	<25	<25	<25	<25	<25
Propionic Acid	mg/l	<1	<1	<1	<1	<1	<1
Pyruvic Acid	mg/l	<10	<10	<10	<10	<10	<10

ng/l = nanograms per liter.

flow line of the plume are plotted in Figure 30-10. The chemical data suggest that several microbial processes are active in the TBA plume including sulfate reduction, manganese reduction, and methanogenesis. More aerobic conditions exist in the plume fringes compared to the plume core. The elevated concentrations of TIC and CO_2 found in the TBA plume under investigation suggest that TBA had been mainly mineralized to CO_2. A minor amount of CH_4 was found in the TBA plume.

ESTIMATION OF NATURAL BIODEGRADATION RATES

Figure 30-11 illustrates TBA concentrations along three ground water flow paths versus distance from presumed source areas (Figure 30-3) in the Plant II Area. TBA concentrations show an approximate log-linear decrease with distance along the portion of the flow paths that are outside of the operational Plant II Area (Figure 30-11). The flow path from well 165 to well 7 shows very little TBA decrease over the first 150 meters (500 feet). This segment of the flow path underlies the operational Plant II Area, so it is possible that there are additional sources of TBA contributing to the shallow ground water plume. There is probably commingling of TBA from suspected source areas upgradient from well 159 (Figure 30-3).

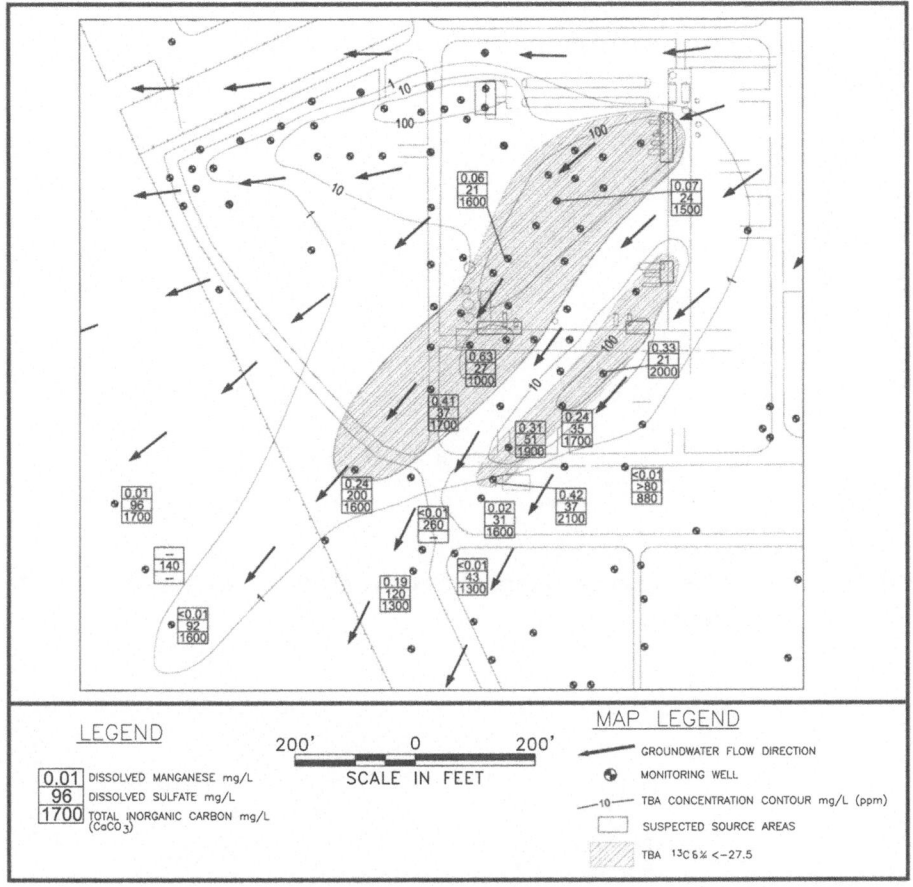

Figure 30-9.   **Key Inorganics in Plant II TBA Plume**

Biodegradation rate constants for TBA in the active front of the Plant II plume were estimated using the approach of Buscheck and Alcantar (1995), which assumes constant source concentrations. In this approach, an equivalent first-order biodegradation rate constant, λ, may be estimated in a steady-state plume using Equation 30-2:

$$\lambda = \frac{v_c}{4\alpha_x}\left(\left[1 + 2\alpha_x\left(\frac{k}{v_x}\right)\right]^2 - 1\right) \tag{30-2}$$

where: $\lambda =$ first-order biological degradation rate constant
$v_c =$ retarded contaminant velocity
$\alpha_x =$ longitudinal dispersivity (feet)
$k/v_x =$ rate of attenuation in contaminant concentration with distance along flow path

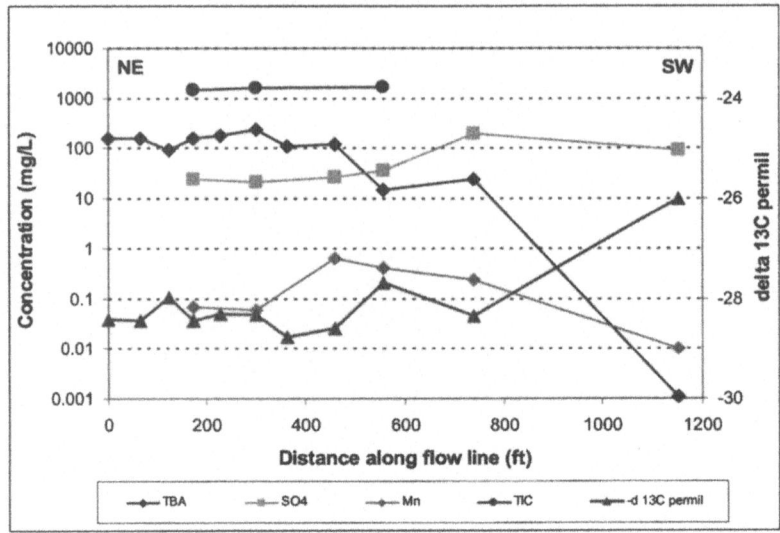

Figure 30-10.   **Variation of Selected Constituents along the Plume Centerline**

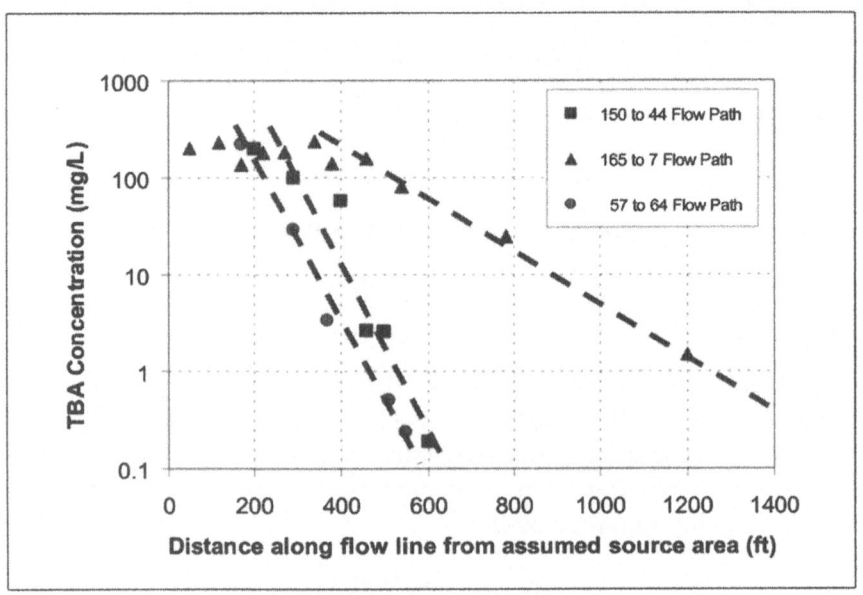

Figure 30-11.   **TBA Concentration Versus Distance along Flow Paths**

The attenuation rate of contaminant concentrations with distance along the flow path (k/v_x in Equation 30-1) can be estimated as the negative slope of a regression of the natural logarithm of contaminant concentrations with distance along the flow path. The retardation factor for TBA was assumed to be one as TBA is miscible with water and is not retarded by adsorption to the aquifer matrix. The contaminant velocity, v_c in equation 30-2, is therefore equal to the ground water flow velocity.

The longitudinal dispersivity, α_x, was estimated using the equation of Xu and Eckstein (1995).

$$\alpha_x = 0.83 * 3.28[\log_{10}(L_p/3.28)]^{2.414} \tag{30-3}$$

where: L_p = plume length (source area to the leading edge of the plume)

The results of the biodegradation rate calculations are summarized in Table 30-2 for the higher value of hydraulic conductivity (and ground water flow velocity) for the S1 sand at the site. Calculations were performed using the highest TBA concentrations reported for the September 2000 to September 2001 time period. Regressions for the linear portions of the three profiles shown in Figure 30-11 were performed on the natural logarithms of the concentration data. The negative slopes of the regressions are reported at the 95% confidence interval in Table 30-2.

Table 30-2. **Parameters and Results of Biodegradation Rate Calculations**

	57 to 64 Flow Path	165 to 7 Flow Path	150 to 44 Flow Path
Total length of plume, Lp (feet)	550	1,200	600
Negative slope of natural log vs distance graph (per foot)[1]	0.018±0.003	0.006±0.001	0.020±0.009
Coefficient of determination (R^2) for regression[1]	0.990	0.994	0.884
TBA retardation factor	1	1	1
Longitudinal dispersivity, α_x (feet)[2]	18.76	26.41	19.54
S1 sand hydraulic gradient, I[3]	0.006	0.005	0.005
S1 sand hydraulic conductivity, K (ft/day)	5.0	5.0	5.0
S1 sand effective porosity, n	0.25	0.25	0.25
Retarded contaminant velocity, V_c (ft/yr)[4]	41	37	40
First-order biodegradation rate constant (per year)[5]	0.97	0.26	1.1
TBA Half-life (years)[6]	0.71	2.65	0.63

Notes: [1] From Figure 30-11 (at 95% confidence interval).
[2] From Equation 30-3.
[3] From Figure 30-2.
[4] V_c = KI/n.
[5] From Equation 30-2.
[6] Half-life = 0.69/biodegradation rate constant.

Calculated first-order biological degradation constants for the 57 to 64 and 150 to 44 flow paths are very similar, at 0.97 and 1.1 per year respectively (Table 30-2). The 165 to 7 flow path has a lower calculated first-order biological degradation constant of 0.26 per year (Table 30-2), possibly due to commingling of two sources of TBA. The range of degradation constants for the three flow paths yield an equivalent half-life for TBA ranging between 0.63 and 2.65 years. The calculated TBA degradation rates for this site are comparable to the low end of the range of degradation rates reported by Kolhatkar *et al.* (2000).

CONCLUSIONS

The investigated TBA plume is stationary or shrinking, indicating attenuation at its leading edge. Several indicators show significant natural attenuation of TBA in the form of biodegradation under aerobic and anaerobic conditions. The plume generally exhibits anaerobic conditions depleted of dissolved oxygen. Redox conditions in the plume range from iron- to sulfate-reducing. Downgradient from source areas, where carbon isotope analyses indicate that degradation is more active, first-order biological degradation constants for TBA have been calculated to range between 0.26 and 1.1 per year, yielding an equivalent half-life for TBA ranging between 0.63 and 2.65 years.

FUTURE WORK

Follow-up research is being planned to better understand the geochemical conditions and the mechanisms of TBA attenuation at the site. Pore water samples from clay units above and below the S1 sand zone will be extracted and analyzed to evaluate the extent of diffusional processes. Microcosm studies are also planned using soil and ground water samples from the site to demonstrate TBA biodegradation under site-specific redox conditions.

REFERENCES

Buscheck, T.E. and Alcantar, C.M. 1995. Regression techniques and analytical solutions to demonstrate intrinsic bioremediation. In: *Intrinsic Bioremediation, Proceedings of the Third International In Situ and Onsite Bioremediation Symposium.* San Diego, California. April 24–27, 1995. Volume 3(1), pp. 109–116. (Hinchee, R.E., Wilson, J.T., and Downey, D.C., Eds.). Columbus, Battelle Publications.

Day, M., Aravena, R., Hunkeler, D., and Gulliver, T. 2002. Application of carbon isotopes to document biodegradation of *tert*-butyl alcohol under field conditions. *Contaminated Soil Sediment and Water.* 2002 Oxygenated Fuels Issue. July/August 2002.

Fortin, N.Y. and Deshusses, M.A. 1999. Treatment of methyl tert-butyl ether vapors in biotrickling filters. 1. Reactor startup, steady-state performance, and culture characteristics. *Environmental Science and Technology.* 33(17), 2980–2986.

Hickman, G.T. and Novak, J.T. 1989. Relationship between subsurface bio-degradation rates and microbial density. *Environmental Science and Technology.* 23(5), 525–532.

Hickman, G.T., Novak, J.T., Morris, M.S., and Rebhun, M. 1989. Effects of site variations on subsurface biodegradation potential. *Research Journal of the Water Pollution Control Federation.* 61(9), 1564–1575.

Hunkeler, D., Butler, B., Aravena, R., Barker, J.F. 2001. Monitoring biodegra-dation of MTBE using compound-specific carbon isotope analysis. *Environmental Science and Technology.* 35(20), 676–681.

Kolhatkar, R., Wilson, J., and Dunlap, L.E. 2000. Evaluating natural biodegra-dation of MTBE at multiple UST sites. In: *Proceedings of the Petroleum Hydrocarbons and Organic Chemicals in Ground Water: Prevention, Detection, and Remediation Conference and Exposition.* Anaheim, California. November 15–17, 2000. pp. 32–49. (Stanley, A., Managing Ed.). Westerville, National Ground Water Association.

Mackay, D., Einarson, M.D., Wilson, R.D., Fowler, B., Scow, K., and Wood, I. 2001. *In Situ* remediation of MTBE at Vandenberg Air Force Base, California. *Contaminated Soil Sediment and Water.* Spring (Special Issue), 43–46.

McDonald, M.G. and Harbaugh, A.W. 1988. A modular three-dimensional finite-difference ground-water flow model. In: *U.S. Geological Survey Techniques of Water-Resources Investigations,* Book 6, Chapter A1, p. 586. Washington D.C., U.S. Department of the Interior.

Mormile, M.R, Liu, S., and Suflita, J.M. 1994. Anaerobic biodegradation of gasoline oxygenates: Extrapolation of information to multiple sites and redox conditions. *Environmental Science and Technology.* 28(9), 1727–1732.

Novak, J.T., Goldsmith, C.D., Benoit, R.E., and O'Brien, J.H. 1985. Biodegra-dation of methanol and tertiary butyl alcohol in subsurface systems. *Water Science Technology.* 27(5), 71–85.

Salanitro, J.P., Johnson, P.C., Spinnler, G.E., Maner, P.M., Wisniewski, H.L., and Bruce, C. 2000. Field-scale demonstration of enhanced MTBE biore-mediation through aquifer bioaugmentation and oxygenation. *Environmental Science and Technology.* 34(19), 4152–4162.

Steffan, R.J., McClay, K., Vainberg, S., Condee, C.W., and Zhang, D. 1997. Biodegradation of the gasoline oxygenates methyl *tert*-butyl ether, ethyl *tert*-butyl ether, and *tert*-amyl methyl ether by propane-oxidizing bacteria. *Applied and Environmental Microbiology.* 63(11), 4216–4222.

Suflita, J.M. and Mormile, M.R. 1993. Anaerobic biodegradation of known and potential gasoline oxygenates in the terrestrial subsurface. *Environmental Science and Technology.* 27(5), 976–978.

USEPA (U.S. Environmental Protection Agency). 1999. *Use of Monitored Natural Attenuation at Superfund, RCRA Corrective Action, and Underground*

Storage Tank Sites. U.S. EPA Office of Solid Waste and Emergency Response Directive 9200.4-17P

Wilke, C. R and Chang, P. 1955. Correlation of diffusion coefficients in dilute solutions. *American Institute of Chemical Engineers Journal*. 1, 264.

Wilson, J. T., Soo Cho, J., Wilson, B.H., and Vardy, J.A. 2000. *Natural Attenuation of MTBE in the Subsurface under Methanogenic Conditions*. U.S. Environmental Protection Agency, Office of Research and Development, Washington, DC, 2000, p 49.

Xu, M. and Eckstein, Y. 1995. Use of weighted least-squares method in evaluation of the relationship between dispersivity and scale. *Ground Water*. 33(6), 905–908.

Yeh, C. K. and Novak, J. T. 1994. Anaerobic biodegradation of gasoline oxygenates in soils. *Water Environment Research*. 66, 744–752.

Zheng, C. 1990. *MT3D, A Modular Three-Dimensional Transport Model for Simulation of Advection, Dispersion, and Chemical Reactions of Contaminants in Ground Water Systems*. Report to Kerr Environmental Research Laboratory, US Environmental Protection Agency, Ada, Oklahoma.

CHAPTER 31

Natural Attenuation of Benzene and MTBE at Four Midwestern U.S. Sites

Joseph Robb, P.G., AMEC Earth and Environmental
Ellen E. Moyer, Ph.D., P.E., Tighe & Bond

This chapter discusses the natural attenuation of benzene and MTBE at four Midwestern U.S. retail gasoline marketing outlets. We chose four sites with many rounds of analytical data that, for confidentiality purposes, are described as Sites A, B, C, and D. This chapter has the following major objectives:

- Assess the natural attenuation of benzene and MTBE using datasets "typical" of gasoline station sites (*i.e.*, data from a limited monitoring well network that are not collected with the primary goal of evaluating natural attenuation);
- Use the Mann-Kendall trend test and the coefficient of variation (CV) test to identify decreasing, stable, or increasing concentration trends at individual wells and, by extension, identify decreasing, stable, or increasing plumes;
- Use Sen's non-parametric trend estimator to quantify the magnitude of concentration trends;
- Demonstrate how trend evaluations can also be used to help identify the need for additional source control measures or site characterization work;
- Identify additional data that would be useful to collect for natural attenuation evaluations at petroleum release sites; and
- Develop practical recommendations for transitioning the focus of site characterization, if warranted, from plume delineation to evaluation of natural attenuation.

The term natural attenuation, as used in this chapter and defined by the USEPA (1999), refers to all of the physical, chemical, or biological processes that act without human intervention to reduce the mass, toxicity, mobility, volume, or concentration of chemicals in soil or ground water (see Chapter 16). The three lines of evidence that may be used to demonstrate natural attenuation include: 1) analytical data showing plume stabilization or a decrease in contaminant mass over time, 2) geochemical data that show conditions are favorable for natural attenuation, and 3) bench-scale microbiological laboratory data.

This evaluation will rely entirely on the first and second lines of evidence to demonstrate natural attenuation. In order to demonstrate stable or decreasing contaminant plumes, time trend evaluations were performed to ascertain the behavior of benzene and MTBE at individual monitoring wells. Trend evaluation results from individual monitoring wells were then interpreted within the hydrogeologic context of each site to describe site-wide trends in plume behavior, or in the case of very limited monitoring well networks, trends in source area concentrations.

TREND ANALYSIS APPROACH

The non-parametric Mann-Kendall test and Sen's non-parametric trend estimator were combined with the CV test to evaluate time trends in benzene and MTBE data. The Mann-Kendall test and Sen's trend estimator are considered well suited to the dataset because they can be used on data that are non-parametric (do not have a specific distribution, such as normal or log normal) and the dataset can contain data at irregularly spaced intervals. In addition, the dataset can contain elevated values compared to the average (outliers) or data reported as below the practical quantitation limit. As with many statistical tests, the validity of the results is increased when the sample size is larger.

Essentially, the Mann-Kendall test is performed as follows:

- The data are listed in the order in which they were collected;
- Each data point is compared to the points that follow in time; and
- The number of times the data increase is compared to the number of times the data decrease. The greater the number of increases or decreases, the more evidence there is for an upward or downward trend.

Sen's estimate of trend is calculated as follows:

- The data are listed in the order in which they were collected;
- Individual slope estimates are made between each data point and all successive data points; and
- A median slope is calculated from the individual slope estimates.

The following guidelines were followed when applying the Mann-Kendall test and Sen's estimate of trend to the dataset and interpreting the results:

- Trend analyses were performed for benzene and MTBE;
- A confidence level of 90% was deemed appropriate for natural atten-uation trend evaluations;
- All non-detect values were assigned one-half of the lowest detection limit in the time series data, as described by Gibbons (1994). This ap-proach removes the bias introduced by detection limits that change over time;
- Mann-Kendall evaluations were not performed for time series data that consisted entirely of points near and below the detection limit, due to the inherent uncertainty of analytical data near the detection limit. For example, a Mann-Kendall evaluation of the series 2 µg/l, 1.5 µg/l, 1 µg/l and 0.5 µg/l would treat each value as different and lower than the preceding value, and result in a determination of a downward trend, even though from a data interpretation perspective these values may essentially be the same number;
- Mann-Kendall trend evaluations were not performed on datasets that exhibited seasonality;
- Trend results for individual wells were grouped together to develop an interpretation of concentration trends across the entire site; and
- Trend results for individual wells were interpreted considering the lo-cation of source area(s), ground water flow directions and approxi-mate dissolved plume locations.

Analytical data from site perimeter monitoring wells often have many consecutive rounds of non-detect data interspersed with several rounds that include detected compounds. Since the non-detect values are all considered equal in the Mann-Kendall evaluation, the result is often a determination of "no trend." However the validity of the "no trend" result is drawn into ques-tion because the Mann-Kendall test does not take into account the variability in the dataset. Since the identification of "no trend" is important for the eval-uation of natural attenuation (*i.e.*, identification of stable plumes or continu-ing sources), the CV was used to measure the variability within each time se-ries dataset. CV is determined by Equation 31-1.

$$CV = \text{Standard Deviation} / \text{Arithmetic Mean} \qquad (31\text{-}1)$$

The CV should be less than or equal to one for a Mann-Kendall "no up-ward or downward trend" result to be considered a stable trend. If the CV is greater than one, the trend is considered unstable. Thus, in order to show a stable plume, one would look for a preponderance of wells, especially at the downgradient edge of a plume, that show no trend in the Mann-Kendall analysis with a CV less than or equal to one. Stable benzene and MTBE con-centrations near the source area indicate an ongoing source.

Sen's estimator of slope was used to quantify downward trends in benzene and MTBE at individual monitoring wells. The guidelines described above for the Mann-Kendall trend test were also applied to the Sen's trend estimator. Slope estimates, calculated in µg/l per year, for individual wells were grouped together to develop an interpretation of concentration trends across the entire site. Slope estimates for benzene were qualitatively compared to MTBE slope estimates at individual monitoring wells and across the entire site.

Although present at each site, toluene, ethylbenzene, and xylenes are not discussed in this chapter. Benzene and MTBE were chosen because benzene is often the risk driver at petroleum release sites and MTBE is often present in high concentrations, due to its relatively high solubility. In addition, for the purposes of comparing MTBE and other gasoline constituents, the fate and transport characteristics of benzene are generally representative of other gasoline constituents such as toluene, ethylbenzene, and xylenes.

GEOCHEMICAL DATA

The identification of significant downward trends provides good evidence that natural attenuation is occurring, but does not illuminate to what extent biological mechanisms are contributing to observed decreases in concentrations. It is now recognized that biological degradation of benzene and MTBE can occur under a variety of TEAPs. Dissolved oxygen and ORP data were available for the four sites considered in this chapter and were qualitatively evaluated to assess the importance of biological degradation. Although the data were insufficient to precisely identify TEAPs, general observations of trends in redox chemistry across the site were made. In general, lower ORP values and depressed dissolved oxygen values in the source area, when compared to background, were considered good evidence of the consumption of oxygen for the biological degradation of benzene and MTBE.

The natural attenuation of benzene and MTBE for four sites is evaluated below. Hydrogeologic conditions, release history, trend evaluation results, and geochemical conditions are discussed for each of the four sites. Site A data are presented and discussed in slightly greater detail than for Sites B, C, and D, in order to demonstrate the methods used to evaluate natural attenuation. At each of the four sites an unspecified amount of soil was removed in the early 1990s during UST upgrades or other subsurface investigation. No further engineered soil or ground water remediation has been performed at these sites.

SITE A

Site A was reported to regulatory authorities in January 1992 when LNAPL was identified in the vicinity of the former dispensing island, canopy footings, and dispenser lines during excavation activities associated with UST

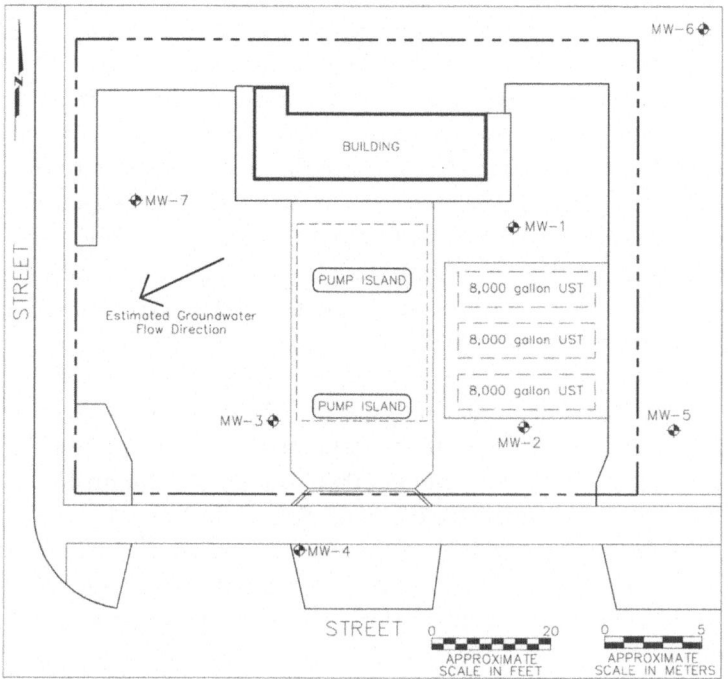

Figure 31-1.   Site A

upgrades. However, no LNAPL has been detected in site monitoring wells. See Figure 31-1 for locations of site features.

Hydrogeology.   The depth to ground water at Site A is 1.5 to 3.7 meters (5 to 12 feet) bgs. The shallow unconfined aquifer consists primarily of unconsolidated glacial deposits described as reddish brown medium stiff clay, containing a trace of silt and fine sand. The average hydraulic conductivity (K), estimated by rising head slug tests performed in site monitoring wells, is 3×10^{-4} cm/s (8×10^{-1} ft/day), and the horizontal hydraulic gradient (i) across the site is approximately 0.09. Based on the equation v = K i / n, where n is effective porosity estimated at 0.25, the average horizontal ground water velocity (v) is estimated to be approximately 34 meters (110 feet) per year. The horizontal ground water flow direction is to the southwest. The dimensions of Site A are 23 meters by 29 meters (75 feet by 95 feet). Thus, based on a release date of at least 9 years ago, and an estimated average ground water velocity of 34 meters (110 feet) per year, dissolved contaminants have had ample opportunity to migrate off-site.

Seasonality.   Between January 1995 and July 1998, increases in ground water elevations generally coincided with increases in dissolved benzene concentrations in source area monitoring wells MW-1, MW-2, MW-3, and MW-4. Since Site A is paved and relatively impervious to infiltration, increases in dissolved benzene concentrations could be due to the seasonal rise of the water table into zones of residual soil contamination. After January 1998, seasonal effects appear to play a smaller role in concentration trends, which indicates a potential depletion of residual soil contamination within the range of seasonal ground water fluctuations. Figure 31-2 presents ground water elevations and the concentrations of total benzene and MTBE at monitoring well MW-1 for the period from January 1995 to November 2000. Similar trends were observed at other source area monitoring wells.

Trends.   Mann-Kendall trend analyses for MTBE were evaluated over 10 sampling rounds from July 1997 to November 2000 and over 7 sampling rounds from April 1998 to November 2000 for benzene. Mann-Kendall trend analyses were not performed for benzene data collected prior to April 1998, or for MTBE data prior to July 1997, due to the presence of seasonal trends (see Figure 31-2). Mann-Kendall trend analyses indicate decreasing trends in benzene and MTBE concentrations at the 90% confidence level in source area monitoring wells MW-1, MW-2, and MW-3. Source area monitoring well MW-4 exhibits a stable trend in MTBE and an unstable trend in benzene

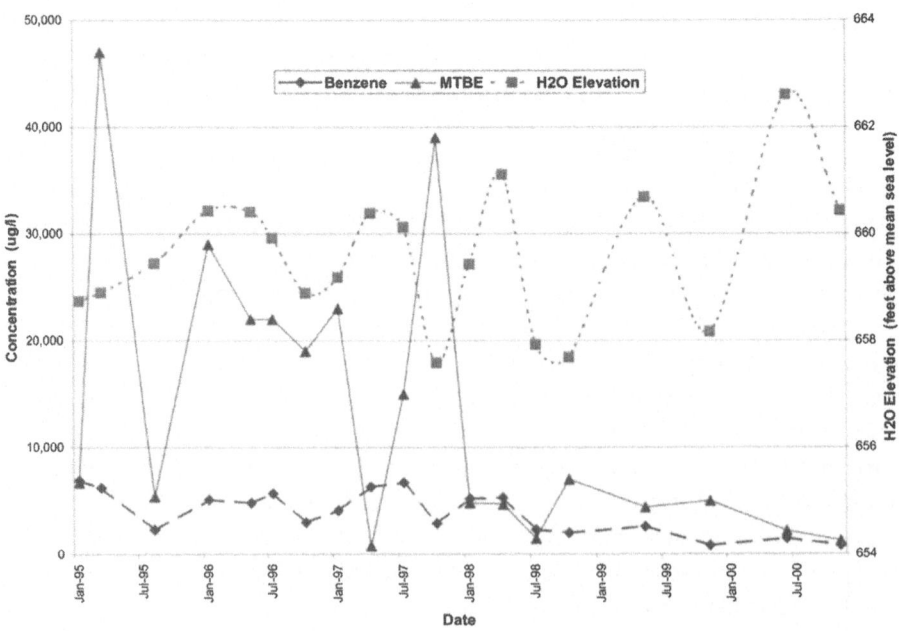

Figure 31-2.   **Concentration Trends and Ground Water Elevations at MW-1 at Site A**

Table 31-1. **Trend Analysis for Benzene and MTBE in Source Area Wells at Site A**

| | Location and Compound | | | | | | | |
| | MW-1 | | MW-2 | | MW-3 | | MW-4 | |
Date	MTBE (μg/l)	Benzene (μg/l)	MTBE (μg/l)	Benzene (μg/l)	MTBE (μg/l)	Benzene (μg/l)	MTBE (μg/l)	Benzene (μg/l)
July 1997	15,000	ND	4,500	ND	190	ND	460	ND
October 1997	39,000	ND	20,000	ND	400	ND	670	ND
January 1998	4,800	ND	50	ND	2,800	ND	2	ND
April 1998	4,700	5,300	2,900	3,400	500	1,400	520	270
July 1998	1,500	2,300	2,100	3,100	590	1,900	180	15
October 1998	7,000	2,000	2,900	3,100	300	1,300	140	4.3
May 1999	4,400	2,600	3,300	2,800	140	760	110	48
November 1999	5,000	820	2,700	1,800	13	780	190	6
June 2000	2,200	1,500	790	1,400	96	670	62	20
November 2000	1,300	850	920	2,100	70	680	150	7
Decreasing Trend at 90% Confidence?	Yes	Yes	Yes	Yes	Yes	Yes	No CV < 1	No CV > 1
Slope (μg/l per year)	−2024	−620	−682	−516	−222	−377	−40	1

ND = No Data

concentrations. Table 31-1 shows benzene and MTBE data from July 1997 to November 2000 at the four source area wells and the results of the trend evaluations.

Decreasing concentrations of benzene and MTBE in wells MW-1, MW-2, and MW-3 are strong evidence of an attenuating source of benzene and MTBE. Where benzene concentrations are highest, (MW-1, MW-2, and MW-3), Sen's slope estimates show benzene is decreasing at rates between 377 μg/l and 620 μg/l per year. At these same wells, MTBE is decreasing at rates between 222 μg/l and 2,024 μg/l per year. The highest estimate of MTBE decline (2,024 μg/l per year at MW-1) may be an overestimate due to a single elevated concentration value in October 1997 and lower values in subsequent sampling rounds. In general, the slope estimates suggest benzene and MTBE concentrations are declining at approximately the same rate in source area wells.

Trend evaluations were not performed for VOC data from perimeter wells MW-5, MW-6, and MW-7 because the data consisted primarily of concentration values near or below the detection limit (see discussion above).

These wells effectively delineate the extent of the dissolved VOC plume to the northeast (hydraulically upgradient) and northwest (cross-gradient).

Geochemical Conditions. Dissolved oxygen and ORP values measured over seven sampling rounds from April 1998 to November 2000 did not show any trends. However, average ORP and dissolved oxygen values over this time period indicate that redox conditions are favorable for the biological degradation of dissolved benzene and MTBE. Table 31-2 provides April 1998 to November 2000 average MTBE, benzene, dissolved oxygen, and ORP data. In general, there is a good correlation between the presence of MTBE and benzene, and low ORP values, relative to background values. In addition, the dissolved oxygen concentration is higher in background well MW-5, and slightly lower in the source area, cross-gradient, and downgradient wells. These results suggest dissolved oxygen is being used as a terminal electron acceptor in the source area.

SITE B

Site B was reported to regulatory authorities in August 1993 when PID field screening identified petroleum-impacted soil in the vicinity of the gasoline transfer lines. No LNAPL was observed during the 1993 excavations, nor has LNAPL been observed in Site B monitoring wells. See Figure 31-3 for locations of site features.

The shallow unconfined aquifer consists primarily of unconsolidated glacial deposits described as dark gray silt with some clay and a trace of fine gravel and coarse sand. The hydraulic conductivity is estimated to be between 10^{-6} and 10^{-4} cm/s (10^{-3} and 10^{-1} ft/day) and the horizontal hydraulic gradient across the site is approximately 0.05. Assuming an effective porosity of 0.25, the average horizontal ground water velocity is estimated to range from 0.06 to 6 meters (0.2 to 20 feet) per year. The depth to ground water ranges from

Table 31-2. **Average MTBE, Benzene, Dissolved Oxygen, and ORP at Site A**

Well ID	Location	MTBE (µg/l)	Benzene (µg/l)	Dissolved Oxygen (mg/l)	Oxidation-Reduction Potential (mV)
MW-1	Source Area	3,729	2,196	3.3	−33
MW-2	Source Area	2,230	2,539	4.3	−48
MW-3	Source Area	283	1,070	4.3	−40
MW-4	Source Area / cross-gradient	193	53	4.3	−27
MW-5	Upgradient	1.3	< 1	5.0	80
MW-6	Upgradient	1.4	< 1	3.7	99
MW-7	Cross- and downgradient	< 1	< 1	4.5	62

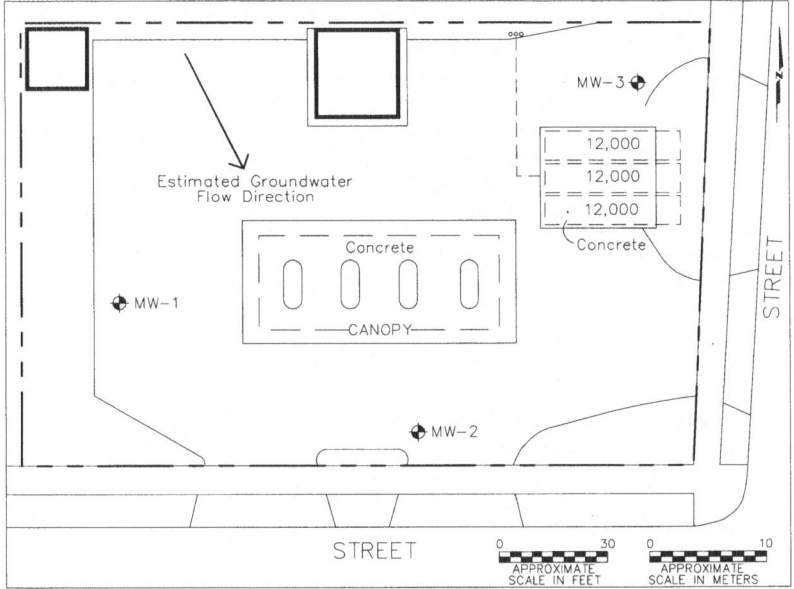

Figure 31-3.   **Site B**

1 to 3.4 meters (3 to 11 feet) bgs, and the horizontal ground water flow direction is to the southeast. The distribution of contaminants at the site is consistent with the timing of the release and the estimated ground water velocity.

Three ground water monitoring wells have been installed at Site B. Table 31-3 provides a description of monitoring well locations relative to the source area, benzene and MTBE concentrations detected historically at each location, statistical trend evaluation results, and average dissolved oxygen and ORP data. Approximately 20 rounds of data were available from April 1994 to November 2000. Data prior to April 1997 appeared to exhibit seasonal trends and were therefore excluded from the trend evaluations.

MTBE exhibits a stable trend at monitoring wells MW-2 and MW-3 and an upward trend at MW-1. Benzene is non-detect at MW-1, exhibits a stable trend at monitoring well MW-2, and an unstable trend at MW-3. The elevated, stable benzene and MTBE concentrations at MW-2, located immediately downgradient of the gasoline-dispensing island, provide strong evidence of a continuing source of petroleum hydrocarbons. The stable trend in MTBE at well MW-3 and upward trend at MW-1 indicate the plume of MTBE may be stable in the northeast, but is expanding in the western part of the site. Elevated MTBE and benzene concentrations at well MW-2 suggest dissolved benzene and MTBE are migrating off-site.

Average dissolved oxygen and ORP values are lowest where benzene and MTBE concentrations are highest (MW-2), while higher average ORP

Table 31-3. **Contaminant Distribution, Trends, and Geochemical Parameters at Site B**

Well ID	Location	April 1997 to November 2000 Benzene and MTBE Concentrations (10 sampling rounds)	Trend Evaluation Results	Average Dissolved Oxygen (mg/l)	Average Oxidation Reduction Potential (mV)
MW-1	40 feet cross-gradient (west) from gasoline-dispensing island	No benzene detected MTBE ranges from 10 to 85 µg/l	Benzene not analyzed (see discussion above) MTBE exhibits upward trend	4.3	78
MW-2	30 feet down-gradient (southeast) of the gasoline-dispensing island	Benzene 400 to 3,000 µg/l MTBE 68 to 5,000 µg/l	Benzene exhibits stable trend, CV < 1 MTBE exhibits stable trend, CV < 1	3.3	−18
MW-3	18 feet cross-gradient (north) of three 12,000-gallon USTs	Benzene ranges from non-detect (ND) to 10 µg/l MTBE ranges from 100 to 1,000 µg/l	Benzene exhibits unstable trend, CV > 1 MTBE exhibits stable trend, CV < 1	4.5	60

and dissolved oxygen values were measured at wells MW-1 and MW-3, which contain little or no benzene and low to moderate concentrations of MTBE. The dissolved oxygen, ORP, benzene, and MTBE results are consistent with the consumption of oxygen for the biological degradation of gasoline constituents.

Overall, results for Site B indicate the following:

- Residual contamination is acting as a continuing source of petroleum hydrocarbons at Site B. Additional source control may be warranted; and
- Biodegradation may be limiting the migration of benzene and MTBE.

Site characterization and development of the natural attenuation argument would be facilitated by installation and sampling of additional monitoring wells.

SITE C

Site C was reported to regulatory authorities in December 1994 when LNAPL was identified in a gasoline transfer line trench excavated for UST upgrades. No LNAPL has been detected in site monitoring wells. See Figure 31-4 for locations of site features.

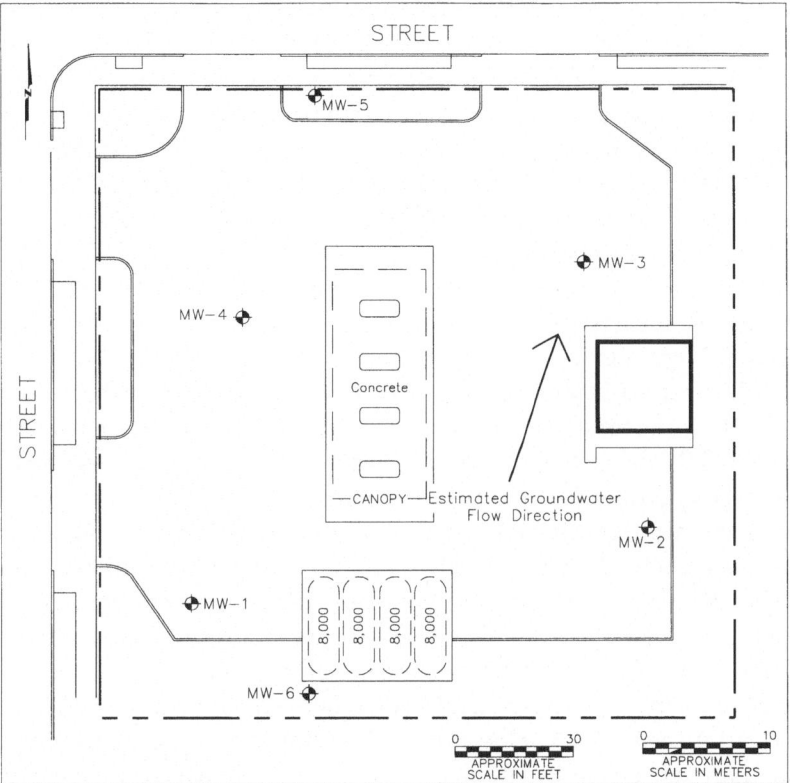

Figure 31-4.   **Site C**

The shallow unconfined aquifer consists primarily of unconsolidated glacial deposits described as gray silty clay with a trace of sand and gravel. The average hydraulic conductivity, based on slug tests performed in site monitoring wells, is estimated to be 9×10^{-8} cm/s (3×10^{-4} ft/day) and the horizontal hydraulic gradient across the site is approximately 0.02. Assuming an effective porosity of 0.25, the average horizontal ground water velocity is estimated to be 3×10^{-3} meters (1×10^{-2} feet) per year. The depth to ground water is 1.5 to 2.4 meters (5 to 8 feet) bgs, and the horizontal ground water flow direction is to the northeast. The shallow depth to ground water and the low permeability of the native aquifer material suggest shallow ground water flow may be influenced by high permeability fill material in utility conduits. Contaminant distribution and the estimated age of the release suggest actual ground water flow rates may be several orders of magnitude greater than 3×10^{-3} meters (1×10^{-2} feet) per year.

Six ground water monitoring wells have been installed at Site C. Between 15 and 17 rounds of benzene and MTBE data were available from August 1995 to November 2000. Concentration data from this time period did not appear to exhibit seasonality and were utilized in the trend evaluations. Table

31-4 provides a description of monitoring well locations relative to the source area, a summary of historical benzene and MTBE detections at each monitoring location, statistical trend evaluation results, and geochemical data.

As summarized in Table 31-4, MTBE and benzene have been consistently detected in all site monitoring wells, except background well MW-2. Where detected, MTBE and benzene usually exhibit downward trends. Where benzene is currently detected (MW-1, MW-4, and MW-6), Sen's slope test estimates benzene is decreasing at rates between 161 µg/l and 1,518 µg/l per year. At three wells where MTBE is currently detected (MW-1, MW-3, and MW-6), Sen's slope test estimates MTBE is decreasing at rates between 47 µg/l and 493 µg/l per year.

MTBE detected in downgradient well MW-5 does not exhibit a downward trend for the time period from April 1995 to November 2000. However, when the MTBE dataset at well MW-5 was divided in two halves, the Mann-Kendall analysis identified an upward trend from April 1995 to October 1997, and a downward trend from October 1997 to November 2000. These results suggest the dissolved plume of MTBE at Site C has been shrinking since October 1997. Recent MTBE concentrations near the property boundary at MW-5 are relatively low (50 µg/l, 75 µg/l) and are similar to MTBE MCLs that have been established by various state regulatory agencies.

At monitoring wells MW-1 and MW-4, benzene appears to be degrading more rapidly than MTBE. However, MTBE appears to be degrading more rapidly than benzene at monitoring well MW-6. Downward benzene trends at wells MW-1, MW-4, and MW-6 indicate the source of benzene is decreasing in strength, and suggest the dissolved plume has receded considerably. The disappearance of benzene at downgradient well MW-5 indicates the dissolved benzene plume no longer extends to the property boundary. Downward trends in MTBE at source area monitoring wells MW-1, MW-3, MW-4, and MW-6 indicate the source of MTBE is also decreasing in strength.

ORP values (presented in Table 31-4) measured during ground water sampling activities suggest biodegradation is limiting the migration of dissolved benzene and MTBE. There is a good correlation between low ORP values, relative to background values, and the presence of benzene and MTBE. For example, monitoring well MW-1 typically has greater than 3,000 µg/l total benzene and MTBE and has an average ORP of --38 mV, while background well MW-2 typically has no benzene or MTBE and has an average ORP of 85 mV. Average dissolved oxygen concentrations range from 2.6 to 3.9 mg/l across the site and show little correlation to the presence of benzene or MTBE.

SITE D

Site D was reported to regulatory authorities in July 1992 when a leak was detected in a premium grade gasoline transfer line. No LNAPL has been

Table 31-4. Contaminant Distribution, Trends, and Geochemical Parameters at Site C

Well ID	Location	April 1995 to November 2000 Benzene and MTBE Concentrations (15–17) sampling rounds)	Trend Evaluation Results	Slope of Downward Trend (µg/l per year)	Average Dissolved Oxygen (mg/l)	Average Oxidation Reduction Potential (mV)
MW-1	40 feet cross-gradient (west) of transfer line trench (suspected source)	Benzene 3,000 to 860 µg/l MTBE 3,100 to 2 µg/l	Decreasing trends in benzene Decreasing trend in MTBE	276 70	3.1	–38
MW-2	Background well	Benzene and MTBE typically non-detect	Trends not evaluated.	NA	2.6	85
MW-3	90 feet downgradient (northeast) of suspected source	Benzene typically non-detect MTBE 730 to 50 µg/l	Benzene not evaluated MTBE exhibits decreasing trend	NA 47	3.9	102
MW-4	70 feet downgradient (northwest) of suspected source	Benzene 12,000 to 2,900 µg/l MTBE 2,900 to < 16 µg/l	Decreasing trend in benzene Decreasing trend in MTBE	1,518 390	2.8	–40
MW-5	115 feet downgradient (north) of suspected source	Benzene 730 µg/l to non-detect MTBE 2,300 to 75 µg/l	Benzene exhibits slight decreasing trend MTBE exhibits unstable trend, CV > 1	1 NA	3.2	51
MW-6	40 feet upgradient (south) of four 8,000-gallon USTs	Benzene 3,100 to non-detect MTBE 9,300 to 12 µg/l	Decreasing trend in benzene Decreasing trend in MTBE	161 493	3.8	52

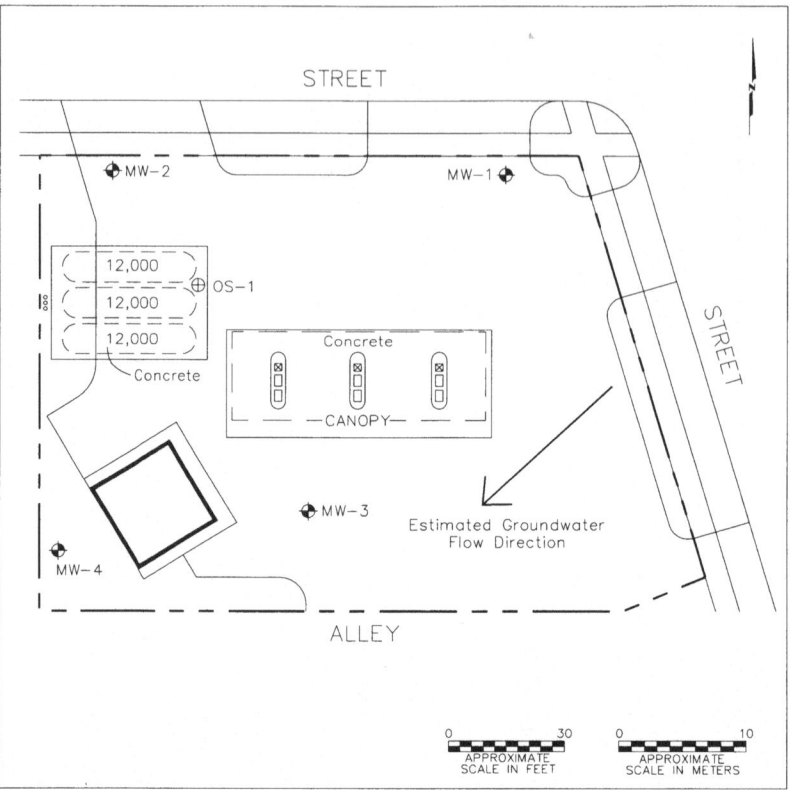

Figure 31-5.   **Site D**

detected in site monitoring wells. See Figure 31-5 for locations of site features.

The shallow unconfined aquifer consists primarily of unconsolidated glacial deposits described as light brown sand and silt overlying brownish gray silty clay with a trace of sand. The depth to the silty clay layer is approximately 1.5 meters (5 feet) bgs and the depth to ground water is 1.2 to 1.8 meters (4 to 6 feet) bgs. Thus, ground water appears to migrate through the thin, more permeable sand and silt layer, and through the underlying silty clay material. The average hydraulic conductivity of the lower unit, based on slug tests performed in site monitoring wells, is estimated to be 5×10^{-7} cm/s (1×10^{-3} ft/day) and the horizontal hydraulic gradient across the site is approximately 0.01. Assuming an effective porosity of 0.25, the average horizontal ground water velocity in the lower unit is estimated to be 6×10^{-3} meters (2×10^{-2} feet) per year. The average ground water velocity in the upper sand and silt layer is expected to be several orders of magnitude greater, in the range of 0.09 to 9 meters (0.3 to 30 feet) per year. The horizontal ground water flow direction is to the southwest. Based on the timing of the petroleum release, and the distribution of dissolved VOCs at Site D, it appears likely some benzene and MTBE migration is occurring in the more permeable sand and silt layer.

Four ground water monitoring wells have been installed at Site D. Table 31-5 provides a description of monitoring well locations relative to the source area, a summary of historical benzene and MTBE detections at each monitoring location, statistical trend evaluation results, and geochemical data. Approximately 20 rounds of benzene and MTBE data were available from April 1994 to November 2000, however, data prior to July 1997 exhibited seasonal trends. Concentration data from after July 1997 appeared to exhibit minimal seasonal variation and, therefore, were utilized in the trend evaluations.

As summarized in Table 31-5, concentrations of benzene compounds at all four site monitoring wells have decreased to near or below the analytical detection limit. Thus, it appears that the plume of benzene has decreased in size and no longer migrates off-site. The detections of MTBE at downgradient monitoring well MW-4 suggest the plume of MTBE may still extend off-site, but the downward trend indicates the plume may be receding. The concentration of MTBE near the property boundary from the most recent round of sampling (November 2000) was relatively low (56 µg/l). The rate at which MTBE is decreasing over time in individual monitoring wells at Site D are similar to those observed at Sites A and C.

In general, the background average dissolved oxygen concentration is high (5.5 mg/l at MW-1), while source area wells (MW-2 and MW-3) exhibit depressed average dissolved oxygen concentrations. The average ORP at source area well MW-3 (–36 mV) is depressed relative to background well MW-1 (110 mV). These trends provide good evidence for the aerobic biodegradation of dissolved petroleum hydrocarbons. Downgradient well MW-4 has relatively high dissolved oxygen and ORP values, which are consistent with the depletion of dissolved petroleum hydrocarbons and the reoxygenation of the aquifer downgradient of the source area. Based on the stoichiometric relationship describing the aerobic biodegradation of benzene and MTBE, 3.1 milligrams of oxygen are consumed during the biodegradation of 1 milligram benzene and 2.75 milligrams of oxygen are consumed during the biodegradation of 1 milligram MTBE. Based on the upgradient dissolved oxygen concentration (5.5 mg/l) and the source area benzene concentrations (less than 1 mg/l), there is more than enough oxygen migrating into this site via ground water migration for the aerobic biodegradation of the remaining dissolved benzene. However, based on the stoichiometric relationship between dissolved oxygen and MTBE, and the concentrations of MTBE in the source area (2.2 mg/l), aerobic biodegradation of MTBE may be oxygen-limited.

CONCLUSIONS

Four petroleum retail marketing outlets with documented releases of petroleum hydrocarbons were evaluated to characterize the natural attenuation of

Table 31-5. Contaminant Distribution, Trends, and Geochemical Parameters at Site D

Well ID	Location	July 1997 to November 2000 Benzene and MTBE Concentrations (17-20 sampling rounds)	Trend Evaluation Results	Slope of Downward Trend (µg/l per year)	Average Dissolved Oxygen (mg/l)	Average Oxidation Reduction Potential (mV)
MW-1	Upgradient—Background	Benzene primarily non-detect MTBE concentrations near and below detection limits	Benzene trends not evaluated MTBE trends not evaluated	NA NA	5.5	110
MW-2	50 feet cross-gradient (northwest) from three 12,000 gal USTs and transfer lines	Benzene concentrations near and below detection limits MTBE 38,000 to 40 µg/l	Benzene trends not evaluated Downward trend in MTBE	NA 994	3.7	100
MW-3	45 feet cross-gradient (southeast) of USTs and transfer lines	Benzene 650 < 0.5 µg/l MTBE 36,000 to 1,700 µg/l	Downward trend in benzene Downward trend in MTBE	10 166	3.2	–36
MW-4	70 feet downgradient (southwest) of transfer lines	Benzene primarily non-detect MTBE 860 to 56 µg/l	Benzene trends not evaluated Downward trend in MTBE	NA 253	5.1	73.7

benzene and MTBE. Where a sufficient number of monitoring wells exist, dissolved plumes were characterized as stable, expanding, or shrinking based on the non-parametric Mann-Kendall trend evaluations. Datasets that did not exhibit upward or downward trends were further characterized as stable or non-stable trends with the CV test. The slopes of downward trends were quantified using Sen's non-parametric indicator of median slope.

The trend evaluations provided a preliminary evaluation of the potential for natural attenuation to play a role in management of each of the four sites. Trend evaluations at Sites A, C, and D provided good evidence of decreasing sources and shrinking or stable benzene and MTBE plumes, while Site B showed evidence of a continuing source and steady-state dissolved plumes of benzene and MTBE. Additional source control measures may be beneficial at Site B.

At Site A, MTBE and benzene concentrations are declining at similar rates. In addition, the downgradient extents of benzene and MTBE appear to be roughly equal at Site A. The comparable behavior of the benzene and MTBE plumes at Site A suggests MTBE and benzene may be attenuating at similar rates. At Sites C and D, the MTBE plume extends beyond the monitoring well network, but the MTBE plumes appear to be shrinking and concentrations at the property boundary are low. The benzene plumes do not appear to extend past the property boundary at Sites C and D. The benzene plumes appear to be attenuating more rapidly than the MTBE plumes at Sites C and D.

In general, average dissolved oxygen and ORP values were lowest where benzene and MTBE concentrations were highest, while higher average ORP and dissolved oxygen values were measured at locations with little or no benzene and low to moderate concentrations of MTBE. The dissolved oxygen, ORP, benzene, and MTBE results are consistent with the consumption of oxygen for the biological degradation of gasoline constituents.

In situ natural attenuation processes include biodegradation, dispersion, dilution adsorption, volatilization, and chemical reactions. However, it should not be necessary to quantify the significance of individual natural attenuation processes in order to show natural attenuation mechanisms are at work. Demonstration of decreasing contaminant concentrations across the site over time should be sufficient to demonstrate the efficacy of natural attenuation, assuming the monitoring well network is adequate in three dimensions and sources have been controlled.

RECOMMENDATIONS

1. Monitoring well networks installed for purposes of plume delineation often are inadequate for purposes of thoroughly evaluating natural attenuation. If preliminary investigations show natural attenuation will play an important role in site management, monitoring wells should be installed

along the axis of the plume, extending from the source area to the downgradient edge of the dissolved plume. Monitoring wells should be placed to provide three-dimensional plume delineation.

2.   All of the monitoring wells at the four sites considered in this chapter were installed as water table wells, with screen intervals bridging the water table. As a result, no information on vertical hydraulic gradients or vertical distribution of contaminants was available. Downward hydraulic gradients could cause downward migration of the dissolved plume, which may not be detected with the conventional monitoring well network. Installation of a shallow-deep well couplet at the downgradient property boundary could strengthen the case for natural attenuation by verifying that the plume is not merely diving beneath the monitoring well network.

3.   Accurate measurement of dissolved oxygen in ground water is useful for an understanding of biological degradation.

4.   The role of biodegradation in reducing contaminant concentrations could be more fully understood by sampling and analyzing for alternate electron acceptors and other natural attenuation parameters including ferrous and ferric iron, sulfate, sulfide, nitrate/nitrite, carbon dioxide, methane, and hydrogen. Samples should be collected at several points in time, including a baseline round. Adequate detection limits and reliable results can usually be achieved at reasonable cost through the use of colorimetric field test kits for ferrous and ferric iron, sulfate, sulfide, nitrate/nitrite, and carbon dioxide.

5.   Demonstration of the production of breakdown products could bolster the case for natural attenuation. Since MTBE is known to biodegrade to TBA as an intermediate product, ground water samples should also be analyzed for this compound. Examination of MTBE/TBA ratios over time at individual monitoring wells could provide supporting evidence of biodegradation.

6.   Reliable estimates of hydraulic conductivity are useful in natural attenuation evaluations. Residual drawdown tests as described by Driscoll (1995) can provide more reliable K estimates than slug tests because they test a larger portion of the aquifer. Residual drawdown tests are not significantly more expensive than slug tests and are cheaper than full-scale pump tests.

REFERENCES

Driscoll, F.G. 1995. *Groundwater and Wells*, Second Edition. St. Paul, U.S. Filter/Johnson Screens.

Gibbons, R.D. 1994. *Statistical Methods for Groundwater Monitoring*. New York, John Wiley & Sons, Inc.

USEPA (U.S. Environmental Protection Agency). 1999. *Use of Monitored Natural Attenuation at Superfund, RCRA Corrective Action, and Underground Storage Tank Sites*. U.S. EPA Office of Solid Waste and Emergency Response Directive 9200.4-17P.

Appendices

Appendix A

MTBE Occurrence in Surface and Ground Water

Edited by James A. M. Thomson, Applied Hydrology Associates, Inc.

INTRODUCTION

This appendix describes the occurrence of MTBE in surface and ground water, including the last ten years of sampling results for the United States Geological Survey (USGS) National Water Quality Assessment (NAWQA) Program. Where benzene, toluene, ethylbenzene, and xylenes (BTEX) data are available, these are also described, but most of the summarized studies only considered methyl tert butyl ether (MTBE). For broader perspective, this appendix also reviews the findings of MTBE sampling programs other than NAWQA in the United States (U.S.), and describes the Environment Agency's study into MTBE occurrence in England and Wales, which provides a valuable counterpoint to the situation in the U.S. The end of the appendix summarizes plume length studies performed in Texas, Florida, California, and North Carolina, and summarizes current conclusions.

In this Appendix, a public water supply (PWS) is defined as a system that serves piped water to at least 25 people or 15 service connections for at least 60 days a year. A community water supply (CWS) is a PWS that serves people year round in their homes (USEPA, 2001).

MTBE AND THE USGS'S NAWQA PROGRAM

David K. Ramsden, Ph.D. and Peter Hicks, URS Corporation

The NAWQA Program. The NAWQA Program is a wide-ranging, systematic water quality survey conducted by the USGS, assessing more than 50 river basins and aquifer systems in nearly all 50 states. The stated program objective is to determine "how and why water quality varies across the nation." To that end, the USGS systematically collects, analyzes, evaluates, and interprets data about water chemistry, hydrology, land use, stream habitat, and aquatic life, using surface water, ground water, sediment, and animal tissue samples from across the U.S. and parts of Canada.

Water quality variations may result from either natural or anthropogenic causes, but attention often focuses on the latter. Parameters include chemicals, microbial indicators, and chemical/physical properties. Table A-1, condensed from the USGS website, summarizes the organization of the study, the groups of parameters monitored, and the planned frequency of monitoring. USGS teams assess different areas of the country using a consistent study design, allowing national evaluations of pesticides, nutrients, volatile organic compounds (VOCs), trace elements, and aquatic ecology in different river and ground water basins. Generally, the NAWQA program attempts a spatial well density of one well per 98 square kilometers (38 square miles) for each location. Monitoring or observation wells that were installed to detect a known or suspected contaminant are avoided, as are wells that are located near roads or highways (Lapham *et al.*, 1995).

The public, Congress, private corporations, and other local, national, and international groups use the gathered data. Data are intended to be used in connection with water resource issues including: agriculture, urbanization, chemical use, drinking water and human health, source water protection, pesticide registration, nutrient enrichment and criteria, fish consumption, beneficial uses, and impaired waters.

Program Status. In 1991, the NAWQA program was initiated in 51 areas, and additional assessment areas are scheduled to be added until 2012. Over the next decade, the USGS will return to 14 major river basins and aquifer systems that were already sampled in the initial stage of the program. Additionally, they will begin sampling two new groups of 14 basins and aquifers, starting in 2004 and 2007, respectively. The new studies will focus on emerging pesticides, indicators of water-borne diseases, and the bioaccumulation of mercury and methyl mercury in aquatic organisms. The original 59 study units from the first cycle of the program covered 65% of the water used for drinking and irrigation nationwide, while the 42 potential study units in the second cycle will cover 60% of the same water uses.

A large existing database, the NAWQA Data Warehouse, is accessible through the Internet (USGS, 2001). Searches can be performed for combinations of locations, analytes, geography, and geopolitical units such as states. The data can be downloaded in a variety of formats that are selectable on the website, including text or spreadsheet formats. The MTBE data are only a small part of the total data available. Consequently, downloads of multiple contaminant results can be slow, depending on the search parameters selected. As data are generated and validated, they become available on the site, though this process may take several years. Consequently, the data statistics or trends may vary from one download to the next, if considerable time passes between data searches.

MTBE Data. MTBE is only one of the many analytes included in the NAWQA program. BTEX compounds are also NAWQA analytes and are the gasoline components most commonly compared to MTBE in ground water or surface water.

Ground Water and Spring Water Data. As of October 17, 2001, the data set includes 4,023 ground water samples (excluding springs) analyzed for MTBE, of which 388 (including estimated values), or about 9.6%, contained detectable concentrations of MTBE. The average concentration of all the MTBE detections, not counting estimated values, was approximately 280 micrograms per liter (μg/l), with a maximum concentration of 23,000 μg/l. (Estimated values are either values near the detection limits, indicating a low level of precision for that analysis, or values much higher than the detection limits, indicating that the value is outside the calibration limits of the instrument [Williamson, 2001]). The MTBE concentrations from 63 estimated detections averaged about 0.09 μg/l. Benzene was detected in 205 samples, or about 5.1% of locations analyzed, with a maximum concentration of 290 μg/l.

These data should not be interpreted to mean that MTBE releases are more common than benzene releases. Two characteristics of benzene, its lower solubility and concentration in gasoline of less than or equal to 2%, combine to limit its practical solubility concentration relative to MTBE. Histograms of MTBE and benzene concentrations for all ground water samples in the database are included as Figures A-1 and A-2. Most of the samples for both analytes were non-detects. A crossplot of the results from ground water samples with reported benzene and MTBE concentrations is provided in Figure A-3.

There were 94 MTBE analyses of spring water in the database. Of these, 91 were non-detects and 3 were detections; none were estimated. There were also 94 benzene analyses of spring water, with 92 non-detects and 2 estimated values. Springs are excluded from the discussion of ground water below.

Surface Water Data. The data set includes 1,515 surface water analyses for MTBE, with 463 detections, 778 non-detects, and 274 estimated detections. Samples included in the "surface water" database query included 15 wet deposition samples (*i.e.*, precipitation samples), 1 artificial sample (no information available), and 1 ground water sample. It was not clear why the latter two were included in the surface water results; this may simply be an error in the database or retrieval from the database. Histograms of MTBE and benzene concentration for all surface water samples in the database are included as Figures A-4 and A-5. The majority of the samples for both analytes were non-detects. A crossplot of the results from surface water samples that reported benzene and MTBE concentrations is provided in Figure A-6.

Table A-1. Summary of NAQWA Study Components*

Study Component	Data Type	Data	Types of Sample Sites	Rationale/Purpose	Number of Sites	Sampling Frequency & Period
Stream Chemistry						
Basic Sites	General Water Chemistry	Concentrations/Seasonal Variation/Annual Loads for: Streamflow, Nutrients, Major ions, Organic Carbon, Suspended Sediment, Water Temperature, Specific Conductance, Alkalinity, pH, Dissolved Oxygen	Basic Fixed Sites	Represent common land uses, as well as basin outflow sites	~10/Study Unit	Monthly + Storms
Intensive Sites	Pesticides	Concentrations/Seasonal Variations for: Basic Site Constituents, 85 Pesticides	Subset of Basic Sites	Represent where land use was most homogeneous + some mixed outflow sites in some study units	~2/Study Unit	Weekly to Monthly: Feb. 1993–Mar. 1994
Synoptic Sites	Water Chemistry + Pesticides	Distribution of: Basic Site Constituents, 85 Pesticides	Basic Sites, Other Sites, and Reference Sites	Spatial distribution of pesticides and nutrients	10–30/Study Unit	1–4/Season
Bed Sediments	Chemistry	Occurrence/Distribution of: Total PCBs, 32 Organochlorine Pesticides, 63 Semivolatile Organics, 44 Trace Elements	Depositional Zones	Most stream sites sampled in other components of study	~20/Study Unit	Once
Fish & Benthics	Chemistry	Occurrence/Distribution in Biota of: Total PCBs, 30 Organochlorine Pesticides, 24 Trace Elements	Most Study Stream Sites	Wherever tissue could be collected	~20/Study Unit	Once

Ground Water Aquifer Survey	Chemistry	Occurrence/Distribution of: Major Ions Nutrients 85 Pesticides	60 Volatile Organic Compounds Dissolved Organic Carbon Radon	Domestic/Public Supply or Monitoring Wells; Varied Depth Wells in Study Unit or Uniform Segment	Represent distribution of chemicals in randomly selected wells in an area	30/Area — 1–2 Areas/Study Unit	Onc
Land Use Effects	Chemistry	Occurrence/Distribution of: Major Ions Nutrients 85 Pesticides	60 Volatile Organic Compounds Dissolved Organic Carbon Radon	Shallow Domestic or Monitoring Wells	Describe effects on shallow ground water of particular pattern of agricultural land use	~30 Wells/1–4 Land Users/Study Unit	Onc

*Adapted from USGS Table on the NAWQA Internet Site (USGS, 2001)

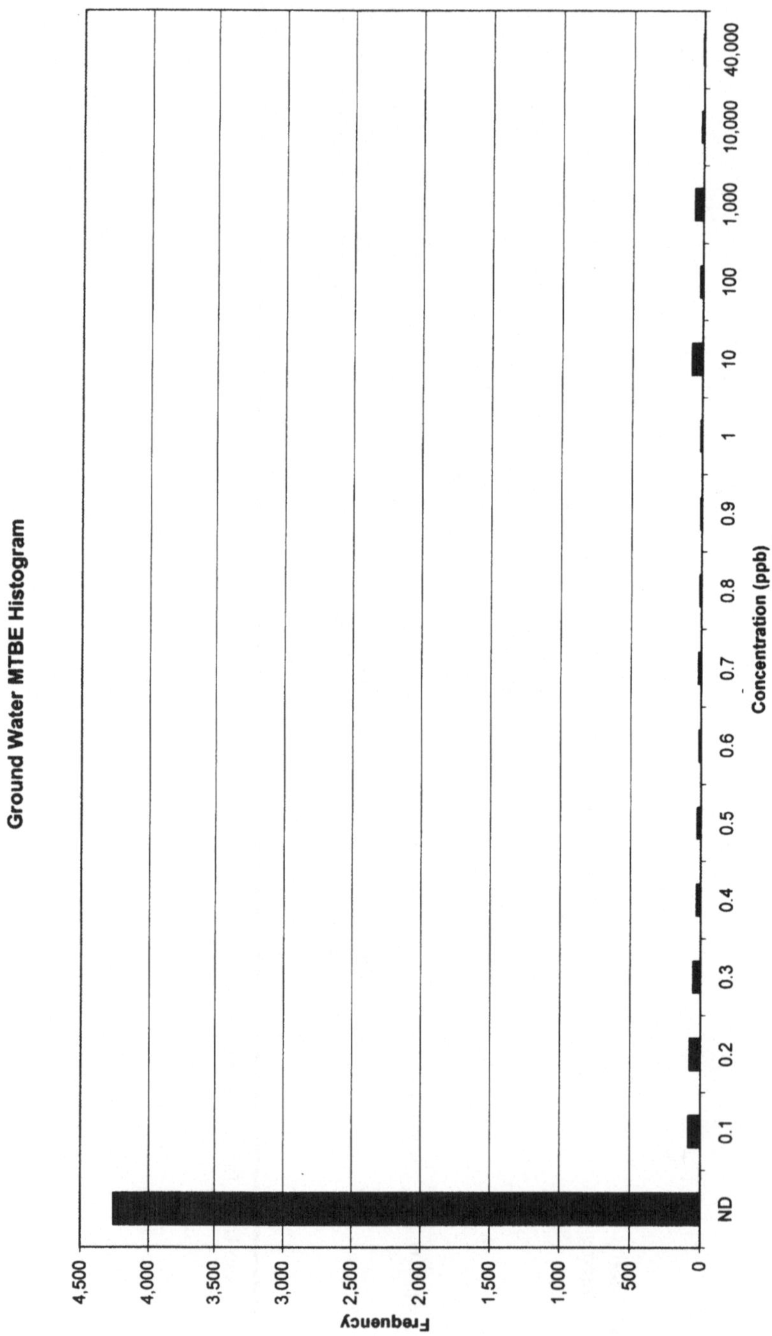

Figure A-1.   Ground Water MTBE Histogram

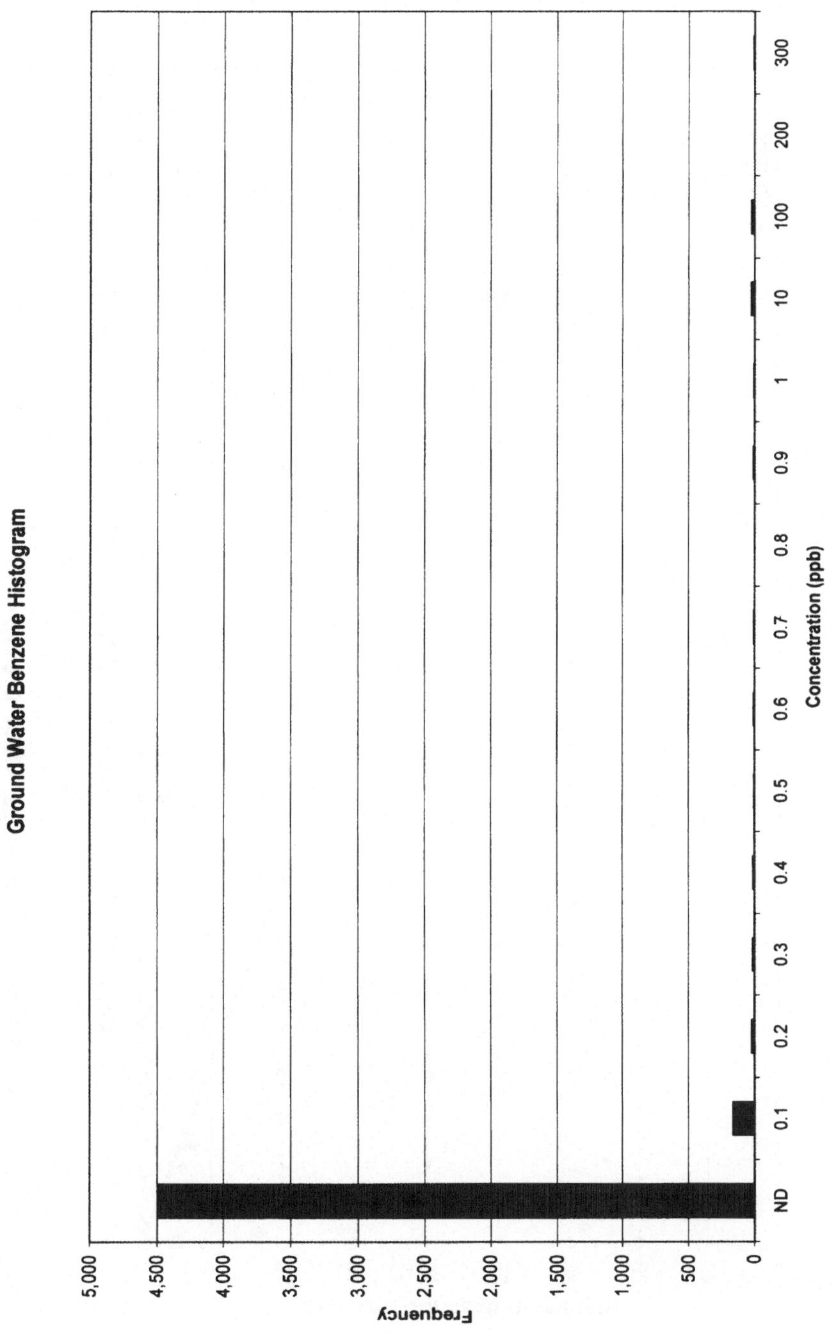

Figure A-2. Ground Water Benzene Histogram

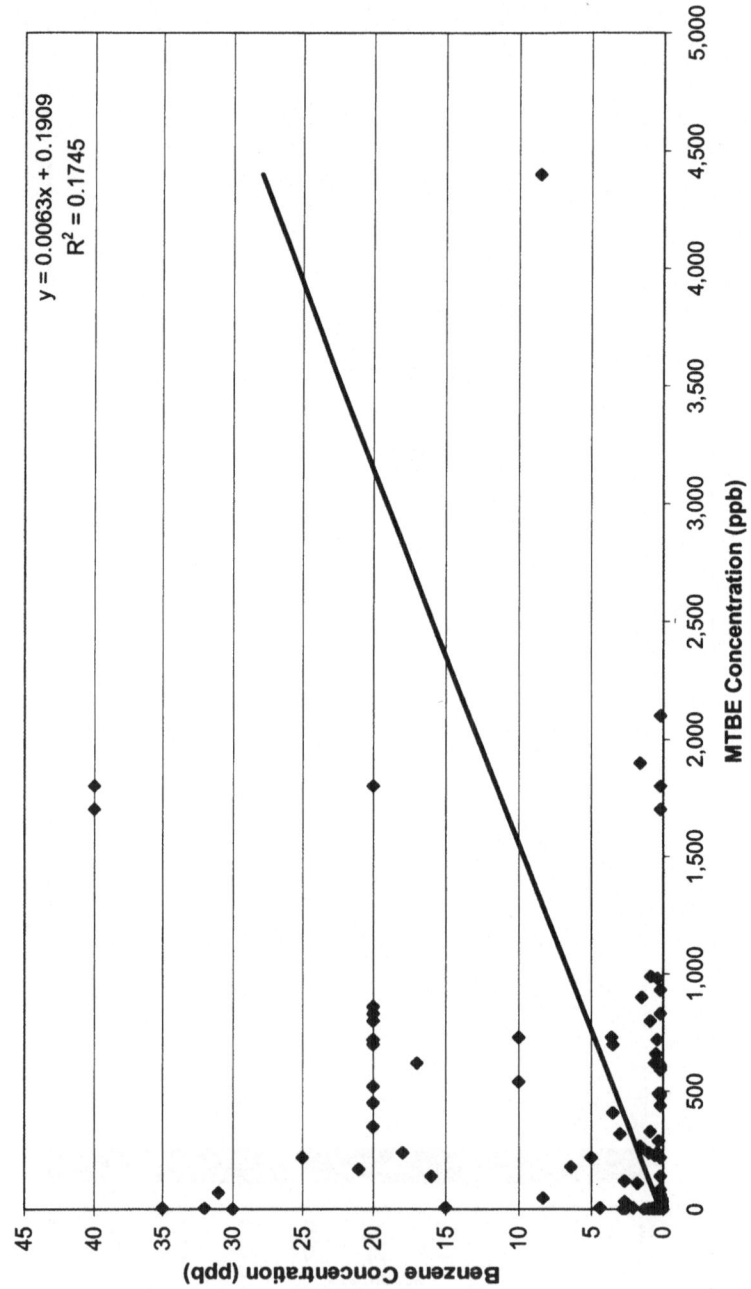

Figure A-3.   Ground Water — Benzene vs. MTBE Crossplot

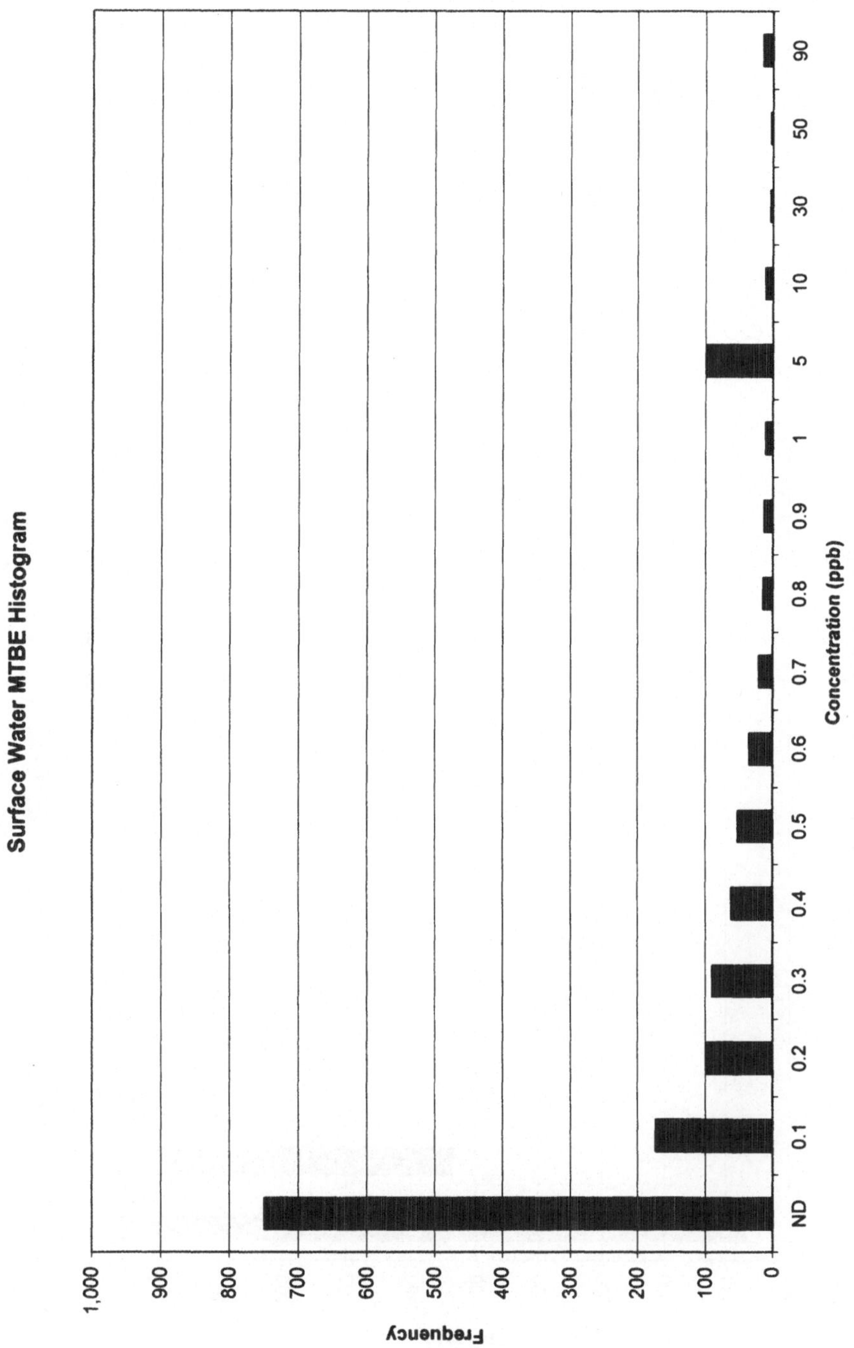

Surface Water MTBE Histogram

Figure A-4. **Surface Water MTBE Histogram**

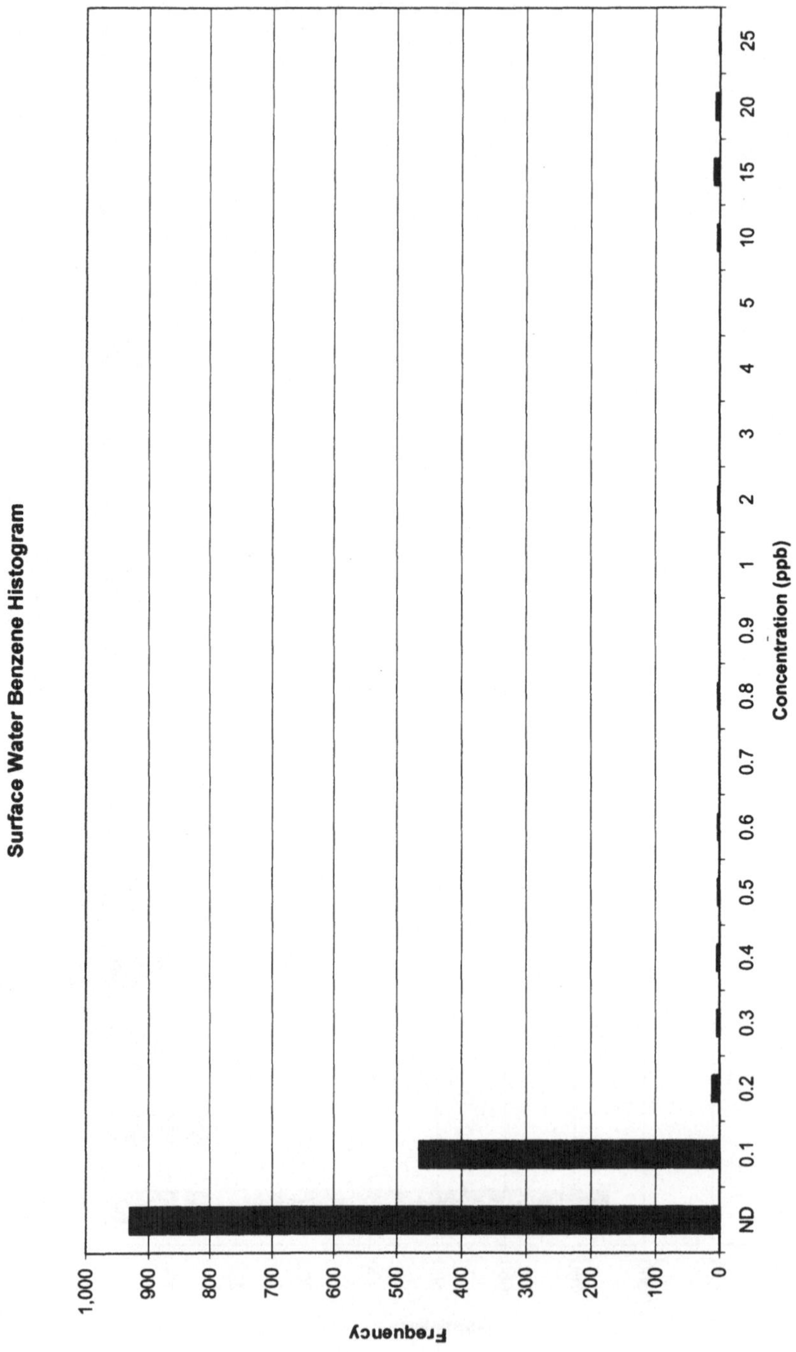

Figure A-5. Surface Water Benzene Histogram

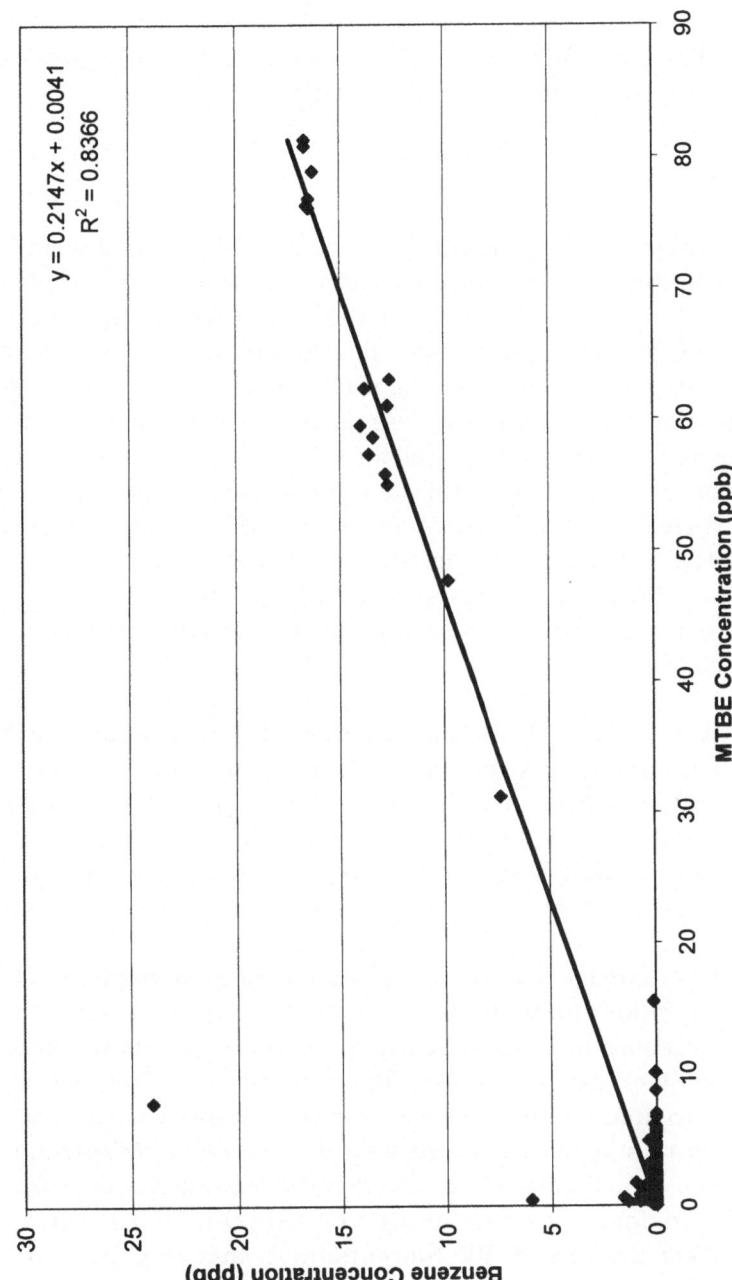

Figure A-6.   Surface Water — Benzene vs. MTBE Crossplot

Table A-1 summarizes the NAWQA study design and the data collection system. Basic sites are analyzed monthly or bimonthly for nutrients primarily with no regularly scheduled pesticide analyses; intensive sites are the same as basic sites except pesticides are analyzed regularly; and synoptic sites are sites sampled occasionally, as needed to fill "spatial gaps" within the basic and intensive sites.

Patterns

Concentrations. Clear concentration trends for MTBE are difficult to discern because of the lack of repeated sampling data from most sample locations. Figure A-7 suggests that, at those locations repeatedly sampled in Colorado, the trend is one of decreasing concentrations of MTBE over time. Assuming that the selection of sample locations eliminates any bias from year to year, and the ratio of detections to analyses is relatively constant from year to year (see Figure A-8), then averaging all the MTBE concentrations above detection limits for each year should indicate whether there is an overall trend of annually decreasing MTBE concentrations. Indeed, as Figure A-9 shows, the average MTBE concentration detected in ground water (excluding springs) declined after the initial two years of the NAWQA study.

Several explanations for the validity of this result being a potential artifact were considered:

- **Were average MTBE concentrations changing because of the changing number of analyses?** As depicted in Figure A-8, the number of analyses remained fairly constant from 1993 to 1998, varying from 500 to 800 analyses per year. The number of analyses reported in the database did decline sharply in 1999; however, significant declines in average MTBE ground water concentrations were evident from 1994 to 1998.
- **Were average MTBE concentrations changing because of the changing ratios of detections to analyses?** For ground water, the ratios of detections to analyses remained the same generally throughout the study to date (Figure A-8). The years 2000 and 2001 had no reported detections but the reported number of analyses is low. The low number of analyses combined with the decreasing MTBE concentration trend and the already low average concentration of detections may account for no detections being reported yet for those years.
- **Were average MTBE concentrations changing because detection limits were changing and skewing the average annual concentrations?** The effect of changing detection limits was evaluated by reviewing the detection limits by year for those samples where MTBE

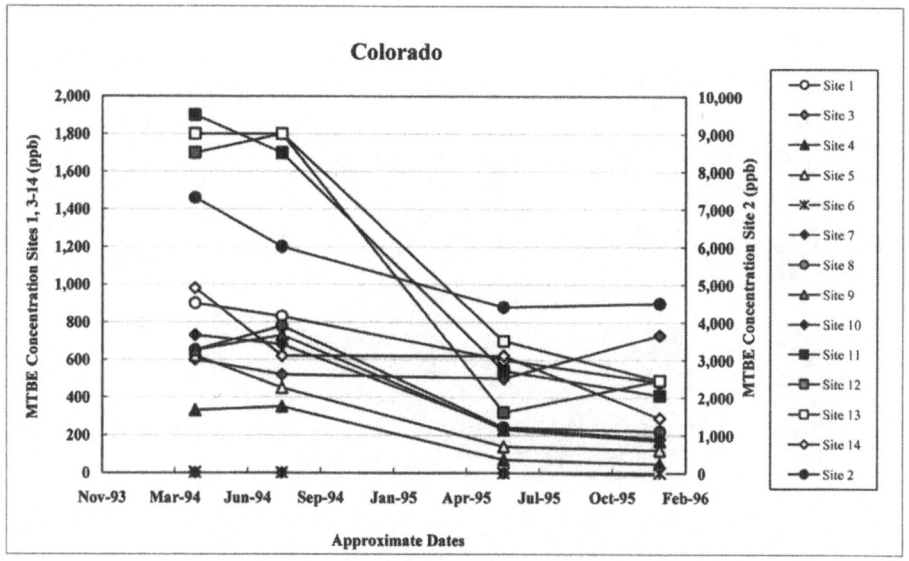

Figure A-7. **Plot of Temporal Sampling and Analysis Results for Multiple Sites in Colorado**
Connecticut and New Jersey also have locations with multiple temporal analyses. These data also generally indicated a decreasing concentration trend over the sampling period.

was not detected. It was assumed that MTBE detections in the same year were analyzed with essentially the same average detection limits. Table A-2 summarizes these results. There is no clear trend for lower (better) detection limits among those samples where MTBE was not detected. Average detection limits were slightly lower for 1996 and 1997 and were then slightly higher in the next three years. However, overall detection limits were about 0.2 µg/l. The small detection limit differences from year to year do not seem to be sufficient to account for the substantial decreases in average MTBE concentrations detected by year.

It is possible that the last few years — 2000, 2001, and possibly 1999 — in which only a few analyses and detections have yet been reported, should not be included in evaluating the ratios over time of detections to analyses. Statistically, fewer analyses and detections make the evaluation more sensitive to "outliers," either exceptionally high or low. Exceptionally high outliers are a larger problem since the upper practical concentration limit is proportionately greater relative to the lower limit, the detection limit. The bounds

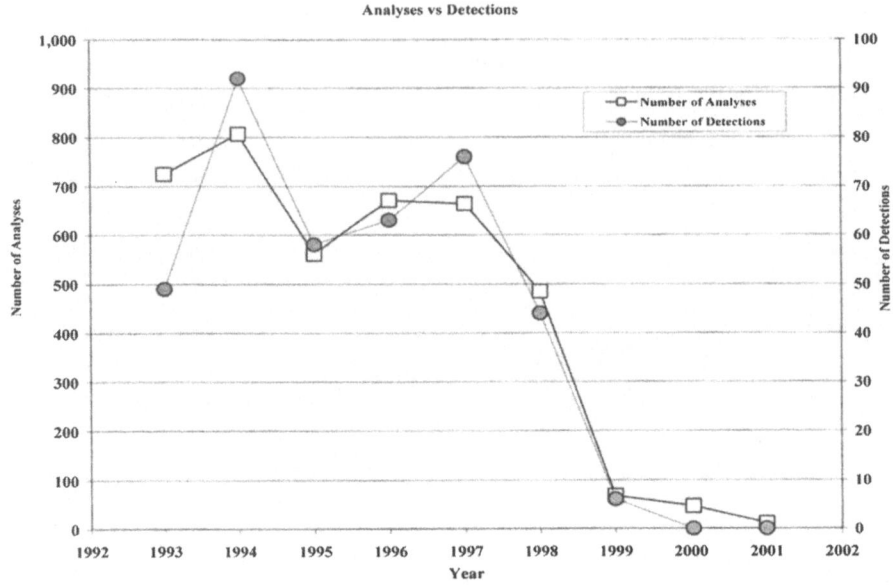

Figure A-8.   **Analyses vs. Detections**
              Graphic comparison of MTBE detections versus analyses by year. MTBE detections generally track the number of analyses by year. This suggests that the USGS has generally achieved the unbiased sampling distribution sought as part of the NAQWA process. Analyses and detections are plotted on axes an order-of-magnitude apart to facilitate visualizing how they track annually. This closeness of the data points illustrates the approximate overall 9-10% detections:analyses ratio.

would thus be further on the upside. As discussed above, there are no reported detections of MTBE in ground water samples in 2000 and 2001 as of the time of this writing. This is probably due to the slow reporting of data from more recent analyses. However, in year 2000 data, 46 analyses were reported with no MTBE detections. When compared to 67 analyses with 6 detections in 1999 for an approximately 9% detection ratio and the average ratio from 1993 through 1999 of about 9.6%, one would have expected about 4 detections in the 46 reported analyses of MTBE in ground water in 2000.

Temporal Changes.   It is difficult to separate temporal and concentration changes since they are usually dependent. The ability to evaluate changes in MTBE temporally appears limited by the NAWQA data set. Data can be recovered nationally or sorted in smaller units by basin, state, county, or year.

Average MtBE Concentrations by Year

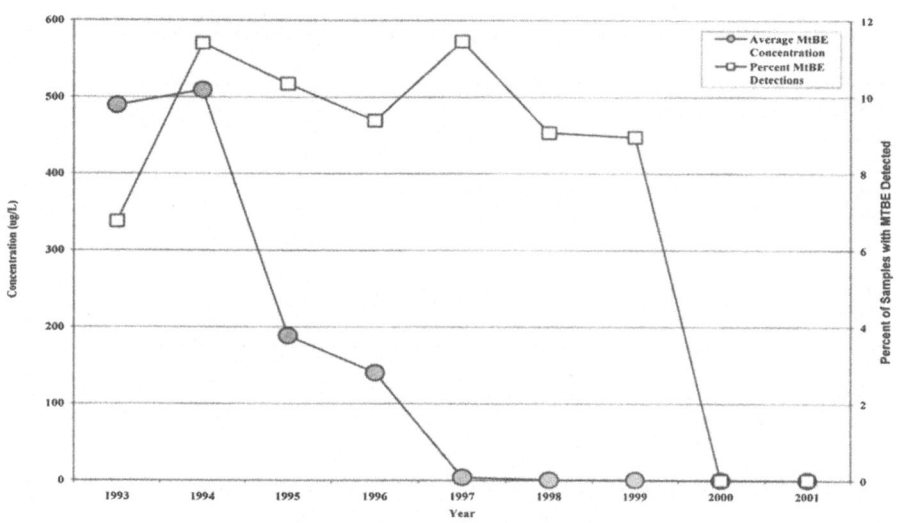

Figure A-9.   **Average MTBE Concentrations by Year**
Average concentration by year of all the ground water de-
tections of MTBE. Assumes that the data are unbiased year
by year based on the random selection of sample locations,
and thus indicative of the trend nationally with any com-
parable large set of randomly selected locations. Indicates
that the average detected concentration of MTBE across
the NAQWA ground water well program has gone down
since the second year of sampling (1994). As seen in Figure
3-8, the detections:analyses ratio has not been appreciably
different from 1993 to 1999 suggesting that decreasing
concentrations are not due to more or fewer analyses or
detections. In 2000 and 2001, the ratio changes but may be
due to the low number of sites reported to date for those
years. Does this reflect later changes in underground stor-
age systems, changing use patterns for MTBE, or possibly
gradual stimulation of MTBE degrading natural systems?

Data can be analyzed in smaller units ("the small picture"); however, there
are few repeated samples at the same locations relative to the total number of
analyses. Alternatively, the data may be analyzed nationwide for changes in
frequency of detections, or changes in concentrations as discussed below. Finally, there may be "the missing picture," which is the large number of sample locations with no MTBE detections and no repeat samples. Temporal data
are critical for evaluating all the parameters since they indicate trends in con-

Table A-2. **Average Detection Limit of MTBE Samples With No Detections, by Year**

	Average Detection Limit (ppb)	Count
1993	0.204	662
1994	0.201	715
1995	0.201	503
1996	0.113	608
1997	0.118	588
1998	0.210	441
1999	0.237	61
2000	0.255	46
2001	0.170	11

tamination or natural attenuation and can indicate whether programs to prevent ground water contamination, such as underground storage tank (UST) improvements, are having an effect that can be projected into the future. The data set contained 201 locations or stations with two or more samples from the same location on different days: 122 locations with 2 samples, 40 locations with 3 samples, 38 locations with 4 samples, and 1 location with 5 samples. The locations with 4 or 5 samples over time were evaluated to determine their patterns of change, *i.e.*, from detection to non-detection or non-detection to detection. Of these 39 sites, 18 had detections for all the samples, 14 had non-detects for all samples, 4 went from detections to non-detects over time, and none of the initial non-detect locations later showed detections. Two sites went from detection to non-detection and back to detection over the course of the sampling. The temporal trends of the 18 multi-sample locations with all samples as detections are discussed below.

The Small Picture.   MTBE data have been collected since at least 1993, but few analyses have been repeated at the same locations. Colorado, New Jersey, and Connecticut each have three or four samples at four or more locations where MTBE was detected. Figure A-7 graphically illustrates data for 13 locations in Colorado, the state with the most multiple temporal analyses with MTBE detections. Visually, there appears to be a temporal trend of decreasing concentrations over about one to one and one-half years for almost all of these ground water sampling locations. At least two locations appear to show an upward trend in the last analysis, and several of the low-concentration locations are difficult to categorize but have generally remained low but with detections. Overall, MTBE concentrations decreased temporally at these Colorado locations as well as in New Jersey and Connecticut, suggesting that this would be the general trend at most of the sites where MTBE was detected if temporal sampling were performed. This would fit with the national trend of MTBE concentrations discussed below. These state results also suggest that the MTBE inputs are not continuing over time at most of these locations. However, since these are single well locations, not plume-oriented well arrays, the significance of such decreases is unclear relative to the specific site and the behavior of any ground water plume at the site. It is not known if these temporal changes are due to active remediation, variation in the ground water flow direction, natural attenuation including biodegradation, simple ground water movement with dilution, removal of the source, or other causes. Since the NAWQA program attempts to randomly select well locations, it is unlikely that numerous wells will be sampled in a small enough area to adequately characterize an MTBE plume.

The Big Picture. A plot of the average detected concentrations by year across the U.S. (Figure A-9), suggests that MTBE concentrations are decreasing. The average detected concentration of MTBE across the NAWQA ground water sampling program has decreased dramatically since the second year of sampling (1994). Annually, MTBE detections and analyses track each other at approximately the same ratio in the ground water data set. As shown in Figure A-8, the detections to analyses ratios are approximately the same for fully reported years, despite considerable annual variation in the number of analyses. Thus, the rate of detections is not significantly increasing or decreasing with time. As seen in Figure A-8 and discussed previously, the detections to analyses ratios are similar from 1993 through 1999, and thus this is unlikely to be causing the declining average concentrations. In 2000 and 2001, the ratio changes, though it may be due to the low number of sites currently reported for those years. Does this detected concentration decline reflect upgraded USTs, changing use patterns for MTBE, or gradual stimulation of natural MTBE-degrading systems? Confirmation of whether these data are predictors of a larger trend across the country remains to be determined by a wider set of data from the same locations within the NAWQA program. One conclusion would be that MTBE is a diminishing problem from a concentration basis across the country.

Spatial Variation. If the detections to analyses ratios do not change annually across the nation, do they change on some other basis in the data set? With this question in mind, the variation in the detections to analyses ratio by basin was evaluated. It was discovered that the basins fall into two categories: those with high ratios of detections to analyses, and those with a considerably lower ratio. Figure A-10 shows the 36 different basins or areas of the study ranked by number of MTBE analyses and then plotted for number of detections of MTBE against that ranking. Three study areas, the South Platte River Basin, the Connecticut/Housatonic/Thames River Basins (Unit MA-

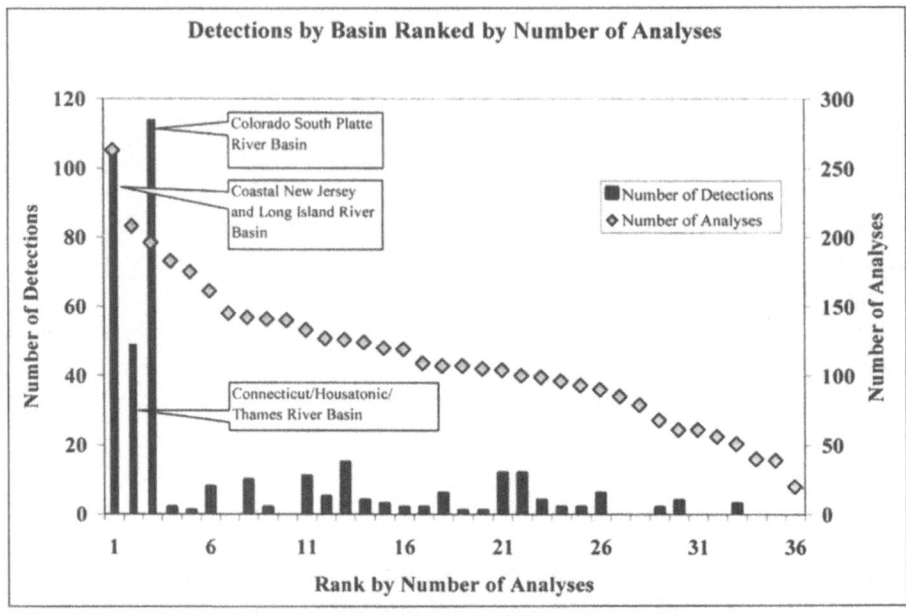

Figure A-10.   **Detections by Basin Ranked by Number of Analyses**
MTBE detections by basins ranked by number of analyses.
This illustrates that the study basins are divided into two
groups: (1) three basins with high numbers of detections,
and (2) the remaining basins with low numbers of detec-
tions. In part this might be attributed to more analyses in
these basins, and these three basins did have more total
analyses performed. However, the grey diamonds, show-
ing the number of analyses in each basin, illustrate that
many other basins also had substantial numbers of analy-
ses, but with appreciably fewer detections. This suggests
that these three basins have substantially higher detec-
tion ratios and thus releases of MTBE. These basins are
identified in the call outs in the graph.

100), and the Coastal New Jersey and Long Island River Basin area are in the
high detections group. All 33 other basins appear to fall into the lower detections group. The three basins with many more detections also have the highest number of analyses. However, as can be seen in Figure A-10 (shaded diamonds), many of the other basins also have nearly as many analyses but
considerably fewer detections. The high-detection basins average a 40.3%
MTBE detections-to-analyses ratio in ground water, while the 33 other basins
or study areas average a 3.6% detections-to-analyses ratio. The three high detection basins combine for 268 detections while the 33 others combine for 120

detections. This analysis suggests that the ground water data set is heavily skewed by the results of these three basins. That is, ignoring the results from the three basins that account for 67% of the total ground water detections, the remaining data indicates a widespread pattern of very few MTBE detections in the remaining basins across the country. This is not to say that the skew is artificial. Rather, the NAWQA MTBE data set indicates that three areas or basins are disproportionately high in MTBE detections relative to the remainder of the country, which has a generally low frequency of MTBE detections. The basins overlap with repeat sample states, but assuming that the selection of resample locations was equally unbiased nationally in all basins or areas, the resample data presumably are not design artifacts in the overall results but rather reflect actual differences from basin to basin.

Why would the three basins described above have a significantly higher detection to analyses ratio than the rest of the basins? A simple map inspection suggests that two of the three units are small in area relative to the number of detections. For example, the Coastal Long Island and New Jersey area is about 15,500 square kilometers (6,000 square miles) in area, but has a high population density. The South Platte River unit is approximately 63,700 square kilometers (24,600 square miles), about four times larger than the Coastal Long Island and New Jersey unit. The Connecticut/Housatonic/ Thames Rivers study area covers a relatively small area compared to other units but both of the high detections, small areal units are generally more densely populated than many other basins. This suggests that the high rate of detection of MTBE could possibly be related to the population density and the density of gasoline service stations. The high detection rates could also relate to UST policies and practices in those basins, length of time or concentrations of MTBE in use, soil and ground water settings in the basins, and many other possible factors.

Conclusions. Relative to MTBE, the following are some conclusions on the data in the NAWQA Data Warehouse as of October 17, 2001.

- The overall ratio of MTBE detections to analyses in ground water has remained relatively constant from 1993 through 1999. The two most recent years, 2000 and 2001, show no detections but have limited reported analyses. The time needed to validate data before placing it into the database may have resulted in incomplete data sets for those years. It will be interesting to see if the relatively constant ratio of about 9.6% detections to analyses continues through the next decade or decreases in response to changes in MTBE use, improved fuel management systems, and similar factors.
- Three basins stand apart by having the largest number of detections without a correspondingly higher numbers of analyses. These three

basins account for about 67% of all the MTBE detections but only 20% of the analyses.

- During the study period, average annual MTBE concentrations in ground water have gone down over the years while the ratio of detections has remained generally constant. This suggests that MTBE concentrations in ground water are diminishing nationwide over time. This parallels on a large scale the results seen in those states with multiple analyses at sample locations with MTBE detections.
- Colorado, New Jersey, and Connecticut, states with three or more temporal MTBE samples from the same locations, show decreasing MTBE concentrations at almost all of the locations. This suggests, as above, that the MTBE concentrations may be declining across the country. Additional data from the same locations may substantiate or dispute this conclusion in the future.

Limitations.   Although NAWQA is an excellent concept, it has some limitations in its usefulness. Discussion of the limitations follows.

Representativeness.   Because NAWQA specifically excludes samples from monitoring or observation wells that were installed to detect a known or suspected contaminant (Lapham *et al.*, 1995), such as wells found at leaking underground storage tank (LUST), Resource Conservation and Recovery Act (RCRA), and Superfund sites, the study serves three purposes extremely well. It provides a picture of national background water quality; identifies areas of previously unknown widespread non-point source contamination; and identifies previously unknown contaminated sites. However, by only sampling in areas with no previously known ground water contamination, it biases the results towards lower average MTBE concentrations and detection frequencies. It is probable that in the second phase of the NAWQA program, when certain sites are resampled, the withdrawal of candidate sample points in hydrocarbon release areas, as these areas are identified, could lead to a decrease in average MTBE concentrations. Additionally, wells located near roads and highways are avoided due to the use of herbicides and road-salt applications along roadsides (Lapham *et al.*, 1995). In eliminating wells near roads and highways, the study has further reduced the likelihood of testing in areas where MTBE and other VOCs are likely to be detected. As the NAWQA program does not sample wells where MTBE or other gasoline compounds are present or expected to be present, it is hard to conclude that the NAWQA MTBE results are truly representative of the nationwide occurrence of MTBE.

Remediation Perspectives.   The data presented in the NAWQA program appear to have little to support site-specific or national remediation perspectives except in possibly predicting whether MTBE contamination will be a diminishing or continuing long-term remediation driver. Because of the needs of the program and the massive study area versus costs, one can expect that the data are best suited for predicting the possible future needs for remediation (or gasoline release prevention) in basins and areas with the greatest number of detections.

Temporal Trends. With only a modest number of repeated temporal samplings at specific sample locations, it is difficult to draw conclusions about temporal changes and trends for contaminants on a smaller, local scale. Some states have multiple repeat samples, while others do not, leaving it currently unclear whether the site-specific declines generally seen at such sites in a few states should be extrapolated to all the states. This prevents evaluation of the frequencies of new releases and evaluation of recontamination at locations where MTBE may have declined, disappeared, or not been present initially. In addition, the three basins with high detection frequencies generally overlap those states with the greatest number of specific locations with repeat sampling. Repeat sampling in limited locations, rather than in all basins or areas, may skew the evaluation of the overall data set. An unbiased research design, including the selection of locations for repeat samples, should eliminate these issues. However, it is not clear what the basis was for choosing the particular locations that were repeatedly sampled.

Program Time-Scale. The extended time for carrying out the study is probably a necessity for cost and personnel issues but clearly is a limiting factor for identifying problems and monitoring the outcome of any remediation actions. The time period required to identify a "new contaminant," incorporate it into the program, achieve adequate coverage, and acquire time sequenced sampling to assess the impact of the contaminant would seem to be long. A response for a "new contaminant" would need a focused, additional study to provide the short-term data needed to determine the extent of any problem. The attempt to initiate the study in so many areas during the initial 10 years and include additional areas in the next 10 years might contribute to the difficulty in determining trends. This has led to "snapshots" across the country, but not to definitive, consistent temporal evaluations at locations within all the states or basins being studied. Consequently, it is uncertain whether chemicals such as MTBE, when detected in a particular year at high frequency or concentration, pose a continuing substantial problem or are transient at most locations. It can be inferred from new data at new locations that there is a continuing problem, or it can be extrapolated that the results from repeatedly sampled locations in a few states or basins are predictive of all the other areas. However, such conclusions cannot be definitively reached without revisiting a geographical range of past locations over a period of time that can show whether the trends are down, up, or stable, nationally or by study area. This is particularly critical since the results of the NAWQA program indicate a general trend nationally of lower MTBE concentrations detected over time. Future decisions on UST programs, gasoline composition, drinking water treatment needs, and similar programs require an understanding of whether concentrations of chemicals such as MTBE are declining due to measures taken to date or due to natural attenuation.

Summary.   Despite the limitations discussed above, the NAWQA program appears to be a useful and needed program to evaluate background water quality across the U.S. It is a very ambitious program that of necessity appears to have made some compromises. From the perspective of those interested in specific contaminants, including MTBE, additional focused studies are needed to gain a better understanding of the extent, fate, and transport of contaminants in ground water and surface water across the nation.

MTBE OCCURRENCE IN THE UNITED STATES

James W. McKinley,   Applied Hydrology Associates, Inc.

National MTBE Survey.   In addition to continuing its NAWQA program, the USGS is also directly involved in other nationwide MTBE studies. The USGS has recently conducted a National MTBE Survey for the American Waterworks Association (AWWA) Research Foundation in cooperation with a team of researchers from the Metropolitan Water District of Southern California and the Oregon Graduate Institute. This study looked at the occurrence of MTBE and other VOCs in reservoirs, rivers, and wells that supply CWSs. The initial findings were reported at the Annual Conference of the AWWA on June 20, 2001, in the fourth and final year of the survey. The study tested 954 randomly selected CWSs, consisting of 579 wells, 171 rivers, and 204 reservoirs, within all 50 states and Puerto Rico (Hirsch, 2001). The study did not include water sources that had been taken off-line due to MTBE detections, and most of the sites in the survey had not previously been tested for MTBE (Clawges *et al.*, 2000).

Preliminary results indicated that MTBE was the second most commonly detected VOC after chloroform, occurring in 9% of all sources sampled at a reporting concentration of 0.2 µg/l. However, MTBE concentrations can reach 1 to 2 µg/l from deposition by precipitation, known as atmospheric washout (Zogorski *et al.*, 1997). The reporting concentrations for the other analytes were not included. MTBE was detected in 14% of surface water sources and 5% of ground water sources, primarily at concentrations below the U.S. Environmental Protection Agency (USEPA) advisory. (The science behind the advisory concentration was outlined in a fact sheet released by the USEPA in

December 1997 [USEPA, 1997]. The fact sheet states that there are no data on human health effects from water containing MTBE. However, concentrations of MTBE below 20 to 40 μg/l would most likely avert any taste or odor effects in drinking water, and would also protect against any potential human health threats. Many states have adopted the lower end of the advisory; when mentioned in this appendix, "USEPA advisory" corresponds to an MTBE concentration of 20 μg/l.)

The frequency of MTBE detection increased with the size of the CWS. MTBE was detected in 4% of systems serving 10,000 people or less, and 15% of systems serving 50,000 people or more. The study concluded that even though MTBE is fairly frequently detected in ground, surface, and drinking water, the concentrations are almost always below the USEPA advisory (Hirsch, 2001).

Northeastern and Mid-Atlantic States. The USGS is also researching concentrations of VOCs, including MTBE, regionally. This is because, even though nationwide detection frequencies and concentrations might be low, there could be certain areas of the U.S. where MTBE is detected at frequencies and concentrations that are significantly higher than national background values. In order to focus on a particular region, the USGS collaborated with the USEPA's Office of Ground Water and Drinking Water to determine the occurrence of MTBE and other VOCs in CWSs for twelve states in the Northeast and Mid-Atlantic regions of the U.S. (Clawges *et al.*, 2000). The states evaluated include Connecticut, Maine, Maryland, Massachusetts, New Hampshire, New Jersey, New York, Rhode Island, Vermont, and Virginia; Delaware and Pennsylvania did not have MTBE data available (Hirsch, 2001). Some of these states were designated reformulated gasoline (RFG) areas and had a longer history of MTBE use than most other states.

Under the Clean Air Act Amendments of 1990, oxygenated gasoline was mandated in areas where concentrations of ozone and carbon monoxide in the atmosphere exceeded air-quality standards. RFG was introduced to help non-attainment states comply with the oxygenate concentrations required by the USEPA; MTBE was the primary oxygenate used in RFG in the Northeast and Mid-Atlantic states (Clawges *et al.*, 2000). As a result, one would expect to see more widespread occurrence of MTBE in this area of the country than in other areas. More importantly, approximately 85 million people nationwide live in RFG areas, and their drinking water supply is split almost equally between surface and ground water sources (Hirsch, 2001). Therefore, it is important to study the occurrence of MTBE in both surface and subsurface water supplies.

From the 10,479 CWSs in the 12-state area that provided VOC data (64 analytes) between 1993 and 1998, the USGS and USEPA randomly selected 2,110 as a representative sample for statistical analysis. This sample of CWSs provides water to about 16% of the total population of the 12-state area. Of the sample, only 1,194 CWSs performed analyses for MTBE, reducing the data set further. Approximately 9% of CWSs had detectable concentrations of MTBE, but less than 1% of tests showed MTBE concentrations at or above the USEPA advisory (Grady and Casey, 1999). Of the CWSs tested, only 2% contained MTBE concentrations equal to or greater than 5 µg/l. The results also indicated that MTBE was detected more frequently in RFG areas (15%) than non-RFG areas (3%) (Clawges *et al.*, 2000). (Note that, due to blending and transport economics, RFG may be used in non-RFG areas.) The Northeastern and Mid-Atlantic States study also included a table summarizing MTBE occurrence in drinking water, by state, from 1978 to 1998. A histogram of this table is included as Figure A-11. Additionally, a histogram of the VOCs with the 10 highest detections to analyses ratios from 1993 to 1998 in the study area is included as Figure A-12 (Grady and Casey, 1999). From the states that provided data, it was extrapolated that a further 180 CWSs in the Northeast and Mid-Atlantic States could contain MTBE concentrations at or above 5 µg/l, and 80 CWSs could exceed 20 µg/l MTBE (Clawges *et al.*, 2000).

Northeast States for Coordinated Air Use Management (NESCAUM). On November 9, 1998, as a follow up to the Northeastern and Mid-Atlantic States

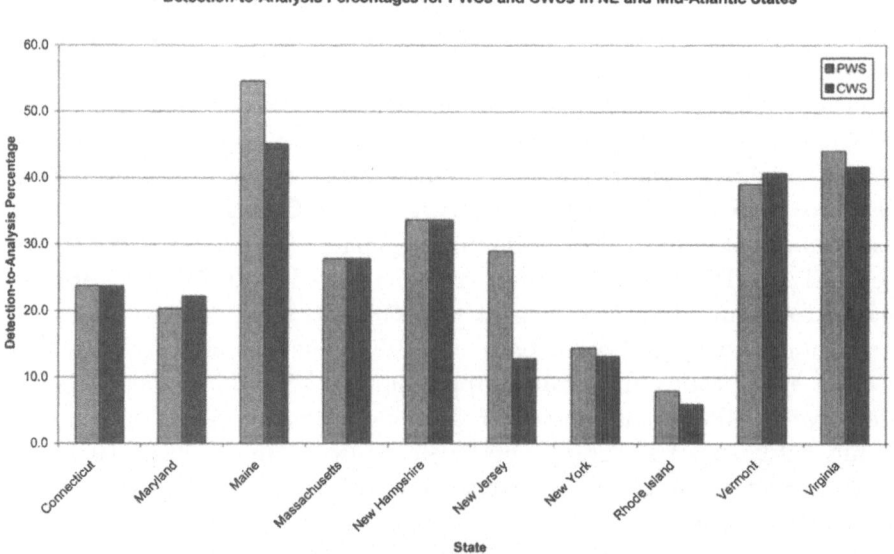

Figure A-11.   **Detection-to-Analysis Percentages for PWSs and CWSs in NE and Mid-Atlantic States**

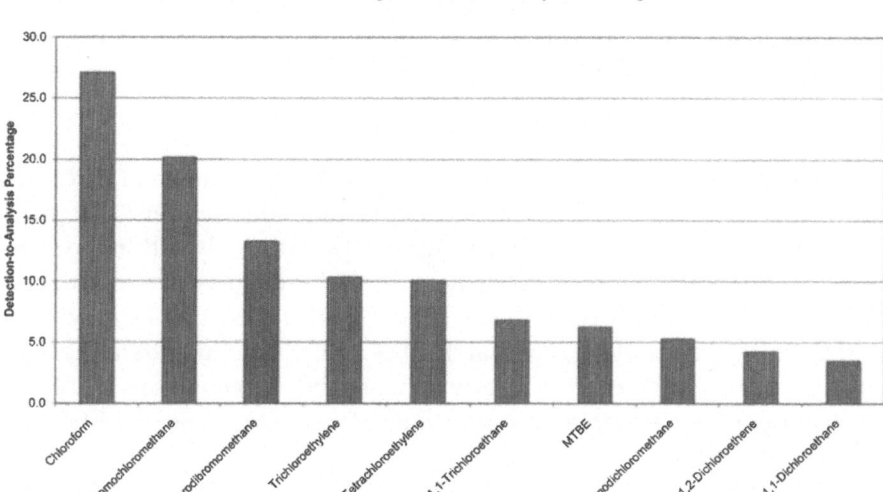

Figure A-12. **Chemicals with the Highest Detection-to-Analysis Percentages**

USGS report, New Hampshire Governor Jeanne Shaheen asked NESCAUM to assess what steps the northeastern states should take to maximize air quality without sacrificing water quality (NESCAUM, 1999). In 1999, NESCAUM released a set of technical reports that outlined the occurrence of MTBE in CWSs in its eight member states, which include Connecticut, Maine, Massachusetts, New Hampshire, New Jersey, New York, Rhode Island, and Vermont. Where MTBE was detected, most concentrations were below the USEPA advisory (Grady and Casey, 1999). The NESCAUM report summarized that at concentrations typically detected in the northeast states (less than 20 µg/l), MTBE does not pose a health problem. However, MTBE could become a threat in the few areas where concentrations exceed 20 µg/l. Since MTBE was one of the most frequently detected VOCs and was cited in the NESCAUM report as being resistant to biodegradation, it was determined that LUSTs could pose a serious health risk. At a minimum reporting concentration of 0.2 µg/l, BTEX compounds were detected at only 12% of the sites that had positive test results for MTBE and infrequently overall (NESCAUM, 1999); MTBE was therefore determined to be the larger threat to human health and the environment. The following section is a state-by-state discussion of the findings of the NESCAUM report as well as individual state studies or AWWA data (Gullick and LeChevallier, 2000) where available. In the AWWA report, surface water samples included raw water intakes and samples taken as part of watershed monitoring programs, while ground water samples included mostly drinking water supply wells, with some monitoring and exploratory wells. For individual states, the AWWA report did not identify whether the wells sampled were water supply, monitoring, or exploratory.

Connecticut. The Connecticut Department of Environmental Protection (CTDEP) is responsible for collecting monitoring data for over 1,000 private wells each year; however, the total number of private wells within the state is currently unknown. More specifically, the CTDEP records all occurrences of MTBE in private wells over the state action level. This action level was lowered from 100 µg/l to 70 µg/l in March 1999. In 1998, before the standard was changed, the CTDEP reported 22 private wells that exceeded the MTBE standard; most of these elevated concentrations were related to nearby LUST sites (NESCAUM, 1999).

The Connecticut Department of Public Health (CTDPH) tested approximately 700 PWSs between 1997 and 1998. Of the PWSs tested in 1997, 30 had positive test results for MTBE, with 4 testing above 10 µg/l, and 1 well testing above the former action limit of 100 µg/l. At this well, MTBE was detected at 210 µg/l, while the benzene concentration was only 1.4 µg/l, below the USEPA drinking water standard of 5 µg/l. Of the PWSs tested in 1998, 45 PWSs had positive test results for MTBE, with 13 having concentrations above 10 µg/l, and 4 having concentrations above 100 µg/l. The highest concentration measured was approximately 17,000 µg/l. The increase in MTBE detections between 1997 and 1990 was thought to have arisen from the increased testing of non-transient non-community water systems (NTNCWSs) by the CTDPH in 1998 (NESCAUM, 1999).

Beginning in January 1997, AWWA started conducting MTBE surveys in various states throughout the U.S. As part of this effort, between 1997 and 1998, water samples were collected from 2 surface water locations and 10 wells across the southwestern, southeastern, and central parts of Connecticut. The wells ranged from 15 to 140 meters (50 to 450 feet) deep; 5 had positive test results for MTBE at low concentrations ranging from 0.5 to 0.8 µg/l. The surface water sources, which included the Mianus River, Palmer Reservoir, Putnam Reservoir, and Dean's Mill Reservoir, all tested negative for MTBE (Gullick and LeChevallier, 2000).

Maine. Maine has had a history of well contamination from small spills of gasoline containing MTBE. In the spring of 1998, three spills occurred that allowed the Maine Department of Environmental Protection to observe how small gasoline releases can result in relatively large, albeit temporary, impacts. Most research performed before 1998 involved large gasoline releases from LUST sites. In May 1998, a water supply well for Whitfield Elementary School had positive test results for MTBE, with a maximum concentration of

900 µg/l. The source of MTBE was identified as a gasoline leak from a car parked less than 30 meters (100 feet) away. Two PWS wells tapping the North Windham aquifer contained MTBE at concentrations of 2 and 4.5 µg/l respectively, reportedly due to a release of probably less than 110 liters (30 gallons) of gasoline combined with several UST overfills from a service station 210 to 330 meters (700 to 1,100 feet) away. Even though the concentrations were below the USEPA advisory, ground water samples collected in the vicinity of the USTs had concentrations of up to 7,140 µg/l for MTBE and 60 µg/l for BTEX (Hunter *et al.*, 1999). The third investigated release occurred at the site of an automobile accident, where 26 to 45 liters (7 to 12 gallons) of gasoline were spilled (Hunter *et al.*, 1999; NESCAUM, 1999). The accident reportedly resulted in MTBE detections at 24 nearby wells, 11 above the USEPA advisory, with a maximum MTBE concentration of 6,500 µg/l. After the spills, data were collected for 10 months at each site. The analyzed data indicated that over the 10-month sampling period, MTBE concentrations decreased by approximately 80% at nearly all monitored wells (Hunter *et al.*, 1999).

Concerned about these small gasoline releases, Maine Governor Angus King directed state health and environmental agencies to conduct a comprehensive MTBE study (Maine, 1998). In 1998, Maine completed a survey of the occurrence of MTBE and other gasoline components in drinking water across the state. In the course of the survey, 951 randomly selected household wells, springs, and lakes, and 793 of the 830 regulated non-transient PWSs were tested. Of the 951 household supplies, 15.8% had positive test results for MTBE, including 1.1% testing above Maine's drinking water standard of 35 µg/l. The State extrapolated the results to the estimated total of 275,000 households statewide with private water supplies, projecting that possibly 1,400 to 5,200 private wells in Maine contain MTBE at concentrations above the State's drinking water standard. Of the sampled household wells, 92.3% were either non-detects for MTBE or had concentrations below 1 µg/l, while 6.6% tested between 1 and 35 µg/l. Other gasoline components were detected infrequently relative to MTBE, and all BTEX concentrations were below drinking water standards. Toluene was the second most commonly detected gasoline component, occurring in 2.1% of all wells tested, while benzene, ethylbenzene, and xylenes were detected in less than 1% of the samples. MTBE detection frequencies correlated with areas of high RFG use, as well as with areas with high population densities (Maine, 1998).

Of the 793 PWSs included in the survey, 16% had positive test results for MTBE, although no samples had MTBE concentrations above the Maine standard of 35 µg/l. Consistent with the household well results, most PWSs (93.9%) were either non-detect for MTBE or had concentrations below 1 µg/l, while 6.1% had concentrations between 1 and 35 µg/l. BTEX components were detected relatively infrequently, with the exception of toluene (detected at 13.1% of PWSs). MTBE detection frequencies correlated with areas of high

RFG use. However, population density did not appear to play as strong a role as it did with the household supply results (Maine, 1998).

Massachusetts. As of the release of the NESCAUM report, Massachusetts did not require the testing and reporting of MTBE concentrations in PWSs. The Massachusetts Department of Environmental Protection (MADEP) has kept a database on MTBE concentrations since 1993, but all the reporting has been voluntary. The approximately 1,600 PWSs in Massachusetts service seven million out of the state's population of nine million people. The MADEP does not maintain records pertaining to private water supply wells.

Of the 1,600 PWSs, 9 reported positive MTBE detections each year from 1993 to 1995. The average concentrations ranged from 1.2 to 2 µg/l, and the maximum concentrations ranged from 68 to 80 µg/l. The number of detections increased from 22 in 1996 to 31 by 1998. However, the highest reported average and maximum concentrations occurred in 1997, at 2 and 690 µg/l respectively. In 1999, the average and maximum concentrations dropped to 1.6 and 230 µg/l respectively, even though the number of detections rose to 33. Since the reporting was voluntary, and MADEP does not know how many supplies were tested but not reported, no accurate conclusions can be drawn from the data (NESCAUM, 1999).

Beginning in January 1991, AWWA collected 38 samples from 13 ground water sources in Massachusetts, along with 1 surface water sample. MTBE was detected at low concentrations in 5 of the ground water sources. Most of the detections could be linked to nearby potential sources, such as gasoline leaks or spills, while none of the detections exceeded 6 µg/l (Gullick and LeChevallier, 2000).

New Hampshire. In 1993, New Hampshire passed a law requiring the testing of approximately 1,125 PWSs for VOCs. By March 1999, New Hampshire reported MTBE detections in 195 PWSs; 24 of these are no longer active. Of the 195 detections, 59 were at concentrations below 0.5 µg/l, 106 were at concentrations between 0.5 and 5 µg/l (96 remain active), 20 were at concentrations between 5 and 15 µg/l (19 remain active), 1 was at a concentration between 15 and 20 µg/l (showing a decreasing trend), 6 were at concentrations between 20 and 40 µg/l (3 remain active), and 3 were at concentrations above 40 µg/l (all are now inactive). Most of these detections were not deemed to be related to a point source release; the use of recreational boats and watercraft on the tested reservoirs could be the main source of MTBE (NESCAUM, 1999).

AWWA collected 16 samples from 11 water supply wells in New Hampshire; no surface water samples were collected. MTBE was detected in only 1 of the samples at a low concentration of 1.4 µg/l (Gullick and LeChevallier, 2000).

New Jersey. In 1996, the New Jersey Drinking Water Quality Institute developed a drinking water standard for MTBE of 70 µg/l and required that all water supplies be routinely monitored for MTBE. As a result, the New Jersey Bureau of Safe Drinking Water has been monitoring for MTBE since 1997. Over the last four years, 400 of 614 public CWSs and 400 of 1,100 NTNCWSs have submitted MTBE results. From July 1997 to September 1998, 85% of CWSs and 84% of NTNCWSs had results below the detection limit for MTBE. Analytical results for the remaining 15% of CWSs and 16% of NTNCWSs ranged between 0.5 and 20 µg/l MTBE. No CWSs tested above 20 µg/l, while only 1 NTNCWS had results between 20 and 70 µg/l, and 1 more NTNCWS had results above 70 µg/l (NESCAUM, 1999). The report concluded that MTBE currently is not a public health concern in New Jersey drinking water supplies.

In 2000, the New Jersey Department of Environmental Protection (NJDEP) released a report on the history of LUST sites and MTBE concentrations within the state. In the northeast, MTBE is by far the most prevalent oxygenate used in RFG, and the NJDEP extrapolated that as much as 187 million gallons of MTBE may be stored in USTs across the state. Of the 2,400 LUST cases within New Jersey, 80% of those had ground water samples with MTBE concentrations exceeding the state standard of 70 µg/l (NJDEP, 2000).

The New Jersey Geological Survey has analyzed data from monitoring wells, PWS wells, and private domestic and non-domestic wells across the state as part of its Ambient Ground Water Quality Network. Since 1995, of the 43 wells tested, 11 (26%) contained detectable MTBE, at concentrations ranging from 0.2 µg/l to 5.4 µg/l. In 1995, 5 of 9 wells had positive test results, with a maximum concentration of 2.3 µg/l. In 1996, 2 of 15 wells in the Barnegut Bay area had positive test results, with a maximum concentration of 1.7 µg/l. In 1997, 4 of 19 wells in the Rancocas area had positive test results, with a maximum concentration of 5.4 µg/l. All detections were well below the USEPA advisory and the New Jersey drinking water standard (NJDEP, 2000).

AWWA concluded that average MTBE concentrations and frequencies of detection were generally higher in New Jersey than in other areas that they had analyzed. RFG use is required year round in New Jersey, and many of the sites that AWWA tested were located in close proximity to LUST sites. Both of these factors may have contributed to the high detection frequencies. AWWA collected 683 samples from 92 ground water sources and 43 samples from 16 surface water sources. Between 1997 and 1998, 13% of the ground water sources and 25% of the surface water sources had one or more samples with positive but low test results for MTBE. In ground water, MTBE concentrations reached a maximum of 11.6 µg/l, and only three sites had concentrations above 1.5 µg/l. The maximum surface water MTBE concentration was 3.3 µg/l; this sample was taken from a highly populated area with many potential areas of non-point source contamination (Gullick and LeChevallier, 2000).

New York.   Before July 1998, New York did not require monitoring or report-ing for MTBE in drinking water supplies. The New York State Department of Health (NYSDOH) now requires testing for supplies servicing over 10,000 people. In addition to the NYSDOH's testing, New York's Wadsworth Labo-ratories conduct testing on small PWSs across the state. In 1997, 17 of 404 small supplies had positive test results for MTBE, with only 1 supply having results over 10 µg/l, at a concentration of 17 µg/l. In 1998, 16 of 381 supplies had positive test results; only 3 had results over 10 µg/l, with a maximum concentration of 33 µg/l (NESCAUM, 1999).

On February 8, 2000, the New York State Department of Environmental Conservation (NYSDEC) released a survey of MTBE concentrations at gaso-line remediation sites across the state. Each year, over 15,000 reports of po-tential gasoline spills and leaks are submitted to NYSDEC. As of the time of the 2000 survey, files remained open on 5,262 sites reported to have the po-tential for elevated MTBE concentrations between July 1978 and September 1998. Of these reported sites, 1,706 were concluded to have MTBE impacts to surrounding ground water. In terms of MTBE concentration, the phrase "im-pact" is not defined. Of the 1,706 impacted sites, 82% had MTBE detections at concentrations above the state action level of 10 µg/l, resulting in impacts to 866 private wells and 47 PWS wells. Since these were all sites that were still being actively remediated, detection frequencies above the state standard were expected to be high. About 70% of these affected supplies were attrib-uted to LUSTs or faulty piping. However, 20% of the affected supplies oc-curred from unknown sources, most likely non-point sources such as surface releases by cars or atmospheric deposition. The survey concluded that New York's experience is typical of northeastern states where the use of RFG is re-quired (NYSDEC, 2000).

AWWA collected 29 samples from 18 surface water sources in southeast-ern New York. Three of these sites had positive test results for MTBE, with a maximum concentration of 0.7 µg/l. Two of the three sites were adjacent to parking lots, and the detected concentrations could have been the result of small gasoline spills from parked cars (Gullick and LeChevallier, 2000).

Rhode Island.   The Rhode Island Department of Health (RIDOH) is responsi-ble for all VOC testing at PWSs within the state. Even though the State does not require specific testing for MTBE, it is reported in nearly all VOC testing. The RIDOH also tests private wells, but only if a gasoline release is suspected. In 1997, 32 out of 143 PWSs had positive test results for MTBE. Average con-centrations were 8 µg/l, with a maximum of 62 µg/l, which is above the state drinking water standard of 40 µg/l. Six private wells had results above 40 µg/l, with a maximum of 400 µg/l. In 1998, 59 out of 197 PWS wells had pos-itive test results, with an average concentration of 5 µg/l and a maximum concentration of 57 µg/l. Three private wells had results above the state standard in 1998, with a maximum concentration of 150 µg/l (NESCAUM, 1999).

Vermont. Of the 76,445 drinking water supplies in Vermont, 1,312 are public and 75,133 are private. Of the 1,312 PWSs, only 650 are required to monitor for VOCs, including MTBE. Even though Vermont is not an RFG area, MTBE concentrations were detected. By March 1999, 95% of the 650 supplies had reported their testing results; 35 had positive test results for MTBE, of which 3 exceeded the state advisory standard of 40 µg/l. Private water supplies in the state are not routinely monitored. Less than 1% of all private wells had been tested by 1999; of these, 175 had positive test results, including 40 with results above 40 µg/l. As is the case with Rhode Island, private wells are only tested when a gasoline release is suspected; therefore statewide occurrences cannot be extrapolated from these data (NESCAUM, 1999).

Midwestern States Study. In October 2001, ENSR International and Applied Hydrology Associates, Inc. released a report on the occurrence of gasoline constituents in the ground water of seven Midwestern states: Colorado, Illinois, Indiana, Kansas, Minnesota, Nebraska, and Wisconsin (ENSR and AHA, 2001). The report analyzed 77 ground water samples collected from 29 PWS systems in Colorado, Nebraska, Minnesota, and Illinois, 231 samples collected from 70 LUST sites in all seven states, and 31 samples collected from intermediate locations between 6 LUST sites and the nearest PWS system. In total, the seven-state study compiled water quality data from 325 wells covering 105 sites. The locations of the LUST and PWS sampling sites are included in Figure A-13 (Gregg *et al.*, 2002). Only two of the PWS samples exceeded standards for any of the gasoline constituents tested (BTEX, ethanol, MTBE, and tert butyl alcohol [TBA]). Both of these sites are located in Hyannis, Nebraska and had reported concentrations of 19 and 170 µg/l benzene. Only four other PWS systems reported any detectable gasoline constituent: 3 µg/l benzene in Minnesota, 3 µg/l MTBE in Illinois, and 0.2 µg/l and 3.6 µg/l MTBE in Colorado. Ethanol and TBA were not detected in any of the PWS samples (ENSR and AHA, 2001).

Of the LUST sites sampled, BTEX and MTBE were detected at frequencies of 86% and 70% respectively. The presence of MTBE was typically correlated with BTEX occurrence, and BTEX concentrations exceeded MTBE plus TBA concentrations at 64% of the sites. Plots comparing concentrations of BTEX versus MTBE plus TBA, benzene versus MTBE, and MTBE versus TBA are included as Figures A-14a, A-14b, and A-14c (Gregg *et al.*, 2002). The spills included in this study were typically from older gasoline releases; this explains the low frequency of ethanol detection (at 2 out of 70 LUST sites) (ENSR and AHA, 2001).

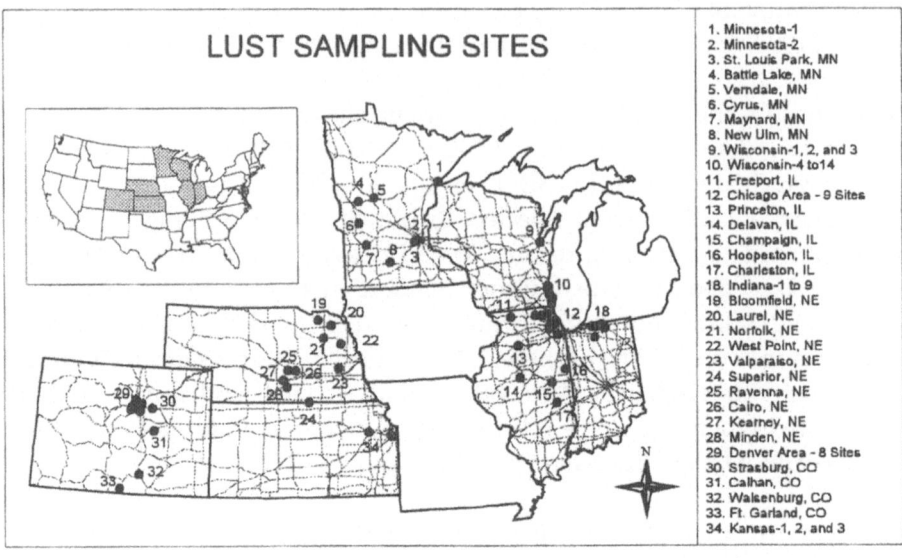

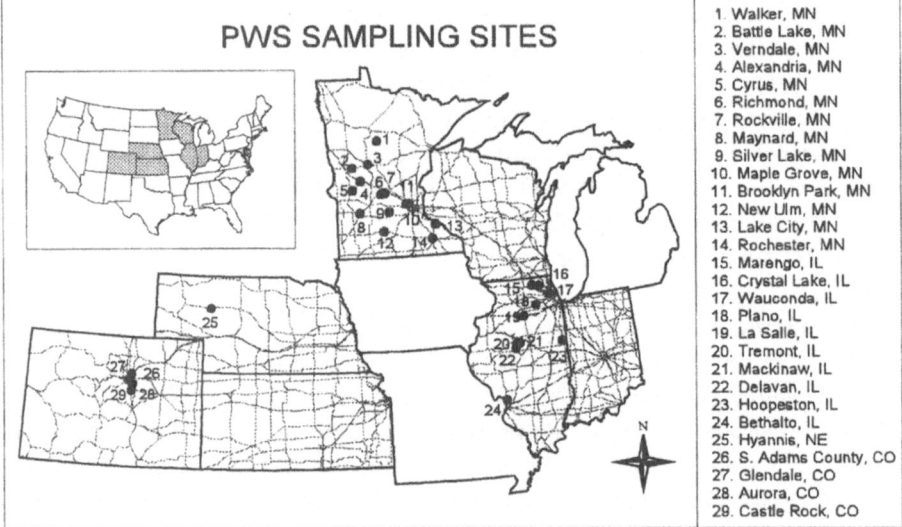

Figure A-13.   **LUST and PWS Sampling Sites**

The following section is a state-by-state discussion of the findings of the PWS section of this report, as well as individual state studies or AWWA results, where applicable.

Colorado.   Colorado was one of the first pilot areas for the "Winter Fuels" program aimed at improving air quality during winter months, and oxygenates have been used in Colorado gasoline since 1988. The Colorado De-

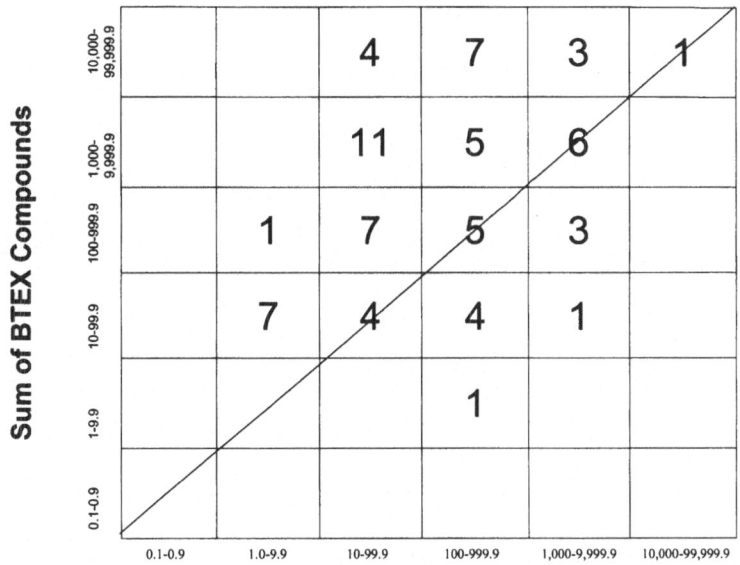

MTBE Plus TBA

Figure A-14a. Comparison of Sum of BTEX versus the Sum of MTBE and TBA Concentrations (ppb) from Midwestern LUST Site Samples

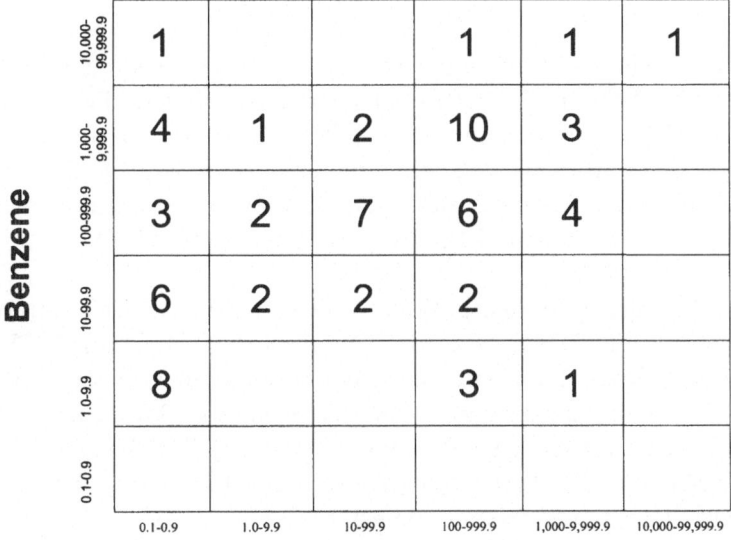

MTBE

Figure A-14b. Comparison of Benzene versus MTBE Concentrations (ppb) from Midwestern LUST Site Samples

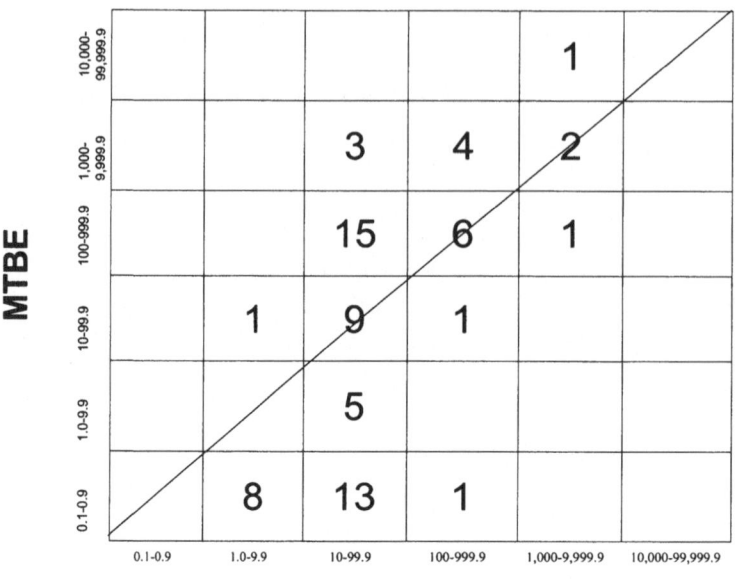

TBA

Figure A-14c.   **Comparison of MTBE versus TBA Concentrations (ppb)
from Midwestern LUST Site Samples**

partment of Public Health and Environment requires the testing of PWSs, but
as of the time of the study, the list of testing parameters did not include
MTBE. In the Midwestern States Study, MTBE was detected at two PWSs in
Colorado, at concentrations well below the USEPA advisory. Both detections
were located in South Adams County, at wells that are located approximately
0.7 miles downgradient of a LUST site (ENSR and AHA, 2001).

Illinois.   In Illinois, the Drinking Water Compliance Assurance Program is responsible for testing PWS systems for VOCs, including BTEX and MTBE. Additionally, the Illinois Environmental Protection Agency began testing for
MTBE in 1993, covering nearly 80% of all PWS systems. Of the 2,379 total PWS
wells that were sampled as a result of the program, 35 (1.4%) had positive test
results for MTBE and 443 (18.6%) had positive test results for BTEX. In the
Midwestern States Study, the only detectable gasoline constituent in Illinois
was MTBE, detected at one PWS at 3 µg/l. However, the nearest LUST site did
not contain significant concentrations of MTBE, so the MTBE source was not
clear (Christer, 2000; Crumly, 2000; IEPA, 2000a and 2000b; McMillan, 2000).
    AWWA collected 88 samples from 14 ground water sources in the north-
central region of Illinois; none of the samples had detectable concentrations

of MTBE. The only surface water sample collected, from the Illinois River, contained no detectable MTBE (Gullick and LeChevallier, 2000).

Indiana. The Indiana Department of Environmental Management is responsible for the testing of every PWS in the state, including analysis for BTEX components. Of the 4,255 PWS systems, 96% are from ground water sources, and the remaining 4% are from surface water sources. In 1999, only 38 of these systems (less than 1%) reported concentrations exceeding any VOC state standard (IDEM, 2000; Loa, 2000; Roeff, 2000).

AWWA collected 243 raw water samples from 99 ground water wells throughout Indiana. MTBE was detected in only one sample at a concentration of 1.5 µg/l, at a well from the west-central part of the state. At that well, however, MTBE was not detected in five prior and two subsequent sampling events. Of the nine surface water sites tested, providing 17 samples, none had positive test results for MTBE (Gullick and LeChevallier, 2000).

Kansas. As early as 1985, the Kansas Department of Health and Environment (KDHE) first started detecting MTBE in ground water samples. Since 1995, all PWS systems have been tested by the KDHE, resulting in MTBE detections at 21 of the 1,027 systems tested over that time period (Hattan, 2000; KDHE, 2000; Plummer, 2000; Spivey, 2000).

Minnesota. The Minnesota Pollution Control Agency has been responsible for MTBE testing at gasoline spill sites since 1989. MTBE was detected at less than 20% of the sampled sites, almost always at low concentrations. The Minnesota Department of Health has sampled PWSs for MTBE and other VOCs since 1989. In 1999, MTBE was detected at five PWS systems, although at concentrations below the USEPA advisory. In the Midwestern States Study, there was only one detection of a gasoline constituent in a Minnesota PWS: benzene at 3 µg/l. The study concluded that this concentration might be related to leaching from an old landfill, rather than the adjacent LUST site (Baker, 2000; MacAfee, 2000; Mandy, 2000).

Nebraska. The Nebraska Department of Health and Human Services (NDHHS) is responsible for testing state PWS systems, although MTBE was not included in its analyte list until February 2000. By October 2000, 1,069 samples had been analyzed; MTBE was detected in 6 of those samples, with concentrations ranging from less than 1 to 5.5 µg/l. In July 2000, the town of Hyannis, Nebraska tested all of its PWS systems and detected benzene at 70 µg/l and MTBE at 2.9 µg/l in one of its wells. As a result, the town stopped using this well. The suspected sources of VOCs were two gas stations, one operating and one abandoned. In the Midwestern States Study, three PWS systems in Hyannis were tested. The well that had been shut down had benzene at a concentration of 19 µg/l in October 2000, but benzene was non-detect in December 2000. However, at another well, located downgradient of the abandoned well, benzene was detected at 170 µg/l. It is suspected that the shutdown of the initial well caused a migration of gasoline components towards the second well (ENSR and AHA, 2001).

Wisconsin.   Since 1982, the Wisconsin Department of Natural Resources (WDNR) has been testing PWS systems for VOCs. Of the 2,230 systems tested since 1982, 5.1% have had positive test results for at least one VOC, most below state standards. In 1998, out of 2,193 wells sampled, 2 exceeded the Wisconsin enforcement standard of 60 µg/l for MTBE, and 3 exceeded the preventive action limit of 12 µg/l. In 1999, 49 (2.2%) out of 2,163 wells had positive test results for MTBE, but none of the concentrations exceeded Wisconsin's state standard or preventive action limit (Damgaard, 2000; Duren, 2000).

In June 2000, the WDNR released the results of its Groundwater Retrieval Network query on the state of Wisconsin's water with respect to MTBE. Of the 71 counties where ground water samples had been collected, 21 had one or more potable water wells at which MTBE had been detected. The earliest PWS data detecting MTBE dated back to 1997. Since 1990, 2,100 private drinking water wells have been sampled, of which 53 had positive test results for MTBE. Of these, 15 exceeded the state standard. More recently, in Freedom, near Outgamie County, 40 out of 70 private wells had positive test results for MTBE. This one town comprises most of the MTBE detection data in Wisconsin. However, a search of gasoline release sites showed that benzene concentrations, rather than MTBE, are most often the driver for remediation. This conclusion is a result of the fact that benzene is a carcinogen and has a lower health standard than MTBE (Pelayo *et al.*, 2000).

Conclusions of the Midwestern States Study.   The final report concluded that gasoline releases have had a minimal impact on PWS systems in the study area, and remediation efforts and natural attenuation at LUST sites appear to have been adequate. In general, BTEX, MTBE, and TBA plumes did not migrate a considerable distance from LUST sites. BTEX components were detected more frequently and at higher concentrations than any of the other constituents sampled (ENSR and AHA, 2001).

Additional State Studies.   In addition to the regional studies mentioned above, some individual states have conducted their own MTBE occurrence studies. The results of these studies are provided in the following sections.

Arizona.   On October 1, 1999, the Arizona Department of Environmental Quality (ADEQ) released a report summarizing the occurrence of MTBE in Arizona. At that time, the city of Phoenix had shut down four shallow production wells as the result of impacts from a LUST site just east of the city (ADEQ, 1999).

Ground water sources provide drinking water for approximately 65% of Arizona's population. During 1998, the Water Quality Division, along with the USGS, performed sampling in the Upper Santa Cruz ground water basin. Sixty samples were collected, mostly from domestic wells, none of which had positive test results for MTBE. In 1998, 58 samples were collected from the Sacramento Valley ground water basin and 28 from the Willcox ground water basin; none had positive test results for MTBE. Due to the high use of motorized watercraft on lakes and ponds in Arizona, the Water Quality Division is planning to implement a testing schedule for surface waters in Arizona.

The ADEQ is not responsible for tracking MTBE concentrations at LUST sites. However, the State Lead Unit of the UST Program does maintain a database of sampling events. These results were included as an appendix to the ADEQ study. MTBE concentrations ranged from 0.93 to 94,000 µg/l, while benzene concentrations ranged from less than 0.5 to 25,000 µg/l. At nearly every site, MTBE concentrations were higher than the corresponding benzene concentrations (ADEQ, 1999).

AWWA collected 122 samples from six wells in Arizona; MTBE was not detected in these samples. All six wells were located near Phoenix in a densely populated area. No surface water samples were collected as part of the study (Gullick and LeChevallier, 2000).

Idaho. In February 1999, the Idaho Division of Environmental Quality (IDEQ) presented a report evaluating the occurrence of MTBE at LUST sites. From October to December 1997, 100 ground water samples were collected at 100 sites where known gasoline releases had occurred. The LUST sites were split into three categories: those where the release had occurred within the last five years, those where diesel was released, and those where the release occurred over five years ago. MTBE was not detected at any of the 12 "diesel" sites. Benzene was detected at 5 sites at a mean concentration of 150 µg/l, toluene at 7 sites at 289 µg/l, ethylbenzene at 6 sites at 188 µg/l, and xylenes at 7 sites at 1,434 µg/l. MTBE was detected at 50% of the "new" release sites and 30% of the "old" release sites, resulting in an overall detection of MTBE at 40% of LUST sites in Idaho. At 56 "new" gasoline release sites, BTEX compounds were detected at mean concentrations of 2,835 µg/l benzene, 5,500 µg/l toluene, 851 µg/l ethylbenzene, and 4,169 µg/l xylene, while MTBE was detected at 2,271 µg/l. At 32 "old" gasoline spill sites, the BTEX components were detected at 599 µg/l benzene, 1,234 µg/l toluene, 357 µg/l ethylbenzene, and 3,213 µg/l xylene, while MTBE was detected at 312 µg/l. Ninety-three percent of benzene detections exceeded the corresponding Idaho risk-based corrective action (RBCA) concentration of 5 µg/l, and 62% of MTBE detections exceeded the corresponding Idaho RBCA concentration of 52 µg/l. Both benzene and MTBE were detected together at 41% of the LUST sites;

when detected together, benzene concentrations were typically higher. As a result of benzene's typically higher concentration relative to its RBCA concentration, the IDEQ study concluded that the benzene concentration is the determining factor for remediation efforts most of the time at LUST sites in Idaho (IDEQ, 1999).

Iowa.   AWWA collected 29 samples from four deep wells and three samples from one surface water source, the Mississippi River. No samples had positive test results for MTBE (Gullick and LeChevallier, 2000).

Kentucky.   AWWA collected 35 samples from eight surface water sites in the east-central part of the state, which included the Kentucky River, Jacobson Reservoir, and Lake Ellerslie. No samples had positive test results for MTBE, and no ground water samples were collected as part of the study (Gullick and LeChevallier, 2000).

Maryland.   In December 2000, the Maryland Department of the Environment's (MDE) Task Force on the Environmental Effects of MTBE released its preliminary report. The study concluded that, even though a major effort has been made to upgrade gasoline USTs across Maryland, gasoline releases from older USTs continue to affect ground water. Maryland has tested PWSs for MTBE since 1995. However, because there is no enforceable federal drinking water standard, testing for MTBE is not compulsory. Maryland currently uses the USEPA advisory of 20 µg/l as guidance for corrective action. Approximately 68% of Maryland's population is served by drinking water supplies originating from surface waters, which generally have been negligibly affected by MTBE. Since the release of the study, two surface water samples had positive test results for MTBE, but at concentrations less than 2 µg/l. Nearly 10% of the population uses wells set in confined aquifers, which are hydraulically separated from LUST releases, 16% are serviced by private wells, and 6% draw their water supply from less protected unconfined aquifers (Maryland Task Force, 2000).

Of the 1,084 PWSs sampled since 1995, 85 systems have had positive test results for MTBE, and only 11 PWSs exceeded the USEPA advisory. Of those 11, 10 now draw their water from alternate sources, while at 1 PWS the MTBE concentration has since dropped below the advisory. In July 1999, the MDE compiled a database of known impacted private water wells throughout the state. As of July 1999, 149 wells had positive detections, and as of the release of the MDE's report in 2000, a total of 267 domestic wells had positive detections for MTBE at some time. However, the study does not indicate how many wells in total were tested for MTBE. Of the 267 wells, 239 are investigation or remediation cases under the Oil Control Program (Maryland Task Force, 2000).

AWWA collected 10 ground water samples from two sites in Maryland; of these, 1 sample contained a trace amount of MTBE (0.5 µg/l). However, 3 more samples were collected at later dates from the same site and MTBE was not detected (Gullick and LeChevallier, 2000).

Michigan.   In March 2000, the Michigan Department of Environmental Quality (MDEQ) compiled a fact sheet on the occurrence of MTBE in Michigan. Michigan does not have an extensive history of using oxygenated fuels. The Michigan Department of Agriculture (MDA) randomly samples gasoline sold across Michigan. In 1998, the MDA determined that only 5% of the gasoline sold in Michigan contained MTBE. Even though MTBE is not a large component of the gasoline sold in the state, MDEQ has identified MTBE as an issue at several LUST sites throughout the state. The MDEQ Storage Tank Division is responsible for documenting all LUST sites as well as BTEX and MTBE concentrations at these sites. MDEQ has established a health-based criterion of 240 µg/l for MTBE, and has adopted an aesthetic (secondary) drinking water standard of 40 µg/l (MDEQ, 2000).

In 1999, the MDEQ launched investigations into the occurrence of MTBE in Michigan's surface waters. No MTBE was detected in any of the samples collected. MDEQ has also analyzed 31,557 water samples from 18,046 community, non-community, and private water wells between October 1987 and September 1999. Of these samples, 903, or 3% of the total number of samples collected (representing 542 locations), had positive test results for MTBE. Of the 3% of results that were positive, 3.1% of the concentrations were greater than 240 µg/l, 9.1% were between 40 and 240 µg/l, and 87.8% were between 1 and 40 µg/l (MDEQ, 2000).

AWWA collected six samples from three wells in the northwest section of Michigan's Upper Peninsula; no samples had positive test results for MTBE. The study area is sparsely populated and is not an RFG area (Gullick and LeChevallier, 2000).

Missouri.   In September 2000, a report outlining the findings of two MTBE meetings in Missouri was released (Missouri, 2000). The Executive Director of the Petroleum Storage Tank Insurance Fund organized the meetings, and contributors included the Department of Health, the Department of Agriculture, the Department of Natural Resources, the Petroleum Storage Tank Insurance Fund, and the Attorney General's Office. The purpose of the meetings was to ensure that threats to human health and the environment from MTBE are minimized. The report summarized that of 5,600 gasoline release sites in Missouri, 4,000 had been remediated. Eighteen of the remaining 1,600 releases had impacted drinking water supplies, resulting in MTBE detections at three PWSs and 28 private wells.

AWWA collected 14 samples from five ground water sources and 7 samples from the Missouri River. None of the surface or ground water samples had positive test results for MTBE. However, the ground water wells were deep, and no MTBE concentrations were expected at those depths (Gullick and LeChevallier, 2000).

New Mexico. AWWA analyzed 28 samples collected from seven ground water wells set in the Ogallala Aquifer along the state's eastern border. The wells ranged from 100 to 120 meters (350 to 400 feet) deep. No samples had positive test results for MTBE; no surface water samples were collected (Gullick and LeChevallier, 2000).

Ohio. AWWA collected 54 samples from 32 ground water sites in north-central Ohio, as well as 4 samples from 2 surface water locations. Of the 32 ground water sites, 15 were at depths ranging from 40 to 80 meters (140 to 270 feet), 15 were at depths ranging from 60 to 160 meters (200 to 540 feet), and 2 were approximately 90 meters (300 feet) deep. None of the ground water or surface water samples had positive test results for MTBE (Gullick and LeChevallier, 2000).

Pennsylvania. AWWA collected 27 samples from four wells, as well as 11 samples from nine surface water locations. None of the 8 surface water samples collected from the western part of the state had positive test results for MTBE. However, the remaining samples were collected near a LUST site at which 3,800 liters (1,000 gallons) of gasoline had been released. Some of this gasoline had entered a creek upstream of the wellfield; MTBE concentrations detected in the creek were 4.4 and 25.1 μg/l. MTBE was detected in samples from wells in the impacted field at concentrations ranging from 0.5 to 14.2 μg/l, and generally showing a decreasing trend over time (Gullick and LeChevallier, 2000).

Virginia. Four ground water samples were collected from two wells in Virginia by AWWA. The wells were approximately 100 meters (350 feet) deep in a metropolitan area; neither well had positive test results (Gullick and LeChevallier, 2000).

Washington. In October 2000, the Washington State Department of Ecology (WADOE) published a study of the occurrence of MTBE in LUST sites in Washington (WADOE, 2000). Based on the USEPA's advisory, Washington established an MTBE action level of 20 μg/l. Oxygenated fuels have not been as widely used in Washington compared to other parts of the U.S. Before WADOE's report, MTBE had been detected in ground water samples in areas of King County, and at LUST sites in Vancouver, Yakima, and Spokane.

The study selected 70 LUST sites, based on representative locations throughout Washington, and analyzed ground water samples for BTEX and

MTBE. Of the 70 sites, 8 (11%) had negative test results for all components, 30 (43%) had positive test results for MTBE, and 26 (37%) of those had MTBE concentrations above 1 µg/l. Of the 62 sites that had positive test results for gasoline components, 24% contained MTBE concentrations above the action level of 20 µg/l, with the highest measured concentration being 7,150 µg/l. The mean concentration of the samples that met or exceeded 1 µg/l for MTBE was 441 µg/l, while the median was 13 µg/l. All of the BTEX components were detected more frequently than MTBE, at 90%, 66%, 61%, and 68% of all sites respectively, and at higher average concentrations: 1,992 µg/l; 3,943 µg/l; 657 µg/l; and 2,251 µg/l respectively. Detections of MTBE and benzene were not well correlated. However, in the instances where MTBE and benzene were both detected, benzene concentrations were higher at 63% of the sites (WADOE, 2000).

West Virginia. AWWA collected 31 surface water samples from 19 locations and 5 ground water samples from 2 wells. The surface water samples were from sites scattered throughout the state, and none had positive test results for MTBE. At one of the two ground water wells, MTBE concentrations were non-detect in 1997, 2.3 µg/l in 1998, and 6.6 µg/l in 1999. The well is approximately 30 meters (100 feet) away from a gasoline station, in an area that could also be subjected to commercial or residential non-point source pollution. As of the release of the study, the actual source of the detected concentrations was not known (Gullick and LeChevallier, 2000).

Conclusions. From the data sets presented above, it appears that MTBE detection frequencies are low in most areas of the U.S. In those areas where detection rates are relatively high compared to the national background, measured concentrations of MTBE are often below the USEPA advisory. Many of the reports listed above did not include BTEX detection data, which would have made for a useful comparison with MTBE detection data.

MTBE OCCURRENCE IN ENGLAND AND WALES

Robert C. Harris and Alwyn J. Hart, Ph.D., Environmental Agency

Fuel Background. The use of ethers in motor fuels in the U.S. has clearly been driven by legislation and concerns over air quality. In the United Kingdom (U.K.) this pressure has not existed, and while much of the unleaded fuel sold does contain MTBE or other ethers, the proportions used are much lower than in the U.S.

The limits for ether content of U.K. gasoline are governed by European Union Directive 85/536/EEC, with a maximum legal concentration of 15% by volume (volume to volume [v/v]). The most common reason for addition of ethers to U.K. fuel is not to increase the oxygen content, since there is no regulation in place for this. Rather, the purpose is to enable refiners to most easily attain octane requirements for their fuel. Some U.K. refineries are able to meet these requirements without the addition of ethers, for example through the use of alkylation plants.

Ether use is typically less than 1% v/v in "unleaded" gasoline (known as premium unleaded petrol) and between 1% and 5% in "super unleaded," although the latter accounts for only a small proportion of the total fuel market. This level of ether use is believed to be the lowest in Europe (Morgenroth and Arvin, 1999).

One final factor to consider is the common practice of fuel exchange between the various refiners. This means that the fuel brand and retail ownership at any particular site cannot be used as a guide to the ether content of fuel sold there.

Water Background.   Ground water accounts for approximately 35% of the public water supply in England and Wales, although geographically the proportion varies greatly. In the densely populated southeast area of England, the proportion rises to around 75%. The Environment Agency is the public body charged with protecting and regulating the water environment, and has set out its approach in a series of policy statements, which in turn rest on assessment tools such as Groundwater Vulnerability maps and Source Protection Zones around major ground water abstractions (Environment Agency, 1998). These large ground water abstractions (almost 2,000) are generally owned by privatized water utility companies and, in most cases, lie within the sandstone or chalk aquifers of England.

There are no specific drinking water standards for MTBE in water, but maximum levels for total hydrocarbons are 10 μg/l.

Study Method.   The Environment Agency (2000) collected data on the occurrence of ethers in ground water from all readily available data sources and archives. This included known spills, leaks, and the results of routine monitoring from the Environment Agency, water companies, major oil companies, and trade associations. The data included information from over 800 site investigations and almost 3,000 samples from public supply and monitoring boreholes. For all sites that had been investigated, information was requested about the site itself (including size, use, etc.); the environmental setting (including aquifer type, ground water vulnerability class, etc.); the analytical methods (including performance characteristics, limit of detection, etc.); and ethers detected (including concentrations, location, plume size, trends, etc.).

While reviewing site investigation reports, particular attention was paid to finding and recording information on the reason for the investigation, including any details of the type and date of the source/spill; maximum concentrations of MTBE or other ethers, TPH, and BTEX; the geology beneath the site; the distance to nearby PWS wells, and any impacts upon those wells; and remedial measures, either on site or at the PWS well.

Each water supply company and Environment Agency region was asked to complete a questionnaire on perceptions of the MTBE problem in relation to PWSs; known incidents involving MTBE or other ethers; and previous investigations and monitoring for ethers and BTEX.

All parties were also asked to supply any data on ether occurrence in ground water, together with relevant background information, including borehole details, *i.e.,* geology and depth of screened section; the sampling methods used and the reason for sampling; the analytical methods (including performance characteristics, limit of detection, etc.); and ethers detected (including concentrations, trends, etc.).

The purpose of collecting so much information about each site was to be able to assess the quality of information presented and to build some confidence that the results really could provide a reasonable picture of ground water contamination.

The major oil companies provided information on 2,069 retail filling stations, transfer depots, and oil terminal sites, which had been investigated for contamination. Analysis for ethers had been carried out at 837 of these sites. A histogram showing the frequencies of concentrations detected at these 837

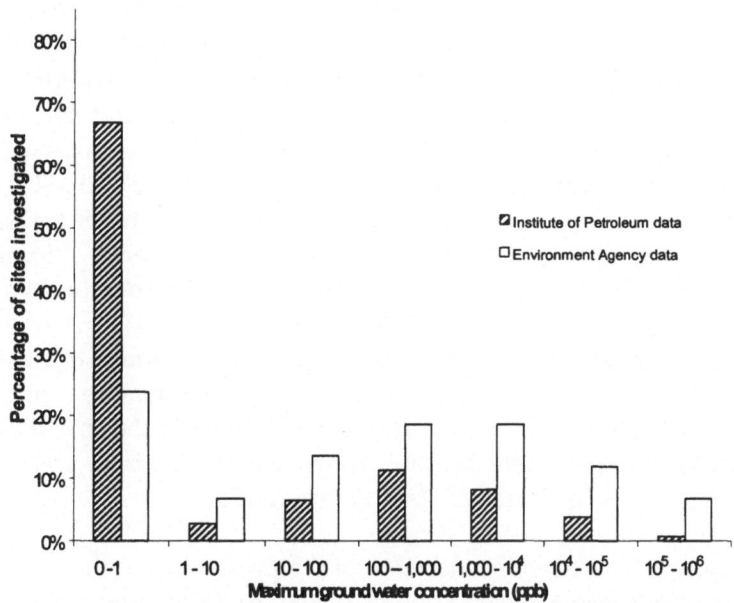

Figure A-15. **Distribution of Maximum MTBE Concentrations in Ground Water from 837 Site Investigations in England and Wales**

sites is shown in Figure A-15. Ethers (primarily MTBE) were detected in ground water or perched water at approximately 29% and 25% of these petrol retail and distribution sites respectively. Forty of the sites with detected MTBE were reported to be located above high vulnerability (high leaching potential of the underlying ground) aquifers (major or minor).

In all, the Environment Agency and the water companies supplied data on almost 3,000 ground water samples from 940 observation boreholes or PWS wells that have been analyzed for ethers. Of the 940, only 255 were regularly analyzed for ethers, of which 32 (12.5%) contained MTBE above the detection limit of 0.1 µg/l, and only 3 contained MTBE at concentrations above the low end of the taste threshold range (taken as 5 µg/l) (McKinnon and Dyksen, 1984). There has only been one case in the U.K. where MTBE has been detected in a municipal water by the public. It was also apparent that the very few wells where such levels of MTBE had been detected were contaminated with a wide range of substances. As a result, MTBE would be unlikely to drive the management or regulatory approach.

Risk Assessment. For the purposes of this project, the location of individual sites was not revealed by the oil companies. Therefore, it was not possible to determine how many of the 40 sites reported to lie on high vulnerability aquifers are within the capture zone of a PWS well and hence to fully assess the risk posed.

It was possible to construct a risk model (Environment Agency, 2000) using data on the location of all retail filling stations (Catalist, 2000) and the 1,944 PWS wells (Environment Agency, 2002). The model was calibrated using known detection rates and abstraction and flow rates for each well along with estimated aquifer properties such as porosity and thickness.

The model predicts that 203 (10%) of the 1,944 PWS boreholes in England and Wales can be expected to contain MTBE, but in only 6 will MTBE exceed the taste threshold of 5 µg/l. A key factor is the proximity of the leak to an abstraction source (Figure A-16) although clearly the influence of local hydrogeological conditions in these fractured consolidated formations will have a major influence.

Using this model, it as also possible to estimate the effects of any increase in MTBE concentration in fuel. Here the results provide more cause for caution, should MTBE concentrations rise to match those of the U.S., the number of detections above the taste threshold may well increase from a handful now to several tens or even hundreds (see Figure A-17).

Conclusions. The overall conclusion of the project is that the ethers do not currently pose a major threat to public water supplies in England and Wales. However, if concentrations in fuel were to rise to 5% or more in the future, the incidence of taste problems may significantly increase.

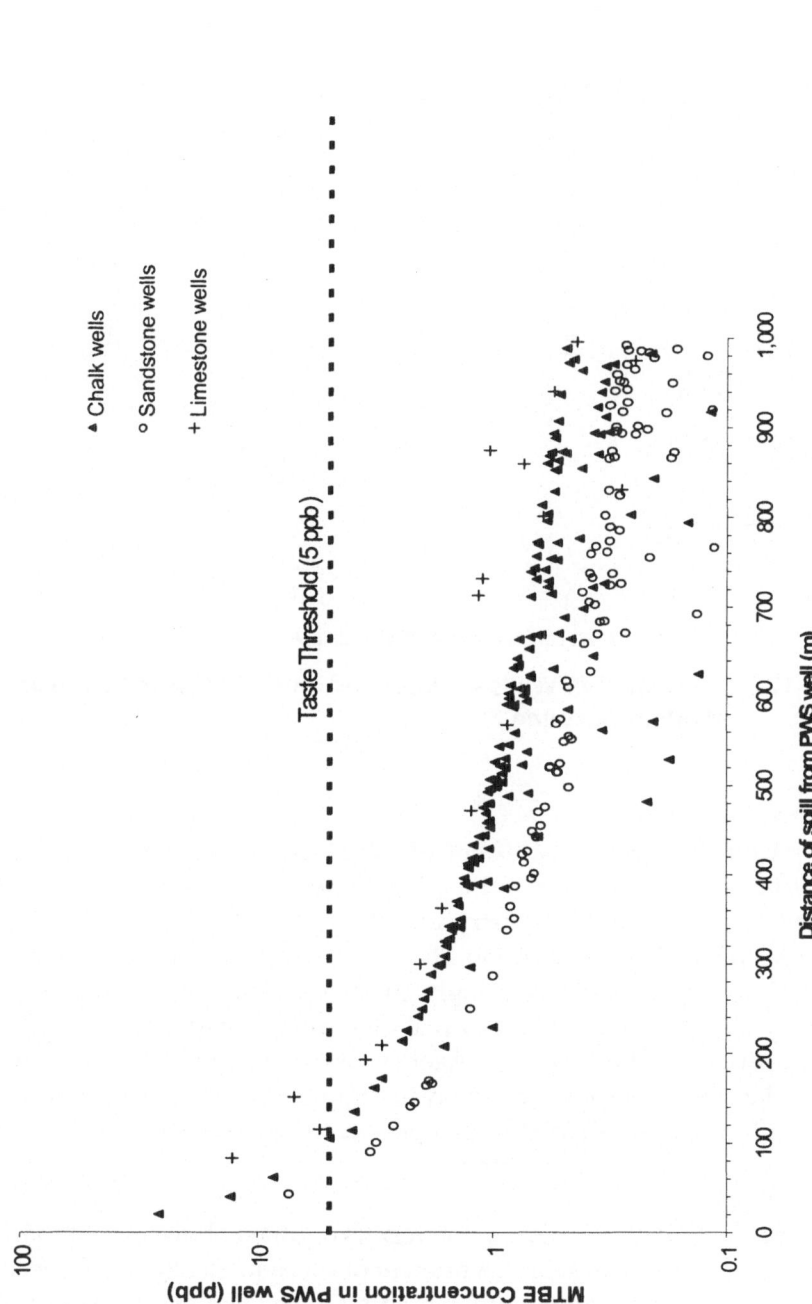

Figure A-16. Modeled relationship between MTBE concentration in PWS wells and distance to spill in England and Wales

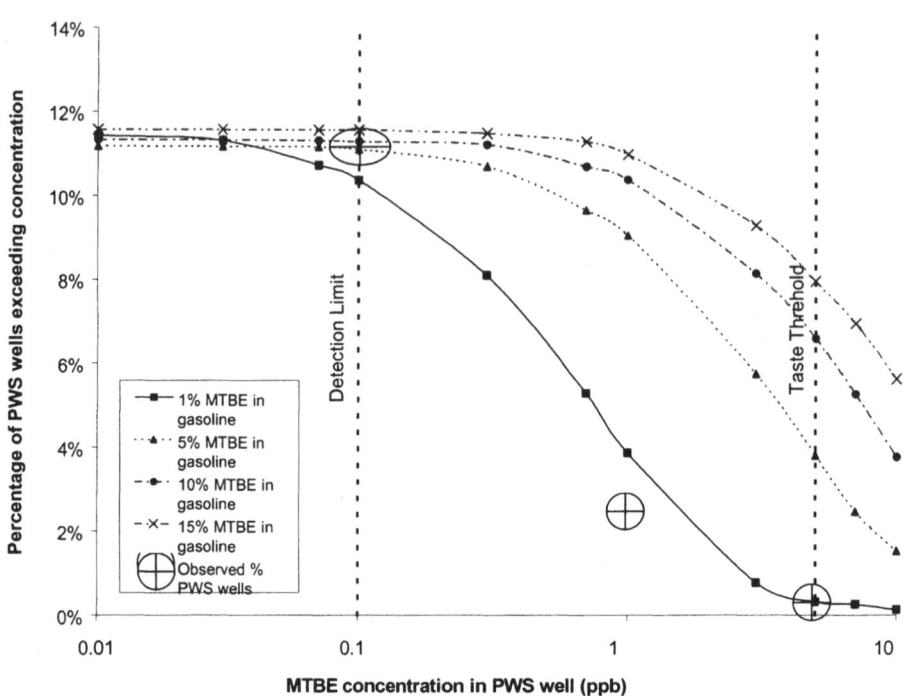

Figure A-17.   **MTBE in PWS Wells — Predicted Outcome from Increasing MTBE in Gasoline**

Impacts.   Until this study, the extent of MTBE contamination of ground water was largely unknown beyond the U.S. However, the prevailing mood was negative, with even oil industry reports raising the possible need to phase out MTBE use in Europe (Anon, 2000). Legislators at both national and European Union levels were increasingly questioning the need for and benefits of MTBE use. This study allowed real data from England and Wales to feed into a European Environment Agency risk assessment of the use of ethers. Similar studies are now underway in a number of European countries. The data thus provide an alternative perspective from the U.S. experience for European policy makers at a key time.

Further Work.   Though reassuring, the results outlined above do not mark an end to Agency interest in MTBE. A number of subsequent research projects are underway to strengthen the understanding of the fate and transport of ethers particularly in consolidated aquifers. These include laboratory based studies on the attenuating properties of U.K. soils and geologies faced with MTBE spills; a site-based study of an MTBE plume in the Chalk aquifer; and further development of stable isotope characterization of MTBE spills.

Monitoring of ground water has recently been reviewed by the Agency and recommendations included: increased monitoring of PWS wells; an expansion of the Environment Agency's monitoring network to include MTBE analysis in all regions; more comprehensive site investigations in areas where PWS wells are vulnerable; and a review of laboratory analytical procedures.

Acknowledgement. The work reported here was funded jointly by the Environment Agency and the Institute of Petroleum. In addition a wide range of other organizations, such as water companies and fuel related trade associations, provided valuable information.

PLUME LENGTH STUDIES (TEXAS, FLORIDA, AND CALIFORNIA)

James W. McKinley, Applied Hydrology Associates, Inc.

On December 15, 2000, the New England Interstate Water Pollution Control Commission (NEIWPCC) released the results from their survey of state experience with MTBE at LUST sites. The survey was performed under a grant by the USEPA's Office of Underground Storage Tanks to determine the role of MTBE at LUST sites in all 50 states. Table A-3 summarizes the survey results.

The above sections have dealt primarily with the frequency of MTBE detections in ground and surface water across the U.S., England, and Wales. However, there have also been a number of studies that summarized the movement and behavior of MTBE in the subsurface. These plume studies, covering Texas, Florida, and California, were conducted to evaluate the fate and transport of MTBE compared with other gasoline components, especially BTEX.

Table A-3. NEIWPCC Survey Results (Frye, 2000)

Does an MTBE-affected site exist with no known release?					
Yes		No		Don't know	
25 states		11 states		14 states	
What percentage of MTBE concentrations exceed 20 ppb at LUST sites?					
0–20%	20–40%	40–60%	60–80%	80–100%	Don't know
7 states	4 states	7 states	10 states	10 states	9 states
What is the average MTBE plume length?					
100 to 250 ft		250 to 500 ft		> 500 ft	Don't know
6 states		8 states		1 state	5 states
How often are MTBE plumes longer than BTEX plumes?					
Most of the time			Some of the time		
15 states			11 states		

Texas. The Texas plume comparison study compiled MTBE and benzene concentration data at 609 LUST sites (Mace and Choi, 1998). The data were adequate for characterizing benzene plume lengths at 289 of the sites and MTBE plume lengths at 99 of the sites. A linearized exponential plume model was used to fit a mathematical description of the plume shape to the observed concentrations. The plume dimensions were then characterized at concentrations of 10, 20, and 100 µg/l for both MTBE and benzene. Ninety-three percent of the LUST sites had detectable MTBE concentrations in the surrounding ground water, with 85% having concentrations above the USEPA advisory.

Ten percent, or 10, of the MTBE plume lengths were zero, due to low or non-detect concentrations, and 10%, or 10, of the plumes extended beyond the edge of the monitoring network. Of the 89 sites with MTBE plumes, most were located in densely populated areas around Dallas and Fort Worth. At the 79 sites where the MTBE plume did not extend beyond the edge of the monitoring well network, the average plume length at a concentration of 10 µg/l was 55 meters (182 feet), 8 meters (27 feet) longer than the corresponding average benzene plume length of 47 meters (155 feet). While the average MTBE plume was longer than the average benzene plume over all LUST sites, at any one site the MTBE plume was longer than the corresponding benzene plume only 56% of the time. Of the 89 characterized sites, several contained a benzene plume and no MTBE plume, which was attributed to older gasoline releases before oxygenated fuels were introduced. At only one site was there a characterized MTBE plume but no benzene plume.

MTBE concentrations at 24% of the wells were non-detects, 9% were decreasing, 50% were stable, 7% were increasing, and 10% were erratic, with no discernable trend. Because 83% of the wells show MTBE plumes to be stable, decreasing, or not detected, the study concluded that MTBE may be naturally attenuating at many of these sites (Mace and Choi, 1998).

Of the 609 LUST sites tested, which included sites where MTBE and benzene plumes had not been characterized, 21% had detectable benzene but no MTBE; whereas 24% of tested sites had detectable MTBE but no benzene. At almost 80% of the sites, MTBE concentrations were detected at higher concentrations than corresponding benzene concentrations. MTBE plume lengths were compared between areas with different hydrogeologic characteristics. The average plume length in the clay-rich Beaumont Formation (53 meters [174 feet]) was smaller than that in the unconsolidated sand-rich Ogallala Formation (74 meters [242 feet]), which is consistent with the relative water velocities in the two aquifers. However, the shallower the depth to ground water, the smaller the plume, which at first appeared contradictory. However, this result is logical, because in this study, the areas with shallower depth to ground water were also generally areas of low permeability soils, slowing vertical migration (Mace and Choi, 1998).

Florida. In Florida, ground water supplies provide drinking water for about 90% of the state's population. The shallow depth to ground water, course-grained soils, high permeability, and low hydraulic gradient are important in understanding the nature of MTBE plumes within the state. Because ground water levels are very shallow and the infiltration rate is high, gasoline spills reach the water table very quickly. However, oxygen is more readily available than in most other settings. The low hydraulic gradients tend to limit the longitudinal spread of any gasoline release relative to the lateral spread, and the coarse-grained soils and high subsurface temperatures allow volatile hydrocarbons to evaporate quickly (Reid *et al.*, 1999).

In the Florida plume comparison study (Reid *et al.*, 1999), performed for British Petroleum (BP), data were reviewed from 149 BP sites throughout Florida. The 149 sites were screened to see if they met several criteria. First, the site must have had a documented gasoline release that contained MTBE. Second, the site must have at least three years of monitoring data for MTBE, so that the historical trend could be characterized. Third, the site must have at least three monitoring wells that detected MTBE, so that the lateral extent of the plume could be characterized. The report pointed out that, with this methodology, results might be biased towards sites that contained larger MTBE plumes, because sites with smaller plumes would be less likely to have the three monitoring wells needed for inclusion. Eighty of these sites met the selection criteria; of these, 55 had sufficient data to compare benzene and MTBE plume lengths. At the 80 study sites, 1,369 monitoring points had been established, at which there were 5,757 samples collected between November 1985 and May 1998. For the 80 sites that were included, a database was created that included concentrations for MTBE and BTEX.

Of the 5,757 samples, 39% detected MTBE and 43% detected benzene. The report explains that this difference could be attributed to the omission of MTBE testing in early samples. The average and maximum MTBE concentrations were 317 and 95,500 µg/l respectively, while the corresponding values for benzene were 365 and 32,000 µg/l. Both benzene and MTBE occurrences were skewed towards lower concentrations.

Unlike the Texas plume comparison study, the data for each site was hand-contoured using the best judgment of the researchers. This approach was taken because of concern that the data points were irregular and might be misinterpreted by a computer model. In the Florida study, using the 10 µg/l contour to define both the MTBE and benzene plumes, MTBE plumes were 34% larger (3,655 square meters [11,985 square feet] versus 2,410 square meters [7,910 square feet]) and 18% longer (40 meters [140 feet] versus 35 meters [115 feet]) than benzene plumes. The relatively greater MTBE plume size compared with length was attributed to the flatter hydraulic gradients in Florida. As a result of the lower gradient, lateral spreading becomes a more dominant transport mechanism than longitudinal spreading. The study then considered the benzene drinking water standard, which is significantly lower than the MTBE advisory. In 96% of the cases, the MTBE plume at a 20 µg/l contour was nearly equivalent in size or shorter than the benzene plume at a 1 µg/l contour. Of all the benzene and MTBE plumes, 4.4% were increasing in size over time, 6.6% were stable, and 89% appeared to be decreasing in size over time (Reid *et al.*, 1999).

California.   In 1998, the State of California conducted its own plume comparison study (Happel *et al.*, 1998). The study examined 236 LUSTs covering 24 counties in California, with 1,858 monitoring wells. Between 1995 and 1996, MTBE was detected at 78% of those sites, at concentrations ranging from several to 100,000 µg/l. Of the 2,297 active PWS wells that were monitored for hydrocarbon impacts, MTBE exceeded 20 µg/l at 0.35%, and benzene exceeded 1 µg/l at 0.42% (Happel *et al.*, 1998).

The California plume comparison study used compiled information from 63 of the LUST sites described above. The plumes were characterized using the 20 µg/l contour for MTBE and the 1 µg/l contour for benzene because of their respective regulatory levels. From the 63 LUST sites, the researchers were able to characterize 50 MTBE and benzene plumes using hand plots and best professional judgment. The study showed that MTBE plumes were equal to or shorter than benzene plumes at 81% of the sites. Ninety percent of the MTBE plumes were shorter than 100 meters (325 feet), while 90% of the benzene plumes were shorter than 120 meters (400 feet). Additionally, MTBE and benzene plume lengths were moderately correlated, but MTBE and benzene concentrations were not correlated. The behavior of the MTBE and benzene plumes was not consistent with the researchers' expectations, since the study had assumed that MTBE was highly recalcitrant and that benzene was naturally attenuated at most spill sites. To investigate these assumptions further, MTBE and benzene occurrence data from 1994 to 1996 were analyzed for 29 sites in San Diego County. The frequency of MTBE and benzene being detected in the same sample dropped from 80% to 60% over the three-year period, while the overall detection frequency for each individual hydrocarbon remained constant (Happel *et al.*, 1998). These data were thought to support the hypothesis that due to MTBE's higher solubility and lower adsorption relative to benzene, MTBE plumes would be more mobile and would gradually dissociate themselves from the corresponding benzene plume over time. At the downgradient wells in the San Diego study, MTBE concentrations were equal to or higher than benzene concentrations in most samples. However, when considering the different state standards, the researchers concluded that benzene was more often the contaminant of concern, even though it was frequently detected at lower overall concentrations than MTBE (Happel *et al.*, 1998). The results of the California plume comparison study appear to support the results of the Florida and Texas studies:

physically, MTBE may be more widespread, but physiologically, benzene is of more concern.

History of MTBE in California. Over recent years, California has received a lot of media attention concerning MTBE, most of it negative. Initial studies, reported in 1995, assumed that detection frequencies would either stay stable or increase as more widespread testing was performed. As a result, there was an initial feeling that MTBE would become a much greater problem than has actually been observed in the last six years. MTBE occurrence in California now appears not to be as widespread as initially thought. On September 5, 2000, the California Department of Health Services (CDHS) estimated that only 0.8% of drinking water sources and 1.9% of PWSs sampled in California in the year 2000 contained detectable MTBE in one or more sample. Of those sources and supplies where MTBE was detected, less than 1% exceeded the state's maximum contaminant level (MCL) of 13 µg/l. The data presented by the CDHS covers systems providing water for 29.9 million people, which comprises nearly 90% of California's population (Williams and Sheehan, 2001).

The CDHS also provided a summary of MTBE detections prior to 2000. From 1995 to 2000, more than 29,000 California drinking water samples, from approximately 4,300 drinking water sources and 1,700 PWSs, were analyzed for MTBE. For that time period, the average detection frequency for MTBE was 1.3% for all samples. The highest detection frequencies occurred in 1995: 3% of sources sampled and 6% of systems sampled, probably due to increased sampling and analysis for MTBE. The frequency dropped to 1.3% of sources sampled and 3% of systems sampled in 1996, and the MTBE detection frequency for all samples has remained below 2% since then (Williams and Sheehan, 2001; Williams et al., 2000).

Even though the detection frequency was generally low over the five-year period, 31 of 58 counties in California had positive test results for MTBE at one time or another. In most counties the frequency was less than 5%, but higher detection frequencies occurred in 5 counties, which represent 9% of California's population (Williams and Sheehan, 2001). The cities of Santa Monica and South Lake Tahoe fall into this group, and the MTBE occurrences there account for much of the media attention given to MTBE in California.

At the end of 1995, MTBE was detected in Santa Monica's water supply wells. Of the city's 11 high volume production drinking water wells, MTBE was detected at 9, which were shut down (Cal/EPA, 1999). In the Charnock well field, where 7 of the wells were located, maximum MTBE concentrations detected were 610 µg/l at the production wells, 17,000 µg/l at the regional monitoring wells, and 230,000 µg/l at nearby LUST source-site monitoring wells (Blue Ribbon Panel, 1999). At the Arcadia well field, which contained two wells, the maximum MTBE concentration (72.4 µg/l) was detected in

1996. At both the Arcadia wells, the MTBE was attributed to nearby LUSTs. The nine abandoned drinking water wells had provided service to over 80% of Santa Monica's residents (Cal/EPA, 1999). As a result of the loss of potable water, the City began purchasing its supplies from the Los Angeles Metropolitan Water District, with Shell, Chevron, and Exxon providing replacement costs for approximately 1,100 hectare-meters (8,900 acre-feet) of water per year (USEPA, 1999). Although identified as an "MTBE spill," it should be considered that the root causes of the shutdown of the Santa Monica wells were (1) the urban growth of both Los Angeles and Santa Monica around the well field locations, combined with (2) an inadequate aquifer or wellhead protection program, and (3) undetected or unremediated leaking gasoline tanks, rather than the presence of MTBE per se.

Beginning in 1996, the South Tahoe Public Utility District began to test for the presence of MTBE. Since MTBE was detected soon thereafter, 13 of the District's 34 drinking water wells were shut down. In eight cases, the wells were taken off-line because MTBE from gasoline spills had reached that location; for the remaining cases, the threat of a nearby plume was the reason for shutting them down. The spills were traced back to leaking tanks and leaks associated with dispensing systems (Dernbach, 2000). On April 27, 2000, the District passed an ordinance banning the sale of gasoline containing more than 0.6% by volume MTBE in the Tahoe basin. The reduction is a marked change from the concentrations of 10 to 12% by volume MTBE used in gasoline prior to the new ordinance (Anon, 2001). As with Santa Monica, although much of the publicity was given to MTBE, it should be considered that the root causes of the loss of drinking water supplies were urban growth, poor wellhead protection, and leaking gasoline tanks.

Fortunately, it seems that the occurrences in Santa Monica and South Lake Tahoe are the exception rather than the rule. According to the CDHS report summarizing the occurrence of MTBE in California drinking water, average MTBE concentrations, for all sources and systems where MTBE was detected, were highest in 1995 and 1996, ranging from 37 to 58 µg/l for sources and 13 to 40 µg/l for systems. These concentrations decreased rapidly in 1997, and remained below 10 µg/l for all sources and systems through 1999. In 2000, however, the average concentration increased slightly to 13 µg/l (12 µg/l for drinking water sources and 15 µg/l for PWSs). Over the five-year period, 95% of drinking water supplies in California had no detectable MTBE. Of the drinking water supplies with detectable MTBE between 1995 and 2000, 73% of samples and 86% of sources and systems had MTBE concentrations below the state's MCL of 13 µg/l (Williams and Sheehan, 2001). The authors point out that these findings are quite different from the media's perspective that MTBE use has resulted in widespread contamination at high concentrations of California's water resources. The study concludes that most of MTBE detections in California are below the MCL, and those that do exceed the standard are generally the result of a LUST or poor management practices. In summary, it was not the use of MTBE that caused the problem, but the fact that gasoline was spilled.

In May 2000, the Santa Clara Valley Water District published a report on the occurrence of MTBE at gasoline spill sites with operating USTs. The report included a summary of 16 case studies at LUST sites with operating USTs where gasoline releases resulted in high concentrations of MTBE in the adjacent ground water. Tanks had been upgraded (in compliance with the Federal UST Regulations and the California Code of Regulations, Title 23 Water, Division 3, Chapter 16, UST Regulations) prior to 1992 at all 16 sites, and at one site the upgraded tank was replaced by a new UST in 1999. The purpose of the study was to determine if the observed concentrations were the result of old gasoline spills or undetected releases from upgraded or replaced tanks that were believed to be less susceptible to releases. If the new or upgraded USTs were the source of the problem, then the study might be able to determine where and why the releases were occurring. The results of the study revealed that 13 (82%) of the sites had previously undetected releases from newer or upgraded UST systems. Two of the 13 newly detected leaks were traced back to definite mechanisms of release: a leak in the primary piping of one UST, and a documented vapor release in the second. Two other cases were the result of undetermined pre-upgrade releases. For the remaining 12 case studies, a specific release mechanism could not be determined. Nevertheless, 12 of the 13 MTBE plumes could be traced back to an origin in the direct vicinity of the UST, while one case study revealed inconclusive evidence. Five of the releases were thought to be related to inadequate sumps around the storage tanks, and seven were thought to be the result of problems with trenches, piping, and dispensers (Tulloch *et al.*, 2000). The study concluded that upgraded or replaced USTs do not offer complete protection from gasoline releases and ground water contamination.

The Santa Clara study also compared MTBE and benzene occurrence. Maximum MTBE concentrations at individual sites ranged from 69,000 to 430,000 µg/l, while corresponding benzene concentrations ranged from nondetect to 23,000 µg/l. Ten of the sites had benzene concentrations below 600 µg/l, and seven had concentrations below 100 µg/l. Of the 16 sites, 14 have or have had an increasing MTBE trend, and half of those exhibited an increase of MTBE concentrations by two or more orders of magnitude since the original MTBE detection (Tulloch *et al.*, 2000). However, since the study was conducted in areas where gasoline was suspected to have leaked from UST sites, it is not surprising that at some point in the historic record, MTBE concentrations increased over a certain period of time. Unfortunately, a historical analysis of benzene concentrations was not included in the study.

As of the release of the District's report, low levels of MTBE had been detected at two PWS wells in Santa Clara County. As a result of these detections, several studies followed, the first being the Free Well Water MTBE Testing Study (Tulloch *et al.*, 2000). The purpose of the study was to test domestic water wells within one-half mile of documented LUST sites. Of the 301 wells that matched the study criteria, 51 were tested. MTBE was detected at 4 of the tested wells, although the highest concentration (9.8 μg/l) was below the state's MCL. The second study was the Focused Groundwater MTBE Monitoring Program (Tulloch *et al.*, 2000). It evaluated MTBE occurrence in Santa Clara County's deeper aquifers in five monitoring areas, each of which was approximately 10 square kilometers (4 square miles). The five study areas contained 104 deep monitoring and supply wells that were sampled every six months starting in 1999. At a 3 μg/l detection limit, there was no MTBE detected in any of the 104 wells (Tulloch *et al.*, 2000). Most MTBE detections historically have occurred in shallower aquifers, so the lack of detections is consistent with historical data.

Because of concerns about MTBE in California, Governor Gray Davis established a panel of University of California (UC) professors and researchers to prepare a comprehensive overview of the human health and environmental effects of MTBE use (Keller *et al.*, 1998; Keller, 1999). The results were released in a November 1998 report titled "Health and Environmental Assessment of MTBE" (Keller *et al.*, 1998) which influenced Davis' decision to phase out the use of MTBE in California by December 31, 2002.

In response to the UC report, Malcolm Pirnie prepared a commentary for the Methanol Institute in August 2001 (Malcom Pirnie, Inc., 2001). Malcolm Pirnie compared predicted and actual MTBE detections in California. The UC report had extrapolated that between 60 and 340 public drinking water supply wells would become impacted by MTBE in the future, resulting in a total of approximately 100 to 400 impacted wells (including the 35 wells that had already been impacted). The UC report had relied on the CDHS's MTBE detection frequency of 1.2% in 2,988 monitored public water wells up to 1998. However, 7,981 additional sources were sampled between March 2000 and June 2001, with an MTBE detection frequency of only 0.15%, bringing the overall detection frequency down to 0.6% (Malcolm Pirnie, 2001). Using this new detection frequency and the total of 10,931 unsampled active PWS wells, the predicted number of future MTBE-impacted public wells is 16. In other words, most MTBE-affected PWSs are already known.

The Malcolm Pirnie report also concluded that the UC researchers had overestimated MTBE impacts to surface waters from recreational vehicles. Since the publication of the UC report, it has been discovered that MTBE volatilizes quite rapidly (in approximately 40 days) when discharged to surface water. Additionally on many of California's drinking water reservoirs, the phase out of 2-stroke boat engines, which previously discharged a considerable amount of BTEX and MTBE to surface waters as unburned fuel, has significantly reduced the frequency and concentrations of MTBE in California surface waters. Based on this new information, the UC's assumption that all impacted recreational water will result in $160 to 200 million of damage is seriously overestimated (Malcolm Pirnie, 2001).

Comparison among Texas, Florida, and California. Studies conducted by the Chevron Research and Technology Company and Integrated Science and Technology compared the results of the plume comparison studies. Buscheck *et al.* (1998) compiled MTBE data for California and Texas. MTBE exceeded concentrations of 1 μg/l at 47% of the operating service station sites investigated in California and 63% of the sites investigated in Texas. Reisinger *et al.* (2000) concluded that because MTBE plumes were generally similar in length to benzene plumes, and because benzene plumes were considered to be biologically controlled, it appears that many of the MTBE plumes are also being naturally biodegraded. The biggest contrast between the Texas and Florida studies was that Texas used the same concentration for MTBE and benzene plume delineations, while Florida took the different drinking water standards into account.

COMPARISON OF PLUME LENGTHS FOR MTBE AND BTEX AT 212 SOUTH CAROLINA SITES

Barbara H. Wilson, Dynamac Corporation

Until recently, it was commonly believed that remediation requirements were different for MTBE and BTEX because MTBE was considered less biodegradable and, therefore, less subject to natural attenuation than the BTEX constituents of fuel spills. Based on its chemistry and biology, MTBE plumes were also thought to recurrently form substantially longer and larger plumes than contamination from BTEX, and indeed, MTBE plumes as long as 1,500 meters (5,000 feet) have been formed from LUSTs (Weaver *et al.*, 1999). Many regulators and others responsible for remediation of fuel spills have been concerned with the potential for the perceived increased MTBE plume lengths (as compared to BTEX plume lengths) to result in increased cleanup costs.

During an evaluation of the cost of MTBE remediation, South Carolina provided an extensive data set from UST cleanups that included the selected remedial technologies, plume length and width, saturated thickness of the impacted aquifer, and concentration information on MTBE and BTEX (Wilson *et al.*, 2001). Information on 144 active corrective action (ACA) and 68 monitored natural attenuation (MNA) sites were provided for a total of 212 sites, with cost information provided for 183 sites.

The 212 plumes were evaluated for equal plume lengths, BTEX greater than MTBE, and MTBE greater than BTEX, and the MTBE and BTEX plume

Table A-4. **Comparison of BTEX and MTBE plume lengths in South Carolina**

	Number	BTEX Plume Length (ft)	BTEX Plume Area (ft²)	MTBE Plume Length (ft)	MTBE Plume Area (ft²)
BTEX=MTBE	117	150 (224)	15,600 (54,568)	150 (224)	16,000 (55,104)
BTEX>MTBE	54	288 (328)	48,100 (88,929)	120 (147)	9,600 (19,113)
BTEX<MTBE	41	170 (179)	14,625 (26,325)	250 (271)	30,000 (44,290)
Sand	121	160 (228)	18,000 (57,092)	150 (197)	15,600 (40,044)
Silt	55	250 (303)	43,200 (62,888)	200 (265)	30,000 (54,758)
Clay	36	128 (193)	13,250 (52,750)	128 (188)	14,300 (39,950)

Median (arithmetic mean)

lengths and areas were also compared in sand, silt, and clay (Table A-4). A frequency distribution of the 212 plumes is presented in Figure A-18.

In the majority of the sites, BTEX plume length was equal to or greater than the MTBE plume length (171 versus 41). The plume lengths ranged from 3 to 400 meters (10 to 1,400 feet) for BTEX and 3 to 400 meters (10 to 1,400 feet) for MTBE. Median values for the 212 BTEX and MTBE plume lengths were 56 meters (185 feet) and 54 meters (178 feet), respectively. Median values for the 212 BTEX and MTBE plume areas were 1,800 square meters (19,800 square feet) and 1,500 square meters (16,200 square feet), respectively. The 90% percentile for the BTEX plume length was 140 meters (480 feet) and 120 meters (400 feet) for the MTBE plume. At 41 sites (18.2%), MTBE plume length was greater than the BTEX plume length with a median value of 75 meters (250 feet) compared to 50 meters (170 feet). For the ACA sites, the median MTBE plume length was 80 meters (260 feet) compared to a median BTEX length of 60 meters (200 feet). The median MTBE plume length was 36 meters (120 feet) compared to 20 meters (68 feet) for the BTEX plume for the MNA sites.

When separated by soil type (sand, silt, or clay), the median plume lengths and plume areas for both MTBE and BTEX plumes were larger in silt than those in sand or clay, although the mean lengths and areas were similar in the three soil types (Table A-4). The larger median length and area values in silt may reflect the presence of gravel stringers that would substantially increase the hydraulic conductivity of the silt layers.

In general, when sorted by soil type, the median plume areas for the ACA sites (2,200 square meters [23,300 square feet] to 4,300 square meters [46,800 square feet]) are approximately an order of magnitude larger than the MNA sites (200 square meters [1,700 square feet] to 450 square meters [4,800 square feet]), indicating that active remediation is chosen for sites that potentially pose a greater risk to a receptor. The median plume lengths and areas for the MNA sites are consistent for both MTBE (length: 11.4 to 24 meters [37.5 to 80 feet], area: 200 to 400 square meters [1,700 to 4,800 square feet]) and BTEX

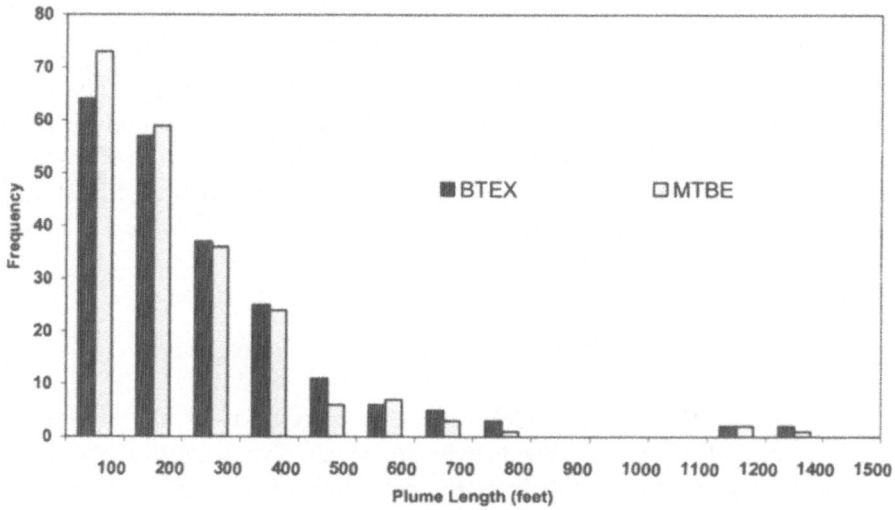

Figure A-18.   **Plume Length Distribution of 212 USTs from South Carolina.**

(length: 12 to 30 meters [40 to 100 feet], area: 200 to 410 meters [1,700 to 4,450 square feet]) in sand, silt, or clay. A similar pattern is observed for the ACA sites for MTBE (length: 30 to 84 meters [100 to 275 feet], area: 2,240 to 3,695 meters [24,150 to 39,775 square feet]) and BTEX (length: 41 to 81 meters [135 to 265 feet], area: 2,200 to 4,300 square meters [23,300 to 46,800 square feet]).

An interesting plume geometry was observed in the ACA sites located in sand, whereby the median widths of the respective MTBE and BTEX plumes, 85 meters (275 feet) and 75 meters (250 feet), were greater than the respective median MTBE and BTEX plume lengths, 60 meters (200 feet) and 70 meters (225 feet). This circular geometry is often observed in floodplain environments and may result from a readily fluctuating ground water flow direction due to a relatively flat hydraulic gradient (Schirmer *et al.*, 2000).

Results from this study are similar to findings of Reid *et al.* (1999), Mace and Choi (1998), and Happel *et al.* (1998) in their assessment of MTBE and benzene plumes in Florida, Texas, and California, respectively. In these studies, the respective median plume lengths were 35, 55, and 31 meters (115, 181, and 101 feet) for Florida, Texas, and California for benzene (at 10 µg/l). Median plume lengths for MTBE (at 10 µg/l) were 43 meters (140 feet) for Florida and 50 meters (163 feet) for Texas. The median plume lengths for BTEX and MTBE in South Carolina were 56 meters (185 feet) and 54 meters (178 feet), respectively.

Natural attenuation of MTBE has been significant at some locations (Wilson *et al.*, 2000; Wilson *et al.*, 1999; Cho *et al.*, 1997), especially under strongly methanogenic conditions (Kolhatkar *et al.*, 2000), and may be a factor in the short plume lengths of the South Carolina data. The concerns of formation of longer MTBE plumes than BTEX plumes based on the chemical and physical characteristics and the perceived biological recalcitrance of the oxygenate were not observed in the data from South Carolina. The relatively modest median values for these plume sets suggest that biological and chemical weathering were sufficient to prevent formation of extensive MTBE and BTEX plumes at many locations.

CONCLUSIONS

James A. M. Thomson,   Applied Hydrology Associates, Inc.

The studies described above represent a very large body of data. However, drawing general conclusions needs to be approached with caution because of variations between studies. Perhaps the most important variable is the purpose of the study or sampling program. The purpose of the sampling program can contain inherent biases towards a particular sample set. Specific program objectives could include one or more of the following:

1. Statistically represent water quality conditions in an aquifer system.
2. Detect previously unknown VOC plumes and attempt to identify their sources.
3. Characterize status of LUST site management.
4. Determine effectiveness of tank upgrade program, remedial response, or natural attenuation, in reducing MTBE occurrence, concentration, or plume size.
5. Identify which chemical should "drive" remediation or the oversight program activities in an administrative area.
6. Compare MTBE fate and transport with benzene. (This is not possible if only MTBE data are reported.)

Other factors that often vary from study to study and from state to state are listed below.

1. Is the study area a designated RFG area or a non-RFG area?
2. If the study area is an RFG area, when was oxygenated fuel introduced, and what type is used (MTBE or ethanol)?
3. If a non-RFG area, is oxygenated fuel used for reasons other than RFG designation?
4. What date was MTBE changed from an optional to a required monitoring analyte? If data that do not include MTBE analyses are mixed with data including MTBE analyses, then as MTBE analyses become more commonly required, this will likely lead to an apparent increase in MTBE detection.
5. For CWSs: do they draw from a ground water or a surface water

source? MTBE breakdown is more rapid and detection frequency is lower in surface water.

6. Are wells to be sampled only selected from known LUST sites? This biases results upwards.

7. Alternatively, are known LUST sites excluded from the sampled population? This biases results downwards.

8. Are "randomly" chosen wells actually non-randomly distributed? For example, wells may follow the position of a specific aquifer formation, which in turn may be associated with urban development.

9. What is the type of well sampled? The type of well sampled affects its "catchment area": samples from large CWS wells represent water drawn from a larger radius than smaller, private domestic supply wells, which in turn sample a larger area than unpumped monitoring wells. However, monitoring wells are typically installed in urban areas, often around gas stations, and therefore have more probability of being in or near a LUST plume.

10. Vertical permeability of the unsaturated zone, affecting the rate of plume development.

11. Lateral permeability and ground water flow rate, affecting the rate of plume growth.

12. Local standards for BTEX and MTBE, and variation in these standards over time.

13. Computer versus hand-drawn plume contours to define plume length.

14. Selection of defining contour for plume length (*e.g.*, 1 µg/l, 20 µg/l, 200 µg/l).

15. Completeness of monitoring network for plume length characterization. (Was vertical delineation complete?)

The broadest conclusions that may be drawn from the U.S. national and state studies are that the threat to water quality posed by MTBE has not proved to be as widespread, persistent, and intractable as initially forecast, and in general it is associated with other gasoline constituents, meaning that the environmental concern is a leaking tank problem rather than an MTBE problem.

Overall, MTBE was detected at higher frequencies and concentrations between 1995 and 1998. If the 1995 to 1998 sampling was truly representative, and if MTBE plumes were rapidly growing, continued sampling would be expected to result in stable or increasing MTBE detection frequencies and concentrations. However, since 1998, results have shown the opposite trend. Detection frequencies and concentrations have reduced since 1998, and in most cases, continue to drop from year to year. This suggests that: (1) the 1995 to 1998 sampling was biased towards sites with higher MTBE concentrations;

(2) the incidence of oxygenated fuel releases is decreasing due to better USTs and improved monitoring; and (3) remedial efforts and natural attenuation are mitigating MTBE impacts to ground water.

REFERENCES

Anon (Anonymous). 2000. MTBE — How should Europe respond? *Petroleum Review.* February 2000, 37–38.

Anon (Anonymous). 2001. *Some FAQ's About MTBE in South Tahoe.* www.tahoe.ceres.ca.gov/stpud/faqsmtbe.html.

ADEQ (Arizona Department of Environmental Quality). 1999. *Report on Methyl Tertiary Butyl Ether.* October 1, 1999. www.adeq.state.az.us/environ/waste/ust/download/1001mtbe.pdf

Baker, M. 2000. Minnesota Drinking Water Protection Program. Personal communication. Provided information on well data. (651-215-0794)

Blue Ribbon Panel (Blue Ribbon Panel on Oxygenates in Gasoline). 1999. *Achieving Clean Air and Clean Water: The Report of the Blue Ribbon Panel on Oxygenates in Gasoline.* EPA-A20-R-99-021. September 15, 1999.

Buscheck, T.E., Gallagher, D.J., Peargin, T.R., Kuehne, D.L., and Zuspan, C.R. 1998. Occurrence and behavior of MTBE in groundwater. In: *Proceedings of The Southwest Focused Ground Water Conference—Discussing the Issue of MTBE and Perchlorate in Ground Water,* Anaheim, California. June 3–4, 1998. National Ground Water Association, p. 2–3.

Cal/EPA (California Environmental Protection Agency). 1999. *Public Health Goal for Methyl Tertiary Butyl Ether (MTBE) in Drinking Water.* Pesticide and Environmental Toxicology Section, Office of Environmental Health Hazard Assessment, California Environmental Protection Agency, Sacramento, California. March 1999.

Catalist. 2000. *Petrol Retail Station Database for the UK.* Catalist, Clifton, Bristol, UK.

Cho, J.S., J.T. Wilson, D.C. DiGiulio, J.A. Vardy, and W. Choi. 1997. Implementation of natural attenuation at a JP-4 jet fuel release after active remediation. *Biodegradation.* 8, 265–273.

Christer, J. 2000. Illinois Environmental Protection Agency. Personal communication to discuss PWS system well sampling. (217-782-8482)

Clawges, R.M., Zogorski, J.S., and Bender, D. 2000. *Abstract; Key MTBE Findings Based on National Water-Quality Monitoring.* Rapid City, South Dakota, USGS.

Crumly, M. 2000. Illinois Environmental Protection Agency, Bureau of Water, Springfield, Illinois. Personal communication to discuss to IEPA VOC and MTBE PWS database. (217-785-0561)

Damgaard, M. 2000. Wisconsin Department of Natural Resources. Personal communication regarding PWS system well data. (608-266-0738)

Dernbach, L.S. 2000. The complicated challenge of MTBE cleanups. *Environmental Science and Technology*. 34(23), 516A–521A.

Duren, S. 2000. Programmer/Analyst, Wisconsin Department of Natural Resources. Personal communication regarding PWS system wells. (608-266-8617)

ENSR and AHA (ENSR International and Applied Hydrology Associates, Inc.). 2001. *Investigation of Selected Gasoline Constituents in Groundwater in Seven States*. Document # 04373-006-200. October 2001.

Environmental Agency. 1998. *Policy and Practice for the Protection of Groundwater*, 2nd Edition. TSO ISBN 0 11 310145 7.

Environmental Agency. 2000. *A Review of Current MTBE Usage and Occurrence in Groundwater in England and Wales*. TSO ISBN 0 11 310181 3.

Environment Agency. 2002. www.envronment-agency.gov.uk/.

Frye, E. 2000. *A Survey of State Experiences with MTBE Contamination at LUST sites-Executive Summary*. The New England Interstate Water Pollution Control Commission. December 15, 2000. www.neiwpcc.org/mtbees. pdf.

Grady, S. J. and Casey, G. D. 1999. *Occurrence and Distribution of Methyl tert-Butyl Ether and Other Volatile Organic Compounds in Drinking Water in the Northeast and Mid-Atlantic Regions of the United States, 1993–1998*. U.S. Geological Survey Water-Resources Investigation Report, 00-4228.

Gregg, W., Moyer, E., Boehm, C., and Day, M. 2002. *Multi-State Survey of Gasoline Constituents in Midwest Groundwater*. Presented at AEHS MTBE Conference, San Diego, California. March 2002.

Gullick, R.W. and LeChevallier, M.W. 2000. Occurrence of MTBE in drinking water sources. *Journal American Water Works Association*. 92:1, 100–113.

Happel, A.M., Beckenbach, E.H., and Halden, R.U. 1998. *An Evaluation of MTBE Impacts to California Groundwater Resources*. Lawrence Livermore National Laboratory, University of California, Livermore, California. UCRL-AR-130897. June 11, 1998.

Hattan, G. 2000. Kansas Department of Health and Environment, Bureau of Environmental Remediation, Topeka, Kansas. Personal communication to discuss PWSs in Kansas. (785-296-5931)

Hirsch, R. M. 2001. *Statement Before the United States House of Representatives, Committee on Energy and Commerce, Subcommittee on Oversight and Investigations*. Associate Director for Water, U.S. Geological Survey, U.S. Department of the Interior. November 1, 2001. sd.water.usgs.gov/nawqa/vocns/USGS_MTBE_testimony.html

Hunter, B., Hahn, B., Shutty, M., and Seaward, P. 1999. *Impact of Small Gasoline Spills on Groundwater*. Preliminary report abstract presented at the Maine Water Conference Meeting. April 15, 1999.

IDEQ (Idaho Division of Environmental Quality). 1999. *An Evaluation of Methyl Tert-Butyl Ether (MTBE) in Groundwater at Leaking Underground*

Storage Tank Sites. Watershed and Aquifer Protection Bureau. February 1999. http://www2.state.id.us/deq/ust/mtbestal.pdf.

IEPA (Illinois Environmental Protection Agency). 2000a. *State MTBE/BTEX Municipal Well Database.*

IEPA. 2000b. Groundwater Monitoring Raw Source Location Report. Division of PWSs.

IDEM (Indiana Department of Environmental Management). 2000. *Indiana Department of Environmental Management 1999 Annual Compliance Report for Indiana PWS Systems.* Drinking Water Branch.

KDHE (Kansas Department of Health and Environment). Kansas Environmental News, April 2000. www.kdhe.state.ks.us/opp/ken/april2000/index.html#mtbe.

Keller, A., Froines, J., Koshland, C., Reuter, J., Suffet, I., and Last, J. 1998. Health and environmental assessment of MTBE, summary and recommendations. In: *Health and Environmental Assessment of MTBE,* Volume 1. UC Toxics Research and Teaching Program, UC Davis. November 1998.

Keller, A. 1999. Health and environmental assessment of MTBE — The California perspective. In: *Proceedings of the American Water Works Association Annual Conference,* Chicago, Illinois. June 20–24, 1999.

Kolhatkar, R., J. Wilson, L.E. Dunlap. 2000. Evaluating natural biodegradation of MTBE at multiple UST sites. In: *Proceedings of the 2000 Petroleum Hydrocarbons and Organic Chemicals in Ground Water-Prevention, Detection, and Remediation.* Houston, Texas.

Lapham, W.W., Wilde, F.D., and Koterba, M.T. 1995. *Ground-Water Data-Collection Protocols and Procedures for the National Water-Quality Assessment Program: Selection, Installation, and Documentation of Wells, and Collection of Related Data.* U.S. Geological Survey Open-File Report 95-398.

Loa, A. 2000. Indiana Department of Environmental Management. Personal communication regarding PWS system well data. (317-308-3283)

MacAfee, P. 2000. Minnesota Drinking Water Protection Program. Provided anecdotal information on ethanol in drinking water and drinking water monitoring. (651-215-0747)

Mace, R.E., and Choi, W.-J. 1998. The size and behavior of MTBE plumes in Texas. In: *Proceedings of the Petroleum Hydrocarbons and Organic Chemicals in Ground Water — Prevention, Detection, and Remediation Conference,* Houston, Texas. November 11–13, 1998. pp. 1–11. (Stanley, A., Ed.) National Ground Water Association and American Petroleum Institute.

Maine. 1998. *The Presence of MTBE and Other Gasoline Compounds in Maine's Drinking Water.* Maine MTBE Drinking Water Study. Bureau of Health, Department of Human Services, Bureau of Waste Management and Remediation, Department of Environmental Protection, Maine Geological Survey, and Department of Conservation. Preliminary Report, October 13, 1998.

Malcom Pirnie, Inc. 2001. *Water Quality Impacts of MTBE: An Update Since the Release of the UC Report.* Prepared for the Methanol Institute, Oakland, California. August 2001.

Mandy, D. 2000. Minnesota Department of Health, Section of Drinking Water Protection. Provided information on PWS systems and testing results.

Maryland Task Force (Maryland Task Force on the Environmental Effects of MTBE). 2000. *December 2000 Preliminary Report.* Maryland State Legislature, House Bill 823. www.mde.state.md.us/was/mtbe_report.pdf.

McKinnon, R. and Dyksen, J. 1984. Removing organics from groundwater through aeration plus GAC. *Journal of the American Water Works Association.* 76(5), 42–47.

McMillan, D. 2000. Illinois Environmental Protection Agency, Bureau of Water, Springfield, Illinois. Personal communication to discuss the PWSs in Illinois. (217-782-2829)

MDEQ (Michigan Department of Environmental Quality). 2000. *MTBE (methyl tertiary-butyl ether) Fact Sheet.* Storage Tank Division. March 2000. www.deq.state.mi.us/std/MTBE.pdf.

Morgenroth, E. and Arvin, E. 1999. *The European perspective to MTBE as an oxygenate in fuels.* International Congress on Ecosystem Health-Managing for Ecosystem Health, Sacramento, California. August 1999, p. 5–20.

NESCAUM. 1999. *RFG/MTBE Finding and Recommendations.* Northeast States for Coordinated Air Use Management, Boston, Massachusetts. August 1999. www.nescaum.org/RFG/RFGPh2.shtml.

NJDEP (New Jersey Department of Environmental Protection). 2000. *MTBE in New Jersey's Environment.* MTBE Work Group. www.state.nj.us/dep/dsr/mtbe/MTBE-NJ.PDF.

NYSDEC (New York State Department of Environmental Conservation). 2000. *Survey of Active New York State Gasoline Remediation Sites with Potential MTBE Contamination.* Division of Environmental Remediation, Bureau of Spill Prevention and Response. Final Report, February 8, 2000.

Pelayo, A., Tobias, J., McKnight, K., and Egre, L. 2000. MTBE in Wisconsin groundwater. In: *Information from Wisconsin Department of Natural Resources Bureau for Remediation and Redevelopment News.* PUB-RR-651, Vol. 10, No. 2, 1–3.

Plummer, D. 2000. Kansas Department of Health and Environment. Personal communication regarding PWS system well data. (785-296-5523)

Reid, J.B., Reisinger, H.J. II, Bartholomae, P.G., Gray, J.C., and Hullman, A.S. 1999. A comparative assessment of the long-term behavior of MtBE and benzene plumes in Florida, USA. In: *Natural Attenuation of Chlorinated Solvents, Petroleum Hydrocarbons, and Other Organic Compounds,* V. 1. p. 97–102. Proceedings of the Fifth International *In Situ* and Onsite Bioremediation Symposium, San Diego, California. April 19–22, 1999. (Alleman, B.C., and Leeson, A., Eds.). Columbus, Ohio, Battelle Publications.

Reisinger H.J., Reid J.B., and Bartholomae P.J. 2000. MTBE and benzene plume behavior a comparative perspective. *Soil, Sediment and Groundwater.* March 2000, 43–46.

Roeff, M. 2000. Indiana Department of Environmental Management. Correspondence regarding PWS system well data. (317-308-3282)

Schirmer, M., Weiss, H., Durrant, G.C., Molson, J.W., and Frind, E.O. 2000. Influence of changing flow directions on plume shape, apparent dispersion and mass loss of biodegradable compounds. In: *Proceedings Groundwater 2000: International Conference on Groundwater Research.* Copenhagen, Denmark. June 6–8, 2000.

Spivey, E. 2000. Kansas Department of Health and Environment. Personal communication regarding PWS system well data. (785-296-6434)

Missouri. 2000. Department of Health, Department of Agriculture, Department of Natural Resources, Petroleum Storage Tank Insurance Fund, and Attorney General's Office. *Report on Inter-Agency Meetings on MTBE.* September 2000. www.pstif.org/apps/mtbe_report.pdf.

Tulloch, C., Crowley, J., and Hemmeter, T. 2000. *An Evaluation of MTBE Occurrence at Fuel Leak Sites with Operating Gasoline USTs.* Santa Clara Valley Water District, San Jose, California. May 2000.

USEPA (United States Environmental Protection Agency). 1997. *Drinking Water Advisory: Consumer Acceptability Advice and Health Effects Analysis on Methyl Tertiary-Butyl Ether.* USEPA Fact Sheet No. EPA-822-F-97-009. December 1997.

USEPA (United States Environmental Protection Agency). 1999. *Unilateral Administrative Order for Water Replacement.* Region IX, USEPA Docket No. RCRA 7003-09-99-0007.

USEPA (U.S. Environmental Protection Agency). 2001. *Where Does My Drinking Water Come From?* www.epa.gov/OGWDW/wot/wheredoes.html.

USGS (United Stated Geological Survey). 2001. *The National Water-Quality Assessment Warehouse.* infotrek.er.usgs.gov/pls/nawqa/nawqa_home.

WADOE (Washington State Department of Ecology). 2000. *Occurrence of Methyl Tertiary-Butyl Ether (MTBE) in Groundwater at Leaking Underground Storage Tank Sites in Washington.* Publication No. 00-09-054. October 2000.

Weaver, J.W., Haas, J.E., and Sosik, C.B. 1999. Characteristics of gasoline releases in the water table aquifer of Long Island. In: *Proceedings of the National Ground Water Association/American Petroleum Institute Conference1999 Petroleum Hydrocarbons Conference and Exposition.* November 17–19, 1999. Houston, Texas.

Williams, P.R.D., Scott, P.K., Sheehan, P.J., Paustenbach, D.J. 2000. A probabilistic assessment of household exposures to MTBE in California drinking water. *Human and Ecological Risk Assessment.* 6(5), 827–849.

Williams, P.R.D. and Sheehan, P.J. 2001. A better perspective on the incidence and implications of MTBE in California's drinking water. *Contaminated*

Soil Sediment and Water. Special Oxygenated Fuels Issue (Spring 2001), 23–28.

Williamson, S. 2001. USGS NAWQA Database Team Leader, personal communication.

Wilson, B.H., Shen, H., Pope, D., and Schemelling, S. 2001. Cost of MTBE remediation. *Bioremediation of MTBE, Alcohols, and Ethers.* The Sixth International *In Situ* and On-Site Bioremediation Symposium. San Diego, California. June 4–7, 2001. (Magar, V.S., Gibbs, J.T., O'Reilly, K.T., Hyman, M.R., and Leeson, A., Eds.). Battelle Press. Columbus, Ohio.

Wilson, J.T., Cho, J.S., Wilson, B.H., and Vardy, J.A. 2000. *Natural Attenuation of MTBE in the Subsurface Under Methanogenic Conditions.* EPA/600/R-00/006.

Wilson, B.H., Shen, J., Cho, J., and Vardy, J. 1999. Use of Bioscreen to evaluate natural attenuation of MTBE. In: *Natural Attenuation of Chlorinated Solvents, Petroleum Hydrocarbons, and Other Organic Compounds.* Proceedings of the Fifth International *In Situ* and On-site Bioremediation Symposium. San Diego, California. April 19–22, 1999. (Alleman, B.C. and Leeson, A., Eds.). Battelle Press. Columbus, Ohio.

Zogorski, J.S., Morduchowitz, A.M., Baehr, A.L, Bauman, B.J., Conrad, D.L., Drew, R.T., Korte, N.E., Lapham, W.W., Pankow, J.F., and Washington, E.R. 1997. Chapter 2: Fuel oxygenates and water quality. In: *Interagency Assessment of Oxygenated Fuels.* Washington D.C., Office of Science and Technology Policy, Executive Office of the President.

ACRONYMS

ACA	active corrective action
ADEQ	Arizona Department of Environmental Quality
AWWA	American Waterworks Association
BTEX	benzene, toluene, ethylbenzene, and xylenes
Cal/EPA	California Environmental Protection Agency
CDHS	California Department of Health Services
CTDEP	Connecticut Department of Environmental Protection
CTDPH	Connecticut Department of Public Health
CWS	community water supply
IDEQ	Idaho Division of Environmental Quality
KDHE	Kansas Department of Health and Environment
LUST	leaking underground storage tank
MADEP	Massachusetts Department of Environmental Protection
MCL	maximum contaminant level
MDA	Michigan Department of Agriculture
MDE	Maryland Department of the Environment
MDEQ	Michigan Department of Environmental Quality

MNA	monitored natural attenuation
MTBE	methyl tert butyl ether
NAWQA	National Water Quality Assessment
NDHHS	Nebraska Department of Health and Human Services
NEIWPCC	New England Interstate Water Pollution Control Commission
NESCAUM	Northeast States for Coordinated Air Use Management
NJDEP	New Jersey Department of Environmental Protection
NTNCWS	non-transient non-community water system
NYSDOH	New York State Department of Health
NYSDEC	New York State Department of Environmental Conservation
PADEP	Pennsylvania Department of Environmental Protection
ppb	parts per billion (equivalent to $\mu g/l$ for dilute solutions)
PWS	public water supply
RBCA	risk-based corrective action
RCRA	Resource Conservation and Recovery Act
RFG	reformulated gasoline
RIDOH	Rhode Island Department of Health
TBA	tert butyl alcohol
U.K.	United Kingdom
U.S.	United States
UC	University of California
USEPA	U.S. Environmental Protection Agency
USFDA	U.S. Food and Drug Administration
USGS	U.S. Geological Survey
UST	underground storage tank
VOC	volatile organic compound
v/v	volume to volume
WADOE	Washington State Department of Ecology WDNR Wisconsin Department of Natural Resources
$\mu g/l$	micrograms per liter (equivalent to ppb for dilute solutions)

Appendix B

Primary Author Contact Information

Name: **Robert M. Cataldo, P.G., LSP**
Affiliation: ENSR International
Address: 2 Technology Park Drive, Westford, MA 01886
Telephone: (978) 589-3000
Fax: (978) 589-3705
Email: rcataldo@ensr.com

Name: **Theodore R. Davis, P.E.**
Affiliation: Southwestern Environmental, Inc.
Address: 16476 Hunters Trail, Montgomery, TX 77356
Telephone: (713) 208-1582
Fax: (936) 449-5087
Email: davist@swbell.net; ted@davismail.com

Name: **Michael J. Day**
Affiliation: Applied Hydrology Associates, Inc.
Address: 950 South Cherry Street, Suite 810, Denver, CO 80246
Telephone: (303) 782-0164 x110
Fax: (303) 782-2560
Email: mday@appliedhydrology.com

Name: **David L. Espy**
Affiliation: ENSR International
Address: 2 Technology Park Drive, Westford, MA 01886
Telephone: (978) 589-3000
Fax: (978) 589-3100
Email: despy@ensr.com

Name: **Kevin T. Finneran, Ph.D.**
Affiliation: GeoSyntec Incorporated
Address: 629 Massachusetts Avenue, Boxborough, MA 01719
Telephone: (978) 263-9588 x246
Fax: (978) 263-9594
Email: kfinneran@geosyntec.com

Name: **Jonathan Greene, P.E.**
Affiliation: Malcolm Pirnie
Address: 1700 West Loop South, Suite 950, Houston, TX 77027
Telephone: (713) 960-7408
Fax: (713) 840-1207
Email: JGreene@PIRNIE.COM

Name: **Evan T. Johnson, P.E., LSP**
Affiliation: Tighe & Bond, Inc.
Address: 53 Southampton Road, Westfield, MA 01085
Telephone: (413) 572-3254
Fax: (413) 562-5317
Email: etjohnson@tighebond.com

Name: **Kara L. Kelley**
Affiliation: Xpert Design & Diagnostics, LLC
Address: 22 Marin Way, Unit 3, Stratham, NH 03885
Telephone: (603) 778-1100
Fax: (603) 778-2121
Email: Kelley@xdd-llc.com

Name: **William B. Kerfoot, Ph.D.**
Affiliation: K-V Associates, Inc.
Address: Madaket Place — Unit B, 766 Falmouth Road, Mashpee,
               MA 02649
Telephone: (508) 539-3002
Fax: (508) 539-3566
Email: wbkerfoot@kva-equipment.com

Name: **Tie Li, Ph.D.**
Affiliation: URS Corporation
Address: 9801 Westheimer, Suite 500, Houston, TX 77042
Telephone: (713) 914-6403
Fax: (713) 789-8404
Email: Tie_Li@urscorp.com

Name: **Ernest E. Lory**
Affiliation: Naval Facilities Engineering Service Center
Address: 1100 23rd Avenue, Port Hueneme, CA 93043
Telephone: (805) 982-5851
Fax: (805) 982-4304
Email: netts@nfesc.navy.mil

Name: **Christopher G. Mariano, P.G., LSP**
Affiliation: ENSR International
Address: 2 Technology Park Drive, Westford, MA 01886
Telephone: (978) 589-3000
Fax: (978) 589-3705
Email: cmariano@ensr.com

Name: **Nancy E. Milkey, P.G.**
Affiliation: Tighe & Bond, Inc.
Address: 53 Southampton Road, Westfield, MA 01085
Telephone: (413) 572-3273
Fax: (413) 562-5317
Email: nemilkey@tighebond.com

Name: **Ellen E. Moyer, Ph.D., P.E.**
Affiliation: Tighe & Bond, Inc.
Address: 53 Southampton Road, Westfield, MA 01085
Telephone: (413) 572-3230
Fax: (413) 562-5317
Email: eemoyer@tighebond.com

Name: **Lee A. Newman, Ph.D.**
Affiliation: University of South Carolina, Norman J. Arnold School of
                Public Health
Address: 800 Sumter Street, Columbia, SC 29208
Affiliation: Savannah River Ecology Laboratory
Address: Savannah River Site, Aiken, SC 29802
Telephone: (803) 777-4795
Fax: (803) 777-3391
Email: newman2@gwm.sc.edu

Name: **Joseph E. O'Connell, ScD. P.E.**
Affiliation: Environmental Resolutions, Inc.
Address: 20372 North Sea Circle, Lake Forest, CA 92630
Telephone: (949) 457-8950
Fax: (949) 457-8956
Email: joconnell@eri-us.com

Name: **Robert J. Pirkle, Ph.D.**
Affiliation: Microseeps, Inc.
Address: University of Pittsburgh, Applied Research Center, 220
             William Pitt Way, Pittsburgh, PA 15238
Telephone: (412) 826-3433
Fax: (412) 826-5245
Email: rpirkle@microseeps.com

Name: **David K. Ramsden, Ph.D.**
Affiliation: URS Houston
Address: 9801 Westheimer, Suite 500, Houston, TX 77042
Telephone: (713) 914-6451
Fax: (713) 789-8404
Email: David_Ramsden@urscorp.com

Name: **Bruce Rittmann, Ph.D.**
Affiliation: Northwestern University, Dept. of Civil and Environmental
             Engineering
Address: 2145 Sheridan Road, Evanston, IL 60208
Telephone: (847) 491-8790
Fax: (847) 491-4011
Email: b-rittmann@northwestern.edu

Name: **Joseph Robb, P.G.**
Affiliation: AMEC Earth and Environmental
Address: 239 Littleton Road, Suite 1B, Westford, MA 01886
Telephone: (978) 692-9090
Fax: (978) 692-9085
Email: jrobb@amec.com

Name: **Gerard E. Spinnler, Ph.D.**
Affiliation: Shell Global Solutions (US) Inc., MTBE Remediation Tech-
             nology Department
Address: Westhollow Technology Center, 3333 Highway 6 South,
             Houston, TX 77082-3101
Telephone: (281) 544-7319
Fax: (281) 544-7261
Email: gerard.spinnler@shell.com

Name: **Robert J. Steffan, Ph.D.**
Affiliation: Envirogen, Inc.
Address: 4100 Quakerbridge Road, Lawrenceville, NJ 08648
Telephone: (609) 936-9300
Fax: (609) 936-9221
Email: steffan@envirogen.com

Name: **Martti R. Suominen**
Affiliation: Neste Marketing Ltd.
Address: P.O.B. 77, 02151 Espoo, Finland
Telephone: 358 50 452 8344
Email: martti.suominen@fortum.com

Name: **Brian D. Symons, P.E.**
Affiliation: The RETEC Group, Inc.
Address: 6800 W. 64th Street, Suite 108, Shawnee-Mission, KS 66202
Telephone: (913) 362-8444
Fax: (913) 362-1044
Email: bsymons@retec.com

Name: **James A.M. Thomson**
Affiliation: Applied Hydrology Associates, Inc.
Address: 950 S. Cherry St., Suite 810, Denver, CO 80246
Telephone: (303) 782-0164
Fax: (303) 782-2560
Email: jthomson@appliedhydrology.com

Name: **Pamela R.D. Williams, Sc.D.**
Affiliation: Exponent
Address: 4940 Pearl East Circle, Ste. 300, Boulder, CO 80301
Telephone: (720) 406-8115
Fax: (303) 444-7528
Email: pwilliams@exponent.com

Name: **Barbara H. Wilson**
Affiliation: Dynamac Corporation
Address: 3601 Oakridge Blvd., Ada, OK 74820
Telephone: (580) 436-6415
Fax: (580) 436-6496
Email: bwilson@dynamac.com

Name: **John T. Wilson, Ph.D.**
Affiliation: USEPA National Risk Management Research Laboratory,
                  Office of Research and Development
Address: P.O. Box 1198, Ada, OK 74820
Telephone: (580) 436-8534
Fax: (580) 436-8703
Email: Wilson.JohnT@epamail.epa.gov

Acronyms

0/00	per mill
$/lb	price per pound
ACA	active corrective action
ADD	average daily dose
AOP	advanced oxidation process
APH	air petroleum hydrocarbon
AQDS	anthraquinone-2,6-disulfonate
AS/SVE	air sparging/soil vapor extraction
AST	aboveground storage tank
ASTM	American Society for Testing and Materials
atm m^3/mol	atmosphere cubic meter per mole
ATSDR	Agency for Toxic Substances and Disease Registry
bgs	below ground surface
BOD	biological oxygen demand
BTEX	benzene, toluene, ethylbenzene, and xylenes
BTU	British thermal unit
Cal/EPA	California Environmental Protection Agency
CA-OEHHA	California Office of Environmental Health Hazard Assessment
CARB	California Air Resources Board
CASWRCB	California State Water Resources Control Board
CDHS	California Department of Health Services
CDPHE	Colorado Department of Public Health and Environment
cfm	cubic feet per minute
CFU	colony forming unit
cm/s	centimeter per second
cm^2/s	square centimeter per second
cmm	cubic meter per minute
CPT	cone penetrometer technology
CSF	cancer slope factor
CV	coefficient of variation
CWS	community water supply

DAI	direct aqueous injection
DAS	diffused aeration stripper
DCA	1,1-dichloroethane
DCE	dichloroethene
DIPE	di-isopropyl ether
DNAPL	dense non-aqueous phase liquid
DPE	dual-phase extraction
dpm	disintegrations per minute
EBCT	empty bed contact time
E-beam	electron beam
E°	oxidation-reduction potential
EPH	extractable petroleum hydrocarbon
ESTCP	Department of Defense's Environmental Security Technology Certification Program
ETBE	ethyl tert butyl ether
EU	European Union
eV	electron volt
FBR	fluidized bed reactor
FID	flame ionization detector
FLDEP	Florida Department of Environmental Protection
ft/day	foot per day
ft²/day	square foot per day
g/day	gram per day
g/eq	gram per equivalent
g/hr	gram per hour
g/kg	gram per kilogram
GAC	granular activated carbon
GC	gas chromatograph(y)
GMP	ground water management permit (New Hampshire)
gpm	gallon per minute
GPR	ground-penetrating radar
HDPE	high density polyethylene
HFM	hollow fiber membrane
HI	hazard index
HIBA	hydroxyisobutyraldehyde; 2- hydroxyisobutyric acid
HQ	hazard quotient
HRT	hydraulic (contact) retention time
HSA	hollow stem auger
IARC	International Agency for Research on Cancer
ICU	internal combustion unit
in Hg	inch of mercury
IPA	isopropyl alcohol
ISCO	in-situ chemical oxidation

ISTD	in situ thermal desorption
KDHE	Kansas Department of Health and Environment
kHz	kilohertz
km	kilometer
kPa	kilopascal
kw	kilowatts
l/day	liter per day
l/min	liter per minute
LADD	lifetime average daily dose
lb/eq	pounds per equivalent
LEL	lower explosive limit
LNAPL	light non-aqueous phase liquid
LOAEL	lowest-observable-adverse-effect level
LPS	low-profile air stripper
LUST	leaking underground storage tank
M	mole per liter
M-s	mole per liter per second
m/z	mass to charge ratio
m^2/g	square meter per gram
m^3/day	cubic meter per day
m^3/hour	cubic meter per hour
MADEP	Massachusetts Department of Environmental Protection
MCL	Maximum Contaminant Level
MCLG	Maximum Contaminant Level Goal
MCP	Massachusetts Contingency Plan
MCRT	mean cell retention time
MEDEP	Maine Department of Environmental Protection
mg/day	milligram per day
mg/kg	milligram per kilogram
mg/kg-day	milligram per kilogram per day
mg/l	milligram per liter
mg/l-h	milligram per liter per hour
mg/m^3	milligram per cubic meter
MHP	2-methyl-2-hydroxy-1-propanol; 2-methyl-1,2-propanediol
MIP	membrane interface probe
MLVSS	mixed liquor volatile suspended solids
mM	millimole per liter
mm Hg	millimeter of mercury
MMP	2-methoxy-2-methyl propionaldehyde
MNA	monitored natural attenuation
MOE	margin of exposure
MPE	multi-phase extraction
MRL	minimum risk level

MS	mass spectrometry
MSD	mass spectral detector
MTBE	methyl tert butyl ether
mV	millivolt
MW	molecular weight
NA	not applicable/not available/not analyzed
NAPL	non-aqueous phase liquid
NAWQA	National Water Quality Assessment
NBVC	Navy Base Ventura County
NESCAUM	Northeast States for Coordinated Air Use Management
NETTS	National Environmental Technology Test Sites
NEX	Naval Exchange
NHAGQS	New Hampshire Ambient Ground Water Quality Standard
NHDPHS	New Hampshire Department of Public Health Services
NJDEP	New Jersey Department of Environmental Protection
NOAEL	no-observed-adverse-effect level
NPDES	National Pollutant Discharge Elimination System
NPDWSA	non-potential drinking water source area
NRC	National Research Council
NTNCWS	non-transient non-community water system
NTP	National Toxicology Program
NYSDEC	New York State Department of Environmental Conservation
OH•	hydroxyl radical
ORC	Oxygen Release Compound®
ORP	oxidation-reduction potential
OSHA	Occupational Safety and Health Administration
OUST	Office of Underground Storage Tanks
PAC	powdered activated carbon
PADEP	Pennsylvania Department of Environmental Protection
PAH	polynuclear aromatic hydrocarbon
PBPK	physiologically based pharmacokinetic
PCB	polychlorinated biphenyl
PCE	perchloroethene
PCL	protective concentration limit
PCR	polymerace chain reaction
PDB	passive diffusion bag
PID	photoionization detector
POTW	publicly owned treatment works
ppb	part per billion
PPE	personal protective equipment
ppm	part per million
ppmv	part per million by volume
psi	pound per square inch

PTS	packed tower stripper
PVC	polyvinyl chloride
PWS	public water supply
RBCA	risk-based corrective action
RCRA	Resource Conservation and Recovery Act
RDX	cyclotrimethylenetrinitramine
RfC	reference concentration
RfD	reference dose
RFG	reformulated gasoline
RI	risk index
RO	reverse osmosis
s	second
SAB	Science Advisory Board
SAVE	spray aeration vacuum extraction
SCAPS	Site Characterization and Analysis Penetrometer System
SCDHEC	South Carolina Department of Health and Environmental Control
scfm	standard cubic foot per minute
scmm	standard cubic meter per minute
SCVWD	Santa Clara Valley Water District
SDWR	secondary drinking water regulation
SER	surfactant-enhanced recovery
SERDP	Strategic Environmental Research and Development Program
SIM	selective ion monitoring
SLR	sludge loading rate
SOD	soil oxidant demand
SPH	separate phase hydrocarbons
SPME	solid phase microextraction
SVE	soil vapor extraction
SVOC	semivolatile organic compound
TAA	tert amyl alcohol
TAME	tert amyl methyl ether
TBA	tert butyl alcohol
TBF	tert butyl formate
TCA	1,1,1-trichloroethane
TCE	trichloroethylene
TEAP	terminal electron accepting process
TIC	total inorganic carbon
TNRCC	Texas Natural Resource Conservation Commission
TNT	trinitrotoluene
TOC	total organic carbon
TPH	total petroleum hydrocarbon

TPHg	total petroleum hydrocarbon-modified for gasoline
U.K.	United Kingdom
U.S.	United States
UC	University of California
uCi	microCurie
UCL	upper confidence limit
ug/kg	microgram per kilogram
ug/kg/day	microgram per kilogram per day
ug/l	microgram per liter
ug/m^3	microgram per cubic meter
USACE	United States Army Corps of Engineers
USCPSC	U.S. Consumer Product Safety Commission
USDOD	U.S. Department of Defense
USDOE	U.S. Department of Energy
USEPA	U.S. Environmental Protection Agency
USFDA	U.S. Food and Drug Administration
USGAO	U.S. General Accounting Office
USGS	U.S. Geological Survey
UST	underground storage tank
UV	ultraviolet
V	volt
VET	vapor extraction test
VOC	volatile organic compound
VPDB	Vienna Peedee Belemnite
VPH	volatile petroleum hydrocarbon

Index